Stories engage students

Every chapter of this text draws students in with a compelling and engaging story that covers the key topics of human biology. This approach allows students to relate research science to their everyday lives, encouraging them to integrate the principles and concepts discussed into their understanding of the world.

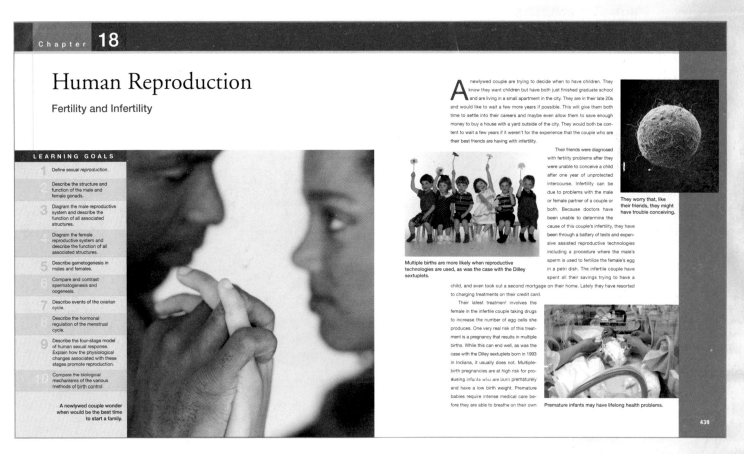

Chapter Opening Stories

Each chapter's opening story is integrated throughout the chapter to help students connect the science they are studying with the story they have read.

The theme of genetic regulation of homeostasis is highlighted throughout

Molecular genetics is introduced early in the text (Chapter 4) so that students learn how genes control biological processes. This allows the authors to emphasize genetic regulation of homeostasis throughout the remainder of the book.

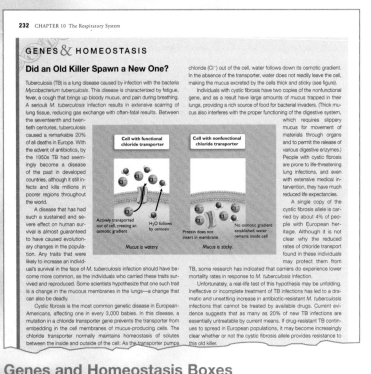

Genes and Homeostasis Boxes

These boxes, which appear in each chapter, show the genetic basis of homeostatic processes.

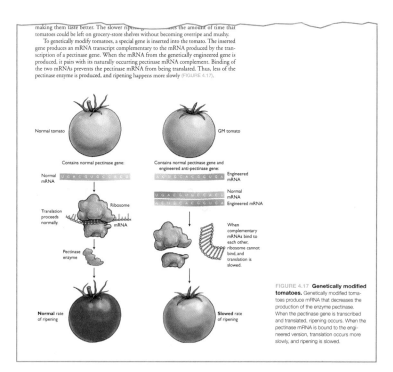

Informative Art

Informative and clearly presented figures illuminate genetic processes and allow visual learners to see and understand anatomy, physiology, and biological processes.

"I feel that Belk/Borden's explanations were very clear, and made me get really interested into the topic. I didn't want to stop reading...I loved the illustrations...I would definitely recommend the book!"

—Mia Scippo, *Monroe Community College*

Critical thinking encourages students to ask "Why?"

Belk and Borden ask students to apply their scientific knowledge in order to critically evaluate the information presented to them.

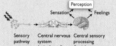

Stop and Stretch

These vignettes help students to pace their reading and ask them to reflect on earlier concepts.

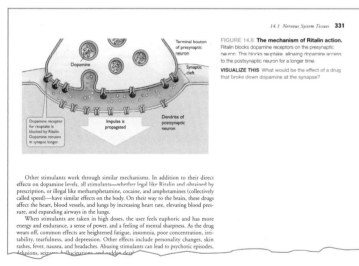

Visualize This

These figures explain a biological process and then encourage students to think through a hypothetical situation involving the same process.

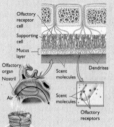

End-of-Chapter Problems

Illustrated chapter summaries, along with extensive problem sets, function as meaningful review as well as to pose critical thinking questions.

Innovative art teaches human biology

Arresting visuals at the opening of chapters summarize each storyline and immediately capture students' interest. Throughout the chapter illustrations and photographs continually build conceptual understanding of the science, often by continuing the opening story. Whenever possible, illustrations reinforce the idea that understanding science relies on careful observation.

> "I like all the figures because I am a visual learner. I think the reading is easier to understand and the figures are more helpful [than my current book]."
>
> —Alicia McDonald, *Ashland University*

Structure of the Heart

In its appearance, the fist-sized heart has only faint similarity to the Valentine's heart. It has a pointed end, called the *apex,* which projects behind the left lung. The heart also has a basic bilateral symmetry, as it consists of two side-by-side muscular pumps. The pumps are completely separated from each other by a muscular wall, or *septum,* and are somewhat independent but also coordinated. One pump, on the right side of the heart, receives oxygen-poor blood from the body and sends it to the lungs, whereas the left pump receives the oxygen-rich blood from the lungs and sends it into general circulation within the body.

The two pumps are each divided into two chambers, a relatively thin-walled but elastic **atrium** (plural: atria) and a thick-walled **ventricle** (FIGURE 9.12). The heart is completely surrounded by the **pericardium**, a membranous sac that holds a small quantity of lubricating liquid. Damage to the heart, including surgery and heart attack, can cause the amount of fluid in the pericardium to increase. As the fluid fills the stiff pericardium, the heart becomes compressed, causing sharp pain.

The chambers of the heart are sealed shut at different stages in its pumping cycle to prevent the backflow, or regurgitation, of blood. The seals take the form of valves that open for blood flow and then close to prevent its passage.

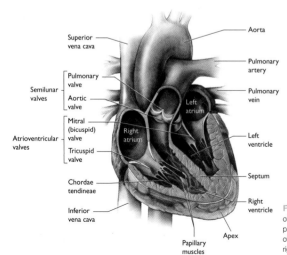

FIGURE 9.12 **The human heart.** The heart consists of four chambers making up two mostly independent pumps. Note that this drawing illustrates the orientation of a person's heart when he is facing you. The left and right sides are relative to that person, not to you.

FIGURE 18.5 **Oogenesis.** (a) One follicle per month goes through steps 1–6 to produce and ovulate an egg cell. (b) Meiosis in human females produces only one fertilizable cell; the others are nutrient-poor polar bodies. If the larger egg cell, with the lion's share of nutrients, is fertilized, it contains all the nutrients and organelles that the fertilized egg (the zygote) will need to serve as the progenitor of all the millions of cells of the human body.

VISUALIZE THIS Would twins produced when ovulation occurs from both ovaries in one month be genetically identical? Why or why not?

Figures with Voice Balloons

These figures present concepts and processes through a paced narrative within the figures to better reveal the biology depicted by the illustration.

Anatomical Figures

These carefully designed figures give students clearly labeled views and information on structures and tissues.

Media provides students with interactive and visual tools for learning

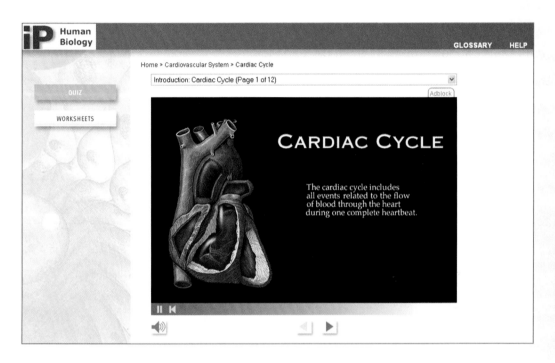

Interactive Physiology® for Human Biology

This companion CD-ROM reinforces readings and lectures with a wealth of outstanding animations, engaging activities, helpful self-testing, and much more. This edition includes a new module on the Immune System along with assignable worksheets to help students gauge their progress and stay on track with their studies.

The Human Biology Place Website

The Human Biology Place student website offers a highly interactive way for students to get more from their studies. Students get started right away with a unique chapter guide highlighting the animations and quizzes they can use to measure their progress and improve their grades. A glossary, flashcards, and a crossword puzzle for every chapter will help students remember and practice their biology vocabulary. A news feed feature will keep your students aware of the most important human biology topics in the news.

www.humanbiology.com

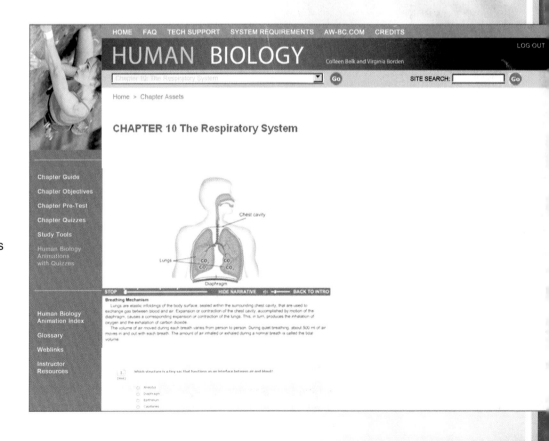

Think INSIDE THE BOX!

A course planning kit that saves you hours of time

NEW! The Teaching Tool Box provides easy access to all teaching resources in one convenient package.

By including all of the instructor supplements (Instructor Resource DVD, Instructor Guide/Test Bank, Computerized Test Bank (TestGen), and Transparency Acetates), *Interactive Physiology* 10-System Suite for Human Biology, plus unique lecture and prep tools (Course-at-a-Glance Guide and Human Biology Support Manual), the Teaching Tool Box helps make prepping for class even easier.

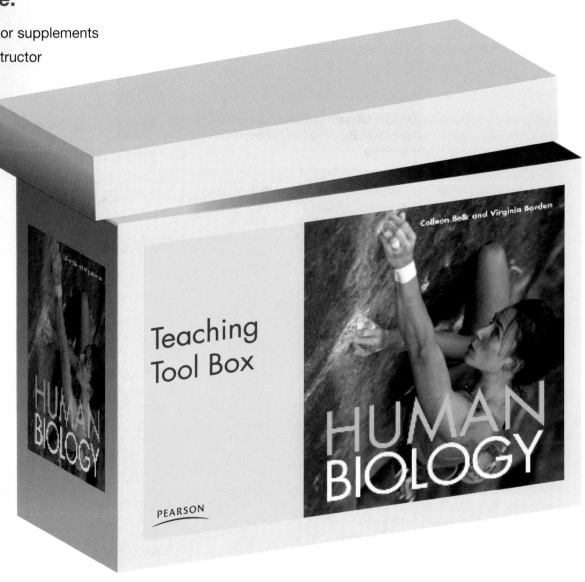

Instructor Resource DVD

This cross-platform DVD organizes all instructor media resources into one convenient, chapter-by-chapter resource package. The IR-DVD includes an image library of all art and tables and selected photos from the book with customizable labels, editable PowerPoint® lecture presentation slides, clicker questions, animations, step-edit art, ABC videos, *Interactive Physiology*® for Human Biology slides, and other resources that help instructors create memorable presentations effortlessly.

Instructor Guide and Test Bank

Instructor Guide by Donald Glassman, *Des Moines Area Community College*, and Test Bank by Jill Feinstein, *Richland Community College*

This excellent, comprehensive resource contains a printed Test Bank with over 1,000 questions, full lecture outlines for each chapter, with in-class and out-of-class activities for the student inserted where appropriate, discussion questions with comprehensive answers that offer information not found in the text, and an in-depth storyline tailored to complement or function as an alternative to the stories that begin each chapter of the main text.

Human Biology Support Manual

This support manual guides novice instructors through lectures for each chapter, suggesting multimedia and print supplements where appropriate to invigorate lectures with additional material. This manual contains sample syllabi for 10 and 16 week courses, teaching tips for new lecturers, a visual walkthrough of lecture outlines for each chapter, and a primer on how to use CourseCompass™.

Course-at-a-Glance Guide

This handy guide to planning a course highlights all the resources available to instructors and students, chapter-by-chapter and week-by-week, following a model of either a 10-week or 16-week syllabus.

Transparency Acetates

This supplement provides full-color acetates, including all the art and tables from the text—almost 500 illustrations in all—in full color, with tabs separating each chapter.

Study Guide

by Catherine Podeszwa

Students will master key concepts and earn a better grade with the helpful study tools found in the study guide. Each chapter includes a chapter summary, key concepts, fill-in-the-blank questions, labeling exercises, crossword puzzles, true/false questions, table completion exercises, plus practice tests to help students understand the material and test their knowledge.

Teaching and learning solutions for successful instructors and students

For Instructors

Teaching Tool Box

978-0-321-59020-6 / 0-321-59020-1
The Teaching Tool Box includes all of the instructor supplements (Instructor Resource DVD, Instructor Guide/Test Bank, Computerized Test Bank (Test Gen), and Transparency Acetates) and unique lecture and prep tools (Course-at-a-Glance guide, Human Biology Support Manual). For more details on these tools please review the two preceding pages.

Computerized Test Bank (Test Gen)

The questions from the printed test bank are also available in this text-specific testing program that allows instructors to view and edit these test questions, add their own questions, and generate tests randomly to create an unlimited variety of quizzes and tests.

CourseCompass™

This nationally-hosted, dynamic, interactive online course management system is powered by Blackboard, the leading platform for Internet-based learning tools. This easy-to-use and customizable program enables professors to tailor content and functionality to meet individual course needs. Every CourseCompass™ course includes a range of preloaded content such as testing and assessment question pools, chapter-level overviews and objectives, interactive animations, and web resources—all designed to help students master core course objectives.

www.pearsonhighered.com/coursecompass

WebCT Premium

This course management system contains a range of pre-loaded content such as testing and assessment question pools, chapter-level overviews and objectives, interactive web-based activities, animations, and web links– all designed to help students master core course objectives.

www.pearsonhighered.com/webct

Blackboard Premium

This course management system contains a range of preloaded content such as testing and assessment question pools, chapter-level overviews and objectives, interactive web-based activities, animations, and web links—all designed to help students master core course objectives.

www.pearsonhighered.com/blackboard

Human Biology Place

This open-access web site includes objective-driven chapter guides to guide student learning, human biology animations with quizzes, extensive chapter quizzes, crossword puzzles, flashcards, glossaries, and a unique news feature allowing students to review and reflect on what's happening in biology news. A password-protected link offers instructor access to a variety of additional instructor resources.

www.humanbiology.com

For Students

Study Guide

Please see preceding page for a complete description.

***Scientific American*, Current Issues in Biology**

These magazines are created specifically for Benjamin Cummings titles, and contain articles from *Scientific American* magazine that present topics directly related to the issues explored in our books. Answers to the questions that follow each article are found on the Companion Website, and accompanying PowerPoint® presentations can be found on the Instructor Resource Center on DVD.

Interactive Physiology® for Human Biology

The Interactive Physiology for Human Biology CD-ROM reinforces tough topics with step-by-step tutorials.

CourseCompass Student Access Kit for Human Biology

Human Biology Place

Please see description under Instructor listings.

What students are saying about Human Biology

"Stories always interest me. It makes it easier for me to get involved in the readings. It helps to see more examples of a concept I'm trying to learn. This book looks superior to my current book. I like how the info is presented."
—Danielle Bergdahl, *Pennsylvania State University Altoona*

"They [the stories] drew me in and made me want to read more…it captivated me and I loved to keep reading."
—Thea Merriweather, *Mott Community College*

"It breaks down scientific knowledge into easily understandable text coupled with informative illustrations and thought provoking questions."
—Jacqueline Suarez, *William Paterson University*

HUMAN BIOLOGY

HUMAN BIOLOGY

Colleen Belk
University of Minnesota–Duluth

Virginia Borden Maier
St. John Fisher College

Benjamin Cummings

San Francisco • Boston • New York
Cape Town • Hong Kong • London • Madrid • Mexico City
Montreal • Munich • Paris • Singapore • Sydney • Tokyo • Toronto

Executive Editor: Gary Carlson
Senior Project Editor: Susan Malloy
Development Manager: Claire Alexander
Development Editor: Donald Gecewicz
Art Development Editor: Jay McElroy
Editorial Assistants: Jessica Young, Kaci Smith
Media Producer: Lucinda Bingham
Managing Editor: Deborah Cogan
Production Supervisor: Beth Masse
Production Management: Martha Emry
Copyeditor: Anita Wagner
Compositor: Pre-Press PMG
Interior Designer: GGS Book Services, Inc.
Cover Designer: Studio A
Illustrators: Imagineering Media Services, Inc.
Art Editor: Elisheva Marcus
Photo Researcher: Yvonne Gerin
Director, Image Resource Center: Melinda Patelli
Image Rights and Permissions Manager: Zina Arabia
Image Permissions Coordinator: Elaine Soares
Manufacturing Buyer: Michael Penne
Marketing Manager: Gordon Lee
Executive Market Development Manager: Josh Frost
Text and Cover Printer: RR Donnelley, Willard
Cover Photo Credit: Getty Images/Jimmy Chin

Library of Congress Cataloging-in-Publication Data

Belk, Colleen M.
  Human biology / Colleen Belk, Virginia Borden.
    p. cm.
  ISBN 0-13-148124-X (student edition)
   1. Human biology—Textbooks. I. Borden, Virginia. II. Title.
QP36.B385 2009
612—dc22   2008016904

Copyright © 2009 Pearson Education, Inc., publishing as Pearson Benjamin Cummings, 1301 Sansome St., San Francisco, CA 94111. All rights reserved. Manufactured in the United States of America. This publication is protected by Copyright and permission should be obtained from the publisher prior to any prohibited reproduction, storage in a retrieval system, or transmission in any form or by any means, electronic, mechanical, photocopying, recording, or likewise. To obtain permission(s) to use material from this work, please submit a written request to Pearson Education, Inc., Permissions Department, 1900 E. Lake Ave., Glenview, IL 60025. For information regarding permissions, call (847) 486–2635.

Many of the designations used by manufacturers and sellers to distinguish their products are claimed as trademarks. Where those designations appear in this book, and the publisher was aware of a trademark claim, the designations have been printed in initial caps or all caps.

ISBN-10: 0-131-48124-X (Student edition)
ISBN-13: 978-0131-48124-4 (Student edition)
ISBN-10: 0-321-57154-1 (Professional Copy)
ISBN-13: 978-0 321-57154-0 (Professional Copy)
2 3 4 5 6 7 8 9 10—DOW—12 11 10

www.pearsonhighered.com

BRIEF CONTENTS

UNIT I	**THE FOUNDATIONS OF LIFE**	
Chapter 1	The Scientific Method *Proven Effective*	1
Chapter 2	The Chemistry of Life *Drink to Your Health?*	26
Chapter 3	Cell Structure and Metabolism *Diet*	54
UNIT II	**GENES AND HOMEOSTASIS**	
Chapter 4	Genes—Transcription, Translation, Mutation, and Cloning *Genetically Modified Foods*	82
Chapter 5	Tissues, Organs, and Organ Systems *Work Out!*	105
UNIT III	**THE HUMAN MACHINE**	
Chapter 6	The Skeletal, Muscular, and Integumentary Systems *Sex Differences in Athleticism*	126
Chapter 7	The Digestive System *Weight-Loss Surgery*	154
Chapter 8	The Blood *Malaria—A Deadly Bite*	178
Chapter 9	The Cardiovascular System *Can We Stop the Number-One Killer?*	198
Chapter 10	The Respiratory System *Secondhand Smoke*	226
Chapter 11	The Urinary System *Surviving the Ironman*	250
UNIT IV	**RESPONDING TO THE ENVIRONMENT**	
Chapter 12	The Immune System *Will Mad-Cow Disease Become an Epidemic?*	272
Chapter 13	Sexually Transmitted Infections *The Cervical Cancer Vaccine*	302
Chapter 14	Brain Structure and Function *Attention Deficit Disorder*	322
Chapter 15	The Senses *Do Humans Have a Sixth Sense?*	352
Chapter 16	The Endocrine System and Stress *Worried Sick*	382
UNIT V	**THE NEXT GENERATION**	
Chapter 17	DNA Synthesis, Mitosis, and Meiosis *Cancer*	404
Chapter 18	Human Reproduction *Fertility and Infertility*	438
Chapter 19	Heredity *Genes and Intelligence*	466
Chapter 20	Complex Patterns of Inheritance *DNA Detective*	486
Chapter 21	Development and Aging *The Promise and Perils of Stem Cells*	510
UNIT VI	**OUR PAST AND FUTURE**	
Chapter 22	Evolution *Where Did We Come From?*	538
Chapter 23	Ecology *Is Earth Experiencing a Mass Extinction?*	562
Chapter 24	Biomes and Natural Resources *Where Do You Live?*	584

CONTENTS

UNIT I THE FOUNDATIONS OF LIFE

Chapter 1
The Scientific Method *Proven Effective* — 1

- **1.1** An Introduction to the Scientific Process 2
  - *The Nature of Hypotheses* 2
  - *Scientific Theories* 3
  - *The Theory of Evolution and the Theory of Natural Selection* 4
  - *The Logic of Hypothesis Testing* 4
- **1.2** Hypothesis Testing 7
  - *The Experimental Method* 7
  - *Controlled Experiments* 8
  - *Minimizing Bias in Experimental Design* 9
  - *Using Correlation to Test Hypotheses* 10
- **1.3** Understanding Statistics 13
  - *What Statistical Tests Can Tell Us* 13
  - *Statistical Significance: A Definition* 13
  - *Factors Influencing Statistical Significance* 15
  - *What Statistical Tests Cannot Tell Us* 16
- **1.4** Evaluating Scientific Information 17
  - *Information from Anecdotes* 17
  - *Science in the News* 17
  - *Understanding Science from Secondary Sources* 19
- Chapter Review 21

Chapter 2
The Chemistry of Life *Drink to Your Health?* — 26

- **2.1** Water: Essential to Life 28
  - *The Building Blocks of Water* 28
  - *The Structure of Water* 29
  - *Water Is a Good Solvent* 30
  - *Water Facilitates Chemical Reactions* 30
  - *Water Is Cohesive* 30
  - *Bottle or Tap?* 33
- **2.2** Acids, Bases, and Salts 33
  - *pH: Measuring the Activity of Ions* 33
  - *Electrolytes* 34
- **2.3** Structure and Function of Macromolecules 35
  - *Carbohydrates* 36
  - *Proteins* 37
  - *Lipids* 40
  - *Nucleic Acids* 43

2.4 Micronutrients 45
   Vitamins 45
   Minerals 47
   Antioxidants 47
   What Are You Paying For? 47
Chapter Review 49

Chapter 3
Cell Structure and Metabolism *Diet*   54

3.1 Food and Energy 56
   ATP Is the Cell's Energy Currency 57

3.2 Cell Structure and Function 59
   Cell Structures 60

3.3 Membrane Structure and Function 64
   Membrane Structure 64
   Transporting Substances Across Membranes 64

3.4 Metabolism—Chemical Reactions in the Body 67
   Enzymes 67
   Cellular Respiration 68
   A General Overview of Cellular Respiration 69
   Glycolysis, the Citric Acid Cycle, Electron Transport, and ATP Synthesis 69
   Calories and Metabolic Rate 74

3.5 Health and Body Weight 74
   Underweight 74
   Obesity 75
Chapter Review 78

UNIT II GENES AND HOMEOSTASIS

Chapter 4
Genes—Transcription, Translation, Mutation, and Cloning
Genetically Modified Foods   82

4.1 What Is a Gene? 84

4.2 Protein Synthesis and Gene Expression 84
   From Gene to Protein 84
   Transcription: Copying the Gene 86
   Translation: Using the Message to Synthesize a Protein 87
   Mutations 88
   Regulating Gene Expression 92

4.3 Producing Recombinant Proteins 93
   Cloning a Gene Using Bacteria 94
   FDA Regulations 96

4.4 Genetically Modified Crops 96
   Potential Benefits of Genetically Modifying Crops 97
   Effect of GMOs on Human Health 98
Chapter Review 100

Chapter 5
Tissues, Organs, and Organ Systems *Work Out!* — 105

5.1 Tissues and Their Functions 106
  Epithelial Tissue 106
  Cell Junctions 107
  Connective Tissue 108
  Muscle Tissue 113
  Nervous Tissue 114

5.2 Body Cavities and Membranes 115
  Body Cavities 115
  Body Membranes 115

5.3 Organs and Organ Systems 116
  Levels of Organization 118
  Interdependence of Organ Systems 118

5.4 Homeostasis 118
  Improving Fitness 119

Chapter Review 121

UNIT III THE HUMAN MACHINE

Chapter 6
The Skeletal, Muscular, and Integumentary Systems
Sex Differences in Athleticism — 126

6.1 The Skeletal System 128
  Bones of the Skeleton 128
  Bone Development, Growth, Remodeling, and Repair 129
  Axial Skeleton: The Central Structure 134
  Appendicular Skeleton 136
  Joints and Movement 138

6.2 The Muscular System 140
  Names and Actions of Skeletal Muscles 140
  Skeletal Muscle Structure 140
  Skeletal Muscle Contraction 142
  Energy Inputs for Muscle Contraction 144

  ■ **GENES & HOMEOSTASIS: MUSCULAR DYSTROPHY** 146

6.3 The Integumentary System 146
  Epidermis 146
  Dermis 147
  Accessory Structures of the Skin 147
  Subcutaneous Layer 148
  What Do Sex Differences Really Mean? 148

Chapter Review 150

Chapter 7
The Digestive System *Weight-Loss Surgery* — 154

7.1 The Digestive Tract 156
  The Wall of the Digestive Tract 157
  The Mouth: The Actions of Teeth and Saliva 157
  The Pharynx and the Esophagus: Transport to the Stomach 159

The Stomach: Digestion in an Acid Bath 159
The Small Intestine: Where Most Digestion Happens 161
Regulation of Digestive Secretions 163
The Large Intestine: Absorption and Elimination 163
Gastric Bypass Surgery: Scaling Back Digestion 165

- **GENES & HOMEOSTASIS: OBESITY** 166

7.2 Three Accessory Organs of the Digestive System 166
The Liver 167
The Gallbladder 168
The Pancreas 169

7.3 Weighing the Risks of Gastric Bypass Surgery 171

Chapter Review 172

Chapter 8
The Blood *Malaria—A Deadly Bite* — 178

8.1 The Constituents of Blood 180
Plasma 180
Formed Elements: The Cellular Portion of Blood 181

8.2 Malaria and the Blood 184
Malaria Infection 185
Anemia and Blood Cell Production 186
Blood Types and Transfusions 186
Recycling Red Blood Cells 189

8.3 Blood Clotting 189
The Clotting Cascade 189
Clotting Disorders 190

- **GENES & HOMEOSTASIS: SICKLE-CELL ANEMIA** 191

8.4 Ending Malaria 192

Chapter Review 193

Chapter 9
The Cardiovascular System *Can We Stop the Number-One Killer?* — 198

9.1 Blood and Lymphatic Vessels: The Circulation Pipes 200
Arteries and Arterioles 200
Capillaries: The Distribution Network 203
Veins: The Path Back to the Heart 204
The Lymphatic System: Draining the Tissues 204
Control of Blood Pressure 205

- **GENES & HOMEOSTASIS: RACE-BASED MEDICINE?** 206

Fixing the Pipes 206

9.2 The Mechanical Heart 209
Structure of the Heart 209
The Cardiovascular Pathway 210
Repairing the Pump 211

9.3 The Electrical Heart 214
The Cardiac Cycle 214
Steadying the Heartbeat 216

9.4 Power for the Heart 217
  Coronary Blood Vessels 217
  Maintaining the Heart's Energy Supply 217
  A Healthy Heart 219
Chapter Review 220

Chapter 10
The Respiratory System *Secondhand Smoke* 226

10.1 Respiratory System Anatomy: The Path of Smoke into the Lungs 228
  Upper Respiratory Tract 228
  Lower Respiratory Tract 229

10.2 Tobacco Smoke and the Respiratory Tract 230
  The Composition of Tobacco Smoke 230

  ■ **GENES & HOMEOSTASIS: DID AN OLD KILLER SPAWN A NEW ONE?** 232
  Smoke Damages the Respiratory System 232

10.3 Inhaling and Exhaling 234
  The Mechanics of Breathing 234
  The Control of Breathing 236
  Smoking and Breathing 237

10.4 Gas Exchange in the Lungs 238
  A Closer Look at Gas Exchange 238
  Smoking and Gas Exchange 240
  Nicotine: Why Tobacco Is Habit-Forming 241

10.5 Beyond the Lungs 242
  The Effects of Smoke on Other Organ Systems 243
  Preventing Smoking-Related Illness 243

Chapter Review 245

Chapter 11
The Urinary System *Surviving the Ironman* 250

11.1 An Overview of the Urinary System 252
  Homeostasis and the Urinary System 252
  Structure of the Urinary System 253
  Urination 254

11.1 Excretion 255
  The Composition of Urine 255
  Urine Formation 256

11.3 Water, pH, and Salt Balance 259
  Hormones and Water Depletion 260
  Countercurrent Exchange in the Kidney 261
  Maintaining Blood pH 263
  Salt Balance: The Right Amount of Sodium 263

  ■ **GENES & HOMEOSTASIS: PLEASE PASS ON THE SALT** 264

11.4 When Kidneys Fail 264
Chapter Review 267

UNIT IV RESPONDING TO THE ENVIRONMENT

Chapter 12
The Immune System *Will Mad Cow Disease Become an Epidemic?* — 272

- 12.1 Infectious Agents   274
  - *Bacteria*   274
  - ■ **GENES & HOMEOSTASIS: MUTATIONS AND THE DEVELOPMENT OF ANTIBIOTIC RESISTANCE**   276
  - *Viruses*   277
  - *Eukaryotic Pathogens*   280
  - *Prions*   280
- 12.2 Transmission of Infectious Agents   283
  - *Direct Contact*   283
  - *Indirect Contact*   283
  - *Vector-Borne Transmission*   283
  - *Ingestion*   284
- 12.3 The Body's Response to Infection: The Immune System   284
  - *First Line of Defense: Skin and Mucous Membranes*   284
  - *Second Line of Defense: White Blood Cells, Inflammation, Defensive Proteins, and Fever*   285
  - *Third Line of Defense: Lymphocytes*   287
  - *Anticipating Infection*   289
  - *Humoral and Cell-Mediated Immunity*   292
  - *There Is No Immune Response to Prions*   294
- 12.4 Preventing an Epidemic of Prion Diseases   295
- Chapter Review   296

Chapter 13
Sexually Transmitted Infections *The Cervical Cancer Vaccine* — 302

- 13.1 The Old Epidemics   304
  - *The Eukaryotes: Pubic Lice and Trichomoniasis*   304
  - *The Bacteria: Chlamydia, Gonorrhea, and Syphilis*   306
  - *The Viruses: Herpes, Hepatitis, and Genital Warts*   310
- 13.2 The New Epidemic—AIDS   310
  - *A Disease of the Immune System*   310
  - *The Course of HIV Infection*   310
  - *Treating HIV Infection*   313
  - *Preventing HIV/AIDS*   315
- ■ **GENES & HOMEOSTASIS: RESISTANCE TO HIV/AIDS**   316
- Chapter Review   317

Chapter 14
Brain Structure and Function *Attention Deficit Disorder* — 322

- 14.1 Nervous System Tissues   324
  - *Neuron Structure*   324
  - *The Creation of Nerve Impulses*   325
  - *Neurotransmitters Carry Signals Between Neurons*   328
  - *Neurotransmitters and Disease*   328
  - *Synaptic Integration*   329
- ■ **GENES & HOMEOSTASIS: A GENETIC LINK TO DEPRESSION**   330
  - *Neurotransmission, ADD, and Ritalin*   330

14.2 The Central Nervous System   334
Spinal Cord   334
The Brain   336
ADD and the Structure and Function of the Brain   339

14.3 The Limbic System and Memory   340
Limbic System Structures   340
Memory   340

14.4 The Peripheral Nervous System   341
The Nerves of the Peripheral Nervous System   341
Somatic System   342
The Autonomic Nervous System   342

14.5 What Causes ADD?   343

Chapter Review   345

Chapter 15
The Senses *Do Humans Have a Sixth Sense?*   352

15.1 Sensing and Perceiving   354
Sensory Receptors   354
Reading and Understanding the Environment   355

15.2 The General Senses   356
Proprioception   356
The Sense of Touch   358
Temperature and Pain   358

15.3 The Chemical Senses   361
Taste   361
Smell   362

15.4 Senses of the Ear   364
Hearing   364
A Sense of Balance   367

15.5 Vision   369
Focusing Light   369
Photoreceptors   372
Vision and Perception   373

■ **GENES & HOMEOSTASIS: COLOR BLINDNESS**   375

15.6 Predicting the Future   376
Understanding Premonitions   376
Expanding the Receptive Field   376

Chapter Review   377

Chapter 16
The Endocrine System and Stress *Worried Sick*   382

16.1 An Overview of the Endocrine System   384
Hormones: Chemical Messengers   384
Endocrine Glands   385
Stress and the Endocrine System   385

16.2 The Endocrine System and Homeostasis   387
The Control Center: The Hypothalamus   387
Turning Down Hormone Release Through Negative Feedback Loops   389

16.3 Other Endocrine Glands 389
   The Pituitary: Regulation of Growth 389
   ■ **GENES & HOMEOSTASIS: SIZE MATTERS** 390
   The Gonads: Sex-Specific Characteristics 390
   The Pancreas: Regulation of Blood-Glucose Levels 391
   The Thyroid and Parathyroid: Metabolism and Development 392
   The Pineal Gland: Hormonal Effects of Light and Darkness 394
   The Thymus: Junction of the Endocrine and Immune Systems 395
   Other Tissues That Produce Hormones 396

16.4 Combating Stress 398

Chapter Review 400

UNIT V THE NEXT GENERATION

Chapter 17
DNA Synthesis, Mitosis, Meiosis *Cancer*   404

17.1 What Is Cancer? 406
   Risk Factors for Cancer 408

17.2 An Overview of Cell Division 409
   DNA Replication 411

17.3 The Cell Cycle and Mitosis 413
   Interphase: Normal Functioning and Preparations 413
   Mitosis: The Nucleus Divides 413
   Cytokinesis: The Cytoplasm Divides 416

17.4 Mutations Override Cell-Cycle Controls 416
   Controls in the Cell Cycle 416
   ■ **GENES & HOMEOSTASIS: INHERITANCE AND CANCER** 417

17.5 Cancer Detection and Treatment 419
   Detecting Cancer 419
   Cancer Treatments: Chemotherapy and Radiation 420

17.6 Meiosis: Making Reproductive Cells 422
   Interphase 425
   Meiosis I 425
   Meiosis II 425
   Crossing Over and Random Alignment 425

Chapter Review 432

Chapter 18
Human Reproduction *Fertility and Infertility*   438

18.1 The Human Reproductive Systems 440
   The Male Reproductive System 440
   The Female Reproductive System 442

18.2 Gametogenesis: Development of Sex Cells 443
   Spermatogenesis: Development of Men's Gametes 444
   Oogenesis: Development of Women's Gametes 446
   ■ **GENES & HOMEOSTASIS: ENDOCRINE DISRUPTORS** 448

18.3 The Menstrual Cycle 450
18.4 The Human Sexual Response 452
18.5 Controlling Fertility 453
   Principles of Fertility Control 453
   Barrier Methods 453
   Hormonal Birth Control 456
   Other Methods of Birth Control 457
   The Future of Birth Control Technology 460
18.6 Health, Lifestyle, and Fertility 460
Chapter Review 461

Chapter 19
Heredity *Genes and Intelligence* 466

19.1 The Inheritance of Traits 468
   Genes and Chromosomes 468
   Producing Diversity in Offspring 468
19.2 Mendelian Genetics: When the Role of a Gene Is Direct 472
   Genotype and Phenotype 472
   Genetic Diseases in Human Beings 473
   Using Punnett Squares to Predict Genotypes of Offspring 474

   ■ **GENES & HOMEOSTASIS: A GENETIC STUTTER** 475

19.3 Quantitative Genetics: When Genes and Environment Interact 476
   Why Traits Are Quantitative 476
   The Heritability of Quantitative Traits 477
19.4 Genes, the Environment, and the Individual 479
   The Use and Misuse of Heritability Calculations 479
   How Do Genes Matter? 479
Chapter Review 482

Chapter 20
Complex Patterns of Inheritance *DNA Detective* 486

20.1 Extensions of Mendelism 488
20.2 Dihybrid Crosses 491
20.3 Sex Determination and Sex Linkage 494
   Chromosomes and Sex Determination 494
   Sex Linkage 494

   ■ **GENES & HOMEOSTASIS: CHANGES TO CHROMOSOME STRUCTURE AND NUMBER** 495

20.4 Pedigrees 498
20.5 DNA Fingerprinting 500
   Copying DNA Through Polymerase Chain Reaction 501
   Size-Based Separation Through Gel Electrophoresis 502
Chapter Review 505

Chapter 21
Development and Aging *The Promise and Perils of Stem Cells* 510

21.1 The Production of Embryonic Stem Cells 512
   Fertilization: Forming the Ultimate Stem Cell 513
   Preembryonic Development 515

21.2 Early Embryonic Development 516

21.3 Organ Formation 517
Cell Migration and Death 517
Early Organogenesis: Development of the Nervous System 518
Later Organogenesis: The Reproductive Organs 519

21.4 Fetal Development and Birth 521
The Purpose of the Placenta in Pregnancy 521
Fetal Circulation 522
Stem Cells and the Fetal Period 523
The Process of Childbirth 524

21.5 Development After Birth 526
Growth and Maturation 526
Puberty 527

■ GENES & HOMEOSTASIS: ARE GIRLS BECOMING WOMEN TOO YOUNG? 528

21.6 Aging 529
Why Do We Age? 529
Effects of Aging 529
Restoring the Brain 531

Chapter Review 532

UNIT VI OUR PAST AND FUTURE

Chapter 22
Evolution *Where Did We Come From?* — 538

22.1 Evidence of Evolution 540
What Is Evolution? 540
Charles Darwin's Revolution 540
Alternative Hypotheses: Scientific and Religious 541
Evidence from Biological Classification 542
Evidence from Homology: Related Species Are Similar 544
Evidence from Biogeography 545
Evidence from the Fossil Record 546

22.2 The Origin of Species 549
Speciation: How One Becomes Two 549
The Theory of Natural Selection 550
Critical Thinking About Natural Selection 551
Evolution: A Robust Theory 552

22.3 Human Evolution 554
Why Human Groups Differ: Selection and Genetic Drift 554
Evolution in the Classroom 557

Chapter Review 557

Chapter 23
Ecology *Is Earth Experiencing a Mass Extinction?* — 562

23.1 Limits to Population Growth 564
Principles of Population Ecology 564
Population Crashes 566

23.2 The Sixth Extinction   567
Measuring Extinction Rates   567
Causes of Extinction   568
The Consequences of Extinction   571

23.3 Saving Species   575
Protecting Habitat   575
Ensuring Adequate Population Size   577
Meeting the Needs of Humans and Nature   578

Chapter Review   580

Chapter 24
Biomes and Natural Resources *Where Do You Live?*   584

24.1 Terrestrial Biomes   586
Forests and Shrublands   588
Grasslands   588
Desert   589
Tundra   589

24.2 Aquatic Biomes   590
Freshwater   590
Saltwater   590

24.3 Human Habitats   591
Energy and Natural Resources   591
Waste Production   592
Climate Change   594
The Future of Our Shared Environment   595

Chapter Review   596

Appendix: Metric System Conversions   A-1

Answers   ANS-1

Glossary   G-1

Index   I-1

Photo Credits   PC-1

PREFACE

Using Narratives to Teach Science

Humans are natural storytellers. Many of the most important lessons we learn about life are taught and reinforced by stories that have been passed on through countless generations. It is clear why storytelling works as a technique for passing on wisdom—people are much more motivated to learn when their own natural curiosity is piqued by the twists and turns of a dramatic story.

Human Biology uses our affinity for stories to teach the latest science in the field of health and human biology. The text engages students in learning this challenging material by using a storyline to present the science in every chapter. Our narrative approach allows us to cover the key areas of human biology, including cell structure and function, classical and molecular genetics, anatomy and physiology, and evolution and ecology in a manner that motivates students to learn. Just as importantly, the storyline allows them to relate research science to their everyday lives, encouraging them to integrate the principles and concepts discussed into their understanding of the world.

The effectiveness of *Human Biology*'s narrative approach is evident in the text's art program as well. The arresting visuals used at the opening of chapters summarize each storyline and immediately capture students' interest. The visual images used throughout the chapters often pick up and continue the story. These images appeal to students with strong visual skills as well, and illustrate that understanding the sciences often relies on careful observation.

The authors know from experience in the classroom that presenting science within a storyline is an effective way to raise student interest and improve their understanding of the material. An engaging textbook helps to support this strategy in the classroom. Because students are more motivated to read, they have a better understanding of the basic science from the textbook. Thus, instructors have the flexibility to use class time more effectively and to focus on developing students' abilities to understand and evaluate information on health, disease, and the history and future of the human population.

Developing Critical Thinking Skills

In addition to chapter narratives *Human Biology* incorporates several other strategies for maximizing student learning, including modeling the process of science, an emphasis on the genetic basis of homeostasis, and a focus on developing students' critical thinking skills.

Modeling the Process of Science

The scientific method is a powerful tool for increasing our understanding of the natural world. Students, however, often feel that "doing science" requires some secret knowledge or special intelligence. In *Human Biology*, the process of science is modeled throughout the text so that students can see how scientists and citizens alike use this process to answer questions.

To this end, the entire first chapter is dedicated to describing how scientific knowledge is acquired, a theme that will continue in following chapters. In each remaining chapter, students will practice evaluating the merit of scientific information using careful systematic reasoning and empirical evidence. For example, Chapter 2 examines the health claims made by various "sports beverages" and provides tools for evaluating those claims. Chapter 4 tells the story of genetically modified organisms and explains to students how scientists discovered and refined techniques of genetic engineering. And Chapter 12 examines the threat posed by mad cow disease and leads students through the process of uncovering the existence of prion infections. Through these narratives, students learn to appreciate how science differs from other ways of understanding the world.

Emphasizing Homeostasis and Its Genetic Basis

One of the remarkable qualities of living organisms is their ability to self-regulate. *Human Biology* highlights the feedback processes that keep systems at equilibrium. Reflecting the modern focus on the genetic basis of homeostatic processes, the text introduces the basics of gene expression in the second unit (Chapter 4) so that students have early exposure to the manner in which genes control biological processes. This early unit also covers homeostasis (Chapter 5) so that students will understand how the systems in the human body collect and react to feedback.

Early coverage and integration of these two important concepts is continued through each of the remaining physiology chapters (Chapters 6–21) in boxed essays entitled **Genes & Homeostasis** that highlight the genetic basis of homeostatic processes. Some of the topics covered in Genes & Homeostasis boxes include muscular dystrophy (Chapter 6), the possible genetic basis of obesity (Chapter 7), the relationship between race and medical conditions (Chapter 9), tuberculosis and its effects on human evolution (Chapter 10), and the decrease in the age of sexual maturity in girls (Chapter 21).

Focusing on Critical Thinking

Our students are operating in a world rife with information from both reliable and unreliable sources. Television, radio, print media, and the Internet present a constant stream of data, articles, and headlines making it difficult for even the savviest student to separate fact from fiction. Helping students apply scientific knowledge in order to critically evaluate information has to be a goal of any up-to-date textbook. Two features found in every chapter provide this opportunity.

We designed **Stop and Stretch** boxes so that students can pace their reading and to underscore the importance of mastering certain scientific concepts. The title, Stop and Stretch, isn't a call for a coffee break. Instead, we want students to slow down, think a bit, and stretch their understanding. Suggested answers to the questions posed in these boxes are found at the end of the book.

In the same vein, we were concerned that students tend to skim the visuals in science textbooks. To encourage students to slow down, assess the information in visuals, and learn to read the text and art program at the same time, we designed a feature called **Visualize This**. These critical thinking and analysis questions appear with many of the key figures in each chapter.

Innovative Chapter Problems Support Learning and Engage Students

Our end-of-chapter exercises are designed with two primary groups of book users in mind. Instructors want a wide array of problems with different setups (from multiple choice to essay), problems that accurately reflect the content of each chapter, as well as problems with interesting scenarios. Students want practice problems as well as other problems that truly engage them, and if answered correctly reflect a degree of comprehension that will allow them to pass tests and feel confident of their mastery of the material.

The first problem set at the end of each chapter is called **Learning the Basics**. These problems test student understanding through familiar formats such as multiple choice, true or false, fill in the blanks, and matching. Students who successfully work through this category will achieve a basic grasp of the topics in the chapter.

The next category of problems is called **Analyzing and Applying the Basics**. These carefully constructed problems present interesting scenarios that ask students to apply their knowledge of the science within a chapter and to think critically about a scientific question. These problems go beyond review to get students to analyze scientific data, and begin to learn how to think like a scientist.

The third category of questions takes the student out of the textbook and classroom and into the world at large. **Connecting the Science** questions ask about the broader implications of science—from ethical issues to personal decisions to political controversies. These questions provide additional support for the goal of connecting the material learned in class to students' daily lives and discussions.

Answers and suggested answers for Learning the Basics, Analyzing and Applying the Basics, and Stop and Stretch questions are found at the end of the book.

Organized to Reflect Modern Biology

The Human Genome Project and innovations in molecular biology have led to an explosion in our understanding of the role of genes in health and disease. Because students now need to understand the relationship between DNA, genes, proteins, and phenotypes when discussing physiology and medicine, we have placed molecular genetics in Chapter 4. We return to the topic of genes and their role in human systems in every subsequent physiology chapter, in our boxed Genes & Homeostasis essays.

In keeping with the text's emphasis on the interrelatedness of systems and levels of organization, we have integrated coverage of muscular and skeletal systems in a single chapter (Chapter 6). This combined chapter contains all of the important information about bone and muscle structure, function, and physiology, but also reflects the close association of these two systems—an association recognized in medicine by the field of orthopedics, the specialty concerned with disorders of the musculoskeletal system.

Comprehensive Supplements and Multimedia Package to Support the Main Text

For the Instructor

Teaching Toolbox (ISBN: 0-321-59020-1) This package includes everything a first-time professor might need to assist him or her in developing a course that will engage students and help them succeed. It also includes materials that experienced professors can use to inject excitement into lectures. The Teaching Toolbox contains all instructor resources listed below in an easy-to-carry container.

Instructor Resource Center on DVD for Human Biology This resource is designed to fully support individual teaching approaches and enhance lectures. The DVD includes the highest-quality multimedia assets available, including all art from the book in labeled and unlabeled JPEG format and label-edit PowerPoint versions, all photos and tables from the text, editable PowerPoint lecture presentation slides, PSR-enabled (clicker) questions, and human biology animations designed especially for your course. BioFlix animations present the most challenging biological content with visually stunning 3-D effects. BLAST! Tutorials provide additional animations and simulations, while Interactive Physiology for Human Biology helps your students visually comprehend complex processes. Videos from ABC News have been specially selected to correlate to the stories in this text and enhance lectures. In addition, each of the stories in the printed supplement *Scientific American*'s Current Issues in Biology are summarized in an individual PowerPoint presentation.

Instructor Guide This manual, written by Donald Glassman of Des Moines Area Community College, contains material that will help instructors craft a course to appeal to students and engage them on many levels. Its contents include full lecture outlines for each chapter, in-class and out-of-class activities for the student, discussion questions whose comprehensive answers offer information not found in the text, and an in-depth alternative storyline tailored to complement or function as an alternative to the stories that begin each chapter of the main text.

Test Bank This printed supplement, by Jill Feinstein of Richland Community College, offers over 1,600 questions that have been rigorously checked for accuracy and cross-referenced to the text. Question types include multiple choice, fill in the blank, short answer, and essay, as well as matching and labeling the art from the main text.

Computerized Test Bank for Human Biology The questions from the printed test bank are also available in this text-specific testing program that allows instructors to view and edit these test questions, add their own questions, and generate tests randomly to create an unlimited variety of quizzes and tests.

Human Biology Support Manual This support manual guides novice instructors through lectures for each chapter, suggesting multimedia and print supplements where appropriate to invigorate lectures with additional material. The manual contains sample syllabi for 10- and 16-week courses, teaching tips for new lecturers, a visual walk-through of lecture outlines for each chapter, and a primer on how to use the on-line course management system CourseCompass.

Course-at-a-Glance Guide for Human Biology This handy guide to planning a course highlights all the resources available to instructors, chapter by chapter and week by week, following a model of either a 10-week or 16-week syllabus.

Transparency Acetates for Human Biology The set of transparencies contains all photos, art, and tables from the book—more than 500 illustrations in all—in full color, with tabs separating each chapter.

Instructor Access Kit for CourseCompass This instructor package offers a nationally hosted course management system with state-of-the-art eLearning tools.

Course Management Systems In addition to CourseCompass, we also offer WebCT and Blackboard support for this course.

For the Student

Study Guide for Human Biology Written by Catherine Podeszwa, this printed supplement appeals to students with a variety of learning styles by offering in-depth, easy-to-follow chapter outlines and questions in an array of formats including crossword puzzles, table completions, art labeling, and sequencing. This guide also contains questions covering the "Roots to Remember" section found in each chapter, which encourages students to apply the tools they learn in this course across disciplines.

***Biological Explorations: A Human Approach* Lab Manual, 6th edition**
This manual, by Stanley E. Gunstream, provides 33 stimulating laboratory exercises for human biology. The level of rigor, user-friendly language, and abundant illustrations

make it ideal for beginning laboratory students. It places emphasis on the scientific method, and the background information that precedes each experiment prepares students for the activities to follow.

The Human Biology Place Web Site (www.humanbiologyplace.com)
This open-access web site includes objective-driven chapter guides to help students direct their learning, human biology animations with quizzes, extensive chapter quizzing, crossword puzzles, flashcards, and glossaries. A password-protected link offers instructor access to a variety of additional instructor resources.

Scientific American **Current Issues in Biology** These magazines are created specifically to accompany Benjamin Cummings textbooks. They contain articles from *Scientific American* magazine that address topics directly related to issues explored in our books. Answers to the questions that follow each article are found on the companion web site. Accompanying PowerPoint presentations can be found on the Instructor Resource Center on DVD.

Interactive Physiology for Human Biology Created specifically for students taking Human Biology, the Interactive Physiology for Human Biology CD helps them grasp tough topics with step-by-step tutorials.

CourseCompass Student Access Kit Provides full student access to an instructor's customized CourseCompass course, including interactive exercises, animations, news, and study tools.

About the Authors

Colleen Belk and Virginia Borden Maier have been colleagues and collaborators for almost 15 years. They have taught both majors and nonmajors biology courses at the University of Minnesota–Duluth and brought the innovations in teaching and learning used in this book to both types of classes. Colleen and Virginia are coauthors of *Biology: Science for Life*, a general biology text for nonmajors, and *Biology: Science for Life Laboratory Manual*.

Acknowledgments

We are indebted to our editor, Gary Carlson. Gary has been a joy to work with—calm, funny, patient, wise, and truly committed to producing an excellent book that meets the needs of students and instructors. We feel fortunate to have had his guidance throughout the project.

We also owe a great deal of thanks to our development editor, Don Gecewicz, a very careful reader and polymath who never failed to offer interesting insights and observations that developed the narratives and spurred critical thinking exercises. In addition to his editing prowess, Don's organizational skills were essential to ensuring that we were on top of all reviewer feedback and to keeping *Human Biology* on track and on time. Special thanks are due to Lisa Tarabokjia, editorial assistant, who managed and organized the reviews, and to Kaci Smith who worked with the supplements authors to develop an unusually effective set of support materials. Susan Malloy, our project editor, kept us on course throughout production and contributed to the design, accuracy, and overall effectiveness of the presentation. Catherine Podeszwa deserves special recognition for her work on the end of chapter "Analyzing and Applying the Basics" questions. Her research and writing allowed the development of questions that cover a wide range of subject areas with significant depth.

We are grateful to the authors of our supplements package, including Donald Glassman, Des Moines Area Community College, author of the Instructor Manual; Catherine Podeszwa, author of the Study Guide; and Jill Feinstein, Richland Community College, author of the Test Bank. In addition, Cherie McKeever, of Montana State University–Great Falls, reviewed final drafts of each chapter in the book and the entire test bank to ensure their accuracy. Clare Hays, of Metropolitan State College of Denver, reviewed most chapters of the book and reviewed the Instructor Manual, offering key insights into its development.

We also appreciate the careful review of the Study Guide for accuracy and usefulness by Celeste Humphrey, of Dalton State College; Mary Celeste Reese, of Mississippi State University; Louis A. Scala, of Passaic County Community College; and Jennifer Warner, of University of North Carolina–Charlotte.

We were very fortunate to have a large number of intelligent and thoughtful reviewers. These close readers were crucial to shaping and improving the text. This book is much better for their efforts. A number of reviewers met with us at a weekend focus group in San Francisco and we are grateful to them for taking time out of their schedules to help us make decisions on features, design, and art and media programs.

Reviewers

Mike Aaron, *Shelton State Community College*

John Aliff, *Georgia Perimeter College*

Laura Ambrose, *Luther College*

Saeid Baki-Hashemi, *Southwest Tennessee Community College*

Tamatha Barbeau, *Francis Marion University*

Erwin Bautista, *University of California–Davis*

Ken Belanger, *Colgate University*

Robert Boyd, *Auburn University*

Mark Condon, *Dutchess Community College*

Karen Crawford, *St. Mary's College (Maryland)*

Susan Crowson, *Framingham State College*

Michael Dann, *Pennsylvania State University*

Christopher Easton, *State University of New York–Potsdam*

Lucinda Elliot, *Shippensburg University*

Beth Erviti, *Greenfield Community College*

Jill Feinstein, *Richland Community College*

Laurine Ford, *Inver Hills Community College*

Glenn M. Fox, *Jackson Community College*

Kathy Germain, *Southwest Tennessee State University*

Eric Gillock, *Fort Hays State University*

Donald Glassman, *Des Moines Area Community College*

Andrew Greene, *Ashland University*

Christine Griffiths, *Nova Southeastern University*

Esther Hager, *Passaic County Community College*

Martin Hahn, *William Patterson University*

Elizabeth Harris, *Duke University*

Clare Hays, *Metropolitan State College of Denver*

Celeste Humphrey, *Dalton State College*

Janice Ito, *Leeward Community College*
Patricia Jaacks, *Jefferson Community College*
R. Kent Johnson, *De Anza College*
Donald Johnston, *Clinton Community College*
Donna Julseth, *Des Moines Area Community College*
Nick Kapp, *Skyline College*
Sherenne D. Kevan, *Wilfrid Laurier University*
Joseph S Keyes, *Sinclair Community College*
Mary Krawczyk, *Arapahoe Community College*
Rene LeBlanc, *Framingham State College*
William J. MacKay, *Edinboro University of Pennsylvania*
Brian R. Maricle, *Fort Hays State University*
Cherie McKeever, *Montana State University–Great Falls*
Deborah Merritt, *Marshall University*
Nancy H. Miller, *Diablo Valley College*
Stacia Moffett, *Washington State University*
Mary Beth Mullen, *Wilkes University*
Glenda H. Orloff, *Berry College*
Phillip Ortiz, *Empire State College*
Thomas Lon Owen, *Northern Arizona University*
Nancy Pencoe, *University of West Georgia*
Mason Posner, *Ashland University*
Katherine Rasmussen, *University of South Dakota*
Mary Celeste Reese, *Mississippi State University*
Laura S. Rhoads, *State University of New York–Potsdam*
Bernice Richards, *Northern Essex Community College*
John Rinehart, *Eastern Oregon University*
Edward Saiff, *Ramapo College of New Jersey*
Hildy Sanders, *Villa Julie College*
Mary Ann Satler, *St. Charles Community College*
Louis A. Scala, *Passaic County Community College*
Emily Schmitt, *Nova Southeastern University*
Beverly Sher, *College of William and Mary*
Dann Siems, *Bemidji State University*
Marc A. Smith, *Sinclair Community College*
Charlotte Spencer, *University of Alberta*
Charles Toth, *Providence College*
Michael Troyan, *Pennsylvania State University*
Jan Trybula, *State University of New York–Potsdam*
Miryam Z. Wahrman, *William Patterson University*
Jennifer Warner, *University of North Carolina–Charlotte*
Nicole Watson, *Minneapolis Community and Technical College*
Robert Wiggers, *Stephen F. Austin State University*
Richard Worthington, *University of Texas–El Paso*

Stephen Wright, *Carson Newman College*

John Zenger, *Brigham Young University–Idaho*

Focus Group Attendees

Jill Feinstein, *Richland Community College*

Donald Glassman, *Des Moines Area Community College*

Celeste Humphrey, *Dalton State College*

Nancy H. Miller, *Diablo Valley College*

Thomas Lon Owen, *Northern Arizona University*

Mary Celeste Reese, *Mississippi State University*

Louis A. Scala, *Passaic County Community College*

Dann Seims, *Bemidji State University*

Jennifer Warner, *University of North Carolina–Charlotte*

Stephen Wright, *Carson Newman College*

Class Testers

Diane Anderson, *Washtenaw Community College*

Dorothy Anderson, *Keystone College*

Isaac Barjis, *New York City College of Technology*

Deborah S. Barry, *Marshall University*

Susan Capasso, *St. Vincent's College*

William Cushwa, *Clark College*

Charles J. Dick, Jr., *Pasco Hernando Community College, North Campus*

Deborah Dodson, *Ohio University*

Douglas P. Easton, *Buffalo State College, SUNY*

Ingebord Eley, *Hudson Valley Community College*

Mary Ethington, *Genesee Community College*

Robert Farrell, *Pennsylvania State University–York*

Brian D. Feige, *Mott Community College*

Steven Fenster, *Ashland University*

Joseph D. Gar, *West Kentucky Community and Technical College*

Andrew Greene, *Ashland University*

Esta Grossman, *Washtenaw Community College*

Celeste Humphrey, *Dalton State College*

Allen Hunt, *Elizabethtown Community and Technical College*

Alison Jassen, *Northwestern Connecticut Community College*

Lori Ann Jukofsky, *Montclair State University*

Gwendolyn M. Kinebrew, *John Carroll University*

Elizabeth F. Kisielewski, *Pennsylvania State University–Harrisburg*

June Klingler, *Phoenix College*

Kathleen Klueber, *Spalding University*

Kent Koerner, *Indiana State University*

Lee Kurtz, *Georgia Gwinnett College*

Suzanne Long, *Monroe Community College*

Kate Lormand, *Arapahoe Community College*
Paul Marshall, *Northern Essex Community College*
Deborah McCool, *Pennslyvania State University–Altoona*
Pamela Monaco, *Molloy College*
John Morris, *New Hampshire Community Technical College–Laconia*
Sarah Myer, *Oral Roberts University*
Deb Olander, *Inver Hills Community College*
Joseph Orgel, *Illinois Institute of Technology*
Thomas Price, *Fairfield University*
Mary Celeste Reese, *Mississippi State University*
Eileen Roark, *Central Connecticut State University*
Nick Roster, *Northwestern Michigan College*
Maureen Sanz, *Molloy College*
Mark Schlueter, *Georgia Gwinnett College*
Maureen Scott, *Norfolk State University*
Terry R. Shank, *Marshall University*
Jennifer Sherwood, *Holy Names University*
Craig M. Story, *Gordon College*
Ken Thomas, *Northern Essex Community College*
Michael Troyan, *Pennsylvania State University–University Park*
Jagan Valluri, *Marshall University*
Jennifer Warner, *University of North Carolina–Charlotte*
Miryam Z. Warhman, *William Paterson University of New Jersey*
M. Eva Weicker, *Alvernia College*
David M. Woods, *Tidewater Community College*
Stephen Wright, *Carson Newman College*
Eugenia Zavras, *Fairfield University*

HUMAN BIOLOGY

Chapter 1

The Scientific Method

Proven Effective

LEARNING GOALS

1. Describe the characteristics of a scientific hypothesis and give examples of ideas that cannot be tested scientifically.
2. Compare and contrast the terms *scientific hypothesis* and *scientific theory*.
3. Distinguish between inductive and deductive reasoning.
4. Why can't the truth of a hypothesis be proven conclusively via deductive reasoning?
5. Describe the features of a controlled experiment and explain how these experiments eliminate alternative hypotheses for the results.
6. How do scientists minimize bias in experimental design?
7. Define *correlation* and explain the benefits and limitations of using this technique to test hypotheses.
8. Describe the information that statistical tests provide.
9. List the factors that can influence whether or not an experiment will return statistically significant results.
10. Summarize the techniques you can use to evaluate scientific information from secondary sources.

How to start a healthy day?

For many of us, the beginning of a semester represents an opportunity for a fresh start, a time to make some resolutions to improve ourselves. Maybe you resolved to adopt healthier personal habits—a better sleep schedule, more exercise, and a healthier diet. Now comes the next step—putting those resolutions into action.

On the first day of classes, you walk into the dining center for a healthy breakfast. What should you choose? A bowl of oatmeal? You have heard oat bran is good for you, but what about all those carbs? How about some protein-rich scrambled eggs? Oh, but they are full of cholesterol. How about a chocolate doughnut? Chocolate is supposed to contain lots of heart-healthy flavonoids, but somehow a doughnut does not seem like a healthy choice. Frustrated, you grab a glass of orange juice. At least the vitamin C will protect you from getting sick—right?

Many people have had this experience. It can be difficult to know how to eat healthy if you follow health news. It seems as though scientists cannot make up their minds about the links between diet and health. One week a news story reports that, for instance, drinking caffeinated coffee carries some health risks, and a week later, another story details the positive effects of caffeine on health. But you also know that we rely on the knowledge gained by science. Information gathered by scientific research has direct effects on our everyday lives—even activities as basic as what we choose to eat. How do we know which ideas are correct?

You can approach questions about eating healthy, about the risk of global warming, about the debate over teaching evolution in schools, and many others with more confidence if you have a solid understanding of the process of science. In this chapter, we introduce you to the process of science and help you understand how to evaluate scientific claims by looking at the relationship between diet and health.

Oatmeal has fiber, which is good. But carbs—aren't they bad?

Eggs have protein—that's good. But cholesterol—isn't that bad?

Orange juice with vitamin C. Is that a good choice? How does anyone know?

1.1 An Introduction to the Scientific Process

The term *science* can refer to a body of knowledge—for example, the science of **biology** is the study of living organisms. Your impression of science may be that it requires near-perfect recall of specific sets of facts about the world. In reality, this goal is impossible, and unnecessary—we do have reference books, after all. The real action in science is not in memorizing what is already known about the world, but in using the process of science to discover something new and unknown about subjects that fascinate us.

In fact, like all humans, you are already a scientist. Think about how you solve many of the dilemmas in your life. Once you identify a problem or difficulty, you think of several possible causes, and then you try to solve it by addressing what you think may be the most likely cause. Suppose that you have been feeling especially tired lately. To regain your energy, you would try to identify the factor that is causing your fatigue. And just as a scientist would do, if your first solution does not work, you would move on to another possible solution. If changing your diet does not result in an improvement in your energy level, you might try changing your work schedule.

This process—making observations of the world, proposing ideas about how something works, testing those ideas, and discarding (or modifying) our ideas in response to the results of a test—is the essence of the **scientific method**. The scientific method allows us to solve problems and answer questions efficiently and effectively. Can we use the scientific method to solve the complicated problem of choosing a healthful diet?

The Nature of Hypotheses

Your food choices are partly based on your ideas about how various foods affect your body's function. When you choose a glass of vitamin C–rich orange juice for breakfast, it may be because you believe the following statement to be true: Vitamin C boosts the function of the immune system and protects us from colds and the flu. This statement is called a hypothesis.

In science, a **hypothesis** is a tentative fact or a proposed explanation for one or more observations. Hypotheses in biology are generated from knowledge about how the body and other biological systems work, experiences in similar situations, familiarity with other scientific research, and logical reasoning. Hypotheses are also shaped by the creative mind (FIGURE 1.1).

The hallmark of science is that hypotheses are subject to rigorous evaluation. Therefore, scientific hypotheses must be **testable**—it must be possible to evaluate a hypothesis through observations of the measurable universe. Not all hypotheses are testable. For

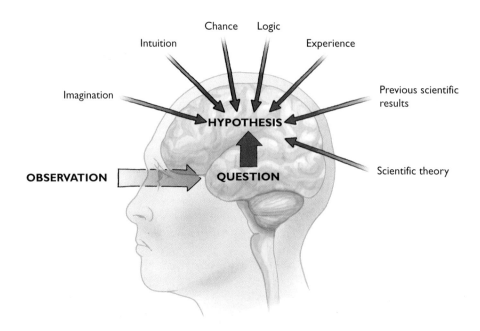

FIGURE 1.1 Hypothesis generation. All of us generate hypotheses. Many different factors, both logical and creative, influence the development of a hypothesis. Scientific hypotheses are both testable and falsifiable.

instance, the statement that "physical health is maintained by a proper balance of psychic energy" is not a scientific hypothesis because psychic energy cannot be seen or measured. "Psychic energy" does not have a material nature and therefore cannot be put to a test. In addition, hypotheses that require the intervention of a supernatural force cannot be tested scientifically. If something is **supernatural**, it is not constrained by the laws of nature, and its behavior cannot be predicted using our current understanding of the natural world.

A scientific hypothesis must also be **falsifiable**; that is, an observation or set of observations could potentially prove it false. The hypothesis that fruit consumption promotes good physical health is falsifiable; the observation that people with high fruit consumption are not measurably healthier than people with low fruit consumption would cause us to reject the hypothesis. Of course, not all hypotheses are proved false, but it is crucial to scientific progress that incorrect ideas are discarded. And discarding faulty ideas can occur only if they are falsifiable. A lack of falsifiability is another reason supernatural explanations are not scientific. Because a supernatural force can cause any possible result of a test, hypotheses that rely on supernatural forces cannot be falsified.

Finally, statements that are value judgments, such as "It is always wrong to eat meat," are not scientific, because different people have different opinions about right and wrong. It is impossible to falsify these types of statements. To find answers to questions of morality, ethics, or justice, we turn to other methods of gaining understanding—such as philosophy and religion.

Consider the following nonscientific hypothesis: "God exists and influences events on Earth." Explain why no observation or set of observations can falsify this hypothesis.

Scientific Theories

Most hypotheses fit into a larger picture of scientific understanding. We can see this relationship when examining how research upended a commonly held belief about diet and health—that chronic stomach and intestinal inflammation is caused by eating too much spicy food. This belief directed the standard medical practice for ulcer treatment. Patients were prescribed drugs that reduced stomach acid levels and advised to avoid eating acidic or highly spiced foods. These treatments were rarely successful, and ulcers were considered chronic, possibly lifelong problems.

In 1982, Australian scientists Robin Warren and Barry Marshall discovered that a bacterium, later named *Helicobacter pylori*, was present in nearly all samples of ulcer tissue that they examined (FIGURE 1.2). From this observation, Warren and Marshall reasoned that infection with *H. pylori* was the cause of most ulcers. Barry Marshall even tested this hypothesis on himself, by consuming a sample of *H. pylori* and subsequently suffering from acute gastric distress.

Warren and Marshall's colleagues were at first unconvinced that ulcers could have such a simple cause. Today, the hypothesis that *H. pylori* infection is responsible for most ulcers is accepted as fact. The primary reasons? (1) All reasonable alternative hypotheses about the causes of ulcers (for instance, consumption of highly spiced foods) have not been supported by numerous experimental tests; and (2) the hypothesis has not been rejected after carefully designed experiments demonstrated that eradication of *H. pylori* in patients cured most ulcers.

The third reason that the relationship between *H. pylori* and ulcers is considered fact is that it conforms to a well-accepted scientific principle, the germ theory of disease. A **scientific theory** is an explanation of a set of related observations based on well-supported hypotheses from several different, independent lines of research. The

(a)

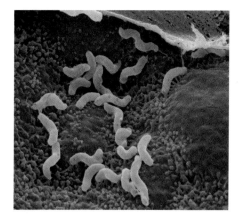

(b)

FIGURE 1.2 **A scientific breakthrough.**
(a) *Helicobacter pylori* (shown in green) (b) Robin Warren and Barry Marshall won the 2005 Nobel Prize in Medicine for their discovery that this organism is the causative agent of nearly all cases of stomach and intestinal ulcers. Their story is a dramatic example of how science progresses by the rejection of incorrect hypotheses.

germ theory arose in the early twentieth century from the accumulated observations of biologists such as Louis Pasteur and Robert Koch. These scientists first noted the relationship between particular diseases and particular microorganisms (that is, organisms too small to be seen with the naked eye). The basic premise of the germ theory is that microorganisms are the cause of many human diseases, through infection.

Pasteur observed that bacteria cause milk to become sour. From this observation, he reasoned that these same types of organisms could injure humans. Koch's demonstration of the link between anthrax bacteria and a specific set of fatal symptoms in mice provided additional evidence for this theory. The germ theory is further supported by the observation that antibiotic treatment that targets particular microorganisms can relieve patients of certain illnesses—as is the case with bacteria-caused ulcers.

In everyday speech, the word *theory* is synonymous with "untested hypothesis." In contrast, scientists use the term when referring to explanatory models about how the natural world works.

The supporting foundation of all scientific theories is multiple hypothesis tests. The germ theory has been supported by numerous tests of hypotheses suggested by the theory—for instance, observations that infectious human diseases ranging from chicken pox to malaria to tuberculosis to yeast infections have all been clearly associated with infection with viruses, protozoans, bacteria, and fungi. The theories of evolution and natural selection, which form a basis for our understanding of all of biology, are also well supported by hypothesis tests and are summarized briefly below.

The Theory of Evolution and the Theory of Natural Selection

The *theory of evolution* states that all modern organisms are descended from a common ancestor that existed in the distant past. Since the appearance of this common ancestor over 3 billion years ago, organisms have been diverging from each other, but they still show evidence of relationship. The evidence for the theory of evolution is thus present in the similarities among living things and in the pattern of shared traits among groups of organisms (FIGURE 1.3).

The evidence for evolution is detailed in Chapter 22; however, basic knowledge of evolutionary relationships helps us understand certain unique human conditions. For instance, the high frequency of back injury in adults is related to the less-than-optimal design of our spine, which was adapted from creatures that walked on all four limbs. Evolutionary relationships are also what make nonhuman animals reasonable substitutes for human subjects when testing the effects of drugs, the outcome of new surgical procedures, and the effect of genes on our health.

A major process by which organisms evolve is natural selection. The *theory of natural selection* is discussed in detail in Chapter 22, but the basic principle is simple: Individual organisms vary from each other and some of these variations increase their chances of survival and reproduction. Any trait that increases survival and reproduction, as long as it has a genetic basis, should become more prevalent in a population as those individuals contribute more offspring to the next generation. In contrast, less successful variants should eventually be lost from the population. The theory of natural selection helps us understand a wide variety of human traits, from our relative hairlessness to our resistance to certain pathogens and susceptibility to other diseases.

Both evolution and natural selection are well-developed theories with abundant observational and experimental support. These theories remain important both because they can help generate hypotheses about human biology and because they have withstood nearly 150 years of hypothesis tests without being disproved.

The Logic of Hypothesis Testing

To understand how hypotheses are tested, let's return to our question about the relationship between diet and health. A student thinking that a chocolate doughnut may be good for him recalls a hypothesis that has received a lot of press recently—that dark

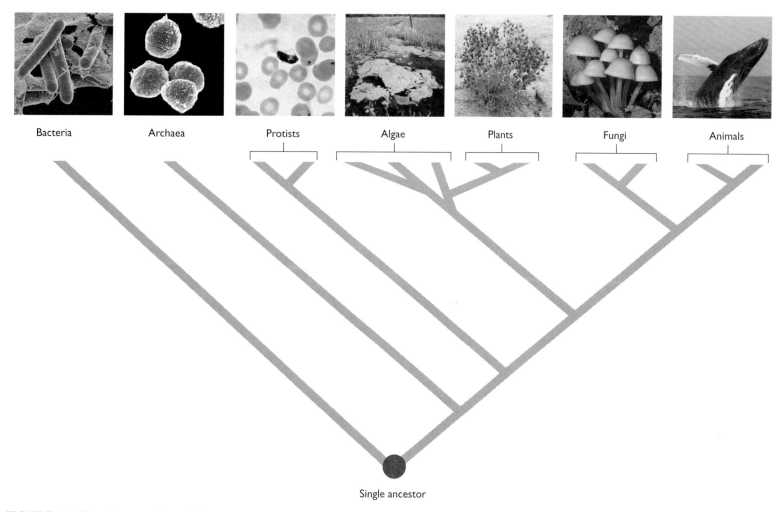

FIGURE 1.3 **The theory of evolution.** All modern organisms derive from a common ancestor that existed billions of years ago.

chocolate can be a health food. Surprisingly, this hypothesis makes logical sense, given a few observations, namely:

1. Dark chocolate is at least 25% cocoa.
2. Cocoa is rich in compounds called flavonoids.
3. Flavonoids are antioxidants, which reduce the risk of damage to cells and tissues.
4. Studies have shown that flavonoid consumption reduces the risk of cardiovascular (that is, heart and blood vessel) disease.

Given these facts, we can state the following falsifiable hypothesis:

Consuming dark chocolate reduces the risk of cardiovascular disease.

The process used to construct this hypothesis is called **inductive reasoning**, which means combining a series of specific observations (here, statements 1–4) to discern a general principle. The set of observations certainly seems like enough information on which to base a decision—start eating dark chocolate if you want to keep your heart and blood vessels healthy. However, a word of caution is in order: Just because an inductively constructed hypothesis makes sense does not mean that it is true. The following example clearly demonstrates this point.

Consider the ancient hypothesis, asserted by Aristotle, that the sun revolves around Earth. Aristotle proposed this hypothesis based on the observation that the sun rose in the east and set in the west every day. For 2,000 years, this hypothesis was considered to be a "fact" by nearly all of Western society. To most people, the hypothesis made perfect sense, especially since it seemed that the Earth was the center of the universe and surrounded by the heavens. It was not until the early seventeenth century that Aristotle's hypothesis was falsified as the result of Galileo's observations of Venus. His measurements helped to confirm the more modern hypothesis, proposed by Nicolaus Copernicus, that Earth revolves around the sun.

So even though the hypothesis about dark chocolate is sensible, it needs to be tested. Hypothesis testing requires a process called **deductive reasoning** or deduction. Deduction involves using a general principle to predict an expected observation. This **prediction** concerns the outcome of an action, test, or systematic investigation. In other words, the prediction is the result we expect from a test of the hypothesis.

Deductive reasoning takes the form of "if/then" statements. If our general principle is correct, then we would expect to observe a specific set of outcomes. A prediction based on the dark-chocolate hypothesis could be: If consuming dark chocolate decreases the risk of cardiovascular disease, then people who eat dark chocolate will experience fewer heart attacks than people who do not eat dark chocolate.

Deductive reasoning is a powerful method for testing hypotheses. However, the structure of an if/then statement means that hypotheses can be clearly rejected if untrue, but almost impossible to prove if true (FIGURE 1.4).

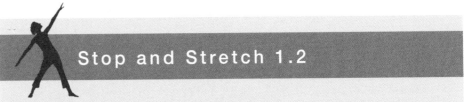

Stop and Stretch 1.2

Consider this if/then statement: If you ate all the chocolate doughnuts, then there won't be any doughnuts left in the box. Imagine I find that the doughnuts are all gone—did I just prove that you ate them?

Consider the possible outcomes of a comparison between people who eat dark chocolate and those who do not. People who eat chocolate may experience more heart attacks than people who do not. They may have the same number of heart attacks as people who avoid chocolate. Or, chocolate-eaters may in fact experience fewer heart attacks. What do these results tell us about the hypothesis?

If people who eat dark chocolate have the same number of heart attacks or more compared to those who do not eat it, the hypothesis that dark chocolate alone provides protection against cardiovascular disease can be rejected. But what if people who eat chocolate do experience fewer heart attacks? In this case, we can only say that the hypothesis has been supported and not disproved. Why? Primarily because there are *alternative hypotheses* that could explain why people with different chocolate-eating habits have different levels of cardiovascular disease.

Consider the alternative hypothesis that lower stress levels reduce the risk of heart disease. And also suppose that people who eat chocolates are more relaxed than those who do not eat them. If both of these hypotheses are true, the prediction that chocolate-eaters have healthier hearts than chocolate-abstainers would be true, but not because the original hypothesis (that chocolate itself reduces the risk of heart attacks) is true. Instead, people who eat chocolate would experience fewer heart attacks than people who do not because they are also less stressed, and low stress means better overall heart health.

A hypothesis that seems true because it has not been rejected by an initial test may be rejected later because of a different test. This is what happened to the hypothesis that vitamin C consumption reduces susceptibility to illness. The argument for the power of vitamin C was popularized in 1970 by Nobel Prize–winning chemist Linus Pauling in his book *Vitamin C and the Common Cold*. Pauling based his assertion—that large doses of vitamin C reduce the incidence of colds by as much as 45%—on the results of a few studies. However, repeated, careful tests of this hypothesis have since failed to support it. In many of the studies Pauling cited, it appears that one or more alternative hypotheses explain the difference in cold incidence between individuals with high–vitamin C diets and those with low–vitamin C diets. Today, most researchers studying the common cold agree that the hypothesis that vitamin C prevents colds has been convincingly falsified.

FIGURE 1.4 **The scientific method.** Tests of hypotheses follow a logical path. This flowchart illustrates the process.

The example of the vitamin C hypothesis illustrates the challenge of communicating scientific information to the general public. You can see why the belief that vitamin C prevents colds is so widespread. A book by a Nobel Prize–winning scientist may have seemed like the last word on the role of vitamin C in preventing illness. However, scientific knowledge progresses by the rejection of incorrect ideas, without regard to who proposed those ideas. After many years of research, it is clear that Pauling's "last word" was wrong.

1.2 Hypothesis Testing

The previous discussion may seem discouraging—how can scientists determine the truth of any hypothesis when there is always a chance that the hypothesis could be falsified? Even if one of the hypotheses about the relationship between diet and health is supported, does the existence of alternative hypotheses mean that we will never know which diet is truly healthiest? The answer is yes—and no.

Hypotheses cannot be proven absolutely true. It is always possible that the true cause of a particular phenomenon may be found in a hypothesis that has not yet been tested. However, in a more practical sense, a hypothesis can be proven beyond a reasonable doubt. That is, when one hypothesis has not been disproved through repeated testing and all reasonable alternative hypotheses have been eliminated, scientists accept that the well-supported hypothesis is, in a practical sense, true.

"Truth" in science can therefore be defined as what we know and understand based on all currently available information. If a hypothesis or theory appears to explain all instances of a particular phenomenon and has been repeatedly tested and supported, it may eventually be accepted as accurate.

One of the most effective ways to test hypotheses is through rigorous scientific experiments.

The Experimental Method

Experiments are sets of actions or observations designed to test specific hypotheses. Generally, an experiment allows a scientist to control some of the conditions that may affect the subject of study. Manipulating the environment allows a scientist to eliminate alternative hypotheses that may explain the result.

Experimentation in science is analogous to what a car mechanic does when he diagnoses a problem. If a car's engine does not turn over, the problem could be the battery. If replacing the battery fails to solve the problem, the next step might be to replace the starter motor. If the mechanic had started by replacing both, he would not know which component had caused the problem, and he would have an unhappy customer who was charged for both parts. Likewise, a scientist who changes many factors and sees a different experimental result cannot determine which of the changes she made was the one that made the difference.

Not all scientific hypotheses can be tested through experimentation. For instance, hypotheses about historical events, such as those concerning how life on Earth originated or the cause of dinosaur extinction, are usually not testable in this way. These hypotheses must instead be tested via careful observation of the natural world. For instance, the examination of fossils and other evidence preserved in rocks allows scientists to test hypotheses regarding the extinction of the dinosaurs.

The observations and information collected by scientists during any type of hypothesis testing are known as **data**. The data are collected on the **variables** of the test, that is, any factor that can change in value under different conditions. In an experimental test, scientists manipulate an **independent variable** (one whose value can be freely changed) in order to measure the effect on a **dependent variable**. The dependent variable may or may not be influenced by changes in the independent variable. Data obtained from well-designed experiments should readily allow researchers to either reject or support a hypothesis. This is more likely to occur if the experiment is controlled.

Controlled Experiments

The most unambiguous support for a theory comes in the form of hypothesis tests called **controlled experiments**. Control has a very specific meaning in science. A **control** for an experiment is a subject similar to an experimental subject, except that the control is not exposed to the experimental treatment. If the control and experimental groups differ at the end of a well-designed test, then any difference between the groups is likely due to the experimental treatment. In other words, the independent variable is the experimental treatment, and the dependent variable is the response of each group to the presence (or absence) of that treatment. A good control eliminates differences between experimental groups in as many independent variables as possible.

Let's return to the hypothesis discussed earlier in the chapter: Consuming dark chocolate reduces the risk of cardiovascular disease. Although the case for this hypothesis is reasonable, it is experimental tests that will allow scientists to prove (or refute) it beyond a reasonable doubt. Scientists at the University of California, San Francisco, recently performed such an experiment. In their experiment, volunteers who consumed chocolate bars containing large amounts of flavonoids had a 21% improvement in one measure of cardiovascular function. The "21% improvement" was in comparison to the control group—volunteers who ate chocolate with few flavonoids.

Control and experimental groups are consciously designed to eliminate as many alternative hypotheses for the results as possible. The first step is to select a pool of subjects in such a way as to eliminate differences between the two groups in a variety of independent variables, such as participants' ages, diets, weight, and overall health. One effective way to minimize differences between the groups is the **random assignment** of individuals to experimental and control groups. For example, a researcher might put all of the volunteers' names in a hat, draw out half, and designate these people as the experimental group and the remainder as the control group. As a result, there is unlikely to be a systematic difference between the experimental and control groups—each group should be a rough cross section of the population in the study. In the chocolate experiment, members of both the experimental and control groups were randomly assigned students and staff at the University of California, San Francisco, who responded to a newspaper ad calling for volunteers.

The second step in designing a good control is interacting with all subjects identically during the course of the experiment. In this study, all participants received the same information about the purported benefits of chocolate consumption, and during the course of the experiment, all participants were given small bars of chocolate to consume each day over the course of two weeks. However, individuals in the control group received chocolate bars that contained very few flavonoids. The difference in flavonoid content is not noticeable to subjects in the experiment, because these chemicals do not noticeably affect taste or texture. Treating all subjects equally ensures that no factor related to the interaction between subject and researcher influences the final results.

The low-flavonoid chocolate in our experiment is equivalent to the "sugar pills" that are given to control subjects when testing a particular drug. Like other intentionally ineffective medical treatments, low-flavonoid chocolates and sugar pills are called **placebos**. Employing a placebo generates only one consistent difference between individuals in the two groups—in this case, the amount of flavonoids they consumed.

In the chocolate study, the data indicated that one measure of cardiovascular function—arterial flexibility—was higher in the experimental group compared to those who received the placebo. Because their study used controls, the researchers can be confident that the groups differed because of the effect of high-flavonoid chocolate. By reducing the likelihood that alternative hypotheses could explain their results, the researchers could strongly infer that they were measuring a real effect of chocolate on cardiovascular health (FIGURE 1.5).

The study described here supports the hypothesis that consuming dark chocolate reduces the risk of cardiovascular disease. However, it is extremely rare that a single experiment will cause the scientific community to accept a hypothesis as "true beyond a reasonable doubt." Only a small number of studies, each using different experimental designs and often evaluating different measures of heart health, have investigated the

Control group	Experimental group
Healthy adult men and women	Healthy adult men and women
Volunteered for study	Volunteered for study
Received **placebo** chocolate	Received **high-flavonoid** chocolate

FIGURE 1.5 **A controlled experiment.** In the chocolate experiment, individuals in both groups were treated identically except for the type of chocolate they consumed.

benefits of dark-chocolate consumption. Some of these studies have shown a positive effect, but others have shown none. In the medical community as a whole, the jury is still out regarding the role of chocolate in a healthy diet. Only through continued controlled tests of the hypothesis will an accurate answer to the question, "Can chocolate be good for your heart?" be discerned.

Minimizing Bias in Experimental Design

Scientists and human research subjects may have strong opinions about the truth of a particular hypothesis even before it is tested. These opinions may cause participants to unfairly influence, or **bias**, the results of an experiment. Minimizing the likelihood of biased experimental results can increase our confidence in a hypothesis test.

One potential source of bias is subject expectation. Individual experimental subjects may consciously or unconsciously display the behavior they feel the researcher expects. For example, an individual who knew she was receiving high-flavonoid chocolate may have felt certain that she was improving her health. This might cause her to relax during the test of artery function, which could improve her results. This potential problem is avoided by designing a *blind experiment*, wherein individual subjects are not aware of exactly what they are predicted to experience. In experiments on drug treatments, this means not telling participants if they are receiving the drug or a placebo.

Another source of bias arises when a researcher makes consistent errors in the measurement and evaluation of results. This phenomenon is called *observer bias*. In the chocolate experiment, observer bias could take various forms. Expecting a particular outcome might lead a scientist to treat patients differently, making those in the experimental group feel more relaxed and those in the control group feel more anxious. These differences in psychological state could influence the volunteers' physiology. Or, if the researcher expected people who ate high-flavonoid chocolate to have improved artery function, she might make small errors in the measurement of that function that influenced the final result.

To avoid the problem of experimenter bias, the data collectors themselves should be "blind." Ideally, the scientist, doctor, or technician applying the treatment does not know which group (experimental or control) any given subject is part of until after all data have been collected and analyzed (FIGURE 1.6). Blinding the data collector ensures that the data are objective, or in other words, without bias.

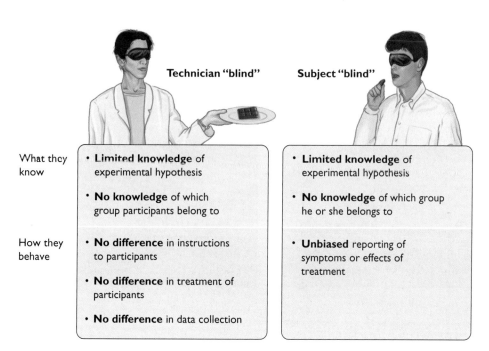

FIGURE 1.6 **Double-blind experiments.**
Double-blind experiments result in more objective data.

We call experiments **double-blind** when both the research subjects and the technicians performing the measurements are unaware of either the hypothesis or whether a subject is in the control or experimental group. Double-blind experiments nearly eliminate the effects of human bias on results. When both researcher and subject have few expectations about the outcome, the results obtained from an experiment should be considered more credible.

Using Correlation to Test Hypotheses

Double-blind, placebo-controlled, randomized experiments represent the gold standard for medical research. However, well-controlled experiments can be difficult to perform when humans are the experimental subjects. The requirement that both experimental and control groups be treated nearly identically means that some people receive no treatment. In the case of healthy volunteers who have little risk of severe cardiovascular problems, the low-flavonoid placebo treatment did not hurt those who received it.

However, placebo treatments are impossible or unethical in many cases. For instance, imagine testing the effectiveness of a birth control drug using a controlled experiment. This would require asking women to take a pill that may or may not prevent pregnancy while not using any other form of birth control!

Experiments on Model Organisms Scientists can use **model organisms** as stand-ins for people when testing hypotheses that might raise practical or ethical concerns if performed on humans. In the case of research on human health and disease, model organisms are typically other mammals. Mammals are especially useful as model organisms in medical research because they are closely related to us evolutionarily. Like us, they have hair and give birth to live young, and thus they share with us similarities in anatomy and function.

The vast majority of animals used in biomedical research are rodents such as rats, mice, and guinea pigs, although some areas of research require animals that are more like humans in size, such as dogs or pigs, or share a closer evolutionary relationship, such as chimpanzees (FIGURE 1.7).

(a) Rat

(b) Dog

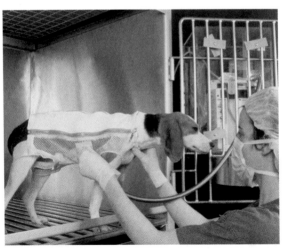

(c) Chimpanzee

FIGURE 1.7 Model organisms in science. (a) The classic "lab rat" is easy to raise and care for and has little genetic diversity to confound the results of an experiment. (b) Dogs have cardiovascular systems very similar to those of humans. (c) Chimpanzees are human beings' closest biological relatives.

The use of model organisms allows experimental testing of drugs and other therapies before these methods are employed on people. Research on model organisms has contributed to a better understanding of nearly every serious human health threat, including cancer, heart disease, Alzheimer's disease, and AIDS. However, ethical concerns about the use of animals in research persist and can complicate such studies. Also, the results of animal studies are not always directly applicable to humans—despite a shared evolutionary history, experimental animals still can have important functional differences from humans. Testing hypotheses about human health in human beings still provides the clearest answer to these questions.

Looking for Relationships Between Factors Another method that scientists can use when controlled experiments on humans are difficult or impossible to perform is to test hypotheses using correlations. A **correlation** is a relationship between two variables. Suggestions about increasing fruit and vegetable consumption to improve heart health are based on a correlation between diet and levels of artery-clogging cholesterol (FIGURE 1.8). Researchers who collected data on dozens of individuals' diets and their cholesterol levels generated this correlation. The researchers had no influence on the diet of the study participants—in other words, this is not a controlled experiment because people were not randomly assigned to different treatments of low or high produce consumption.

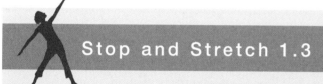

Stop and Stretch 1.3

Even though the researchers didn't manipulate the independent variable in the study comparing fruit intake and cholesterol levels, there are still an independent variable and a dependent variable. What are they?

Let's examine the results of this study, presented in Figure 1.8. The horizontal axis of the graph, or ***x*-axis**, illustrates the independent variable, a scale of fruit and vegetable consumption—from low on the left edge of the scale to high on the right. The vertical axis of the graph, the ***y*-axis**, is a scale for the dependent variable, which is the average level of LDL cholesterol measured in blood samples of the participants in each category. LDL (low-density lipoprotein) cholesterol is commonly known as

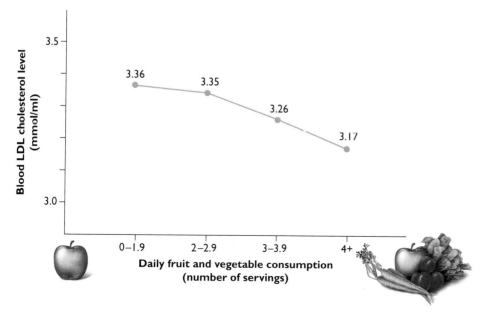

FIGURE 1.8 **Correlation between diet and cholesterol level.** This graph summarizes the results of an experiment that compared levels of "bad" LDL cholesterol in groups of individuals with different self-reported fruit and vegetable consumption.

VISUALIZE THIS Imagine that researchers divided the category of individuals who ate 4+ servings of fruits and vegetables into additional groups, say individuals who ate 4–4.9 servings, those who ate 5–5.9 servings, and those who ate 6+ servings. If the correlation still held, what would the graph look like with these additional data points?

"bad cholesterol" because it can accumulate in and clog blood vessels, and thus it is a good proxy for heart health. The line connecting the four points on the graph illustrates the correlation between these two factors. Because the line falls to the right, there is a *negative correlation* between these two factors. In other words, these data tell us that as fruit and vegetable consumption increases, LDL cholesterol levels decline. But does this relationship mean that eating fruits and vegetables causes lower LDL cholesterol levels?

To conclude that produce consumption reduced cholesterol, we require the same assurances that are given by a controlled experiment. In other words, we must assume that the individuals measured for the correlation are similar in every way, except for their fruit and vegetable intake. Is this a good assumption? Not necessarily. Most correlations cannot control for alternative hypotheses, and thus other independent variables could be determining the results. For example, people who eat more fruits and vegetables may exercise more than people who eat less. Also, people who eat more fruits and vegetables may eat less fat. These differences among people who have different diets may also influence their cholesterol level (FIGURE 1.9). Therefore, even with a strong correlation between the two variables, we cannot strongly infer that fruit consumption alone causes a decrease in blood cholesterol. The same complication occurs with a *positive correlation*, where both variables increase together.

As you can see, it is difficult to demonstrate a cause-and-effect relationship between two variables simply by showing a correlation between them. In other words, correlation does not equal causation. For example, a commonly cited correlation exists between fat consumption and cancer rates. It is true that when comparing populations of different countries, as the percentage of fat in the diet increases, so does the cancer rate. But conflicting results among studies of this question have caused scientists to seriously question this relationship. Instead, both fat consumption and cancer rates may be related to wealth—wealthier individuals can afford diets higher in fat, may be exposed to more cancer-causing compounds, and are less likely to die from infectious diseases.

Correlational studies are the main tool of *epidemiology*, the study of the distribution and causes of diseases. The data about fruit and vegetable consumption and cholesterol level came from one type of epidemiological study—a cross-sectional survey. In this type of survey, many individuals are both tested for the presence of a particular condition and asked about their exposure to various factors. The limitations of cross-sectional surveys include the effect of subject bias and poor recall on the part of survey participants, in addition to all of the problems associated with interpreting correlations. Table 1.1 provides an overview of the other strategies epidemiologists employ to study the links between what we eat and our health.

(a) Does fruit and vegetable consumption cause low cholesterol?

(b) Or do factors associated with fruit and vegetable consumption cause low cholesterol?

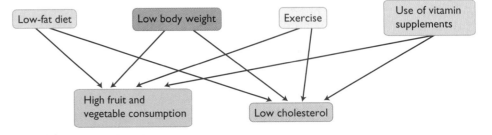

FIGURE 1.9 Correlation does not signify causation. A correlation typically cannot eliminate all alternative hypotheses.

TABLE 1.1 | Types of Epidemiological Studies

Name	Description	Pros	Cons
Ecological studies	Examine specific human populations for unusually high levels of various diseases (e.g., documenting a "cancer cluster" around an industrial plant).	Inexpensive and relatively easy to do	Unsure if exposure to environmental factor is actually related to onset of the disease
Cross-sectional surveys	Question individuals in a population to determine amount of exposure to an environmental factor and whether disease is present or not.	More specific than ecological study	Expensive Subjects may not know exposure levels Cannot control for other factors that may be different among individuals in survey Cannot be used for rare diseases
Case-control studies	Compare exposures to specific environmental factors between individuals who have a disease and individuals matched in age and other factors but who do not have a disease.	Relatively fast and inexpensive Best method for rare diseases	Does not measure absolute risk of disease as a result of exposure Difficult to select appropriate controls to eliminate alternative hypotheses Examines just one disease possibly associated with an environmental factor
Cohort studies	Follow a group of individuals, measuring exposure to environmental factors and disease prevalence.	Can determine risk of various diseases associated with exposure to particular environmental factor	Expensive and time-consuming Difficult to control for alternative hypotheses Not feasible for rare diseases

1.3 Understanding Statistics

During a review of scientific literature on diet and health, you may hear that a particular food item has a significant effect on some measure of health. This statement might cause you to sit up and take notice—after all, "significant" means important or meaningful. But in science, this isn't always the case. In scientific studies, "significance" has a different meaning. To understand this difference, we need a basic understanding of statistics.

What Statistical Tests Can Tell Us

Statistics is a specialized branch of mathematics used in the evaluation of data; the word is also used to describe the numerical data that have been collected.

A hypothesis test typically relies on a small subgroup, or **sample**, of a population. Statistics can be employed to summarize characteristics of the sample—for instance, the average blood cholesterol level of volunteers in the study. Statistical tests can then be used to extend the results from a sample to the entire population. When scientists conduct an experiment, they hypothesize that there is a true, underlying effect of their experimental treatment on the entire population. An experiment on a sample of a population can only estimate this true effect because a sample is always an imperfect snapshot of an entire population.

Statistical tests help scientists evaluate, given the number of individuals sampled and the variation among individuals, how likely it is that their snapshot provides an accurate picture of the whole population. Imagine trying to determine the average hair length of all women students in a class of 400 by taking a snapshot of a random subset of 10 women.

If hairstyles were identical among the women in the snapshot, you could reasonably assume that the average hair length in the class is approximately equal to the average hair length in the snapshot. However, you are much more likely to see that women in the snapshot have a variety of hairstyles. When you calculate the average hair length in the snapshot, it is unlikely to be identical to the average hair length you would calculate if you measured everyone in the entire class.

If the women in the snapshot exhibited a wide variety of hairstyles, from crew cuts to long braids, it would be more difficult to determine whether the average hair length in the snapshot is at all close to the average for the class. Your snapshot could, by chance, contain the one woman in the class with extremely long hair, causing the average length in this sample to be much longer than the average length for the class (FIGURE 1.10).

In the experiment with the high- and low-flavonoid chocolates, researchers observed a 21% increase in artery flexibility in the experimental group compared to the controls. By taking the variation in the results in the experimental and control groups into account, statistical tests tell us the likelihood that this improvement reflects the true effect of chocolate on cardiovascular health.

Statistical Significance: A Definition

A **statistically significant** result is one that is very unlikely to be due to chance differences between the experimental and control groups. We can explore the role that statistical tests play in determining significance by evaluating more closely the results of the chocolate study. When the data from this experiment were summarized, researchers observed that the 11 members of the experimental group had experienced a 12% increase in their arterial flexibility over the two weeks of the study, while the 10 members of the control group had experienced a 9% decline in flexibility. On the surface, this result supports the hypothesis. However, recall the snapshot of women's hair length. A statistical test is necessary because of the effect of chance.

The effect of chance on experimental results is known as **sampling error**—more specifically, sampling error is the difference between a sample of a population and the population as a whole. In any experiment, the group of individuals assigned to the experimental treatment and the group assigned to the control treatment will differ from each other in random ways. Even if there is no true effect of an experimental treatment, the results observed in the experimental and control groups will never be exactly the same.

FIGURE 1.10 **Statistics evaluate a snapshot of the population.** (a) If the snapshot, or sample, shows little variation among individuals, the average hair length in the snapshot is likely to be similar to the average hair length in the class. (b) If the snapshot shows large variation among individuals, it is difficult to determine whether or not the average hair length in the class is similar to the average of the sample.

(a) Sample is small but uniform, and thus likely to be a good representation of the class.

(b) Sample is small and variable, and thus may not be a good representation of the class.

For example, we know that people naturally differ in their cardiovascular function. If we give high-flavonoid chocolate to one volunteer and placebo to another, it is likely that the two volunteers will have different values for arterial flexibility. But even if the chocolate-eater had better artery function than the placebo-eater, you would probably say that the test did not tell us much about our hypothesis. The chocolate-eater might just have had greater arterial flexibility to begin with.

Now imagine that we had five volunteers in each group and saw a difference, or that the difference was only 5% instead of 21%. How would we determine if the chocolate had an effect? Statistical tests allow researchers to look at their data and determine how likely it is—that is, the **probability**—that the result is due to sampling error. In this experiment, a statistical test distinguished between two possibilities for why the experimental group had greater arterial flexibility: The difference was either more likely due to the effect of chocolate or more likely due to a chance difference from the control group. The test results indicated that there was a low probability, less than 1 in 3,000 (0.027%), that the experimental and control groups were so different simply by chance. In other words, the difference between the two groups is statistically significant and was very likely caused by chocolate consumption (FIGURE 1.11).

The amount of variability in a sample is often expressed as the **standard error**. The standard error is used to generate the **confidence interval**—the range of values that has a high probability (usually 95%) of containing the true population average. Put simply, the average of a sample is unlikely to be exactly the average of the population. The average plus the standard error represents the highest likely value for the population average, and the average minus the standard error represents the lowest likely average value for the population. The confidence interval provides a way to express how much sampling error is influencing the results.

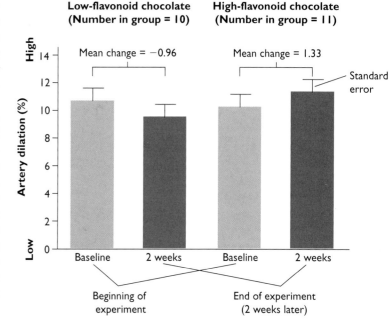

FIGURE 1.11 Interpreting data from graphs. This figure summarizes the results of the study discussed in the chapter.

VISUALIZE THIS Describe in words the results summarized by this graph.

Source: Engler, M. B., et al. 2004. Flavonoid-Rich Dark Chocolate Improves Endothelial Function and Increases Plasma Epicatechin Concentrations in Healthy Adults. *Journal of the American College of Nutrition* 23(3): 197–204.

Stop and Stretch 1.4

Opinion polling before a recent election indicated that candidate A was favored by 47% of likely voters and candidate B was favored by 51% of likely voters. The standard error was 3%. Why did reporters refer to this poll as a "statistical tie"?

Factors Influencing Statistical Significance

One characteristic of experiments influencing the power of statistical tests is **sample size**—the number of individuals in the experimental and control groups. If a treatment has no effect, a small sample size might mislead researchers into thinking that it does; this was the case with the vitamin C hypothesis described earlier. Subsequent tests with larger sample sizes, encompassing a wider variety of individuals with different underlying susceptibilities to illness, allowed scientists finally to reject the hypothesis that vitamin C prevents colds and the flu.

Conversely, if the effect of a treatment is real, but the sample size of the experiment is small, a single experiment may not allow researchers to determine convincingly that their hypothesis has support. In the same chocolate study discussed above, the experimental group experienced a 5% decline in LDL cholesterol levels while the placebo group experienced an 8% increase, for a total difference equaling a 13% improvement in cholesterol status for the flavonoid group. Because only 21 volunteers participated in the experiment, this smaller difference between the groups was not statistically significant.

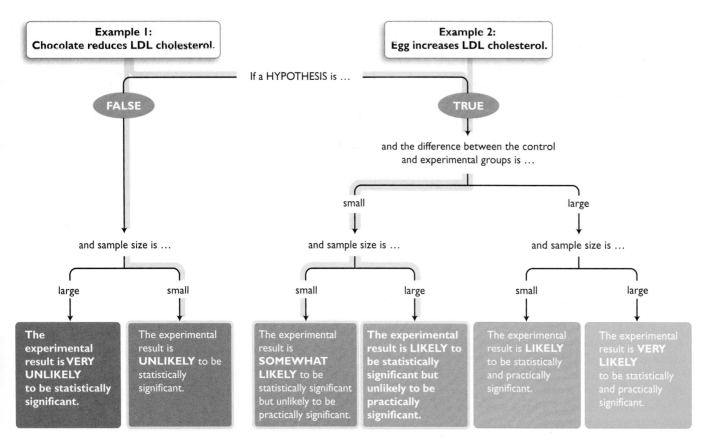

FIGURE 1.12 Factors that influence statistical significance. In Example 1, the true effect of chocolate on cholesterol is unknown because the sample size was so small that the result was ambiguous. The data indicated that the effect of chocolate on LDL cholesterol may be small, or the difference may have been due entirely to chance. In Example 2, a large sample size allowed researchers to perceive a very small effect.

VISUALIZE THIS The Nurse's Health Survey, which has followed nearly 80,000 women for over 20 years, has found that a diet low in refined carbohydrates such as white flour and sugar, but relatively high in vegetable fat and protein, cuts heart disease risk by 30% relative to women with a diet high in refined carbohydrates. How does this result map out on the flowchart here?

The true effect of high-flavonoid chocolate on cholesterol levels is still unknown, but this result does not rule out a possible positive effect.

The more participants there are in a study, the more likely it is that researchers will see a true effect of an experimental treatment, if one exists. If the sample size is large, any difference between an experimental and control group, no matter how small, is more likely to be statistically significant. For example, a study that pooled data from more than 3,500 subjects found that increasing egg intake from four per week to six per week increases LDL cholesterol levels by less than 3%. In this case, a large sample size allowed researchers to see that egg consumption has a real effect on blood cholesterol—but contrary to common belief, this effect is relatively tiny.

The relationship between hypotheses, experimental tests, sample size, and statistical and practical significance is summarized in FIGURE 1.12.

There is one final caveat to this discussion. A statistically significant result is typically defined as one that has a 5% probability or less of being due to chance alone. If all scientific research uses this same standard, as many as 1 in every 20 statistically significant results (that is, 5% of the total) is actually reporting an effect that is not real. In other words, some statistically significant results simply represent a surprisingly large difference between the experimental and control group that occurred only by chance.

An experiment with a statistically significant result will still be considered to support the hypothesis it was meant to test. However, the small probability that the results are due to chance explains why one supportive experiment is usually not enough to convince all scientists that a particular hypothesis is accurate. Even with a statistical test indicating that the result had a likelihood of less than 0.05% of occurring by chance, we should begin to feel assured that dark chocolate improves cardiovascular function only after locating additional tests of this hypothesis that give similar results.

What Statistical Tests Cannot Tell Us

All statistical tests operate with the assumption that the experiment was designed and carried out correctly. In other words, a statistical test evaluates the chance of sampling error, not observer error. A statistically significant result should never be taken as the last word on an experimentally tested hypothesis—an examination of the experiment itself is required.

In the chocolate experiment, the experimental design minimized the likelihood that alternative hypotheses could explain the results by randomly assigning subjects to treatment groups, using an effective placebo, and blinding both the data collectors and the subjects. These factors increase our confidence that the results of the experiment represent a real effect of chocolate on cardiovascular function.

Statistical significance is also not equivalent to significance as we usually define the term, that is, as "meaningful or important." A very large sample size can reveal a very small effect, as described earlier in the study reporting the effects of egg consumption on cholesterol level. Unfortunately, experimental results reported in the news often use the term *significant* without making this distinction. Understanding that problem, as well as other misleading aspects of how science is sometimes presented, will enable you to better use scientific information.

1.4 Evaluating Scientific Information

The previous sections should have given you some insight into why definitive scientific answers to our questions are slow in coming. However, a well-designed experiment can certainly allow us to approach the truth.

Looking critically at reports of experiments can help us make well-informed decisions about actions to take. Most of the research on diet and health is first published as primary sources, written by the researchers themselves and peer reviewed by other experts within the scientific community (FIGURE 1.13). The process of **peer review**, in which other scientists critique the results and conclusions of an experiment before it is published in a professional journal, helps increase confidence in scientific information. Peer-reviewed research articles in journals such as *Science*, *Nature*, the *Journal of the American Medical Association*, and hundreds of others represent the first and most reliable sources of current scientific knowledge.

Unfortunately, evaluating the hundreds of scientific papers that are published weekly is a task no one of us can perform. Even if we focused only on a particular field of interest, the jargon used in many scientific papers is a significant barrier to the nonexpert public. Instead of reading the primary literature, most of us receive our scientific information from secondary sources such as books, news reports, and advertisements. How can we evaluate information from these sources?

Information from Anecdotes

Information about diet and health is often in the form of **anecdotal evidence**—meaning that the advice is based on one individual's personal experience. A friend's enthusiastic plug for a herbal supplement that she felt helped her is an example of a *testimonial*—a common form of anecdote. Advertisements that use a celebrity to pitch a product "because it worked for them" are classic forms of testimonials. Although anecdotes may indicate that a product or treatment has merit, only well-designed tests can help determine its safety and efficacy.

Science in the News

Popular news sources provide a steady stream of health information. However, stories about scientific research often do not contain information about the adequacy of controls, the number of subjects, the experimental design, or the source of the scientist's funding. How can anyone evaluate the quality of research that drives headlines about the benefits of eating chocolate or the risks of consuming a particular dietary supplement?

FIGURE 1.13 Primary sources: publishing scientific results. Both before and after publication in a scientific journal, a paper is reviewed by other scientists who evaluate the research presented in the paper and the researchers' conclusions.

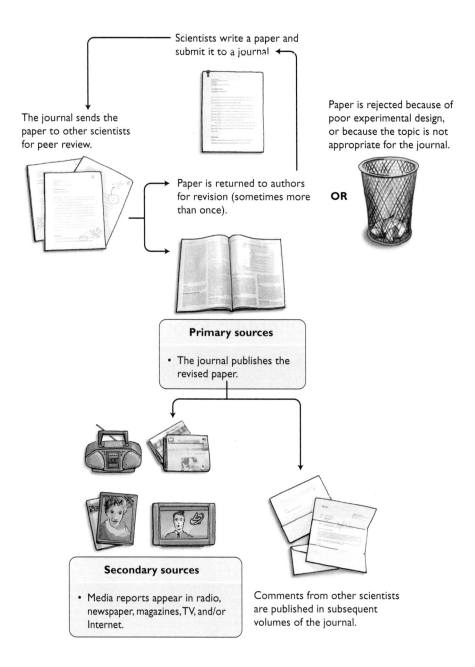

First, you must consider the source of media reports. News organizations are more reliable reporters of fact than entertainment tabloids, and news organizations with science writers should be better reporters of the substance of a study than those without. Television talk shows, which have to fill airtime, regularly have guests who promote a particular health claim. Too often these guests may be presenting information that has not been subjected to peer review and is based on anecdotes or an incomplete summary of the primary literature.

Paid advertisements are a legitimate means of disseminating information. However, claims in advertising should be very carefully evaluated. Advertisements of over-the-counter and prescription drugs must conform to rigorous government standards regarding the truth of their claims—claims which have been tested through controlled studies. However, lower standards apply to advertisements for herbal supplements, many health food products, and diet plans. Be sure to examine the fine print, because advertisers often are required to clarify the statements made in their ads.

Another commonly used source for health information is the Internet. As you know, anyone can post information on the Internet. Typing in "diet and health" on a standard web search engine will return millions of web pages—from highly respected academic and government sources to small companies trying to sell their products, or individuals who have strong, but completely unsupported, ideas. It can be difficult to determine the reliability of a well-designed web site. Here are some things to consider when using the web as a resource for health information:

1. Choose sites maintained by reputable medical establishments, such as the National Institutes of Health (NIH) or the Mayo Clinic.
2. It costs money to maintain a web site. Consider whether the site seems to be promoting a product or agenda. Advertisements for a specific product should alert you to a web site's bias.
3. Check the date when the web site was last updated, and see if the page has been updated since its original posting. Science and medicine are disciplines that must frequently incorporate new data into hypotheses. A reliable web site will be updated often.
4. Determine if unsubstantiated claims are being made. Look for references, and be suspicious of any sites that describe studies that are not from peer-reviewed journals.

Understanding Science from Secondary Sources

Once you are satisfied that a media source is relatively reliable, you should examine the scientific claim that it presents. Use your understanding of the process of science and of experimental design to evaluate the story and the science. Does the story about the claim present the results of a scientific study, or is it built around an untested hypothesis? Is the story confusing correlation with causation? Does it seem that the information is applicable to nonlaboratory situations, or is it based on results from preliminary or animal studies?

Look for clues about how well the reporters did their homework. Scientists usually discuss the limitations of their research in their papers. Are these cautions noted in an article or television piece? If not, the reporter may be overemphasizing the applicability of the results.

Then, note if the scientific discovery itself is controversial. Does it reject a hypothesis that has long been supported? Does it concern a subject that is controversial in human society (like the origin of racial differences or homosexuality)? Might it lead to a change in social policy? In these cases, be extremely cautious. New and unexpected research results must be evaluated in light of other scientific evidence and understanding. Reports that lack comments from other experts may omit important problems with a study or fail to place it in context with other research. Table 1.2 provides a checklist of questions to answer as you evaluate a news report about a recent scientific development.

Finally, the news media generally highlight only those stories that editors and producers find newsworthy. News organizations are also more likely to report a study that supports a hypothesis rather than one that gives less supportive results, even if both types of studies exist.

Even with well-researched stories, you may still find situations where reports on several scientific studies seem to give conflicting and confusing results. Either the reporter is not giving you enough information, in which case you may want to read the researchers' papers yourself, or the researchers themselves are just as confused as you are. Such confusion is the nature of the scientific process. Early in our search to understand a phenomenon, many hypotheses are proposed and discussed. Some are tested and rejected immediately, and some are supported by one experiment but later rejected by more thorough experiments. Only by clearly understanding the process and pitfalls of scientific research can you distinguish "what we know" from "what we don't know."

TABLE 1.2 | A Guide for Evaluating Science in the News

Questions	Possible answers	
	Preferred answer ☑	**Raises a red flag** ☑
1. What is the basis for the story?	Hypothesis test ❏	Untested assertion ❏ *No data to support claims in the article.*
2. What is the affiliation of the scientist?	Independent (university or government agency) ❏	Employed by an industry or advocacy group ❏ *Data and conclusions could be biased.*
3. What is the funding source for the study?	Government or nonpartisan foundation ❏	Industry group or other partisan source (with bias) ❏ *Data and conclusions could be biased.*
4. **If the hypothesis test is a correlation:** Did the researchers attempt to eliminate reasonable alternative hypotheses?	Yes ❏	No ❏ *Correlation does not equal causation. One hypothesis test provides poor support if alternatives are not examined.*
If the hypothesis test is an experiment: Is the experimental treatment the only difference between the control group and the experimental group?	Yes ❏	No ❏ *An experiment provides poor support if alternatives are not examined.*
5. Was the sample of individuals in the experiment a good cross section of the population?	Yes ❏	No ❏ *Results may not be applicable to the entire population.*
6. Was the data collected from a relatively large number of people?	Yes ❏	No ❏ *Study is prone to sampling error.*
7. Were participants blind to the group they belonged to and/or to the "expected outcome" of the study?	Yes ❏	No ❏ *Subject expectation can influence results.*
8. Were data collectors and/or analysts blinded to the group membership of participants in the study?	Yes ❏	No ❏ *Observer bias can influence results.*
9. Did the news reporter put the study in the context of other research on the same subject?	Yes ❏	No ❏ *Cannot determine if these results are unusual or fit into a broader pattern of results.*
10. Did the news story contain commentary from other independent scientists?	Yes ❏	No ❏ *Cannot determine if these results are unusual or if the study is considered questionable by others in the field.*
11. Did the reporter list the limitations of the study or studies he or she is reporting on?	Yes ❏	No ❏ *Reporter may not be reading study critically and could be overstating the applicability of the results.*

For each question, check the appropriate box. Note the number of red flags raised by the time you have reached this point. Some of the issues raised may be more serious than others but, in general, the fewer the red flags, the more thorough the report and the more reliable the scientific study.

Chapter REVIEW

ROOTS TO REMEMBER

The following roots of words come mainly from Latin and Greek and will help you to decipher terms:

bio- means life.

deduc- means to reason out, working from facts.

induc- means to rely on reason to derive principles (and also to cause to happen).

hypo- means under, below, or basis.

-ology means the study of, or branch of knowledge about.

KEY TERMS

1.1 An Introduction to the Scientific Process
biology *p. 2*
deductive reasoning *p. 6*
falsifiable *p. 3*
hypothesis *p. 2*
inductive reasoning *p. 5*
prediction *p. 6*
scientific method *p. 2*
scientific theory *p. 3*
supernatural *p. 3*
testable *p. 2*

1.2 Hypothesis Testing
bias *p. 9*
control *p. 8*
controlled experiment *p. 8*
correlation *p. 11*
data *p. 7*
dependent variable *p. 7*
double-blind *p. 10*
experiment *p. 7*
independent variable *p. 7*
model organism *p. 10*
placebo *p. 8*
random assignment *p. 8*
variable *p. 7*
x-axis *p. 11*
y-axis *p. 11*

1.3 Understanding Statistics
confidence interval *p. 15*
probability *p. 15*
sample *p. 13*
sample size *p. 15*
sampling error *p. 14*
standard error *p. 15*
statistical significance *p. 14*
statistics *p. 13*

1.4 Evaluating Scientific Information
anecdotal evidence *p. 17*
peer review *p. 17*

SUMMARY

1.1 The Process of Science

- Science is a process of testing hypotheses—statements about how the natural world works. Scientific hypotheses must be testable and falsifiable (pp. 2–3; Figure 1.1).

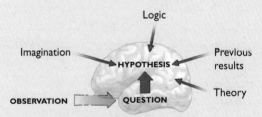

- A scientific theory is an explanation of a set of related observations based on well-supported hypotheses from several different, independent lines of research. The theories of evolution and natural selection, which concern the relationships among modern organisms and how they developed, are two theories that form the basis of biology (pp. 3–4).

- Hypotheses are often generated via inductive reasoning, whereby scientists infer a general principle through observing a number of specific events (p. 5).

- Hypotheses are tested through the process of deductive reasoning, which allows researchers to make specific predictions about expected observations (p. 6).

1.2 Hypothesis Testing

- Absolutely proving hypotheses is impossible. However, well-designed scientific experiments can allow researchers to strongly infer that their hypothesis is correct (p. 7).

- Controlled experiments test hypotheses about the effect of experimental treatments by comparing a randomly assigned experimental group with a control group. Controls are individuals who are identical to the experimental group in all respects, except for application of the treatment (p. 8).

- Bias in scientific results can be minimized with double-blind experiments that keep subjects and data collectors unaware of which individuals belong in the control or experimental group (p. 9).

- In situations where performing controlled experiments on humans is considered unethical, scientists sometimes employ model organisms, such as other mammals (p. 10).

- Some hypotheses can be tested using correlation, in which scientists look for a relationship between two factors. A correlation can show a relationship between two factors, but it does not eliminate all alternative hypotheses (p. 11–12; Figure 1.9).

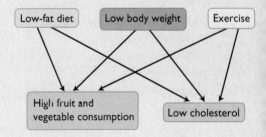

- Epidemiology, the study of the causes of human diseases, relies on different types of correlational studies (pp. 12–13).

1.3 Understanding Statistics

- Statistics help scientists evaluate the results of their experiments by determining whether results appear to reflect the true effect of an experimental treatment on a sample of a population (p. 14).

- A statistical test indicates the role that chance, or sampling error, plays in the experimental results. A statistically significant result is one that is very unlikely to be due to chance differences between the experimental and control group (pp. 14–15; Figure 1.10).

- Even when an experimental result is highly significant, hypotheses are tested multiple times before scientists come to consensus on the true effect of a treatment (pp. 16–17).

1.4 Evaluating Scientific Information

- Primary sources of information are experimental results published in professional journals and peer reviewed by other scientists before publication (p. 17).

- Anecdotal evidence is an unreliable means of evaluating information, and media sources are of variable quality; distinguishing between news stories and advertisements is important when evaluating the reliability of information. The Internet is a rich source of information, but users should look for clues to a particular web site's credibility (pp. 17–19).

- Stories about science should be carefully evaluated for information on the actual study performed, the universality of the claims made by the researchers, and other studies on the same subject. Sometimes confusing stories about scientific information are a reflection of controversy within the scientific field itself (pp. 19–20).

LEARNING THE BASICS

1. A scientific hypothesis is _____.
   a. an opinion
   b. a proposed explanation for an observation
   c. a fact
   d. easily proved true
   e. an idea proposed by a scientist

2. A good scientific hypothesis must be _____.
   a. testable
   b. true
   c. falsifiable
   d. a and c are correct
   e. a, b, and c are correct

3. Which of the following is an example of inductive reasoning?
   a. All cows eat grass.
   b. My cow eats grass and my neighbor's cow eats grass: therefore all cows probably eat grass.
   c. If all cows eat grass, when I examine a random sample of all the cows in Minnesota, I will find that they all eat grass.
   d. Cows may or may not eat grass, depending on what type of farm they live on.

4. For the hypothesis "Drinking ginger tea is an effective treatment for nausea," which of the following is a prediction based on the hypothesis?
   a. People who drink ginger tea recover from nausea more quickly than people who not drink ginger tea.
   b. People who do not drink ginger tea have more severe nausea than people who do drink ginger tea.
   c. The bacteria responsible for food poisoning cannot live in ginger tea.
   d. People who drink ginger tea feel healthier than do people who do not drink ginger tea.
   e. Consuming ginger tea causes vomiting.

5. If I perform a hypothesis test where I demonstrate that the prediction I made above is true, I have _____.
   a. proved the hypothesis
   b. supported the hypothesis
   c. not falsified the hypothesis
   d. b and c are correct
   e. a, b, and c are correct

6. A scientific theory is _____.
   a. based on unsupported hypotheses
   b. equivalent to a hypothesis
   c. an untested idea about "how the world works"
   d. a basic principle that informs our understanding of the world
   e. more than one of the above is correct

7. Control subjects in an experiment _____.
   a. should be similar in most ways to the experimental subjects
   b. should not know if they are in the control or experimental group
   c. should have the same interactions with the researchers as the experimental subjects
   d. help eliminate alternative hypotheses that could explain experimental results
   e. all of the above

8. Which of the following scenarios describes the random assignment of individuals in an experiment?
   a. a lottery in which people who drew odd numbers are assigned to the experimental group and where even numbers are assigned to the control group
   b. a survey that places all cat-lovers in one group and all cat-haters into another group
   c. a health clinic that assigns people to the experimental or treatment group based on the time of day they visited the clinic
   d. a pharmaceutical researcher chooses which patients will receive the experimental drug and which will receive a placebo

9. An experiment in which neither the participants in the experiment nor the technicians collecting the data know which individuals are in the experimental group and which ones are in the control group is known as _____.
   a. controlled
   b. biased
   c. double-blind
   d. falsifiable
   e. unpredictable

10. All of the following can be effective model organisms for studies on human health *except* _____.
    a. dogs
    b. rats
    c. guinea pigs
    d. chimpanzees
    e. all of the above are appropriate model organisms

11. A relationship between two factors, for instance between outside temperature and the number of people with active colds in a population, is known as a(n) _____.
    a. significant result
    b. correlation
    c. hypothesis
    d. alternative hypothesis
    e. experimental test

12. When the temperature rises in urban areas, the number of violent incidents in the city increases. This is an example of _____.
    a. negative correlation
    b. positive correlation
    c. hypothesis
    d. observer bias
    e. controlled experiment

13. When the temperature rises in urban areas, the number of violent incidents in the city increases. From this statement you can conclude _____.
    a. urban residents hate hot weather
    b. there is a correlation between hot weather and violence
    c. high temperatures make routine incidents become violent
    d. a and c are correct
    e. a, b, and c are correct

14. Statistical tests tell us _____.
    a. if an experimental treatment showed more of an effect than would be predicted by chance
    b. if a hypothesis is true
    c. if an experiment was well designed
    d. if the experiment suffered from any bias
    e. how similar the sample was to the population from which it was drawn

15. The effect of chance on the results of a hypothesis test is known as _____.
    a. bias
    b. poor control
    c. statistical insignificance
    d. placebo
    e. sampling error

16. Which of the following factors affects the likelihood that an experimental result will be statistically significant?
    a. the size of the sample
    b. the actual effect of the experimental treatment
    c. the variability of the sample
    d. how large the effect of the experimental treatment is
    e. all of the above

17. The process of submitting a scientific paper to other scientists in the field for critique before publishing it in a journal is known as _____.
    a. collaboration
    b. correlation
    c. peer review
    d. anecdotal

18. A primary source for learning about scientific results is _____.
    a. the news media
    b. anecdotes from others
    c. articles in peer-reviewed journals
    d. the Internet
    e. all of the above

19. Which of the following is the least reliable source of scientific information?
    a. a web site maintained by the Mayo clinic
    b. a paper published in a peer-reviewed journal
    c. the *New York Times* science section
    d. an advertisement containing a testimonial from a famous actor
    e. all sources are equally reliable

20. **True or False?** To be a scientist, one must memorize a diverse set of very specific facts about a particular area of study, for instance, about all living things _____.

ANALYZING AND APPLYING THE BASICS

1. In an experiment on the effect of vitamin C on reducing the severity of cold symptoms, college students visiting their campus health service with early cold symptoms either received vitamin C or were treated with over-the-counter drugs. Students then reported on the length and severity of their colds. The timing of dosages and the type of pill were different, so both the students and the clinic health providers knew which treatment students were receiving. This study reported that vitamin C significantly reduced the length and severity of colds experienced in this population. Why might this result be questionable, given the experimental design?

2. Brain-derived neurotrophic factor (BDNF) is a substance produced in the brain that helps nerve cells (neurons) to grow and survive. BDNF also increases the connectivity of neurons and improves learning and mental function. A 2002 study that examined the effects of intense wheel-running on rats and mice found a positive correlation between BDNF levels and running distance. What could you conclude from this result?

3. Samuel George Morton, who published data in the 1840s, reported differences in brain size among human races. His research indicated that Europeans had larger brains than did Native Americans and Africans. His measures of brain size were based on skull volume, calculated by packing individual skulls with mustard seed and then measuring the volume of the seeds they contained. When the biologist Stephen Jay Gould reexamined Morton's data in the 1970s, he found that Morton had systematically erred in his measurements, consistently underestimating the size of the African and Native American skulls. According to Gould, Morton did not realize that he was affecting his own results to support his hypothesis that Europeans had larger brains than the other groups. How could Morton have designed his experiment to minimize the effect of this bias on his results?

Use the following information to answer questions 4 and 5.

Skin care products often make claims about their ability to improve skin tone or reduce wrinkles. The following information was found on the packaging of two skin care products:

Product 1: A dietary supplement claims to have a "revolutionary biovitamin complex" that enhances the skin's support system, removes toxins from the skin, and improves the "clarity of the epidermis." The following disclaimer is on the package: "These statements have not been reviewed by the Food and Drug Administration."

Product 2: A serum designed to reduce puffiness and wrinkles around the eyes claims an "excellent hydrating effect: +47% after one hour, +30% after 6 hours." The serum claims a "reduction of wrinkles up to 86%" on volunteers participating in a study conducted by an "independent laboratory certified by the French Ministry of Health."

4. Which product's claims are more likely to be true? Explain your answer.

5. What other information would you need in order to evaluate the claims made by product 2?

Use the following information to answer questions 6–8.

FIGURE 1.14 shows the survival of mice on a normal diet, a high-calorie diet, and a high-calorie diet that was supplemented with the chemical resveratrol.

6. What conclusion can you draw from the data on the graph?

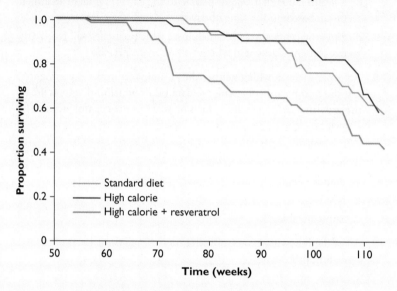

FIGURE 1.14 **Effect of resveratrol on life span in mice**
Source: Baur, J. A., et al. 2006. Resveratrol Improves Health and Survival of Mice on a High-Calorie Diet. *Nature* (444): 337–342.

7. The following excerpt is part of an article in the *New York Times* that was based on the resveratrol research partially summarized by the Figure 1.14 graph. The article begins:

> Can you have your cake and eat it? Is there a free lunch after all, red wine included? Researchers at the Harvard Medical School and the National Institute on Aging report that a natural substance found in red wine, known as resveratrol, offsets the bad effects of a high-calorie diet in mice and significantly extends their lifespan.
>
> Their report, published electronically yesterday in *Nature*, implies that very large daily doses of resveratrol could offset the unhealthy, high-calorie diet thought to underlie the rising toll of obesity in the United States and elsewhere, if people respond to the drug as mice do.
>
> Resveratrol is found in the skin of grapes and in red wine and is conjectured to be a partial explanation for the French paradox, the puzzling fact that people in France enjoy a high-fat diet yet suffer less heart disease than Americans.

What does the beginning of this article imply about the relationship of diet to human health?

8. Later in the *New York Times* article on resveratrol, the author states:

> The mice were fed a hefty dose of resveratrol, 24 milligrams per kilogram of body weight. Red wine has about 1.5 to 3 milligrams of resveratrol per liter, so a 150-lb person would need to drink 750 to 1,500 bottles of red wine a day to get such a dose.

How does the information in the paragraph above relate to the suggestion made by the article's author in question 7?

CONNECTING THE SCIENCE

1. Much of the research on prevention and treatment of the common cold is performed by scientists employed or funded by drug companies. Often these companies do not allow scientists to publish the results of their research for fear that competitors at other drug companies will use this research to develop a new drug before they do. Should our society allow scientific research to be owned and controlled by private companies?

2. Should society restrict the kinds of research performed by government-funded scientists? For example, many people believe that research performed on tissues from human fetuses should be restricted. These people believe that such research would justify abortion. If most Americans feel this way, should the government avoid funding this research? Are there any risks associated with prohibiting public funding of certain research topics?

Chapter 2

The Chemistry of Life

Drink to Your Health?

LEARNING GOALS

1. Describe the components of an atom and how they interact.
2. State several properties of water that make it a good solvent.
3. Compare and contrast covalent, ionic, and hydrogen bonds.
4. Describe the pH scale and what makes a substance an acid or a base.
5. Describe how buffer systems in the human body can help maintain homeostasis.
6. List the building-block molecules of a carbohydrate, a protein, and a fat.
7. Discuss the manner in which proteins fold.
8. Describe the process of hydrogenation.
9. Make a rough sketch of the structure of a DNA molecule, and label each of the following: a nucleotide, a sugar, a nitrogenous base.
10. Compare and contrast the classes of micronutrients: vitamins, minerals, and antioxidants.

What is the best way to rehydrate your body?

Sports drinks, energy drinks, fitness waters, and vitamin waters have become a common sight in the beverage aisle of most grocery stores and line the coolers of convenience stores. The first to hit the shelves were sports drinks, like Gatorade, containing small amounts of sugar and something called "electrolytes." The orange Gatorade bucket became ubiquitous on football sidelines and baseball dugouts, bolstering the idea that these drinks were a true advance over plain tap water for rehydrating the body after a workout.

The success of Gatorade spawned a long line of imitators. Soon, sports drinks were joined on the shelves by energy drinks, which contain caffeine and much more sugar. One of the most common energy drinks, Red Bull, sponsors many athletes and competitions, including downhill skier Daron Rahlves, golfer Chris DiMarco, a major league soccer team, and some extreme sports events. Marketers of these drinks are now sending the message that these drinks can actually improve your athletic performance.

Energy drinks have not gone over as well with older, more calorie-conscious buyers, who tend to gravitate toward the bottled waters advertising purity. However, an even more recent trend appealing to these buyers has been fitness or vitamin waters, which are low-sugar drinks containing artificial sweeteners, flavorings, vitamins, and minerals. Many of these drinks are sold as special blends of supplements targeted at individuals with different needs. Now, instead of simply maintaining good health, water is being sold as a type of drug that makes you feel better.

Is it possible to drink your way to better athletic performance or enhanced health? This chapter will attempt to help you determine if these drinks serve a purpose or are simply a clever ploy by marketers to convince you to purchase these products.

What about bottled water or fitness waters supplemented with vitamins and minerals?

Are sports drinks the way?

Are these drinks better than tap water? How can we know?

2.1 Water: Essential to Life

Human beings can survive for several weeks with no nutrition other than water. Without water, though, human survival is limited to just a few days. A decrease in the body's optimal water level, called **dehydration**, can lead to impaired physical and mental abilities, cramps, and heat exhaustion. Large water deficits can result in hallucinations, heat stroke, and death.

Every day, we lose about 3 liters of water as sweat, and in urine and fecal waste. To avoid the negative health impacts of dehydration, this water must be replenished. We can replace some water through consumption of food. A typical adult obtains about 1.5 liters of water per day from food consumed, leaving a deficit of about 1.5 liters (about six 8-ounce glasses) that one must drink to replenish. A person who exercises needs to consume more water to compensate for the water that is lost as sweat.

The Building Blocks of Water

Water is made up of two elements: hydrogen and oxygen. **Elements** are the fundamental forms of matter—the simplest chemical substances, composed of atoms that cannot be broken down by normal physical means such as boiling.

Atoms are the smallest units that have the properties of any given element. Ninety-two natural elemental atoms have been described by chemists, and several more have been created in laboratories, but only a handful are found in living organisms (Table 2.1). Each element has a one- or two-letter symbol: H for hydrogen, O for oxygen, and Ca for calcium, for example.

Atoms are composed of subatomic particles called **protons**, **neutrons**, and **electrons**. Protons have a positive electric charge; these particles and the uncharged neutrons make up the **nucleus** of an atom. All atoms of a particular element have the same number of protons, giving the element its **atomic number**. The negatively charged electrons are found outside the nucleus in an "electron cloud." Electrons are attracted to the

TABLE 2.1 | Elements Found in Humans

Element	Symbol	% of Weight	Functions
Oxygen	O	65	Component of water and most other complex molecules.
Carbon	C	18	Backbone of many biological molecules
Hydrogen	H	10	Part of many biological molecules and water
Nitrogen	N	3	Part of proteins
Calcium	Ca	2	Constituent of bone
Phosphorus	P	1	Part of nucleic acids, cell membranes, and bone
Potassium	K	0.3	Essential for nerve function
Sulfur	S	0.2	Component of proteins
Sodium	Na	0.1	Essential for nerve action; important electrolyte
Chlorine	Cl	0.1	Component of digestive acid
Magnesium	Mg	Trace	Muscle contraction
Iron	Fe	Trace	Helps carry oxygen in blood

positively charged nucleus (FIGURE 2.1). A *neutral atom* has equal numbers of protons and electrons. **Ions** do not have an equal number of protons and electrons. In this case, the atom is not neutral and has an electrical charge.

An atom's **mass number** is the sum of the numbers of protons and neutrons in its nucleus. **Isotopes** are versions of a chemical element containing the same number of protons but different numbers of neutrons in the nucleus. For example, carbon comes in several isotopes: Carbon-12, containing 6 protons and 6 neutrons, is the most common isotope. Carbon-14, with 6 protons and 8 neutrons, may be familiar to you as the isotope used to determine the age of ancient human artifacts.

Some isotopes are unstable and break down over time. As isotopes break down, they release energy in the form of rays and subatomic particles. Unstable isotopes that break down are called **radioactive isotopes**. Small amounts of certain radioactive isotopes can be ingested by humans to detect changes in the body's organs and tissues without having to undergo an exploratory surgery. For example, if a person drinks a solution containing a very small amount of radioactive iodine, it becomes concentrated in the thyroid gland. An image of the thyroid can then be used to assess this organ's health. Small cancers that have not spread to other locations in the body can be treated through the use of radiation from radioisotopes to destroy cancer cells.

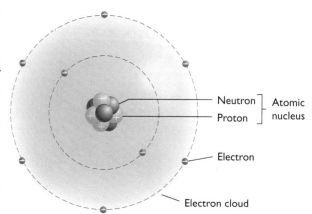

FIGURE 2.1 Atomic structure of oxygen. An oxygen atom contains a nucleus made up of 8 protons and 8 neutrons. Orbiting electrons surround the nucleus. Although the number of particles within each atom differs, all atoms have the same basic structure.

The Structure of Water

The chemical formula for water is H_2O, indicating that it contains two hydrogen atoms and one oxygen. Water, like other **molecules**, consists of two or more atoms joined by chemical bonds. A molecule can be composed of the same or different atoms. For example, a molecule of oxygen consists of two oxygen atoms joined to each other. If the atoms differ, as is the case when two hydrogen atoms and one oxygen atom form water, the substance can be referred to as a **compound**. The atoms of molecules and compounds are held together by chemical bonds.

The ability of hydrogen and oxygen, or other elements, to make chemical bonds depends on the atom's electron configuration. The electrons in the electron cloud that surrounds the atom's nucleus have different energy levels based on their distance from the nucleus. The first energy level, or **electron shell**, is closest to the nucleus, and the electrons located there have the lowest energy. The second energy level is a little farther away, and the electrons located in the second shell have a little more energy. The third energy level is even farther away, and its electrons have even more energy, and so on.

Each energy level can hold a specific maximum number of electrons. The first shell holds two electrons, and the second and third shells each hold a maximum of eight. Electrons fill the lowest energy shell before advancing to fill a higher energy-level shell. Atoms with the same number of electrons in their outermost energy shell, called the **valence shell**, exhibit similar chemical behaviors. When the valence shell is full, the atom will not normally form chemical bonds with other atoms. Atoms whose valence shells are not full of electrons do have the ability to combine via chemical bonds.

Atoms with four or five electrons in the outermost valence shell tend to share electrons to complete their valence shells. When atoms share electrons, a type of bond called a **covalent bond** is formed. Covalent bonds hold the oxygen and hydrogen atoms together in a water molecule (FIGURE 2.2). In standard chemical notation, a short line indicating a shared pair of electrons symbolizes a covalent bond. Atoms can share two or even three valence electrons with each other—for example, the oxygen gas we breathe (O_2) consists of two oxygen atoms held together by a double covalent bond. Double bonds are symbolized by two horizontal lines (FIGURE 2.3).

Atoms with one, two, or three electrons in their valence shell tend to lose electrons and, therefore, become positively charged ions, whereas atoms with six or seven electrons in the valence shell tend to gain electrons and become negatively charged ions. Positively and negatively charged ions associate via an ionic bond, which has some differences from a covalent bond.

Ionic bonds form between charged atoms attracted to each other by similar, opposite charges. For example, a sodium atom forms an ionic bond with a chlorine atom to

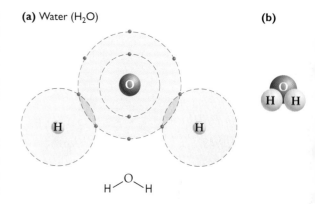

FIGURE 2.2 Covalent bonding in water. A covalent bond forms when two atoms share a pair of electrons. (a) In water, oxygen shares one pair each with two hydrogen atoms. The structure of water using more typical chemical notation is H_2O. (b) This space-filling model is another way to depict the structure of water, without showing the electrons.

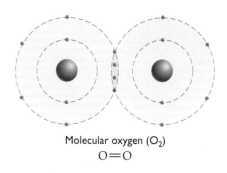

FIGURE 2.3 Double covalent bonds. Molecular oxygen consists of two oxygen atoms sharing two pairs of electrons to form a double bond.

FIGURE 2.4 Ionic reaction. During the formation of sodium chloride, an electron is transferred from the sodium atom to the chlorine atom. Note that both atoms now have complete valence shells.

VISUALIZE THIS Why does this ionic reaction result in the addition of a positive charge to sodium and a negative charge to chloride?

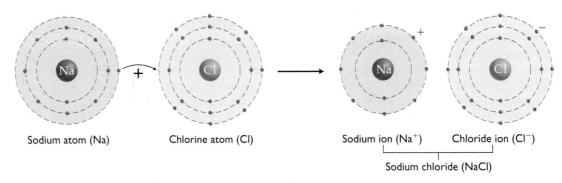

Sodium atom (Na)   Chlorine atom (Cl)   Sodium ion (Na$^+$)   Chloride ion (Cl$^-$)

Sodium chloride (NaCl)

Stop and Stretch 2.1

Examine the electron configuration of the carbon and oxygen atoms shown here. What sort of bond are these atoms likely to form with each other, and why?

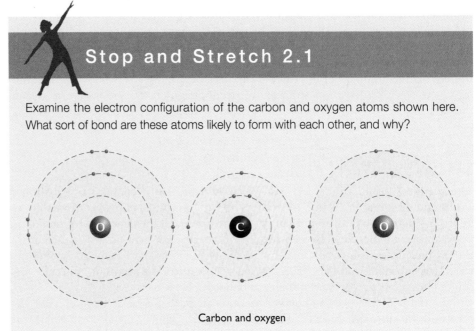

Carbon and oxygen

produce table salt (sodium chloride) when the sodium atom gives up an electron and the chlorine atom gains one (FIGURE 2.4). More than two atoms can be involved in an ionic bond. For instance, calcium will react with two chlorine atoms to produce calcium chloride (CaCl$_2$). This is because calcium has two electrons in its valence shell—when it loses these it has two more protons than neutrons, giving it a double positive charge. Each chlorine atom, with seven valence electrons, picks only one more electron to have a stable outer shell and a single negative charge. Thus, two chlorine ions will be attracted to a single calcium ion.

Ionic bonds are about as strong as covalent bonds. They can be more easily disrupted, however, when mixed with certain liquids containing electrical charges. Water is one liquid that causes ions in molecules to *dissociate*, or fall apart.

Water Is a Good Solvent

Water has the ability to dissolve a wide variety of substances. A substance that dissolves when mixed with another substance is called a **solute**. When a solute is dissolved in a liquid, such as water, the liquid is called a **solvent**. Once dissolved, components of a particular solute can pass freely throughout the water, making a chemical mixture or **solution**. Supplements added to flavored water and sports drinks, from sugar to salt to artificial flavors, are invisible to us when we drink them. This is a testament to the effectiveness of water's dissolving power.

Water is a good solvent because it is **polar**, meaning that different regions, or poles, of the molecule have different charges. The polarity arises because oxygen is more attractive to electrons, that is, it is more **electronegative**, than most other atoms, including hydrogen. As a result of oxygen's electronegativity, electrons in a water molecule spend more time near the nucleus of the oxygen atom than near the nuclei of the hydrogen atoms. With more negatively charged electrons near it, the oxygen in water carries a partial negative charge, symbolized by the Greek letter delta, δ^-. The hydrogen atoms thus have a partial positive charge, symbolized by δ^+ (FIGURE 2.5). Any molecule consisting of charged atoms will dissolve in water because water molecules can surround individual parts of the molecule (FIGURE 2.6). Molecules that dissolve in water in this way are called **hydrophilic**.

When atoms of a molecule carry no partial charge they are said to be **nonpolar**. Nonpolar molecules such as oil do not contain charged atoms and are referred to as **hydrophobic** because they do not easily mix with water.

Water's ability to dissolve other compounds makes it a good medium for carrying materials into and around the body—in fact, the liquid portion of your blood is a chemical mixture consisting of primarily water and dozens of dissolved substances. Some of the symptoms of dehydration occur because decreased amounts of water in the blood result in decreased blood volume and pressure, resulting in less oxygen and fewer nutrients being delivered to all parts of the body, including the brain.

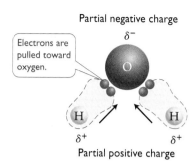

FIGURE 2.5 Polarity in a water molecule. Water is a polar molecule because oxygen pulls electrons closer to its nucleus than hydrogen does. Because the atoms in the molecule do not share electrons equally, different parts of the molecule have different partial charges. Partial charges are symbolized by the Greek letter delta (δ).

Water Facilitates Chemical Reactions

Because it is such a powerful solvent, water can facilitate **chemical reactions**, which are changes in the chemical composition of substances. Solutes in a mixture, called **reactants**, can come in contact with each other, permitting the modification of chemical bonds that occur during a reaction. The molecules formed as a result of a chemical reaction are known as **products**. For example, when carbon dioxide (CO_2) is dissolved in water, it reacts with some of the water molecules to form carbonic acid, in the following chemical reaction:

$$H_2O + CO_2 \longrightarrow H_2CO_3$$

In this equation, a mixture of the reactants water and carbon dioxide yields the product carbonic acid. Chemical reactions are typically written in this fashion, with the arrow standing for "yields." Notice that on both sides of the arrow, the number of atoms of each type is the same: three oxygen, two hydrogen, and one carbon. All chemical equations are balanced in this manner. Atoms do not disappear from the equation, nor are new atoms added.

Manufacturers of *carbonated* water encourage the reaction described above by injecting carbon dioxide at a high pressure into water. The carbon dioxide remains in the solution until the pressure is released (that is, until the can or bottle is opened), when it is allowed to bubble out of the solution. Because carbonation results in the formation of a weak acid, carbonated waters have an extra "bite" that many people enjoy.

Water Is Cohesive

Water molecules tend to orient themselves so that the hydrogen atom (with its partial positive charge) of one molecule is near the oxygen atom (with its partial negative charge) of another molecule (FIGURE 2.7a). The weak attraction between hydrogen atoms and oxygen atoms in adjacent molecules forms a **hydrogen bond**. Hydrogen bonding is a type of weak chemical bond that forms when a partially positive hydrogen atom is attracted to a partially negative atom. Hydrogen bonds can be *intra*molecular, involving different regions within the same molecule, or they can be *inter*molecular, between different molecules, as is the case in hydrogen bonding between different water molecules. FIGURE 2.7b shows the hydrogen bonding that occurs between water molecules in liquid form. Table 2.2 outlines the characteristics of covalent, ionic, and hydrogen bonds.

The tendency of like molecules to stick together is called **cohesion**. Cohesion is much stronger in water than in most liquids as a result of hydrogen bonding and is an important property of many biological systems. For instance, many plants depend on cohesion to help transport a continuous column of water from the roots to the leaves.

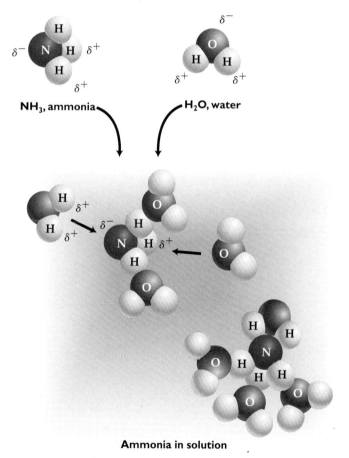

FIGURE 2.6 Water as a solvent. Charged molecules easily dissolve into water because the opposite charged ends of water molecules surround a charge, separating it from its neighbors. In this example, a charged compound is dissolving in water.

FIGURE 2.7 Water is a cohesive substance. (a) Water molecules stick together because the partial negative charge of one molecule is attracted to the partial positive charge of another, forming a hydrogen bond. (b) Each water molecule can form multiple hydrogen bonds with other molecules in a solution.

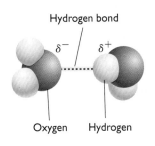

(a) Bonds between two water molecules

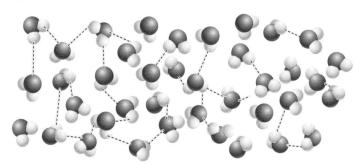

(b) Bonds between many water molecules

When heat energy is added to water, its initial effect is to disrupt the hydrogen bonding among water molecules. Therefore, this heat energy can be absorbed without changing the temperature of water. Only after the hydrogen bonds have been broken can added heat increase the temperature. In other words, the initial input of energy is absorbed. Water's ability to absorb a lot of energy before becoming hot is the reason that water can be used to cool a hot engine—or internal organs.

Inside cells, water helps to stabilize temperatures even though chemical reactions in cells are constantly producing heat. When enough heat energy is present, such as when you are running on a warm summer day, hydrogen bonds between water molecules do break apart. The liquid water *evaporates* when molecules escape into the air. This cools the skin and the blood circulating below it. As sweat, made primarily of water, is moved to the surface of the body, it absorbs heat produced by the body tissue and evaporates from the skin. As a result, body temperature declines, preventing the body from overheating (FIGURE 2.8). Dehydration is dangerous because a decline in body water levels prevents sweating, allowing body temperature to rise to harmful levels.

TABLE 2.2 | Main Chemical Bonds Found in Biological Molecules

Type of Bond	Characteristics	Example
Covalent	Atoms joined together by shared electrons	
Ionic	Atoms of opposite charges are attracted to each other	
Hydrogen	Weak bond joins hydrogen in one polar molecule with electronegative atom in another polar molecule	

Bottle or Tap?

Bottled water and carbonated water can replace fluid as well as tap water can. So is there any reason to use these products? Many Americans apparently think so: Sales of bottled water are close to 8 billion gallons annually.

Some consumers choose bottled water because of concerns about the quality of tap water. The Food and Drug Administration (FDA) sets high standards for bottled water. However, this agency generally uses the same standards applied by the Environmental Protection Agency (EPA) to tap water, meaning that water from both sources is equally clean. In fact, nearly 40% of bottled waters are actually derived directly from municipal tap water. Bottlers may, however, distill or filter this water to remove molecules that impart a smell or taste that consumers find objectionable. These treatments also remove fluoride, an element that has been added to municipal water supplies and that helps to prevent tooth decay by killing bacteria in the mouth. Some dentists have become concerned that replacing children's tap water with bottled water could contribute to greater risk of cavities. However, too little research has been performed to be able to support or reject this hypothesis.

Most consumers choose bottled water for its convenience and portability, but at a cost. The billions of water bottles used each year require 1.5 million barrels of oil to produce. Although the bottles are recyclable, 86% of the bottles go to landfills each year (FIGURE 2.9). The energy required to transport bottled water far exceeds the energy required to clean an equivalent amount of tap water.

Plain or carbonated water is not the only beverage touted as "healthy" on the store shelves. Many sports drinks contain a number of additives that influence their characteristics. In the next section, we explore the chemical effects of adding various solutes to water and examine how they affect the quality of this essential drink.

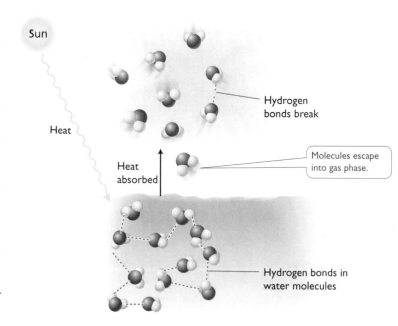

FIGURE 2.8 Water absorbs heat. Hydrogen bonds break as they absorb heat. Only after the hydrogen bonds are disrupted can water change from liquid to gas and evaporate. In this manner, large amounts of heat can be transferred, cooling objects in contact with the water.

2.2 Acids, Bases, and Salts

Some of the additives found in sports drinks change the chemistry of the water they are added to. Two important factors that are often modified are the water's acidity and its salt content.

pH: Measuring the Activity of Ions

Hydrogen is a simple atom—just a single proton orbited by a single electron. When this electron is shared with another, more electronegative atom in a molecule, the hydrogen atom has a tendency to dissociate, or break apart, from the molecule. The resulting ion, really just a lone proton, is designated as H^+, indicating that it has a positive charge resulting from the loss of its negatively charged electron. When a hydrogen ion dissociates from a water molecule, it leaves its electron with its companion OH^- ion (called a hydroxyl). These ions can react with other charged molecules and help to bring them into the water solution.

Represented as a chemical formula, the dissociation of water is written as

$$H_2O \rightleftharpoons H^+ + OH^-$$

Note that the "yields" arrows in this equation point both ways. This indicates that the process is constantly reversing under normal conditions. In a glass of pure water, molecules are constantly dissociating and reassociating, but the number of H^+ and OH^- ions remains relatively constant.

FIGURE 2.9 A hidden cost of bottled water. The production of containers for bottled water requires 1.5 million barrels of oil a year. Most of these bottles are not recycled.

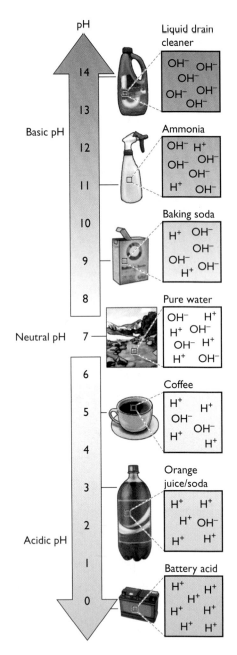

FIGURE 2.10 What the pH scale measures. The pH scale is a measure of hydrogen ion concentration. Each pH unit actually represents a tenfold (10×) difference in the concentration of H^+ ions. For example, a substance with a pH of 5 has 100 times more H^+ ions than a substance with a pH of 7. Water has a pH of 7 and is therefore neutral; that is, it has as many H^+ ions as OH^- ions. The pH of most cells is very close to 7.

VISUALIZE THIS How would the pH of pure water change if a small amount of ammonia was added to it? How much more acidic than coffee is soda?

The **pH scale** is a measure of the relative percentage of H^+ ions in a solution (FIGURE 2.10) and ranges from 0 to 14. Because there are equal numbers of H^+ and OH^- ions in pure water, it is neutral, which on the pH scale is 7. An **acid** is a compound that donates H^+ ions to a solution, causing the pH level to decline, while a **base** decreases H^+ concentration (primarily by releasing OH^- ions to bond with them), causing pH levels to rise. As we described earlier, when carbon dioxide is added to water, the result is the production of carbonic acid—this leads to a decline in pH.

Some of the flavorings added to bottled waters serve to lower their pH, notably lemon and lime juice. Although these flavorings deliver a pleasant taste, the resulting acidity of the drink can be destructive to tooth enamel, causing higher rates of tooth decay. When these drinks also contain sugar, which encourages bacterial growth, they have the potential to become quite damaging to teeth.

Homeostasis, the maintenance of stable internal conditions, will be covered in detail in Chapter 5. However, the maintenance of blood pH is a good example of homeostasis in humans and other living organisms and can serve as an introduction to the topic.

To maintain normal body function, the pH of body fluids must remain within a narrow range. In any solution, pH is stabilized by a pair of **buffers**, which can take up excess H^+ and OH^- ions. Blood pH in a healthy person is around 7.4. Carbonic acid, a normal constituent of blood, dissociates into bicarbonate (HCO_3^-) and re-forms constantly. This helps stabilize pH. When hydroxide ions are added to blood, the following reaction occurs:

$$OH^- + H_2CO_3 \rightleftharpoons HCO_3 + H_2O$$

Hydroxide ion + Carbonic acid $\rightleftharpoons$ Bicarbonate ion + Water

When hydrogen ions are added to the blood, this reaction occurs:

$$H^+ + HCO_3^- \rightleftharpoons H_2CO_3$$

Hydrogen ion + Bicarbonate ion $\rightleftharpoons$ Carbonic ion

Electrolytes

You have probably noticed that sweat has a salty taste. The sweat excreted onto your skin contains a number of ions, including sodium and potassium, which impart this flavor. These ions are also known as electrolytes.

Electrolytes form when substances dissociate into ions in solution. Electrolytes play important roles in the body, including the conduction of electrical impulses along nerves and through muscle. The loss of electrolytes through sweating can lead to some of the symptoms of dehydration, including nausea, dizziness, and fatigue. Many sports drinks contain added electrolytes, as do intravenous solutions used in hospitals.

It is possible to replace sodium and potassium lost through sweat by modifying your diet. Adding a little table salt to foods will replace the lost sodium. Many fruits and vegetables are rich in potassium. Because you can replace electrolytes by eating a healthful diet, it is not necessary for most people to consume sports drinks, even when exercising. However, for athletes who engage in intense exercise, beverages that contain electrolytes are better than water at replacing essential body fluids lost through sweat. Electrolyte replacement is important for anyone who is sweating heavily or who is engaging in endurance events.

Electrolytes are added to water in the form of **salts**, which are the product of reactions between acids and bases and which maintain their structure thanks to ionic bonds. Salts include any compound composed of positive and negative ions that result in a neutral product. For example, the electrically neutral table salt (NaCl) consists of Na^+ and Cl^- ions.

2.3 Structure and Function of Macromolecules

Although carbon is only one of many elements that living organisms require, it makes up the majority of their mass. In fact, the branch of chemistry concerned with carbon-containing compounds is called **organic chemistry**—"organic" referring to life. Carbon is an ideal element for the foundation of life's chemistry because its outer valence shell contains only four electrons—permitting bonding with up to four other elements. Like a Tinkertoy connector, carbon's multiple sites for connections allow carbon-containing molecules to take an almost infinite variety of shapes including rings, chains, and branched chains (FIGURE 2.11).

Many of the additives in sports drinks are organic, from sugars to proteins to essential fats. Although these substances all contain a carbon backbone, their configurations and additional elements give them very different properties. The organic chemicals found in living organisms come in four broad categories: carbohydrates, lipids, proteins, and nucleic acids. Most of these nutrients can be obtained from one's diet, and various health drinks often contain one or more chemicals as additives.

Structurally, all of these molecules are composed of subunits that are joined together and are called **macromolecules**. Individual subunits, called **monomers**, are joined to one other to produce multiunit **polymers**.

The building or synthesis of polymers involves a chemical reaction called **dehydration synthesis**, because water is released when an OH^- ion is removed from one monomer and an H^+ ion is removed from another monomer to join the subunits together by covalent bond (FIGURE 2.12a). Breaking polymers apart into their component monomers involves the reverse reaction, called **hydrolysis**. During hydrolysis, water is added across the chemical bond holding the monomers together, thereby releasing them (FIGURE 2.12b).

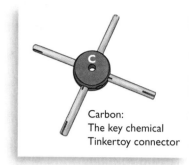

Carbon: The key chemical Tinkertoy connector

Carbon dioxide (CO_2)

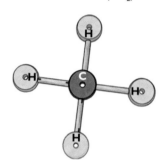

Methane (CH_4)

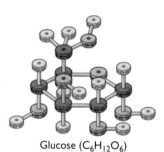

Glucose ($C_6H_{12}O_6$)

FIGURE 2.11 Carbon, the chemical Tinkertoy connector. Because carbon can connect with up to four other elements, carbon-containing compounds can be very diverse in shape.

VISUALIZE THIS In what ways is the carbon molecule different from a Tinkertoy connector?

(a) Dehydration reaction in the synthesis of a polymer

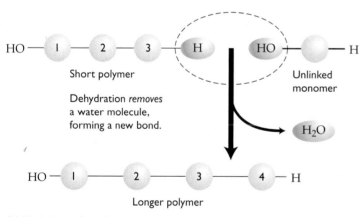

(b) Hydrolysis of a polymer

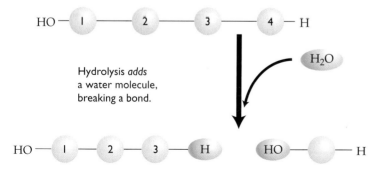

FIGURE 2.12 Dehydration synthesis and hydrolysis.
(a) Dehydration synthesis reactions remove water to join subunits together.
(b) Hydrolysis reactions add water across chemical bonds to break apart more complex substances into their individual subunits.

Carbohydrates

Carbohydrates are organic compounds that are commonly called sugars. Endurance athletes such as runners, cyclists, and swimmers often supplement their carbohydrate intake by ingesting carbohydrate-rich energy drinks. Let's learn about the structure and function of carbohydrates before analyzing the merits of this practice.

Carbohydrates play structural roles and serve as the major source of chemical food energy in cells. Energy is stored in the chemical bonds between the carbon, hydrogen, and oxygen atoms that make up carbohydrate molecules. The sugar glucose has six carbon, twelve hydrogen, and six oxygen atoms. The molecular formula for glucose is $C_6H_{12}O_6$. Glucose is a simple sugar, or *monosaccharide*, which consists of a single ring-shaped structure (FIGURE 2.13a). Disaccharides are two rings joined together by dehydration synthesis. Lactose, found in milk, is a *disaccharide* composed of glucose and another monosaccharide called galactose. The sugar you bake with or sprinkle into coffee is sucrose, a disaccharide composed of glucose and fructose (FIGURE 2.13b).

Polymers of sugar monomers are called **polysaccharides** (FIGURE 2.13c). Plants use tough polysaccharides of glucose in their cell walls as a sort of structural skeleton. This molecule, cellulose, is the most abundant carbohydrate on Earth. The external skeletons of insects, spiders, and lobsters are composed of polysaccharides, and the cell walls that surround bacterial cells are rich in structural polysaccharides. Plants, such as potatoes, wheat, rice, and corn, store excess carbohydrates as polymers of *starch*. Animals store excess carbohydrates as *glycogen* in muscles and the liver. Both starch and glycogen are polymers of glucose (FIGURE 2.14).

When multiunit sugars are composed of many different branching chains of sugar monomers, they are called *complex carbohydrates*. Complex carbohydrates, such as those found in fruits and vegetables, are often involved in storing energy for later use (Table 2.3). The body digests complex carbohydrates more slowly than it does simpler sugars, because

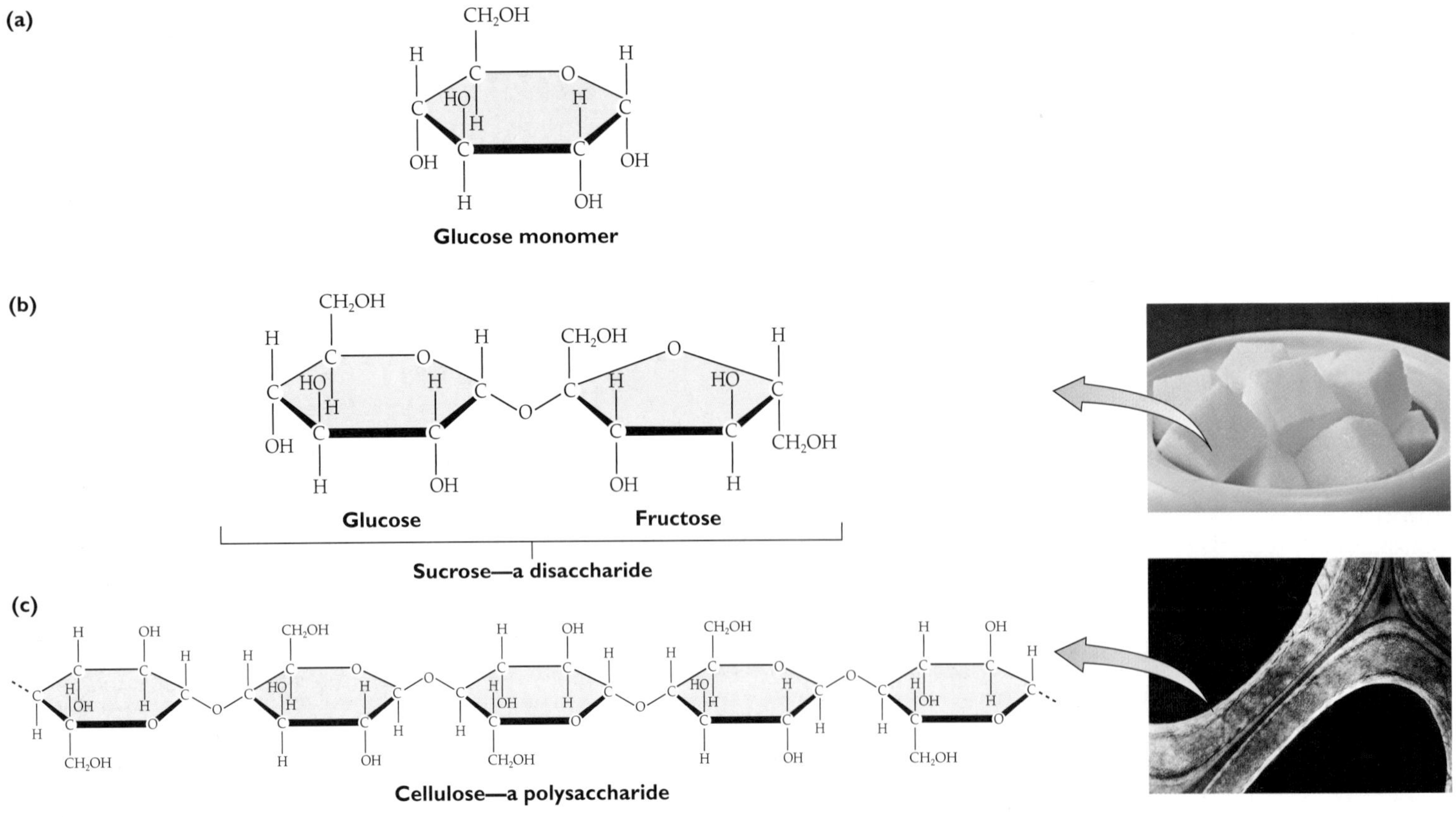

FIGURE 2.13 Carbohydrates. (a) Monosaccharides, such as the glucose molecule shown here, are individual sugar molecules. (b) Disaccharides are two monosaccharides joined together. (c) Polysaccharides are long chains of sugars joined together. The monosaccharide glucose and the disaccharide sucrose are important sources of energy, while cellulose plays a structural role in plant cell walls.

TABLE 2.3	Complex Versus Simple Carbohydrates	
	Simple Carbohydrates	**Complex Carbohydrates**
Characteristics	Digested quickly	Digested slowly
	Simple mono- or disaccharides	Polysaccharides
	Do not contain additional nutrients (that is, vitamins and minerals)	Often contain additional nutrients
Examples	Table sugar, fruit, fruit juice, milk, yogurt	Vegetables, bread, pasta, nuts, beans

complex carbohydrates have more chemical bonds to break. Endurance athletes load up on complex carbohydrates for several days before a race to increase the amount of easily accessible energy that they can draw on during competition.

Dietary **fiber**, also called roughage, is composed mainly of cellulose, which humans cannot digest into component monosaccharides. For this reason, dietary fiber passes through the digestive system and into the large intestine. Some fiber is digested by bacteria living there, and the remainder gives bulk to the feces. Fruits and vegetables tend to be rich in dietary fiber. Although fiber is not digested, it is still an important part of a healthful diet. Fiber helps protect blood vessels and may also decrease your risk of various cancers by reducing the amount of time potentially harmful substances remain in the large intestine.

A diet rich in complex carbohydrates from fruits, vegetables, and grains not only provides you with all the carbohydrate you should need for daily activities but has the added benefit of providing the vitamins, minerals, and fiber that come with these foods. Many sports drinks supply carbohydrates along with electrolytes and water. The additional carbohydrate is probably unnecessary except in cases of prolonged, intense exercise. In fact, in some cases, the added sugars increase the calorie count of these seemingly "healthy" drinks well beyond the calories in more traditional soft drinks (FIGURE 2.15).

Proteins

Your body requires **proteins** for a wide variety of processes. In fact, proteins serve so many important functions that you will learn about them in detail in several later chapters. For example:

- Proteins called *enzymes* help speed up chemical reactions inside cells and serve as components of cell membranes (Chapter 3).
- Proteins help regulate gene expression (Chapter 4).
- Proteins are important structural components of cells. Some cells, such as animal muscle cells, are composed almost entirely of proteins (Chapter 6).
- Proteins can function as chemical messengers (called *hormones*) that course throughout an organism's body (Chapter 16).

Proteins are large molecules made of monomer subunits called **amino acids** (Table 2.4). There are 20 commonly occurring amino acids. Your body is able to synthesize many of them. Those your body cannot synthesize are called **essential amino acids** and must be supplied by the foods you eat. Like carbohydrates, amino acids are made of carbons, hydrogens, and oxygens. In addition, amino acids contain nitrogen as part of an *amino group* ($-NH_2^+$) attached to one end of the amino acid. On the other end of the amino acid is a *carboxyl group* (FIGURE 2.16a). Lastly, each amino acid has its own particular *side group*. Side groups of amino acids give

Potatoes contain **starch**.

Animal muscle contains **glycogen**.

FIGURE 2.14 Stored carbohydrates. All cells contain carbohydrate, protein, and fat, but some have more than others. When carbohydrate is in excess, plants store the excess as starch, and animals store the excess carbohydrates as glycogen. The glucose molecules are abbreviated as hexagons here.

FIGURE 2.15 Which drink is healthier?

VISUALIZE THIS Compare and contrast the nutrition labels on an energy drink and a soda.

each individual amino acid its particular chemical properties (FIGURE 2.16b). Side groups can also interact to affect the three dimensional structure of the protein (FIGURE 2.16c).

Long polymers of amino acids are sometimes called *polypeptides*, so named because the covalent bond joining adjacent amino acids is a **peptide bond**. The precise sequence of amino acids in a polypeptide chain is called a protein's **primary structure** (FIGURE 2.17a). The linear sequence of amino acids making up the primary structure is

TABLE 2.4 | Amino Acids*

Nonpolar Amino Acids	Negatively Charged Amino Acids	Polar Amino Acids
Valine	Aspartic acid	Threonine
Leucine	Glutamic acid	Glycine
Isoleucine		Serine
Methionine	**Positively Charged Amino Acids**	Cysteine
Phenylalanine	Arginine	Asparagine
Tryptophan	Histidine	Glutamine
Alanine	Lysine	Tyrosine
Proline		

*Essential amino acids are highlighted in yellow.

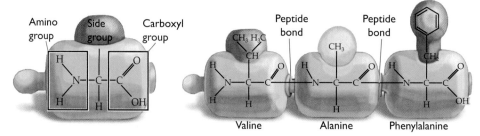

FIGURE 2.16 Amino acids, peptide bonds, and proteins. (a) All amino acids have the same backbone but different side groups. (b) Amino acids are joined together by dehydration synthesis to form covalent bonds called peptide bonds. Long chains of these are called polypeptides. (c) Polypeptide chains fold upon themselves

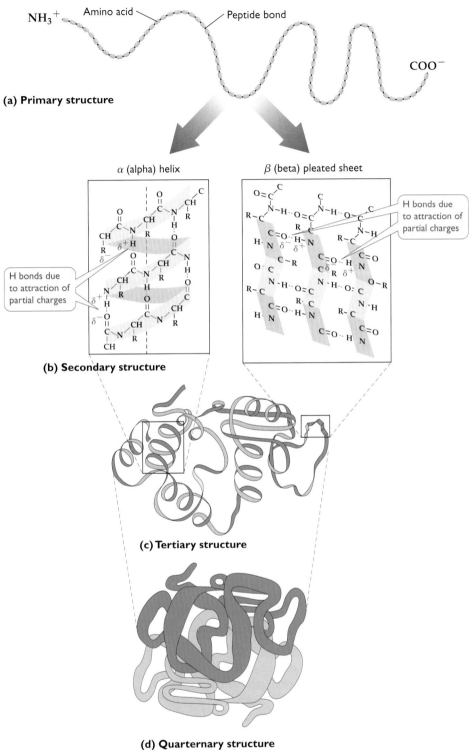

FIGURE 2.17 **Levels of protein organization.**
(a) The primary level of protein structure is its linear sequence of amino acids. (b) Hydrogen bonds between amino acids cause the formation of alpha helices and beta-pleated sheets. (c) A protein further folds due to interactions between side groups to give tertiary structure. (d) Some proteins have a quaternary structure, which arises when two or more polypeptides join to form a single protein.

held together by peptide bonds. Different polypeptides are composed of different linear orders of the 20 amino acids.

The **secondary structure** of a protein comes about when the polypeptide folds on itself in a regular, repeated fashion. These foldings can produce helical structures called *alpha helices* or accordion-like pleats called *beta-pleated sheets*. Secondary structures form when hydrogen bonds form between the amino group of one amino acid and the carboxyl group of another (FIGURE 2.17b).

In addition to the peptide bonds that form the primary structure and the hydrogen bonds that form secondary structures, the side groups of amino acids can also interact with each other. Side groups interact to give the protein its globular three-dimensional shape or **tertiary structure** (FIGURE 2.17c). Due to the differing chemical makeup of each amino acid side group, all of the different types of chemical bonds—hydrogen, ionic, and covalent—are involved in the formation of tertiary structure. Each polypeptide contains a unique sequence of amino acids and thus has a unique tertiary structure.

Certain proteins are made up of more than one polypeptide chain. In this case, the three-dimensional structure created by the interaction of these chains is called the **quaternary structure** (FIGURE 2.17d), but the principle is the same—a protein's unique shape confers its specialized chemical properties. The three-dimensional structure of a protein is also influenced by environmental conditions, such as pH and temperature, which can affect the interactions among amino acid side chains. For example, heating a protein can break the bonds maintaining its structure. A protein so treated is said to have been *denatured*. Adding heat to a protein-rich egg denatures the proteins in the egg white. This change is visible when the egg white transforms from a clear gel-like substance into a harder white substance during cooking.

Protein-rich foods include beef, poultry, fish, beans, eggs, nuts, and dairy products such as milk, yogurt, and cheese. Any protein that you consume is first digested into its component amino acids before entering the bloodstream for use in building the proteins your body needs. Digestion also acts on most protein enzymes that are marketed as supplements to your own enzymes, such as superoxide dismutase (SOD) and protease—these proteins cannot pass into your bloodstream undigested, so their purchase is most likely a waste of money. Enzymes that are functional in the digestive system itself, either in the stomach or the intestines, may provide benefits to some people as supplements. One example is lactase, which breaks down milk sugars in the intestine, relieving lactose-intolerant individuals from intestinal gas and abdominal discomfort.

A few energy drinks contain added proteins, but more common are those that contain specific amino acids. One highly touted additive is taurine, an amino acid that is not essential but can be synthesized from the essential amino acid methionine. Taurine is often advertised as an energy booster, but there is no scientific evidence that its supplementation alone reduces fatigue or increases endurance. Other commonly added amino acids include proline and tryptophan. In individuals with well-balanced diets, there is little need for amino acid supplementation. In a few athletes who are actively building muscle—such as bodybuilders—protein and amino acid supplementation may be useful to increase muscle mass. For most of us, excess protein will simply be used as an alternative energy source to power our activities or will be stored as fat.

Lipids

Lipids are partially or entirely hydrophobic substances made primarily of hydrogens and carbons. There are three different kinds of lipids, each with its own unique structure and function.

Fats The body uses fats as a source of energy. Gram for gram, fats contain a little more than twice as much energy as carbohydrates or protein. Most of this energy is stored in the hydrogen-carbon (*hydrocarbon*) bonds of the fat.

The structure of a **fat** includes *glycerol*, a three-carbon molecule of alcohol—meaning that it contains a hydroxyl group (−OH). Three long, hydrogen-and-carbon-rich (hydrocarbon) chains are attached to this glycerol molecule (FIGURE 2.18a). Like the hydrocarbons present in gasoline, these hydrocarbons can be burned to produce energy. The long hydrocarbon chains are called the **fatty acid** tails of the fat. Your body can synthesize most of the fatty acids it requires. The two that cannot be synthesized, *omega-3* and *omega-6*, are called **essential fatty acids**. Like essential amino acids, essential fatty acids must be obtained from the diet.

Omega-3 fatty acids have gained credibility as a healthful supplement since the 1970s when scientists showed that native Inuit in Greenland had lower rates of heart disease than other people living in the same country. Scientists linked the better heart health of the Inuit to their increased consumption of fish, whale, and seal, which are rich in omega-3 fatty acids. Research since that time supports the heart health benefits of omega-3 fatty acids, and most doctors now recommend that people eat two servings

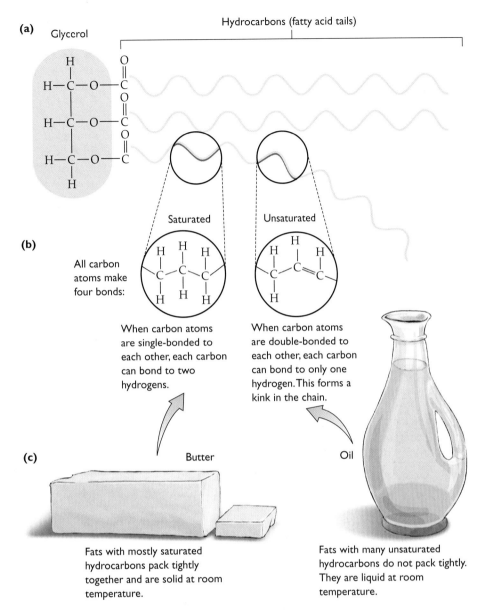

FIGURE 2.18 **Fats.** (a) Fatty acids can differ in length and the number and placement of double bonds. (b) Unsaturated fats have carbon-carbon double bonds. (c) The degree of saturation of a fat affects its consistency.

of fish every week. Other foods containing omega-3 acids are walnuts, flaxseed, and canola or soybean oil. Many foods are now fortified with omega-3, but few drinks are—primarily because it is difficult to combine the hydrophobic fatty acid with water and because it often has a "fishy" taste that consumers find objectionable. If these challenges are overcome by manufacturers, omega-3-enhanced sports and energy drinks are likely to appear.

The fatty acid tails of a fat can differ in the number and placement of double bonds. When the carbons of a fatty acid are bound to as many hydrogens as possible—that is, they do not double bond with other carbons—the fat is said to be a **saturated fat** (saturated in hydrogens). When there are carbon-to-carbon double bonds, the fat is not saturated in hydrogens, and it is an **unsaturated fat** (FIGURE 2.18b). The more double bonds, the higher the degree of unsaturation. When a fat contains many unsaturated carbons, it is referred to as *polyunsaturated*. In general there are fewer negative health effects associated with unsaturated fats than with saturated fats.

The double bonds in unsaturated fats make the structures kink instead of lying flat. This form prevents adjacent fats from packing tightly together, so unsaturated fat tends to be liquid at room temperature. The vegetable oils used in cooking are examples of unsaturated fats. In fact, plants tend to be richer in unsaturated fats than animals. Saturated fats, with their absence of carbon-to-carbon double bonds, pack tightly together. This is why saturated fats, such as butter, are solid at room temperature (FIGURE 2.18c).

42 CHAPTER 2 The Chemistry of Life

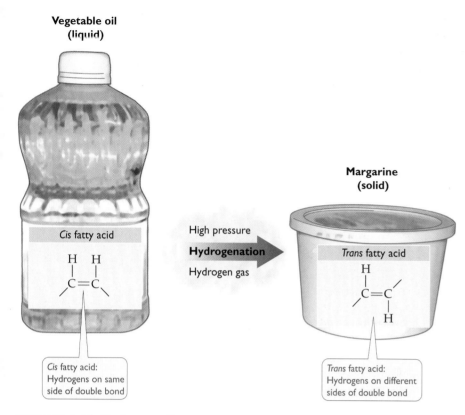

FIGURE 2.19 **Hydrogenation.** Adding hydrogen gas to vegetable oil forces the addition of hydrogen atoms to the fatty acid chains, some of which are incorporated in the unnatural *trans* configuration. The addition of hydrogen atoms means that there are fewer double bonds, so liquids can be solidified by this process.

Commercial food manufacturers sometimes add hydrogen atoms to unsaturated fats by combining hydrogen gas with vegetable oils under pressure. This process, called *hydrogenation*, increases the fat's level of saturation. Hydrogenation retards spoilage and solidifies liquid oils, thereby making food seem less greasy. Margarine is often vegetable oil that has undergone hydrogenation.

When hydrogen atoms are on the same side of the carbon-to-carbon double bond, they are said to be in the *cis* configuration. Naturally occurring unsaturated fats have their hydrogen atoms in the *cis* configuration. When hydrogen atoms are on opposite sides of the double bond, they are said to be in the *trans* configuration (FIGURE 2.19). Such **trans fats** are not found in nature. During the process of hydrogenation, about two-thirds of the hydrogens in the molecules become converted to the *trans* configuration.

Although the results of definitive studies are not yet in, the potential health risks of consuming foods rich in *trans* fatty acids, common in fast foods, include increased risk of clogged arteries, heart disease, and diabetes. Because fats contain more stored energy per gram than carbohydrates and proteins do, and because excess saturated fat intake is associated with several diseases, nutritionists recommend that you limit the amount of fat—especially saturated and hydrogenated fat—in your diet to less than 30%. Foods that are rich in fat include meat, milk, cheese, vegetable oils, and nuts.

Phospholipids   **Phospholipids** are similar to fats except that each glycerol molecule is attached to two fatty acid chains (not three, as you would find in a dietary fat). The third bond in a phospholipid is to a phosphate functional group ($-PO_4$). The phosphate group together with the glycerol molecule produces a hydrophilic head, and the two fatty acid tails are hydrophobic (FIGURE 2.20). Phospholipids often have an additional head group attached to the phosphate, which also confers unique chemical properties on the individual phospholipid. When mixed with water, phospholipids will cluster together with their hydrophilic heads facing out and the hydrophobic tails hidden inside. Because of their ability to form membranes around compartments, phospholipids are important constituents of cells.

Steroids   **Steroids** are composed of four fused carbon-containing rings, with variable side groups that give the molecules their unique properties. *Cholesterol* (FIGURE 2.21) is a steroid that you have probably heard about. The main function of cholesterol in animal cells is to help maintain the fluidity of membranes.

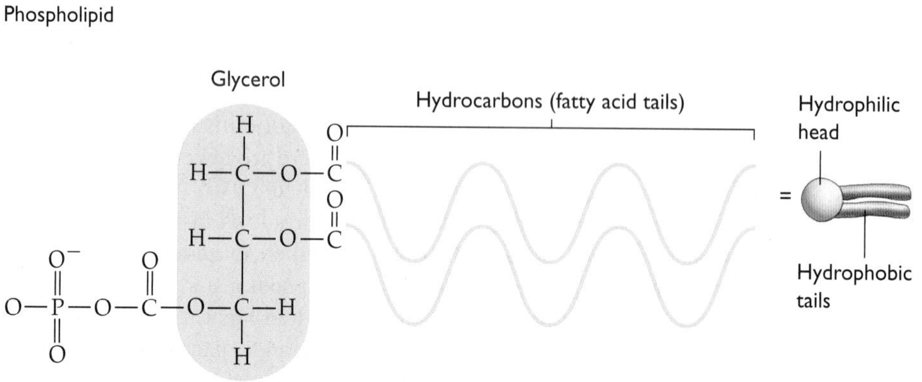

FIGURE 2.20 **Phospholipids.** Phospholipids are composed of a glycerol backbone with two fatty acids attached and one phosphate head group. They have hydrophilic heads and hydrophobic tails. They are constituents of cell membranes.

Because lipids like cholesterol are not soluble in aqueous (water-based) solutions, cholesterol is carried throughout the body, attached to proteins in structures called lipoproteins. **Low-density lipoproteins (LDLs)** have a high proportion of cholesterol (in other words, they are low in protein). LDLs distribute both the cholesterol synthesized by the liver and the cholesterol derived from diet throughout the body. LDLs are also important for carrying cholesterol to cells, where it is used to help make plasma membranes and hormones.

High-density lipoproteins (HDLs) contain more protein than cholesterol. HDLs scavenge excess cholesterol from the body and return it to the liver, where it is used to make bile and helps digest fats. The cholesterol-rich bile is then released into the small intestine, and from there much of it exits the body in the feces. The LDL/HDL ratio is an index of the rate at which cholesterol is leaving body cells and returning to the liver. A lower ratio is healthier, indicating that more cholesterol is being excreted than is allowed to travel in the bloodstream.

Your physician can measure your cholesterol level by determining the amounts of LDL and HDL in your blood. If your total cholesterol level is over 200 or your LDL level is above 100 or so, your physician may recommend that you decrease the amount of cholesterol and saturated fat in your diet. This may mean eating more plant-based foods and less meat, since plants do not have cholesterol. Saturated fat is thought to raise cholesterol levels by stimulating the liver to step up its production of LDLs and slowing the rate at which LDLs are cleared from the blood.

Cholesterol is not all bad; in fact, some cholesterol is necessary—it is present in cell membranes to help maintain their fluidity, and it is the building block, or *precursor*, for other steroids, such as the hormones estrogen and testosterone. You do, however, synthesize enough cholesterol so that you do not need to obtain much from your diet.

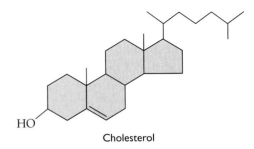

FIGURE 2.21 Steroids. Some steroids are a type of lipid composed of four fused rings. Cholesterol is a steroid common in animal cell membranes.

Nucleic Acids

Nucleic acids get their name from the fact that these molecules are highly concentrated in the nucleus of a cell, a structure whose primary function is to store these molecules. The most familiar of the molecules in this category, **deoxyribonucleic acid (DNA)**, serves as the primary storehouse of genetic information in nearly all living organisms. Genes are long stretches of DNA that, along with proteins, make up structures called chromosomes. Chromosomes are copied to pass on to new cells. In the case of sex cells, fertilization of an egg cell by a sperm cell brings the chromosomes of two parents together.

FIGURE 2.22 shows the three-dimensional structure of a DNA molecule and zooms inward to the chemical structure. You can see that DNA is composed of two curving strands that wind around each other to form a double helix. These strands are made up of long strings of monomers called **nucleotides** joined to each other along the length of the helix by covalent bonds. Each nucleotide is made up of a sugar, a phosphate, and a nitrogen-containing structure called a **nitrogenous base**. The sugar in DNA is called deoxyribose. The nitrogenous bases of DNA have one of four different chemical structures, each with a different name: *adenine* (A), *guanine* (G), *thymine* (T), and *cytosine* (C).

Each strand of the helix consists of a series of sugars and phosphates alternating along the length of the helix, which is called the **sugar-phosphate backbone**. The phosphate groups give the backbone a negative charge, which some proteins can bind to in order to turn specific genes on or off.

The strands of the helix align so that the nucleotides face "up" on one side of the helix and "down" on the other side of the helix. For this reason, the two strands of the helix are said to be **antiparallel**. Note in Figure 2.22 that one end of the DNA molecule has a phosphate group attached to a carbon (designated the 5′ carbon to refer to its position) and that the other end of the same strand has a hydroxyl group attached to the 3′ carbon.

Nitrogenous bases in the two strands of the helix form hydrogen bonds with each other across the width of the molecule. On a DNA molecule, an adenine (A) on one strand always hydrogen bonds with a thymine (T) on the opposite strand. Likewise, guanine (G) always hydrogen bonds with cytosine (C). The term *complementary* is used to describe these pairings. For example, A is complementary to T, and C is complementary

44 CHAPTER 2 The Chemistry of Life

FIGURE 2.22 DNA structure. (a) DNA is a double-helical structure composed of sugars, phosphates, and nitrogenous bases. (b) Each strand of the helix is composed of repeating units of sugars and phosphates, making the sugar-phosphate backbone, and nitrogenous bases. (c) A phosphate, a sugar, and a nitrogenous base constitute the structure of a nucleotide.

VISUALIZE THIS What property of the base pairs allows the width of the helix to remain constant?

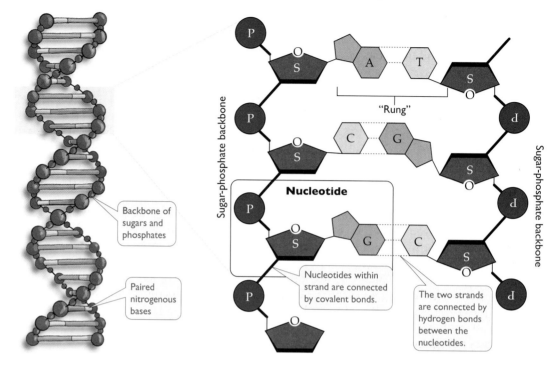

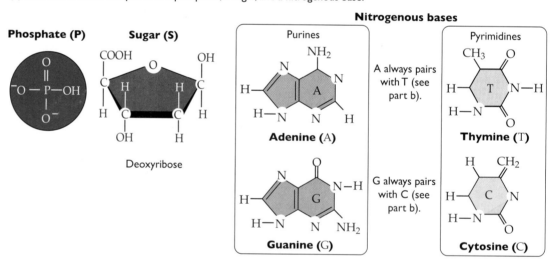

to G. The overall structure of a DNA molecule can be likened to a rope ladder that is twisted, with the hand rails of the ladder composed of the sugar-phosphate backbone, and the rungs of the ladder composed of the paired nitrogenous bases.

As a result of the base-pairing rule (A pairs with T; G pairs with C), the width of the DNA helix is uniform. There are no bulges or dimples in the structure of the DNA helix, because A and G, called *purines*, are structures composed of two rings, and C and T are single-ring structures called *pyrimidines*. A purine always pairs with a pyrimidine and vice versa, so there are always three rings across the width of the helix. A-to-T base pairs have two hydrogen bonds holding them together. G-to-C pairs have three hydrogen bonds holding them together.

Cells also contain various forms of the nucleic acid **ribonucleic acid (RNA)**. RNA differs from DNA in that it is single stranded, contains the sugar ribose in its backbone, and contains the nitrogenous base uracil (U) instead of thymine. RNA is also involved in the genetic work of the cell—its specific functions will be explored in detail in Chapter 4.

DNA or RNA that is ingested is broken down by the stomach into its component nucleotides. There is little reason to ingest either additional DNA or RNA in waters or other forms of supplements. Unless you have a rare genetic disease that negatively affects nucleotide synthesis, your body can make all the nucleotides it needs.

Stop and Stretch 2.2

DNA is found in every cell of our bodies. Sequences of DNA called genes encode biological characteristics. If the same sequences of DNA are present in every nonsex cell of our bodies, why might two cells differ from each other? For instance, if an eye cell and a liver cell contain the same sequences of DNA, why do they differ in both structure and function?

2.4 Micronutrients

In addition to carbohydrates, proteins, lipids, and nucleic acids, your body requires a number of additional substances to maintain proper function. Substances such as vitamins and minerals are required in very small amounts by the body and are called **micronutrients**. They are neither destroyed during use nor burned for energy. Many sports and energy drinks include these vitamins and minerals, both to promote health and to distinguish themselves from their competitors. Fitness waters in particular are sold as superior ways to rehydrate because they contain vitamin or mineral supplements.

Vitamins

Vitamins are organic substances, most of which the body cannot synthesize. They often function as *coenzymes*, substances that help enzymes and thus speed up the body's chemical reactions. When a vitamin is not present in sufficient quantities, deficiencies can affect every cell in the body because many different enzymes, all requiring the same vitamin, are involved in many different body functions. In addition, some vitamins help with the absorption of other nutrients. For example, vitamin C increases the absorption of iron from the intestine. Some vitamins may even help protect the body against cancer and heart disease and may slow the aging process.

The only vitamin human cells can synthesize is vitamin D, also called calcitriol. However, sunlight is required for vitamin D synthesis, so people living in northern regions can develop deficiencies in this nutrient. Vitamin D deficiency can lead to disrupted bone growth (FIGURE 2.23). To stave off widespread deficiency, several foods are fortified with vitamin D—most notably milk and other dairy products. Vitamin K, which is required for proper blood clotting, is not manufactured by human cells, but is produced by the bacteria living in our large intestines from undigested food waste. All other vitamins must be supplied by the foods that you eat or through dietary supplements.

Vitamins may be either hydrophilic, in which case they can be added to fitness waters, or hydrophobic. Because water-soluble vitamins are excreted and not stored, these nutrients are more likely than fat-soluble vitamins to be the source of dietary deficiencies. The symptoms of deficiency of the major hydrophilic vitamins are detailed in Table 2.5. As we have seen with other nutrients, if your diet is healthful, it is unnecessary to supplement your daily intake with the extra vitamins in fortified water. The benefits to a healthy individual of these added nutrients are minimal.

Hydrophobic vitamins (A, D, E, and K) are carried into the body in fat and can remain there. As a consequence, individuals are unlikely to experience deficiencies of these nutrients unless they are chronically undernourished. On the other hand, it is possible to have an excess of fat-soluble vitamins stored in body tissues. The symptoms of fat-soluble vitamin overdose range from mild to severe and are detailed in Table 2.6 alongside the symptoms of vitamin deficiency.

FIGURE 2.23 **Vitamin D deficiency.** Lack of vitamin D during growth leads to rickets, which is characterized by weak, malformed bones. Exposure to sunshine ensures that adequate vitamin D can be produced by the body, but many consumer products include vitamin D as insurance against this disease.

46 CHAPTER 2 The Chemistry of Life

TABLE 2.5 | Water-Soluble Vitamins

- Small organic molecules
- Will dissolve in water
- Cannot be synthesized by body
- Supplements packaged as pressed tablets
- Excesses usually not a problem because water-soluble vitamins are excreted in urine, not stored

Vitamin	Sources	Functions	Effects of Deficiency
Thiamin (B_1)	Pork, whole grains, leafy green vegetables	Required component of many enzymes	Water retention and heart failure
Riboflavin (B_2)	Milk, whole grains, leafy green vegetables	Required component of many enzymes	Skin lesions
Folic acid	Dark green vegetables, nuts, legumes (dried beans, peas, and lentils), whole grains	Required component of many enzymes	Neural-tube defects, anemia, and gastrointestinal problems
B_{12}	Chicken, fish, red meat, dairy	Required component of many enzymes	Anemia and impaired nerve function
B_6	Red meat, poultry, fish, spinach, potatoes, and tomatoes	Required component of many enzymes	Anemia, nerve disorders, and muscular disorders
Pantothenic acid	Meat, vegetables, grains	Required component of many enzymes	Fatigue, numbness, headaches, and nausea
Biotin	Legumes, egg yolk	Required component of many enzymes	Dermatitis, sore tongue, and anemia
C	Citrus fruits, strawberries, tomatoes, broccoli, cabbage, green pepper	Synthesis of particular tissues; improves iron absorption	Scurvy and poor wound healing
Niacin (B_3)	Nuts, leafy green vegetables, potatoes	Required component of many enzymes	Skin and nervous system damage

TABLE 2.6 | Fat-Soluble Vitamins

- Small organic molecules
- Will not dissolve in water
- Cannot be synthesized by body (except vitamin D)
- Supplements packaged in oily gel caps
- Excesses can cause problems since fat-soluble vitamins are not excreted readily

Vitamin	Sources	Functions	Effects of Deficiency	Effects of Excess
A	Leafy green and yellow vegetables, liver, egg yolk	Component of eye pigment; important for immunity	Night blindness, scaly skin, skin sores, and blindness	Drowsiness, headache, hair loss, abdominal pain, and bone pain
D	Milk, egg yolk	Helps calcium be absorbed and increases bone growth	Bone deformities	Kidney damage, diarrhea, and vomiting
E	Dark green vegetables, nuts, legumes, whole grains	Required component of many enzymes	Birth defects, anemia, and gastrointestinal problems	Fatigue, weakness, nausea, headache, blurred vision, and diarrhea
K	Leafy green vegetables, cabbage, cauliflower	Helps blood clot	Bruising, abnormal clotting, and severe bleeding	Liver damage and anemia

Minerals

Minerals are substances that do not contain carbon but are essential for many cell functions. Because they lack carbon, they are said to be *inorganic*. Minerals are important for proper fluid balance, muscle contraction, conduction of nerve impulses, and building bones and teeth. Calcium, chlorine, magnesium, phosphorus, potassium, sodium, and sulfur all are minerals.

Like most vitamins, minerals aren't synthesized in the body and must be supplied through your diet. Many fitness waters contain minerals in the form of electrolytes. Table 2.7 lists the various functions of minerals that your body requires and what happens when there is a deficiency or an excess of certain minerals.

Antioxidants

Several vitamins and minerals belong to a class of substances called **antioxidants**, chemicals thought to play a role in the prevention of many diseases including cancer and heart disease. Antioxidants protect cells and tissues from damage caused by highly reactive substances that are generated by normal cell processes. These highly reactive substances, called *free radicals*, have an incomplete electron shell. Free radicals are thus prone to *oxidize*, that is, remove electrons from, other molecules in the body. Antioxidants bind free radicals and prevent them from damaging DNA and other macromolecules. Antioxidants are abundant in fruits and vegetables, nuts, and grains (Table 2.8).

What Are You Paying For?

With all the dietary supplements available in the soft drink aisle, you might wonder if it is possible to eat an unhealthy diet but maintain your health by using the appropriate fitness and sports drinks. In some cases, the supplements available in these drinks can fill an important gap in nutrition—for example, vegetarians may be deficient in vitamin B_{12} and those who drink little milk may be deficient in calcium. However, eating a wide

TABLE 2.7 | Minerals

- Will dissolve in water
- Inorganic elements
- Cannot be synthesized by body
- Supplements packaged as pressed tablets

Mineral	Sources	Functions	Effects of Deficiency	Effects of Excess
Calcium	Milk, cheese, dark green vegetables, legumes	Bone strength, blood clotting	Stunted growth, osteoporosis	Kidney stones
Chloride	Table salt, processed foods	Formation of acid in stomach	Muscle cramps, reduced appetite, poor growth	High blood pressure
Magnesium	Whole grains, leafy green vegetables, legumes, dairy, nuts	Required component of many enzymes	Muscle cramps	Neurological disturbances
Phosphorus	Dairy, red meat, poultry, grains	Bone and tooth formation	Weakness, bone damage	Impaired ability to absorb nutrients
Potassium	Meats, fruits, vegetables, whole grains	Water balance, muscle function	Muscle weakness	Muscle weakness, paralysis, and heart failure
Sodium	Table salt, processed foods	Water balance, nerve function	Muscle cramps, reduced appetite	High blood pressure
Sulfur	Meat, legumes, milk, eggs	Components of many proteins	None known	None known

TABLE 2.8 | Antioxidants

- Protect cells from damage caused by free radicals
- Thought to have a role in disease prevention

Antioxidant	Source
Beta-carotene	Foods rich in beta-carotene are orange in color; they include carrots, cantaloupe, squash, mangoes, pumpkin, and apricots. Beta-carotene is also found in some leafy green vegetables such as collard greens, kale, and spinach.
Lutein	Lutein, which is known to help keep eyes healthy, is also found in leafy green vegetables such as collard greens, kale, and spinach.
Lycopene	Lycopene is a powerful antioxidant found in watermelon, papaya, apricots, guava, and tomatoes.
Selenium	Selenium is a mineral (not an antioxidant) that serves as a cofactor for many antioxidant enzymes, thereby increasing their effectiveness. Rice, wheat, meats, bread, and Brazil nuts are major sources of dietary selenium.
Vitamin A	Foods rich in vitamin A include sweet potatoes, liver, milk, carrots, egg yolks, and mozzarella cheese.
Vitamin C	Foods rich in vitamin C include most fruits, vegetables, and meats.
Vitamin E	Vitamin E is found in almonds, many cooking oils, mangoes, broccoli, and nuts.

variety of whole foods such as fruits, vegetables, and grains provides you with a better chance of achieving a healthful diet because they contain a variety of the micronutrients your body needs. An orange, for example, provides vitamin C, beta-carotene (an antioxidant and precursor to vitamin A), calcium, fiber, and other nutrients. Fitness water supplemented with vitamin C lacks these other healthy components.

In some cases, drinking sports drinks during exercise can be counterproductive. Many sports drinks are high in calories. For an individual who is exercising to lose or maintain weight, these drinks undermine the goal. Some of these drinks have so many additives that their solute concentration is higher than that of blood, that is, they are *hypertonic* to blood. Hypertonic drinks can draw water from body tissues as they pass through the digestive system, actually increasing the risk of dehydration. These sports drinks are only appropriate for athletes who need to load up on carbohydrates before an event.

Rehydration is most effective when a drink is *hypotonic* to the blood, containing a lower concentration of solutes. Another additive to watch out for is caffeine, a drug that actually increases water excretion from the kidneys.

The health claims made by makers of fitness waters and sports drinks may also be misleading. Before a prescription or nonprescription drug is released to the public, the U.S. Food and Drug Administration requires scientific testing to prove its safety and effectiveness. However, the same is not true of dietary supplements, such as taurine or superoxide dismutase added to drinks. Thus, claims on the labels of fitness waters and energy drinks about the efficacy of their "formula" have not necessarily been proven. Although some of the claims might end up having merit, there is no guarantee that they will be of benefit—or that they will not cause harm if used over a long period.

As with any product that makes claims about its benefits, you should carefully evaluate the claims of these drinks before deciding whether to buy them. Use what you know about the process of science and the importance of well-controlled studies in your evaluation. If you don't have time or a desire to do the research yourself, consult your doctor or a trustworthy web site using the guidelines presented in Chapter 1. Armed with sufficient information, you can more assuredly drink to your health.

Chapter REVIEW

ROOTS TO REMEMBER

The following roots of words come mainly from Latin and Greek and will help you to decipher terms:

hydro- means water.
-lysis means to break apart.
-mer means subunit.
macro- means large.
micro- means small.
mono- means one.
-philic means loving.
-phobic means fearing.
poly- means many.

KEY TERMS

2.1 Water: Essential to Life
atomic number *p. 28*
atom *p. 28*
chemical reaction *p. 31*
cohesion *p. 31*
compound *p. 29*
covalent bond *p. 29*
dehydration *p. 28*
electronegative *p. 30*
electron *p. 28*
electron shell *p. 29*
element *p. 28*
hydrogen bond *p. 31*
hydrophilic *p. 31*
hydrophobic *p. 31*
ion *p. 29*
ionic bond *p. 29*
isotope *p. 29*
mass number *p. 29*
molecule *p. 29*
neutron *p. 28*
nonpolar *p. 31*
nucleus *p. 28*
polar *p. 30*
product *p. 31*
proton *p. 28*
radioactive isotope *p. 29*
reactant *p. 31*
solute *p. 30*
solution *p. 30*
solvent *p. 30*
valence shell *p. 29*

2.2 Acids, Bases, and Salts
acid *p. 34*
base *p. 34*
buffer *p. 34*
electrolyte *p. 34*
homeostasis *p. 34*
pH scale *p. 34*
salt *p. 34*

2.3 Structure and Function of Macromolecules
amino acid *p. 37*
antiparallel *p. 43*
carbohydrate *p. 36*
dehydration synthesis *p. 35*
deoxyribonucleic acid (DNA) *p. 43*
essential amino acid *p. 37*
essential fatty acid *p. 40*
fat *p. 40*
fatty acid *p. 40*
fiber *p. 37*
high-density lipoprotein (HDL) *p. 43*
hydrolysis *p. 35*
lipid *p. 40*
low-density lipoprotein (LDL) *p. 43*
macromolecule *p. 35*
monomer *p. 35*
nitrogenous base *p. 43*
nucleic acid *p. 43*
nucleotide *p. 43*
organic chemistry *p. 35*
peptide bond *p. 38*
phospholipid *p. 42*
polymer *p. 35*
polysaccharide *p. 36*
primary structure *p. 38*
protein *p. 37*
quaternary structure *p. 40*
ribonucleic acid (RNA) *p. 44*
saturated fat *p. 41*
secondary structure *p. 39*

KEY TERMS (continued)

steroid *p. 42*
sugar-phosphate backbone *p. 43*
tertiary structure *p. 40*

trans fat *p. 42*
unsaturated fat *p. 41*

2.4 Micronutrients
antioxidant *p. 47*
micronutrient *p. 45*

mineral *p. 47*
vitamin *p. 45*

SUMMARY

2.1 Water: Essential to Life

- Atoms are the smallest units of matter, and they are made up of protons, neutrons, and electrons (p. 28).

- A water molecule is a chemical compound of two hydrogren atoms and one oxygen atom (p. 29).

- Chemical bonding depends on an element's electron configuration. Atoms whose outermost energy level, or valence shell, is not full will combine to form molecules. When the electrons are shared, the chemical bond is called covalent. When ions that have lost or gained valence shell electrons are held together by mutual attraction, the result is an ionic bond (p. 29).

- Water is a good solvent because it is a polar molecule, having a partial negative charge on one pole and a partial positive charge on the other. Water molecules cluster around other charged atoms and molecules, dissolving them (p. 30; Figure 2.5).

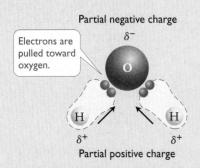

- Water facilitates chemical reactions by allowing dissolved reactants to come in contact with each other (p. 31).

- The polarity of water permits the attraction between the hydrogen atoms and the oxygen atom in adjacent molecules, forming a hydrogen bond. As a result, water is cohesive and slow to change form—making it a good vehicle for absorbing and removing heat (p. 31).

2.2 Acids, Bases, and Salts

- Ions are charged particles formed when an atom loses or gains one or more electrons (p. 33).

- When a hydrogen atom loses its electron it becomes a hydrogen ion, which is a bare proton (p. 33).

- The pH scale is a measure of the relative percentages of H^+ ions in a solution and ranges from 0 (acidic) to 14 (basic) (p. 34; Figure 2.10).

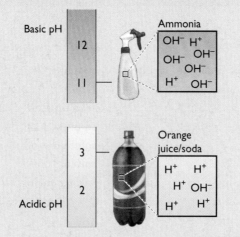

- Salts are produced by the reaction of an acid with a base and are kept together via ionic bonds (p. 34).

2.3 Structure and Function of Macromolecules

- Organic compounds have carbon atoms comprising their skeleton. Carbon can covalently bond to four other atoms, making them a very flexible skeleton (p. 35).

- Macromolecules (polymers) are composed of subunits (monomers) joined together by dehydration synthesis (pp. 35–36).

- Carbon-containing molecules show molecular diversity and, as the basis of life's chemistry, are called organic molecules (p. 35; Figure 2.11).

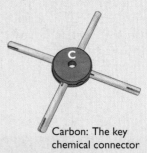

Carbon: The key chemical connector

- Carbohydrates serve as the major source of energy in cells. Carbohydrates can be single-unit monosaccharides or multiple-unit polysaccharides (pp. 36–37).

- Proteins play enzymatic, transport, gene expression, structural, and chemical messenger roles in cells. They are composed of amino acid monomers arranged in different orders. Interactions among the amino acid side chains produce the three-dimensional shapes of proteins, leading to their unique functions (pp. 37–40).

- Lipids are hydrophobic and come in three different forms. Fats are composed of glycerol and three fatty acids. Fats store energy, and they pad and insulate bodies. Phospholipids are composed of glycerol, two fatty acids, and a phosphate group. They are important structural components of cell membranes. Steroids are composed of four fused carbon-containing rings. Cholesterol is a steroid found in most animal cell membranes and helps maintain fluidity. Other steroids function as hormones (pp. 40–43).

- Nucleic acids are polymers of nucleotides, each of which is composed of a sugar, a phosphate, and a nitrogen-containing base (A, C, G, or T) (p. 43).

- DNA, the genetic material of humans, is double stranded and helical. RNA is involved in carrying the message of the DNA and is single stranded (p. 44).

2.4 Micronutrients

- Micronutrients are dietary substances required in very small amounts (p. 45).
- Vitamins are organic micronutrients, most of which the body cannot synthesize. Many vitamins serve as coenzymes to help enzymes function properly (pp. 45–47).
- Minerals are micronutrients that do not contain carbon but are essential for many cell functions (p. 47).
- Antioxidants are chemicals that scavenge free radicals from cells, preventing the damage these compounds can cause to other macromolecules (p. 47; Table 2.8).

Lycopene

Beta-carotene

LEARNING THE BASICS

1. Water _____.
   a. is a good solute
   b. contains two oxygen atoms for every one hydrogen atom
   c. functions as an enzyme
   d. makes hydrogen bonds with other molecules
   e. has an acidic pH

2. Electrons _____.
   a. are negatively charged
   b. comprise the nucleus along with neutrons
   c. are attracted to the negatively charged nucleus
   d. have the least energy if located farthest from the nucleus
   e. participate in hydrogen bonding but not ionic or covalent bonds

3. When an atom loses an electron, all of the following statements about it are true *except* _____.
   a. it would be signified by the symbol X^+, where X is the chemical symbol for the atom
   b. it has an overall positive charge
   c. it is called an ion
   d. it can dissolve in water
   e. it will form covalent bonds with other positive ions

4. Water is polar because _____.
   a. it has fewer electrons than other molecules
   b. it forms a solid easily at room temperature
   c. it dissolves other polar molecules
   d. oxygen is strongly electronegative, pulling electrons away from hydrogen
   e. it forms hydrogen bonds with other water molecules

5. Water absorbs excess heat energy because _____.
   a. it is almost always cooler than body temperature
   b. hydrogen bonds between water molecules require lots of energy to disrupt
   c. hydrogen has room in its valence shell for excess electrons
   d. it can form bonds with energy particles, separating them from each other
   e. it is rarely found in gas form

6. When added to water, sodium hydroxide (NaOH) dissociates into Na^+ and OH_2^-. You would expect sodium hydroxide to _____ pH.
   a. decrease
   b. increase
   c. have no effect on

7. Carbohydrates _____.
   a. store energy
   b. can be monomers or polymers
   c. are found in plant and animal cells
   d. all of the above

8. Fiber _____.
   a. is a complex carbohydrate
   b. is composed, in part, of cellulose
   c. is difficult for humans to digest
   d. may prevent certain cancers
   e. all of the above are correct

9. Which of the following terms is least like the others?
   a. monosaccharide
   b. phospholipid
   c. fat
   d. steroid
   e. lipid

10. Different proteins are composed of different sequences of _____.
    a. sugars
    b. glycerols
    c. fats
    d. amino acids

11. Essential amino acids _____.
    a. all have the same side group
    b. are only found in the proteins of meat
    c. are amino acids that your body can't synthesize and that must be obtained from the diet
    d. are found only in plant foods
    e. are more likely to be found in processed foods than in whole foods

12. Lipids are generally nonpolar, which means that _____.
    a. they do not dissolve in water
    b. they have partial charges that are distributed unevenly around the molecule
    c. they are hydrophilic
    d. they do not contain double bonds
    e. their electron shells are full

13. The macromolecules that contain genetic information include _____.
    a. proteins
    b. nucleic acids
    c. lipids
    d. carbohydrates
    e. all of the above contain genetic information

14. Examples of micronutrients include all of the following except _____.
    a. RNA
    b. vitamin B_{12}
    c. calcium
    d. vitamin D

15. Fruits and vegetables are generally more healthful than dietary supplements because _____.
    a. they contain fiber
    b. they contain antioxidants
    c. they contain many different vitamins and minerals
    d. they are whole, unprocessed foods
    e. all of the above

16. A molecule of sucrose contains 12 atoms of carbon, 22 atoms of hydrogen, and 11 atoms of oxygen. The formula for sucrose is _____.

17. Monomers are joined into multiunit _____ by the process of _____ synthesis.

18. Micronutrients include _____ and _____.

19. **True or False?** Bottled water is almost always cleaner than tap water.

20. **True or False?** Our bodies can manufacture nearly all necessary vitamins and minerals from simpler components.

ANALYZING AND APPLYING THE BASICS

1. Any molecule containing oxygen can be polar. The structure of methanol (CH_3OH) is shown in FIGURE 2.24. Which part of this molecule will have a partial negative charge, and which will have a partial positive charge?

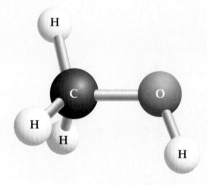

FIGURE 2.24 **Methanol.**

2. Carbon dioxide can be dissolved in water at high pressure but is released as gas when the pressure is released. Water boils at a lower temperature at high altitudes, where air pressure is lower. Given these observations, how do you think differences in pressure affect conversions between liquids and gases?

3. Cells in the lining of the stomach excrete hydrochloric acid (HCl), which dissociates into H^+ and Cl^-. How might this affect the pH of the stomach juices? How does a change in pH facilitate the digestion of proteins into amino acids?

4. Energy bars like PowerBar are high-carbohydrate and high-protein foods packaged in a portable and flavorful form. Under what conditions of physical activity and diet does it make sense for an individual to add these bars to his or her diet?

5. Glucose and fructose have the same molecular formula but different chemical properties. Why might this be?

6. Use the base-pairing rules to predict the sequence of DNA nucleotides on the strand complementary to the sequence 5'-AACGATCCG-3'.

Questions 7–9 concern the research that established that DNA is the molecule of heredity.

7. In 1928, a scientist named Frederick Griffith performed an experiment using bacteria and mice. When he injected mice with a disease-causing, or virulent, strain of the bacteria *Streptococcus pneumoniae*, they died. When he injected another group of mice with a non-disease-causing, or avirulent, strain of the same bacteria, the mice did not die. Griffith noted that the disease-causing strains of bacteria were coated in polysaccharides. To determine whether the virulent factor was in the polysaccharide coat surrounding the bacteria, Griffith heated the virulent strain, knowing that the heat would kill the bacterial cell but would not affect the polysaccharide coat. What result did he predict if the virulence was contained in the polysaccharide?

8. Next, Griffith injected another group of mice with heat-killed virulent cells and live nonvirulent bacterial cells. Remarkably, the mice died. From this experiment, Griffith deduced that the dead virulent strain somehow passed its genetic material to the living nonvirulent strain.

In the early 1940s, Oswald Avery and Maclyn McCarty began to try to determine the nature of the substance being passed between strains of Griffith's bacteria. The scientists isolated the substance that was being passed and dubbed it the "transforming principle" because it transformed the properties of one strain of bacteria into another. When this substance was treated with enzymes that inactivate proteins and those that destroy lipids, transformation still occurred. What did this result tell them about the chemical nature of the transforming principle?

9. Avery and McCarty's results indicated that nucleic acids must be the genetic material. Other scientists were skeptical. In 1952, Alfred Hershey and Martha Chase performed an experiment that allowed them to determine whether protein or DNA was the hereditary molecule.

   Hershey and Chase used a virus that infects bacteria. This virus was composed of only DNA and a coating of proteins. During an infection, the viruses hijack the bacterial cell and direct the synthesis of more viruses. Since the virus only contained DNA and protein, Hershey and Chase wanted to determine whether it was the viral DNA or viral proteins that were able to direct the synthesis of new viruses. In other words, which chemical was acting as the hereditary material?

   The scientists grew some viruses in liquid that contained radioactive phosphorus and others in liquid that contained radioactive sulfur. The radioactive phosphorus labels only DNA, and radioactive sulfur labels only proteins. Viruses with radioactive DNA and viruses with radioactive proteins were then allowed to infect bacterial cells. Hershey and Chase allowed these viruses to infect the cells, then washed them to remove all external materials. Bacteria in both cultures were then measured for radioactivity. Only the bacteria grown with the viruses with radioactive DNA showed any trace of radioactivity. How does this result confirm that it was the DNA that contained the transforming principle?

10. The energy drink Red Bull is banned in several European countries due, in part, to concerns over its high caffeine content. Excess caffeine consumption can cause anxiety, heart palpitations, irritability, and difficulty sleeping. Many people use high-caffeine energy drinks as a mixer for alcohol. This practice has many health care practitioners worried. Why might mixing alcohol with energy drinks have even more negative health consequences than mixing alcohol with more conventional mixes such as pop, water, or juice?

CONNECTING THE SCIENCE

1. Water's characteristic as an excellent solvent means that many human-created chemicals (including some that are quite harmful) can be found in bodies of water around the globe. How would our use and manufacture of harmful chemicals be different if most of these chemicals could not be dissolved and diluted in water but instead accumulated where they were produced and used?

2. Select one fitness water or sports or energy drink and research whether or not the claims it makes on its label are backed up by scientific evidence. Use the strategies for interpreting scientific evidence that you learned about in Chapter 1.

3. The packaging and marketing of many of the drinks discussed in the chapter are meant to evoke good health and high energy. Do you think the packaging and claims are misleading? Whose responsibility is it to ensure that consumers are purchasing products that are truly healthful for them—the manufacturers, who could disclose more information, or the consumers themselves?

Chapter 3

Cell Structure and Metabolism

Diet

LEARNING GOALS

1. What macromolecules provide the main dietary sources of energy?
2. What is the difference between a calorie and a Calorie?
3. How do macromolecules store energy?
4. How does ATP energize cellular reactions?
5. Describe the structure and function of the nucleus, ribosomes, the ER, mitochondria, and the Golgi apparatus.
6. How do enzymes speed up the rate of metabolic reactions?
7. How do active and passive transport differ?
8. What kinds of substances can pass freely through the membrane? What types require help to pass through membranes?
9. How are cellular respiration and breathing respiration related?
10. What is the chemical equation of cellular respiration?

The ideal body image is unattainable for most people.

People come in all shapes and sizes, yet attractiveness tends to be more narrowly defined by the images of men and women we see in the popular media. Nearly all media images equate attractiveness and desirability with a very limited range of body types.

The ideal body type for men has long meant being tall, broad-shouldered, and muscular with so little body fat that every muscle is visible. For many men, no amount of diet and exercise will ever allow them to attain this body type.

Current standards for female beauty are equally unforgiving and include small hips; long, thin limbs; and an absence of fat anywhere on the body except for the breasts. For most females, this body type is not only unattainable but is actually at odds with their biology.

Into this clash between body image and human biology steps the average college student—worried about appearance, yet also trying to find time to study, exercise, and socialize. To complicate matters, the student is now making all of his or her own decisions about food, often on a limited budget.

Making choices that help maintain a healthy body weight is not always easy. Typical meal-plan choices at college dining centers are often greasy and full of calories—and available in unlimited portions. Making healthful choices is further complicated by the presence of conveniently located fast-food restaurants that serve time-pressed students inexpensive foods in "supersized" portions.

The combination of unrelenting pressure to be thin and a glut of readily available but unhealthful foods can lead students to establish (or continue) poor eating habits that continue far beyond college life. If these eating habits persist, they can result in lifelong battles with obesity or self-inflicted starvation.

Making good decisions about how much and what to eat requires an understanding of the role of food in the body.

Most college students are, for the first time, making all their own choices about food.

Making unhealthy choices now can create habits that lead to future health problems associated with anorexia . . .

. . . and obesity.

3.1 Food and Energy

The food that organisms ingest provides building-block molecules that can be broken down and used as raw materials for growth, for maintenance and repair, and as a source of energy. Foods that organisms require to fulfill these functions include proteins, carbohydrates, and fats (FIGURE 3.1). You learned about the structure and function of these macromolecules in Chapter 2, so the discussion in this chapter will focus on the roles these nutrients play in the diet.

Energy is stored in the chemical bonds of foods. When energy-rich bonds between the carbon, hydrogen, and oxygen atoms of these macromolecules are broken, the energy released can be converted into a form that the body can use.

Excess energy released from food can be stored as glycogen in the muscles and liver. When muscle and liver cells are filled with glycogen, additional energy is stored as fat. Fat is often found in association with muscles, which then have easy access to the stored energy. Humans store fat just below the skin, to help cushion and protect vital organs, to insulate the body from cold weather, and to stockpile energy in case of famine (FIGURE 3.2).

The body tissue where fat is stored is called **adipose** tissue. Adipose tissue is found in the abdomen, buttocks, and hips—parts of the body that contribute greatly to body image. Human males tend to store fat in their abdomens, making it difficult for many men, even those with low body fat levels, to achieve the kind of abdominal-muscle definition seen in some athletes. Human females store fat in the

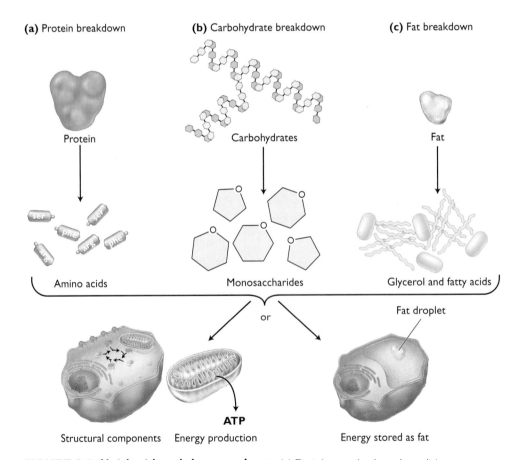

FIGURE 3.1 Nutrient breakdown and use. (a) Proteins are broken down into component amino acids, which can be used to produce structural components of cells or stored as fat. (b) Polysaccharides are broken down into component monosaccharides and used in structural components of cells, burned for energy, or stored as fat. (c) Fats are broken down into glycerol and fatty acids, which can be used to build structural components of cells, burned for energy, or stored.

adipose tissue of their buttocks, hips, thighs, and breasts, leading to a naturally more curvy appearance that is at odds with the stick thin body type displayed by most fashion models. To maintain a constant amount of body fat, energy intake must closely parallel energy expenditure.

To meet the energy demands of daily living, the energy stored in the chemical bonds of food is not used directly. Instead, food energy is converted into a high-energy chemical that can power most cellular activities.

ATP Is the Cell's Energy Currency

The chemical that cells use to power their activities is called **adenosine triphosphate**, or **ATP**. ATP can supply energy to cells because it stores energy obtained from the movement of electrons from food molecules in its own bonds.

ATP is a special type of nucleotide. Recall from Chapter 2 that nucleotides consist of a nitrogenous, or nitrogen-containing, base—adenine (A), guanine (G), cytosine (C), thymine (T), or uracil (U)—plus a sugar and a phosphate group (made up of the elements phosphorus and oxygen). ATP is a nucleotide *tri*phosphate. It contains the nitrogenous base adenine, the sugar ribose, and not one but three phosphates (FIGURE 3.3). Each phosphate in the series of three is negatively charged. These negative charges repel each other, which contributes to the stored energy in this molecule.

ATP behaves much like a coiled spring. To understand this, think about loading a dart gun. Pushing the dart into the gun requires energy from your arm muscles, and much of the energy you exert will be stored in the coiled spring inside the dart gun (FIGURE 3.4a). When you shoot the dart gun, the energy is released from the gun and used to perform some work—in this case, sending a dart through the air. Likewise, energy captured from the degradation of food molecules is stored in the phosphate bonds of an ATP molecule. Releasing a phosphate group from ATP liberates this energy for the cell to use to perform work (FIGURE 3.4b). After the removal of a phosphate group,

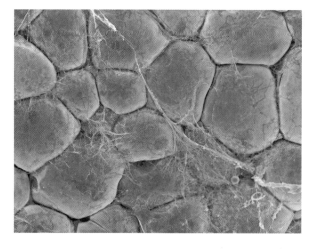

FIGURE 3.2 **Fat storage.** Humans store fat in adipose cells under the skin.

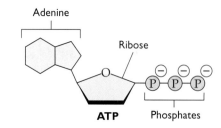

FIGURE 3.3 **The structure of ATP.** ATP is a nucleotide with a total of three phosphate groups. Notice that the phosphate groups are all negatively charged.

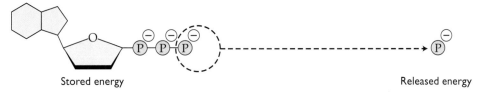

FIGURE 3.4 **Stored energy.** (a) A dart gun uses energy stored in the coiled spring and supplied by the arm muscle to perform the work of propelling a dart. (b) ATP provides energy stored in its bonds to perform cellular work.

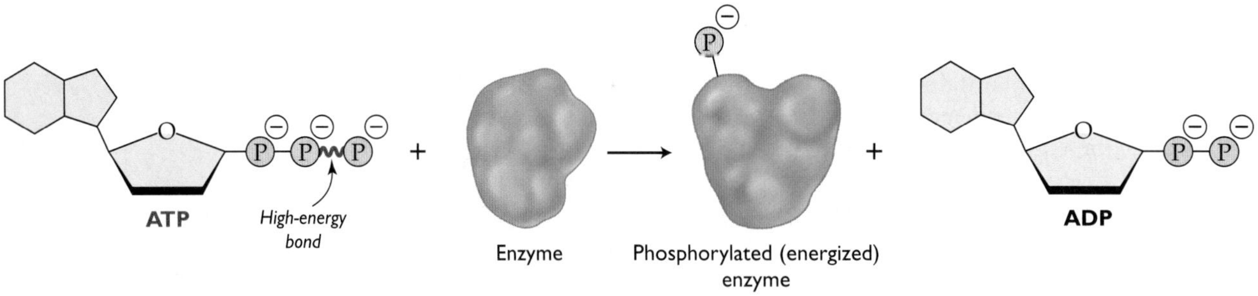

FIGURE 3.5 Phosphorylation. The terminal phosphate group of an ATP molecule can be transferred to another molecule, in this case an enzyme, to energize it. When ATP loses a phosphate, it becomes ADP. The enzyme that gained the phosphate group becomes energized.

ATP is converted into **adenosine diphosphate (ADP)**, which has two phosphates (hence *di*phosphate instead of *tri*phosphate).

The phosphate group that is removed from ATP can be transferred to another molecule. Thus, one way for ATP to energize other compounds is through **phosphorylation**, which means the addition of a phosphate. When a molecule, say, an enzyme, needs energy, the phosphate group is transferred from ATP to the enzyme, and the enzyme now carries the energy it needs to perform its job (FIGURE 3.5).

The energy released by the removal of the outermost phosphate of ATP can be used to help cells perform many different kinds of work. ATP helps power *mechanical work* such as the movement of a whiplike flagellum to help propel a sperm cell, *transport work* such as the movement of substances across membranes during active transport, and *chemical work* such as the making of complex molecules from simpler ones (FIGURE 3.6).

Thus far we have seen that food energy can be stored as glycogen in liver and muscle or as fat in the body, or converted into the energy stored in ATP and used by

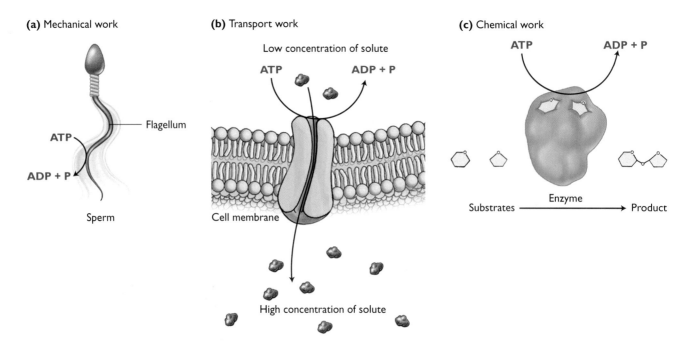

FIGURE 3.6 ATP and cellular work. ATP powers many different types of work in a cell.

Stop and Stretch 3.1

Consider the structures of a dietary fat, a carbohydrate, and a protein, illustrated here. Are all the elements required to produce fat found in carbohydrate and protein?

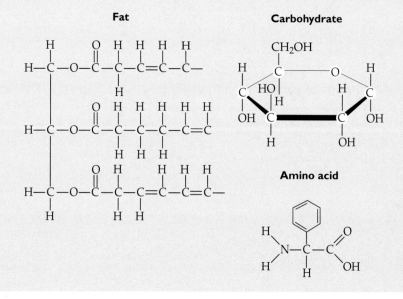

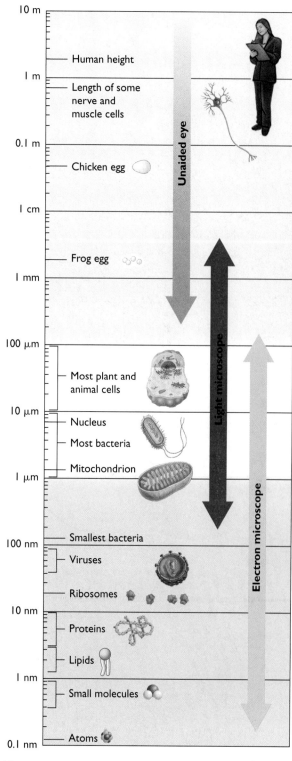

cells. To understand how a cell determines whether to store energy or use it to synthesize ATP, it is necessary to become acquainted with some of the structures inside cells. Then we will revisit the ATP molecule to learn how cells use food energy to produce it.

3.2 Cell Structure and Function

Cells are the fundamental structural unit of life. All living things are made of cells. Some cells, such as a frog's egg cell, can be seen with the naked eye, but most cells are so small that they must be viewed through a light microscope or electron microscope.

Light microscopes use lenses and light rays to produce an image that the human eye can see. *Electron microscopes* use magnetic lenses and electrons to produce an image that is projected onto film or a computer screen that humans can view. Transmission electron microscopes produce two-dimensional images, and scanning electron microscopes produce three-dimensional views.

Due to the very small nature of most cells, the units used to describe their measurement may be unfamiliar to you. While you have probably heard of a meter (m), you may not have heard of a nanometer (1×10^{-9} m). This notation means that there are 1,000,000,000 (1 billion) nanometers in a meter. FIGURE 3.7 shows that most cells and subcellular structures measure between 1 and 100 micrometers (a micrometer is 1×10^{-6} m, or one-millionth of a meter, abbreviated 1 μm). For perspective, the period at the end of this sentence is approximately 300 micrometers (or 300,000 nanometers) in diameter.

Measurements
1 centimeter (cm) = 10^{-2} meter (m) = 0.4 inch
1 millimeter (mm) = 10^{-3} m
1 micrometer (μm) = 10^{-3} mm = 10^{-6} m
1 nanometer (nm) = 10^{-3} μm = 10^{-9} m

FIGURE 3.7 Cell size. Most cells are in the 1- to 100-μm range and are therefore visible only under a microscope.

CHAPTER 3 Cell Structure and Metabolism

FIGURE 3.8 Ratios of surface area to volume. As cells increase in size, the amount of surface area available for nutrient exchange does not increase proportionately. The cell here has doubled in size while its ratio of surface area to volume has decreased.

VISUALIZE THIS How would the ratio of surface area to volume change if a cuboidal cell went from having 1-mm sides to 3-mm sides?

Stop and Stretch 3.2

If you were sitting at a desk that was approximately 1 meter long (about 3.3 feet), how many 100-micrometer cells could be lined up single file from one end of the desk to the other?

The small size of cells is a function of the ratio of surface area to volume. Nutrients enter cells and wastes exit by crossing the outer surface of the cell. The greater the surface area, the higher the rate of nutrient and waste exchange. As a cell's volume increases, the demand for nutrients increases but the surface area does not increase proportionately (FIGURE 3.8). Therefore, it tends to work out better for cells to stay small.

In spite of their small size, a lot is going on inside cells. When it comes to energy utilization, each living cell is like a tiny factory whose goal is to efficiently break apart the incoming nutrients, and then send them off to other work areas for more specialized processing.

Cell Structures

Structures inside cells help cells perform many different functions. They also help a cell maintain its size and shape. Subcellular structures called **organelles** are to cells as organs are to the body. Each organelle performs a specific job required by the cell, and all organelles work together to keep an individual cell healthy and to produce the raw materials that the cell needs to survive.

Nucleus The **nucleus** is a spherical structure that houses the DNA and serves as the cell's control center. The nucleus is surrounded by two membranes, together called the *nuclear envelope*. The nuclear envelope is studded with *nuclear pores* that regulate traffic into and out of the nucleus. Inside the nucleus is *chromatin*, composed of DNA and proteins. The *nucleolus* is where ribosomes are produced. The fluid inside the nucleus is called the *nucleoplasm* (FIGURE 3.9).

Cytosol Between the nucleus and the outer boundary of the cell lie organelles surrounded by the **cytosol**, a watery matrix containing salts and many of the enzymes

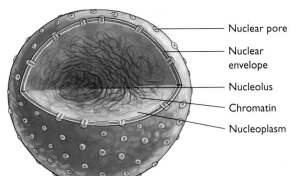

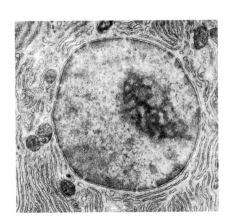

FIGURE 3.9 Nucleus. The nucleus is surrounded by two membranes, together called the nuclear envelope, which is perforated by protein pores that regulate traffic into and out of the nucleus. Ribosomes are built in the darkly staining nucleolus. The chromatin inside the nucleus consists of DNA and proteins.

VISUALIZE THIS Locate the structures labeled in the drawing on the micrograph.

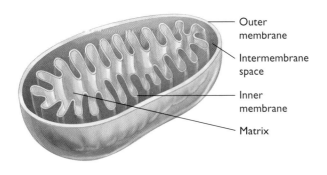
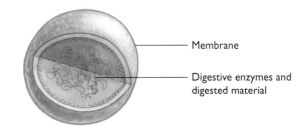

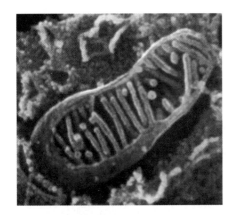

FIGURE 3.10 Mitochondrion. The inner and outer membranes of the mitochondrial envelope are separated by the intermembrane space. The matrix is the semifluid medium inside this organelle.

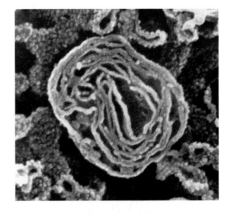

FIGURE 3.11 Lysosome. Lysosomes are membrane-enclosed sacs of digestive enzymes. Inside this lysosome is a partially digested organelle.

required for cellular reactions. The cytosol surrounds the organelles. The term **cytoplasm** includes the cytosol and organelles.

Some subcellular structures, including mitochondria and lysosomes, are involved in the breakdown of nutrients to release their energy.

Mitochondria Mitochondria (FIGURE 3.10) are energy-harvesting organelles surrounded by two membranes. The *inner* and *outer mitochondrial membranes* are separated by the *intermembrane space*. The highly convoluted inner membrane carries many of the proteins involved in producing ATP. The fluid *matrix* inside of the mitochondrion is the location of many of the reactions necessary to release energy from nutrients.

Lysosomes Lysosomes (FIGURE 3.11) are also involved in the release of energy from nutrients. A lysosome is a membrane-enclosed sac of enzymes that degrade proteins, carbohydrates, and fats. Lysosomes roam around the cell and engulf nutrients as well as dead and dying organelles.

Other subcellular structures are involved with synthesizing and processing proteins.

Ribosomes Ribosomes (FIGURE 3.12) are workbenches where amino acids are joined together to produce proteins. Ribosomes are built in the nucleus and shipped out through nuclear pores to the cytosol. They can be found floating in the cytosol or tethered to a membranous structure called the endoplasmic reticulum.

Endoplasmic Reticulum The **endoplasmic reticulum (ER)** (FIGURE 3.13) is a large network of membranes that begins at the nuclear envelope and extends into the cytosol. ER with ribosomes attached is called *rough ER*. Proteins synthesized on rough ER will be secreted from the cell or will become part of the plasma membrane. ER without ribosomes attached is called *smooth ER*. The function of the smooth ER depends on cell type but includes tasks such as detoxifying harmful substances and synthesizing lipids.

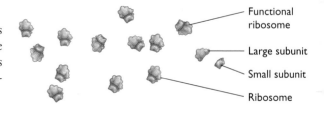

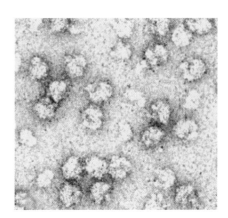

FIGURE 3.12 Ribosomes. A functional ribosome consists of a large and a small subunit. Each subunit is composed of RNA and protein.

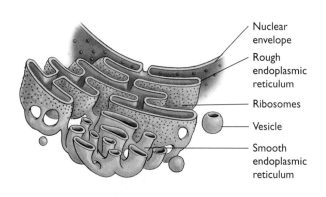
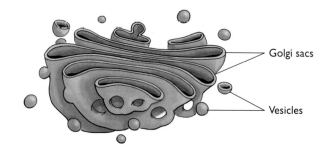
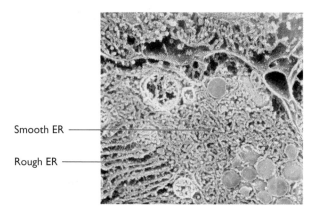
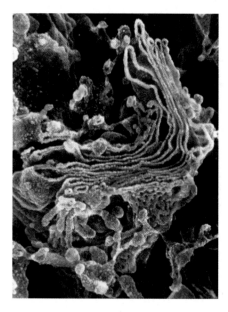

FIGURE 3.13 Endoplasmic reticulum. The endoplasmic reticulum is a series of membranous sacs. ER with ribosomes attached is rough ER while the more tubular ER is smooth ER. Vesicles bud out from the ER and carry proteins to the Golgi apparatus.

FIGURE 3.14 Golgi apparatus. Vesicles arrive at the Golgi membranes carrying cargo to be modified and sorted and sent to its next destination.

VISUALIZE THIS Locate a vesicle budding off the Golgi in the micrograph.

Golgi Apparatus The **Golgi apparatus** (FIGURE 3.14) is a stack of membranous sacs. *Vesicles*, which are membrane sacs, pinch off from the ER and travel to fuse with the Golgi apparatus and empty their protein contents. The proteins are then modified, sorted, and sent to the correct destination in new transport vesicles that bud off from the Golgi apparatus.

Other subcellular structures help cells divide, maintain their shape, and move.

Centrioles **Centrioles** (FIGURE 3.15) are barrel-shaped structures composed of structural proteins called microtubules. *Microtubules* are rigid structures and with the help of other proteins facilitate chromosome movement during cell division. Centrioles are involved in microtubule formation during cell division and the formation of cilia and flagella.

Cytoskeleton The **cytoskeleton** (FIGURE 3.16) is a network of filaments and tubules found in the cytoplasm. Cytoskeletal elements are protein fibers that help a cell maintain its structure and help it move. The cytoskeleton in the cell is similar to the bony skeleton in humans in that it provides structural support and facilitates movement.

FIGURE 3.17 shows a drawing of an animal cell with its full complement of organelles and other structures. All of these subcellular components work together to break apart nutrients and recycle them for different uses. However, before your cells can use the nutrients you ingest to build structural components of cells, to generate ATP, or to convert to fat for storage, the nutrients must first gain access to the cell by crossing the outer boundary of the cell.

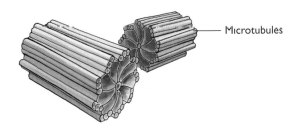
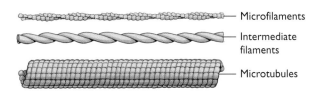
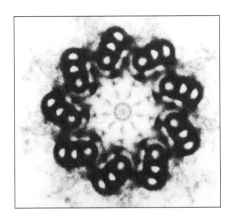
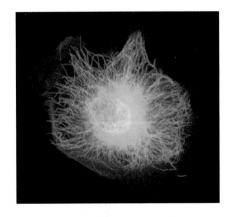

FIGURE 3.15 Centrioles. Centrioles are involved in microtubule formation. Microtubules are found in cilia and flagella as well as throughout the cell. The micrograph shows a cross-section through a centriole.

FIGURE 3.16 Cytoskeletal elements. The three types of cytoskeletal elements are microfilaments, intermediate filaments, and microtubules. Cytoskeletal elements help the cell maintain its shape. Microtubules are stained green and microfilaments are orange in the micrograph.

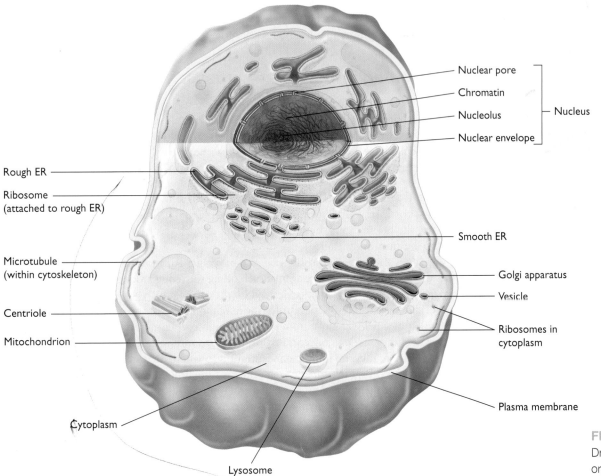

FIGURE 3.17 Animal cell. Drawing of an animal cell showing organelles and structures.

3.3 Membrane Structure and Function

The outermost boundary of the cell is called the **plasma membrane**. The plasma membrane isolates the cell's contents from the environment and serves as a barrier that determines which substances are allowed into and out of the cell.

Membrane Structure

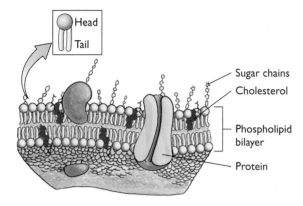

FIGURE 3.18 **Plasma membrane.** The plasma membrane is a bilayer of phospholipids with proteins embedded. It can have carbohydrate chains attached to its extracellular surface that aid in cell-to-cell recognition.

The plasma membrane and the membranes that surround and make up some organelles are composed, in part, of phospholipids, which you learned about in Chapter 2. Phospholipids are lipids that have two fatty acid tails and one phosphate head group. When phospholipid molecules are placed in a watery solution, such as in a cell, they orient themselves so that their hydrophilic heads are exposed to the water and their hydrophobic tails are away from the water. In cells, they cluster into a form called a **phospholipid bilayer** in which the tails of the phospholipids interact with themselves and exclude water, while the heads maximize their exposure to the surrounding water both inside and outside of the membrane (FIGURE 3.18).

The bilayer of phospholipids has many different proteins embedded in the membrane or attached to its surface. These proteins can carry out enzymatic functions, or anchor the cytoskeleton to the plasma membrane. Some membrane proteins called *transport proteins* help move substances from one side of the membrane to the other. Membrane proteins can also serve as receptors for substances outside the cell, enabling the communication of external conditions to the inside of the cell.

In addition to phospholipids and proteins, human cell membranes also contain cholesterol. Cholesterol helps maintain the fluidity of membranes by preventing phospholipids from packing too tightly.

Short carbohydrate chains can be attached to the phospholipids, forming *glycolipids*. Likewise, *glycoproteins* are proteins with short carbohydrate chains attached. These carbohydrates serve as cell markers that are particular to a given individual and cell type.

Crossing the plasma membrane is easier for some substances than others.

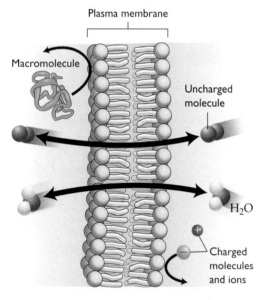

FIGURE 3.19 **Selective permeability.** Biological membranes allow some substances to pass but prevent others from doing so.

Transporting Substances Across Membranes

Cell membranes are **selectively permeable** in the sense that they allow some substances to cross and prevent others from crossing (FIGURE 3.19). This characteristic allows cells to maintain a degree of independence from the surrounding solution.

Some substances can cross the membrane without any help from transport proteins or any energy input. In other situations, energy is required for transport to occur.

Passive Transport: Diffusion, Facilitated Diffusion, and Osmosis Imagine opening a bottle of your favorite carbonated beverage. The hissing sound you hear when removing the bottle cap is the sound of carbon dioxide gas trickling out of the beverage and into the surrounding air.

Why does the carbon dioxide leave the beverage? All molecules contain energy that makes them vibrate and bounce against each other, scattering around like billiard balls during a game of pool. In fact, molecules will bounce against each other until they are spread out over all the available volume. Part of the reason that the carbon dioxide leaves the bottle is because molecules will move from their own high concentration (in the bottle) toward a region where they are less concentrated (in the surrounding air). This movement of molecules from where they are in high concentration to where they are in low concentration is called **diffusion**. This movement of molecules is sometimes described as movement *down* a concentration gradient and does not require an input of outside energy; it is spontaneous.

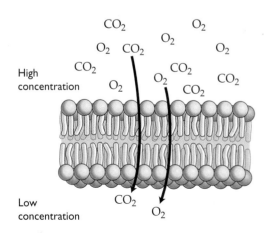

FIGURE 3.20 **Diffusion.** Diffusion of molecules across the plasma membrane occurs with the concentration gradient and does not require energy from the cell. Small hydrophobic molecules, carbon dioxide, and oxygen molecules can diffuse across the membrane.

Diffusion also occurs in living organisms. When substances diffuse across the plasma membrane, we call the movement **passive transport**. Passive transport does not require an input of energy from the cell. The structure of the phospholipid bilayer that composes the plasma membrane prevents many substances from diffusing across it. Only very small, hydrophobic molecules are able to cross the membrane by diffusion. In effect, these molecules dissolve in the membrane and slip from one side to the other (FIGURE 3.20).

Because fats are hydrophobic, they can cross the membrane. Carbon dioxide can cross the membrane unaided, as can oxygen molecules.

Hydrophilic molecules and charged molecules such as ions are unable to diffuse across the hydrophobic core of the membrane. Instead, these molecules can be transported across membranes by proteins embedded in the lipid bilayer. When this type of transport does not require an input of energy from the cell it is called **facilitated diffusion**. As with passive transport, facilitated diffusion moves substances down their concentration gradient. Facilitated diffusion is so named because the specific membrane proteins are helping, or facilitating, the diffusion of substances across the plasma membrane (FIGURE 3.21).

The permeability of the plasma membrane to water presents a problem for many organisms. In environments where the water contains low levels of solutes, water from outside the cell will flow into the cell, causing it to burst.

The movement of water across a membrane is a type of passive transport called **osmosis**. Like other substances, water moves from its own high concentration to its own low concentration.

Body fluids are typically **isotonic** to cells. *Iso* means "same" and *tonic* refers to a property called tonicity. **Tonicity** is the concentration of solute in a solution. Therefore, having an isotonic solution in contact with body cells means there is an equal concentration of solutes and water on both sides of the cell's membrane, allowing a cell to maintain its size and shape. When a cell is in a **hypertonic** solution, the solution contains a higher concentration of dissolved solute than the cell, so water leaves the cell by osmosis, causing the cell to shrivel (FIGURE 3.22a). In an isotonic solution, water crosses the

FIGURE 3.21 Facilitated diffusion. Facilitated diffusion is the diffusion of molecules through protein channels. The molecules' movement takes place with their concentration gradient and does not require an input of the cell's energy.

(a) Hypertonic solution

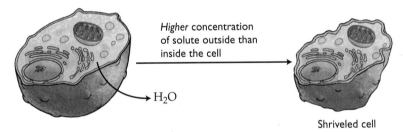

(b) Isotonic solution

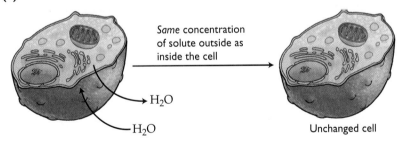

(c) Hypotonic solution

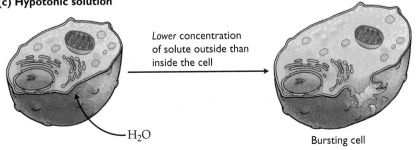

FIGURE 3.22 Osmosis. Osmosis is the movement of water in response to a concentration gradient. Water moves toward a region that has more dissolved solute. (a) When a cell is placed in a hypertonic environment, water will leave the cell and it will shrink. (b) In an isotonic environment, water leaves and enters the cell in equal amounts. (c) In a hypotonic environment, water enters the cell and the cell can burst.

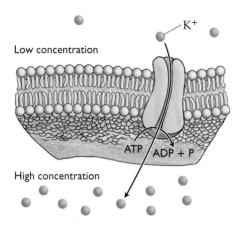

FIGURE 3.23 Active transport. Active transport moves substances against their concentration gradient and requires energy (ATP) to do so.

membrane in both directions but there is no net movement of water (FIGURE 3.22b). If a cell is placed in a **hypotonic** solution the concentration of dissolved solute is greater inside the cell than outside and water will enter the cell by osmosis, causing it to burst (FIGURE 3.22c).

Stop and Stretch 3.3

Intravenous solutions that are given to patients in hospitals are usually isotonic to their cells. Why is it important to use isotonic solutions when replacing fluids intravenously?

Active Transport: Pumping Substances Across the Membrane In some situations, a cell will need to maintain a concentration gradient. For example, nerve cells require a high concentration of certain ions inside the cell and other ions outside the cell to transmit nerve impulses. To maintain this difference in concentration across the membrane an input of energy is required.

Active transport is transport that uses proteins, powered by ATP, to move substances up, or against, a concentration gradient (FIGURE 3.23). Proteins that actively transport substances are often referred to as "pumps" because they pump substances against concentration gradients. One type of pump that is commonly active in nerve and muscle cells, and which you will learn more about in Chapter 14, pumps sodium ions out of the cell and potassium ions into the cell.

Exocytosis and Endocytosis: Movement of Large Molecules Across the Membrane Larger molecules are often too big to diffuse across the membrane or to be transported through a protein, regardless of whether they are hydrophobic or hydrophilic. Instead, they must be moved around inside membrane-bounded vesicles that can fuse with membranes. **Exocytosis** (FIGURE 3.24a) occurs when a membrane-bounded vesicle, carrying some substance, fuses with the plasma membrane and secretes its contents into the exterior of the cell. **Endocytosis** (FIGURE 3.24b) occurs when a substance is brought into the cell and the plasma membrane buds inward, bringing the substance with it. Endocytosis and exocytosis can also be used to perform the bulk transport of many molecules into or out of the cell at one time.

Now that we see how nutrients can get into cells, we are ready to learn how the body determines whether to use the energy stored in nutrients immediately or to store the energy for later use.

(a) Exocytosis

(b) Endocytosis

FIGURE 3.24 Movement of large substances. (a) Exocytosis is the movement of substances out of the cell. (b) Endocytosis is the movement of substances into the cell.

Stop and Stretch 3.4

Consider red blood cells placed in solutions of various concentrations of dissolved solute. For each of the following images, determine whether the red blood cell has been placed in a hypertonic, hypotonic, or isotonic solution.

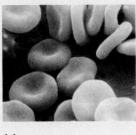 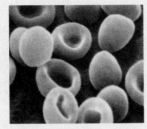 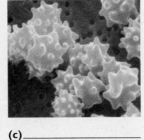

(a) _____ (b) _____ (c) _____

3.4 Metabolism—Chemical Reactions in the Body

Metabolism is a general term used to describe all of the chemical reactions occurring in the body. When too much food is consumed, metabolic reactions convert carbohydrates and proteins into fats and store them, along with any excess fat consumed.

The amount of fat that a given individual will store depends partly on how quickly or slowly he or she breaks down food molecules into their component parts.

Enzymes

All metabolic reactions are regulated by proteins called **enzymes** that speed up, or **catalyze**, the rate of reactions. Enzymes help synthesize polymers from monomers and help hydrolyze macromolecules into their component parts. When enzymes help your body break down the foods you ingest, they help liberate the energy stored in the chemical bonds of foods. The speed and efficiency of many different enzymes will lead to an overall increase or decrease in the rate at which a person can metabolize food. Thus, when you say that your metabolism is slow or fast, you are actually referring, in part, to the speed at which enzymes catalyze chemical reactions in your body.

To break chemical bonds, molecules must absorb energy from their surroundings, often by absorbing heat. Heating chemicals may speed up a reaction, but in living tissue, heating cells to an excessively high temperature can damage or kill them. For this reason, enzymes help catalyze the body's chemical reactions without requiring heat for the reactants to break their chemical bonds. By decreasing the energy required to start the reaction, enzymes allow chemical reactions to occur more quickly.

Activation Energy The energy required to start the metabolic reaction serves as a barrier to catalysis and is called the **activation energy**. If not for the activation energy barrier, many of the chemical reactions in cells would occur relentlessly, whether the products of the reactions were needed or not. Because most metabolic reactions have to surpass the activation energy barrier before proceeding, they can be regulated by the presence or absence of enzymes. Enzymes catalyze metabolic reactions by decreasing the activation energy barrier. How do they do this?

The chemicals that are metabolized by an enzyme-catalyzed reaction are called **substrates**. One way enzymes can decrease the activation energy barrier is by binding to

FIGURE 3.25 Enzymes. The enzyme sucrase is splitting the disaccharide sucrose into its monosaccharide subunits, glucose and fructose. The enzyme can then be recycled to perform the same reaction again and again.

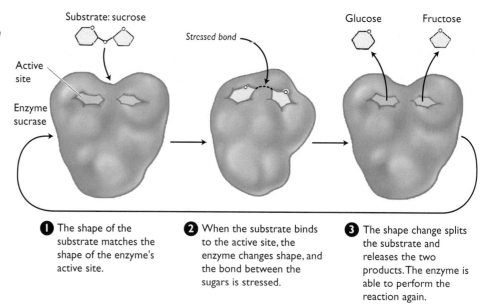

① The shape of the substrate matches the shape of the enzyme's active site.

② When the substrate binds to the active site, the enzyme changes shape, and the bond between the sugars is stressed.

③ The shape change splits the substrate and releases the two products. The enzyme is able to perform the reaction again.

their substrate and placing stress on its chemical bonds, decreasing the amount of initial energy required to break the bonds. The region of the enzyme where the substrate binds is called the enzyme's **active site**. Each active site has its own shape and chemistry. When the substrate binds to the active site, the enzyme changes shape slightly in order to envelop the substrate. This shape change by the enzyme in response to substrate binding is called **induced fit**, because the substrate induces the enzyme to change shape to conform to the substrate's contours. When the enzyme changes shape, it places stress on the chemical bonds of the substrate, making them easier to break. In this manner, an enzyme can help convert the substrate to a reaction product and then resumes its original shape so that it can perform the reaction again (FIGURE 3.25).

Different enzymes catalyze different reactions due to a property called **specificity**. Enzymes are usually named for the reaction they catalyze and end in the suffix *-ase*. For example, sucrase is the enzyme that breaks down table sugar (sucrose). The specificity of an enzyme is the result of its shape and the shape of its active site. Different enzymes have unique shapes because they are composed of amino acids in varying sequences. The 20 amino acids, each with its own unique side group, are arranged in a distinct order for each enzyme, producing enzymes of all shapes and sizes, each with an active site that can bind with its particular substrate.

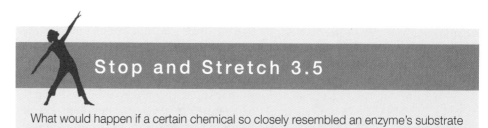

What would happen if a certain chemical so closely resembled an enzyme's substrate that it could bind to the active site in place of the actual substrate?

Many enzyme-catalyzed reactions are required to break nutrients down and utilize their energy to synthesize ATP.

Cellular Respiration

The main function of **cellular respiration** is to convert the energy stored in chemical bonds of food into energy that cells can use. Energy is stored in the electrons of chemical bonds, and when bonds are broken, electrons can be moved from one molecule to another.

Cells are constantly producing ATP. During this process, cells produce carbon dioxide and use oxygen to produce water. Because some of the steps in cellular respiration require oxygen, they are said to be **aerobic** reactions, and cellular respiration is called **aerobic respiration**. When you breathe in, oxygen enters your lungs where it diffuses into the blood for transport to cells. The carbon dioxide that is a waste product of cellular respiration must be excreted from your body; it is exhaled during breathing (FIGURE 3.26).

Most foods can be broken down and their energy used to produce ATP as they are routed through a complex pathway. Because carbohydrate metabolism follows the longest pathway, we will first look at glucose metabolism and then see where other macromolecules like proteins and fats feed into the pathway.

A General Overview of Cellular Respiration

The equation for glucose breakdown is

$$C_6H_{12}O_6 + 6O_2 \longrightarrow 6CO_2 + 6H_2O$$
Glucose plus oxygen yields carbon dioxide plus water.

Glucose is an energy-rich sugar, but the products of its metabolism—carbon dioxide and water—are energy poor. The energy released during the conversion of glucose to carbon dioxide and water is used to synthesize ATP.

Many of the chemical reactions in this process occur in mitochondria through a series of complex reactions that break apart the glucose molecule. The carbon and oxygen atoms that make up the original glucose molecule are released from the cell as carbon dioxide. The hydrogen atoms present in the original glucose molecule combine with oxygen to produce water (FIGURE 3.27). Gaining an appreciation for *how* this happens requires a more in-depth look.

Glycolysis, the Citric Acid Cycle, Electron Transport, and ATP Synthesis

To harvest energy from glucose, the 6-carbon glucose molecule is first broken down into two 3-carbon **pyruvic acid** molecules. This part of the process actually occurs in the fluid cytosol, rather than in an organelle, and is called **glycolysis** (FIGURE 3.28). Glycolysis does not require oxygen but does produce a small amount of ATP.

As glucose is converted to pyruvic acid, several electrons are stripped from the molecule. If these electrons simply floated around in a cell they would damage it; instead they are carried by molecules called electron carriers. One of the electron carriers used by cellular respiration is a chemical called **nicotinamide adenine dinucleotide (NAD)** (FIGURE 3.29). NAD^+ picks up 2 hydrogen atoms and releases 1 positively charged proton. Each **hydrogen atom** is composed of 1 negatively charged electron and 1 positively charged proton. When NAD^+ picks up 2 hydrogen atoms (each with 1 proton and 1 electron), it uses 1 proton and 2 electrons, releasing the remaining

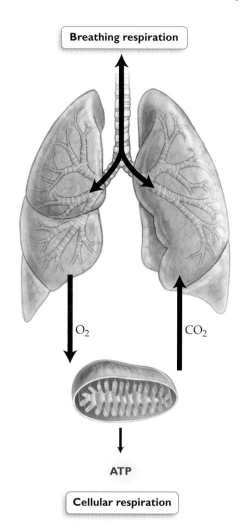

FIGURE 3.26 **Respiration.** (a) When you inhale, you bring oxygen from the atmosphere into your lungs. This oxygen is delivered through the bloodstream to tissues that use it to drive cellular respiration. (b) The carbon dioxide produced by cellular respiration is released from cells and diffuses into the blood and to the lungs. Carbon dioxide is released from the lungs when you exhale.

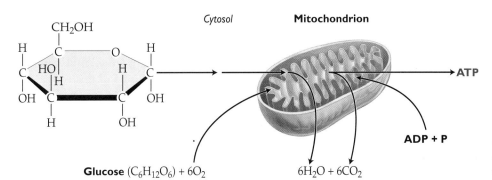

FIGURE 3.27 **Overview of cellular respiration.** The breakdown of glucose by cellular respiration requires oxygen and ADP plus a phosphate. The energy stored in the bonds of glucose is harvested to produce ATP (from ADP and P), releasing carbon dioxide and water.

FIGURE 3.28 Glycolysis. Glycolysis occurs in the cytosol and does not require oxygen. Glycolysis is the enzymatic conversion of glucose into two pyruvic acid molecules. The pyruvic acid molecules move into the mitochondrion for further breakdown. Small amounts of NADH and ATP are made during glycolysis.

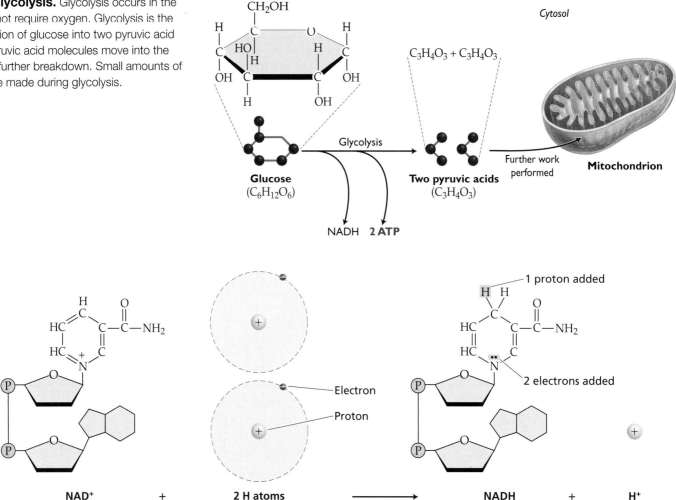

FIGURE 3.29 Nicotinamide adenine dinucleotide (NAD). NAD is a dinucleotide in the sense that it contains two sugars, two phosphates, and the nitrogenous base adenine. This molecule can pick up a hydrogen atom along with its electrons. Hydrogen atoms are composed of one negatively charged electron that circles the one positively charged proton. When NAD^+ encounters two hydrogen atoms (from food), it utilizes each hydrogen atom's electron and only one proton, thus releasing one proton.

proton. NAD^+ serves as a sort of taxicab for electrons. The empty taxicab (NAD^+) picks up electrons. The full taxicab (NADH) carries electrons to their destination, where they are dropped off, and the empty taxicab returns for more electrons. Thus NAD^+ is converted to NADH, which travels to the destination and releases a hydrogen atom and an electron.

A careful look back at Figure 3.28 shows that the conversion of glucose into 2 pyruvic acid molecules results in the loss of 4 hydrogen atoms. These hydrogen atoms and their associated electrons are picked up by NAD^+ to produce NADH.

After glycolysis, the pyruvic acid is *decarboxylated* (loses a carbon dioxide molecule), and the 2-carbon fragment that is left is further metabolized inside the mitochondrion.

Once the pyruvic acid is inside the mitochondrion, the energy stored in its bonds is converted into the energy stored in the bonds of ATP. The first step of this conversion is called the *citric acid cycle*.

Citric Acid Cycle The **citric acid cycle** is a series of reactions catalyzed by eight different enzymes located in the matrix of each mitochondrion. The citric acid cycle breaks down the remains of a carbohydrate, harvesting its electrons and releasing carbon dioxide into the atmosphere (FIGURE 3.30). These reactions are a cycle because

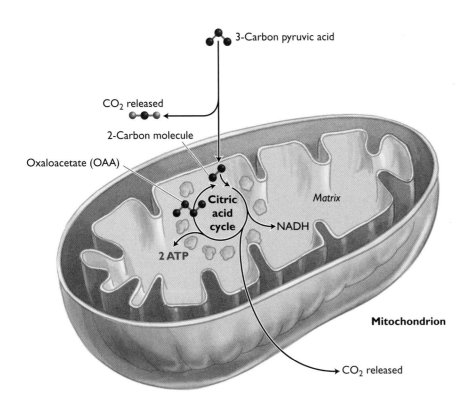

FIGURE 3.30 **The citric acid cycle.** The three-carbon pyruvic acid molecules generated by glycolysis are decarboxylated, leaving a two-carbon molecule that enters the citric acid cycle within the mitochondrial matrix. The two-carbon fragment reacts with a four-carbon OAA molecule and proceeds through a stepwise series of reactions that results in the production of more carbon dioxide and regenerates OAA. NADH and ATP are also produced.

every turn of the cycle regenerates the first reactant in the cycle. Therefore, the first reactant in the cycle, a 4-carbon molecule called oxaloacetate (OAA), is always available to react with incoming 2-carbon fragments. The citric acid cycle also produces some ATP and NADH. The NADH produced deposits its electrons at the top of the electron transport chain.

The Electron Transport Chain The **electron transport chain** is a series of proteins embedded in the inner mitochondrial membrane that moves electrons from one protein to another. The electrons are pulled toward the bottom of the electron transport chain by oxygen in the matrix of the mitochondrion. Because oxygen is very electronegative, or electron-loving, it pulls electrons toward itself. Each time an electron is picked up by a protein or handed off to another protein, the protein moving it changes shape. This shape change facilitates the movement of protons from the matrix of the mitochondrion to the intermembrane space. Therefore, as proteins in the electron transport chain are moving electrons down the electron transport chain toward oxygen, they are also moving hydrogen ions across the inner mitochondrial membrane and into the intermembrane space.

The result is a decrease in the concentration of hydrogen ions in the matrix and an increase in the concentration of hydrogen ions within the intermembrane space. Whenever a concentration gradient exists, molecules will diffuse from an area of high concentration to an area of low concentration. Because charged ions cannot diffuse across the hydrophobic core of the membrane, they escape through a protein channel in the membrane called **ATP synthase**. This enzyme uses the energy generated by the rushing hydrogen ions to synthesize ATP from ADP and phosphate, much as water rushing through a mechanical turbine can be used to generate electricity. The electrons that were pulled down the electron transport chain then combine with the oxygen at the bottom of the chain and 2 hydrogen atoms in the matrix to produce water (FIGURE 3.31).

Overall, the two pyruvic acids produced by the breakdown of glucose during glycolysis are converted into carbon dioxide and water. Carbon dioxide is produced when it is removed from the pyruvic acid molecules during the citric acid cycle, and water is formed when oxygen combines with hydrogen atoms at the bottom of the electron transport chain. A summary of the process is shown in FIGURE 3.32.

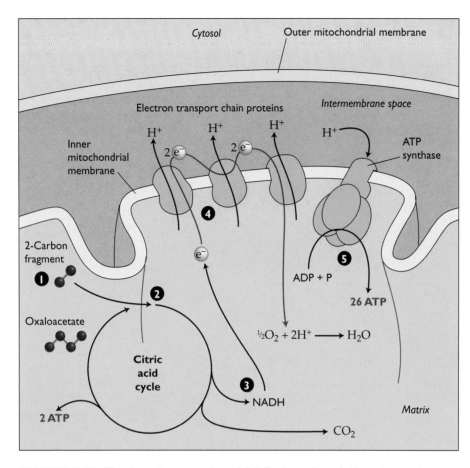

FIGURE 3.31 Electron transport and ADP phosphorylation. Energy from electrons added to the top of the electron transport chain is used to produce ATP.

VISUALIZE THIS What would happen to electron transport and ATP production if oxygen were not available to serve as the final electron acceptor?

Imagine placing a tiny tube in the inner mitochondrial membrane that would allow substances to freely diffuse across the membrane. What effect might this have on ATP synthesis?

Metabolism of Proteins and Fats Proteins and fats are broken down and their subunits merge with the carbohydrate breakdown pathway. FIGURE 3.33 shows the points of entry for proteins and fats.

Excess protein is not stored by the body. Protein is broken down into component amino acids, which are then used to synthesize new proteins. Protein may also be broken down to produce energy; however, this process takes place only when fats or carbohydrates are unavailable. The first step in producing energy from the amino acids of a protein is to remove the nitrogen-containing amino group of the amino acid. Amino groups are then converted to a compound called urea, which is excreted in the urine. The carbon, oxygen, and hydrogen atoms remaining after the amino group is removed undergo further breakdown and eventually enter the mitochondria, where they are fed through the citric acid cycle and produce carbon dioxide, water, and ATP.

Enzymes in adipose cells break the bonds holding glycerol and fatty acids together, freeing the fatty acids. Next the fatty acids are hydrolyzed into even smaller units that can undergo cellular respiration. Most cells will break down fat only when carbohydrate supplies are depleted.

Fermentation Aerobic respiration is one way for organisms to generate energy. It is also possible for cells to generate energy in the absence of oxygen, a process called **anaerobic respiration**. Muscle cells normally produce ATP by aerobic respiration. However, if oxygen supplies diminish, as would be the case with intense exercise, cells get most of their ATP from glycolysis, which does not require oxygen. When glycolysis happens without aerobic respiration, the cells run low on NAD^+, which is converted into NADH by glycolysis only.

The cells use a process called **fermentation** to regenerate NAD^+. No usable energy is produced by fermentation. Fermentation simply recycles NAD^+. Fermentation cannot, however, be used for very long since one of the by-products of this reaction leads to the buildup of a compound called lactic acid. Lactic acid is produced by the actions of NADH, which has no place to dump its electrons during fermentation since there is no electron transport chain and no oxygen to accept the electrons. Instead, NADH gets rid of its electrons by giving them to the pyruvic acid produced by glycolysis (FIGURE 3.34). Adding electrons to pyruvic acid produces lactic acid, which is transported to the liver where liver cells use oxygen to convert it back to pyruvic acid.

This requirement for oxygen, to convert lactic acid to pyruvic acid, explains why you continue to breathe heavily even after you have stopped working out. Your body needs to supply oxygen to your liver for this conversion, sometimes referred to as "paying back your oxygen debt." The accumulation of lactic acid also explains the phenomenon called "hitting the wall." Anyone who has ever felt as though their legs were turning to wood while

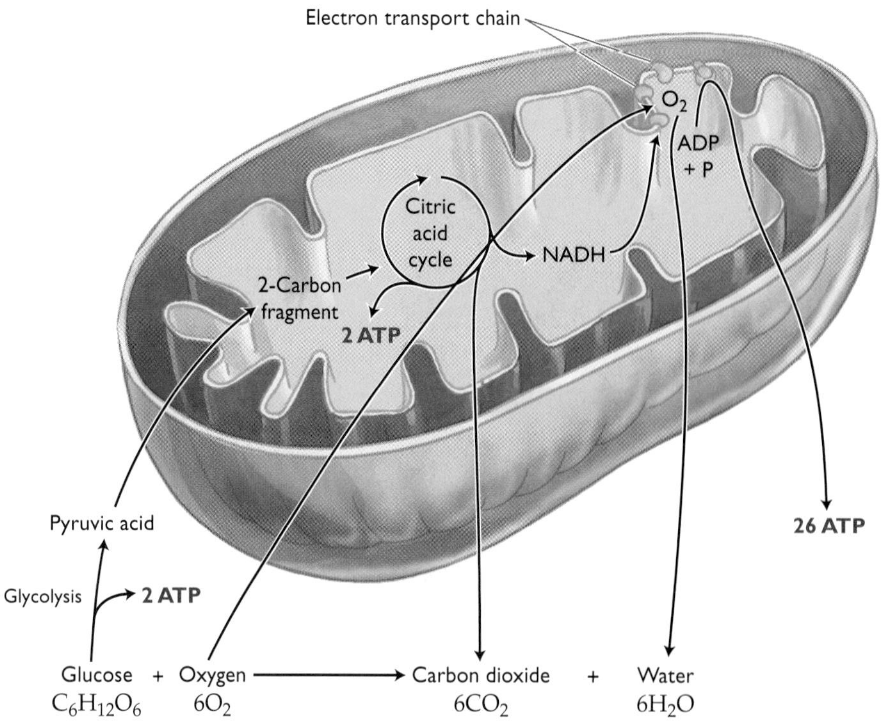

FIGURE 3.32 Summary of cellular respiration. This figure shows the inputs and outputs of cellular respiration.

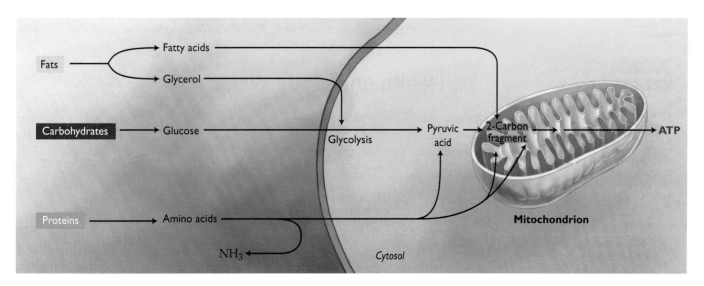

FIGURE 3.33 Metabolism of other macromolecules. Carbohydrates, proteins, and fats can all undergo cellular respiration. They feed into different parts of the pathway. Any available carbohydrates are metabolized before fat and dietary protein. Muscle proteins can be degraded and fed into this pathway in cases of starvation.

Human muscle

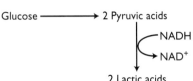

FIGURE 3.34 Fermentation. Fermentation recycles NAD⁺ by removing electrons from NADH and giving them to pyruvic acid, producing lactic acid.

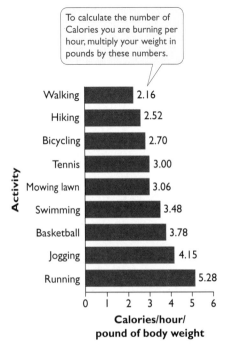

FIGURE 3.35 Energy expenditures for various activities. This bar graph can help you determine how many Calories you burn during certain activities.

running or biking knows this feeling. When your muscles are producing lactic acid by fermentation for a long time, the oxygen debt becomes too large, and muscles shut down until the rate of oxygen consumption outpaces the rate of oxygen utilization.

Now that you have learned how cells use food to make ATP, we can consider why it is that some people store more energy than others.

Calories and Metabolic Rate

Energy is measured in units called calories. A **calorie** is the amount of energy required to raise the temperature of 1 gram of water by 1 degree Celsius. In scientific literature, energy is usually reported in kilocalories, and 1 kilocalorie equals 1,000 calories of energy. However, in physiology—and on nutritional labels—the prefix *kilo-* is dropped, and a kilocalorie is referred to as a **Calorie** (with a capital *C*). Calories are consumed to supply the body with energy to do work, including maintaining body temperature. When the supply of Calories is greater than the demand, the excess Calories can be stored by the body as fat.

A person's **metabolic rate** is a measure of his or her energy use. This rate changes according to the person's activity level. For example, people require less energy when asleep than when exercising. The **basal metabolic rate** represents the resting energy use of an awake, alert, but sedentary, person. The average basal metabolic rate is 70 Calories per hour, or 1,680 Calories per day. However, this rate varies widely among individuals because many factors influence each person's basal metabolic rate: exercise habits, body weight, age, genetics, and sex. The healthfulness of your diet can also affect metabolism, because an enzyme that is missing its vitamin coenzyme (found in fruits, vegetables, and grains) may not be able to perform at its optimal rate.

Exercise requires energy, which allows you to consume more Calories without having to store them. As for body weight, a heavy person uses more Calories during exercise than a thin person does. FIGURE 3.35 shows the number of Calories used per hour for various activities based on body weight. As you age, your metabolism and level of activity tend to slow down, so older people need fewer calories than younger people.

As you will see in the next chapter, genes code for the production of enzymes, so people with different genetic makeups will have different metabolic rates. Males require more Calories per day than females do because testosterone, a hormone produced in larger quantities by males, increases the rate at which fat breaks down. Men also have more muscle than women, which requires more energy to maintain than fat does. Both of these factors lead to females storing more body fat than males. On average, healthy women have 22% body fat, and healthy men have 14%.

3.5 Health and Body Weight

To maintain essential body functions, women need at least 12% body fat but not more than 32%; for men, the range is between 3% and 29%. If a person is not within these ranges, problems with fertility in females or obesity in both sexes can result.

Underweight

Self-starvation, or **anorexia**, is quite common on college campuses as women try to attain a body type that is at odds with their biological need to store fat.

In females, very low body fat can cause **amenorrhea**, or lack of menstruation. This occurs when the protein **leptin**, which is secreted by fat cells, signals the brain that there is not enough body fat to support a pregnancy. Hormones (such as estrogen) that regulate menstruation are inhibited, and menstruation ceases. Lack of menstruation can be permanent and causes sterility in women who are underweight.

The damage done by the lack of estrogen is not limited to the reproductive system—bones are affected as well. Estrogen secreted by the ovaries during the menstrual cycle acts on bone cells to help them maintain their strength and size. Anorexics reduce the development of dense bone and put themselves at a much higher risk of breaking their weakened bones, in a condition called **osteoporosis**.

Starvation can also starve heart muscles to the point that altered rhythms develop. Blood flow is reduced, and blood pressure drops so much that the little nourishment present cannot get to the cells.

Some women concerned about storing too much fat allow themselves to eat—sometimes very large amounts of food (called binge-eating)—but prevent the nutrients from being turned into fat by purging themselves, often by vomiting. Binge-eating followed by purging is called **bulimia**.

Besides experiencing the same health problems that anorexics face, bulimics can rupture their stomachs through forced vomiting. They often have dental and gum problems caused by stomach acid being forced up into their mouths during vomiting, and they can become fatally dehydrated.

As you have seen, not having enough body fat causes many problems, particularly for women. Storing too much body fat also causes health problems for both sexes.

Obesity

One in four Americans is considered to be obese. Obesity results in many health problems including high blood pressure, heart attack, stroke, and joint problems. An additional risk of being overweight is the increased risk of one form of diabetes. Diabetes is discussed fully in Chapter 7. For now, it is enough to know that one form of diabetes, called type 2 diabetes, is associated with obesity and that diabetes results in an increase in blood-glucose levels that can cause excessive thirst, weakness, confusion, convulsions, coma, and even death. Approximately 90% of the 21 million Americans with diabetes have this obesity-induced form of diabetes.

Stop and Stretch 3.7

One of the earliest signs that type 2 diabetes might be developing is increased thirst. Based on what you know about blood-glucose levels in people with diabetes and tonicity in cells, explain why this symptom might be common among diabetics.

Although being overweight in general can cause many health problems, the location of the fat also matters. People who store fat on their upper bodies are more at risk than people who store fat on their buttocks and thighs. Men tend to store more weight in their abdomens than women, and overconsumption of alcohol can increase fat accumulation around the stomach (beer belly) in both males and females. Measuring the circumference of your waist (just above your belly button) can help determine whether you are at a higher risk for the obesity-related health problems. A waist circumference greater than 40 inches in males and 35 inches in females indicates increased risk for diabetes, high blood pressure, and cardiovascular disease.

Another way to measure weight and health risk, the **body mass index (BMI)**, can be used to help assess weight-related health risks. BMI uses both height and weight to determine a value that correlates an estimate of body fat with the risk of illness and death.

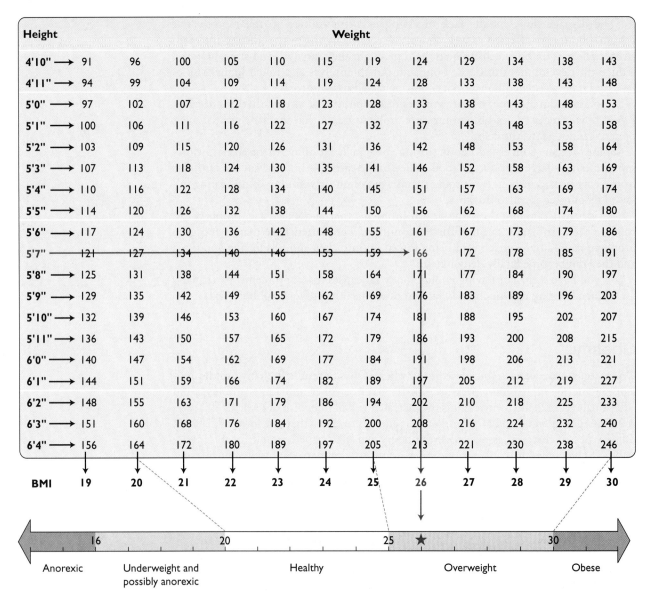

FIGURE 3.36 BMI. The BMI chart uses both height and weight to determine a numeric value that correlates an estimate of body fat with risk of illness and death for both men and women. A person who is 5′ 7″ and weighs 166 pounds has a BMI of 26 and would be considered overweight.

VISUALIZE THIS We know that women who are underweight have many health problems and would be likely to suffer health consequences. Why might this correlation between underweight and illness also be seen in males even though there are not really any health risks for males who are underweight?

A person's BMI is obtained by dividing their weight (in kg) by their height (in meters squared). The chart (FIGURE 3.36) displays the results of that calculation for a range of weights and heights. BMI measures were first used by insurance companies in the 1940s to help evaluate whether a potential client might become a costly user of health insurance. However, BMI does not account for differences in frame size, gender, or muscle mass. In fact, studies show that as many as one in four people may be misclassified by BMI tables, because this measurement provides no means to distinguish between lean muscle mass and body fat. For example, an athlete with a lot of muscle will weigh more than a similarly sized person with a lot of fat, because muscle is heavier than fat.

If your BMI does not fall within the healthy range (BMI of 20–25), focusing on getting the right number of calories from healthy sources along with getting 30 min-

FIGURE 3.37 USDA Food Guide Pyramid. This newly designed food guide stresses the importance of physical activity. The size of each triangle represents the relative proportion of your diet that should be composed of each food group.

VISUALIZE THIS What does the unlabeled yellow triangle represent? Why do you suppose this triangle is smaller than the others?

utes of exercise most days is a safer, more realistic path to health than trying to attain a particular body type (FIGURE 3.37).

If your BMI falls within the healthy range, and your waist circumference is normal, you should still focus on eating healthful foods and getting exercise, but you probably have no reason to worry about health risks from being underweight or obese. Even though you might not look like a supermodel, movie star, or professional athlete, you are likely to be storing the right amount of body fat.

Chapter REVIEW

ROOTS TO REMEMBER

The following roots of words come mainly from Latin and Greek and will help you to decipher terms:

a- means without, or a lack of.

-ase is a common suffix in names of enzymes.

cyto- and **-cyte** mean cell or a kind of cell.

endo- means inside.

hemo- relates to blood.

-ic is a suffix commonly used to name acids.

iso- means the same or equal.

-plasm means a fluid.

KEY TERMS

3.1 Food and Energy
adenosine diphosphate (ADP) *p. 58*
adenosine triphosphate (ATP) *p. 57*
adipose *p. 56*
phosphorylation *p. 58*

3.2 Cell Structure and Function
centriole *p. 62*
cytoplasm *p. 61*
cytoskeleton *p. 62*
cytosol *p. 60*
endoplasmic reticulum (ER) *p. 61*
Golgi apparatus *p. 62*
lysosome *p. 61*
mitochondria *p. 61*
nucleus *p. 60*
organelle *p. 60*
ribosome *p. 61*

3.3 Membrane Structure and Function
active transport *p. 66*
diffusion *p. 64*
endocytosis *p. 66*
exocytosis *p. 66*
facilitated diffusion *p. 65*
hypertonic *p. 65*
hypotonic *p. 65*
isotonic *p. 65*
osmosis *p. 65*
passive transport *p. 64*
phospholipid bilayer *p. 64*
plasma membrane *p. 64*
selectively permeable *p. 64*
tonicity *p. 65*

3.4 Metabolism—Chemical Reactions in the Body
activation energy *p. 67*
active site *p. 68*
aerobic *p. 69*
aerobic respiration *p. 69*
anaerobic respiration *p. 73*
ATP synthase *p. 71*
basal metabolic rate *p. 74*
calorie *p. 74*
Calorie *p. 74*
catalyze *p. 67*
cellular respiration *p. 68*
citric acid cycle *p. 70*
electron transport chain *p. 71*
enzyme *p. 67*
fermentation *p. 73*
glycolysis *p. 69*
hydrogen atom *p. 69*
induced fit *p. 68*
metabolic rate *p. 74*
metabolism *p. 67*
nicotinamide adenine dinucleotide (NAD) *p. 69*
pyruvic acid *p. 69*
specificity *p. 68*
substrate *p. 67*

3.5 Health and Body Weight
amenorrhea *p. 74*
anorexia *p. 74*
body mass index (BMI) *p. 75*
bulimia *p. 75*
leptin *p. 74*
osteoporosis *p. 75*

SUMMARY

3.1 Food and Energy

- Carbohydrates, proteins, and fats store energy in their chemical bonds (p. 56).
- Cells use a chemical called ATP as their energy currency (pp. 57–58; Figure 3.3).

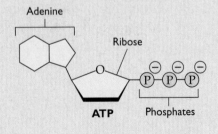

- Fat is stored in adipose tissue (p. 56).

3.2 Cell Structure and Function

- All living things are composed of cells (p. 59).
- Organelles perform many coordinated functions in cells (pp. 60–63; Figure 3.10).

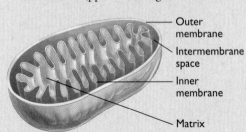

- The nucleus houses the DNA and serves as the control center of the cell (p. 60).
- Mitochondria are kidney-bean-shaped, double-membrane-bounded organelles that help convert the energy stored in food molecules into ATP (p. 61).
- Lysosomes help break down ingested food before it is sent to the mitochondria (p. 61).
- Ribosomes have two subunits. Each subunit is composed of RNA and protein. A functional ribosome serves as a workbench where proteins are assembled (p. 61).

- Rough endoplasmic reticulum is a series of membranous sacs with ribosomes attached (p. 61).
- Smooth endoplasmic reticulum is membranous and tubular shaped, does not have ribosomes attached, and is involved in detoxification and other metabolic functions (pp. 61–62).
- Proteins are modified and sorted in the sacs of the membranous Golgi apparatus (p. 62).
- Centrioles are involved in moving genetic material when a cell divides (pp. 62–63).
- The cytoskeleton helps give the cell shape and can help a cell move (pp. 62–63).

3.3 Membrane Structure and Function

- Nutrients move across the plasma membrane, which functions as a semipermeable barrier that allows some substances to pass and prevents others from crossing (p. 64; Figure 3.19).

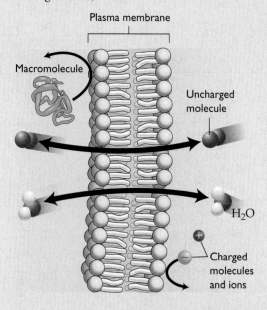

- The plasma membrane is composed of two layers of phospholipids, in which are embedded proteins and cholesterol. Carbohydrate chains can bind to proteins and lipids on the outer surface of the membrane (p. 64).
- Passive transport mechanisms include simple diffusion, osmosis, and facilitated diffusion (diffusion through proteins). Passive transport always moves substances with their concentration gradient and does not require energy (pp. 64–65).
- Osmosis, the diffusion of water across a membrane, helps cells maintain their shape (p. 65).
- Cells in an isotonic environment maintain their shape, whereas those in a hypertonic environment shrivel. Cells in a hypotonic environment expand and may burst (p. 65).
- Active transport is an energy-requiring process that occurs when proteins in cell membranes move substances against their concentration gradients (p. 66).
- Larger molecules move into and out of cells enclosed in membrane-bounded vesicles (p. 66).

3.4 Metabolism—Chemical Reactions in the Body

- The chemical reactions that occur in cells to build up or break down macromolecules are called metabolic reactions (p. 67).
- Metabolism is governed by enzymes. Enzymes are proteins that catalyze specific cellular reactions, first by binding the substrate to the enzyme's active site. This binding causes the enzyme to change shape (induced fit), placing stress on the bonds of the substrate and thereby lowering the activation energy (pp. 67–68; Figure 3.25).

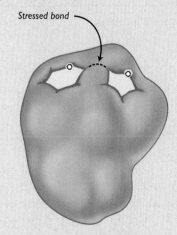

- As sugars, proteins, and fats go through cellular respiration, energy stored in their chemical bonds is released and used to synthesize ATP (p. 69).
- Cellular respiration begins in the cytosol, where a 6-carbon sugar is broken down into two 3-carbon pyruvic acid molecules during the anaerobic process of glycolysis. The pyruvic acid molecules then move across the two mitochondrial membranes and into the matrix of the mitochondrion, where the citric acid cycle strips them of carbon dioxide and electrons. The electrons are carried by electron carriers to the inner mitochondrial membrane; there, they are added to a series of proteins called the electron transport chain. At the bottom of the electron transport chain, electronegative oxygen pulls the electrons toward itself. As the electrons move down the electron transport chain, the energy that they release is used to drive protons into the intermembrane space. Once there, the protons rush through the enzyme ATP synthase and produce ATP from ADP and phosphate. When electrons reach the oxygen at the bottom of the electron transport chain, they combine with the oxygen and hydrogen ions to produce water (pp. 69–73).
- Energy is measured in units called Calories (p. 74).
- Metabolic rate is a measure of energy use (p. 74).

3.5 Health and Body Weight

- Being either underweight or overweight can result in serious health consequences (pp. 74–75).
- Waist circumference and BMI measurements are ways to determine whether one's body fat level is healthful (pp. 75–76).
- Eating right and exercising are the best ways to achieve a healthy body weight (p. 77; Figure 3.37).

LEARNING THE BASICS

1. ATP _____.
   a. is composed of adenine, a sugar, and three phosphates
   b. stores energy that can be used by cells
   c. is produced during cellular respiration
   d. can be used to perform many types of cellular work
   e. all of the above are true

2. ATP energizes cellular activities by _____.
   a. releasing heat
   b. acting as an enzyme
   c. direct transfer of a phosphate group
   d. releasing electrons to drive reactions

3. Which of the following pairs is mismatched?
   a. nucleus : DNA
   b. lysosome : digestive reactions
   c. mitochondria : protein synthesis
   d. cytoskeleton : microtubules
   e. cell membrane : phospholipid bilayer

4. Which of the following are made of two subunits and are composed of RNA and protein?
   a. Golgi
   b. mitochondria
   c. chloroplasts
   d. ribosomes
   e. ER

5. What membrane surface molecules serve as molecular identification tags?
   a. phospholipids
   b. fatty acids
   c. proteins
   d. cholesterol
   e. carbohydrates

6. A red blood cell placed in a hypotonic environment will _____.
   a. retain its shape
   b. shrivel
   c. burst

7. Which of the following is a *false* statement regarding enzymes?
   a. Enzymes are proteins that function as catalysts.
   b. Enzymes display specificity for their substrates.
   c. Enzymes provide the energy for the reactions they catalyze.
   d. The activity of enzymes can be influenced by their environment.
   e. An enzyme may be used many times over.

8. Water moves by osmosis from a _____.
   a. hypertonic solution to a hypotonic cell interior
   b. hypotonic solution to a hypertonic cell interior
   c. solution of greater solute concentration to one of lesser concentration
   d. region where there is less water to a region where there is more

9. Transport proteins in the plasma membrane are required for _____.
   a. diffusion
   b. osmosis
   c. facilitated diffusion
   d. active transport
   e. more than one of the above

10. Which of the following would have the most difficult time crossing a cell membrane unaided?
    a. carbon dioxide
    b. oxygen
    c. a small nonpolar molecule
    d. a polar molecule

11. Which of the following is a correct statement regarding diffusion?
    a. It requires an expenditure of ATP energy by the cell.
    b. It is a passive process.
    c. It occurs when molecules move from a region of their own lower concentration to one of their own higher concentration.
    d. It requires membrane proteins.

12. Which of the following would not be found in a cell membrane?
    a. phospholipids
    b. proteins
    c. cholesterol
    d. nucleotides

13. A substance moving across a membrane against a concentration gradient is moving by _____.
    a. passive transport
    b. osmosis
    c. facilitated diffusion
    d. active transport
    e. diffusion

14. A cell that is placed in a hypertonic solution of salty seawater will _____.
    a. take sodium and chloride ions in by diffusion
    b. move water out of the cell by active transport
    c. use facilitated diffusion to break apart the sodium and chloride ions
    d. lose water to the outside of the cell via osmosis

15. Match each organelle or structure with its function.
    _____ detoxification   a. ribosome
    _____ modifies proteins   b. nucleolus
    _____ where ribosomes are made   c. smooth ER
    _____ helps a cell maintain its shape   d. Golgi apparatus
    _____ where proteins are made   e. cytoskeleton

16. **True or False?** Proteins and carbohydrates can be converted into fats.

17. **True or False?** A person's metabolic rate is affected by his or her age, sex, and genetic makeup.

ANALYZING AND APPLYING THE BASICS

1. Two people with very similar diets and similar exercise levels have very different amounts of body fat. Why might this be the case?

2. Brown fat, or brown adipose tissue, is found in human infants, young mammals, and adult mammals that live in cold climates. This type of tissue often forms a layer around vital organs. Brown fat differs from white fat in that it contains more nerve cells, more blood vessels, and greater numbers of mitochondria. What is the likely purpose of brown fat? Explain your answer.

3. Maintaining a low-calorie diet, around 30–50% below normal, has been shown to increase life expectancy in yeast, worms, flies, and rats. A 2002 study determined that yeast can produce 30% more generations when maintained in a 0.5% glucose solution versus a 2% glucose solution.

    An increase in life span appears to be linked to an increase in enzyme production by the *SIR2* gene. This enzyme production is stimulated by NAD^+, which is produced in abundance when yeast cells are starved. As NAD^+ increases, yeast cells switch from fermentation to cellular respiration. Why can this switch occur?

4. The CB1 receptor is located mainly in neurons in the brain. Certain neurotransmitters produced by the body can bind to this receptor and induce appetite. Cannabis also shows an affinity for the CB1 receptor, triggering excessive hunger in some marijuana users. A French drug company has developed a drug that blocks the CB1 receptor, thus reducing appetite. Compare the interaction between the appetite-suppressant drug and the CB1 receptor to the interactions between enzymes and substrates.

Questions 5–8 refer to the following information.

The human body maintains energy homeostasis (a balance between energy intake and energy use) using a variety of hormones that interact with the brain. The following table shows hormones that affect appetite.

Hormone	Produced by	Effect
Cholecystokinin	Gastrointestinal tract	Causes sense of fullness after meal
Ghrelin	Stomach	Stimulates appetite when stomach is empty
Insulin	Pancreas	Decreases appetite
Leptin	Fat tissue	Decreases appetite
Melanocortin	Hypothalamus	Decreases appetite
NPY	Hypothalamus	Stimulates appetite
PYY(3-36)	Colon	Inhibits eating for up to 12 hours

5. Using the table above, describe how hormonal feedback systems may help the body to maintain energy homeostasis before and after a meal.

6. As a person loses weight, fat cells shrink and less leptin is secreted. This sends a signal to the hypothalamus to increase production of NPY. As NPY production increases, melanocortin production decreases. What effects would these changes in hormone levels have on the person who is trying to lose weight?

7. The *ob/ob* mouse is a laboratory strain that lacks the leptin-producing gene. This strain of mouse eats all the time and is three times heavier than normal mice. When *ob/ob* mice were injected with leptin, fat stores decreased in the mice and their body weights became normal. Could injections with leptin help obese humans to obtain normal weights? Explain your answer.

8. Considering the information above and in the chapter, what might be some practical, drug-free ways to lose weight?

CONNECTING THE SCIENCE

The New York City Board of Health recently adopted the nation's first major municipal ban on the use of artificial trans fats in restaurant cooking.

1. If New York bans trans fats in restaurant cooking, should state health regulators ban it in commercially prepared/packaged food like Oreos, Chips Ahoy, or Lays potato chips?

2. Do you think banning trans fats is the same as banning smoking (eliminating a dangerous substance)? If so, should alcohol be banned?

3. The Chicago City Council recently banned restaurants from serving foie gras, a pate made of goose or duck liver. To fatten the birds and enlarge their livers before slaughter, the birds are force-fed through a tube placed down their throat for several weeks. Do you think the City of Chicago was ethically correct in banning restaurants from serving foie gras?

Chapter 4

Genes—Transcription, Translation, Mutation, and Cloning

Genetically Modified Foods

LEARNING GOALS

1. Describe the structure and function of a gene.
2. Describe the synthesis of a protein from transcription through translation.
3. Which organelles and structures are involved in transcription and translation?
4. Compare and contrast mRNA, tRNA, and rRNA.
5. Describe the structure and function of transfer RNA (tRNA).
6. How do mutations affect gene expression?
7. Explain how two different genes would differ from each other.
8. Describe what makes any two proteins different from each other.
9. How is gene expression regulated?
10. Describe the process of cloning a gene using bacterial cells.

Many activists are concerned about the effects of producing and eating genetically modified foods.

Emotions run high in the debate over *genetically engineered* foods. These foods, also called *genetically modified* or *GM* foods, have had one or more of their genes modified, or a gene from another organism inserted alongside their normal complement of genes. People on both sides of the issue are making their cases dramatically.

Demonstrators dressed in biohazard suits tossed crackers, cereal, and pasta into a garbage can in front of a supermarket while shareholders inside voted on whether to remove genetically modified foods from the store shelves or not. Late one summer night, a group calling itself Seeds of Resistance hacked down a plot of genetically modified corn at the University of Maine. Other detractors destroyed the world's first outdoor trial of genetically engineered coffee.

In response to concerns over genetically modified foods, some parents refuse to send their kids to day-care centers that use milk from cows treated with growth hormone, and most natural food co-ops won't sell the so-called frankenfoods.

Is milk from hormone-treated cows safe?

On the other side of the issue are proponents of genetically modifying foods who believe the technology holds tremendous promise. Advocates claim that genetically modified foods have the potential to help wipe out hunger and cure human disease. They see a future in which all children will be vaccinated through the use of edible vaccines and common human diseases like heart disease will be prevented when people begin to eat genetically modified animals such as pigs that produce omega-3 fatty acids. Proponents even see environmental benefits because some genetically modified crops allow farmers to use far smaller amounts of chemicals that damage the environment.

Is it ethical to genetically modify animals?

Between these contrasting views, it is hard to find the middle ground. How can the average consumer determine whether it is okay to purchase and consume these foods? To answer this question for yourself, you must first gain a basic understanding of what genes are, what genes do, and what it means to modify a gene.

Will genetic technologies help decrease farmers' reliance on pesticides?

4.1 What Is a Gene?

Genes are segments of DNA that carry information about traits that can be passed from parents to offspring. You learned in Chapter 2 that DNA (deoxyribonucleic acid) is located within the nucleus and composed of the nucleotides adenine (A), guanine (G), cytosine (C), and thymine (T).

The function of genes is to serve as instructions for making polypeptides. Individual polypeptides can fold on themselves to produce a functional protein, or several polypeptides can join together to produce the functional protein. The sequence of nucleotides in the DNA can be thought of as a blueprint that directs the cell to build a particular polypeptide. In earlier chapters, you learned about proteins in membranes that help to transport substances, and about enzymes, which are proteins that help to catalyze metabolic reactions. Genes composed of DNA code for both of those types of proteins, and genes can code for many other proteins required by cells. Proteins give cells—and by extension, entire individuals—nearly all of their genetically determined characteristics.

Genes encode proteins. Long stretches of genes are packaged to produce **chromosomes**. A chromosome typically carries thousands of genes. Human chromosomes are linear structures composed of DNA wrapped around various proteins. An organism's full set of chromosomes is called its **genome**.

A more thorough understanding of genes and their role in heredity will be detailed in subsequent chapters. For now, it is enough to know the following:

- Genes are composed of DNA.
- Genes encode proteins.
- Genes are located on chromosomes.

Stop and Stretch 4.1

Assume that a particular protein is composed of two different polypeptides folded together. How many genes encode this protein?

To genetically modify foods, scientists can move a gene known to produce a certain protein from one organism to another. Genetically modifying foods can also involve making changes in the amount of protein a gene produces. Regulating the amount of protein available for the cell to use is also referred to as regulating *gene expression*.

4.2 Protein Synthesis and Gene Expression

Let's first examine how cells normally use DNA instructions to produce proteins and how gene expression is controlled. Then we can turn our attention to how scientists manipulate these processes to genetically modify foods.

From Gene to Protein

Protein synthesis involves using the instructions carried by a gene to build a particular polypeptide or protein. As described earlier, genes do not build proteins directly. Instead, genes carry the instructions that dictate how a protein should be built. Understanding protein synthesis requires that we review a few basics about DNA, proteins, and RNA.

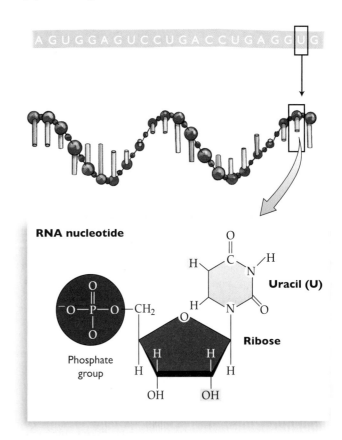

FIGURE 4.1 DNA and RNA. (a) DNA is double stranded. Each DNA nucleotide is composed of the sugar deoxyribose, a phosphate group, and a nitrogen-containing base (A, G, C, or T). (b) RNA is single stranded. RNA nucleotides are composed of the sugar ribose, a phosphate group, and a nitrogen-containing base (A, G, C, or U).

VISUALIZE THIS Point out the structural difference between the sugar in DNA and RNA, and between the nitrogenous bases uracil and thymine.

First, DNA is a double-stranded polymer of nucleotides. The nitrogenous bases of nucleotides hydrogen bond with each other based on complementarity (A to T, and C to G). Second, proteins are large molecules composed of amino acids. Each protein has a unique function dictated by its particular structure. The structure of a protein is the result of the order of amino acids that constitute it, because the chemical properties of amino acids cause a protein to fold in a particular manner.

Before a protein can be built, a copy of instructions carried by a gene has to be made. When the gene is copied, the copy is made up not of DNA (deoxyribonucleic acid) but of **RNA (ribonucleic acid)**. Therefore, it is important to understand the differences between DNA and RNA.

RNA is also a polymer of nucleotides. A nucleotide is composed of a sugar, a phosphate group, and a nitrogen-containing base. Whereas the sugar in DNA is deoxyribose, the sugar in RNA is ribose. Also, RNA has the nitrogenous base uracil (U) in place of thymine. One final difference between RNA and DNA is that RNA is usually single stranded, not double stranded like DNA (FIGURE 4.1).

When a cell requires a particular protein, a strand of RNA is produced using DNA as a guide or template. RNA nucleotides are able to make base pairs with DNA nucleotides. C and G make a base pair, and U pairs with A.

The RNA copy then serves as a blueprint that tells the cell which amino acids to join together to produce a protein. Thus, the flow of genetic information is from DNA to RNA to protein (FIGURE 4.2).

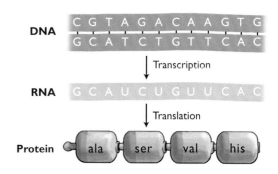

FIGURE 4.2 The flow of genetic information. Genetic information flows from a gene (DNA) to an RNA copy of the DNA gene, to the amino acids that are joined together to produce the protein coded for by the gene.

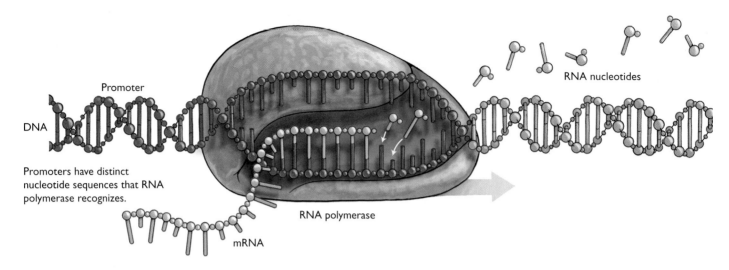

FIGURE 4.3 Transcription. After locating the promoter, the enzyme RNA polymerase ties together nucleotides within the growing RNA strand as they bind to their complementary base on the DNA. Only when a complementary base pair is made between DNA and RNA does the polymerase add an RNA nucleotide to the growing strand. Complementary bases are formed via hydrogen bonding of A with U and G with C. Note that only one strand of the DNA is used as a template for the synthesis of the mRNA. When the RNA polymerase reaches the end of the gene, the mRNA transcript is released.

VISUALIZE THIS Based on what you know about the size of various nitrogenous bases (the purines A and G are composed of two ringed structures, while the pyrimidines C and T are composed of single rings), propose a sequence of RNA that is being synthesized in this illustration.

How does this flow of information actually take place in a cell? Going from gene to protein involves two steps. The first step, called **transcription**, involves producing the RNA copy of the required gene. In the same way that a transcript of a speech is a written version of the words spoken by the speaker, transcription inside a cell is a process that produces a copy with the RNA nucleotides substituted for DNA nucleotides. The second step, called **translation**, involves decoding the copied RNA sequence and producing the protein for which it codes. In the same way that a translator helps determine the meaning of words in two different languages, translation in a cell involves moving from the language of nucleotides (DNA and RNA) to the language of amino acids and proteins.

Transcription: Copying the Gene

Transcription occurs in the nucleus and involves the copying of a DNA gene into RNA. The copy is synthesized with the help of an enzyme called **RNA polymerase**.

To begin transcription, the RNA polymerase binds to a nucleotide sequence at the beginning of every gene, called the **promoter**. Once the RNA polymerase has located the beginning of the gene by binding to the promoter, the enzyme rides along the DNA, unzipping the double helix. Once the helix is unwound, hydrogen bonding occurs between the DNA gene and the complementary RNA nucleotides (C : G and A : U). The RNA polymerase then ties together the nucleotides one after another along the sugar-phosphate backbone (FIGURE 4.3).

To tie adjacent nucleotides together, the RNA polymerase performs a dehydration synthesis reaction, resulting in the formation of a covalent phosphodiester bond between the nucleotides. This process results in the production of a single-stranded RNA molecule called **messenger RNA (mRNA)**, which carries the message of the gene that is to be expressed. A newly synthesized mRNA transcript consists of *introns*, or intervening sequences, which correspond to sequences of DNA that don't code for proteins, and *exons*, which carry the protein-building instructions. Introns are spliced out of the transcript before it leaves the nucleus to be translated.

FIGURE 4.4 The structure of a ribosome. Ribosomes are composed of two subunits. Each subunit, in turn, is composed of rRNA and protein. Ribosomes are the site of protein synthesis.

Translation: Using the Message to Synthesize a Protein

The second step in going from gene to protein requires that the mRNA be used to produce the actual protein that the gene encodes. The decoding of the gene occurs in the cytoplasm during the process of translation. For translation to occur, a cell needs the following: mRNA, a supply of amino acids, structures called ribosomes, and transfer RNA molecules.

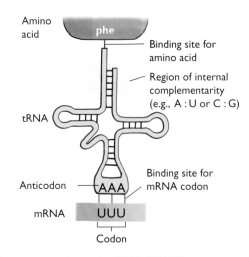

Ribosomes Ribosomes are subcellular, globular structures (FIGURE 4.4) that are composed of another kind of RNA called **ribosomal RNA (rRNA)**, which is wrapped around many different proteins. Each ribosome is composed of two subunits—one large and one small. When a ribosome is assembled in this fashion, the mRNA can be threaded through it. In addition, the ribosome is able to bind to structures called **transfer RNA (tRNA)** that carry amino acids.

Transfer RNA (tRNA) Transfer RNA (FIGURE 4.5) is yet another type of RNA found in cells. An individual transfer RNA molecule carries one specific amino acid and interacts with mRNA to place the amino acid in the correct location of the growing polypeptide.

As mRNA moves through the ribosome, small sequences of nucleotides are exposed. These sequences of mRNA, called **codons**, are three nucleotides long and encode a particular amino acid. Transfer RNAs also have a set of three nucleotides, which will bind to the codon if the right sequence is present. These three nucleotides at the base of the tRNA are therefore called the **anticodon**, and they complement a codon on mRNA. The anticodon on a particular tRNA binds to the complementary mRNA codon. In this way, the codon calls for the incorporation of a specific amino acid.

When a tRNA anticodon binds to the mRNA codon, a covalent peptide bond is formed between amino acids, again by dehydration synthesis. The ribosome adds the amino acid that the tRNA is carrying to the growing chain of amino acids that will eventually constitute the finished protein. The transfer RNA functions as a sort of cellular translator, fluent in both the language of nucleotides (its own language) and the language of amino acids (the target language).

To help you understand protein synthesis, let us consider its similarity to an everyday activity such as baking a cake (FIGURE 4.6). To bake a cake, you would consult a recipe book (genome) for the specific recipe (gene) to make your cake (protein). You may copy the recipe (mRNA) out of the book so that the original recipe (gene) does not become stained or damaged. The original recipe (gene) is left in the

FIGURE 4.5 Transfer RNA (tRNA). Transfer RNA molecules are composed of an RNA strand that has regions of internal complementarity. Each transfer RNA has a characteristic three-nucleotide anticodon that binds to the mRNA codon. Each tRNA also carries the amino acid corresponding to the mRNA codon to which it binds.

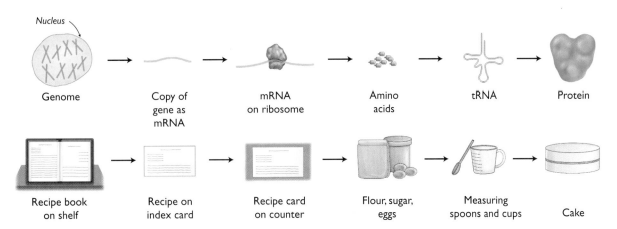

FIGURE 4.6 Cake baking and protein synthesis. The process of protein synthesis can be likened to cake baking.

book (genome) on a shelf (nucleus), so that you can make another copy when you need it. The original recipe (gene) can be copied again and again. The copy of the recipe (mRNA) is placed on the kitchen counter (ribosome) while you assemble the ingredients (amino acids). The ingredients (amino acids) for your cake (protein) include flour, sugar, butter, milk, and eggs. The ingredients are measured in measuring spoons and cups (tRNAs) that are dedicated to one specific ingredient. The measuring spoons and cups bring the ingredients to the kitchen counter. Like the amino acids that are combined in different orders to produce a specific protein, the ingredients in a cake can be used in many ways to produce a variety of foods. The ingredients (amino acids) are always added according to the instructions specified by the original recipe (gene).

Inside of cells, the sequence of bases in the DNA dictates the sequence of bases in the RNA, which in turn dictates the order of amino acids that will be joined together to produce a protein. Protein synthesis ends when a codon that does not code for an amino acid, called a **stop codon**, moves through the ribosome. When a stop codon is present in the ribosome, no new amino acid can be added, and the growing protein is released. Once released, the protein folds up on itself and moves to where it is required in the cell or moves to the Golgi apparatus for modification. A summary of the process of translation is shown in FIGURE 4.7.

The process of translation allows cells to join amino acids in the sequence coded by the gene. Scientists can determine the sequence of amino acids that a gene calls for by looking at a chart called the **genetic code**.

Genetic Code The genetic code shows which mRNA codons code for which amino acids (FIGURE 4.8). Note that there are 64 codons, 61 of which code for amino acids. Three of the codons are stop codons that occur near the end of the mRNA. Because stop codons do not code for an amino acid, protein synthesis ends when a stop codon enters the ribosome. In the table, you can see that the codon AUG functions both as a start codon (and thus is found near the beginning of each mRNA) and as a codon dictating that the amino acid methionine (met) be incorporated into the protein being synthesized. Methionine is often excised later in the process.

Notice also that the same amino acid can be coded for by more than one codon. For example, the amino acid threonine (thr) is incorporated into a protein in response to the codons ACU, ACC, ACA, and ACG. The fact that more than one codon can code for the same amino acid is referred to as *redundancy* in the genetic code.

There is, however, no situation where a given codon can call for more than one amino acid. For example, AGU codes for serine (ser) and nothing else. Therefore, there is no *ambiguity* in the genetic code as to what amino acid any codon will call for. The genetic code is also universal in the sense that different organisms, with very few exceptions, use the same codons to specify the same amino acids. The *universality* of the genetic code allows a gene from one organism to produce the same protein if inserted into another organism. Universality is why a protein normally produced in one organism can be inserted into and expressed by another organism for the purposes of genetic engineering.

Changes to the DNA can result in the production of a polypeptide that does not function as it should.

Mutations

Changes to the DNA sequence, called **mutations**, can affect the order of amino acids incorporated into a protein during translation. Mutations to a gene can alter the order of the amino acids of the encoded protein.

Mutations can result in either a nonfunctional protein or a protein different from the one previously called for. If this protein does not have the same amino acid composition, it

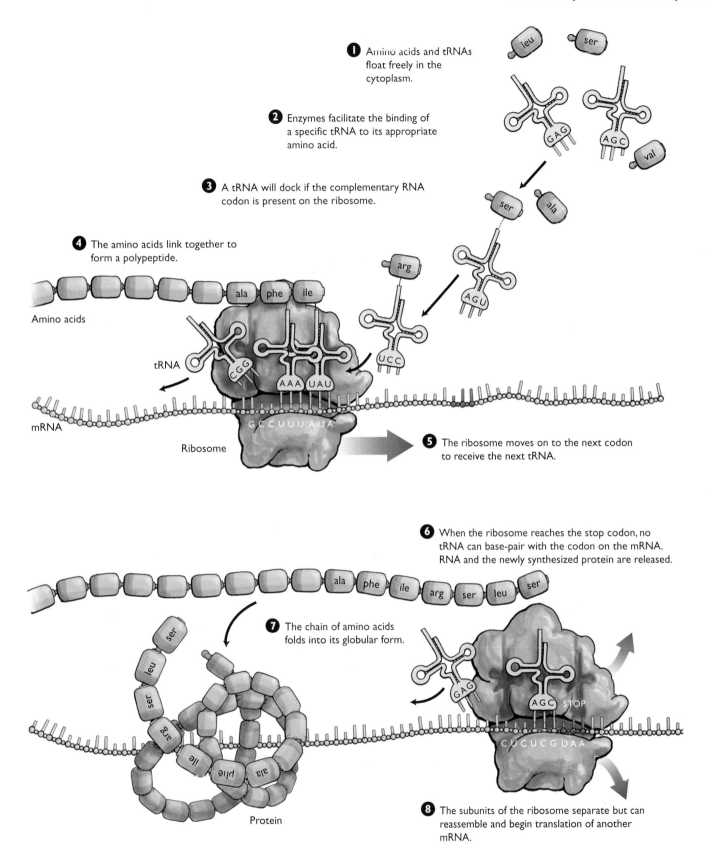

FIGURE 4.7 Translation. During translation, mRNA directs the synthesis of a protein. The mRNA codon that is exposed in the ribosome binds to its complementary tRNA molecule, which carries the amino acid coded for by the DNA gene. When many amino acids are joined together, the required protein is produced. When the translation machinery reaches a stop codon, the newly synthesized protein is released into the cytoplasm.

CHAPTER 4 Genes—Transcription, Translation, Mutation, and Cloning

	Second base			
First base	**U**	**C**	**A**	**G**
U	UUU, UUC Phenylalanine (phe) UUA, UUG Leucine (leu)	UCU, UCC, UCA, UCG Serine (ser)	UAU, UAC Tyrosine (tyr) UAA **Stop codon** UAG **Stop codon**	UGU, UGC Cysteine (cys) UGA **Stop codon** UGG Tryptophan (trp)
C	CUU, CUC, CUA, CUG Leucine (leu)	CCU, CCC, CCA, CCG Proline (pro)	CAU, CAC Histidine (his) CAA, CAG Glutamine (gln)	CGU, CGC, CGA, CGG Arginine (arg)
A	AUU, AUC, AUA Isoleucine (ile) AUG Methionine (met) **Start codon**	ACU, ACC, ACA, ACG Threonine (thr)	AAU, AAC Asparagine (asn) AAA, AAG Lysine (lys)	AGU, AGC Serine (ser) AGA, AGG Arginine (arg)
G	GUU, GUC, GUA, GUG Valine (val)	GCU, GCC, GCA, GCG Alanine (ala)	GAU, GAC Aspartic acid (asp) GAA, GAG Glutamic acid (glu)	GGU, GGC, GGA, GGG Glycine (gly)

Third base: U, C, A, G

FIGURE 4.8 The genetic code. The genetic code shows which amino acid is coded for by each mRNA codon. Look at the left-hand side of the table for the base of the first nucleotide in the codon. There are four rows, one for each possible RNA nucleotide—A, C, G, or U. By looking at the intersection of the second-base columns at the top of the table and the first-base rows, you can then narrow your search for the codon to four different codons. Finally, the base of the third nucleotide in the codon on the right-hand side of the table determines the amino acid that a given mRNA codon codes for. Note that three codons, UAA, UAG, and UGA, do not code for an amino acid. These three are stop codons. The codon AUG is a start codon, found at the beginning of most protein-coding sequences.

VISUALIZE THIS What codons on DNA would code for the incorporation of histidine?

may not be able to perform the same function (FIGURE 4.9). Single nucleotide changes or deletions are called **point mutations**. A **missense mutation** can occur when a point mutation results in the substitution of one amino acid for another, often with serious impacts on the structure and function of the protein. For example, you will learn in Chapter 8 that the substitution of a single nucleotide in the gene that encodes hemoglobin compromises the ability of this protein to carry oxygen, resulting in sickle-cell disease. Another type of point mutation, the **nonsense mutation**, changes the codon from one that codes for an amino acid to a stop codon. The presence of a stop codon in the mRNA will lead to premature termination of translation, and an abbreviated version of the protein will be produced.

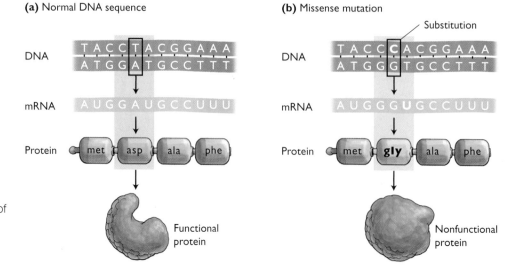

FIGURE 4.9 Mutation. A single nucleotide change from the normal sequence (a) to the mutated sequence (b) can result in the incorporation of a different amino acid. If the substituted amino acid has chemical properties different from those of the original amino acid, then the protein may assume a different shape and thus lose its ability to function properly.

In some cases, a mutation has no effect on a protein. Such cases may occur when changes to the DNA result in the production of the mRNA codon that codes for the same amino acid as was originally called for. Due to the redundancy of the genetic code, a mutation that changes the mRNA codon from ACU to ACC will have no impact because both of these codons code for the amino acid threonine. This is called a **neutral mutation** (FIGURE 4.10). In addition, mutations can result in the substitution of one amino acid for another with similar chemical properties, which may have little or no effect on the protein.

Inserting or deleting a single nucleotide can have a severe impact because the addition (or deletion) of a nucleotide can change the groupings of nucleotides in every codon that follows. Changing the triplet codon groupings is called altering the **reading frame**. All nucleotides located after an insertion or deletion will be regrouped into different codons, producing a **frameshift mutation**. For example, inserting an extra letter "H" after the fourth letter of the sentence, "The dog ate the cat," could change the reading frame to the nonsensical statement, "The dHo gat eth eca t." Inside cells, this shift in the frame often results in the incorporation of a stop codon and the production of a shortened, nonfunctional protein.

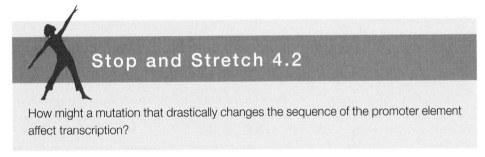

Stop and Stretch 4.2

How might a mutation that drastically changes the sequence of the promoter element affect transcription?

All cells in all organisms are capable of undergoing protein synthesis, with different cell types selecting different genes from which to produce proteins.

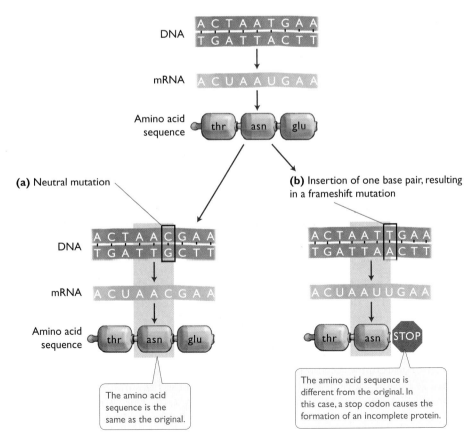

FIGURE 4.10 Neutral and frameshift mutations. (a) Neutral mutations are changes to the DNA that, due to the redundancy of the genetic code, do not change which amino acid is called for. (b) Frameshift mutations involve the insertion (or deletion) of a nucleotide and can result in the production of mRNA that produces the wrong protein or one that terminates translation too soon.

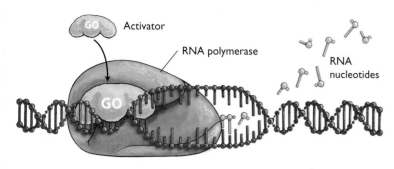

FIGURE 4.11 Regulation of gene expression by regulating transcription. Gene expression can be regulated by activation, during which activator proteins help RNA polymerase bind the promoter. This is a common method of gene regulation in humans.

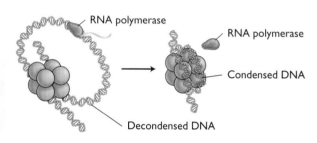

FIGURE 4.12 Regulating gene expression by chromosome condensation. RNA polymerase can transcribe decondensed chromosomes more readily than condensed chromosomes.

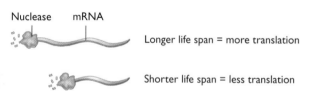

FIGURE 4.13 Regulating gene expression by regulating mRNA degradation. The longer a particular mRNA is present in the cell, the more times it will be translated and the higher the number of proteins that will be produced.

Regulating Gene Expression

Each cell in your body, except sperm or egg cells, has the same complement of genes you inherited from your parents. Yet each cell expresses only a small percentage of those genes. In other words, different cell types transcribe and translate different genes. For example, because your liver and pancreas each perform specific functions, the cells of your liver turn on or express one suite of genes and the cells of your pancreas, another. Turning a gene on or off, or modulating that gene more subtly, is called regulating gene expression. The expression of a given gene is regulated so that it is turned on and turned off in response to the cell's needs, in order to maintain homeostasis. In other words, cells regulate which individual genes are expressed under a given set of circumstances rather than simply flipping a switch to turn all genes on or off at once. One way to regulate gene expression is to regulate the rate of transcription.

Regulation of Transcription Gene expression is most commonly regulated by controlling the rate of transcription. Regulation of transcription can occur at the promoter, which you will recall is the sequence of nucleotides adjacent to a gene to which the RNA polymerase binds to initiate transcription. When a cell requires a particular protein, the RNA polymerase enzyme binds to the promoter for that particular gene and transcribes the gene.

Transcription can be controlled via the actions of proteins called **activators** that help the RNA polymerase bind to the promoter, thus facilitating gene expression (FIGURE 4.11). The rate at which the polymerase binds to the promoter is also affected by substances that are present in the cell. For example, the presence of alcohol in a liver cell might result in increased transcription of a gene involved in the breakdown of alcohol. It is also possible to prevent transcription from occurring by blocking the promoter region. When certain repressor proteins bind to the promoter, the polymerase cannot bind to initiate transcription. This method of regulation is more common in bacterial cells than it is in human cells, but does occur in both. Other methods by which cells regulate gene expression follow.

Regulation by Chromosome Condensation For a gene to be transcribed in humans, the chromosome in the region of the gene must first open up, or decondense, so that the RNA polymerase can access the gene (FIGURE 4.12).

When chromosomes condense, the DNA is tightly wound around chromosomal proteins. Condensation prevents RNA polymerase from gaining access to the DNA and thus prevents transcription and gene expression. Chromosomes may decondense one or more segments at a time. For example, chromosomes in developing egg cells decondense many segments at a time so that the rapidly developing egg cell can be supplied with all the proteins it needs.

Regulation by mRNA Degradation Cells can also regulate the expression of a gene by regulating how long a messenger RNA is present in the cytoplasm. Enzymes called *nucleases* roam the cytoplasm, cutting RNA molecules by binding to one end and breaking the bonds between nucleotides. If a particular mRNA has a long "tail" of extra nucleotides, it will survive longer in the cytoplasm and be translated more times (FIGURE 4.13). All mRNAs are eventually degraded in this way. Otherwise, once a gene had been transcribed one time, it would be expressed forever.

Regulation of Translation It is also possible to regulate many of the steps of translation. For example, the binding of the mRNA to the ribosome can be slowed or hastened, as can the movement of the mRNA through the ribosome. Likewise, a cluster of many ribosomes, a **polyribosome**, can produce several proteins from a single mRNA (FIGURE 4.14).

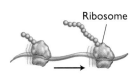

FIGURE 4.14 Regulating gene expression by regulating translation. A polyribosome can increase the rate of translation.

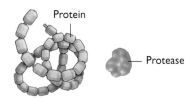

FIGURE 4.15 Regulating gene expression by regulating protein degradation. The activity of proteases affects the length of time proteins persist in a cell.

Regulation of Protein Degradation Once a protein is synthesized, it will persist in the cell for a characteristic span of time. Like the mRNA that provided the instructions for its synthesis, the life of a protein can be affected by *protease* enzymes inside the cell that degrade the protein (FIGURE 4.15). Speeding up or slowing down the activities of these enzymes can change the amount of time that a protein is able to be active inside a cell.

Now that you have an understanding of how cells synthesize proteins and how this synthesis is regulated, we can turn our attention to how scientists manipulate gene expression and regulation to genetically modify foods.

4.3 Producing Recombinant Proteins

One of the earliest genetic modifications was the alteration of a protein involved in milk production in cows. During the early 1980s, genetic engineers at Monsanto Company began to produce *recombinant bovine growth hormone (rBGH)* in their laboratories. rBGH is a protein that has been produced by genetically engineered bacteria. (Gene abbreviations are italicized when referring to the actual gene and not italicized when referring to the protein the gene encodes.) In this case, the *BGH* gene was removed from its original location in the cow genome and placed into a bacterial cell, producing a new combination of genes in the bacterium. Any time an organism is modified to produce a new complement or arrangement of genes, the organism is said to have undergone **recombination**. Bacteria will transcribe the cow gene and, due to the universality of the genetic code, translate the gene and produce the same protein as a cow would.

Before the advent of genetic technologies, growth hormone was extracted from the pituitary glands of slaughtered cows and then injected into live cows. Growth hormones act on many different organs to increase the overall size of the body and, in cows, to increase milk production. It is also possible to obtain human growth hormone from the pituitary glands of human cadavers. When human growth hormone is injected into humans who have a condition called *pituitary dwarfism*, their size increases. However, harvesting growth hormone from the pituitary glands of cows and humans is laborious, and many cadavers are necessary to obtain small amounts of the protein.

Genetic engineers at Monsanto realized that genetic engineering would allow them to control the synthesis of this protein by engineering bacteria to produce large quantities of bovine growth hormone in the laboratory. They could also override the regulatory controls on the expression of this gene by injecting any amount they wished into dairy cows. By doing so, scientists could cause an increase in milk production.

The first step in the production of the rBGH protein is to transfer the *BGH* gene from the nucleus of a cow cell into a bacterial cell. Bacteria are single-celled organisms that reproduce very rapidly. They can thrive in the laboratory if they are allowed to grow in a liquid broth containing the nutrients necessary for survival. Bacteria with the *BGH* gene can serve as factories to produce millions of copies of this gene and its protein product. Making many copies of a gene is called **cloning** the gene.

Cloning a Gene Using Bacteria

The following three steps are involved in moving a *BGH* gene into a bacterial cell (FIGURE 4.16).

Step 1. Remove the Gene from the Cow Chromosome The gene is sliced out of the cow chromosome on which it resides by exposing the cow DNA to enzymes that cut DNA. These enzymes, called **restriction enzymes**, act like highly specific molecular scissors. Restriction enzymes cut DNA only at specific sequences, called palindromes. *Palindromes* are words or expressions that read the same backwards and forwards. For example, the word *racecar* is a palindrome. An example of a molecular palindrome is shown here:

Note that the bottom middle sequence is the reverse of the top middle sequence.

Many restriction enzymes cut the DNA at these palindromic sites in a staggered pattern, leaving unpaired bases called *sticky ends*:

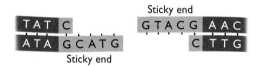

The unpaired bases form bonds with any complementary bases with which they come in contact. The enzyme selected by the scientist cuts on both ends of the *BGH* gene but not inside the gene:

The unpaired bases can bind to unpaired bases of a gene that has been cut with the same enzyme.

Because different individual restriction enzymes cut DNA only at specific points, scientists need some information about the entire suite of genes present in a particular organism to determine which cutting sites for a restriction enzyme surround the gene of interest. Cutting the DNA generates many different fragments, only one of which will carry the gene of interest.

Step 2. Insert the *BGH* Gene into the Bacterial Plasmid Once the gene is removed from the cow genome, it is inserted into a bacterial structure called a **plasmid**. A plasmid is a circular piece of DNA that normally exists separate from the bacterial chromosome and can replicate independently of the bacterial chromosome. Think of the plasmid as a ferry that carries the gene into the bacterial cell where it can be replicated.

To incorporate the *BGH* gene into the plasmid, the plasmid is cut with the same restriction enzyme used to cut the gene. Cutting both the plasmid and gene with the same enzyme allows the "sticky ends" that are generated to base-pair with each other (A to T and G to C). When the cut plasmid and the cut gene are placed together in a test tube, they re-form into a circular plasmid with the extra gene incorporated.

The bacterial plasmid has now been genetically engineered to carry a cow gene. At this point, the *BGH* gene is referred to as the *rBGH* gene. The *r* is used to indicate that this product is genetically engineered, or recombinant, because it has been removed from its original location in the cow genome and recombined with the plasmid DNA.

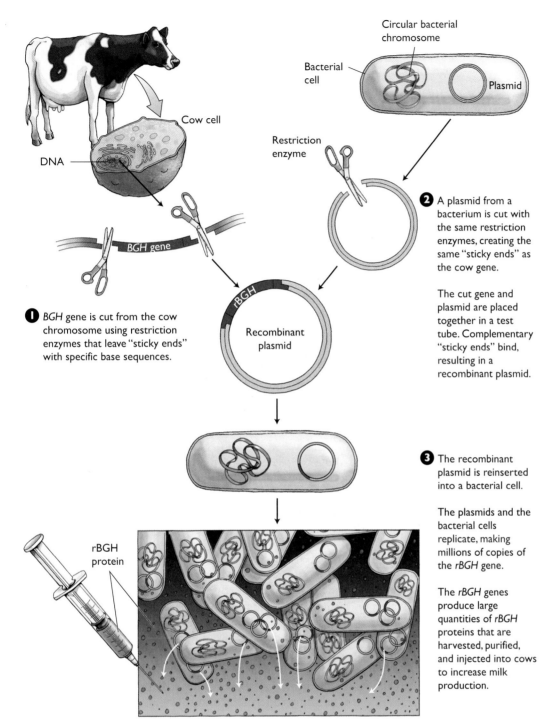

FIGURE 4.16 **Cloning genes using bacteria.** Bacteria can be used as factories for the production of human or other animal proteins.

VISUALIZE THIS What happens if the cut plasmid shown in step 2 does not bind to the *rBGH* gene?

Step 3. Insert the Recombinant Plasmid into a Bacterial Cell The recombinant plasmid is now inserted into a bacterial cell. Bacteria can be treated so that their cell membranes become porous. When they are placed into a suspension of plasmids, the bacterial cells allow the plasmids into the cytoplasm of the cell. When bacteria replicate so does the plasmid, making thousands of copies of the *rBGH* gene. Using this procedure, scientists can grow large amounts of bacteria capable of producing BGH.

Once scientists successfully clone the *BGH* gene into bacterial cells, the bacteria produce the protein encoded by the gene. When a gene from one organism is incorporated into the genome of another organism, a **transgenic organism** is produced. A transgenic organism is more commonly referred to as a **genetically modified organism (GMO)**.

Bacteria can be genetically engineered to produce many proteins of importance to humans. For example, bacteria are now used to produce the clotting protein missing from people with hemophilia, as well as human insulin for people with diabetes and human growth hormone for people with some types of growth deficiencies.

Why do scientists need to have some information about an organism's genome to clone a particular gene?

Once scientists had produced millions of rBGH-producing bacteria, they were able to break open the bacterial cells, isolate the BGH protein, and inject it into cows. Close to one-third of all dairy cows in the United States now undergo daily injections with recombinant bovine growth hormone. These injections increase the volume of milk that each cow produces by around 20%.

Before marketing the recombinant protein to dairy farmers, the Monsanto Company had to demonstrate that its product would not be harmful to cows or to humans who consume the cows' milk. This involved obtaining approval from the U.S. Food and Drug Administration (FDA).

FDA Regulations

The FDA is the agency of the U.S. federal government charged with ensuring the safety of all domestic and imported foods and food ingredients (except for meat and poultry, which are regulated by the U.S. Department of Agriculture). The manufacturer of any new food that is not *generally recognized as safe (GRAS)* must obtain FDA approval before marketing its product. Adding substances to foods also requires FDA approval, unless the additive is GRAS.

According to both the FDA and Monsanto, there is no detectable difference between milk from treated and untreated cows and no way to distinguish between the two. Even if there were increased levels of rBGH in the milk of treated cows, there should be no effect on the humans consuming the milk, because we drink the milk and do not inject it. Drinking the milk ensures that any protein in it will be digested by the body, just like any other protein that is present in food. Therefore, in 1993, the FDA deemed the milk from rBGH-treated cows as safe for human consumption.

In addition, since the milk from treated and untreated cows is indistinguishable, the FDA does not require that milk obtained from rBGH-treated cows be labeled in any manner. Vermont is the only state that requires labeling of rBGH-treated milk. However, many distributors of milk from untreated cows label their milk as "hormone free," even though there is no evidence of the hormone in milk from treated cows. There is, however, some evidence that cows treated with rBGH are more susceptible to certain infections than untreated cows. Due in part to concerns about these animals' welfare, Europe and Canada have banned rBGH use.

The rBGH story is a little different from that of genetically modified crop foods because rBGH protein is produced by bacteria and then administered to cows. When foods are genetically modified, the genome of the food itself is altered.

4.4 Genetically Modified Crops

Unless you eat only certified organic foods, you have probably eaten lots of genetically modified foods, likely without even realizing it. Before trying to understand whether modified foods are more or less healthful than unmodified foods, we will first try to understand why crops are modified to begin with.

Potential Benefits of Genetically Modifying Crops

Crop plants can be genetically modified to increase their shelf life. Tomatoes were the first genetically engineered fresh produce, and they became available in American grocery stores in 1994. These tomatoes were engineered to soften and ripen more slowly. An enzyme called *pectinase* causes the ripening process in some produce, including tomatoes. This enzyme breaks down pectin, a naturally occurring substance found in plant cells. When the enzyme pectinase is active, it helps break down the pectin, and the produce softens. A longer ripening time means that tomatoes could stay on the vine longer, thus making them taste better. The slower ripening also increases the amount of time that tomatoes could be left on grocery-store shelves without becoming overripe and mushy.

To genetically modify tomatoes, a special gene is inserted into the tomato. The inserted gene produces an mRNA transcript complementary to the mRNA produced by the transcription of a pectinase gene. When the mRNA from the genetically engineered gene is produced, it pairs with its naturally occurring pectinase mRNA complement. Binding of the two mRNAs prevents the pectinase mRNA from being translated. Thus, less of the pectinase enzyme is produced, and ripening happens more slowly (FIGURE 4.17).

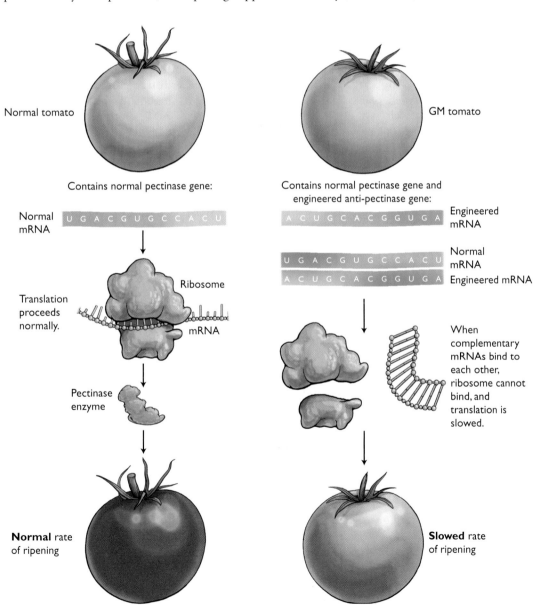

FIGURE 4.17 **Genetically modified tomatoes.** Genetically modified tomatoes produce mRNA that decreases the production of the enzyme pectinase. When the pectinase gene is transcribed and translated, ripening occurs. When the pectinase mRNA is bound to the engineered version, translation occurs more slowly, and ripening is slowed.

FIGURE 4.18 Golden rice. Golden rice has been genetically engineered to produce beta-carotene. Beta-carotene makes the rice yellow as compared to unmodified rice.

Crop plants are genetically modified to increase the yield an individual farmer can obtain. For example, engineering crops to be resistant to pesticides and herbicides allows farmers to apply these chemicals to their fields and kill pests and weeds that would not carry these resistance genes. The farmers would not harm the crop plants, thereby increasing yield. This may also allow farmers to use less of these chemicals, which are known to be toxic to humans.

Scientists can also increase the nutritional value of crops. Some genetic engineers have placed a gene that causes the synthesis of beta-carotene in rice, a staple food for many of the world's people. Scientists hope that the engineered rice will help decrease the number of people who become blind in less-developed nations due to vitamin A deficiency. In many of these countries, the traditional diet lacks beta-carotene, which is required to synthesize vitamin A, a vitamin needed for proper vision. Therefore, eating this genetically modified rice, called golden rice, would offer a reliable source of vitamin A (FIGURE 4.18). However, golden rice has never been approved for human consumption.

Although increasing shelf life, yield, and nutritional value may help feed many more people, questions about risks to human health have arisen about genetic technologies.

Effect of GMOs on Human Health

Most of us are already eating GM foods. Scientists estimate that over half of all foods in U.S. markets contain at least small amounts of GM foods. Over 80% of all soybeans grown in the United States are genetically modified for herbicide resistance. Soybean-based ingredients, including oil and flour, are often produced from genetically modified plants and appear as one or more ingredients in many different processed foods.

Close to 40% of the U.S. corn crop is genetically modified to produce its own pesticide against certain species of caterpillar. Because GM corn is not separated from non-GM corn by farmers or food processors and because many processed food ingredients are corn based, including corn starch and corn syrup, GM corn is thought to be present in most of our processed foods.

Most of the canola oil in the United States is extracted from GM rapeseed plants, which are engineered for herbicide resistance. Canola oil is used in many different products, including vegetable oil, salad dressing, margarine, fried foods, chips, cookies, and pastries.

Genetically modified cotton varieties resistant to caterpillars now account for over 70% of the cotton crop. Although cotton is more often used for clothing than for foods, cottonseed oil is used in cooking oils, salad dressing, peanut butter, chips, crackers, and cookies.

Of the 12 different GM plants approved for production and consumption, 8 are not commonly grown. Very few farmers are growing GM potatoes, squash, papaya, tomato, sugar beets, rice, flax, and radicchio, likely in response to consumer fears about the health consequences of eating these foods. Products that do not contain GMOs are often labeled to promote that fact (FIGURE 4.19).

How Are GM Foods Evaluated for Safety? The FDA becomes involved in testing the GM crop only if the food from which the gene comes has never been tested or when there is reason to be concerned that the newly inserted gene may encode a protein that will prove to be a toxin or an allergen.

Allergies are a serious problem for the close to 8% of Americans who experience reactions to foods. Symptoms of food allergy range from a mild upset stomach to sudden death. A person who knows he must avoid eating peanuts may not know to avoid a food that has been genetically modified to contain a peanut gene that may cause a reaction—although no such food currently exists. Recently, a project to genetically modify peas to be pest resistant was canceled after tests showed that the peas caused an allergic reaction in mice.

If the gene being shuffled from one organism to another is not known to be toxic or cause an allergic reaction, the FDA considers the GM food to be effectively equivalent to the foods from which it was derived; that is, the GM food is GRAS. If a modified crop contains a gene derived from a food that has been shown to cause a toxic or allergic reaction in humans, it must undergo testing before it can be marketed.

This method of determining potential hazards worked well in the case of a modified soybean that carried a gene from the Brazil nut. This engineering was done in an effort to increase the protein content of soybeans. Brazil nuts were known to cause allergic reactions in some people, so modified beans were tested and did indeed cause an allergic reaction in susceptible people. The product was withdrawn, and no one was harmed.

In evaluating potential harmful effects, scientists focus on the protein produced by the modified plant and not the actual gene that is inserted. This focus on the protein is because the gene itself is digested and broken down into its component nucleotides when it is eaten and therefore will not be transcribed and translated inside human cells.

While there is hope that genetic engineers will be able to help solve hunger problems by making farming more productive, there are also concerns about any negative health effects of GM foods. It remains to be seen whether genetic engineering will constitute a lasting improvement to agriculture.

FIGURE 4.19 "No GMO" labeling. Many food manufacturers and consumers consider the use of unmodified foods to be a selling point for their products.

Stop and Stretch 4.4

Genetic modifications have occurred over the last several thousand years due to farmers' use of selective breeding techniques—breeding those cattle that produce the most milk or crossing crop plants that are easiest to harvest. How is this artificial selection different from genetic engineering techniques used today? (And how, in some ways, might these organisms also be considered "genetically modified"?)

Chapter REVIEW

ROOTS TO REMEMBER

The following roots of words come mainly from Latin and Greek and will help you to decipher terms:

chromo- means color.

-ic is a common ending of acids.

nucleo- or **nucl-** refer to a nucleus.

-ose is a common ending for sugars, such as ribose, sucrose, and fructose.

-some or **-somal** relate to a body, whether the whole body (somatic) or a small body (ribosome).

KEY TERMS

4.1 What Is a Gene?
chromosome *p. 84*
gene *p. 84*
genome *p. 84*

4.2 Protein Synthesis and Gene Expression
activator *p. 92*
anticodon *p. 87*
codon *p. 87*
frameshift mutation *p. 91*
genetic code *p. 88*
messenger RNA (mRNA) *p. 86*
missense mutation *p. 90*
mutation *p. 88*
neutral mutation *p. 91*
nonsense mutation *p. 90*
point mutation *p. 90*
polyribosome *p. 92*
promoter *p. 86*
protein synthesis *p. 84*
reading frame *p. 91*
ribosomal RNA (rRNA) *p. 87*
ribosome *p. 87*
RNA (ribonucleic acid) *p. 85*
RNA polymerase *p. 86*
stop codon *p. 88*
transcription *p. 86*
transfer RNA (tRNA) *p. 87*
translation *p. 86*

4.3 Producing Recombinant Proteins
cloning *p. 93*
genetically modified organism (GMO) *p. 95*
plasmid *p. 94*
recombination *p. 93*
restriction enzyme *p. 94*
transgenic organism *p. 95*

SUMMARY

4.1 What Is a Gene?

- Genes are segments of DNA that carry information about traits that can be passed from parents to offspring.
- DNA (deoxyribonucleic acid) is located within the nucleus and composed of the nucleotides adenine (A), guanine (G), cytosine (C), and thymine (T) (p. 84).
- Different genes are composed of nucleotides placed in different orders (p. 84).
- Genes are located on chromosomes (p. 84).
- Genes carry instructions for synthesizing proteins (p. 84; Figure 4.2).

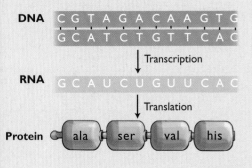

- An organism's full set of chromosomes is called its genome (p. 84).

4.2 Protein Synthesis and Gene Expression

- Protein synthesis involves the processes of transcription and translation (pp. 85–86).
- Transcription occurs in the nucleus of human cells when an RNA polymerase enzyme binds to the promoter, located at the start site of a gene, and makes the messenger RNA (mRNA) copy of the DNA gene. The main differences between RNA and DNA are that the sugar in RNA is ribose (not deoxyribose) and the nitrogenous bases are adenine, guanine, cytosine, and uracil (no thymine in RNA) (p. 86).
- Translation occurs in the cytoplasm of human cells and involves mRNA, ribosomes, and tRNA. mRNA carries the code from the DNA, and ribosomes are the site where amino acids are assembled to synthesize proteins. Transfer RNA (tRNA) carries amino acids. The anticodon on a tRNA molecule binds to an mRNA segment called a codon and peptide bond formation occurs (p. 87).
- A particular tRNA carries a specific amino acid. Each tRNA has its unique anticodon that binds to the codon and carries instructions for its particular amino acid (p. 87, Figure 4.5).

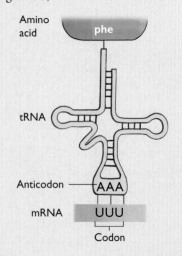

- The amino acid coded for by a particular codon can be determined by looking at the genetic code (p. 90).

- The flow of genetic information is from the DNA sequence to the mRNA transcript to the encoded protein (p. 90).

- Mutations are changes to DNA sequences that can affect protein structure and function. Neutral mutations are changes to the DNA that do not result in a different amino acid being incorporated. Insertions or deletions of nucleotides can result in frameshift mutations that change the protein drastically (pp. 90–91).

- A given cell type expresses only a small percentage of the genes that an organism possesses (p. 92).

- Human cells turn the expression of a gene up or down by increasing transcription in several ways: through the use of proteins that stimulate RNA polymerase binding; by varying the time that DNA spends in the uncondensed, active form; by altering the mRNA life span; by slowing down or speeding up translation; and by affecting the protein life span (p. 92–93).

4.3 Producing Recombinant Proteins

- Modern genetic engineering techniques enable scientists to produce recombinant protein in the lab by placing a gene for the protein into plasmids, which clone the gene by making millions of copies of it as they replicate themselves inside their bacterial hosts. Bacteria can then express the gene by transcribing the DNA to produce the mRNA copy and translating the mRNA into a protein (pp. 93–96; Figure 4.16).

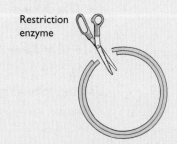

4.4 Genetically Modified Crops

- Crop plants are genetically modified to increase their shelf life, yield, and nutritional value (p. 97; Figure 4.17).

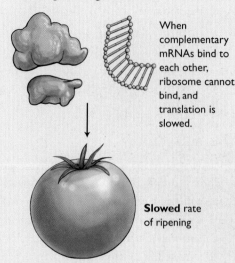

- Although there have been no documented incidents of negative health effects from GM food consumption, there is concern that some GM foods may cause allergic reactions or become toxins (pp. 98–99).

LEARNING THE BASICS

1. List the order of nucleotides on the mRNA that would be transcribed from the following DNA sequence: CGATTACTTA.

2. Using the genetic code (Figure 4.8), list the order of amino acids encoded by the following mRNA nucleotides: CAACGCAUUUUG.

3. Select the correct ending for this statement: Ribosomes are composed of two subunits, _____.
    a. one composed of tRNA and one of rRNA
    b. each composed of tRNA and rRNA
    c. one composed of protein and one composed of rRNA
    d. each composed of protein and rRNA
    e. each composed of protein only

4. How do RNA and DNA differ?
    a. RNA is single stranded and DNA is double stranded.
    b. RNA contains the sugar ribose and DNA contains deoxyribose.
    c. RNA contains uracil and DNA contains thymine.
    d. All of the above are differences between DNA and RNA.

5. Transcription _____.
    a. synthesizes new daughter DNA molecules from an existing DNA molecule
    b. makes an RNA copy of a gene that is to be expressed
    c. pairs thymine (T) with guanine (G)
    d. occurs on ribosomes

6. Transfer RNA (tRNA) _____.
    a. carries monosaccharides to the ribosome for synthesis
    b. is made of messenger RNA
    c. has an anticodon region, which is complementary to the mRNA codon
    d. is the site of protein synthesis

7. During the process of transcription, _____.
    a. DNA serves as a template for the synthesis of more DNA
    b. DNA serves as a template for the synthesis of RNA
    c. DNA serves as a template for the synthesis of proteins
    d. RNA serves as a template for the synthesis of proteins

8. Translation results in the production of _____.
    a. RNA
    b. DNA
    c. protein
    d. individual amino acids
    e. transfer RNA molecules

9. The RNA polymerase enzyme binds to _____, initiating transcription.
   a. amino acids
   b. tRNA
   c. the promoter sequence
   d. the ribosome

10. A particular triplet of bases in the coding sequence of DNA is TGA. The anticodon on the tRNA that binds to the mRNA codon is _____.
    a. TGA
    b. UGA
    c. UCU
    d. ACU

11. Look again at the genetic code in Figure 4.8. A mutation that changes the DNA from AAA to AAC would result in
    a. a missense mutation
    b. a nonsense mutation
    c. a neutral mutation
    d. a frameshift mutation

12. A particular protein is 300 amino acids long. Which of the following could be the number of nucleotides in the DNA that codes for this protein?
    a. 3
    b. 100
    c. 300
    d. 900

13. Plasmids are _____.
    a. linear RNA molecules
    b. cellular genomes
    c. viruses composed of RNA
    d. circular extrachromosomal bacterial DNA

14. Genetically modified tomatoes might ripen more slowly due to the _____.
    a. decreased concentration of pectinase
    b. faster transport time
    c. decreased tRNA concentration
    d. absence of the pectinase gene

15. **True or False?** A mutation is a change to the DNA sequence of a gene.

16. **True or False?** The universality of the genetic code explains why a codon that codes for alanine in humans would also code for alanine in bacteria.

17. **True or False?** A tRNA that carries lysine should have the same anti-codon as a tRNA that carries phenylalanine.

Use the following terms to answer questions 18–20.

a. plasmid   c. cloning   e. amino acid
b. sticky end   d. transcription   f. carbohydrate

18. _____ Single-stranded regions of non-hydrogen-bonded nitrogenous bases produced by cutting DNA with restriction enzymes

19. _____ Making many exact copies of a gene

20. _____ The subunit of a protein

ANALYZING AND APPLYING THE BASICS

1. Take another look at the genetic code (Figure 4.8). Do you see any similarities between codons that code for the same amino acid? Based on this difference, why might a mutation that affects the nucleotide in the third position of the codon be less likely to affect the structure of the protein than a mutation that affects the codon in the first position?

2. Genes encode RNA polymerase molecules. What would happen to a cell that has undergone a mutation to its RNA polymerase gene?

3. Place a box around the six-base-pair site at which a restriction enzyme would be most likely to cut:

   ATGAATTCCGTCCG
   TACTTAAGGCAGGC

4. How can inserting a gene that helps prevent infestation by a certain pest actually result in the increased resistance of pests to that gene's product?

Questions 5 and 6 refer to the following information.

In a landmark paper, Nirenberg et al. (1965) explored the RNA code and its relationship to amino acids. They used triplet codon sequences to direct the binding of radioactively labeled (C^{14}) RNA to ribosomes. High binding activity (a higher number) indicated a relationship between the type of RNA and the triplet codon. Some of these results are shown in Table 4.1.

Table 4.1 Specificity of Trinucleotides for C^{14}-sRNA

RNA Codon	C^{14}-Arg	C^{14}-Asp	C^{14}-His	C^{14}-Gly	C^{14}-Met
AUG	−0.11	−0.06	−0.02	−0.17	1.00
CAU	−0.31	−0.01	0.52	−0.13	−0.04
CGC	1.63	0.02	−0.03	−0.02	0.01
GAU	−0.14	1.29	−0.03	−0.23	−0.09
GGU	−0.33	0.01	−0.03	3.04	−0.08

5. According to Table 4.1, what sequence of codons would produce the amino acid sequence his-met-asp-his?

6. Use Figure 4.8 to determine what amino acids the following codon sequence codes for.

   CGCACCCGAGGCGGCGCAGUC

7. tRNAs are also coded for by genes (even though they are not proteins). What might happen if the gene that encodes a tRNA undergoes a mutation in what will be transcribed into the anticodon?

CONNECTING THE SCIENCE

1. The first "test-tube baby," Louise Brown, was born over 30 years ago. Sperm from her father was combined with an egg cell from her mother. The fertilized egg cell was then placed into the mother's uterus for the period of gestation. At the time of Louise's conception, many people were concerned about the ethics of scientists performing these in vitro fertilizations. Do you think human cloning will eventually be as commonplace as in vitro fertilizations are now? Why or why not?

2. It is now possible for corporations to patent seeds to which they have made small genetic manipulations. Once they are patented, farmers must pay for the use of these seeds. Some companies have even engineered "terminator" genes that prevent farmers from using seeds produced by plants grown in their own fields. If it took hundreds of thousands of years of evolution to produce a seed with a particular genetic makeup, do you think companies should have the right to change one or a few genes and obtain a patent?

3. Coffee plants are being modified so that all the coffee beans can be harvested at once. Traditionally grown coffee is more time-consuming to harvest and thus more expensive than engineered coffee. The result may be the disappearance of small family plantations and serious economic disruption to the communities in the developing world that rely on coffee for trade. Do you think laws should be enacted to protect the family farmer in this situation?

Chapter 5

Tissues, Organs, and Organ Systems

Work Out!

LEARNING GOALS

1. Compare and contrast the structure and function of the epithelia.
2. Compare and contrast exocrine and endocrine glands.
3. Identify the structure and function of tight junctions, gap junctions, and adhesion junctions.
4. List the major differences between the different types of connective tissues.
5. How do the three separate types of muscle tissue differ from each other?
6. How do neurotransmitters help cells communicate?
7. List the major organs found in ventral and dorsal body cavities.
8. List the five types of body membranes covered in this chaper.
9. Define the term *organ*.
10. Give an example of how two different organ systems work together.

The components of physical fitness include strength,

How will I find time to work out today? Maybe I can skip working out today and use the time to study. I can always exercise this weekend. Or should I skip the party tonight and get up early tomorrow morning and work out before class?

This internal battle is familiar to anyone who is concerned about his or her physical fitness. The intense demands on the time of most college students, working adults, and those raising families make it hard to find time to exercise. Yet we know that eating right is not enough. To be healthy one has to eat right *and* exercise.

Exercise provides many health benefits. In addition to lowering your risk of heart disease and obesity, considerable evidence shows that exercise will help prevent the onset of diabetes and may help prevent certain cancers. In addition, many studies show that exercising has a positive impact on mental health. Most people who exercise regularly feel better, have more energy, and even get a better night's sleep.

How do you know whether you are getting enough exercise to reap these benefits? Is your workout routine helping you become physically fit? Physical fitness has many components, including muscle strength and endurance, aerobic fitness, flexibility, and body fat percentage.

aerobic fitness,

Strength is the ability of muscles to exert force during exercise or daily activity. We know that daily activities are less tiring if you are fit. Fit people have more endurance than unfit people and become fatigued less easily.

Aerobic fitness refers to the ability of the heart, lungs, and muscles to use oxygen efficiently during exertion. Aerobic exercises typically are rhythmic exercises that use large muscle groups and force the heart and lungs to work harder than they would at rest. Such exercise strengthens the heart and other muscles and increases the ability of the lungs to take in oxygen to deliver to the rest of the body. Aerobic exercises include running, bicycling, stair climbing, swimming, and any exercise that makes you feel winded.

Flexibility is the range of motion of the joints. Flexible joints are not stiff, instead bend easily. Unless people work at staying flexible, their range of motion decreases with age and their joints

endurance,

flexibility, and core conditioning.

105

become less pliable. Maintaining a flexible and strong spine often involves conditioning with exercises that strengthen the core muscles including the abdominal muscles involved in supporting the spine.

Another facet of physical fitness, body fat percentage, is the percentage of body fat to lean (nonfat) tissues such as bone, muscle, organs, and blood. If you weigh 200 pounds and have 25% body fat, your body consists of 50 pounds of fat and 150 pounds of lean body mass. Exercise can help you increase your lean muscle mass and decrease your body fat.

To learn how to increase your muscle strength, endurance, aerobic fitness, and flexibility, and how to decrease your body fat percentage, let us examine how exercise affects the different types of body tissues.

5.1 Tissues and Their Functions

Groups of similar cell types that perform a common function make up a **tissue**. The human body has four main types of tissue: epithelial, connective, muscle, and nervous.

Epithelial Tissue

Epithelial tissues, or **epithelia**, are tightly packed sheets of cells that cover organs and outer surfaces and line hollow organs, vessels, and body cavities (FIGURE 5.1).

The sheets of epithelial tissue are usually not attached on all surfaces, as in other tissue types. Instead, epithelial tissues are anchored on one face of the tissue but free on the other tissue face. The free tissue, called the *apical* surface, can be exposed to body fluids or the external environment. For example, the epithelial tissue that lines blood vessels is attached to the wall of the blood vessel, but its apical surface is exposed to the bloodstream. The epithelium that constitutes the outer layer of skin, called the *epidermis*, is anchored to the underlying tissue but has a surface exposed to air. This pattern of attachment to only one face is also true of the epithelia that line other parts of the body, including respiratory surfaces, the digestive tract, and the urinary and genital tracts.

Epithelial cells function in protection, absorption, and also secretion of various substances. Epithelial cells in the skin help protect the body from injury, ultraviolet light, water loss, and disease-causing organisms. Epithelial cells that line blood vessels and intestines absorb nutrients and water. However, the epithelium is *avascular* in that it does not have its own blood supply.

Groups of epithelial cells that are specialized to secrete a substance are called **glands**. Outgrowths of epithelial tissue function as glands that secrete substances such as mucus, oils, and sweat. FIGURE 5.2 shows examples of the two categories of glands found in the human body. **Exocrine glands** secrete their products into hollow organs or ducts, or onto surfaces such as the skin. The glands in your mouth that secrete saliva, the glands in your stomach that secrete acid, and those in your skin that produce sweat are all examples of exocrine glands. Take a look at your arms when you exercise on a hot, humid day to see sweat-producing exocrine glands in action. Evaporation of sweat cools the body, allowing it to function more efficiently and work longer in warm conditions. Sweat also removes metabolic waste products. **Endocrine glands** secrete substances called hormones, which modify the functions of other organs, into the bloodstream. We will cover hormones and endocrine glands in detail in Chapter 16.

Epithelial tissues are classified according to their shape and the number of cell layers. Differently shaped epithelial cells have different structures and functions (FIGURE 5.3). *Simple tissue* types are composed of one cell layer, but *stratified tissue* types are composed of more than one layer of cells.

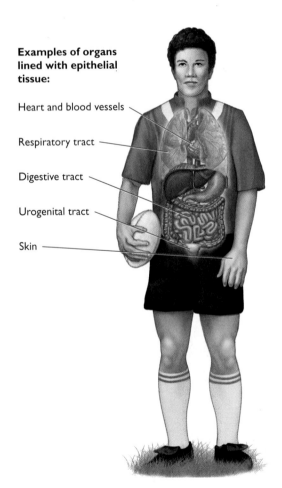

FIGURE 5.1 **Epithelial tissues.** Epithelia line organs including the heart and blood vessels, respiratory tract, and digestive and urogenital tracts.

Examples of organs lined with epithelial tissue:
- Heart and blood vessels
- Respiratory tract
- Digestive tract
- Urogenital tract
- Skin

Squamous epithelium consists of flattened cells. These "squashed" cells form the outer surface of the skin and line the inner surfaces of some vessels and organs. These cells' function is protection and nutrient exchange. The walls of the smallest blood vessels, called *capillaries*, are simple squamous epithelium. Because these vessels are only one cell layer thick, nutrient exchange between the blood and tissues is enhanced. The thin walls of the capillaries facilitate delivery of oxygen and nutrients at all times, but the rate at which this occurs increases during exercise. When squamous epithelium is composed of multiple layers it is called *stratified squamous* epithelium. Such stratified squamous epithelium is located in areas of the body subject to mechanical stress or abrasion, such as the skin.

Cuboidal epithelium is composed of tightly packed cells that are neither flat nor tall. Cuboidal epithelium lines kidney tubules and is found on the surface of ovaries. Most glands are made of cuboidal epithelium. In addition to protecting other cells and tissues, these cells secrete and absorb water and small molecules.

Columnar epithelium is composed of tall, column-shaped cells. The nucleus in these cells is usually located at the bottom of the cell, near the attached surface. These cells line the intestine and other parts of the digestive system, parts of the reproductive system, and most of the respiratory system. The function of the cells of the columnar epithelium is to secrete and absorb substances.

Beneath the cells of an epithelial tissue is a layer of noncellular material called the **basement membrane**. The basement membrane provides structural support to the epithelial cells and joins the epithelium to underlying tissues. The polysaccharides and proteins that make up the basement membrane are secreted by the epithelial cells. The basement membrane provides chemicals and growth factors that are important for healing damage to the epithelium.

Epithelial tissues constantly shed dead cells, which are replaced when cells copy themseves during the process of cell division (Chapter 17). Dead skin cells can be rubbed off by clothing, and cells that line the mouth and intestines are scraped off by food and digestive chemicals. These cells are replaced by others beneath them. For the skin, the result is a wholly new epidermis every four to six weeks. The highly regenerative property of epithelia is also displayed by the epithelium of the stomach lining, which is replaced once or twice a week.

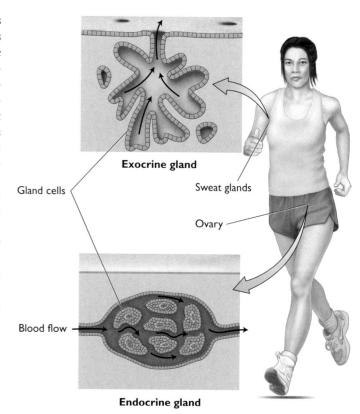

FIGURE 5.2 Glandular epithelia. Glandular epithelia secrete a product. Endocrine glands secrete products into the bloodstream. Exocrine glands secrete products into hollow organs or ducts.

Cell Junctions

Epithelial and other cells can be joined to each other by several different types of **intercellular junctions**. These junctions are composed mainly of proteins and serve to anchor these and other types of cells to each other and to allow cells to interact. Cells that have channels between them can communicate with each other.

Tight junctions form tight, impermeable barriers between the plasma membranes of adjacent cells (FIGURE 5.4a). Since some substances must be prevented from passing through any small spaces between cells (acids from the digestive system, for example), these junctions help limit the flow of chemicals in and between tissues. **Adhesion junctions** or *desmosomes* are junctions composed of filaments that join cells to each other (FIGURE 5.4b). These filaments are flexible enough to allow some movement by adjacent cells. For example, adhesion junctions in the epithelium of the skin allow the skin to move. **Gap junctions** occur when two adjacent plasma membranes are joined via channels made of proteins (FIGURE 5.4c). These connecting channels permit the passage of water, ions, and other substances between cells. Gap junctions in the heart allow the substances that cause contraction to flow between cells, ensuring rhythmic contractions. When you work out, your heart beats more quickly to move more oxgygen and nutrients to your tissues.

To obtain oxygen and nutrients, the epithelium must be next to an area of connective tissue.

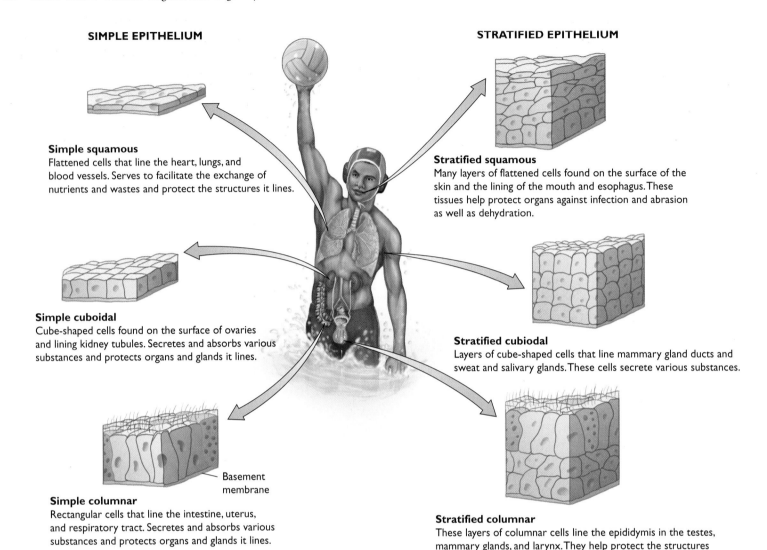

FIGURE 5.3 Types of epithelial tissues. Simple epithelial tissues—squamous, cuboidal, and columnar—are one cell layer thick. Stratified epithelium can be many layers thick.

Connective Tissue

Connective tissues function, as their name implies, to form connections. They usually bind organs and tissues to each other. In connective tissue, the cells are not tightly packed as in epithelium. The spaces in between the cells are filled with an amorphous, noncellular **ground substance** along with protein fibers that collectively constitute the **matrix**.

In some areas, the connective tissue is highly cellular; in others, the protein fibers or ground substance of the matrix is the main component. The ground substance of the matrix varies in consistency from solid to semifluid to fluid. The consistency of the matrix is a function of the types of fibers and ground substances present. Blood, for example, has a liquid matrix. Adipose or fat tissue has a thick fluid, or viscous, matrix. Bone has a solid matrix. Specific compositions and functions of many of the different types of connective tissue follow.

Loose Connective Tissue Loose connective tissue (FIGURE 5.5a), also called *areolar* connective tissue, is the most widespread connective tissue in animals. It connects epithelia to underlying tissues, holds organs in place, and acts as padding under the skin and elsewhere. It allows some organs, such as the lungs and bladder, to expand. The tissue is called loose connective tissue because of the loose weave of its fibers. The cells of loose connective tissue, called **fibroblasts**, synthesize and secrete proteins into the matrix. The matrix of loose connective tissue is its chief structural feature.

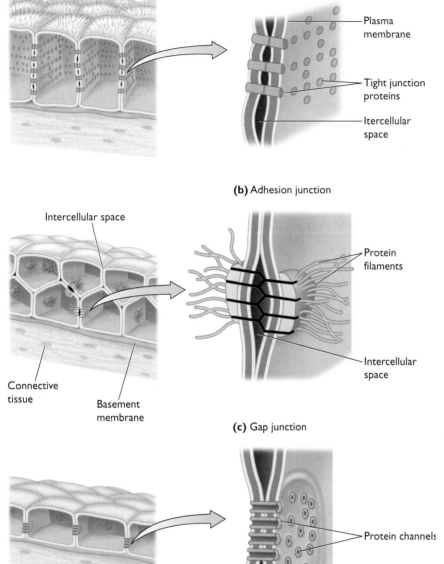

FIGURE 5.4 **Junctions between cells.** (a) Proteins between cells at a tight junction block leakage between adjacent cells. (b) Adjacent cells adhere to each other via the filaments of proteins present at adhesion junctions. (c) Cylindrical proteins span plasma membranes of adjoining cells at a gap junction, allowing quick movement of substances between cells.

Two proteins are important constituents of this matrix: collagen and elastin. **Collagen fibers** give connective tissue the ability to bear mechanical stress without breaking, a property called *tensile strength*. **Elastin fibers** allow connective tissue to stretch without breaking, a property called elasticity. Collagen fibers are like strong guitar strings, while elastin fibers resemble stretchy rubber bands.

The degradation of elastin in skin that occurs with age and exposure to sunlight and cigarette smoke causes a decrease in the elasticity of the skin, leading to wrinkles. Stretch marks occur when elastin and other fibers are damaged, resulting in visible scars.

Exercise may lessen some of the effects of aging on connective tissues. As we age, the fibroblasts that synthesize and secrete collagen become less productive. Any exercise that increases your heart rate will facilitate the transport of nutrients to connective tissues. The influx of nutrients from exercise helps fibroblasts receive the nutrients they need to keep producing these fibers efficiently. In addition, exercise increases the production of substances by the fibroblasts that help promote the growth of new blood

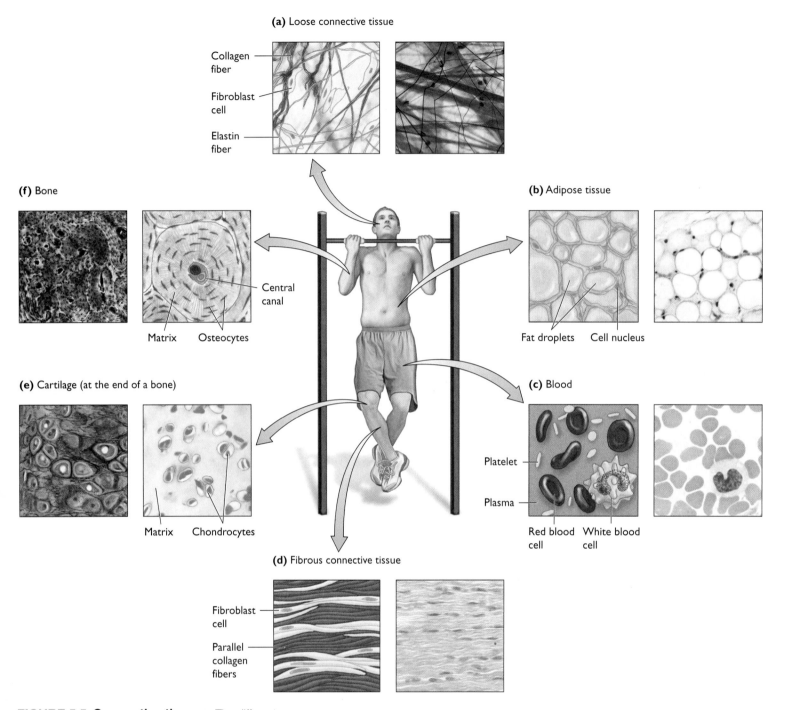

FIGURE 5.5 Connective tissues. The different types of connective tissue are each composed of cells and a specialized matrix.

VISUALIZE THIS Structures are labeled in the art in this figure. Can you identify the same structures in the corresponding micrographs?

vessels. Therefore, exercise helps pump more blood and also increases the number of vessels available for nutrients to travel to their destinations.

Adipose Tissue Adipose (fat) **tissue** (FIGURE 5.5b) connects the skin to underlying structures and, as you learned in Chapter 3, insulates and protects organs. Adipose tissue is found beneath the skin and around organs such as the kidneys and heart. Cells are the main constituent of this connective tissue. Only a small amount of a soft matrix is associated with adipose tissue.

Adipose cells, or **adipocytes**, are specialized for the synthesis and storage of energy-rich reserves of fat (lipids). A fat droplet fills the cytoplasm of these cells. The droplet shrinks when the fat is used for energy. Adipocytes also produce a number of chemicals, called *adipokines*, that can influence disease. Adipokines trigger changes that can cause heart disease, stroke, and type 2 diabetes.

All forms of exercise burn calories that might otherwise be stored on the body as fat. During intense exercise, your body works harder and burns more calories. Even after you stop exercising, your body burns calories at a slightly increased rate for several hours.

The American Council of Sports Medicine suggests that the typical range of body fat in fit women is 21–24% and in fit men 14–17%. The range in trained female athletes is 14–20% and in trained male athletes 6–13%. Keep in mind, though, that it is far more important to focus on attaining overall body fitness than to become obsessed with any one measure of fitness, such as body fat percentage. In fact, most experts agree that it is more healthful to be overweight and fit than to be thin and out of shape.

Blood Blood (FIGURE 5.5c) is a connective tissue with a liquid matrix. Blood circulates throughout the body through blood vessels, transports oxygen and nutrients to cells, and carries away waste products. The bloodstream allows distant parts of the body to be in communication with each other through the sharing of transported substances.

The cellular component of blood tissue includes red cells, which carry oxygen; white cells, which help fight infection; and cell fragments called platelets, which function in clotting. Blood cells are suspended in a protein-rich fluid matrix called *plasma*. When the clotting proteins are removed from the plasma, the remaining liquid is called *serum*. The blood will be covered in detail in Chapter 8.

When you breathe more quickly and deeply during aerobic exercise, you maximize the amount of oxygen and other nutrients your blood can bring to your tissues. Regular exercise can produce a long-lasting increase in the number of red blood cells and in total blood volume, both of which increase the ability of the body to transport nutrients to cells even when you are not working out. In addition, blood vessels dilate as they move more oxygen to your muscles, and over time, more small blood vessels will develop. This development of more and bigger blood vessels is part of the reason why some fit individuals have prominent blood vessels under the surface of their skin.

Dense Fibrous Connective Tissue **Dense fibrous connective tissue** occurs in two forms. Dense irregular fibrous connective tissue has bundles of collagen fibers that are randomly oriented, as in the layer of skin beneath the epidermis called the *dermis*. Dense regular tissues have collagen fibers in parallel. This occurs in the connective tissue that forms the *tendons*, which join muscles to bones, and the *ligaments*, which connect bones to each other. The densely packed parallel collagen fibers of the matrix are the most conspicuous feature of this tissue (FIGURE 5.5d). As in loose connective tissue, the cells of this tissue are called fibroblasts.

Exercise can help strengthen and stabilize joints, preventing injury to ligaments and tendons. However, some sports injuries that result from overuse or physical trauma can damage tendons and ligaments.

Tendonitis is an inflammation of the tendon caused by repetitive motion. Healthy tendons are tough but flexible bands of tissue that glide smoothly as muscles contract. Tendons are often covered with protective sheaths filled with lubricant, which helps them withstand exercise. If these sheaths become inflamed, movement becomes painful. *Achilles tendonitis* is inflammation surrounding the tendon of the heel. *Tennis elbow* is tendonitis of the elbow.

Ligaments can be torn when the direction of movement is changed suddenly. These so-called pivot shift injuries are common in basketball and soccer. A ligament can also be torn by physical trauma, such as a hard blow of a helmet hitting a knee during a football tackle. Once a ligament is torn, its ability to connect bones to each other is compromised and the joint can become unstable. When a joint is destabilized, damage to cartilage that protects that joint surface can also happen.

Cartilage Cartilage is a type of connective tissue composed of cells called **chondrocytes** that secrete the dense matrix that surrounds the cells (FIGURE 5.5e). The matrix is rich in collagen and other structural proteins. Cartilage provides flexible support for the ears and nose and allows for shock absorption. The high water content of the matrix also enhances the cushioning function of cartilage. At joints, the ends of bones are covered in cartilage, which permits smooth gliding where joint surfaces contact each other. There are no blood vessels in cartilage, which is part of the reason why injured cartilage does not heal well.

Hyaline cartilage has a matrix that contains only collagen fibers. Hyaline cartilage is found in the nose, at the ends of long bones, and in the ribs. The skeleton of a fetus is made of hyaline cartilage, which is later replaced by bone. This smooth tissue is often found in locations of the body where friction occurs.

Elastic cartilage contains more elastic fibers and is more flexible. Elastic cartilage is found in regions where flexible cartilage is required. You can bend your outer ear because it is composed of elastic cartilage.

Fibrocartilage has a matrix packed with bundles of strong collagen fibers. Fibrocartilage can withstand tension and pressure and acts as a cushion in the disks between vertebrae in the back and also in the knee joint.

Stop and Stretch 5.1

Some athletes experience a loss of cartilage in their knee joint due to injury or wear and tear. When the cartilage that covers the bones wears down enough to allow direct contact between the heads of the bones that form this joint, a painful condition called *osteoarthritis* results.

Early treatments of osteoarthritis include injecting steroids such as cortisone into the knee joint. Steroids help prevent inflammation caused by the bones rubbing together and can help decrease pain. Unfortunately, these steroid injections can cause serious side effects and their pain-relieving effects are transient. Scientists are trying to grow cartilage cells in the lab to provide cartilage replacements. Another treatment for osteoarthritis is a total knee replacement—the heads of the long bones of the knee are replaced by a metal prosthesis. However, knee replacement must be delayed as long as possible because the prosthesis will eventually wear out. Progression of osteoarthritis seems to be delayed by strength training. Why might strength training help delay the progression of osteoarthritis? What types of exercises would be best for a person with ligament or cartilage damage who wants to prevent the onset of this condition?

Bones **Bones** make up the skeleton and are connected to each other at the joints. In addition to being a framework for the body, bones support and protect other tissues and organs. Marrow, inside of the bone, produces blood cells.

Bone is a rigid type of connective tissue whose branched cells, called **osteocytes**, secrete substances that harden into a solid matrix of collagen and calcium and other minerals (FIGURE 5.5f). Bone serves as a reservoir of calcium and minerals that the body can use if dietary levels of these substances are low. Calcium is needed by the body for nerve impulses, muscle contraction, and many other functions, in addition to forming the hard part of bones. Canals inside bones house nerves and blood vessels. Bone structure will be covered in more depth in Chapter 6.

Bone is a living tissue that changes in response to other tissues. Bone health is improved by weight-bearing exercise such as running and weight lifting. Strength training and weight-bearing exercise help preserve bone mass and may also help increase the density of bones. Strong bones help prevent *osteoporosis*, or degeneration of bone tissue with age. Weight-bearing exercise helps prevent osteoporosis because bone cells grow and divide under physical stress. Because bone mass begins to decrease beginning around age 30 especially for women, exercising while you are younger helps build up bone mass to a level high enough to prevent or postpone debilitating bone losses.

Whereas connective tissue serves to support and connect, muscle tissue enables body movement. Muscle is the most abundant type of tissue in humans.

Muscle Tissue

Muscle is a highly specialized tissue capable of contracting and helping the body move. **Muscle tissue** is composed of bundles of long, thin, cylindrical cells called muscle fibers. Muscle fibers contain specialized proteins, **actin** and **myosin**, that cause the cells to contract when signaled by nerve cells.

Muscle tissue is distinguished according to whether its function requires conscious thought—such as walking—and is thus **voluntary**, or its function requires no conscious thought and is therefore **involuntary**, such as the beating of the heart.

Muscle tissue is also differentiated by the presence of bands that look like stripes under a microscope, called **striated muscle**, or the lack of obvious bands, called **smooth muscle** tissue. The striated appearance is due to the banding pattern formed by actin and myosin deposits in the cell. Muscle tissue that lacks these striations also contains actin and myosin deposits, just not in a banded pattern. Humans have three types of muscle tissue: cardiac, skeletal, and smooth.

Cardiac Muscle **Cardiac muscle** is found only in the wall of the heart. This involuntary, striated tissue undergoes rhythmic contractions to produce the heartbeat. Cardiac muscle cells are highly branched and interwoven (FIGURE 5.6a). They have gap junctions that allow a contraction signal to be propagated among many muscle cells at once, producing a coordinated heartbeat to help pump blood throughout the body. Cardiac muscle cells do not reproduce themselves after birth. If a cardiac muscle cell dies, a neighboring cell may be able to take its place but the dead cell is not replaced. This is why damage to the heart can result in permanent disability.

Regular aerobic exercise increases the strength of the heart muscle, allowing more blood to be pumped with each beat. The heart then becomes not only bigger and stronger but also more efficient. Therefore, as your heart fitness increases, you can do more work with less effort. For this reason, trained athletes tend to have lower resting heart rates than other people.

Skeletal Muscle **Skeletal muscle** is located in muscles that attach to bone and produces all the movements of body parts in relation to each other. Skeletal muscle is responsible for voluntary movements such as walking. In FIGURE 5.6b, note the presence of multiple nuclei in the cells in this tissue.

Exercise causes an increase in the size of skeletal muscle cells, not an increase in the number of cells. This is because skeletal muscles do not reproduce after birth. When a tissue increases in size without increasing the number of cells, this is called **hypertrophy**. In contrast, the wasting away of a tissue is called **atrophy**.

Increasing the strength of skeletal muscles helps maintain correct posture and can prevent back pain. The stretching and strength-enhancing effects of yoga and Pilates help maintain skeletal muscle strength and joint flexibility. To increase strength with weight lifting, you should work the muscle harder than it is accustomed to and gradually increase the resistance. For increasing endurance, concentrate on lifting lighter weights for more repetitions.

Smooth Muscle Smooth muscle is not striated (FIGURE 5.6c). This involuntary muscle type is composed of spindle-shaped cells that form the musculature of internal organs, blood vessels, and the digestive system. Smooth muscle contracts more slowly than skeletal muscle, but it can remain contracted for a long time. Consider the difference between the contractions of the smooth muscle that lines blood vessels to move blood as opposed to the contraction of the biceps of the upper arm. Contractions of the smooth muscles of the blood vessels are involuntary and are held for long periods of time. Contractions of the biceps are voluntary and occur more quickly.

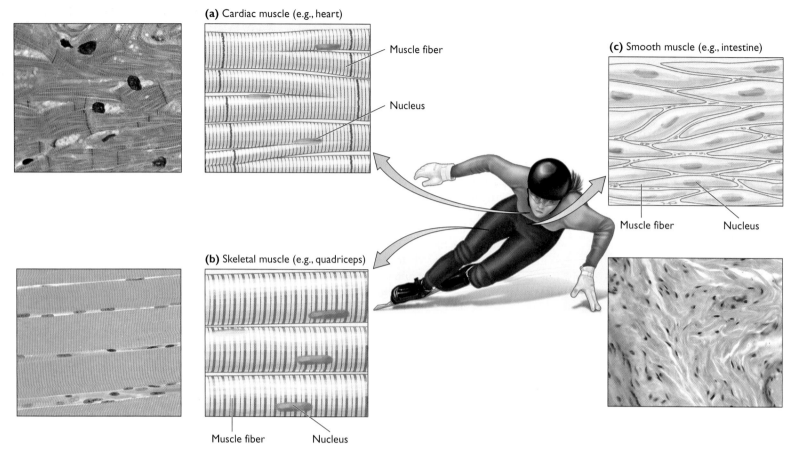

FIGURE 5.6 Muscle tissue. Cardiac (a) and skeletal (b) muscles are striated, while smooth muscle (c) is not.

VISUALIZE THIS What causes the striped pattern seen in cardiac and skeletal muscle?

Nervous Tissue

Nervous tissue is composed, in part, of cells called **neurons** that can conduct and transmit electrical impulses. The main function of nervous tissue is to help the body sense stimuli, process the stimuli, and transmit signals to and from the brain to the rest of the body to respond to stimuli. For example, if a soccer ball is suddenly kicked in a new direction, a soccer player can stop running and change direction without much thought. This action requires the player's body to recognize the change in direction and immediately coordinate a response to it. Nervous tissue makes such quick reaction possible. The brain, the spinal cord, and nerves connected to them contain nervous tissue, which is covered in greater detail in Chapter 14.

In addition to neurons, nervous tissue contains support cells called **neuroglia** (FIGURE 5.7). These cells are common components of nervous tissue and take up most of the volume of nervous tissue. Although neuroglia were once thought to function only as a glue that binds and supports neurons, scientists are now finding many different functions for these cells. **Astrocytes** are star-shaped neuroglia that bring nutrients to neurons. Others, called **microglia**, remove foreign substances from the nervous tissue. **Oligodendrocytes** and **Schwann cells** are neuroglial cells that synthesize a coating called myelin that insulates neurons, helping to increase the speed at which nerve impulses can travel.

Nerve cells communicate with each other via the secretion of chemicals called **neurotransmitters**. Some neurotransmitters are affected by exercise. Exercise produces *endorphins*, which are neurotransmitters that produce feelings of well-being. Endorphins are the source of a natural "high" experienced by many runners and other athletes. Exercise may also help to fight depression by increasing the concentration of certain neurotransmitters associated with positive emotions. Recent studies also suggest that neurotransmitters released during exercise may help control anxiety.

You have seen that the functioning of many different tissues can be affected positively by exercise. Let's now turn our attention to the body cavities and membranes that are also important components of a healthy body.

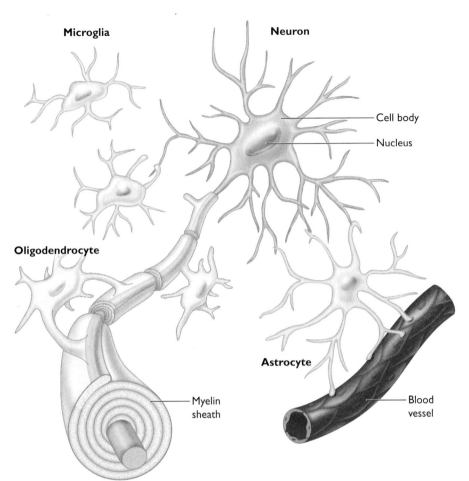

FIGURE 5.7 **A neuron and neuroglia.** Microglia in the brain become mobile in response to inflammation and remove debris. Substances entering neurons from blood pass through astrocytes located between neurons and capillaries. Oligodendrocytes form the layers of protective coating (myelin) that surround nerve fibers in the brain and spinal cord.

5.2 Body Cavities and Membranes

The human body is divided into two main cavities, each of which has a particular position and is further subdivided into smaller, more specialized cavities that house particular organs and glands.

Body Cavities

The two main cavities are the ventral cavity and the dorsal cavity (FIGURE 5.8). Consisting of the *thoracic* (chest) and *abdominal* cavities, the **ventral cavity** is at the front of the body. The thoracic cavity contains the heart and lungs, and the abdominal cavity contains the abdominal organs including the stomach, liver, spleen, gall bladder, intestines, urinary bladder, and internal reproductive organs. A horizontal sheet of muscle called the **diaphragm** separates the thoracic cavity from the abdominal cavity.

The **dorsal cavity** is composed of the cranial cavity and the vertebral canal. The cranial cavity is inside the skull and houses the brain. The vertebral canal is formed by vertebrae and contains the spinal cord.

Body Membranes

Body membranes line internal surfaces of organs and body cavities. These sheetlike membranes can have different structures, but they all function to protect and/or lubricate organs.

Epithelial membranes are composed of a sheet of epithelium bound to underlying connective tissue. They line body surfaces, cavities, ducts, and tubes. Epithelial membranes include mucous, serous, and cutaneous membranes.

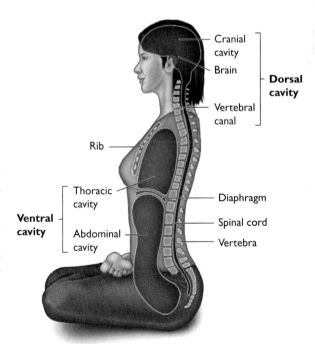

FIGURE 5.8 **Body cavities.** The dorsal ("toward the back") body cavity contains the cranial cavity and vertebral canal. The brain is in the cranial cavity, and the spinal cord is in the vertebral canal. The ventral ("toward the front") cavity is divided by the diaphragm into the thoracic and abdominal cavities.

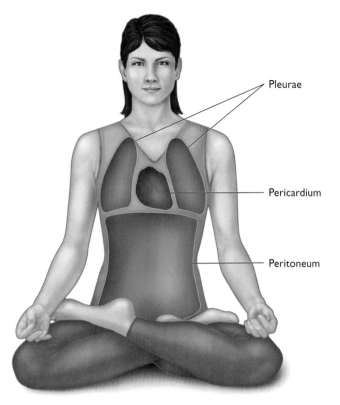

FIGURE 5.9 Serous membranes. The locations of several serous membranes are shown above.

Mucous membranes are the pink, moist, epithelial membranes that line the tubes of the digestive, respiratory, reproductive, and urinary systems. Mucous membranes are able to absorb substances, and many can also secrete protective substances. For instance, nasal membranes are mucous membranes that secrete mucus to protect the body from bacterial and viral infections. When you have a cold, more mucus is secreted partly in response to this invasion. Mucus in the digestive system protects the walls of the stomach and small intestine from acidic digestive juices.

Serous membranes are double-layered membranes that line the thoracic and abdominal cavities and the organs contained therein. They are composed of epithelium and loose connective tissue. They secrete watery fluid between the layers that keeps the membranes lubricated. Serous membranes support internal organs and provide a lubricated smooth surface to prevent chafing between adjacent body structures. The serous membranes are named according to their location (FIGURE 5.9). **Pleural membranes** are a type of serous membrane that lines the thoracic cavity and lungs. The **pericardium** is a membranous sac that surrounds the heart. The **peritoneum** is the membrane that lines the abdominal cavity and covers its organs. A double layer of peritoneum, called the *mesentery*, attaches organs in the abdominal cavity to the abdominal wall.

The **cutaneous membrane** is a type of epithelial membrane that forms the relatively hard, dry skin.

Some membranes are not composed of epithelium. These include meningeal and synovial membranes.

Meningeal membranes are found within the dorsal cavity. Made of connective tissue, these membranes cover the brain and spinal cord.

Stop and Stretch 5.2

The meningeal membranes that protect the brain can become swollen if the brain is injured. Such injuries, called concussions, are common among football, hockey, and soccer players. When the soft tissue of the brain is slammed against the hard skull, bruising of the brain and swelling of surrounding membranes can occur. An athlete who has been diagnosed with a concussion should not play again until the concussion has completely healed. A second impact to the brain before it is completely healed can result in permanent disability and death. Why might a second injury to the brain be more severe than an initial injury, even though the impact is not as powerful?

Synovial membranes are also composed of connective tissue. These membranes line the capsule formed by the ligaments of synovial joints, like the knee. Synovial membranes secrete synovial fluid into the joint cavity to lubricate joints. When the joint capsule is damaged by injury, its ability to stay lubricated decreases. Athletes with knee injuries are sometimes given injections of synthetic synovial fluid to help lubricate the joint.

5.3 Organs and Organ Systems

Many structures in the body are assembled from different tissue types. **Organs** are structures composed of two or more tissues packaged together, working in concert to carry out the organ's specific function. When many organs interact to perform a common function, these organs are said to be part of an **organ system**. Human organs and organ systems will be covered in detail in subsequent chapters. For now, Table 5.1 briefly outlines the functions and component organs involved in each organ system.

TABLE 5.1 Organs and Functions of the Human Organ Systems

Organ System		Function	Organ System		Function
Digestive	Esophagus, Stomach, Liver, pancreas, gallbladder, Small and large intestine	Ingests and breaks down food so that it can be absorbed by the body Chapter 7	**Cardiovascular**	Blood vessels, Heart	Enables the transport of nutrients, gases, hormones, and wastes to and from cells of the body Chapter 9
Urinary	Kidney, Ureter, Bladder, Urethra	Eliminates liquid wastes; regulates water balance Chapter 11	**Endocrine**	Pituitary gland, Thyroid, parathyroid, Thymus, Gonads, others	Secretes hormones into bloodstream for regulation of body activities Chapter 16
Respiratory	Trachea, Bronchi, Lung	Enables gas exchange, supplying blood with oxygen and removing carbon dioxide Chapter 10	**Nervous**	Brain, Spinal cord, Nerves	Senses environment; communicates with and activates other parts of the body Chapters 14 and 15
Skeletal	Cartilage, Bone	Provides mechanical support for the body; stores minerals; produces red blood cells Chapter 6	**Lymphatic and Immune**	Thymus, Lymph nodes, Lymphatic vessels, Spleen	Protects against infections Chapter 12
Muscular	Skeletal muscles	Enables movement, posture, and balance via contraction and extension of muscles Chapter 6	**Reproductive—Female**	Ovary, Uterus, Cervix, Vagina	Produces eggs, and supports the development of offspring Chapter 18
Integumentary	Hair, Nails, Skin	Protects body from environment, injury, and infection; stores fat Chapter 6	**Reproductive—Male**	Prostate, Testicle, Penis	Produces and delivers sperm and associated fluids Chapter 18

All of the organ systems in a body work together to form the functional organism. The coordination of activity among tissues, organs, and organ systems in an organism can be likened to the way the engine (organ system) of a motor vehicle (organism) functions through the combined actions of its different parts, such as the metal, nylon, and plastic that make up the car (tissues) and valves and pistons (organs). Many different systems in the vehicle, such as the electrical system (organ system) and the fuel-injection system (organ system), work together to keep the vehicle (organism) functioning.

Levels of Organization

Living things display increasingly complex levels of organization, from cells, to tissues, to organs, to organ systems, to an entire organism (FIGURE 5.10). Body parts at each level act in concert to regulate the organism's functions.

For example, the regulation of a bicyclist's body temperature while riding through different weather conditions requires several levels of organization. The brain (an organ) sends signals through the cells of the nervous system (an organ system) to the epidermis (tissue) that result in changes in individual units (cells) that make up sweat glands and blood vessels, resulting in an overall change in the amount of sweat secreted and in the dilation of blood vessels.

Interdependence of Organ Systems

We have learned that during exercise, your heart must beat quickly to deliver oxygen more effectively to the rest of your body. Therefore, the force of each beat increases to optimize blood flow to your muscles and back to your lungs. Stronger heart contractions also help increase the amount of oxygen your lungs can deliver to your bloodstream to nourish cells. This integration of the cardiovascular and respiratory systems demonstrates another property of organ systems, which is that different organ systems rely on each other. Without the cardiovascular system to move oxygen through the bloodstream, the respiratory system would have little effect. In fact, the functions of these two systems are so closely integrated that they are often called the *cardiorespiratory system*. As we go on to study other organ systems, you will see many more examples of the interdependence of different organ systems.

It's not just organ systems that interact to help keep the body functioning properly. In the next section you will learn how cells, tissues, organs, and organ systems all work together to maintain bodily functions.

5.4 Homeostasis

The ability to maintain relatively constant internal conditions even under extreme situations is called **homeostasis**. Maintaining this steady state requires the pooled effort of the cells, tissues, organs, and organ systems. Each cell engages in basic metabolic activities that will ensure its own survival. Cells of a tissue perform activities that contribute to the proper functioning of organs. Organs perform required functions alone, or in conjunction with organ systems, to contribute to the maintenance of a stable body.

Humans must maintain heart rate, blood pressure, water and mineral balance, temperature, and blood-glucose levels within a narrow range for survival. Keeping physical and chemical properties within tolerable ranges relies on feedback from one system to another. **Feedback** is information that is sent to a control center, such as the brain, which in turn directs a cell, tissue, or organ to respond by turning up or turning down a given process. Feedback mechanisms apply to almost every physiological process and can have a negative or positive effect.

Negative feedback occurs when the product of a process inhibits the process. It applies in cases of certain *variables* that must be closely controlled, such as body temperature or blood-calcium levels. A deviation away from the *set point* that the body is trying

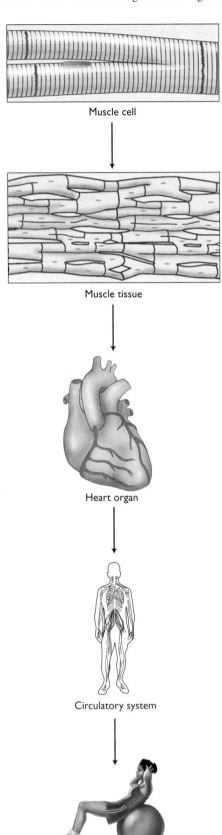

FIGURE 5.10 **Levels of organization.** Cells give rise to tissues. Tissues with a common function make up organs. Organs working together form an organ system, and a whole organism often has many organ systems that communicate with each other.

to maintain is detected by a *sensor*. If deviation will result in the loss of homeostasis, a *control center* activates *effectors* in an attempt to restore homeostasis. If the effectors are successful and the set point is restored, the control center will signal the effectors to shut off. Thus, negative feedback is self-regulating and functions to maintain homeostasis.

As an example, **thermoregulation**—the regulation of body temperature—occurs by negative feedback, much the way that temperature is controlled in a house. A thermostat in a room functions as both an information sensor and a control center. When the thermostat registers a temperature below a set point, it triggers the furnace (effector) to turn on. The output of the furnace then warms the house. When the temperature rises past the set point, the thermostat senses that change and cuts off power to the furnace.

Similarly, when the body temperature increases above normal (37°C, or 98.6°F), a sensor in the brain is activated. This sensor sends signals to the body to dilate blood vessels near the skin and to activate sweat glands, allowing heat to escape. As a result, the body temperature cools. When body temperature returns to the set point, the control center directs blood vessels near the surface of the skin to constrict and also inhibits sweating (FIGURE 5.11). If body temperature continues to drop, the control center triggers muscles to begin to shiver, generating heat via metabolism. These negative feedback mechanisms keep our body temperatures within a very narrow range, even when swimming in 18°C (65°F) water or running a marathon in 40°C (105°F) air temperatures.

Positive feedback occurs when the product of the process intensifies the process. Sensors detect some change away from homeostasis, and the control center activates effectors to amplify the change. The response of positive feedback is different because the purpose of positive feedback is different. Positive feedback is usually used to accomplish some event such as blood clotting or childbirth. Also, positive feedback mechanisms have natural limits. They proceed until the event is over. For example, as the bladder fills with urine, nerve cells that measure stretch are stimulated. This information triggers contraction of the muscles of the bladder, leading to discomfort and a feeling of needing to urinate. As urine is released from the bladder, it triggers additional contractions, ensuring that the bladder empties completely (FIGURE 5.12). The production of breast milk is also under positive feedback control because suckling by an infant stimulates more milk production.

Most homeostatic regulation is via negative feedback because positive feedback amplifies a response and can actually drive the body away from homeostasis.

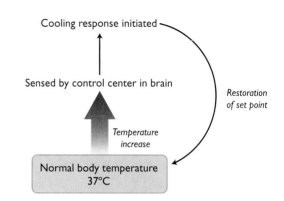

FIGURE 5.11 **Negative feedback.** Regulation of body temperature within a narrow range relies on a negative feedback system in which the brain triggers a response that helps the body dissipate heat when body temperature rises. When these changes allow body temperature to return to normal, the response is stopped.

VISUALIZE THIS Identify the control center, sensor, effector, and set point in this example.

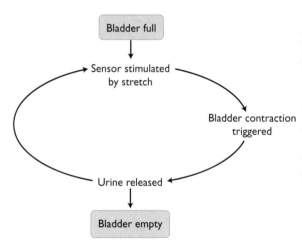

FIGURE 5.12 **Positive feedback.** The presence of urine in the bladder triggers a response that causes the bladder to contract. Release of a little urine triggers further contractions until the bladder is completely empty.

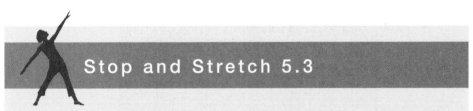

Stop and Stretch 5.3

During labor, the pregnant mother's uterus, which houses the developing fetus, begins to undergo strong rhythmic contractions. The hormone oxytocin causes uterine contractions, which force the child's head against the bottom of the uterus, called the cervix. Stretching of the cervix stimulates the release of more oxytocin, increasing the strength of the contractions and resulting in even more pressure on the cervix. Is this an example of positive or negative feedback?

When a person is not in good shape, or is injured, many tissues, organs, and systems of the body can be affected and homeostasis can be harder to attain. Exercising can help improve overall fitness and decrease the likelihood of injury, but only if it is done safely.

Improving Fitness

To work toward attaining overall body fitness, most experts recommend at least 30 minutes of aerobic exercise on most days and muscle-strengthening exercises (including stretching, core strengthening, and flexibility exercises) two or three times per week.

To exercise effectively and safely, you should learn to measure your heart rate. Measuring your heart rate during exercise allows you to determine if you are working out too intensely (which is dangerous for your heart) or not hard enough (which is less productive).

To measure your resting heart rate, count the number of pulses at any place on the body at which an artery is close to the surface and a pulse can be felt, such as the radial artery of the wrist (FIGURE 5.13). You can count for 60 seconds or count for 15 seconds and multiply by 4. The typical adult has a resting heart rate of 60–80 beats per minute (bpm), whereas a trained athlete might have a heart rate closer to 40 bpm. As your fitness level increases, your resting heart rate should decrease.

Your maximum heart rate is the highest number of beats per minute that your heart should reach while exercising. Maximum heart rate can be estimated by subtracting your age from 220. Your target range during exercise can be calculated by multiplying your maximum heart rate first by 60% (0.60), then by 80% (0.80). For example, a twenty-year-old would have a maximum heart rate of 200 bpm. While exercising, this twenty-year-old should try to keep to a heart rate between 120 and 160 bpm. As a person increases in fitness, that person will be able to train for longer periods of time and more intensely without increasing his or her heart rate above the top of the target heart zone. You may have seen a chart displaying the target heart rate for various age groups at your local gym or fitness center (FIGURE 5.14).

Next time you are trying to decide whether or not to exercise, remember that in addition to many health benefits provided by exercise, people who exercise have more energy and tire less easily. Even though exercising takes time, you get the time back when you sleep better or get work done more efficiently.

FIGURE 5.13 **Measuring your heart rate.** To measure your heart rate, press on the carotid artery of your neck. Count the number of beats for 15 seconds and multiply that number by 4.

HEART RATE TARGET
(10 Second Count)

AGE	55%	60%	70%	80%	85%
15	19	21	24	27	29
20	18	20	23	27	28
25	18	19	23	26	28
30	17	19	22	25	27
35	17	19	22	25	26
40	17	18	21	24	26
45	16	18	20	23	25
50	16	17	20	23	24
55	15	17	19	22	23
60	15	16	19	21	23
65	14	16	18	21	22
70	14	15	18	20	21
75	13	15	17	19	21
80	13	14	16	19	20
85	12	14	16	18	19
90	11	13	15	17	18

FIGURE 5.14 **Target heart zone.** This chart shows target heart rates of people of various age ranges.

Chapter REVIEW

ROOTS TO REMEMBER

The following roots of words come mainly from Latin and Greek and will help you to decipher terms:

adipo- is from the Latin word for fat, and the Greek root for fat, also commonly seen, is **lip-** or **lipo-**.

cut- or **cuta-** relate to the skin.

cyto- and **-cyte** mean cell or a kind of cell.

dia- and **trans-** mean through.

epi- means on or upon.

homeo- means the same.

lig- comes from a Latin verb meaning to join or connect.

neuro- means nerve.

oligo- means few.

osteo- relates to bone.

peri- means around or surrounding.

pleur- and **pleura-** relate to the ribs and flanks.

thermo- relates to warmth and heat.

-ula (**-ule** and **-uscle**, too) is an ending meaning something small, such as a tub(e)ule.

KEY TERMS

5.1 Tissues and Their Functions
actin *p. 113*
adhesion junction *p. 107*
adipocyte *p. 110*
adipose tissue *p. 110*
astrocyte *p. 114*
atrophy *p. 113*
basement membrane *p. 107*
blood *p. 111*
bone *p. 112*
cardiac muscle *p. 113*
cartilage *p. 111*
chondrocyte *p. 111*
collagen fiber *p. 109*
columnar epithelium *p. 107*
cuboidal epithelium *p. 107*
dense fibrous connective tissue *p. 111*
elastin fiber *p. 109*
endocrine gland *p. 106*
epithelia *p. 106*
epithelial tissue *p. 106*
exocrine gland *p. 106*
fibroblast *p. 108*
gap junction *p. 107*
gland *p. 106*
ground substance *p. 108*
hypertrophy *p. 113*
intercellular junction *p. 107*
involuntary muscle *p. 113*
loose connective tissue *p. 108*
matrix *p. 108*
microglia *p. 114*
muscle tissue *p. 113*
myosin *p. 113*
nervous tissue *p. 114*
neuroglia *p. 114*
neuron *p. 114*
neurotransmitter *p. 114*
oligodendrocyte *p. 114*
osteocyte *p. 112*
Schwann cell *p. 114*
skeletal muscle *p. 113*
smooth muscle *p. 113*
squamous epithelium *p. 107*
striated muscle *p. 113*
tight junction *p. 107*
tissue *p. 106*
voluntary muscle *p. 113*

5.2 Body Cavities and Membranes
cutaneous membrane *p. 116*
diaphragm *p. 115*
dorsal cavity *p. 115*
meningeal membrane *p. 116*
mucous membrane *p. 116*
pericardium *p. 116*
peritoneum *p. 116*
pleural membrane *p. 116*
serous membrane *p. 116*
synovial membrane *p. 116*
ventral cavity *p. 115*

5.3 Organs and Organ Systems
organ *p. 116*
organ system *p. 116*

5.4 Homeostasis
feedback *p. 118*
homeostasis *p. 118*
negative feedback *p. 118*
positive feedback *p. 119*
thermoregulation *p. 119*

SUMMARY

5.1 Tissues and Their Functions

- Tissues are composed of similar cell types that perform a common function (p. 106).

- Epithelia line and cover organs, vessels, and body cavities. They are a tightly packed, avascular tissue with one free surface (pp. 106–107).

- Glands, formed of epithelial cells specialized to secrete a substance, function in protection, secretion, and absorption. Endocrine glands secrete hormones into the bloodstream. Exocrine glands secrete their products, including saliva, sweat, and stomach acid, into hollow organs or ducts, or onto skin or other surfaces (p. 106).

- Epithelial tissue is classified according to shape and number of cell layers. Squamous epithelial cells are flattened cells. Cuboidal epithelial cells are roughly cube shaped. Columnar epithelial cells are column shaped. Simple tissue is one cell layer thick, while stratified tissue has many layers (p. 107; Figure 5.3).

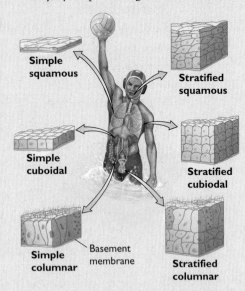

- The noncellular basement membrane is located beneath the cells of an epithelial tissue and functions to provide support to the epithelial cells and join the epithelium to underlying connective tissue (p. 107).

- Junctions between cells serve to anchor cells to each other and allow cells to interact (p. 107).

- Connective tissues bind tissues and organs to each other. The different types of connective tissues are each composed of a characteristic cell type, embedded in a characteristic matrix (p. 108).

- Loose connective tissue connects epithelia to underlying tissues and holds organs in place. The cellular component is mostly fibroblasts. The matrix is rich in the proteins collagen and elastin, which provide this tissue with tensile strength and elasticity (p. 108–109).

- Adipose tissue connects the skin to underlying structures and insulates and protects organs. Cells of this tissue (adipocytes) synthesize and store fat and have very little matrix (p. 110–111).

- Blood is a type of connective tissue that transports oxygen and nutrients to body cells. Blood cells have a liquid matrix called plasma (p. **xx**).

- Dense fibrous connective tissue forms the tendons and ligaments. The cells of this tissue are fibroblasts, and the matrix is rich in collagen (p. 111).

- Cartilage is a flexible, shock-absorbing tissue composed of cells called chondrocytes that secrete a dense, collagenous matrix (p. 111).

- Bone tissue provides support for the body. The cells of this tissue are called osteocytes, and the matrix is rich in collagen and minerals including calcium (p. 112).

- Muscle tissues are composed of fibers that contract and conduct electrical impulses. The three types of muscle tissue are skeletal, smooth, and cardiac (p. 113).

- Cardiac muscle forms the heart. Such muscle is involuntary and striated (p. 113).

- Skeletal muscle is attached to bones, responsible for voluntary movements, and striated (p. 113).

- Smooth muscle forms the musculature of internal organs and blood vessels. Smooth muscle is involuntary and not striated (p. 113).

- Nervous tissue, found in the brain, spinal cord, and nerves, is composed of neurons and supporting neuroglia. Nervous tissue senses stimuli and transmits signals throughout the body (p. 114).

5.2 Body Cavities and Membranes

- The ventral cavity of the human body includes the thoracic and abdominal cavities. The thoracic cavity is separated from the abdominal cavity by the diaphragm. The dorsal cavity consists of the cranial cavity and vertebral canal (p. 115; Figure 5.8).

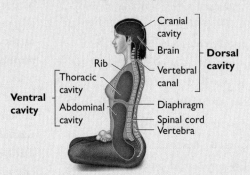

- Body cavities and surfaces are lined with membranes that protect and lubricate the structures they line (p. 115).

- Epithelial membranes include mucous, serous, and cutaneous membranes. Meningeal and synovial membranes are composed of connective tissue (p. 116).

5.3 Organs and Organ Systems

- Organs are groups of tissues working in concert (p. 116).

- Organ systems are suites of organs working together to perform a function or functions (p. 116–118; Figure 5.10).

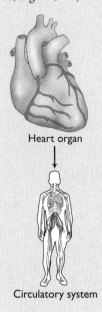

5.4 Homeostasis

- Cells, tissues, organs, and organ systems all work together to maintain the constant internal environment required for homeostasis (p. 118).

- Negative feedback helps keep the internal environment within certain stable limits. When negative feedback occurs, the product of a process slows or stops a process (p. 118–119; Figure 5.11).

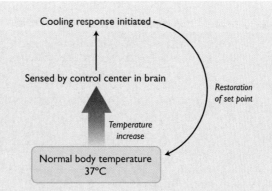

- Positive feedback brings about rapid change in the same direction as the stimulus (p. 119).

LEARNING THE BASICS

1. Epithelia _____.
   a. are loosely packed tissues that hold organs in place
   b. can be free-floating tissues such as blood
   c. line organs and cavities and have one surface exposed
   d. include the tissues that hold joints together
   e. are tissues that conduct nerve impulses

2. Connective tissues include _____.
   a. cartilage
   b. blood
   c. fat
   d. bone
   e. all of the above

3. Muscle tissues _____.
   a. contain actin and myosin, whether or not they are striated
   b. that facilitate movements requiring conscious thought are involuntary
   c. increase in cell number with exercise
   d. in the heart are smooth
   e. all of the above

4. Chemicals secreted by neurons that allow these cells to communicate with each other are called _____.
   a. neuroglia
   b. astrocytes
   c. oligodendrocytes
   d. neurotransmitters

5. Exocrine and endocrine glands are primarily composed of which type of tissue?
   a. connective
   b. adipose
   c. gap junctions
   d. epithelial

6. Which of the following types of intercellular junctions allows materials to pass between cells?
   a. tight junction
   b. gap junction
   c. adhesion junction
   d. more than one of the above

7. What type of tissue has collagen and elastin fibers?
   a. nervous tissue
   b. adipose tissue
   c. connective tissue
   d. muscle tissue

8. Actin and myosin are proteins found in which type of tissue?
   a. nervous tissue
   b. adipose tissue
   c. connective tissue
   d. muscle tissue

9. A matrix substance is found in which tissue type?
   a. connective
   b. epithelial
   c. nervous
   d. muscle

10. Skeletal muscle _____.
    a. is striated
    b. is under voluntary control
    c. has many nuclei
    d. all of the above are true

11. Cardiac muscle _____.
    a. is found only in the heart
    b. is under voluntary control
    c. has cells with many nuclei
    d. cells are shaped like cubes
    e. all of the above

12. Smooth muscle _____.
    a. is striated
    b. is under voluntary control
    c. is composed of highly branched and interwoven cells
    d. forms much of the musculature of internal organs

13. Both cardiac and skeletal muscle _____.
    a. cells contain one nucleus
    b. are under involuntary control
    c. cells are shaped like cubes
    d. are striated

14. Tendons _____.
    a. connect muscles to bones
    b. are composed of loose connective tissue
    c. are composed of smooth muscle
    d. all of the above are true

15. Ligaments _____.
    a. connect bones to each other
    b. contain fibroblasts
    c. are composed of dense fibrous connective tissue
    d. all of the above are true

Choose the correct term for each definition in items 16–20.

a. cartilage
b. smooth
c. fibroblast
d. ventral cavity
e. cuboidal epithelium
f. ligament
g. dorsal cavity
h. astrocyte
i. cardiac
j. tendon
k. squamous epithelium
l. basement membrane

16. _____ Flattened epithelial cells
17. _____ Connective tissue that covers joint surfaces
18. _____ Houses the cranial cavity and vertebral canal
19. _____ Cells that synthesize connective tissue fibers
20. _____ Type of involuntary muscle sheets that line the stomach

ANALYZING AND APPLYING THE BASICS

1. List the major ways in which muscle, nervous, and connective tissues respond to exercise.

2. The interior of the small intestine is lined with columnar epithelial cells. How would the shape of these cells enhance the function of the epithelium of the small intestine?

3. Retinoids are compounds derived from vitamin A that are used in many acne treatments. Besides reducing acne and its effects, retinoids help to reduce the fine wrinkles and coarse skin texture associated with sun damage. Propose a mechanism that would allow retinoids to improve the condition of the damaged skin.

4. How do the organs in the thoracic cavity work together to keep the body functioning properly?

5. Pericarditis is a disease in which inflammation of the pericardium squeezes the heart and causes chest pain. Pericarditis can result from an infection, heart attack, spreading cancer, or injury. What physical change in the pericardium likely causes the symptoms of pericarditis?

6. Vital lung capacity is the maximum amount of air that can be exhaled after your deepest breath. You can get a rough measure of vital lung capacity by inflating a round balloon. FIGURE 5.15 shows lung capacity in liters compared with balloon diameter in centimeters.
   Suppose that you measured your vital lung capacity every month for the next 12 months. How would the diameter of the balloon change if you used those 12 months to train for a marathon? Explain your answer.

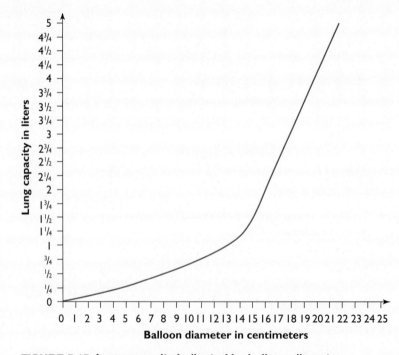

FIGURE 5.15 Lung capacity indicated by balloon diameter.

7. Explain how your organ systems work together to allow you to play tennis.

CONNECTING THE SCIENCE

1. Bone marrow, the spongy tissue inside bones that helps produce blood cells, can be removed and used to help cancer patients replace infection-fighting white cells lost to chemotherapy. Because the recipient of a bone marrow transplant might reject the tissue if genetically determined markers on the surface of cells differ, siblings can be a good source of bone marrow. Bone marrow can be collected from the back of your pelvic bone with a needle and syringe. Should parents be allowed to require their minor children to donate bone marrow to their siblings with cancer?

2. Humans have two kidneys but can survive with one. Therefore live donors can donate a kidney. Kidney transplants are most effective when a relative donates a kidney because two related individuals will have more cell surface markers in common than would two unrelated individuals. However, people do donate kidneys to unrelated individuals that share enough markers for the transplant to be accepted. Many people on waiting lists for kidney transplants will spend years undergoing time-consuming dialysis treatments, and many will die before a matched kidney becomes available. The desperation of this situation has led some people to attempt to buy kidneys from donors willing to sell them. One person even attempted to sell a kidney on eBay. Should private citizens be able to sell their own organs? Why or why not?

3. Most countries have banned the sale of kidneys, but there are brokers who will, for tens of thousands of dollars, attempt to locate a donor. These black market kidneys tend to come from very poor people in less developed nations who are desperate for money to help themselves and their families survive. Should private citizens be allowed to travel to other countries to purchase organs?

Chapter 6

The Skeletal, Muscular, and Integumentary Systems

Sex Differences in Athleticism

LEARNING GOALS

1. List the types of tissues found in the skeletal system.
2. Describe the structure and functions of bones.
3. Compare the activities of osteoclasts and osteoblasts.
4. List the bones that compose the pelvic girdle.
5. How do osteocytes, which are embedded in solid bone, receive nutrients?
6. Describe the structure and functions of muscles.
7. What are antagonistic muscle pairs?
8. Describe the mechanism of muscular contraction.
9. List the functions of the integumentary system.
10. Describe the structure and functions of the skin.

Many young athletes dream of playing professional sports.

The scene on the baseball field is one that many young athletes spend their summers dreaming about. For this baseball pitcher, the dream has come true. On a warm Friday evening late in July, the sun is beginning to set over left field in a beautiful, open-air, brick baseball stadium. On the mound, a 23-year-old lefthander stands motionless, squinting to better see the catcher's signals.

The scoreboard above the centerfield wall shows that it is the top of the sixth inning, and the Duluth-Superior Dukes are leading the Sioux Falls Canaries by a score of 2 to 0. The Dukes pitcher has given up only three hits this game, but the Canaries may be putting together a rally, with one runner on base and their number-three hitter at bat.

The pitcher glances at the runner on second base, turns to face the batter, and in one strong, fluid motion, rears back and hurls the ball across the inside corner of home plate. "Strike three. The batter is out," yells the home-plate umpire.

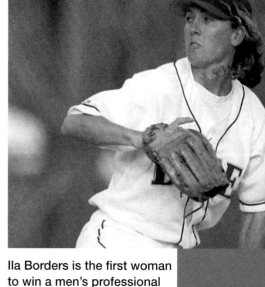
Ila Borders is the first woman to win a men's professional baseball game.

The stadium erupts as the fans jump to their feet cheering wildly. The pitcher walks off the mound toward the dugout to a standing ovation. For the next half-inning, the hometown fans chant the pitcher's name in unison, until the southpaw finally comes out of the dugout to salute the cheering fans. When the pitcher raises the black-and-purple Dukes baseball cap toward the fans behind home plate, her long hair spills across the back of her uniform, obscuring her name and number.

The pitcher is Ila Borders, the first woman in the starting lineup of a men's professional baseball game and, on that warm July evening, the first woman to win a men's professional baseball game. This accomplishment was so impressive that *Sports Illustrated* honored Borders as one of the top 100 female athletes in history. The uniform and glove that served her in this historic victory adorn the prestigious Baseball Hall of Fame in Cooperstown, New York. National television stations ESPN and CNN broadcast footage of her pitching that night.

Annika Sorenstam has had success in golf tournaments against men.

Danica Patrick competes against men as a racecar driver.

Will Ila Borders be the first of many women to play men's professional baseball? Will women infiltrate professional football and hockey leagues next? Will the successes of women golfers and auto racers be replicated in other sports as opportunities increase for women to compete? Or, are the bodies of women and men simply constructed so differently that men and women will be able to compete against each other only rarely?

To answer these questions, we have to delve into the study of sex differences—average differences between males and females. Sex differences are measured using average differences because, for most traits, there is a fair amount of overlap between the sexes. As you will learn, these differences can be biological, such as differences in muscle mass, or cultural, such as differences in opportunities to practice and exposure to high-quality coaching.

6.1 The Skeletal System

The skeletal system protects internal organs, serves as an anchor for muscular attachments, aids in movement, stores minerals, and produces blood cells. This internal framework is composed mainly of connective tissues.

The connective tissues of the skeletal system include cartilage, ligaments, tendons, and bone. Cartilage protects the ends of bones from degradation. Because it has no nerves, it can absorb shock without causing pain. Ligaments and tendons are the dense fibrous connective tissues that bind bones to each other and connect muscles to bones, respectively. Chapter 5 covered cartilage, ligaments, and tendons in detail and introduced bone tissue. This section focuses on the structure and function of bones.

Bones of the Skeleton

Bones are hard body tissues that provide support for the actions of muscles and protect the soft parts of the body. Bones come in a variety of shapes and sizes. *Long bones* are so named because they are longer than they are wide. *Short bones*, such as some of the bones found in the feet and hands, are cube shaped or roughly as long as they are wide. *Flat bones*, including some of the skull bones, are platelike with broad surfaces. *Round bones*, such as the kneecap (patella), are circular in shape. *Irregular bones* can have many different shapes to allow connection with other bones.

The ability of bones to protect organs and tissues is a result of bone's rigidity. Bones are firm and strong because they contain hardened mineral salts, such as calcium phosphate. The minerals stored in bones can be liberated when needed by other parts of the body. In addition to the nonliving minerals found in bone, bones contain living cells, nerves, and blood vessels. Tissue in the interior of bones also helps produce blood cells.

Structurally, bone tissue can be tightly or loosely packed. **Compact bone** forms the hard outer shell of bones and is composed of cylindrical structures called *osteons*. Bone cells called **osteocytes** are found inside cavities called **lacunae**, which are arranged in concentric circles around a central canal (FIGURE 6.1a). Lacunae are connected to each other and to the central canal by even smaller canals called *canaliculi* through which osteocyte cytoplasm can extend (FIGURE 6.1b). Nutrients diffuse through blood vessels located in the central canal to osteocytes located nearby. Osteocytes located farther away from the central canal obtain nutrients when they are passed from cells closer to the central canal through gap junctions. Waste products produced by osteocytes diffuse back to the central canal and are removed via blood vessels.

Loosely packed **spongy bone** is the porous, honeycomb-like bone found on inner surfaces and at the ends of long bones. Spongy bone is less dense than compact bone but is still very strong. Support structures called *trabeculae* function as tiny beams to strengthen the bone (FIGURE 6.1c). Trabeculae enable bone to absorb force from multiple directions, which is why they are found at the ends of long bones. In some types of bones, the spaces between trabeculae are filled with **red bone marrow**, where blood cells are produced. A layer of tough fibrous connective tissue rich in blood vessels and

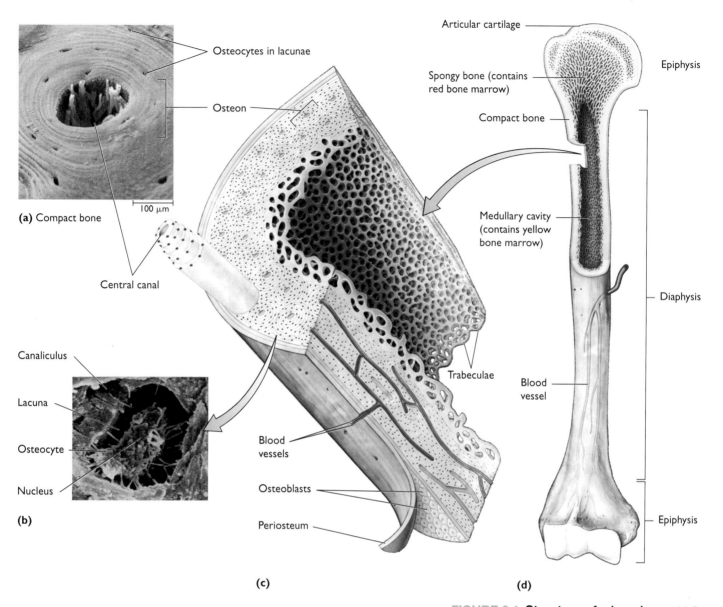

FIGURE 6.1 **Structure of a long bone.** (a) Compact bone forms the hard outer shell of bones. Cylindrical osteons are the subunits, each with a central canal. Osteocytes are found inside cavities called lacunae that surround the central canal. (b) Canaliculi connect lacunae to each other and to the central canal. (c) Spongy bone contains supportive trabeculae. The periosteum is a layer of tough fibrous connective tissue rich in blood vessels and nerves. (d) The epiphysis forming the bulbous tip of a long bone is covered in cartilage, as are all other articular surfaces. The diaphysis, or shaft, of the long bone is composed of compact bone. The medullary cavity is filled with fatty yellow bone marrow.

nerves, called the **periosteum** (FIGURE 6.1c), surrounds the bone except on surfaces that are jointed.

The bulbous spongy bone at the tip of a long bone is called the **epiphysis** (FIGURE 6.1d). The epiphysis is covered with cartilage to prevent the ends of bones at a joint from rubbing against each other. The shaft of a long bone is called the **diaphysis**. The walls of the diaphysis are composed of compact bone. The cavity enclosed by the compact bone is called the **medullary cavity**, which is filled with **yellow bone marrow** that is mainly fat.

Bones at joint surfaces that articulate or move against each other are covered with specialized cartilage called *articular cartilage*. Articular cartilage is composed of chondrocytes embedded within an extracellular matrix. When the extracellular matrix is rich in hyaluronic acid, it is called *hyaline cartilage*.

Bone Development, Growth, Remodeling, and Repair

Early bone development occurs as a fetus grows in the uterus. Once developed, bones are not inert. Instead, the living tissues that make up bones are able to grow and repair themselves.

Bone Development Long bones first begin to form when an embryo is about 6 weeks old. The bones that are visible on a sonogram (FIGURE 6.2) were originally composed of tough hyaline cartilage. As the bony skeleton begins to develop, the cartilage dissolves and is replaced with bone. This process, called *ossification*, is illustrated in FIGURE 6.3.

When the fetus is a few months old, cartilage-producing cells called **chrondroblasts** die and the cartilage matrix they secreted begins to dissolve. This dissolution makes room for the development of blood vessels. Once the periosteum develops, it produces bone-forming cells called **osteoblasts** that travel through the newly formed blood vessels to the developing shaft of the bone. Osteoblasts secrete collagen, which strengthens the developing bone. In addition to secreting collagen, osteoblasts secrete enzymes that promote the crystallization of mineral salts of calcium phosphate, producing *hydroxyapatite*, the main salt found in bone. As the concentration of hydroxyapatite increases around the osteoblasts, the cell eventually becomes trapped in its own secretions, forming lacunae.

Bone Growth Bones lengthen throughout childhood into late adolescence or early adulthood. The **growth plate**, or **epiphyseal plate**, is within the epiphyseal region (FIGURE 6.4a) and contains four layers (FIGURE 6.4b). The *resting zone* serves to attach the epiphyseal plate to the bony tissue of the epiphysis. The next layer, the *proliferating zone*, is undergoing cell division to produce new chondroblasts. In the *degenerating zone*, cartilage cells are dying off, and it is in the fourth layer, the *ossification layer*, that bone is forming.

The epiphyseal plate is not replaced with bone until early adulthood. Around the time a child goes through adolescence, sex hormones stimulate both osteoblasts and chondroblasts, producing a growth spurt. Osteoblasts grow faster than the chondroblasts, so the

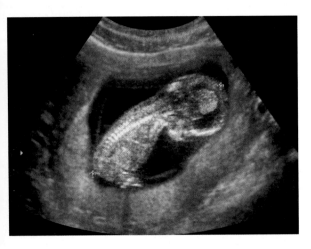

FIGURE 6.2 Fetal sonogram. The skeleton in this fetus is undergoing a process in which the cartilage is being replaced by bone.

FIGURE 6.3 Bone development. After a few months of fetal development, ossification begins and the cartilaginous skeleton is replaced with bone. Blood vessels soon infiltrate the developing bone.

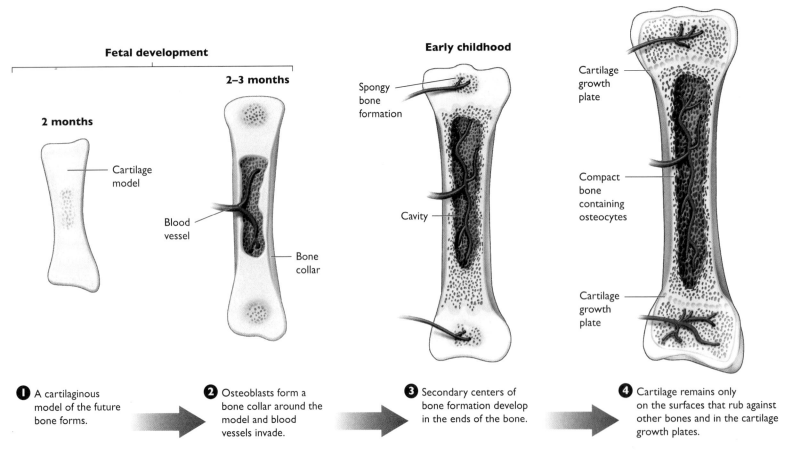

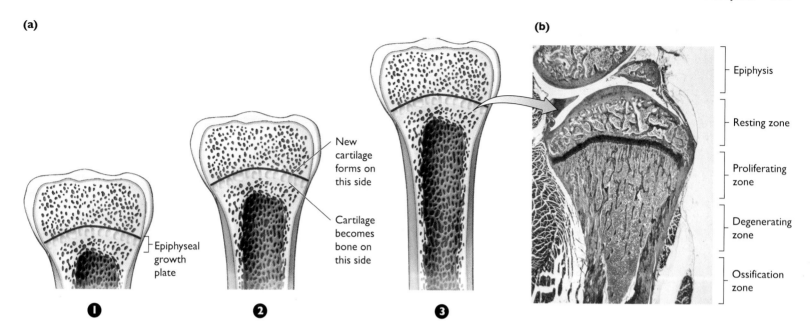

FIGURE 6.4 **Bone growth.** (a) New cartilage is produced on the outer surface of the epiphyseal plate. Cartilage on the inner surface is converted to bone. (b) Chondrocytes produce new cartilage at the resting zone within the epiphysis, and the cartilage becomes bone at the ossification layer near the diaphysis.

bone growth eventually overtakes the growth of the cartilage and calcifies it, closing the epiphyseal plate. When the epiphyseal plates close, no more bone growth will occur. Epiphyseal plates in women close around age 18 and in men around age 20.

From infancy to adulthood, humans grow in overall size, but a larger proportion of growth occurs in the extremities, or arms and legs, as compared to the torso. As evidence of this pattern, consider how short toddlers' legs look in relationship to their bodies. A growth spurt occurs during adolescence, with a larger proportion of growth occurring in the extremities than in the torso. Puberty occurs later and lasts longer in boys than in girls, and bone growth continues until later in life in boys than in girls. Because of this, the average adult man has longer arms and legs than the average adult woman. Overall, the average man is 15 cm (5.9 inches) taller than the average woman.

The differences between the skeletons of women and men contribute to different ways of performing certain physical activities. Because women generally have long torsos relative to their leg length, their center of gravity (the point on the body where the weight above equals the weight below) is lower than in men. Individuals with a lower center of gravity are better able to maintain balance. Competitive gymnastics recognizes this difference in center of gravity—women, but not men, compete on the balance beam apparatus. Ice skating, ballet, and other forms of dance also require a strong feel for balance, and the lower center of gravity may enhance a woman's ability to stay on her feet.

Men, with their generally longer legs and arms and higher center of gravity, have more difficulty balancing, but they bring more power to activities that rely on the lever action of their extremities. A longer arm or leg transmits more force. Therefore, men typically have faster and stronger slap shots in hockey, stronger kicks in soccer, and faster swings and throws in baseball. Longer legs also take longer strides, which means that men will usually be faster runners than women.

Once formed, bones are not static structures. Bones are living tissues that constantly undergo a process called bone remodeling.

Bone Remodeling Every year, approximately 10% of the human skeleton gets broken down and replaced. **Osteoclasts** are specialized bone cells that break bone down. They dissolve hydroxyapatite and collagen fibers. When osteoclasts digest bone, it is reabsorbed by the body as the minerals stored in the bone are released into the bloodstream.

Osteoblasts secrete bone-building collagen and thereby help bone tissue to regenerate itself. Maintaining the delicate homeostasis between bone breakdown and regeneration is integral to maintaining strong, healthy bones.

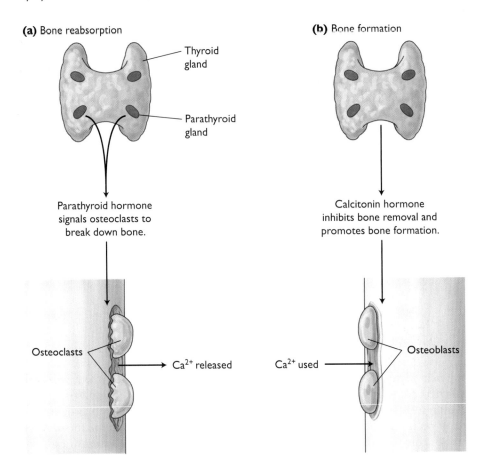

FIGURE 6.5 Bone remodeling. Bone is in a constant state of remodeling, whereby old bone is removed by osteoclasts and new bone is laid down by osteoblasts. (a) Bone removal is stimulated by parathyroid hormone, resulting in the release of calcium. (b) Bone formation is stimulated by the hormone calcitonin, and results in the uptake of calcium.

Weight-bearing exercise and dietary calcium contribute to the maintenance of bone health. Weight-bearing exercise triggers bone building; therefore strength and thickness of bones increase with exercise. Your body needs calcium to help blood clot, muscles contract, and nerves fire, and to facilitate many reactions catalyzed by enzymes. When dietary calcium is low, calcium must be supplied by the bones, and its removal weakens them.

Calcium is delivered to the bloodstream in response to the effects of **parathyroid hormone (PTH)**, which is produced by the parathyroid glands of the neck (FIGURE 6.5a). Release of this hormone signals the kidneys to decrease the amount of calcium secreted in the urine and stimulates the breakdown of bone, so that stored calcium is released into the blood. The thyroid gland secretes another hormone, **calcitonin**, which decreases the concentration of calcium within the blood when the level rises too high (FIGURE 6.5b). Therefore, the opposing actions of PTH and calcitonin keep blood calcium levels within the limited range required for homeostasis.

Osteoblasts and osteoclasts are also involved in maintaining calcium levels. When calcium levels in the blood are high, osteoblasts remove calcium from the blood and use it to make new bone. When calcium levels in the bloodstream are low, osteoclasts break down bone and release calcium into the bloodstream.

Estrogen impedes the activities of osteoclasts. Estrogen levels decline in older women; therefore osteoclast activity increases and the homeostatic balance between bone breakdown and bone building can be altered. If bone breakdown outpaces building, a bone-weakening condition called *osteoporosis* can result (FIGURE 6.6). This condition increases the likelihood of bone breakage.

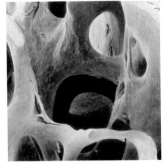

(a) Healthy bone   (b) Osteoporotic bone

FIGURE 6.6 Normal and osteoporotic bone. Healthy bone tissue (a) becomes damaged and weakened when bone resorption outpaces bone deposition, as in osteoporotic bone (b).

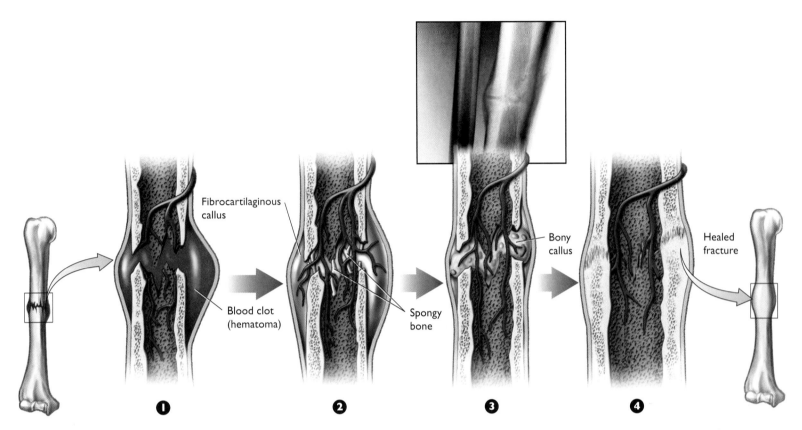

FIGURE 6.7 Steps in the repair of a fracture.
Bone fracture damages blood vessels, and a hematoma is produced. Fibroblasts migrate to the hematoma and lay down cartilage, producing a fibrocartilaginous callus. Osteoclasts remove broken and dead fragments of bone, and osteoblasts convert the callus into a bony callus. Bone remodeling eventually fully heals the fracture.

VISUALIZE THIS What feature of a remodeled bone might make it unusual for a subsequent break to occur at the same site?

Bone Repair A fractured or broken bone undergoes a step-by-step process to repair itself (FIGURE 6.7). When a bone is fractured, damage to the blood vessels supplying the bone produces a mass of clotted blood in the space between the broken bones. This mass of clotted blood is called a *hematoma* and occurs within the first eight to ten hours after fracture. The first step in repairing this damage begins within the next few days as fibroblasts migrate to the area and help lay down cartilage between the broken ends. The resulting structure is the *fibrocartilaginous callus*. This callus is present between the bones for several weeks and can be felt as a hard raised protrusion at the fracture point. Next, osteoclasts arrive and remove the broken and dead fragments of the original bone and break down the hematoma. Osteoblasts lay down new bone, producing the *bony callus*. Lastly, osteoblasts and osteoclasts remodel the bony callus into bone closely resembling the original form. However, the resemblance is not perfect. An X-ray can reveal bone breaks that occurred many years ago.

Stop and Stretch 6.1

A *compound* fracture is a fracture that pierces the skin. When a fracture is *impacted*, the ends of bones are wedged into each other. When a bone is twisted to the point of fracture, ragged edges produce what is called a *spiral* fracture. Why might these types of fractures be more difficult to repair and take longer to heal than a fracture that results in a clean break?

134 CHAPTER 6 The Skeletal, Muscular, and Integumentary Systems

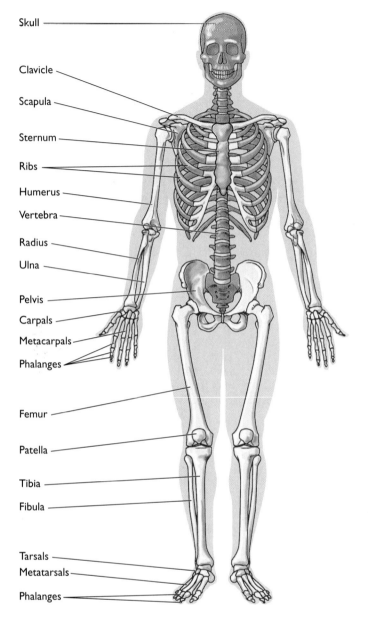

FIGURE 6.8 **The human skeleton.** The human skeleton can be divided into those bones that support the trunk (shaded red in the figure), called the axial skeleton, and those that compose the limbs, called the appendicular skeleton. Cartilage between joints is shaded blue.

The adult human skeleton is composed of 206 bones. The 206 bones are organized into the axial and the appendicular skeletons (FIGURE 6.8). The **axial skeleton** supports the trunk of the body and consists largely of the bones making up the skull, vertebral column (spine), ribs, and sternum (breastbone). The **appendicular skeleton** is composed of the bones of the hip, shoulder, and limbs.

Axial Skeleton: The Central Structure

Skull The human **skull** protects the brain and gives structure to the face. The *braincase*, or **cranium**, forms all the upper portion of the skull, with the bones of the face situated beneath it. The cranial bones are flat bones that enclose and protect the brain.

The cranium protects the sense organs for sight, hearing, smell, and taste. The cranium consists of a relatively few large bones (FIGURE 6.9a). The **frontal bone** forms the forehead region and shows a clear sex difference, being generally more rounded in females and with a less-pronounced ridge above the eyes. The cranium also consists of the **sphenoid**, which extends across the floor of the cranium and attaches to all other cranial bones; the two **temporal bones**, near the ears; an **occipital bone** that forms the base of the skull; and two **parietal bones**, each with a roughly square outline that together form most of the side walls of the cranium. The **ethmoid bone**, found in front of the sphenoid, also helps form the orbits of the eyes and the nasal septum, which separates the nasal cavities.

The largest part of the skeleton of the face is formed by the two **maxillae** (FIGURE 6.9b). Each maxilla helps compose the upper jaw of one side. The lower jaw, a single bone called the **mandible**, is larger in males than in females, giving males a more prominent jawline. The maxillae help form the eye sockets and allow the opening for the nose between them, just below the **nasal bones**. Sockets for teeth are located at the lower part of each maxilla. The **zygomatic bones** form cheekbone protuberances. An opening formed between the temporal and zygomatic bones is the site of attachment for jaw muscles. This opening is larger in males to allow for the connection of thicker muscles to support the larger lower jaw in males. Air pockets within bones of the face form **sinuses** (FIGURE 6.10). Skeletal differences in the skull do not affect athleticism but do contribute to the facial feature differences seen between men and women.

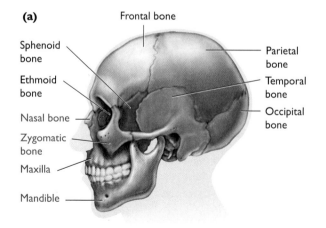

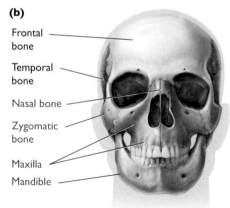

FIGURE 6.9 **Skull bones.** Side (a) and frontal (b) views of the bones that compose the cranium and face. Cranial bones are labeled in black and facial bones in red.

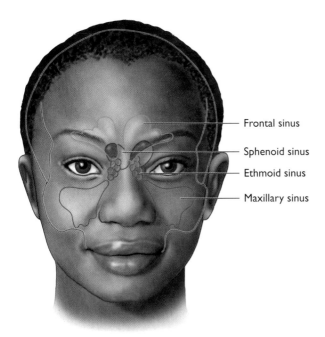

FIGURE 6.10 **Sinuses.** Air-filled cavities in the bones of the skull are called the sinuses.

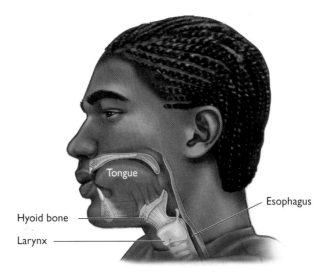

FIGURE 6.11 **Hyoid bone.** The location of the hyoid bone, involved in swallowing.

Hyoid Bone The **hyoid bone** is located at the root of the tongue in the front of the neck (FIGURE 6.11). It attaches to the temporal bones by muscles and ligaments and to the larynx, or voice box, by a membrane. Since these connections are not made by bones, the hyoid bone does not *articulate*, that is, form a joint, with other bones. Instead, it has anchoring and supportive functions. The hyoid bone anchors the tongue and serves as the site of attachment of muscles required for swallowing. During swallowing, muscles attached to the hyoid bone elevate it and the floor of the mouth simultaneously. This action forces the tongue upward and food backward.

Vertebral Column The **vertebral column** is not so much a straight column as a spring that takes the shape of the letter S (FIGURE 6.12). This springy column is composed of 33 individual bones called **vertebrae**, which together house and protect the spinal cord. Nerves branch from the spinal cord to control skeletal muscles. Located between vertebrae are **intervertebral disks**. These resilient fibrocartilaginous pads cushion and protect every vertebra.

The *cervical vertebrae* are located in the neck region, the *thoracic vertebrae* in the upper back region, and the *lumbar vertebrae* in the small of the back. At the end of the vertebral column are the *sacrum* and *coccyx*. In addition to protecting the spinal cord, the vertebral column also serves as an anchoring point for muscles at the *spinous processes*, and for the ribs at the *rib facets*.

Rib Cage The **rib cage**, or thoracic cage, consists of the thoracic vertebrae, the 24 curved **ribs** (12 pairs), and the **sternum**, or breastbone (FIGURE 6.13). The first seven pairs of ribs are attached to the breastbone by hyaline cartilages called *costal cartilages*. Of the remaining five pairs of ribs, the first three have costal cartilages connected to the cartilage above them. The last two pairs, the floating ribs, have cartilages that end in the muscle in the abdominal wall. The lack of direct attachment of the lower five pairs of ribs eases expansion of the lower part of the rib cage and movement of the diaphragm.

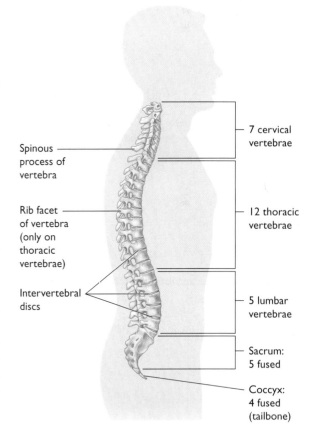

FIGURE 6.12 **Vertebral column.** The 7 cervical vertebrae are located in the neck region. The 12 thoracic vertebrae are located in the chest region. The 5 lumbar vertebrae are located in the small of the back. The sacrum is composed of 5 fused vertebrae in adults, and the coccyx (tailbone) is composed of 4 fused vertebrae. Muscles that move the vertebral column anchor at the spinous processes, and ribs attach at the rib facets.

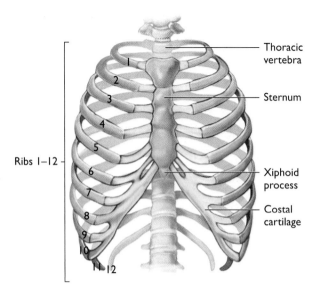

FIGURE 6.13 Rib cage. The rib cage protects the heart and lungs. It is composed of the 12 thoracic vertebrae, 12 pairs of ribs, costal cartilage, and the sternum.

VISUALIZE THIS Can you find the floating ribs in this drawing?

Appendicular Skeleton

The appendicular skeleton consists of the bones within the pelvic or hip girdle, and the pectoral or shoulder girdle, as well as the limbs attached to them. Girdles connect the arms and legs to the axial skeleton by bones or muscles.

Pelvic Girdle The **pelvic girdle**, also called the bony pelvis, (FIGURE 6.14) is a basin-shaped complex of bones that connects the trunk of the body to the legs. The paired hip bones connect to each other in front and to the five fused vertebrae forming the sacrum in back. The hip bones are composed of three fused bones each: the blade-shaped **ilium**, which flares to produce the width of the hips, the **ischium**, behind and below, and the **pubis**, in front. Each ischial bone has a prominence, or tuberosity, and these tuberosities or "sit bones" are what the body rests on when seated. The two pubic bones join at a cartilaginous joint called the *pubic symphysis*. Each of the bones of the pelvic girdle contributes a part of the **acetabulum**, a cup-shaped socket that, along with the femur (thighbone), forms the hip joint.

A ring shape is formed by the bones of the pelvic girdle. The ring made by the pelvic girdle forms the pelvic inlet, or birth canal, in females. The round pelvic inlet that is typical of most women (FIGURE 6.15a) is produced by a bony pelvis that is flatter and broader than a man's pelvis, which has an inlet that resembles an elongated oval (FIGURE 6.15b). A flatter pelvis also requires that the bony pelvis be tipped forward to bring the hip bones to the front of the body; this is achieved by the curvature of the lower spine. The spinal curvature and greater pelvic tilt in individuals with broad pelvises elevates the buttocks and gives a curvy appearance to the profile of women. Conversely, the male pelvis lowers the buttocks and gives a flat appearance to a man's profile. Differences in the architecture of the bony pelvises of males and females affect the geometry of the legs.

In the upper leg, the **femur** (thighbone) has two large processes, the *greater* and *lesser trochanters*, which are places of attachment for thigh and buttock muscles and hip flexors. At the distal end (the end more distant from the body center), the femur has *medial* and *lateral condyles* that articulate with the lower leg. In the lower leg are the **tibia** (shinbone) and the smaller, non-weight-bearing **fibula**. The **patella**, a round bone, is the kneecap.

When you bring your feet together, your femurs extend diagonally to your knees from where they attach at the hips. A broader female pelvis results in femurs farther away from each other at the point of attachment than femurs attached to a narrower pelvis. Therefore, the *Q angle* (formed between the kneecap, the femur, and the line of the tendon from kneecap to shinbone) increases as the breadth of the bony pelvis

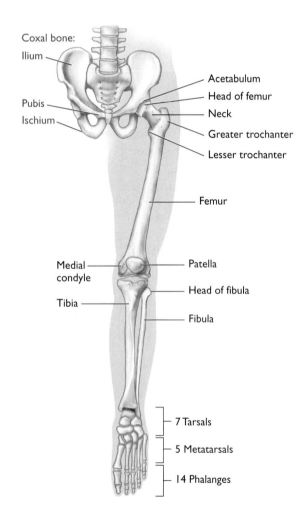

FIGURE 6.14 Pelvic girdle and lower limb bones. Names of bones in the pelvic girdle are in red and those in the lower limbs are in black.

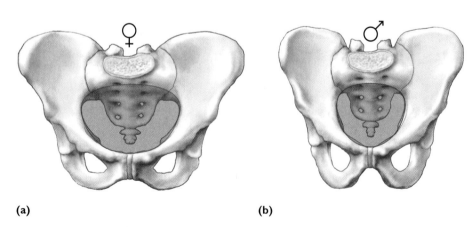

FIGURE 6.15 Bony pelvis. The (a) female and (b) male pelvic inlets differ in shape. The female pelvic inlet evolved to allow the passage of a baby's head during childbirth.

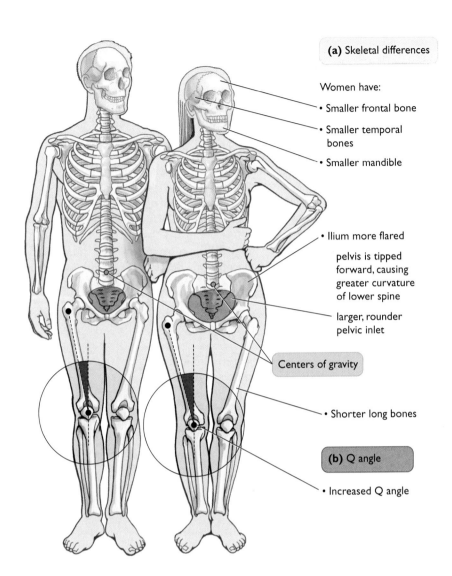

(a) Skeletal differences

Women have:
- Smaller frontal bone
- Smaller temporal bones
- Smaller mandible
- Ilium more flared
- pelvis is tipped forward, causing greater curvature of lower spine
- larger, rounder pelvic inlet

Centers of gravity

- Shorter long bones

(b) Q angle

- Increased Q angle

FIGURE 6.16 Sex differences in the skeleton. Human male and female skeletons are hard to distinguish. (a) Consistent differences are found in the frontal bone, temporal bones, mandible, and the pelvic bones. The female pelvic inlet is larger and rounder than the male pelvic inlet. (b) Q angles in males are typically smaller than Q angles in females. (One side of the Q angle goes up along the femur. The other side is a line extended up from the patellar tendon, which connects the kneecap to the tibia.)

increases. FIGURE 6.16 illustrates this along with the other sex differences associated with the human skeleton.

An increased Q angle has been thought to be a sex-influenced risk factor for knee injury. For instance, knee injuries among basketball players are much more common in women, who generally have greater Q angles than men do. However, many scientists who study sports injuries are quick to point out that social differences in early physical activity and conditioning may have prevented development of the leg-muscle strength that supports the knee-stressing movements typical of basketball. To discern the effects of greater Q angles accurately, studies must compare men and women who have followed similar conditioning and training regimens.

In the foot, the seven **tarsals** form the ankle and heel, and five **metatarsals** form the arches, which together give the foot strength and enable it to act as a lever. The distal ends of the metatarsals form the ball of the foot. The **phalanges** are the toe bones of the foot (see Figure 6.14).

Pectoral Girdle The components of the girdle of the upper extremity, the **pectoral girdle** (FIGURE 6.17), are the shoulder blade, or **scapula**, and the collarbone, or **clavicle**. The head of the **humerus**, the long bone of the upper arm, fits into the **glenoid cavity**, a depression in the scapula. The range of motion of the pectoral girdle, and of the scapula in particular, is greater than that of the pelvic girdle. Along with this greater facility of motion comes a greater risk of dislocation. For this reason, of all joints of the body, the shoulder is most often the site of dislocation. Another injury common to athletes of both

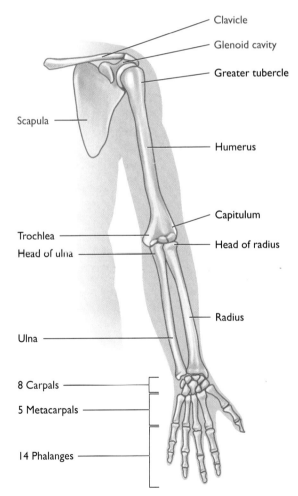

FIGURE 6.17 Pectoral girdle and upper limb bones. Names of bones and some of their features are red for the pectoral girdle and black for the upper limbs.

sexes involves the *rotator cuff.* Rotator cuff injuries are rips and tears to the tendons that extend to the humerus from four small muscles that originate in the scapula.

The forearm, like the lower leg, has two long bones, the **radius** and the **ulna**. Two protuberances of the humerus, the *trochlea* and the *capitulum*, articulate with the ulna and radius, respectively. A large projection at the back of the ulna, called the *olecranon*, forms the lump at the back of the elbow and fits into a depression of the humerus. The radius is shorter than the ulna and has as its most distinctive feature a thick, disk-shaped head. Although attached to the ulna, the head of the radius is free to rotate. As the head rotates, the shaft and outer end of the radius can swing in an arc, allowing the hand an extensive range of motion.

The skeleton of the wrist consists of eight small **carpal bones**, which are arranged in two rows of four each. From these the **metacarpals** and phalanges fan out to form the scaffold for the palm and fingers.

The hand is able to make a wide range of fine movements. The thumb in particular aids these movements in its *opposable* motion. Because the thumb can be brought across, or opposed to, the palm and thus can touch the tips of the fingers, humans can handle a wide variety of instruments with great precision. The next section will help you gain an appreciation for some of the types of movements other joints can undergo.

Joints and Movement

Bones are joined to each other at joints. Some joints are immovable or have little range of motion. For instance, the joints between the cranial bones, called *sutures*, are immovable, and the cartilaginous joints between vertebrae undergo very limited movements. Other joints allow a large range of motion. In the **synovial joints**, bones are separated from each other by a fluid-filled cavity (FIGURE 6.18). The ligaments that hold the joints in place form a *capsule* around the joint. The joint capsule is lined with a synovial membrane that secretes synovial fluid to lubricate the joint. Articular surfaces in these joints are covered with cartilage.

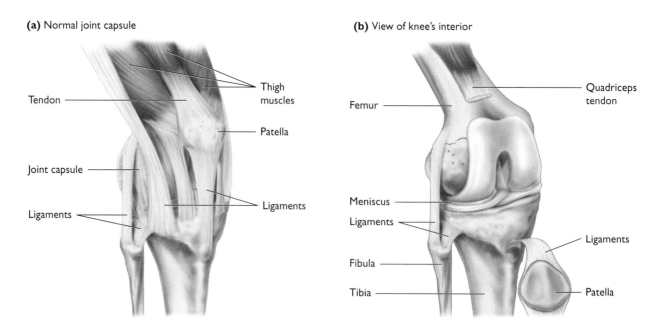

FIGURE 6.18 **A synovial joint.** Synovial joints are freely movable joints. A tough capsule, lined inside with a synovial membrane and the fluid it secretes, envelops the joint. (a) This view of the knee shows the muscles, tendons, and ligaments in their normal positions surrounding the joint capsule. (b) The muscle and some of the capsule are removed to show more of the knee's interior.

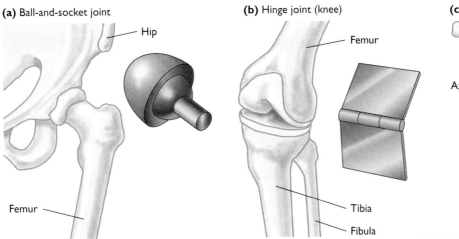
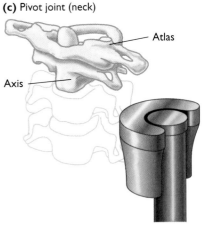

FIGURE 6.19 **Types of joints.** (a) The hip is an example of a ball-and-socket joint. The ball-shaped end of the femur fits into a cup-shaped socket, allowing a wide range of motion. (b) The hinge joint of the knee allows movement in a single plane. Hinge joints are formed when a projection on one bone fits into a depression within another. (c) The pivot joint of the neck allows for freedom of movement of the head.

Synovial joints are categorized by the types of motion they allow. **Hinge joints** such as those in the knee and elbow allow back-and-forth movement similar to opening and closing a door. The **pivot joint** in the neck helps the head to turn from side to side. **Ball-and-socket joints** in the hips and shoulders enable arms and legs to move in three dimensions (FIGURE 6.19).

Some of the movements permitted by synovial joints are illustrated in FIGURE 6.20. **Flexion** decreases the joint angle and **extension** increases the joint angle. Hyperextension is the extension beyond the normal anatomical position. **Adduction** is the movement of a body part toward the midline. **Abduction** is the movement of a body part away from the midline. **Rotation** is the movement of a body part around its own axis and **circumduction** is the movement of a body part in a wide circle.

While differences in skeletal structure between an average male and female can favor females in sports where a lower center of gravity is helpful, the male body has a clear advantage in sports requiring greater strength. A taller, heavier skeletal frame supports more muscle, and muscle mass does affect athleticism.

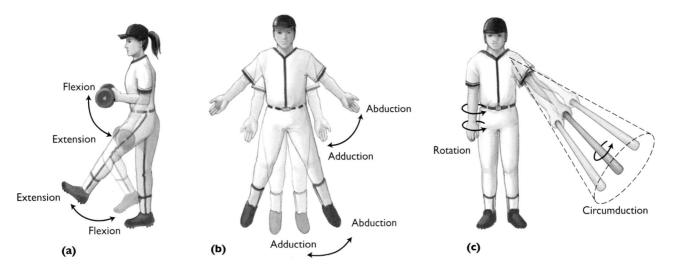

FIGURE 6.20 **Joint movements.** (a) Joint angles decrease during flexion and increase during extension. (b) Adduction is when body parts move toward the middle of the body, and abduction is when body parts move away from the midline. (c) Rotation moves a body part around its own axis, and circumduction moves a body part so that a cone shape is outlined.

6.2 The Muscular System

A major function of the muscular system is to help the body move. The body moves when skeletal muscles shorten, or contract. Skeletal muscle is one of three types of muscle in humans. The other two types are cardiac and smooth muscle. The structure and function of the different types of muscle cells were covered in Chapter 5. The role of skeletal muscles in movement will be addressed in detail here.

Most skeletal muscle attaches to bones at tendons and produces voluntary body movements by interacting with the skeleton. The **origin** of a muscle is on a stationary bone, and the **insertion** is on the bone that moves. Skeletal muscle is nourished by blood vessels that supply oxygen and nutrients, and muscle contraction is controlled by the nervous system.

FIGURE 6.21 outlines one type of muscle movement. Many movements, such as the ability to move the arm to pitch a baseball, require that two muscle groups perform opposite tasks. Contraction of the biceps muscle shortens that muscle while the triceps muscle is relaxed. Likewise, when the biceps muscle is relaxed, the triceps muscle is contracted. If both muscles contracted at once, the arm would be rigid. Therefore, smooth body movement requires that one muscle contracts as the opposing muscle relaxes. Such pairs of muscles, in which each muscle works with a muscle of the opposite effect, are called **antagonistic muscle pairs**.

Names and Actions of Skeletal Muscles

FIGURE 6.22 shows the locations and functions of the major human skeletal muscles. The names of muscles are based on various characteristics such as size (the gluteus *maximus* is large); shape (the *deltoid* is shaped like the Greek letter delta, Δ); location (the latissimus *dorsi* is located at the back or dorsally); number of attachments (the biceps *brachii* has two origins); or action (the *adductor* longus adducts the thigh, or moves the thigh toward the body).

Skeletal Muscle Structure

Recall from Chapter 5 that skeletal muscle cells are cylindrical in shape, striated, and have many nuclei. The microscopic structure of a skeletal muscle involves smaller and

FIGURE 6.21 Antagonistic muscle pairs. Movements are often produced through the actions of two opposing muscles. (a) Contraction of the biceps flexes the forearm, and (b) contraction of the triceps extends the forearm. Tendons attach muscles to bones.

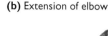

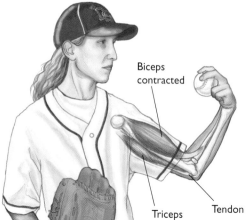
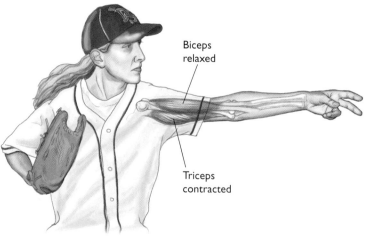

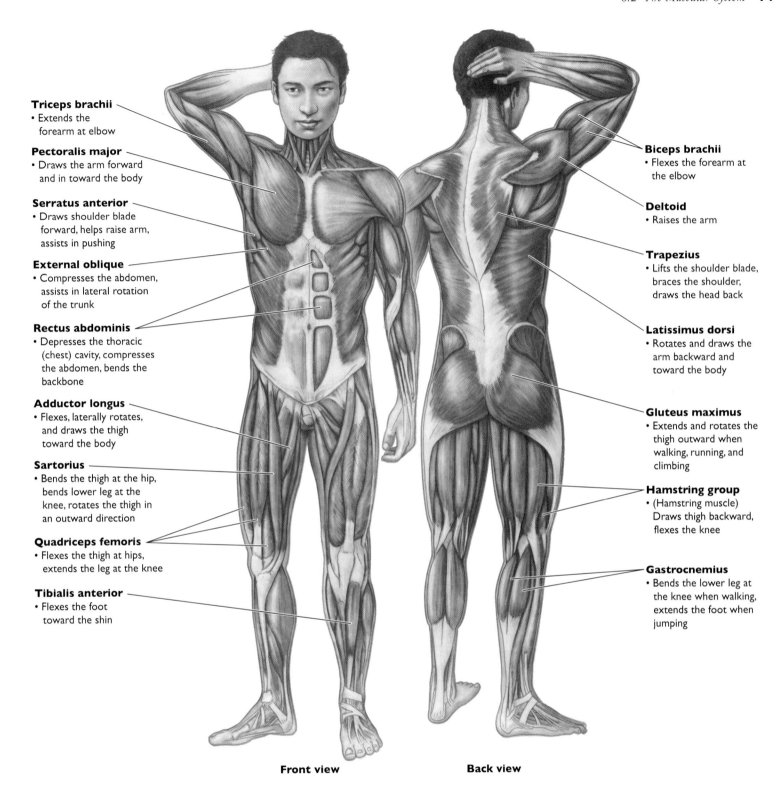

FIGURE 6.22 Human skeletal muscles. Major human skeletal muscles and their functions are depicted.

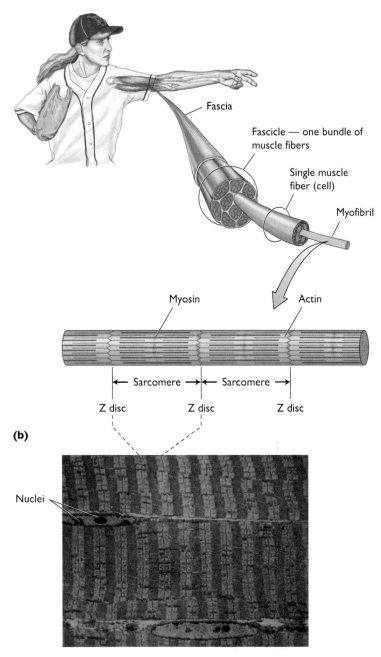

FIGURE 6.23 **Skeletal muscle structure.** (a) Muscle cells (fibers) are arranged in parallel bundles inside the muscle's fascia. The myofibrils inside a muscle cell are linear arrangements of sarcomeres. Sarcomeres are bounded by Z discs. (b) This photomicrograph shows a skeletal muscle cell with its myofibrils in register, producing the characteristic banded or striated pattern.

smaller parallel arrays of filaments (FIGURE 6.23a). A muscle contains bundles of parallel **muscle fibers**. Each muscle fiber is a single cell. Bundles of skeletal muscle fibers are called **fascicles**. Muscles are covered with connective tissue **fascia** that extends beyond the muscle and becomes its tendon to anchor the muscle to the bone.

Within each muscle fiber is a parallel array of threadlike filaments called **myofibrils**. Each myofibril is a linear arrangement of **sarcomeres**, which are the unit of contraction of a muscle fiber. A sarcomere is composed of thin filaments of the protein **actin** and thick filaments of the protein **myosin**. Each myofibril may contain thousands of sarcomeres aligned end to end. The sarcomere is the region between two dark lines, called **Z discs**, in the myofibril. Bands of all myofibrils in a cell are aligned together, or in register, like pipes of the same length stacked neatly with their ends aligned. This pattern gives the skeletal muscle its characteristic striped appearance (FIGURE 6.23b).

Other cellular organelles such as mitochondria are located in the cytoplasm, which in muscle cells is called the **sarcoplasm**. The sarcoplasm also contains glycogen, the storage form of sugar required for muscle contraction. Muscle cells are red because they are rich in the red pigment *myoglobin*, which binds oxygen required for muscle contraction.

Skeletal Muscle Contraction

When a shortstop moves to field a ball, his bones move because the skeletal muscles attached to them shorten as each sarcomere shortens. The contraction of an entire skeletal muscle occurs via the simultaneous shortening of all the sarcomeres in its cells. This contraction is stimulated when a skeletal muscle is activated by a nerve.

Nerves Activate Skeletal Muscles Skeletal muscles are stimulated to contract by special nerve cells called **motor neurons**. These neurons secrete a neurotransmitter called *acetylcholine*. Neurotransmitters are chemicals released by nerve cells that act on another nerve cell or a muscle cell. After the neurotransmitter initiates a response, it must be removed or it will continue to act indefinitely. An enzyme found between nerve and muscle, acetylcholinesterase, breaks down the acetylcholine, stopping the excitatory signal.

A nerve fiber, together with all the muscle fibers it innervates, is called a **motor unit**. The junction between a motor neuron and a skeletal muscle is called the **neuromuscular junction**. When an impulse from the motor neuron arrives at the neuromuscular junction, acetylcholine is released from the nerve and diffuses across the junction to the muscle cell. When acetylcholine binds to the muscle cell membrane, the muscle cell transmits that impulse down the muscle plasma membrane, called the **sarcolemma** in muscle cells. The sarcolemma has extensions called **T tubules** that penetrate the cell and contact the **sarcoplasmic reticulum** (a modified endoplasmic reticulum), which stores calcium ions (FIGURE 6.24).

The arrival of an electrical impulse triggers the release of calcium ions into the cell, where calcium comes in contact with the myofibrils and initiates contraction. Therefore, activation by a nerve increases the concentration of calcium near the contractile proteins. When calcium is absent, no contraction can occur. Likewise, when a muscle cell is no longer stimulated by a nerve, contraction stops.

Calcium initiates contraction through the actions of two other proteins associated with the actin filament (FIGURE 6.25). Threads of **tropomyosin** wind around the actin filament, and **troponin** is attached to the tropomyosin. Calcium ions released by the sar-

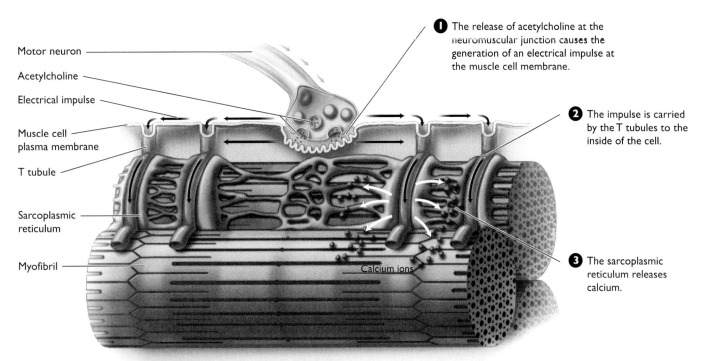

FIGURE 6.24 Nerve activation of a muscle. Nerve activation leads to calcium release.

coplasmic reticulum associate with troponin. This binding changes the shape of troponin. The shape change is relayed to the tropomyosin thread underneath the troponin, causing the tropomyosin thread to shift its position. This exposes the binding sites for myosin.

The actual contraction of a sarcomere takes place through a series of steps outlined below.

Sliding Filaments Shorten the Sarcomere Sarcomere shortening (contraction) involves coordinated sliding and pulling motions within the sarcomere (FIGURE 6.26). Actin filaments are anchored to the Z discs at the ends of a sarcomere. They overlap a parallel, stationary set of myosin molecules that are not attached to Z discs. During contraction, actin filaments slide over the fixed myosin filaments, pulling the ends of the sarcomere with them. The sarcomere ends are brought toward the center when the head of the myosin molecule attaches to binding sites on the actin molecules and pulls toward the center. Contraction of the sarcomere can be likened to the hand-over-hand action of pulling a rope. The stationary hands (myosin) contract and release the sliding rope (actin).

The **sliding-filament model**, or mechanism, involves step-by-step contractions of the sarcomere. First, the high-energy compound ATP binds to the myosin head, causing the myosin head to detach from the binding site on the actin filament. Second, the breakdown of ATP provides energy to the myosin head, and it changes shape or "cocks," like a trigger pulled back, forming a cross bridge between the actin filament and the myosin molecule. Third, the myosin head binds to another site on actin and pulls it along in a short, powerful stroke. Then it releases the actin and reattaches farther along. Ultimately this process serves to pull the Z discs closer together, shortening the sarcomere.

The Genes & Homeostasis feature (on page 146) discusses a condition called muscular dystrophy that arises when muscle cells break down as a result of the forces of contraction.

Testosterone released by the testes at puberty increases the size of muscle fibers. Muscle fibers contain testosterone receptors in their cytoplasm. The presence of testosterone stimulates the cells to increase in mass—in part by retaining dietary nitrogen, which is used to produce more muscle proteins. Therefore, males tend to have greater muscle mass than females. In sports that require brute strength, such as football, men have an advantage, but in sports like baseball, where success depends more on agility and hand-eye coordination, the effect of sex differences is reduced.

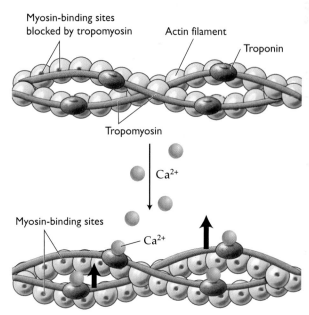

FIGURE 6.25 Calcium binds troponin. Calcium binding to troponin causes the underlying tropomyosin to move. This exposes the sites for myosin to bind to actin, as required for subsequent steps in muscle contraction.

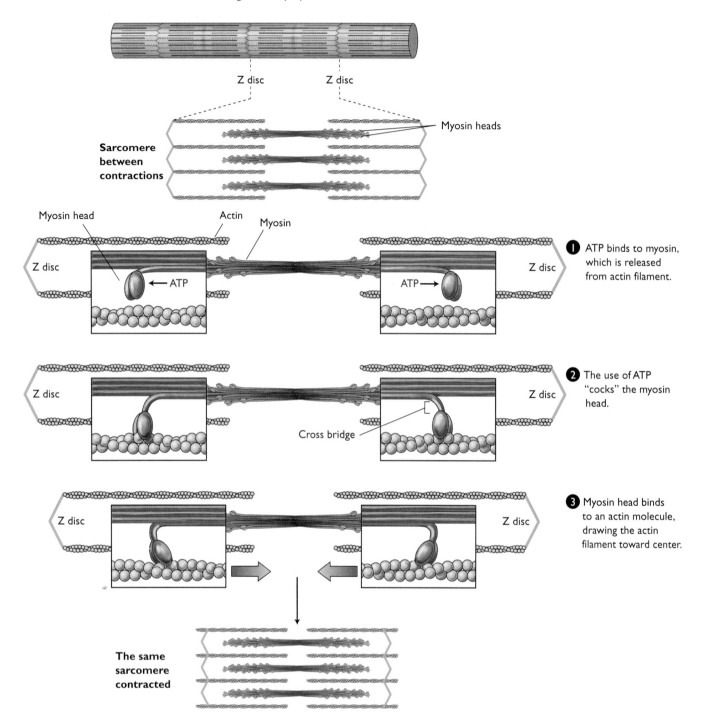

FIGURE 6.26 **Muscle contraction.** The sarcomere of a myofibril consists of actin molecules attached to the Z disc, plus myosin molecules (step 1). Using energy from ATP, the myosin head binds to actin and pulls it toward the center (step 2). Since pulling is happening at both ends of the sarcomere, the sarcomere shortens, allowing contraction of the muscle cell and movement of the muscle (step 3).

Energy Inputs for Muscle Contraction

Every movement of your body requires the use of muscles, which in turn require energy. Keeping skeletal muscles supplied with energy is a bit more complex than supplying energy to other tissues. A resting muscle cell only contains enough ATP to sustain a few seconds of activity. After that, other mechanisms are called upon.

In Chapter 3 you learned that cellular respiration produces ATP. However, this method of ATP production cannot keep up with the demands of an active muscle. To supplement the production of ATP by cellular respiration, skeletal muscle cells evolved a mechanism for synthesizing ATP from another compound, **creatine phosphate**, which has as a part of its structure a high-energy phosphate group (FIGURE 6.27a). This phosphate can be transferred to ADP, producing ATP. Because a resting muscle has much more creatine

Stop and Stretch 6.2

Skeletal muscles can be of different varieties, based on how quickly they use ATP to produce a contraction. Slow-twitch fibers break down ATP and cause contractions more slowly, while fast-twitch fibers break down ATP and cause contractions more quickly.

Slow-twitch fibers contain many mitochondria and are well supplied with blood vessels. They also store oxygen in myoglobin. The presence of myoglobin and many blood vessels give a red color to these muscle fibers, so they are sometimes called "red" muscle. Because they store oxygen, they have a reduced need for oxygen from the bloodstream, enhancing endurance.

Fast-twitch fibers have fewer mitochondria and blood vessels, and they are sometimes called "white muscle." Contractions by fast-twitch muscle are rapid and powerful but cannot be sustained for long. Fast-twitch fibers are more often used to power brief, high-intensity activities such as sprinting.

Do you think athletes who are successful at different sports may differ in their percentages of fast- and slow-twitch muscle fibers? For instance, would world-class marathoners be more likely to have a higher than average percentage of fast-twitch or slow-twitch fibers in their legs? How about world-class weight lifters?

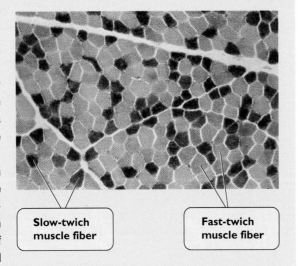

Slow-twich muscle fiber

Fast-twich muscle fiber

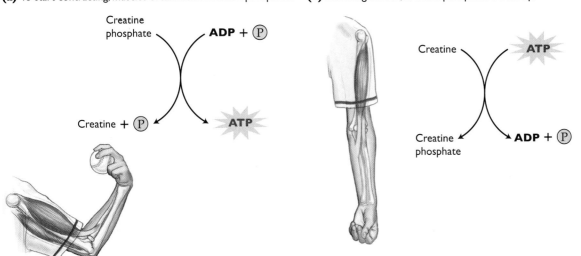

(a) To start contracting, muscles break down creatine phosphate.   **(b)** In resting muscle, creatine phosphate is built up.

FIGURE 6.27 Muscle contraction requires energy. (a) After exhausting the supply of ATP, muscle cells break down stored creatine phosphate to phosphorylate ADP, producing ATP. (b) Resting muscles synthesize creatine phosphate and store it for later use. (Note: The circled P stands for a phosphate group, as in Figure 3.3.)

GENES & HOMEOSTASIS

Muscular Dystrophy

Muscular dystrophy (MD) is a progressive disease that causes muscle wasting. A progressive disease is one in which the symptoms become increasingly more severe. Over time, people with MD become weaker and weaker, leading to problems with balance and coordination.

Mutations to one gene, called the dystrophin gene, are responsible for many cases of MD. When normal, this gene codes for the dystrophin protein, which is very common in skeletal muscle. This protein helps anchor muscle cells to the extracellular matrix that helps to stabilize the cells.

Versions, or alleles, of this gene that are mutant produce proteins that are not able to help stabilize muscle cells. The muscle cells gradually break down as they go through their normal contractions and relaxations, leading to muscle wasting as muscle fibers are replaced with scar tissue. Mutations to different regions of the gene can result in the production of proteins that are less able or more able to perform their anchoring and stabilizing functions. This is why two people with muscular dystrophy may have differing levels of severity of symptoms. For instance, for some people the disease is mild and progresses slowly. For others severe muscle wasting occurs, resulting in confinement to a wheelchair.

You may be aware that this condition is more common in males than in females. This is because the gene that encodes the dystrophin protein is located on the X chromosome. Because females have two X chromosomes in all of their body cells, they have two chances to have a normal allele. Males, with their one X and one Y chromosome, have only one opportunity to receive the normal allele.

---

phosphate than ATP, this process is used to produce ATP during rapid, intense exercise. In addition, this reaction occurs in the midst of the sliding filaments, so the ATP produced can immediately be used to fuel contraction. Hydrolysis of creatine phosphate along with cellular respiration can keep a cell supplied with ATP as long as oxygen is available. Therefore, the stored creatine only holds the cell over for a few additional seconds, until blood cells can deliver more oxygen. If oxygen supplies become diminished, fermentation (Chapter 3) takes over. Once the muscle is at rest, creatine phosphate is rebuilt by the steady source of ATP produced by cellular respiration (FIGURE 6.27b).

A highly trained person, male or female, actually has more mitochondria in his or her muscle cells than someone who does not train. With more mitochondria, more ATP is produced and there is less reliance on creatine phosphate and fermentation for supplying muscles with energy.

Even though the mechanism of muscle contraction works the same way in males and females, males have more muscle mass due to the actions of testosterone. Further, even if a man and a woman had similar amounts of muscle mass, the man would look more muscular because women store more body fat directly underneath their skin.

6.3 The Integumentary System

The **integumentary system**, composed of the skin and accessory organs, protects the body from invasion by harmful pathogens, chemicals, and ultraviolet light. It also prevents water loss from the body and helps regulate temperature.

The skin has two regions, the superficial layer, or epidermis, and the inner layer of dermis. Many different glands originate as outgrowths of these regions. The subcutaneous layer, or hypodermis, is found between the skin and underlying muscle or bone (FIGURE 6.28).

Epidermis

The superficial **epidermis** is the outer, nonvascular layer of skin composed of stratified squamous epithelial cells. The epidermis is composed of many thin, hard layers of cells. The outer 25–30 layers are dead cells. Living cells below this layer actively divide and create new layers that get pushed toward the surface. When the migrating cells get too

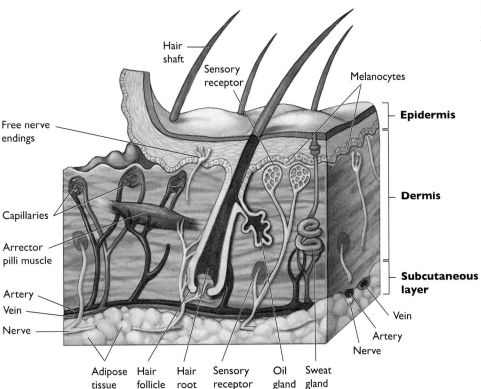

FIGURE 6.28 **Skin anatomy.** The epidermis and dermis constitute the skin. The subcutaneous layer is the hypodermis.

far away from the blood supply in the dermis, they flatten and begin to degenerate. These cells produce *keratin*, a waterproof protein. Specialized cells called **melanocytes** produce the pigment melanin, which absorbs ultraviolet light and gives different humans different skin colors.

Dermis

The **dermis**, located beneath the epidermis, is composed of dense fibrous connective tissue. The dermis is rich in collagen and elastin, and it contains blood vessels that nourish the skin. The dermis also contains sweat glands, hair follicles, and specialized sensory receptors for touch, pain, pressure, and temperature.

Accessory Structures of the Skin

Nails, hair, and the oil and sweat glands are accessory structures formed as outgrowths of the skin.

Nails Nails develop in pocketlike folds of the skin near the tips of fingers and toes. Nails grow from the white, half-moon region of epithelial cells at the base of the nail. These cells produce keratin as they grow out over the nail bed. The cuticle is a fold of skin that covers the nail root.

Hair Hairs, produced only by mammals, start their formation as cylindrical pegs that begin in the dermis and perforate the epidermis. Hair grows in length when new cells are produced by cell division of the epidermal cells that form the root of the hair. Like nails, hair cells produce keratin as they are pushed farther from the root. Contraction of *arrector pili* muscles in the dermis causes the hair to raise up and goose bumps to develop when we are cold.

FIGURE 6.29 Swimmer in streamline position. Fat makes a woman's body smoother and more buoyant.

Sebaceous Glands **Sebaceous glands** are oil-producing glands that grow out from the epithelium that surrounds a hair. The waxy secretions produced by these glands lubricate the hair and the skin.

Sweat Glands **Sweat glands** are present in all regions of the skin. They help regulate body temperature, because sweat absorbs body heat and carries it away as the sweat evaporates. The mammary glands, peculiar to mammals, are specialized sweat glands.

Underneath the skin but above muscle and bone is an additional layer of structures that support the skin and connect it to muscle and bone.

Subcutaneous Layer

The **subcutaneous layer**, or **hypodermis**, is composed of loose connective tissue and adipose tissue. This layer is not actually part of the skin but underlies the skin. Because women store more body fat than men, this layer is thicker in women than in men and tends to give women's bodies and muscles a smoother appearance.

In addition to an absolute difference in the amount of body fat stored, there is also a difference in the location of body fat storage between males and females. At puberty, young women begin to store fat in their torso, abdomen, hips, thighs, and buttocks. This body fat, in addition to skeletal differences, gives women's bodies their curvy shape. Men usually carry most of their fat in the abdomen.

Some evidence suggests that women may metabolize fat differently than men do. Females seem to utilize more fat to produce energy than do males, who tend to use the body's stored sugars more readily. In women, this reliance on fat has the effect of slowing down glucose metabolism, meaning that more sugars are available for prolonged exercise. This difference may result in a greater tolerance for endurance events among women. In fact, women win endurance events such as ultra-marathons as often as men do despite differences in musculature.

The extra body fat that gives women increased endurance may also be a physical advantage in long-distance swimming events. Fat increases buoyancy and enables women to maintain the most energy-conserving streamlined position with less effort (FIGURE 6.29). In addition, fat not only provides increased insulation, which slows the loss of body heat, but also stores energy that can be converted to ATP during endurance events.

What Do Sex Differences Really Mean?

Average differences between males and females can tell us something about athleticism, but they are certainly not the whole story.

To understand the whole story of athleticism and gender, we must evaluate the impact that our culture has on intensifying average differences in athletic performance. Strong cultural forces often determine the amount of time athletes spend practicing and the sports they choose to play, as well as the amount of support they receive for doing so.

Growing up in the 1980s, Ila Borders was prevented from registering for Little League Baseball when she first wanted to play at age 10. In college, where she was the first woman to be given a baseball scholarship, and the only woman on the team, she was taunted by her own teammates as well as by members of the opposing team. In a television interview with Mike Wallace of CBS's *60 Minutes*, Borders spoke of how her own teammates would throw baseballs at her when she had her back turned and how players from opposing teams yelled obscenities at her.

Physical activity is important to overall health, regardless of gender. Studies have shown that children of both sexes who participate in sports are less likely to drop out of high school, smoke, or drink alcoholic beverages.

Children who exercise are also more likely to continue exercising in adulthood than children who do not exercise. In a country where the vast majority of adults do not get the recommended amount of exercise, encouraging everyone to become more athletic makes sense. When good exercise habits are carried into adulthood, risks of heart disease, obesity, diabetes, and many cancers decrease. An additional benefit is lowered cholesterol, and studies suggest that exercise may decrease anxiety and depression. Sex differences that lead to differences in strength, speed, balance, and endurance are meaningless for recreational athletes. What really matters to most athletes is finding an activity they enjoy, so that they can keep healthy for as long as possible.

Stop and Stretch 6.3

Average differences hide the fact that there are a wide *range* of values for each of these characteristics influencing athleticism, and the ranges for males and females typically overlap a great deal. Consider the example of body fat: The average healthy woman has 22% of her body weight in fat, and the average healthy man has 14% body fat. This seems like a large difference, but most males and females fall in the same range (see figure). Healthy women have from 12% to 32% body fat, and the range is 3–29% for men. Most males and females fall within the body-fat range of 12–29%.

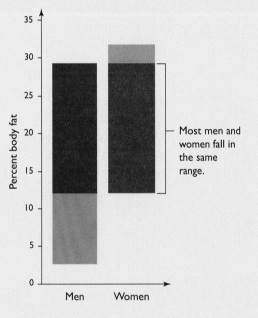

Is it possible to determine a person's gender based on their body-fat percentage? If all you knew about a person was that his or her body-fat percentage was 18%, would you be able to predict whether the person was male or female? Likewise, could you predict a person's body-fat percentage solely on the basis of his or her gender? Look again at the bar graph. The range of body-fat percentages for normal women shows a 20 percentage-point difference from lowest to highest. For men, the range is 26 points. The average difference between men and women is only 6 points. What does this average difference tell us about the differences *within* a group versus the differences *between* the two groups?

Chapter REVIEW

ROOTS TO REMEMBER

The following roots of words come mainly from Latin and Greek and will help you to decipher terms:

-blast refers to an early formation, or to a place where something forms.

-clast is from a Greek verb meaning to break or to break down.

derm is a root derived from the Greek word for skin (the Latin root is **cut-** or **cutan-**).

melan- and **melano-** mean black.

my- or **myo-** means muscle.

-oma is an ending that indicates a swelling, a pathology, or a tumor.

osteo- relates to bones (the Latin form is **os-** or **ossi-**).

pector- relates to the chest.

sarc- or **sarco-** means body.

sub- is the Latin for under (the Greek is **hypo-**).

-ule is an ending for something small, as in tubule.

KEY TERMS

6.1 The Skeletal System
abduction *p. 139*
acetabulum *p. 136*
adduction *p. 139*
appendicular skeleton *p. 134*
axial skeleton *p. 134*
ball-and-socket joint *p. 139*
calcitonin *p. 132*
carpal bone *p. 138*
chrondroblast *p. 130*
circumduction *p. 139*
clavicle *p. 137*
compact bone *p. 128*
cranium *p. 134*
diaphysis *p. 129*
epiphyseal plate *p. 130*
epiphysis *p. 129*
ethmoid bone *p. 134*
extension *p. 139*
femur *p. 136*
fibula *p. 136*
flexion *p. 139*
frontal bone *p. 134*
glenoid cavity *p. 137*
growth plate *p. 130*
hinge joint *p. 139*
hyoid bone *p. 135*
humerus *p. 137*
ilium *p. 136*
intervertebral disk *p. 135*
ischium *p. 136*
lacunae *p. 128*
mandible *p. 134*
maxillae *p. 134*
medullary cavity *p. 129*
metacarpal *p. 138*
metatarsal *p. 137*
nasal bone *p. 134*
occipital bone *p. 134*
osteoblast *p. 130*
osteoclast *p. 131*
osteocytes *p. 128*
parathyroid hormone *p. 132*
parietal bone *p. 134*
patella *p. 136*
pectoral girdle *p. 137*
pelvic girdle *p. 136*
periosteum *p. 129*
phalanges *p. 137*
pivot joint *p. 139*
pubis *p. 136*
radius *p. 138*
red bone marrow *p. 128*
rib *p. 135*
rib cage *p. 135*
rotation *p. 139*
scapula *p. 137*
sinus *p. 134*
skull *p. 134*
sphenoid *p. 134*
spongy bone *p. 128*
sternum *p. 135*
synovial joint *p. 138*
tarsal *p. 137*
temporal bone *p. 134*
tibia *p. 136*
ulna *p. 138*
vertebra *p. 135*
vertebral column *p. 135*
yellow bone marrow *p. 129*
zygomatic bone *p. 134*

6.2 The Muscular System
actin *p. 142*
antagonistic muscle pair *p. 140*
creatine phosphate *p. 144*
fascia *p. 142*
fascicle *p. 142*
insertion *p. 140*
motor neuron *p. 142*
motor unit *p. 142*
muscle fiber *p. 142*
myofibril *p. 142*
myosin *p. 142*
neuromuscular junction *p. 142*
origin *p. 140*
sarcolemma *p. 142*
sarcomere *p. 140*
sarcoplasm *p. 142*
sarcoplasmic reticulum *p. 142*
sliding-filament model *p. 143*
tropomyosin *p. 142*
troponin *p. 142*
T tubule *p. 142*
Z disc *p. 142*

6.3 The Integumentary System
dermis *p. 147*
epidermis *p. 146*
hair *p. 147*
hypodermis *p. 148*
integumentary system *p. 146*
melanocyte *p. 147*
nail *p. 147*
sebaceous gland *p. 148*
subcutaneous layer *p. 148*
sweat gland *p. 148*

SUMMARY

6.1 The Skeletal System

- The skeletal system provides support and protection, allows for movement, and stores minerals and fat. Some bones produce blood cells (p. 128).

- Hardened mineral salts give bones their strength and can be released into the blood as needed. In addition to the nonliving minerals found in bone, bones contain living cells, nerves, and blood vessels (p. 128; Figure 6.1).

- The hard outer shell of bones, compact bone, is composed of cylindrical osteons. Osteocytes embedded in lacunae surround an osteon's central canal. The shaft, or diaphysis, of a bone is compact bone. Spongy bone is the porous, honeycomb-like inner bone found in flat bones and epiphyses of long bones. The epiphysis at the end of long bones contains spongy bone. The periosteum, rich in blood vessels and nerves, surrounds the bone (pp. 129–130).

- Bones develop when the prenatal cartilaginous skeleton is dissolved and replaced with bone cells. Osteoblasts secrete collagen fibers that strengthen the developing bone and facilitate the production of hydroxyapatite. Bone growth occurs from the outside of the epiphyseal plate (pp. 130–131).

- Bones are remodeled throughout one's lifetime. In a cycle of homeostasis, osteoclasts dissolve and release the calcium, helping regulate the level of blood calcium, and osteoblasts rebuild the bone. These cells also help repair fractures (pp. 131–132).

- The 206 bones of the human skeleton are organized into the axial and the appendicular skeletons. The axial skeleton supports the trunk of the body and consists of the skull, hyoid bone, vertebral column, ribs, and sternum. The appendicular skeleton is composed of the bones of the pelvic and pectoral girdles and the lower and upper limbs (pp. 134–138).

- Joints are regions where bones come together. Different types of joints allow for different kinds of movements. The synovial joints are freely movable and permit many types of movements (pp. 138–139).

6.2 The Muscular System

- The three types of muscles are smooth, cardiac, and skeletal. Skeletal muscle cells are cylindrical, striated, and have many nuclei. They form the skeletal muscles that attach to the skeleton. Skeletal muscle is voluntary and contraction is controlled by the nervous system (p. 140).

- Many muscle movements are coordinated by the combined actions of antagonistic pairs of muscles. Each muscle in the antagonistic pair has an opposite effect (p. 140).

- A whole skeletal muscle contains bundles of muscle fibers, or fascicles, and attaches to bones by tendons (pp. 140–142).

- Muscle fibers are bundles of parallel myofibrils. Myofibrils are linear arrays of sarcomeres, which are the unit of muscle contraction (p. 142; Figure 6.23).

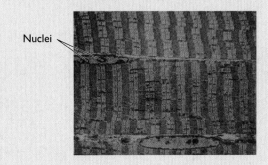

- The sarcolemma is the plasma membrane of a muscle fiber, and the sarcoplasmic reticulum is the endoplasmic reticulum. The sarcolemma forms T tubules that penetrate the cell and contact parts of the sarcoplasmic reticulum that serve as storage sites for calcium ions. Calcium causes changes in troponin and tropomyosin, and thus acts as the trigger for muscle contraction (pp. 142–143).

- Sarcomeres are composed of actin and myosin. Actin filaments attached to the ends of a sarcomere are pulled toward the middle of the sarcomere by the movement of myosin heads. These myosin heads attach to the actin filaments, pull them toward the center of the sarcomere, release the filaments, and reattach farther down the filament (pp. 143–144).

6.3 The Integumentary System

- The skin functions as a protective barrier against abrasion, bacteria, ultraviolet light, and dehydration. It also helps regulate body temperature (p. 146).

- The skin consists of the outer epidermis and the underlying dermis. Most epidermal cells secrete keratin and are pushed to the surface, giving skin its tough outer layer. Melanocytes in the dermis produce pigment that gives color to the skin. The dermis is the dense fibrous connective tissue layer containing epidermally derived glands, hair follicles, nerves, blood vessels, and sensory receptors. The adipose-containing subcutaneous layer lies below the dermis (pp. 146–148; Figure 6.28).

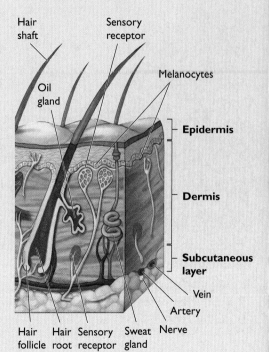

LEARNING THE BASICS

1. Spongy bone is found _____.
   a. in the periosteum
   b. inside lacunae
   c. at the ends of bones
   d. in the diaphysis

2. Bones of the limbs form part of the _____.
   a. appendicular skeleton
   b. axial skeleton
   c. spongy bone
   d. compact bone

3. Osteoblasts _____.
   a. are found in the marrow of bone
   b. regulate bone deposition
   c. regulate bone reabsorption
   d. are more active when calcium is low

4. Which bone in the skull is movable?
   a. frontal
   b. mandible
   c. maxilla
   d. temporal
   e. occipital

5. Which of the following is not a classification of bone shape?
   a. curved
   b. short
   c. flat
   d. round
   e. long

6. Which of the following is not a bone of the appendicular skeleton?
   a. scapula
   b. humerus
   c. clavicle
   d. radius
   e. vertebra

7. **True or False?** Bones can store minerals and fat.

8. **True or False?** The term *phalanges* refers to bones in both the fingers and toes.

9. Match each bone to its location.
   _____ clavicle   a. shoulder blade
   _____ frontal bone   b. forehead
   _____ zygomatic bone   c. cheekbone
   _____ humerus   d. collarbone
   _____ scapula   e. upper arm bone

10. Match each bone or bone group to the feature that appears on it.
    _____ femur   a. olecranon
    _____ scapula   b. acetabulum
    _____ ulna   c. trochanter
    _____ ribs   d. costal cartilage
    _____ pelvic girdle   e. glenoid cavity

11. The antagonistic pairs of arm and leg muscles _____.
    a. allow healthy muscles to compensate for injured ones
    b. allow muscles to produce opposing movements
    c. allow different types of rotations of joints
    d. allow myofibrils to fire in response to different stressors

12. During muscle contraction, _____.
    a. the actin heads pull the sarcomere closed
    b. myosin attaches to actin and pulls it toward the center of the sarcomere
    c. myofibrils shrink due to the actions of testosterone
    d. muscle fibers contract under the actions of testosterone

13. Which of the following is a true statement regarding muscle contraction?
    a. Actin filaments are stationary, and myosin heads can move.
    b. Z discs are pulled toward the center of the sarcomere.
    c. Actin is used up during this process.
    d. No ATP is required for movement to occur.

14. Muscle fibers _____.
    a. are single cells
    b. are aligned in parallel rows
    c. are bundled together to form fascicles
    d. contain myofibrils
    e. all of the above are true

15. The sarcolemma is _____.
    a. the site of storage of calcium ions
    b. found only in cardiac muscle
    c. the plasma membrane of a muscle cell
    d. the endoplasmic reticulum of a muscle cell

16. Label the diagram of a muscle fiber using the following terms: myofibril, sarcomere, sarcoplasmic reticulum, sarcolemma, T tubule

17. A lack of calcium would cause _____.
    a. a strong contraction
    b. an absence of contraction
    c. the sarcolemma to contract
    d. T tubules to regenerate

18. Melanocytes _____.
    a. are more prevalent in light skin
    b. are only found in the hypodermis
    c. produce a pigment that absorbs ultraviolet light
    d. are sebaceous glands

19. **True or False?** The subcutaneous layer lies between the dermis and epidermis and contains accessory organs of the skin.

20. **True or False?** The dermis is a region of connective tissue that contains sensory receptors, nerve endings, and blood vessels.

ANALYZING AND APPLYING THE BASICS

1. Propose a mechanism for bone loss during menopause, keeping in mind the decrease in hormone levels that accompanies the cessation of menstruation.

2. What steps could a perimenopausal woman (between the ages of 35 and 45) take now to protect herself from bone loss during menopause?

3. Shattered or diseased bone may need replacement with a bone graft. In some cases, marine coral is used as a bone graft substitute. The structure and calcium content of some species of coral are compatible with human bone. Coral also does not carry the risks of disease or rejection sometimes encountered with human bone grafts. When implanted, the coral graft allows bone cells to attach and generate new bone. Would the coral in this type of graft remain intact after implantation in a human body? Explain your answer.

4. Why are muscle fibers and their filaments aligned in the same direction?

5. Hypogonadism in males is the inability of the testicles to produce testosterone or sperm. How would untreated hypogonadism affect muscle development during puberty?

6. During exercise, sweat glands work to dissipate heat and control the body's core temperature. Although women have a higher number of sweat glands than men, several studies have shown that men produce more sweat. Men also begin sweating when their core is at a lower temperature. Discuss how these differences between men and women could affect core body temperature and overall athletic performance.

7. The outer protective layers of epidermis are composed of dead, flattened cells. As the body interacts with its environment, these cells are rubbed off. In order for the body to remain protected, what must be true about cells in the lower layers of the epidermis?

CONNECTING THE SCIENCE

1. When it comes to sex differences, the overlap between genders can be quite significant, even in terms of anatomical differences. For instance, the female and male pelvises have different shapes, which can affect how the legs are attached. Yet, according to the classic study on pelvises, only 40–50% of women have a female pelvis as defined in anatomy books. An android (male type) of pelvis appears in 33% of women. How is our ability to make generalizations about the sexes affected by such differences?

2. Provide several examples of cultural influences that sway our views of biological differences between males and females.

Chapter 7

The Digestive System

Weight-Loss Surgery

LEARNING GOALS

1. List the organs of the digestive tract and describe the functions of each.
2. List the different types of teeth and indicate their function.
3. How and where does peristalsis occur?
4. Discuss the mechanism by which food is swallowed and describe how food is prevented from entering the trachea.
5. Identify the secretions produced by the stomach and explain how they aid in digestion.
6. Explain what factors enable the small intestine to serve as the major site of digestion and absorption.
7. Describe the structure and function of the liver.
8. How does the gallbladder aid digestion?
9. In what ways does the pancreas function as both an exocrine gland and an endocrine gland?
10. Explain how homeostasis of blood glucose levels is maintained.

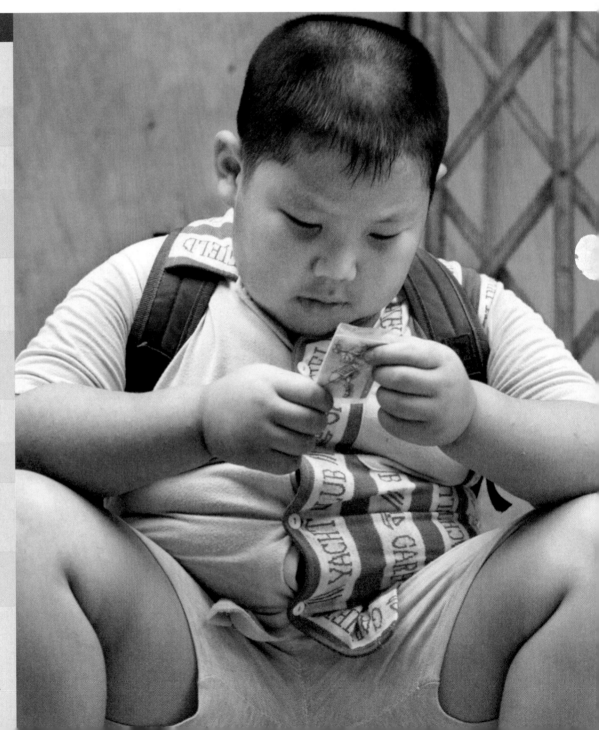

Obesity is on the rise.

We are getting fatter. Obesity has become one of the most common, and dangerous, health problems in the United States. In the past three decades, the number of obese people in the U.S. has doubled, and this trend shows no signs of reversing anytime soon. Most estimates suggest that one in three U.S. adults is obese. And it is not just the adults who are gaining an alarming amount of weight—so are children. Approximately 15% of six- to eleven-year-olds are obese, and studies show that obese children will likely become obese adults.

Along with obesity come severe health consequences including increased risk of diabetes, high blood pressure, joint problems, and certain cancers. In addition, a social stigma befalls the obese. Overweight people are stereotyped as lazy and lacking in self-control. They suffer higher rates of employment discrimination, lower college acceptance rates, and are less likely to marry than normal-weight individuals. The compromised health and quality of life experienced by the obese have led many to turn to what seems like a drastic measure—surgery to help them lose weight. Over the last decade, the rate of weight-loss surgery has increased 10-fold.

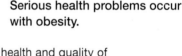

Serious health problems occur with obesity.

Many obese people are going under the knife to help them lose weight.

The results of weight-loss surgery are stunning. Within the first two years of surgery, most people lose 50–60% of their excess weight, and the majority are able to keep the pounds off. Overall health and quality of life also improve. However, as with all surgeries, there are risks, many of which are increased in the obese because their health is already compromised. Infections at the site of the incision, blood clots, and adverse reactions to anesthesia are common. The scariest, however, is the risk of death. Around one of every 100 or so people who lie down on the operating table for this surgery die, either during the operation or within the next 30 days. Those who survive are at lifelong risk for nutritional deficiencies and other surgery-induced complications.

Do the health risks of being obese outweigh the risks associated with weight-loss surgery? To answer this question, you should have an understanding of how the digestive tract normally functions and how its functions are modified by this surgery.

Though often successful, this surgery is also very risky.

7.1 The Digestive Tract

The digestive system consists of organs and glands that work together to break foods into their component parts for reassembly into forms that the body can use, or for use in generating energy. The energy stored in the chemical bonds of food can only be harvested after food is broken down by hydrolysis into successively smaller units. This breakdown occurs as food is passed through a long tube with openings at the mouth and anus, called the **alimentary canal** or **digestive tract** (FIGURE 7.1).

Digestion is the breakdown of food into substances the body can absorb, and involves both mechanical and chemical processes. *Mechanical digestion* includes the crushing, grinding, and churning of food. *Chemical digestion* includes the chemical actions of acids and enzymes. As food is mechanically broken apart into smaller and smaller pieces, more and more surface area is exposed to chemicals and enzymes, and digestion is facilitated.

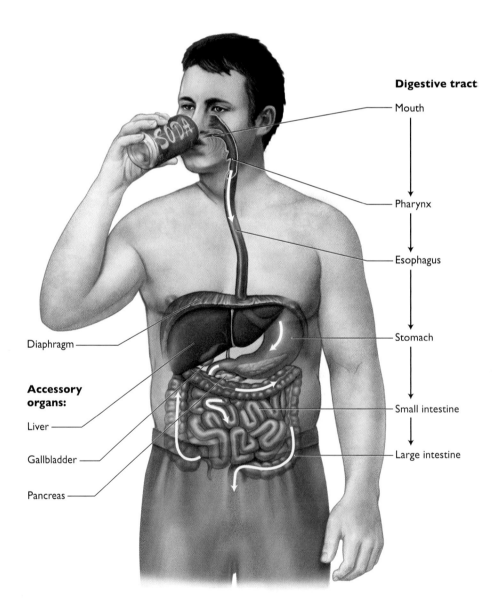

FIGURE 7.1 The digestive system. The digestive system consists of the digestive tract (alimentary canal) and accessory organs. Food moves through the alimentary canal following the path shown.

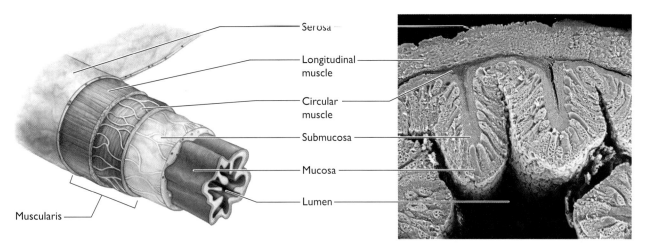

FIGURE 7.2 **The wall of the digestive tract.** Many different tissue types and layers are found in the wall of the digestive tract. The micrograph shows the various layers.

The Wall of the Digestive Tract

From the throat to the anus, the wall of the digestive tract is suited to some part of the job of moving, digesting, and absorbing food. By looking first at the general structure of the wall of the digestive tract, we can save time later by focusing on how it differs in specific organs.

The wall of the digestive tract is composed of several layers (FIGURE 7.2). The *mucosal layer* or *mucosa* lines the central cavity or *lumen* of the digestive tract. The mucosa is a mucous membrane layer composed of epithelia surrounded by connective tissue and then smooth muscle, so it is equipped to reduce and withstand friction. Epithelial cells in the mucosa secrete digestive enzymes, and specialized *goblet* cells secrete mucus.

Outside the mucosal layer is the *submucosa,* a layer composed of connective tissue and containing blood vessels. Lymph nodes in the submucosa are called *Peyer's patches* and, like all lymph nodes, aid in preventing infection.

The *muscularis,* the layer surrounding the submucosa, is composed of an inner circular muscle that encircles the gut and an outer, longitudinally oriented muscle.

A final *serosa* layer makes up the outer surface of the tract and consists of a thin layer of squamous epithelium. The serosa secretes fluids that keep the outer surface of the intestines moist and helps the organs of the abdominal cavity slide against each other.

The Mouth: The Actions of Teeth and Saliva

The breakdown of food begins in the **mouth**, or *oral cavity,* where chewing fractures food into tiny pieces. The teeth grind food, which increases the surface area exposed to enzymes within the mouth and stomach.

The roof of the mouth, or palate, separates the nasal cavity from the oral cavity (FIGURE 7.3a). The palate consists of an anterior **hard palate** containing several bones and a posterior **soft palate** composed of skeletal muscle. The soft palate ends in a fleshy projection called the *uvula,* which rises up to close off the opening to the nasal cavity during swallowing.

The surface of the tongue is covered with sensory receptors called **taste buds** that help you taste the food you eat. The function of the taste buds is described in detail in Chapter 15. Inside the tongue is skeletal muscle that shapes food into a ball called a *bolus* and pushes the bolus toward the back of the throat. **Tonsils** are masses of lymphatic tissue located in the back of the mouth on either side of the tongue that help protect the body against infection by invading microorganisms.

Teeth (FIGURE 7.3b) help us chew food into smaller pieces to both increase the surface area exposed to digestive substances and to ease swallowing. **Incisors** are specialized

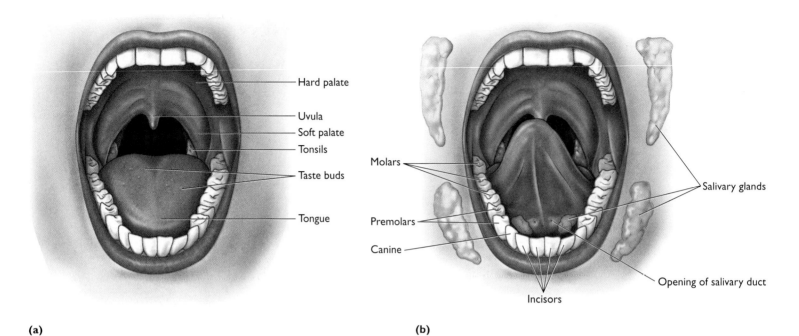

(a)

(b)

FIGURE 7.3 The oral cavity. (a) Food is shaped into a bolus in the oral cavity. (b) Food is ground into smaller pieces by the 32 adult teeth.

VISUALIZE THIS Identify the incisors, canines, premolars, and molars on the top row of teeth. How many of each type of tooth are present in the mouth to give 32 total teeth? Identify the parotid, sublingual, and submandibular salivary glands in this drawing.

for biting off pieces of food. **Canine teeth**, or *cuspids*, are sharp enough to help rip food apart. **Premolars** (also called *bicuspids*) are used for tearing apart food. **Molars** have broad surfaces that enhance grinding. The last molars are sometimes called the *wisdom teeth*. These molars erupt in early adulthood. Because the last molars may grow in crookedly and crowd other teeth or may fail to emerge completely and are thus *impacted*, they often are removed.

Children have 20 teeth that begin to emerge around age two. These are gradually replaced by adult teeth. Typically, adults have a total of 32 teeth.

Enzymes present in the saliva are secreted from three pairs of exocrine glands, called *salivary glands*. The largest pair of salivary glands, the *parotids*, are located in the area of the upper cheeks, just in front of the ears. The *sublinguals* are located beneath the tongue. The *submandibular* glands lie beneath the floor of the oral cavity. Salivary ducts carry saliva to the mouth from these glands. Saliva contains the enzymes salivary amylase and lingual lipase, which begin the breakdown of starch and lipids, respectively, within the mouth. Saliva also moistens food, making it easier to swallow, and kills bacteria in the mouth.

Structurally, each tooth has a visible region called the **crown** and a region below the gums called the **root** (FIGURE 7.4). The crown has a layer of **enamel**, a hard outer shell composed of calcium and phosphate compounds; a living layer of **dentin**, which is bonelike in structure; and an inner **pulp cavity** containing nerves and blood vessels. (When a *root canal* is performed, a dentist removes diseased nerves and blood vessels from the pulp cavity.) Each tooth sits in the jawbone in a socket lined with the *periodontal membrane*, which attaches the tooth to the bony socket.

Bacteria in the mouth subsist on food left between the teeth. Acids produced as by-products of bacterial metabolism can dissolve tooth enamel to cause *dental caries*, or *cavities*. When left untreated, cavities can deepen, eroding the dentin and pulp cavity, which can be very painful. As teeth decay, the surrounding gum tissue can become inflamed, a condition called *gingivitis*. If inflammation of the gums spreads to the periodontal membrane, *periodontitis* can result in the loss of bone and loosening of the teeth from the jawbone. Periodontitis is the main cause of tooth loss in adults. Stimulation of the gums by flossing can help prevent this condition. Flossing can also remove bacteria and food remnants between teeth, thereby decreasing risk of infection.

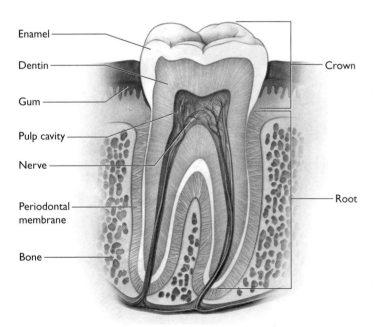

FIGURE 7.4 The structure of a tooth. The crown of the tooth is above the gums and the root lies below the gums.

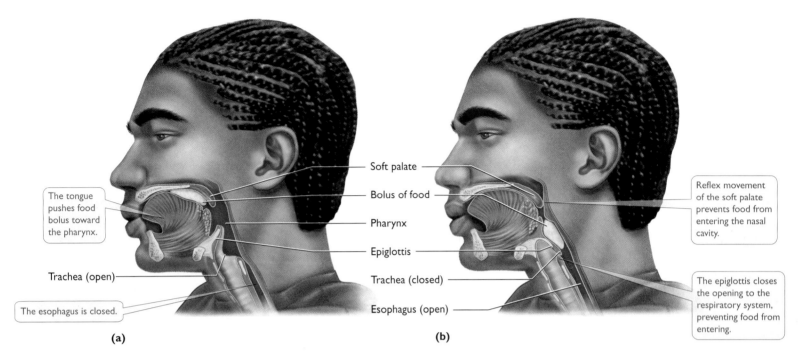

FIGURE 7.5 **Swallowing.** Swallowing begins as the tongue shapes the ball of food into a bolus and pushes it into the pharynx. (a) Before swallowing, the esophagus is closed, and the trachea and upper pharynx above the soft palate are open. (b) During swallowing, the soft palate rises, which closes off the passage to the nasal cavity, and the epiglottis closes off the trachea and opens the esophagus.

The Pharynx and the Esophagus: Transport to the Stomach

When you swallow chewed food (FIGURE 7.5), it moves from your mouth to your **pharynx**. The pharynx, forming the back of your throat, branches into the *trachea* (or windpipe, which leads to the lungs) and the **esophagus** (a long muscular tube that leads to the stomach). Therefore, food and air passages meet in the pharynx. A flap of cartilage called the **epiglottis** keeps swallowed food from entering the trachea (FIGURE 7.5b). If food does enter the trachea ("goes down the wrong way" or "wrong pipe"), coughing will usually force it out.

No chemical digestion occurs in the esophagus. The esophagus functions simply to transport the bolus of food to the stomach. The movement of food is accelerated by **peristalsis**, rhythmic waves of contractions in the muscularis layer. Peristalsis occurs throughout the digestive tract from the esophagus on, providing a constant movement of material through the digestive system.

Movement of food and secretions is also controlled by bandlike muscles called **sphincters** that encircle tubes and act as valves. When a sphincter contracts, the tube it encircles is closed off. When it relaxes, the tube is allowed to open. At the juncture where the esophagus joins the stomach, the *gastroesophageal sphincter* relaxes to allow food to pass into the stomach and contracts to prevent the acidic contents of the stomach from moving backward into the esophagus.

Heartburn, also known as *gastroesophageal reflux disease (GERD)*, is a condition in which acidic digestive juices from the stomach do flow backward (reflux) into the esophagus, causing inflammation and irritation of the cells that line this tube. Many people experience periodic bouts of heartburn, but in chronic sufferers the gastroesophageal sphincter fails to prevent backflow. Symptoms include a burning pain in the upper abdomen and a sour taste in the back of the throat. Over-the-counter antacids can be used to treat sporadic bouts of heartburn. Chronic heartburn should be medically treated, because open sores called *ulcers* can develop in the esophagus from exposure to acid.

The Stomach: Digestion in an Acid Bath

The **stomach** (FIGURE 7.6a) is a large muscular sac located on the left side of the body beneath the diaphragm. The typical stomach expands when full to accommodate several liters of food and drink and empties in two to six hours, depending on what was eaten.

FIGURE 7.6 The stomach. (a) The wall of the stomach is thick and allows for expansion after ingestion of food and drink. (b) This photomicrograph shows the gastric glands that line the stomach. These glands secrete mucus and a gastric juice that aids in protein digestion. (c) A bleeding ulcer in the stomach lining.

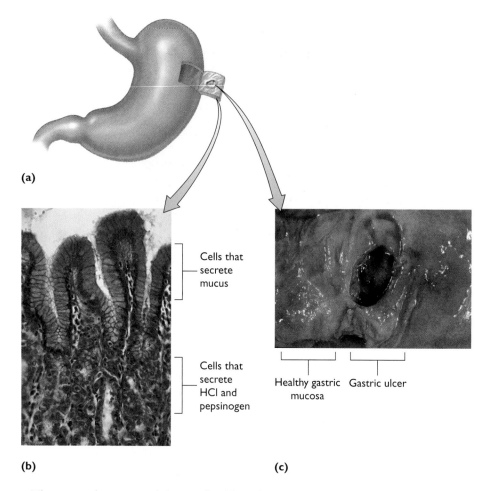

The stomach stores and digests food but does not absorb most food. Caffeine, aspirin, and alcohol are absorbed directly from the stomach. Drinking alcohol when your stomach is full will prevent you from feeling the effects of alcohol as quickly as if you drink alcohol when your stomach is empty. In a full stomach, food can absorb some of the alcohol, leaving less to be absorbed into the bloodstream.

The stretching of the stomach as food enters it serves as a signal for the peristalsis that began in the esophagus to increase in intensity. Stomach contractions force its contents against the walls, and like the kneading of dough, the contents are mixed together into a paste-like consistency. This mixing continues the process of mechanical digestion.

Chemical digestion in the stomach begins when *gastric juice* is secreted by cells of the gastric glands embedded in the stomach wall (FIGURE 7.6b). Gastric juice contains hydrochloric acid (HCl). HCl causes the stomach to have a highly acidic pH (around pH 2), which helps kill most bacteria that may be present in the food. HCl also breaks down connective tissue in meats you have eaten. Also present in gastric juice is *pepsinogen,* a molecule that is modified to form the enzyme **pepsin** once exposed to HCl in the stomach. Once in this active form, pepsin can digest proteins. A thick layer of protective mucus secreted by columnar epithelial cells lining the stomach prevents the stomach itself from being digested.

If HCl does penetrate the mucus, the wall of the stomach can break down and a stomach, or *peptic,* ulcer can develop (FIGURE 7.6c). It used to be thought that spicy food or stress alone caused ulcers, but it is now known that peptic ulcers are associated with infection by the bacterium *Helicobacter pylori.* Treatment with antibiotics can kill the bacteria causing the infection, and antacids can reduce damage caused by stomach acids, allowing the ulcers to heal.

Partially digested food and gastric juice mix to produce a creamy mass called **chyme**. Chyme is slowly squirted into the small intestine by the *pyloric sphincter,* a small circular muscle located at the junction between the stomach and small intestine. As the acidic chyme enters the small intestine, its acidity is neutralized by basic secretions from the pancreas to prevent damage to the intestinal wall.

The Small Intestine: Where Most Digestion Happens

The tubelike **small intestine** of an adult is about 6 meters long, or close to 20 feet. In spite of its tremendous length, the small intestine is named for its small diameter as compared to the diameter of the large intestine.

Recall that carbohydrate and lipid digestion begin in the mouth and protein digestion begins in the stomach. Even though digestion begins earlier, the small intestine is the main site of chemical digestion. The small intestine is also the primary site for absorption of nutrients into the bloodstream.

The small intestine consists of three distinct regions. The **duodenum**, closest to the stomach and about 10 inches long, is the site of most digestion. Absorption occurs in the adjacent **jejunum** and **ileum**. The ileum, at around 11.5 feet, is considerably longer than the jejunum at 8.2 feet.

The wall of the small intestine is folded on itself (FIGURE 7.7a), and the folds themselves are covered with fingerlike projections called **villi** (singular: villus). The multicellular villi enhance absorption of the products of digestion by increasing the small intestine's surface area (FIGURE 7.7b). At the center of each villus are capillaries and a

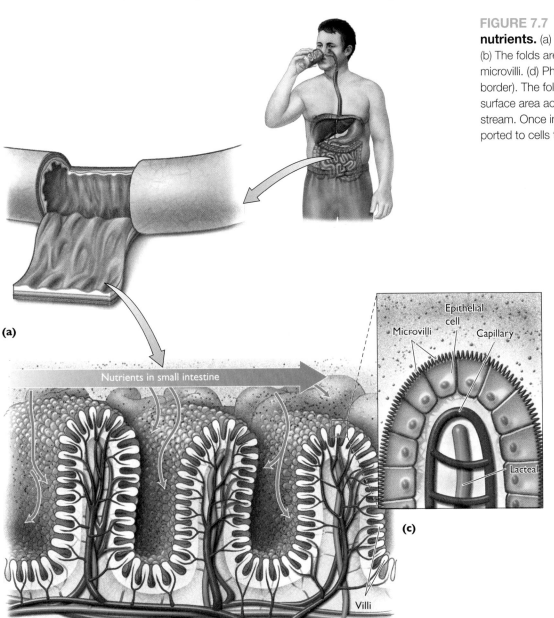

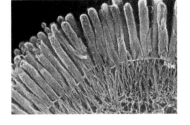

FIGURE 7.7 **The small intestine absorbs nutrients.** (a) The wall of the small intestine is folded. (b) The folds are covered in villi. (c) Villi are covered in microvilli. (d) Photomicrograph of intestinal microvilli (brush border). The foldings of the small intestine increase the surface area across which nutrients can enter the bloodstream. Once in the bloodstream, nutrients can be transported to cells throughout the body.

lacteal (lymph capillary) that transport nutrient subunits to larger blood and lymphatic vessels.

Surface area inside the small intestine is increased even further because each epithelial cell of a villus is covered with small projections of the plasma membrane, called **microvilli** (FIGURE 7.7c). Microvilli give the mucosal surface a velvety appearance sometimes called the *brush border* (FIGURE 7.7d). Together, the intestinal folds, villi, and microvilli increase the surface area of the small intestine about 500 times, to a total area that is larger than that of a tennis court. Increasing surface area allows for increased absorption of nutrients across cell membranes.

Once digested, nutrients are absorbed across the intestinal wall into the circulatory and lymphatic systems. From there, the individual subunits of nutrients enter cells where they can be used as building blocks for the synthesis of cellular structures or be metabolized to produce energy (FIGURE 7.8).

Most of the enzymes that facilitate digestion in the small intestine are produced by accessory organs, described in the next section of the chapter. Cells in the small intestine do make the enzyme lactase, which hydrolyzes lactose, a common sugar found in milk, into monosaccharides that can be absorbed by the bloodstream.

Lactose intolerance is a common dietary problem caused by a genetic deficiency in the production of lactase. People with lactose intolerance are unable to digest large amounts of lactose. When not enough lactase is available to digest lactose, bacteria in the intestine metabolize the sugar, producing lactic acid. The buildup of lactic acid causes bloating, gas, cramps, and diarrhea. Lactose intolerance can be treated by taking dietary supplements that contain the lactase enzyme or by limiting consumption of dairy products.

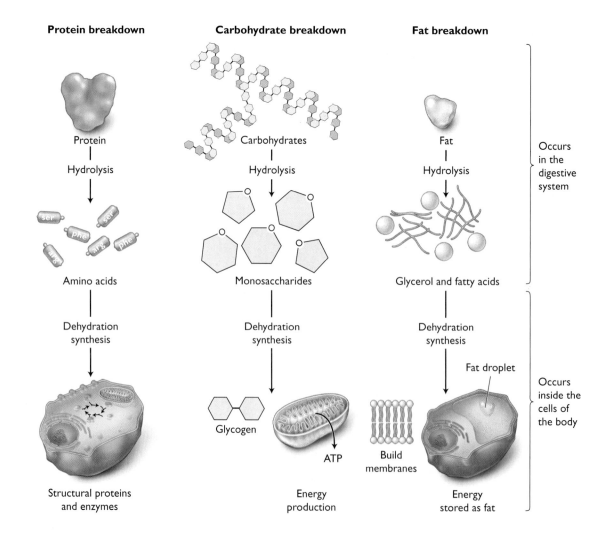

FIGURE 7.8 You are what you eat. Food is digested into subunits of nutrients that are used to synthesize cellular structures and to help generate energy in the form of ATP.

Stop and Stretch 7.1

Children are rarely lactose intolerant, but many people become so as they age. Devise a genetic hypothesis to explain why more adults than children are lactose intolerant.

Regulation of Digestive Secretions

The secretion of digestive juices is regulated hormonally (FIGURE 7.9). After a meal, the stomach produces the hormone *gastrin*, which circulates through the bloodstream and stimulates the upper part of the stomach to produce more gastric juices, thus facilitating digestion.

The duodenum of the small intestine secretes the hormones *secretin* and *cholecystokinin (CCK)*. HCl from the stomach stimulates the release of secretin, and partially digested proteins and fats stimulate the release of CCK. Once in the bloodstream, secretin and CCK cause the pancreas and gallbladder to increase their output of digestive juices.

The Large Intestine: Absorption and Elimination

Materials that are not absorbed by the small intestine are passed into the **large intestine**. The large intestine is shorter than the small intestine (approximately 1.5 meters, or 5 feet, in length) but is wider in diameter and has a smooth inner wall. The primary functions of the large intestine are the absorption of water, salts, and some vitamins and the formation, storage, and lubrication of **fecal matter** until it can be expelled. Fecal matter (feces) is composed largely of water along with fiber, bacteria, and other indigestible materials.

The large intestine consists of four regions: the cecum, colon, rectum, and anus (FIGURE 7.10). The **cecum** is the first part of the large intestine. It is shaped somewhat like a pouch, and a small fingerlike projection called the **appendix** dangles from it. The appendix is a vestigial organ, meaning that it has lost much of its function during evolution. In mammals that eat large amounts of plant matter, this organ houses bacteria that aid in the digestion of this tough material. In humans, the appendix may play a role in fighting infection.

Appendicitis is inflammation of the appendix, caused by its infection. When a person has an attack of appendicitis, the appendix may be surgically removed to prevent it from bursting and releasing bacteria into the abdominal cavity.

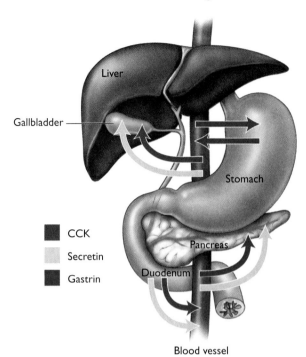

FIGURE 7.9 **Hormonal control of the digestive system.** After eating, the hormone gastrin enters the bloodstream, which stimulates the stomach to produce gastric juice. Secretin and CCK from the small intestine cause the pancreas and gallbladder to increase their secretion of digestive substances.

Stop and Stretch 7.2

No known side effects are associated with a missing appendix after its removal. How does this observation support the hypothesis that the appendix is vestigial? Does the lack of noticeable side effects "prove" that the appendix is nonfunctional?

The **colon** connects to the cecum and continues vertically up the right side of the abdomen (ascending colon), horizontally across to the left side (transverse colon), vertically down the left side (descending colon), and then loops (at the sigmoid flexure, or sigmoid colon) to join the terminal portion of the large intestine, the **rectum**. The small and large intestine are sometimes referred to together as the *bowels*.

FIGURE 7.10 **The large intestine.** The anatomy of the large intestine includes the cecum, colon (ascending, transverse, descending, and sigmoid), rectum, and anus.

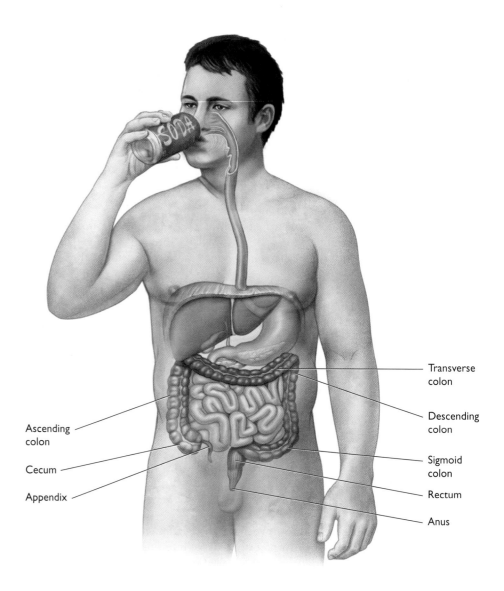

The colon can be the source of many debilitating diseases and conditions. *Polyps* are small growths that develop from the epithelial lining. These can be detected by a small, flexible, lighted microscope called an *endoscope*. An endoscope can be inserted into a hollow organ such as the rectum or colon to provide a view of its interior. Polyps are typically removed to prevent them from developing into cancers.

Colitis and *Crohn's disease* are called inflammatory bowel diseases because they stem from chronic inflammation of the large intestine. These conditions result in abdominal pain and bloody diarrhea and can cause fatigue and weight loss. Colitis usually affects only the innermost lining of the large intestine and rectum, whereas Crohn's disease can occur anywhere in the large intestine and is more likely to spread deep into the layers of affected tissues. It is not known what causes these conditions. Treatments include pharmaceutical agents that decrease inflammation, and sometimes the surgical removal of the affected part of the colon.

Diverticulosis is a disorder that occurs when small pouches, called *diverticula,* arise in the wall of the large intestine. When fecal matter is pushed into the pouches, a bulge in the wall of the large intestine can occur. This condition can progress to *diverticulitis* when the feces-filled sacs become infected and inflamed, causing abdominal pain. If diverticula burst, massive blood loss can occur.

When the undigested fibrous material that makes up fecal material builds up in the rectal cavity, nerves in the rectal wall are stimulated. These nerves send a message to the brain making the individual conscious of the need to eliminate this solid waste, that is, to defecate. Fecal material exits the body through the sphincter at the end of the digestive tract, the **anus**.

Hard, dry stools that are difficult to eliminate cause the bloating and pain in the abdomen associated with *constipation*. Exercise, eating high-fiber foods, and drinking lots of water can help treat and prevent constipation. The loose, watery stools and abdominal pain that accompany *diarrhea* can be caused by intestinal infection or by stress. When pathogenic bacteria, viruses, or other parasites are present in the intestine, the intestinal wall becomes irritated and peristalsis increases. The increased rate of peristalsis causes less water to be absorbed by fecal matter. Stress can also drive the intestines out of homeostasis because the nervous system stimulates the intestinal wall and thereby increases peristalsis.

Now that you have an understanding of the digestive tract we can turn our attention to the manner in which this is altered during gastric bypass surgery.

Gastric Bypass Surgery: Scaling Back Digestion

For some reason, people who eat well past the point of meeting the body's basic needs are not responding to normal homeostatic mechanisms that tell them to stop eating when they are full. Gastric bypass surgery involves decreasing the size of the stomach so that it will be full even when very small amounts of food are consumed.

Gastric bypass surgery is designed to allow food to bypass the major sites of digestion and absorption, the stomach and duodenum (FIGURE 7.11). To bypass these organs, a surgeon creates a small pouch from the upper portion of the stomach. This tiny pouch, about the size of a walnut, serves as the new stomach, now so small that it feels full after ingesting just a few ounces of food. The duodenum, the first part of the small intestine, is also bypassed, further decreasing digestion and absorption, and the small pouch is attached instead to the jejunum. The bypassed portion of the stomach is not removed during this surgery. Instead, the stomach is sewn shut and remains in the abdomen.

Many complications can arise after gastric bypass surgery. Infection at the site of the surgery is always a risk, as with any surgery. Likewise, if either the new or the original stomach does not heal properly, leakage of stomach acids into the abdomen can corrode abdominal organs. Many individuals who have had this surgery cannot eat sweet foods without feeling weak or faint. Others may suffer from a condition called *dumping syndrome* in which the contents of the small stomach pouch move too rapidly into the small intestine, causing nausea, weakness, sweating, and diarrhea.

In many cases, nutritional deficiencies develop as a result of limited nutrient absorption. For example, iron and vitamin deficiencies may cause *anemia*, a reduction

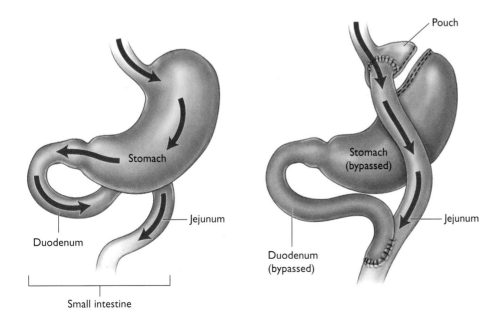

FIGURE 7.11 **Gastric bypass surgery.** Before gastric bypass, food enters the stomach and moves to the duodenum of the small intestine (left). After this surgery, food is redirected and bypasses most of the stomach and the duodenum, flowing instead into the jejunum, which limits the absorption of nutrients (right).

GENES & HOMEOSTASIS

Obesity

Storing fat on the body is thought to have evolved as a mechanism to protect early humans against times of famine. Today, most Americans don't have to worry about the scarcity of food. In fact, overabundance of food is more of a concern. When overeating is combined with a lifestyle that is far more sedentary than that of our ancestors, obesity is the expected result. However, not all people who overeat and are inactive become obese. Are these differences between individuals the result of genetic differences?

It is thought that around 1% of all cases of obesity are caused by mutations to single genes. Scientists studying obesity now know that mutations to any of six different individual genes can result in obesity. The most common of these mutations involves a gene that codes for a protein called the melanocortin-4 receptor. Stimulation of the normal version of this receptor—found in the hypothalamus of the brain, a region of the brain known to control eating behaviors—inhibits eating. When the gene that encodes this receptor is mutated, the receptor is not able to be stimulated and eating continues well past the point of satiety.

In the laboratory, scientists are able to disrupt the gene that encodes this receptor in mice. Mice with this nonfunctional version of the receptor do not respond to signals to stop eating. When these mice are given a chemical substance that can bind to the receptor, they do stop eating. If scientists were able to devise a pharmaceutical agent to stimulate the mutated receptor in humans, obesity caused by this mutation might be prevented.

The remaining cases of obesity are caused by a combination of poor eating habits and/or mutations to many different genes or other poorly understood factors.

in the number of oxygen-carrying red blood cells. One particular type of anemia, *pernicious anemia,* is more common after gastric bypass surgery. The cells that line the stomach produce a protein called intrinsic factor that helps the body absorb vitamin B_{12}. Blood and nerve cells require vitamin B_{12} to function properly, and deficiencies result in fatigue, shortness of breath, tingling sensations, and diarrhea. Even though the stomach is not removed, its ability to produce intrinsic factor is compromised after this bypass surgery.

The Genes & Homeostasis feature discusses some of the genetic causes of obesity.

Stop and Stretch 7.3

Ghrelin is a hormone produced by cells that line the stomach and is thought to trigger appetite. Scientists attempting to discover whether individuals who have had gastric bypass surgery have lowered ghrelin levels have tested the levels before and after patients had gastric bypass surgery. Scientists also tested ghrelin levels in non-obese patients who were having abdominal surgery for other reasons. Why was it important to test ghrelin levels in the non-obese patients as well as the obese patients?

7.2 Three Accessory Organs of the Digestive System

The liver, gallbladder, and pancreas are considered to be accessory organs of the digestive system because they are located outside the alimentary canal yet produce or secrete substances required for digestion. FIGURE 7.12 shows how the *common hepatic duct, common bile duct,* and *pancreatic duct* transport secretions to the duodenum of the small intestine from the liver, gallbladder, and pancreas, respectively.

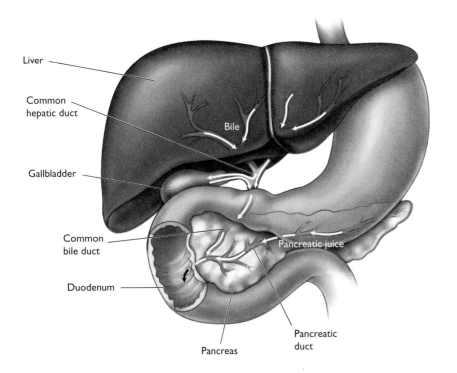

FIGURE 7.12 **Accessory organs of the digestive tract and their associated ducts.** The common hepatic duct transports bile from the liver to the gallbladder. The common bile duct transports bile from the gallbladder to the duodenum of the small intestine. The pancreatic duct transports digestive enzymes from the pancreas to the duodenum.

The Liver

The **liver** is a large, reddish-brown organ found on the right side of the abdominal cavity below the diaphragm. It is about the size of a football, weighs around 1.8 kilograms (4 pounds), and is divided into four lobes. Lobes are further subdivided into **lobules** (FIGURE 7.13). Cells of the lobules secrete substances into the blood and filter substances out of the blood. Three structures are found between lobules: (1) a bile duct that takes bile away from the liver, (2) a branch of the hepatic artery that brings oxygen-rich blood to the liver, and (3) a branch of the hepatic portal vein that transports nutrients from the intestines. A central vein inside the lobule receives blood that has percolated through the lobule.

The liver produces **bile**, a yellowish-green by-product of blood metabolism. Bile has a yellow-green color because it contains a pigment called **bilirubin**, which is produced by the breakdown of hemoglobin when red blood cells are recycled. Bile contains salts that help dissolve fats. Bile salts cause **emulsification**, a process that prevents fats from clumping up together and helps mix them with water. Fats that have been emulsified are broken into smaller particles and thus have more exposed surface area than does a large clump of fat. Bile produced by the liver is secreted through the hepatic ducts and into the small intestine. Thus, secretion of bile into the small intestine facilitates digestion because smaller, emulsified fats are easier for enzymes to digest.

Hepatic portal veins carry nutrients from capillary beds in the intestines to capillary beds in the liver, where the nutrients can be further metabolized or stored. Fats entering the liver from the intestine are converted into fatty acids and carbohydrates and transported by the blood to the tissues. Glucose monomers are joined to each other to produce a polymer of glucose called glycogen. The liver removes glucose from the blood when glucose levels are high, such as after a meal, and stores them as glycogen. When a meal has not been eaten recently, blood-glucose levels fall, and the liver hydrolyzes bonds between glucose molecules in glycogen and releases glucose monomers into the bloodstream. In this manner, glucose homeostasis is maintained.

In addition to its digestive functions, the liver also performs filtering. As blood flows through the liver, it reaches liver cells called *hepatocytes*. In addition to producing bile and performing other functions for the liver, hepatocytes remove toxic materials, dead cells, pathogens, drugs, and alcohol from the bloodstream.

FIGURE 7.13 Liver lobule. Liver lobules filter substances from the blood for processing by liver cells. The bile duct transports bile out of the liver. Each lobule has a large central vein that drains the liver of blood. A branch of the hepatic artery brings oxygen-rich blood to each lobule. A branch of the hepatic portal vein transports nutrients from the intestines.

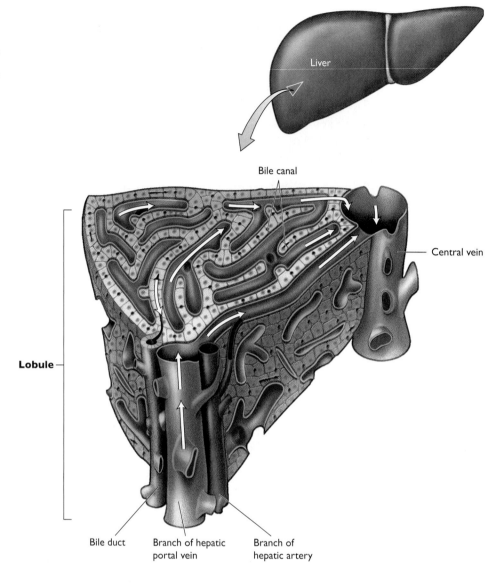

In fact, the liver gets its color from the pint of blood that it is typically filtering. Each pint of blood being filtered by the liver contains substances from the food you eat and drink as well as substances you breathe in and absorb through your skin. The liver metabolizes toxins and secretes their by-products into the bloodstream, where they are carried to your kidneys to be excreted in urine or to the intestines, where they will be excreted in feces.

The liver also manufactures many chemicals, including the cholesterol found in cell membranes. The liver manufactures blood plasma proteins, such as clotting factors and proteins that carry oxygen in the blood. It also synthesizes some immune proteins that help fight infection and plays a role in the removal of dead or dying red blood cells.

The liver is a unique organ in that it can retain functions even if a large majority of its cells are diseased. It can regenerate itself within a few weeks. However, in spite of its remarkable regenerating abilities, the liver can be damaged past the point of repair. Alcohol and drug abuse and infection by certain viruses can cause irreparable liver disease called *cirrhosis* in which healthy liver tissue is replaced with scar tissue.

The Gallbladder

The **gallbladder** is a greenish sac attached to the bottom of the liver. The gallbladder stores excess bile produced by the liver. It also concentrates bile. When needed, bile is released through the common bile duct and into the duodenum of the small intestine to help emulsify any fats present. Salts contained in bile sometimes come out of solution and form crystals that enlarge to form *gallstones* in the gallbladder or its associated ducts. This painful condition often requires that the gallbladder be surgically removed.

The Pancreas

The **pancreas**, located behind the stomach, is a long, irregularly shaped gland. Many of the digestive enzymes used in the small intestine are produced and secreted by the pancreas. The pancreas produces and secretes **pancreatic amylase**, which breaks down starches; **lipase**, which helps to metabolize lipids; and *proteases* such as **trypsin**, which help digest proteins.

As you learned earlier, when acidic chyme along with proteins and fats enter the duodenum, secretin and CCK levels increase. When these hormones are passed from the bloodstream to the pancreas, pancreatic cells respond by producing digestive enzymes along with water and **sodium bicarbonate (NaHCO$_3$)**, which flow together into the intestine. Sodium bicarbonate neutralizes acidic chyme from the stomach. This is helpful because pancreatic enzymes function optimally at a more neutral pH than does pepsin.

The pancreas also secretes the hormones **insulin** and **glucagon** directly into the blood to regulate blood-glucose levels. Insulin is secreted when blood-glucose levels are high and enables sugar to enter cells. Glucagon is released when blood glucose is low, causing the liver to break down glycogen and release it into the blood (FIGURE 7.14). A delicate balance between insulin and glucagon secretion facilitates homeostasis of blood glucose levels. Difficulty maintaining this blood-sugar balance is the hallmark of **diabetes mellitus**, a disease that can be strongly correlated with obesity.

Diabetes: Sugar, Obesity, and a Malfunctioning Pancreas Disrupt Homeostasis During digestion, the body hydrolyzes the complex carbohydrates found in foods such as bread, pasta, fruits, and vegetables, releasing individual glucose subunits. Glucose is absorbed into the bloodstream but cannot enter the cells without insulin, which triggers cells to take up glucose. After eating, when the amount of blood sugar increases, so does insulin production. This extra insulin serves to unlock cells so that more sugar can enter, providing a source of energy for the body and maintaining the level of sugar in the bloodstream within a healthful range. FIGURE 7.15 illustrates how blood-glucose homeostasis is altered in diabetics.

There are two forms of diabetes. **Type 1**, or **insulin-dependent diabetes mellitus (IDDM)**, usually arises in childhood. People with type 1 diabetes do not produce enough

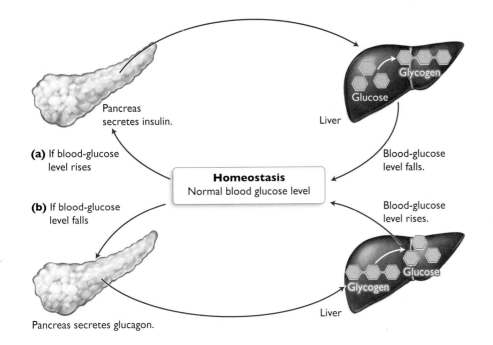

FIGURE 7.14 Regulation of blood-glucose levels. (a) When intake of glucose is high, such as after a meal, the amount of glucose in the blood increases. In response to the increase in blood glucose, the pancreas secretes the hormone insulin, which helps glucose enter cells where it is stored as glycogen. (b) When the blood-glucose level is low, the pancreas secretes the hormone glucagon, which stimulates the liver to break down glycogen and release glucose units into the bloodstream.

VISUALIZE THIS Is blood-glucose regulation an example of positive or negative feedback?

FIGURE 7.15 Diabetes. Lowered insulin secretion or lack of responsiveness to insulin causes difficulty in regulating blood-sugar levels.

VISUALIZE THIS How would blood-glucose homeostasis be disrupted if the cells of the pancreas stopped functioning at an optimal level?

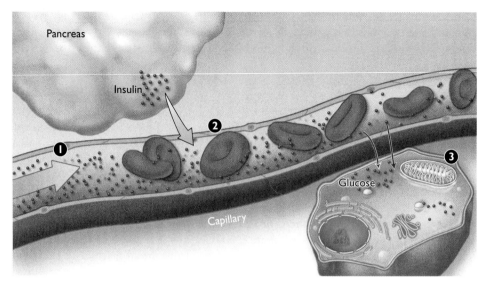

① Glucose is higher following a meal.

② When blood glucose is high, the pancreas secretes insulin into the bloodstream. A person with diabetes might not produce enough insulin, so sugar stays in the blood longer.

③ Insulin triggers the cells of the body to take up glucose. The cells of a person with diabetes might not respond to insulin. Therefore insulin stays in the blood rather than fueling cellular activities.

insulin, because their immune systems mistakenly destroy their own pancreatic cells. Because the body cannot produce insulin, daily injections of the hormone are required. Type 1 diabetes is not correlated with obesity.

Type 2, or **non-insulin-dependent diabetes mellitus** (NIDDM), usually occurs after 40 years of age, and is strongly correlated with obesity. Although it is thought of as an adult disease, type 2 diabetes appears to be increasing among young people at a rate that parallels childhood obesity. In fact, over a quarter of cases are now diagnosed before age 40. Type 2 diabetes arises either from decreased pancreatic secretion of insulin or from reduced responsiveness to insulin by target cells. When target cells, such as muscle cells, become resistant to insulin, the hormone no longer functions as a key that unlocks the door for sugar. As a result, sugar accumulates in the bloodstream. No one knows for certain why obesity leads to insulin resistance, although excess weight and inactivity greatly increase the chances of this occurring.

Stop and Stretch 7.4

One hypothesis for how obesity leads to type 2 diabetes involves cell-signaling molecules called *adipokines*. *Central body obesity,* excess weight in the abdomen, is directly correlated with both obesity and with an increase in the circulating level of one adipokine in particular, *resistin*. Based on its name, resistin, can you propose a mechanism by which this adipokine might increase the likelihood of diabetes?

Whatever the cause, the rise in obesity has led to a tremendous increase in the number of people with type 2 diabetes. Current estimates are that fully 7% of the U.S. population now suffers from this disease. People with type 2 diabetes who cannot maintain homeostasis via changes to their diet and exercise regimens must inject themselves with insulin to be able to control blood-glucose levels. Failure to control these levels can have negative effects. Low blood sugar causes shakiness, weakness, hunger, dizziness, nausea, slurred speech, drowsiness, and confusion. High blood sugar causes the blood to become thick and syrupy, producing excessive thirst, weakness, confusion, convulsions, coma, and death.

Nerve damage occurs in half of all people with diabetes. Scientists think that excess sugar damages the walls of capillaries that supply the nerves. This deterioration can cause damage to the sensory nerves of the legs, resulting in tingling, numbness, burning, or pain, and can lead to loss of feeling in the affected limbs. This puts the diabetic at risk for serious injury or infection that may require amputation, especially in the feet.

Diabetes also damages the kidneys, increasing the likelihood of a need for dialysis and possibly a kidney transplant. Most people with type 2 diabetes will also experience a deterioration of the blood vessels of the retina, resulting in impaired vision or blindness.

The majority of people with type 2 diabetes who undergo gastric bypass surgery restore their body's ability to control blood-glucose levels, and decrease the debilitating effects of this disease.

7.3 Weighing the Risks of Gastric Bypass Surgery

Because gastric bypass surgery helps people to lose weight and keep it off, and because the surgery alleviates the health problems associated with diabetes, it might seem that all obese people should have gastric bypass surgery. However, the benefits must be balanced against the risks of surgery, which are serious. As you learned in the chapter opening, the risk of death from this type of surgery is quite high.

Another way to measure the risks of surgery is to compare the risk of death from surgery to the risk of death from obesity. Obese people who have the surgery have a lower risk of death *over the long term* than obese people who don't have the surgery.

A retrospective study calculated the risk of death for 3,300 obese people who underwent surgery and 60,000 obese people who did not. This study showed that 16% of obese people who did not have gastric bypass surgery died during the 15 previous years, compared with 12% of those who did have the surgery. This study, the results of which appeared in the November 2004 issue of the *Journal of the American College of Surgeons*, also showed that the health benefits of this surgery are more pronounced in younger patients. When the patient is under 40 years of age, 3% of those who had the bypass died compared to a 14% death rate for those under 40 who did not have the surgery. Therefore, the limited data so far suggest that although gastric bypass surgery is risky, the benefits may outweigh the risks for some obese people. However, it is important to remember that gastric bypass surgery is an extreme treatment for a problem that might be solved with less invasive measures such as diet and exercise.

Stop and Stretch 7.5

A study reported in the September 2004 *Annals of Surgery* followed more than 5,000 obese people who did not have gastric bypass surgery and 1,000 obese people who did have a gastric bypass. Five years after the study began, researchers showed that 0.68% of those who had a gastric bypass had died, compared to 6% of those who did not have the bypass. Note that the results of this study differ somewhat from the *Journal of the American College of Surgeons* study mentioned in the preceding paragraph. How would you summarize the results of these two studies? Is the information presented in one study enough information upon which to base one's decision?

Chapter REVIEW

ROOTS TO REMEMBER

The following roots of words come mainly from Latin and Greek and will help you to decipher terms:

ante- and **anterior** mean before, forward, or in front.

-ase is a common ending for enzymes.

dent- is from the word for tooth.

dia- means through (dia-betes means to pass through).

gast- and **gastro-** mean stomach.

glot- and **glos-** refer to the tongue.

gluc- or **gluco-** relate to sugar (as do **glyc-** and **glyco-**)

hep- and **hepato-** relate to the liver.

-itis means a swelling or inflammation.

lip- and **lipo-** refer to fat.

pep- (as in peptic and pepsin) refers to digestion.

post- and **posterior** mean after or behind.

KEY TERMS

7.1 The Digestive Tract
alimentary canal *p. 156*
anus *p. 164*
appendicitis *p. 163*
appendix *p. 163*
canine tooth *p. 158*
cecum *p. 163*
chyme *p. 160*
colon *p. 163*
crown *p. 158*
dentin *p. 158*
digestion *p. 156*
digestive tract *p. 156*
duodenum *p. 161*
enamel *p. 158*
epiglottis *p. 159*
esophagus *p. 159*
fecal matter *p. 163*
hard palate *p. 157*
ileum *p. 161*
incisor *p. 157*
jejunum *p. 161*
large intestine *p. 163*
microvillus (microvilli) *p. 162*
molar *p. 158*
mouth *p. 157*
pepsin *p. 160*
peristalsis *p. 159*
pharynx *p. 159*
premolar *p. 158*
pulp cavity *p. 158*
rectum *p. 163*
root *p. 158*
small intestine *p. 161*
soft palate *p. 157*
sphincter *p. 159*
stomach *p. 159*
taste bud *p. 157*
tonsil *p. 157*
tooth (teeth) *p. 157*
villus (villi) *p. 161*

7.2 Three Accessory Organs of the Digestive System
bile *p. 167*
bilirubin *p. 167*
diabetes mellitus *p. 169*
emulsification *p. 167*
gallbladder *p. 168*
glucagon *p. 169*
insulin *p. 169*
lipase *p. 169*
liver *p. 167*
lobules *p. 167*
pancreas *p. 169*
pancreatic amylase *p. 169*
sodium bicarbonate (NaHCO$_3$) *p. 169*
trypsin *p. 169*
type 1 insulin-dependent diabetes mellitus (IDDM) *p. 170*
type 2 non-insulin-dependent diabetes mellitus (NIDDM) *p. 170*

SUMMARY

7.1 The Digestive Tract

- The digestive tract, or alimentary canal, is a long tube that begins at the mouth and ends at the anus (p. 156; Figure 7.1).

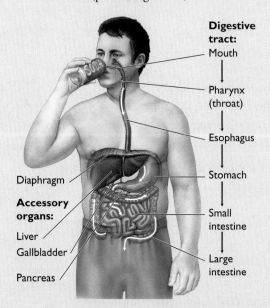

- Initial digestion involves mechanical and chemical breakdown of food. The chemical breakdown of starch and lipids begins in the mouth as the salivary glands release salivary amylase and lingual lipase. The mechanical breakdown of food by the teeth increases the surface area exposed to these and other secretions (p. 156).

- The tongue forms the bolus of food that is pushed into the pharynx. The epiglottis ensures that food leaving the pharynx is routed to the esophagus for transport to the stomach (p. 157).

- Food is pushed through the digestive tract by the wavelike muscular movements of peristalsis (p. 159).

- Ringlike sphincter muscles contract and relax to help regulate the movement of food and secretions through the alimentary canal (p. 159).

- The chemical breakdown of protein begins in the stomach. Churning of the stomach aids in mechanical digestion, while pepsin and acids carry out chemical digestion (p. 160).

- Partially digested food and gastric juice (chyme) slowly exit the stomach and enter the small intestine, where most digestion and absorption occur (p. 160).

- The small intestine consists of the duodenum, ileum, and jejunum. Its lining contains three features that substantially increase the surface area over which absorption can occur: folds, multicellular villi, and microvilli (pp. 161–162).

- The large intestine consists of the cecum, colon, rectum, and anus. It functions to absorb water, salts, and vitamins and to store and lubricate fecal matter until it is eliminated from the body (pp. 163–164).

7.2 Three Accessory Organs of the Digestive System

- The liver, gallbladder, and pancreas are accessory organs that secrete substances into the digestive tract that aid in digestion (p. 166; Figure 7.12).

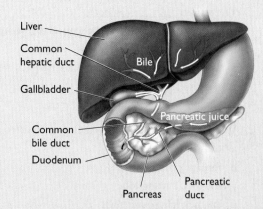

- The liver is the body's major chemical processing center. It absorbs and metabolizes nutrients and helps regulate blood-glucose levels. The liver separates nutrients from wastes and produces many substances required for proper body function including bile, cholesterol, and blood proteins (pp. 166–168).

- The gallbladder stores and concentrates bile for release into the small intestine as needed (p. 168).

- The pancreas secretes digestive enzymes into the small intestine to facilitate digestion, and sodium bicarbonate to neutralize acidic chyme. The pancreas also secretes hormones that regulate blood-glucose levels (p. 169).

LEARNING THE BASICS

1. The mouth _____.
   a. performs mechanical and chemical digestion
   b. contains a total of two salivary glands
   c. contains the tongue, which is a tendon
   d. has a soft palate composed of several different bones
   e. has a hard palate composed of muscle

2. Which of the following associations is incorrect?
   a. tonsils: taste
   b. molars: grinding food
   c. incisors: cutting food
   d. enamel: hard outer shell
   e. pulp cavity: nerves and blood vessels

3. The appendix is attached to the _____.
   a. small intestine
   b. gallbladder
   c. cecum
   d. pharynx
   e. trachea

4. The esophagus is a muscular tube that _____.
   a. stores undigested remnants of food
   b. absorbs nutrients
   c. breaks down proteins
   d. transports food from the pharynx to the stomach
   e. secretes pepsin

5. The large intestine _____.
   a. stores undigested remnants of food
   b. is the major site of nutrient absorption
   c. breaks down proteins
   d. is a passageway from the pharynx to the stomach
   e. secretes pepsin

6. Tracing the path of food as it passes through the digestive tract, which of the following is the correct order?
   a. mouth, esophagus, stomach, pharynx, small intestine
   b. mouth, pharynx, esophagus, anus, large intestine
   c. mouth, pharynx, large intestine, small intestine
   d. mouth, pharynx, stomach, esophagus, anus
   e. mouth, pharynx, esophagus, stomach, small intestine

7. Which of the following is *not* secreted by the pancreas?
   a. lipase
   b. trypsin
   c. pancreatic amylase
   d. sodium bicarbonate
   e. pepsin

8. Which association is incorrect?
   a. mouth: digestion of starches
   b. liver: production of bile
   c. small intestine: digestion and absorption of nutrients
   d. stomach: food storage
   e. large intestine: filtering blood

9. Which association is incorrect?
   a. colon: the small intestine
   b. liver: produces blood proteins
   c. pharynx: forms the back of the throat
   d. pancreas: produces alkaline secretions and digestive enzymes
   e. gallbladder: stores and concentrates bile

10. The lymphatic vessel within an intestinal villus that aids in the absorption of fats is a _____.
    a. microvillus
    b. lipase
    c. lacteal
    d. cecum
    e. bilirubin

11. Which of the following is not a part of the small intestine?
    a. duodenum
    b. jejunum
    c. ileum
    d. cecum

12. The small intestine _____.
    a. has a smooth interior lining
    b. is larger in diameter than the large intestine
    c. possesses microvilli that are covered with villi
    d. possesses villi that contain capillaries and a lacteal
    e. All of the above are true.

13. The large intestine _____.
    a. has a highly folded inner lining
    b. is shorter than the small intestine
    c. is the major site of digestion and absorption
    d. is also called the cecum
    e. All of the above are true statements

14. Which of the following liver components removes toxic materials, dead cells, pathogens, drugs, and alcohol from the bloodstream?
    a. lobe
    b. gallbladder
    c. lobule
    d. portal
    e. hepatocyte

15. Which of the following is not a function of the liver?
    a. filtering toxins
    b. destroying damaged blood cells
    c. storing glycogen
    d. manufacturing blood proteins
    e. storing bile

16. Bile _____.
    a. is stored in the pancreas
    b. helps break down glycogen
    c. emulsifies fats
    d. removes water from indigestible materials in the large intestine
    e. all of the above

17. Match each digestive system structure with its function.
    _____ liver
    _____ large intestine
    _____ stomach
    _____ small intestine
    _____ salivary gland

    a. where most digestion occurs
    b. protein digestion begins here
    c. formation and storage of feces
    d. secretes enzymes that begin starch digestion in the mouth
    e. receives blood carrying absorbed nutrients

18. Match each digestive system hormone or enzyme to its function.
    _____ amylase
    _____ lipase
    _____ sodium bicarbonate
    _____ glucagon
    _____ trypsin

    a. hormone secreted by the pancreas
    b. protein-digesting enzyme
    c. fat-digesting enzyme
    d. starch-digesting enzyme
    e. neutralizer of acidic chyme

19. Use the following list of terms to label the diagram (FIGURE 7.16).
    Pancreas   Diaphragm
    Large intestine   Pharynx
    Liver   Esophagus
    Gallbladder   Small intestine
    Stomach

20. **True or False?** Peristalsis takes place from the beginning of the esophagus to the anus.

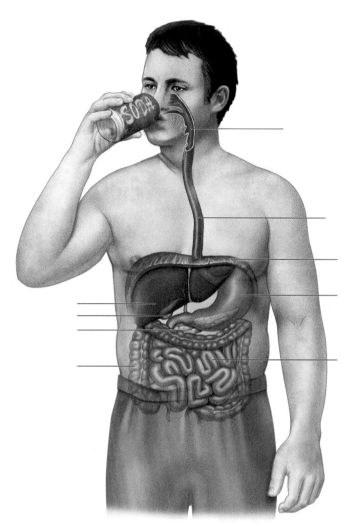

FIGURE 7.16 **Label the diagram.**

ANALYZING AND APPLYING THE BASICS

Questions 1–3 refer to the following information.

Dental erosion is the loss of hard parts of the tooth, such as enamel, through chemical processes not involving bacteria. In 1991, a study in Finland examined factors promoting dental erosion. The following table shows the results of the study.

Risk Factor	Risk of Erosion (Number of times greater than that of patients without risk factor)
Vomiting (weekly or more often)	31
Other gastric symptoms (weekly or more often)	10
Low flow of saliva (≤ 0.1 ml/min)	5
Soft drinks (4–6 per week or more)	4
Sports drinks (weekly or more)	4

1. Considering the data in the table, what chemical property increases risk of dental erosion?

2. If risk factors were additive, what would be the risk of dental erosion for a person who drank a soft drink once a day and a sports drink twice a week?

3. Bulimia is an eating disorder in which episodes of binge-eating are followed by purging—usually by vomiting. Propose a hypothesis that would explain why a bulimic would be at increased risk for dental erosion.

4. Students studying protein digestion designed an experiment using four different test tubes as shown in **FIGURE 7.17**. Each test tube contained a chunk of meat. The other contents are listed in the figure. Rank the test tubes in order, from the greatest amount of digestion to the least, and explain your rankings.

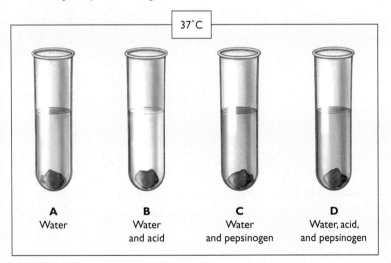

FIGURE 7.17 Digestion experiment.

5. It is known that type 1 diabetes involves the destruction of beta cells in the pancreas, which reduces insulin production. Research indicates that proinsulin, the precursor to insulin, is involved in the destruction of beta cells. However, scientists have not discovered the precise mechanism for this destruction.

   One research group proposes a mechanism related to the folding of proinsulin. During proinsulin production in the endoplasmic reticulum of beta cells, the protein folds on itself to form a tertiary structure. Researchers found high levels of misfolded proinsulin in mice prone to diabetes. These scientists hypothesize that misfolded proteins could stress the cleanup mechanisms in beta cells, eventually leading to cell death. Propose a treatment strategy if this were the mechanism of beta-cell destruction.

Questions 6 and 7 refer to the following information.

Acetaminophen is a common pain reliever. Taken at the recommended dose of 1,000 milligrams (mg) in a 24-hour period, it is safe and effective for adults. However, when more than 4,000 mg is taken within a 24-hour period, acetaminophen can damage the liver. When combined with alcohol, as little as 2,000 mg in 24 hours can cause liver damage, via the following steps:

   A. Acetaminophen is normally processed through two complex pathways (sulfation and glucuronidation) before it is eliminated by the liver. These pathways produce a product that is harmless to the human body.

   B. When too much acetaminophen is taken, the normal pathways become overwhelmed. Any excess acetaminophen is then processed by the cytochrome P-450 system. The pathway produces NAPQI, a potentially toxic compound.

   C. NAPQI is processed by the glutathione system, to become a harmless compound.

   D. However, if the glutathione system is overwhelmed, NAPQI can build up in the liver and cause damage.

6. Chronic alcohol use can reduce the concentration of glutathione in the liver. Alcohol can also make the cytochrome P-450 system more active. Explain how these effects from alcohol would increase the toxicity of acetaminophen to the liver.

7. Which dosage level of acetaminophen might be more toxic: a dose of 4,500 mg over a 24-hour period, or a single dose of 3,000 mg? Explain your answer.

CONNECTING THE SCIENCE

1. Should insurance companies, or the government, help pay the costs of gastric bypass surgery? Why or why not?
2. Why do you think obesity has increased so much in the past few decades?
3. Is the correlation between resistin and central body obesity enough evidence to prove that this adipokine is responsible for causing diabetes? Why or why not?

Chapter 8

The Blood

Malaria—A Deadly Bite

LEARNING GOALS

1. Summarize the major functions of blood in the body.
2. List the constituents of both the liquid and solid portions of the blood and describe their functions.
3. Explain how homeostasis of the oxygen-carrying capacity of the blood is maintained.
4. Describe the ABO and Rh blood type systems and explain the consequences of a blood type mismatch between donors and recipients.
5. Describe the fate of red blood cells when they die.
6. Describe how the circulatory system minimizes blood loss from a damaged vessel, and summarize the process of blood clotting.

The parasites carried by these creatures threaten the survival of millions of people every year.

It sounds like a plot from a science fiction movie. Under cover of darkness, strange-looking airships filled with invaders seek human hosts. The invaders use the human bodies as nurseries, feeding off the blood of their victims and causing serious illness, even death. If this were the plot of a typical Hollywood movie, the take-charge military officer and the brilliant and attractive scientist would team up to save the world from this vicious foe.

Unfortunately, the scenario is not fiction, and the invader remains among us. That invader? *Plasmodium*: the single-celled creatures that cause malaria, creatures that are transmitted by mosquitoes active during twilight and night. The victims? Three to five hundred million people in the tropics and subtropics every year.

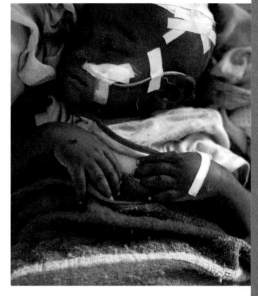

The victims of malaria are mostly poor children in the developing world.

In the period after World War II, scientists and the United Nations tried to fill the hero's role in the story of malaria—they set as their goal the eradication of this scourge. By the 1970s, the disease had become nearly nonexistent in the United States, Europe, and the more developed countries of the subtropics. However, unlike smallpox and polio, which yielded to the advances of scientific knowledge, malaria remained stubbornly difficult to completely eliminate.

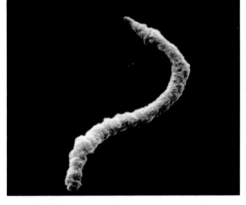

Scientists continue to seek tools to defeat this deadly foe . . .

The campaign to eradicate the disease fizzled in the 1980s, and malaria continued to wreak havoc on poor populations in Africa and South America. Dedicated health scientists and nongovernmental organizations were stymied by a lack of interest and funding from governments, foundations, and private corporations. Inattention to the disease has allowed death rates due to malaria to double over the past 20 years. Current projections indicate that without major changes in our approach to the disease, the death rate will double again by 2027.

Malaria may have remained a seemingly unsolvable problem for years to come if not for new heroes in the story—Bill and Melinda Gates. Bill Gates is the founder of Microsoft Corporation and the world's wealthiest man,

. . . aided by some unlikely allies, including the world's richest man.

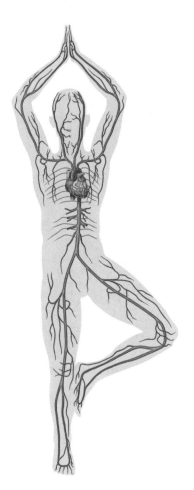

FIGURE 8.1 **The circulatory system.** This organ system is made up of pathways for fluid connected to a pump, the heart. The connective tissue in this system is the blood.

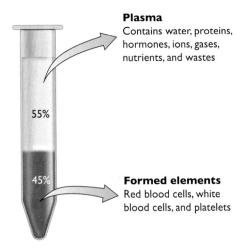

FIGURE 8.2 **Constituents of blood.** In a given volume of blood, 55% is liquid and 45% is made up of formed elements—cells and cell fragments. The solid and liquid portions can be separated from each other by spinning a tube of blood in a centrifuge, causing the formed elements to settle at the bottom.

with a net worth of at least $56 billion. The charitable foundation started by Gates and his wife—which doubled in size thanks to contributions from investor Warren Buffett, the world's second-wealthiest man—has pledged more than $500 million to malaria research. The pledge has stimulated new efforts from governments and pharmaceutical firms.

"Gates has just electrified the field," says Patrick Duffy, a researcher at the Walter Reed Army Institute of Research, summarizing the research community's response to these new funds. Just one measure of the impact of this pledge is that the Gates Foundation provided more than one-third of the global funding for malaria research in 2004.

The Gates Foundation is interested in tackling malaria despite the failure of initial eradication attempts because it is a devastating disease that is both treatable and preventable. An increased understanding of the biology of malaria gained as a result of these grants will reduce the death rate—currently 2,000 to 3,000 children every day. In this chapter, we examine the effects of malaria on its target, the blood, and look at the possibilities of a future where this disease takes much less of a toll.

8.1 The Constituents of Blood

The **circulatory system** (FIGURE 8.1) is the organ system that ties all other organ systems together. Chapter 9 covers the system in detail—in this chapter we are concerned with its principle connective tissue, **blood**.

Blood's primary function is the transportation of materials, such as nutrients from the digestive system and oxygen from the lungs, to all cells. Blood also transports waste away from these cells, carrying carbon dioxide and metabolic wastes to sites where they can be eliminated from the body. Other functions of blood include the regulation of body temperature, water volume, and pH, and providing and distributing materials that fight infection. Blood is so essential to maintaining homeostasis that loss of less than one-third of its volume from an otherwise healthy adult will cause death.

A liquid so rich in nutrients presents a tempting food source for other organisms. The organisms that cause malaria use particular cells in the blood as their food source. And the way malaria is transmitted relies on another organism that dines on human blood.

Malaria is a vector-borne disease, meaning that it is not transmitted directly from one individual to the next, but instead is carried and passed on by another species. The *vector* in this case is a mosquito. Not all mosquitoes can transmit malaria—only those species that are in the genus *Anopheles*. Inside both the mosquito and the human body, species of *Plasmodium* that cause malaria act as **parasites**, benefiting from the resources of the host while providing no benefit in return.

As is true of all mosquito species, the only *Anopheles* mosquitoes that bite are adult females, who use human blood to provide energy for their developing eggs. When a mosquito drinks blood from a human, she is ingesting a mixture of cells and liquids. Of the 5 liters (11 pints) of blood in the vascular system of an adult, about 2.75 liters (55%) are liquid. The remaining 45% consists of **formed elements**, that is, cells and cell fragments (FIGURE 8.2).

Plasma

The yellowish liquid portion of the blood is a solution called **plasma** (FIGURE 8.3). Plasma is approximately 90% water by volume. The remaining 10% consists of hundreds of dissolved substances, including nutrients and wastes, gases, salts, and proteins. Plasma is the vehicle for distribution of drugs throughout the body. This liquid also dissolves and transports chemical messengers, such as hormones, that trigger responses far from their initial release.

About 60% of the protein in plasma is made up of **albumins** manufactured by the liver from amino acids supplied by our diets. Albumins often function as transport proteins, carrying materials that are not soluble in water, such as steroid hormones and fats. One important role of albumin is to maintain tissue water balance; albumin increases the solute concentration of the blood, thus preventing the osmotic flow of water from the blood to the solute-rich body tissues.

Stop and Stretch 8.1

The child in this photo is showing signs of kwashiorkor. The name for this condition comes from the Ghanian word for "rejected one," reflecting that it often appears in toddlers shortly after being weaned from protein-rich breast milk. The swollen belly characteristic of this condition occurs because water accumulates in abdominal tissues. Use your understanding of the role of albumin to describe the pathway by which a lack of protein causes this symptom.

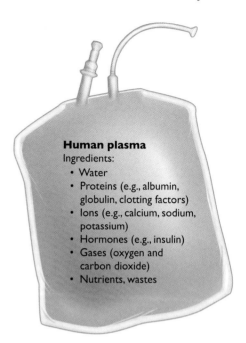

FIGURE 8.3 **Plasma.** Plasma can be collected from donors in whole blood, then separated out by a centrifuge. Plasma is given to patients with bleeding disorders to provide clotting proteins, and to individuals with serious burns to provide extra antibodies.

Plasma also contains proteins called **globulins**. Globulins include transport proteins that carry insoluble or very small molecules throughout the body. **Antibodies**, proteins that attack foreign proteins and infectious organisms, are also globulins. Antibodies are produced by a type of white blood cell, one of the formed elements of the blood. The production and function of antibodies are detailed in Chapter 12, on the immune system.

The remaining plasma proteins are associated with the process of blood clotting. When blood is removed from the body, the clotting proteins will coagulate the formed elements into a gel, leaving a liquid called **serum**—the plasma minus its clotting factors.

Formed Elements: The Cellular Portion of Blood

The solid portion of the blood is primarily made up of red blood cells (giving blood its bright red color) with much smaller numbers of white blood cells and platelets. All of these formed elements are produced through the process of *hematopoiesis* from constantly dividing, undifferentiated **stem cells** in the bone marrow of large bones (FIGURE 8.4).

Red Blood Cells Red blood cells, or **erythrocytes**, are uniquely adapted to their primary task—shuttling oxygen from the lungs to the rest of the body. When they have completed their development, these cells have little internal structure. In other words, they lack a nucleus and other organelles. Without these organelles, erythrocytes are flexible enough to squeeze through the smallest blood vessels. At maturity, red blood cells are essentially disk-shaped, membrane-bounded bags packed with oxygen-carrying proteins. Red blood cells are also small and pinched in the middle, which makes their surface area large relative to their volume, ensuring rapid diffusion of oxygen into and out of the cells.

FIGURE 8.4 **Blood cells arise in the bone marrow.** The inner chamber of larger bones contains stem cells that give rise to the blood cells.

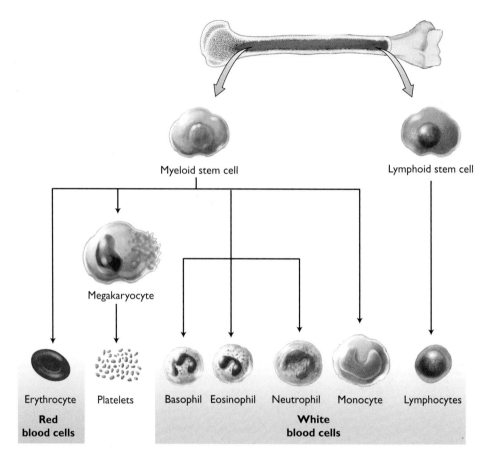

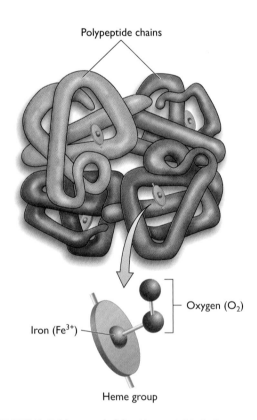

FIGURE 8.5 **Hemoglobin.** Hemoglobin is the blood's oxygen shuttle, picking up oxygen in the lungs and unloading it in active tissues. The protein is made up of four chains (shown here in different colors), each with an active site called a heme group; the heme group binds oxygen in regions where oxygen is abundant and releases it in tissues where oxygen levels are low.

Erythrocytes make up the overwhelming majority of cells in whole blood. A single cubic millimeter of blood—less than a drop—contains 4–6 million red blood cells. The large number of erythrocytes in blood contributes to its thick consistency.

The color of a red blood cell is due to the 250–300 million oxygen-carrying **hemoglobin** proteins it contains (FIGURE 8.5). A hemoglobin molecule is made up of four chains, each of which contains a single iron atom. The iron atom, contained within a structure called a *heme group*, can bind to a single oxygen molecule (O_2). One hemoglobin molecule can thus carry four O_2 molecules. With oxygen bound to it, the molecule is called *oxyhemoglobin* and has a bright red color.

The binding of O_2 to hemoglobin occurs only under certain conditions—that is, when the relative amount of oxygen in the environment is high, and when the pH is not very low. In conditions of low oxygen and low pH, such as in metabolically active tissues that are using up oxygen and producing waste acid, hemoglobin releases oxygen molecules. When this occurs, the protein, now known as *deoxyhemoglobin*, reverts to a purple tone. The blood in your veins is rich in deoxyhemoglobin, which is why veins under the skin appear blue or purple.

After releasing oxygen, heme groups will bind carbon dioxide under low oxygen and pH conditions. In fact, hemoglobin is responsible for transporting about 25% of the CO_2 in our blood. Oxygen and carbon dioxide uptake and excretion are discussed in more detail in Chapter 10, on the respiratory system.

As detailed below, *Plasmodium* causes a dramatic loss of red blood cells during the course of an infection. The result of this loss is a decline in the oxygen-carrying capacity of the blood, causing many of the symptoms of malaria. Control of *Plasmodium* in the body is the responsibility of another class of formed elements, the white blood cells.

White Blood Cells **Leukocytes**, or white blood cells, are significantly larger than red blood cells and contain all of the organelles found in most other body cells. These cells make up the smallest fraction of formed elements in a volume of blood—only about 5,000–11,000 per cubic millimeter, or 1% of the total volume. Yet white blood cells are crucial to immune system function, attacking invading organisms as well as removing toxins, wastes, and damaged cells throughout the body.

White blood cells have traditionally been classified by their appearance: those with large membrane-bounded cavities, or *vesicles*, that are easily seen under a light microscope, and those without. Large vesicles appear as granules in the cells, and thus cells containing them are called the *granular leukocytes*. The *agranular leukocytes* are the remaining cells, containing small vesicles that are not readily visible (Table 8.1). However, it may be more useful to group leukocytes into two categories by their functions: those that participate in the general nonspecific response to pathogens and those that respond to specific pathogens.

As we will discover in Chapter 12, the nonspecific immune response is immediately mobilized in response to the presence of pathogens. The white blood cells involved in this response include eosinophils, basophils, neutrophils, and monocytes. The first two of these make up a relatively small percentage of white blood cells. **Eosinophils** (2–4% of white blood cells) mainly function to defend against large pathogens such as tapeworms and flukes. **Basophils** (<1% of white blood cells) secrete histamine, a chemical that triggers inflammation, in response to injury. The more numerous **neutrophils** and some **monocytes** are phagocytic, which means that they surround and engulf invaders to destroy them. Neutrophils (60% of white blood cells) target the smaller, more common invaders such as bacteria, while monocytes that have differentiated into phagocytic **macrophages** engulf damaged body cells and larger invaders.

The response of monocytes to malaria infection causes the disease's classic symptoms: fever, chills, and sweating. After a monocyte encounters *Plasmodium*, it releases a number of cell-signaling proteins called *cytokines* to stimulate further immune response.

TABLE 8.1 | White Blood Cells and Their Functions

	Cell Type	Description	Function	
Granulocytes	Eosinophil	Stain readily with red eosin. Bilobed nucleus.	Surround large parasitic cells and organisms and secrete digestive enzymes to destroy them. Mediate allergic reactions.	**Nonspecific immune response**
	Basophil	Stain readily with basic blue. Bilobed nucleus.	Release histamine to initiate inflammatory response.	
	Neutrophil	Do not stain readily. Multilobed nucleus.	Surround and engulf foreign cells.	
Agranulocytes	Monocyte	Large cells, do not stain readily.	Engulf foreign cells, dead or damaged self cells, other debris.	
	Macrophage	Derived from monocytes. Irregular outline with many "arms."	Same as monocytes, but activities take place primarily in body tissues.	
	B lymphocyte	Very large, round nucleus.	Produce antibodies to specific toxins, cells, viruses, and other invaders.	**Specific immune response**
	T lymphocyte	Very large, round nucleus.	Target and destroy foreign cells and abnormal host cells.	

One of these cytokines triggers a dramatic increase in body temperature. The fever that results may slow down the growth of the malaria parasite, and it increases the metabolism of immune system cells, allowing them to work faster.

Some of the cytokines released by monocytes trigger increased activity in lymphocytes, the white blood cells involved in the specific immune response. The two types of lymphocytes are named for the organs they mature in: **B lymphocytes** mature in bone and **T lymphocytes** complete their development in the thymus, a small organ behind the sternum. Both types of lymphocytes respond to unique chemical markers found on invading organisms by producing a specific response. B lymphocytes produce protein antibodies that can attach to the surface of a particular invader and disable it. T lymphocytes can differentiate into a variety of types of killer cells targeted toward a particular type of pathogen or disabled cell. Lymphocyte function will be explored more thoroughly in Chapter 12.

Stop and Stretch 8.2

Leukemia is cancer of the blood and occurs when blood stem cells divide uncontrollably, creating immature white blood cells that clog blood vessels and interfere with normal body functions. Explain why one treatment for leukemia is the destruction of bone marrow and its replacement with marrow from a healthy donor.

Plasmodium has several strategies for evading both the nonspecific and specific portions of the immune response. Malaria's ability to avoid white blood cells, detailed in section 8.2, is what makes this disease so devastating and difficult to control.

Platelets The third category of formed elements in the blood is the **platelets**, membrane-bounded cell fragments without nuclei that play an essential role in the process of blood clotting (discussed in section 8.3). Platelets are derived from **megakaryocytes** in the bone marrow. These small fragments are relatively abundant in the blood—each cubic millimeter contains roughly 400,000 platelets—but they are short-lived, lasting only three to five days. Megakaryocytes produce nearly 200 billion new platelets every day.

Platelets appear to play a key role in one of the most serious complications of malaria infection, a condition called cerebral malaria. In the next section, we examine the course of malaria infection and its effects on the constituents and roles of blood in the body.

8.2 Malaria and the Blood

Malaria may be among the oldest human diseases, likely having evolved along with our species. It appears to have become more dangerous during the human transition to an agricultural way of life 5,000 to 10,000 years ago. Five-thousand-year-old Chinese texts advised men to arrange remarriages for their wives before they left for known malarial regions. Egyptian mummies from 2000 B.C.E. show convincing evidence of chronic malaria infection.

By the period of European exploration, malaria was a global disease, causing perhaps 10% of all deaths. Although malaria is not as prevalent today as it was in the eighteenth and nineteenth centuries, the human toll of malaria is now estimated to be between 750,000 and 3 million deaths per year. Nearly 80% of these deaths are of children under

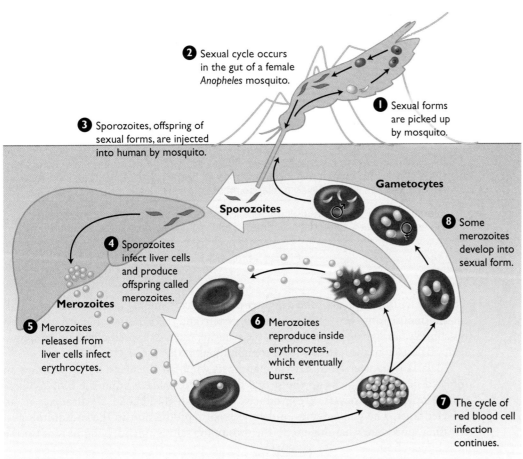

FIGURE 8.6 **The life cycle of *Plasmodium falciparum*.** The reproductive cycle of this deadly parasite is extremely complex. Three other species of **Plasmodium** cause malaria and all have a similar life cycle, although some, such as **P. vivax**, have the ability to remain dormant in liver cells for many months.

the age of five, nearly all of them in Africa. Understanding how malaria kills is central to understanding how it may be controlled.

Malaria Infection

The several species of *Plasmodium* that cause malaria infect erythrocytes (see Table 8.2 on the next page). The life cycle of the most deadly of these parasites, *Plasmodium falciparum*, is described in FIGURE 8.6. As you can see from the figure, the life cycle of a malaria parasite is complicated and consists of several distinct life stages.

P. falciparum in the salivary gland of a mosquito are in a form called sporozoites. When an infected mosquito bites an individual, a small number of sporozoites are injected into the bloodstream. Within seconds, these cells enter human liver cells. In the liver, undetected by the immune system, a single sporozoite divides hundreds of times to produce about 30,000 offspring, in a form called merozoites. Eventually, the liver cell is so packed with merozoites that it bursts, releasing a shower of infectious particles into the body.

This huge pulse of parasites stimulates the powerful immune system response from monocytes, leading to an hours-long bout of fever, chills, and sweating. At this stage, the parasites can only survive the immune system's efforts to destroy them by rapidly attaching to and entering red blood cells. In essence, the parasites hide, and thus the immediate immune response shuts off.

Once inside an erythrocyte, a single parasite can divide several times. After about two days, the thousands of infected erythrocytes burst open, releasing millions of new parasites.

TABLE 8.2 | Forms of Malaria

Parasite	Geographical Range	Characteristics of Infection
Plasmodium falciparum	Widespread in the tropics and subtropics.	Infects any red blood cell. Fever may last for several days. The species most likely to cause severe disease, and the only one likely to cause mortality.
P. vivax	Widespread, except in Africa. Most common form in temperate areas.	Infects immature red blood cells. Fever spikes every other day. Some parasites take up long-term residence in the liver, causing a relapse months to years later.
P. ovale	Primarily western tropical Africa. Not present if P. vivax is present.	Infects immature red blood cells. Fever spikes every other day during infection. Causes only mild symptoms, including anemia. Some parasites take up long-term residence in the liver, causing a relapse months to years later.
P. malariae	Widespread but spotty distribution in the tropics.	Infects only older, dying red blood cells. Symptoms may be moderate to severe. Fever spikes every three days. Untreated infections may persist for decades.

The fever returns until all of the daughter parasites have entered new erythrocytes. This cat-and-mouse game between the immune system and the parasite can continue almost indefinitely, until drug treatment kills the parasite, the host's specific immune response finally defeats it, or the host dies from complications of the disease.

Malaria is transferred to new human hosts when an *Anopheles* mosquito bites an infected person and picks up the sexual form of *Plasmodium* from the blood. These forms, called gametocytes, are produced in a small number of erythrocytes in an infected host. The gametocytes undergo sexual reproduction in the mosquito's gut, releasing sporozoites that migrate to the mosquito's salivary gland, lying in wait for the next human host.

Anemia and Blood Cell Production

Continued rounds of *Plasmodium* reproduction and cell destruction can cause a dramatic decline in the number of erythrocytes in an infected person. This decline in *hematocrit*, the percentage of red blood cells in the blood, results in **anemia**. The symptoms of anemia include fatigue, weakness, and breathlessness—all of these symptoms result from the blood's reduced capacity for oxygen delivery.

Low oxygen levels caused by a decline in red blood cells trigger the release of the hormone **erythropoietin** from the kidneys. Erythropoietin stimulates erythrocyte production by the bone marrow. As with most aspects of homeostasis, this is a negative feedback loop—as hematocrit levels rise, more oxygen is delivered to the kidneys and erythropoietin production is shut off (FIGURE 8.7).

In many children and pregnant women infected with *P. falciparum*, the erythropoietin feedback system cannot make up for the loss of red blood cells during the infection. Nearly half of the deaths due to malaria are a direct result of severe anemia.

Blood Types and Transfusions

Mild anemia caused by inadequate diet, such as a deficiency in iron or in vitamin B_{12} (another nutrient required for red blood cell production), can be overcome by modifying one's food intake. In contrast, severe anemia such as that caused by severe malaria is usually treated by **blood transfusion**, wherein whole blood from a donor is provided to the patient. In severe cases of malaria, where 5% or more of a patient's red blood cells are infected, an exchange transfusion may be performed: Infected blood is drawn from the patient as uninfected donor blood is supplied.

It is also possible to transmit malaria by blood transfusion, if the donor is carrying the malaria parasite. Some forms of malaria can lie dormant in the liver for months or years, releasing sporozoites only rarely. As a consequence, survivors of malaria, and even individuals who have visited malaria-prone regions, are not permitted to donate blood for some time—until it is clear that they are parasite-free.

A successful blood transfusion requires a match in **blood type** between donor and recipient. The surfaces of red blood cells are studded with a variety of sugars and proteins that allow the cells to exchange materials with the body. The most important of these surface features in medical practice are the ABO and Rh factors.

In the **ABO system**, blood type is determined by a single gene that comes in three distinct forms, called *alleles*. One allele codes for an enzyme resulting in the "A" sugar being produced, another codes for "B" sugar production, and a third codes for no sugar production. Although all three types of alleles are present in most human populations, a given individual carries only two alleles. The various combinations of alleles can result in four blood types in the population; A, B, AB (where both sugars are produced), and O (where no sugar is produced).

Physicians must take ABO blood groups into account when performing blood transfusions. This is because individuals produce an immune response, in the form of anti-A or anti-B antibodies, against those sugars that they do not carry on their own red blood cells. The foreign sugars are thus called **antigens**, because they are substances that provoke an immune response.

When foreign red blood cells are introduced into an individual's bloodstream, a *transfusion reaction* occurs. First, the recipient's antibodies bind to the foreign antigens on the donated blood. The blood cells tagged by antibodies then form clumps in a process called **agglutination**. Tagged cells attacked by the recipient's immune system break open, releasing their hemoglobin. The free hemoglobin then clogs the tubules in the kidneys, possibly resulting in death (FIGURE 8.8).

Transfusions between individuals with the same blood type are generally trouble-free, since the antigen profile of the donor and recipient are identical. Transfusions between blood types are sometimes possible, if the transfusion consists of only blood cells, minus the plasma containing the donor's antibodies. Table 8.3 shows compatible and incompatible recipients and donors in the ABO blood system. Note that this table illustrates that individuals with AB blood type make no antibodies to any ABO blood types and so can receive any of them via transfusion, while individuals with type O blood have no ABO antigens and so their blood will not agglutinate in type A, B, or AB individuals.

Another molecule on the surface of red blood cells is called the **Rh factor**. Someone who is Rh positive has the Rh protein on their red blood cells, while

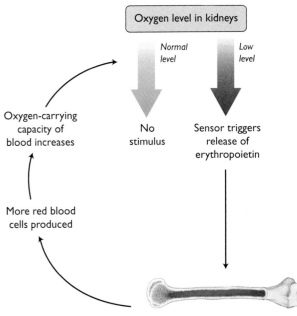

FIGURE 8.7 Homeostasis of blood oxygen. Low oxygen levels in blood reaching the kidney trigger the release of the hormone erythropoietin. Erythropoietin stimulates red blood cell production, increasing the blood's oxygen-carrying capacity. Once oxygen levels return to the set point, erythropoietin production is turned off.

VISUALIZE THIS Blood doping is a practice that elite endurance athletes sometimes engage in, although it is banned by most professional sports organizations. Doping consists of injecting EPO, a synthetic version of erythropoietin, to induce excess red blood cell production. How would this practice enhance the performance of an endurance athlete? Why might blood doping be harmful to the athlete?

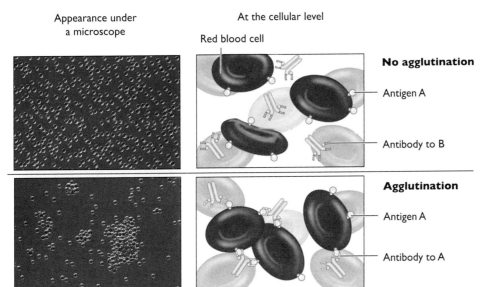

FIGURE 8.8 Blood types and antibodies. Blood mixed with serum containing antibodies to the surface antigens on its red blood cells clumps, or agglutinates. If a person receives an incompatible blood type, these clumps will lead to kidney damage and may cause death.

TABLE 8.3 | Donor/Recipient Relationships in the ABO Blood System

Recipient Blood Type	Antibodies Produced	Donor Blood Type				
		AB	A	B	O	
AB	None	No reaction	No reaction	No reaction	No reaction	Universal recipient
A	Anti-B	Agglutination	No reaction	Agglutination	No reaction	
B	Anti-A	Agglutination	Agglutination	No reaction	No reaction	
O	Anti-A, Anti-B	Agglutination	Agglutination	Agglutination	No reaction	
					Universal donor	

someone who is Rh negative does not. Physicians must also take Rh status into account when performing transfusions—anyone can receive Rh⁻ blood, but Rh⁺ blood transfused into an Rh⁻ recipient may result in agglutination. The ABO and Rh systems are typically referred to together—for instance, A⁺ (type A, Rh positive), or AB⁻ (type AB, Rh negative).

Stop and Stretch 8.3

When individuals have lost a large amount of blood, they can receive transfusions of a sugar solution that has the same solute concentration as plasma as a short-term replacement before their blood is typed and matched. However, in special cases the patient may need red blood cells. In these emergencies the blood of choice is O⁻. Explain why packed O⁻ red blood cells are the safest emergency transfusion.

Another situation where Rh-type status matters is during pregnancy. An Rh⁻ mother and an Rh⁺ father can produce an Rh⁺ baby. Because the blood supply of mother and baby are separate during development, this baby's blood often does not initiate an immune response—until birth, miscarriage, or elective abortion, when the mother becomes exposed to the Rh⁺ cells. As a result of this exposure, the mother will make anti-Rh antibodies to the Rh antigen.

In a subsequent pregnancy with an Rh⁺ baby, these antibodies can attack the fetus's blood, causing destruction of its blood cells and resulting in *hemolytic disease*. Hemolytic disease causes anemia and may lead to brain damage and even death in these babies. An Rh incompatibility between mother and child can be resolved by injecting Rh⁻ women with anti-Rh antibodies (an injection that goes by the trade name RhoGAM). The antibodies destroy the baby's blood cells in the mother's bloodstream before she has an opportunity to mount her own immune defense (FIGURE 8.9). This can occur either within a few days of giving birth to an Rh⁺ baby, or in the last trimester of pregnancy. Interestingly, the Rh⁻ trait is most common in European populations and is nearly absent in African, Asian, and American Indian groups.

Differences among individuals and populations in their blood types may reflect their evolutionary history. For example, one blood type that is common in West Africans but rare in other population groups is Duffy negative, a name that reflects the absence of the **Duffy antigen** on the surface of red blood cells. Louis Miller, a scientist at the National Institutes of Health, connected this fact to the observation that West Africans are nearly immune to malaria caused by *Plasmodium vivax* (although they are susceptible to other malaria-causing organisms). Miller's research demonstrated that the Duffy antigen is essential for *P. vivax* invasion into erythrocytes.

The increased survival of individuals who were Duffy negative must have resulted in natural selection for this blood type in regions where *P. vivax* is common. African-Americans are routinely screened for the Duffy antigen, to ensure that Duffy-negative individuals do not receive blood containing the antigen.

Recycling Red Blood Cells

Over time, red blood cells lose their flexibility and can no longer easily fit through the smallest blood vessels. Old and damaged red blood cells are carried by the circulatory system to the liver and *spleen*, a lymph system organ that lies just to the left of the liver in the abdominal cavity.

Within the liver and spleen, dying red blood cells are engulfed by macrophages, after which the components of the cell are recycled. Iron is transported to bone marrow for incorporation into new hemoglobin molecules. The other proteins and lipids in the cell are disassembled and carried in the bloodstream to sites of active protein or fat synthesis. The heme groups of the erythrocyte's hemoglobin molecules are transported to the liver for processing and excretion. The final product of heme degradation, a yellowish chemical called *bilirubin*, is excreted in bile, modified by bacteria in the intestines to a brown-colored pigment, and passed out of the body in feces. Under healthy conditions, individual erythrocytes live for about four months and are recycled at a rate of 200 billion cells per day.

During *Plasmodium* infection, the recycling of erythrocytes may be many times higher than normal. The subsequent increase in the number of dying red blood cells in the spleen, along with increased immune system activity, causes this organ to become painfully enlarged. Bilirubin production also becomes greater than the rate of excretion. The chemical thus accumulates in blood plasma, causing skin, mucous membranes, and even the whites of the eyes to turn yellow, a condition called **jaundice**. Jaundice is typically a sign of severe liver disease, such as hepatitis and cirrhosis, but in the case of malaria it is simply a side effect of high rates of erythrocyte breakdown.

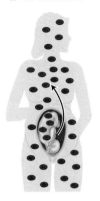 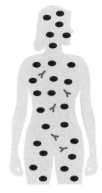 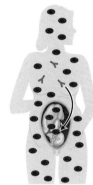

During birthing, a small amount of the baby's Rh⁺ blood enters the Rh⁻ mother's circulation.

As a result the woman develops antibodies against the Rh antigen.

When the woman becomes pregnant with a second Rh⁺ child, the now abundant antibodies cross the placenta and destroy the fetus's red blood cells, leading to hemolytic disease.

FIGURE 8.9 Rh incompatibility. An Rh⁻ mother will produce antibodies to the Rh antigen in late pregnancy or after delivery of an Rh⁺ child. In a subsequent Rh⁺ pregnancy, her immune system can attack the developing fetus's blood, leading to severe depletion of erythrocytes in the newborn.

VISUALIZE THIS How would this figure be different if it illustrated the effect of an anti-Rh antibodies (RhoGAM) injection on the pregnant mother?

8.3 Blood Clotting

The vessels of the circulatory system extend throughout the body—no cell is more than 0.1 mm (the thickness of a sheet of paper) away from the bloodstream. The 160,000 km (100,000 miles) of thin-walled blood vessels in the body are susceptible to damage. One of the most important roles of the blood is to limit loss of fluids from these vessels following an injury.

The Clotting Cascade

Platelets contain enzymes and other proteins that are crucial to **hemostasis**, the processes that stem the flow of blood out of damaged blood vessels. Immediately after a rupture occurs, muscles in the walls of blood vessels contract to restrict the flow of blood, a process known as *vascular spasm*. Collagen proteins in the blood vessel walls trigger changes in platelets, making them swell and become sticky. Within seconds,

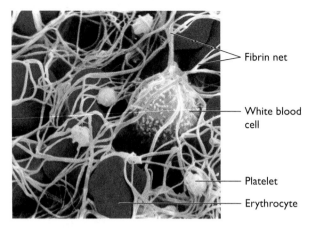

FIGURE 8.10 A blood clot. A blood clot consists of a net of protein fibers that trap platelets and red blood cells, forming a temporary patch over a damaged blood vessel.

these sticky platelets become attached to the ends of the damaged tissue, forming a temporary plug to help restrict blood flow.

A platelet plug may be effective at stopping blood from flowing from a small area of damage, but larger ruptures require a more solid patch. A **blood clot** consists of a net made up of the protein **fibrin**, which forms a screen over the damaged area (FIGURE 8.10). Blood cells and platelets become trapped in this screen, making a gel-like barrier to further blood flow. The fibrin threads of the clot form at the conclusion of a complex pathway that involves at least 12 clotting factors.

When blood vessel damage occurs, nearby cells release the clotting factor *thromboplastin*. This protein initiates a series of reactions that eventually produce an enzyme called *prothrombin activator*. Prothrombin activator converts the plasma protein *prothrombin* into the enzyme **thrombin**. Thrombin in turn catalyzes the conversion of another inactive plasma protein, **fibrinogen**, into fibrin. The main events of blood clot formation are summarized in FIGURE 8.11.

As a blood clot solidifies, the platelets within it begin to contract, pulling the damaged vessel walls closer together. Cell division in these walls and in surrounding tissues eventually seals the cut, and the fibrin patch dissolves. Clots can form within blood vessels as well as on their surfaces. A blood clot that remains in place is called a **thrombosis**, whereas one that travels in the bloodstream is called an **embolism**.

Clotting Disorders

A deficiency of platelets, called *thrombocytopenia*, can occur for a number of reasons, including diseases of the bone marrow. Low platelets are also a sign of serious malaria infection, as platelets get trapped in the enlarged spleen. The danger of low platelets is that uncontrolled bleeding, or **hemorrhage**, will occur, leading to **shock**, that is, inadequate blood flow to the major organs, and even death. Uncontrolled bleeding can also occur if any of the steps of the clotting sequence fail to occur properly. **Hemophilia** is an inherited condition in which mutations in one or more clotting factor genes lead to poor or no blood clot formation. Chapter 20 examines the genetics of the most common form of hemophilia in detail.

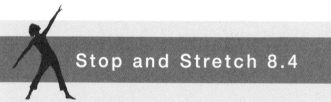

Stop and Stretch 8.4

When a clot forms within a blood vessel, it can block blood flow. Given your understanding of the functions of the circulatory system described at the beginning of the chapter, what is the likely consequence of thrombus or embolus formation?

FIGURE 8.11 Blood clotting. The process of fibrin formation, essential to the formation of a blood clot, requires a cascade of events.

VISUALIZE THIS Disseminated intravascular clotting (DIC) is a condition that occurs when the clotting cascade is inappropriately initiated in blood vessels throughout the body. The most severe side effect of DIC is uncontrolled bleeding when an actual injury does occur. Why might this be?

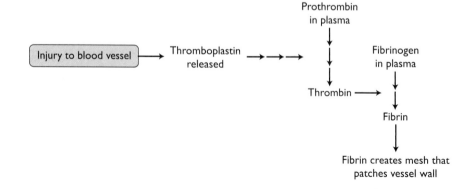

The most deadly complication of malaria is the formation of blood clots in the vessels supplying the brain. These clots do not form as a result of the clotting cascade, but instead from changes to the erythrocytes themselves.

When malaria parasites begin to mature inside a red blood cell, the cell begins to change in form. In particular, the cell surface becomes knobby and covered with a sticky protein. As a result, infected cells begin to adhere to the inner surface of blood vessels. Accumulations of cells become blockages, reducing gas and nutrient exchange between blood and nearby tissues. Oxygen levels decline and acid levels increase in nearby tissues as a result of the use of anaerobic respiration to supply energy.

In the brain, these effects lead to the symptoms of *cerebral malaria*: dizziness, confusion, convulsions, coma, and even death. See the Genes & Homeostasis feature for a description of a genetic mutation that reduces the risk of malaria infection but which interestingly carries a similar risk of clot formation.

GENES & HOMEOSTASIS

Sickle-Cell Anemia

The most common inherited blood disorder in the United States is **sickle-cell anemia**, occurring in about 72,000 individuals. Nearly all of the affected individuals are blacks with ancestry tracing to West Africa.

Individuals with sickle-cell anemia are disposed to experiencing sickling crises, during which blood cells deform from the flexible, rounded shape into stiff crescents (sickle shapes; see photo). The deformed cells cannot pass easily through the smallest blood vessels and thus form clogs, or thrombi, throughout the affected body part. The result is oxygen deprivation in the affected organ, severe pain, and possibly tissue death. Sickled cells also have much shorter lives than normal red blood cells, so anemia results when erythrocyte production cannot keep up with destruction. Until effective treatment (including oxygen therapy) became more available, individuals with sickle-cell anemia often died within the first three years of life.

The change in cell shape during a sickling crisis is due to the hemoglobin molecules within a cell forming long, rigid chains when oxygen or water levels are very low within the cell. These conditions may occur during exercise, illness, or dehydration. In the most common form of sickle-cell anemia, the abnormal hemoglobin results from a single amino acid change in each chain of a hemoglobin molecule. The amino acid change arises from a mutation in a single base pair of the hemoglobin gene. Individuals with sickle-cell anemia carry two copies of this mutated allele and, therefore, cannot make "normal" hemoglobin.

The fact that sickle-cell anemia is found almost exclusively in West Africans led scientists to speculate that, like lack of the Duffy antigen, the mutation must confer some protection from malaria. In fact, individuals with either one or two copies of the sickle-cell allele are resistant to *P. falciparum* malaria. When erythrocytes infected by *P. falciparum* form a clot in small blood vessels, the subsequent oxygen deprivation causes the mutant hemoglobin in these cells to form rigid crystals. The SHFF hemoglobin molecules apparently pierce the *P. falciparum* cell membrane, killing the parasite.

Even though sickle-cell anemia is deadly, the fact that a single copy of the sickle-cell allele is protective against malaria infection allows the sickle-cell allele to be maintained in human populations in malaria-prone areas. Because individuals with one copy of the allele have higher rates of survival than individuals with no copies, they contribute more children to the next generation. Half of the children produced by these individuals will carry the sickle-cell allele, so the allele remains common despite the serious disease it can cause.

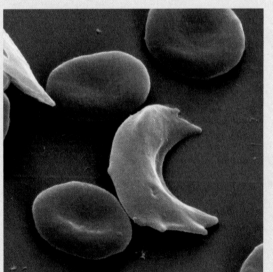

Normal and sickled blood cells.

8.4 Ending Malaria

The primary treatment for malaria infection is chloroquine, a cheap, widely available drug. Unfortunately, in many regions, notably large parts of Africa and Southeast Asia, this drug has become less effective in the past 30 years because *P. falciparum* has evolved resistance to it. Alternative therapies for drug-resistant malaria are more expensive or may have severe side effects, limiting their use in impoverished countries hard hit with malaria outbreaks. In some areas where chloroquine treatment had been abandoned for decades because of widespread resistance, the parasite has reverted to a nonresistant form; however, readoption of chloroquine treatment is likely to result in the redevelopment of resistance. Because of the difficulties inherent in treating malaria, much of the current research focuses on prevention of the disease.

Several of humanity's most deadly diseases have been tamed as a result of widespread **immunization**, also known as vaccination. This technique, discussed in Chapters 12 and 13, involves introducing portions of infectious organisms so that the immune system will develop antibodies to the invader. An immunized individual will be resistant to infection for as long as the immune response lasts. Most successful vaccines have been produced against disease agents such as simple viruses, like those that cause smallpox and polio, or bacteria, such as the species that causes tetanus.

Because malaria is caused by a complex organism with a complicated life cycle, the disease has frustrated control by a vaccine. But a vaccine is not out of the question—in fact, the Gates Foundation has pledged over $250 million to the Malaria Vaccine Initiative since 1999, a funding increase that has spurred increased activity on this front. In 2004, researchers from the pharmaceutical firm GlaxoSmithKline reported that a tested vaccine had reduced malaria incidence among immunized children by 58%. Although the vaccine was not completely effective, this drug trial represented the first major progress toward the elimination of malaria through immunization.

Even the most optimistic scenario predicts that an effective vaccine will not be widely available in malaria-prone regions until 2015. In the meantime, strategies to reduce the transmission of *Plasmodium* represent the best hope for reducing malaria's toll. Some of these strategies are high tech, including the development of new treatment drugs and better drug delivery systems, which decrease the risk that mosquitoes will pick up infected blood. The Gates Foundation has pledged $165 million toward these efforts.

However, some strategies are elegantly simple and low cost—including reducing the amount of mosquito breeding areas near human habitation. This requires draining or filling small ponds and preventing the accumulation of standing water in containers such as old tires. Treating house interiors with mosquito-control chemicals can reduce the risk of bites as well. One of the easiest and lowest-cost preventive measures is the use of insecticide-coated bed nets (FIGURE 8.12), especially surrounding individuals most at risk for severe disease, such as small children. Research has indicated that treated bed nets alone can reduce the risk of malaria by 25% or more. These results have inspired a number of grassroots efforts, including "Nothing but Nets," a program spearheaded by *Sports Illustrated* magazine writer Rick Reilly that encourages sports fans to donate $10 to purchase mosquito nets for a single family in malaria-prone areas.

The story of malaria control will probably not have a Hollywood ending in which efforts of a few maverick scientists and military commanders save the day. Malaria is so complex that many different, interacting factors must be employed for long-term reasonable control of the disease to come about. But the lesson of the first eradication effort is telling—if we were to give up our efforts to control malaria now, we would doom millions of people to severe illness and death. With efforts from real-life heroes like the Gateses, Warren Buffett, Rick Reilly, and thousands of hard-working scientists, medical personnel, and donors of single bed nets, the ending of the story about our struggle with this ancient killer could be a happy one.

FIGURE 8.12 A simple way to reduce malaria. Insecticide-treated bed nets, used over the sleeping areas of young children and pregnant women, are an effective way to reduce the risk of malaria transmission.

Chapter REVIEW

ROOTS TO REMEMBER

The following roots of words come mainly from Latin and Greek and will help you to decipher terms:

a- or **an-** are prefixes for without or lacking.

-cyte and **cyto-** refer to a cell.

-emia is the ending for conditions or diseases of the blood.

erythro- means red.

hemo- and **hemato-** relate to blood.

leuko- means white.

-phage and **phago-** come from the verb meaning to eat.

-poiesis means to make or the making of.

-rrhage (also **-rrhagia**) comes from the verb to burst or to flow excessively.

thromb- and **thrombo-** mean clot.

KEY TERMS

8.1 The Constituents of Blood
albumin *p. 181*
antibody *p. 181*
B lymphocyte *p. 184*
basophil *p. 183*
blood *p. 180*
circulatory system *p. 180*
eosinophil *p. 183*
erythrocyte *p. 181*
formed elements *p. 180*
globulin *p. 181*
hemoglobin *p. 182*
leukocyte *p. 183*
macrophage *p. 183*
malaria *p. 180*
megakaryocyte *p. 184*
monocytes *p. 183*
neutrophil *p. 183*
parasite *p. 180*
plasma *p. 180*
platelet *p. 184*
serum *p. 181*
stem cells *p. 181*
T lymphocyte *p. 184*

8.2 Malaria and the Blood
ABO system *p. 187*
agglutination *p. 187*
anemia *p. 186*
antigen *p. 187*
blood transfusion *p. 186*
blood type *p. 187*
Duffy antigen *p. 189*
erythropoietin *p. 186*
jaundice *p. 189*
Rh factor *p. 187*

8.3 Blood Clotting
blood clot *p. 190*
embolism *p. 190*
fibrin *p. 190*
fibrinogen *p. 190*
hemophilia *p. 190*
hemorrhage *p. 190*
hemostasis *p. 189*
shock *p. 190*
sickle-cell anemia *p. 191*
thrombin *p. 190*
thrombosis *p. 190*

8.4 Ending Malaria
immunization *p. 192*

SUMMARY

8.1 The Constituents of Blood

- Blood is the connective tissue in the circulatory system. Blood's primary functions are materials transport, regulation of body temperature and water volume, and providing constituents of the immune system (p. 180).

- Blood consists of a liquid portion, called plasma, and a solid portion made up of formed elements (p. 180).

- Plasma carries nutrients and gases, helps maintain water balance with body tissues, and contains proteins that transport other molecules, attack invaders, and prevent blood loss (pp. 180–181).

- All blood cells derive from stem cells in the marrow of long bones (p. 181).

- The primary function of erythrocytes is carrying oxygen, bound to the protein hemoglobin, from the lungs to other body organs (pp. 181–182; Figure 8.5).

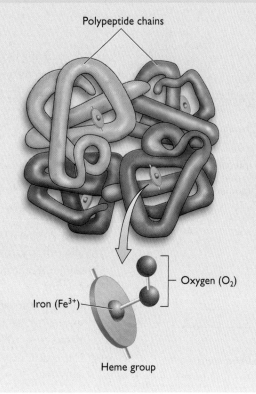

- The blood contains a variety of leukocytes, including those that function in the nonspecific immune response targeted at all invading organisms, and those that are targeted toward specific pathogens (pp. 183–184).

- Platelets are cell fragments that function in blood clotting (p. 184).

8.2 Malaria and the Blood

- The malaria parasite kills red blood cells, leading to anemia. Under normal conditions, a negative feedback loop consisting of signals of blood oxygen levels and a hormone produced by the kidney help to maintain adequate erythrocyte levels (pp. 185–186; Figure 8.7).

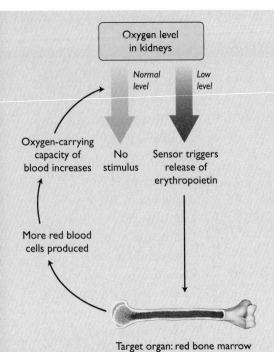

- An incompatibility in Rh blood type between mother and fetus can lead to anemia in the newborn. This problem can be prevented by administering anti-Rh antibodies to the mother during pregnancy (pp. 187–188).
- Old and damaged red blood cells are recycled by the spleen and liver. One waste product of hemoglobin breakdown is bilirubin, a yellow pigment that can accumulate in the blood if red blood cells are dying at a high rate or the liver is diseased. Excess bilirubin results in jaundice (p. 189).

8.3 Blood Clotting

- After a blood vessel is damaged, its walls spasm to restrict blood release. Platelets in the area become sticky and fuse to the damaged tissue ends, forming a temporary plug. If the area of damage is large, a clot consisting of a fibrin net and trapped platelets and red blood cells forms over the break (pp. 189–190).
- The production of fibrin from its precursor fibrinogen in the plasma requires a cascade of events. The first step of this cascade is the release of thromboplastin from the damaged tissue (p. 190; Figure 8.11).

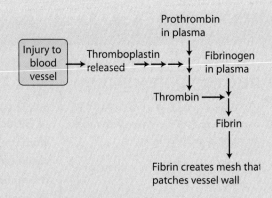

8.4 Ending Malaria

- The organisms that cause malaria are becoming resistant to the most common and effective drugs. As a result, new approaches, including high-tech development of a vaccine and low-tech preventive measures such as bed nets, must be employed to win the battle with this disease (p. 192).

- Blood can be replaced by transfusion from a donor. Because blood cells contain cell surface molecules that can stimulate an immune response, it is important to match the donor's and recipient's blood types (pp. 186–187).

LEARNING THE BASICS

1. All of the following are functions of the blood *except* _____.
    a. transportation of oxygen to the tissues
    b. prevention of fluid loss from blood vessels
    c. production of hormones to trigger blood cell production
    d. maintenance of body water balance
    e. transport of cells and proteins that help fight infections

2. All of the following are normal constituents of blood *except* _____.
    a. serum
    b. cells
    c. water
    d. proteins
    e. *Plasmodium*

3. The role of albumins in plasma is primarily _____.
    a. antibiotic
    b. maintaining water balance between blood and body cells
    c. participation in the clotting cascade
    d. determining blood type
    e. triggering the production of additional red blood cells

4. The stem cells for all formed elements in the blood are found in _____.
    a. the bone marrow
    b. the spleen
    c. the thymus
    d. the blood itself
    e. all of the above

5. The shape of an erythrocyte _____.
    a. allows it to easily escape the blood vessels
    b. minimizes the number of hemoglobin molecules it can contain
    c. makes it especially susceptible to damage
    d. maximizes the surface area for oxygen exchange with the tissues
    e. is irregular, making it well suited for malaria parasite infection

6. All of the following leukocytes are involved only in the nonspecific immune response *except* _____.
   a. monocytes
   b. B lymphocytes
   c. eosinophils
   d. neutrophils
   e. macrophages

7. The function of platelets is to _____.
   a. participate in blood clotting
   b. transport oxygen in the blood
   c. produce antibodies
   d. engulf invading cells
   e. produce bilirubin

8. Which of the following characteristics are shared by erythrocytes and platelets?
   a. contain hemoglobin
   b. lack nuclei
   c. live for less than six months
   d. b and c are correct
   e. a, b, and c are correct

9. Which of the following cells are destroyed by the malaria parasite?
   a. T lymphocytes
   b. platelets
   c. eosinophils
   d. monocytes
   e. erythrocytes

10. The symptoms of anemia _____.
    a. are fatigue, weakness, and breathlessness
    b. can be relieved in some cases by improving the diet
    c. indicate that the hematocrit may be low
    d. result from the blood's reduced capacity for oxygen delivery
    e. all of the above are correct

11. All of the following are true of the protein erythropoietin *except* _____.
    a. it is produced by the kidney
    b. it triggers the production of red blood cells
    c. it is a hormone
    d. it can cause the agglutination of blood cells
    e. it is part of a negative feedback loop

12. An individual's blood type is determined by _____.
    a. the percentage of red blood cells in whole blood
    b. sugars and other substances on the surface of red blood cells
    c. the production of antibodies by leukocytes
    d. the likelihood of blood clots forming under certain conditions
    e. whether the mother received RhoGAM before the individual was born

13. Individuals with blood type O _____.
    a. can only receive blood from other type O individuals
    b. can receive blood from type A, B, AB, and O individuals
    c. can donate blood to type A, B, AB, and O individuals
    d. a and c are correct
    e. b and c are correct

14. An individual with Duffy-negative blood _____.
    a. was born in Africa
    b. is protected from some types of malaria
    c. should not receive a transfusion of Duffy-positive blood
    d. b and c are correct
    e. a, b, and c are correct

15. Jaundice results from _____.
    a. accumulation of the hemoglobin waste product, bilirubin
    b. accumulation of dying red blood cells in the spleen
    c. an overactive liver
    d. reduced recycling of red blood cells

16. All of the following help reduce blood loss from a damaged vessel *except* _____.
    a. spasms in the vessel walls that help pinch the vessel shut
    b. the production of fibrin net
    c. increased stickiness of platelets, causing them to adhere to the site of the cut
    d. a rapid decline in the number of platelets
    e. release of thromboplastin

17. Put the steps of the clotting cascade in order:
    I. Conversion of prothrombin to thrombin
    II. Catalysis of fibrinogen to fibrin by thrombin
    III. Production of the enzyme prothrombin activator
       a. I, II, III
       b. I, III, II
       c. III, I, II
       d. III, II, I
       e. II, III, I

18. Hemophilia occurs _____.
    a. when a portion of the clotting cascade is nonfunctional
    b. when platelet levels are low
    c. when clots randomly occur in blood vessels
    d. as a result of cerebral malaria
    e. as a result of severe infection

19. All of the following would help reduce the number of deaths due to malaria, *except* _____.
    a. use of insecticide-treated bed nets
    b. the use of chemicals that reduce the activity of the immune system
    c. development of a malaria vaccine
    d. control of mosquito populations near human habitation
    e. more effective antimalaria drugs and better drug distribution

20. **True or False?** Malaria is affecting fewer and fewer people every year.

ANALYZING AND APPLYING THE BASICS

1. The retinoblastoma (Rb) protein and macrophages are key components in the process of maturation of erythrocytes. To begin the process, Rb protein appears to regulate macrophage maturation. Mature macrophages then bind to immature red blood cells, called erythroblasts. When mature, erythroblasts become erythrocytes, which separate from the macrophage to become individual cells within the bloodstream.

   How would a genetic defect that causes a deficiency in Rb protein potentially affect a developing organism?

Questions 2–4 refer to the following information.

Porphyria is a rare disease that affects heme production. People suffering from certain forms of porphyria have anemia and skin that is sensitive to light. In fact, the fictional Dracula may have been inspired in part by the symptoms of porphyria.

Recent studies have indicated that some forms of porphyria may result from the faulty transport of the molecule porphyrin into mitochondria. Porphyrin is made inside red blood cells. This molecule must travel into the mitochondria of the blood cell, where it binds to iron to make heme. Since both porphyrin and mitochondria are negatively charged, porphyrin must be actively transported across the mitochondrial membrane by a transporter protein. This protein, called ABCB6, may be dysfunctional in some people suffering from porphyria.

2. Would iron supplements help treat anemia in a person suffering from porphyria? Explain your answer.

3. The fictional Dracula survived by drinking other people's blood. Could some porphyria patients be treated with blood transfusions?

4. How would untreated porphyria affect the person who is suffering from it?

5. Over time, men may accumulate too much iron in their tissues. Some scientists hypothesize that an excess of iron in the blood can lead to cardiovascular disease. This hypothesis is based on studies that found that iron increases oxidation of cholesterol in blood, which may damage the arteries. A Finnish study seems to support the hypothesis. It found that male blood donors who gave blood once a year or more—thereby removing some of the iron in their systems—had a significantly lower risk (88%) of having a heart attack.

   Why are women less likely to build up excess iron in their blood?

6. In 2004, the COX2 inhibitor Vioxx (rofecoxib) was pulled off the market when it was found to dramatically increase the risk of heart attacks and strokes in users. COX2 inhibitors suppress the production of the protein prostacyclin. Prostacyclin causes pain and inflammation, so COX2 inhibitors are used to treat arthritis and other diseases that cause chronic inflammation. But studies using mice have also shown that prostacyclin helps to suppress platelet activity.

   Explain how the action of a COX2 inhibitor could potentially lead to heart attack or stroke.

7. In severe malaria, infected red blood cells clump together and stick to the sides of blood vessels. The drug heparin has anticoagulant properties that minimize the stickiness of these infected cells. Why would an injury be particularly dangerous to someone with severe malaria who is taking heparin?

Questions 8 and 9 refer to the following information.

Glucose-6-phospate dehydrogenase (G6PD) deficiency is the most common enzyme-deficiency disease in the world. It occurs most often in males who have African, Middle Eastern, or Southeast Asian ancestry. The most severe effects of G6PD deficiency are on erythrocytes, which are poorly protected against damage by oxidizing chemicals (that is, chemicals that remove electrons from other molecules) in the absence of G6PD. Under any physical condition that increases the number of oxidizing chemicals, such as infection or the intake of certain drugs, red blood cells will be destroyed, causing anemia.

8. Devise a hypothesis to explain why G6PD deficiency may be more common in areas where malaria is common.

9. Describe how you could test your hypothesis.

10. Hemoglobin C is a type of hemoglobin common in residents of West Africa. This form of hemoglobin has been found by some studies to offer protection against malaria without the deleterious effects of hemoglobin S—the form that causes sickle-cell anemia. A study of 4,348 subjects in Burkina Faso, in West Africa, found the results in Table 8.4.

Table 8.4 Hemoglobin Alleles and Malaria

	Number of Patients with Each Genotype Frequency						
	n	AA	AC	AS	CC	SC	SS
Healthy Subjects	3,513	2,333	763	335	58	23	1
Subjects with Severe Malaria	359	290	63	4	1	1	0
Subjects with Noncomplicated Malaria	476	381	74	19	0	2	0

n indicates the number of subjects in each group.

In the table, AA represents an individual who has two genes for hemoglobin. AC represents an individual with one allele for hemoglobin C and one normal allele. AS represents an individual with one allele for hemoglobin S and one normal allele. CC represents an individual with two alleles for hemoglobin C. SC represents an individual with one allele for hemoglobin S and one for hemoglobin C. SS represents an individual with two alleles for hemoglobin S.

Calculate the percentage of healthy subjects compared to the total subjects for each genotype frequency. Which alleles appear to offer resistance to malaria in this study? Explain your answer.

CONNECTING THE SCIENCE

1. Malaria remains one of the ten most deadly diseases in the world, but it affects primarily children in Africa, Asia, and South America. Even though malaria is not a risk to the population in the United States, do you think the U.S. government should continue to fund research on malaria? Why or why not?

2. Although Bill and Melinda Gates have been very generous in their grants to malaria research, to other global health issues (including HIV/AIDS), and to education, they still have an enormous personal fortune. Many other billionaires and multimillionaires do not share the Gateses' commitment to using some of their wealth to solve pressing global problems. Some critics have argued that, rather than relying on the philanthropy of the very wealthy to solve these problems, governments should heavily tax these fortunes and use the money to provide benefits that the population as a whole agrees upon. What do you think?

Chapter 9

The Cardiovascular System

Can We Stop the Number-One Killer?

LEARNING GOALS

1. List the primary components of the circulatory system and describe the function of each.
2. Compare and contrast the structure and function of arteries, arterioles, capillaries, and veins.
3. How are materials exchanged between body and blood in a capillary bed?
4. Summarize how blood pressure is regulated in the body.
5. Describe the structure of the heart, including the four chambers and the valves.
6. How does blood move through the double circulation system?
7. Describe the heart's electrical system, including the structures involved and the path and timing of the electrical signal.
8. Illustrate the cardiac cycle.
9. Describe the coronary circulation.
10. Summarize the various ways that components of the cardiovascular system can fail and describe the medical tools for repairing those failures.
11. List the five steps for maintaining good heart health.

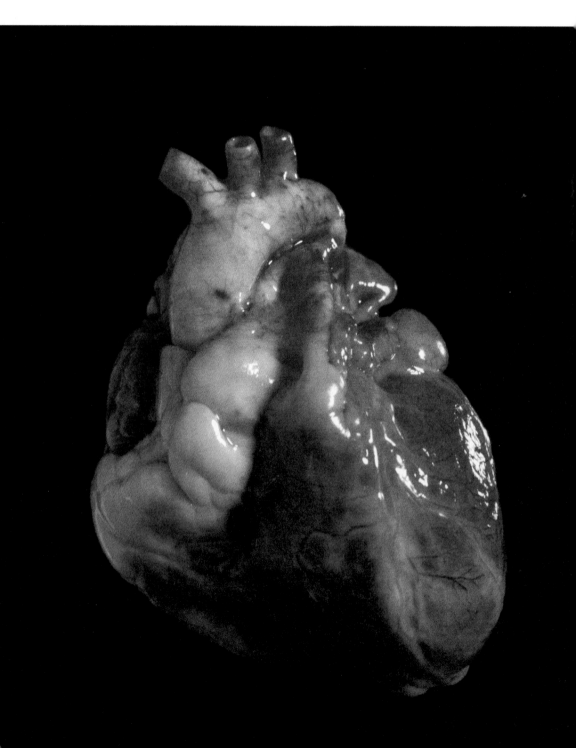

This organ is a killer.

The numbers are sobering: one-third of all U.S. deaths; nearly 700,000 Americans per year; more than cancer, accidents, and AIDS combined.

Diseases of the heart and cardiovascular system are the number-one—by a long shot—cause of death in the United States. And cardiovascular disease doesn't just cause mortality, but morbidity (illness) as well. The American Heart Association estimates that more than 71 million Americans, almost 25% of the population, are living with one or more forms of diseases of this system, including high blood pressure, heart attacks, chest pain, and stroke. Heart disease accounts for more than 6 million hospitalizations every year.

Millions of Americans suffer from heart disease.

You would not expect such a simple organ system to cause so much trouble. In principle, the blood vessels appear to be simply a piping system, and the heart is just a pump. Like a fuel pump that supplies gasoline to a car engine or a pump that draws water up into a town's water tower, the heart has a reservoir that fills with liquid and then is compressed to force the liquid out. The rhythm of the heart pump is driven by an electrical system, as in many mechanical pumps. A supply of richly oxygenated blood powers the heart, just as all mechanical pumps have a source of power.

But the heart is just a simple pump.

If you can repair a car's fuel pump, why not just repair heart pumps, and if you can patch a leaky pipe, why not just replace damaged blood vessels?

In reality, despite being responsible for so many deaths, the heart and to some extent the blood vessels are more amenable to repair than nearly any other organ system in the body. Scientists and doctors have devised an array of tools and techniques to keep damaged and failing cardiovascular systems in working order. These advances have had a major impact on the health and survival of individuals with cardiovascular disease. To appreciate the medical fixes of this organ system and consider whether or not its malfunctioning will remain our number-one killer, we need an understanding of how this remarkable machine works.

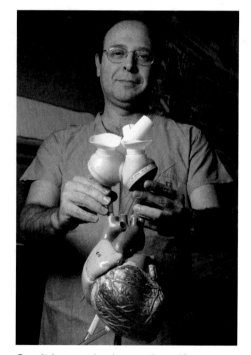

Can it be repaired or replaced?

9.1 Blood and Lymphatic Vessels: The Circulation Pipes

The system of tubes that carries blood to and from the heart is generally referred to as the *vascular system*. The **cardiovascular system** consists of the vascular system, the heart, and the blood that travels through these organs. The cardiovascular system along with the lymphatic system (described below) make up the *circulatory system,* which is responsible for moving fluids, nutrients, hormones, salts, and dissolved gases around the body and exchanging these substances with body tissues (FIGURE 9.1).

Arteries and Arterioles

Components of the vascular system, broadly termed **blood vessels**, include arteries, veins, and capillaries (FIGURE 9.2). **Arteries** carry blood from the heart, and **veins** bring it back. **Capillaries** are the tiny, thin-walled blood vessels that create a net of channels between the smallest arteries and the smallest veins.

The largest arteries have thick, elastic walls made up of smooth muscle and connective tissue. These walls balloon out as contraction of the heart causes a mass of blood to flow into the system; then they snap back to resting size once the blood passes by. The wave of blood is called a **pulse**. You can measure your heart rate by feeling the pulse as it passes through an elastic artery close to the surface of the skin, such

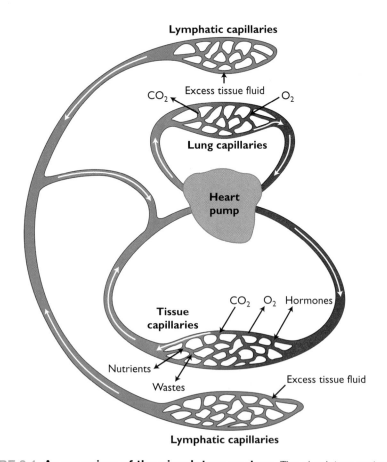

FIGURE 9.1 **An overview of the circulatory system.** The circulatory system is made up of the cardiovascular system, containing the heart and blood vessels, and the lymphatic system, which drains fluid from the tissues and returns it to the vascular system. In this figure and throughout, red indicates oxygenated blood and blue indicates deoxygenated blood.

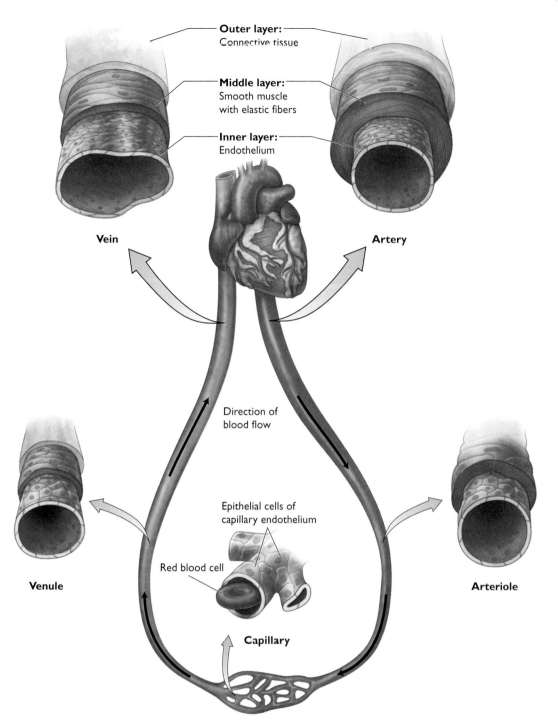

FIGURE 9.2 The vascular system. Arteries carry blood away from the heart, and veins carry blood to the heart. These two types of blood vessels are connected by nets of capillaries. The arteries, veins, and venules consist of three tissue layers, arterioles are made up of two, and capillaries have only a single layer.

202 CHAPTER 9 The Cardiovascular System

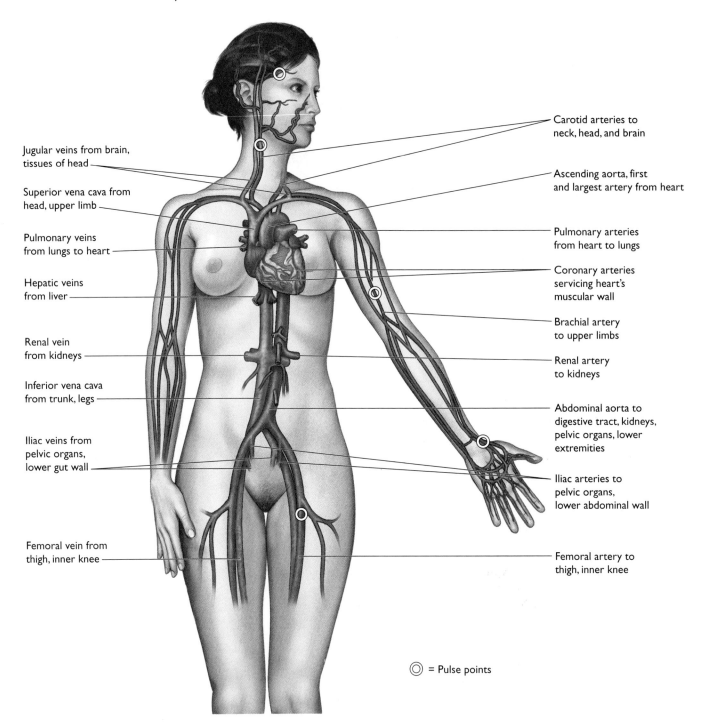

FIGURE 9.3 Pulse points. The major arteries that pass near the surface of the skin permit the measurement of heart rate as the pulse of blood passes through them. As you can see from this figure, the major veins run essentially parallel to the major arteries.

as near the *carotid arteries* in the neck or the radial arteries supplying the hand (FIGURE 9.3).

The smallest arteries are called **arterioles**, and, because they do not have a connected tissue sleeve, they lack the elasticity of the major arteries. These vessels are encircled by smooth muscle that can regulate the size of the vessels. Arterioles are not simple pipes, but a bit more like the nozzle at the end of a water hose—tightening a nozzle restricts the flow and raises the water pressure, while loosening it allows more water to pass at lower pressure. Similarly, *vasoconstriction*, which reduces arteriole diameter, reduces flow and increases blood pressure, whereas *vasodilation*, the relaxing of arterioles, increases flow and decreases pressure (FIGURE 9.4).

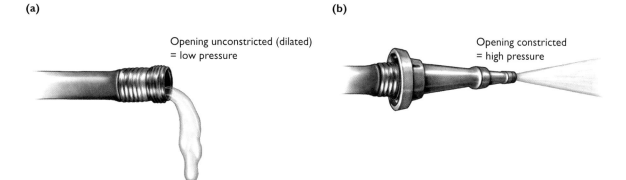

FIGURE 9.4 Arteriole diameter and blood pressure. (a) When arterioles are dilated, blood flows through at lower pressure, equivalent to water flowing out of the end of a hose. (b) Constriction of the arteriole increases the pressure of blood, just as constriction of the nozzle at the end of the hose puts pressure on the water, made obvious by the increase in the distance the water spray travels.

Blood flows from arterioles to the capillaries, the tiny blood vessels in which materials exchange occurs.

Capillaries: The Distribution Network

If the vascular system is analogous to a community's water system, the arteries represent the main pipes from the water treatment plant, and the capillaries represent the individual pipes running to and from each household and place of business.

A typical capillary has thin walls made up of a single layer of *endothelium,* that is, simple squamous epithelial cells. The thin endothelium allows gases, nutrients, wastes, and other small molecules to diffuse rapidly into and out of the bloodstream. Motile white blood cells called *macrophages* can also squeeze between the epithelial cells lining the capillaries to patrol body tissues. Some capillaries have gaps or *pores* between cells that allow larger molecules, water, and even some cells to be exchanged with certain tissues.

With a diameter of 0.8 millimeter, a capillary is not much larger than a single red blood cell. This narrow diameter requires that red blood cells pass through in single file, slowing down the flow, much as a three-lane highway merging to a single lane will cause a traffic slowdown (FIGURE 9.5). The low flow rates through capillaries allow enough time for diffusion and active transport to make needed exchanges between the tissues and bloodstream.

The exchange of gases, nutrients, and wastes between the bloodstream and body tissues occurs only within **capillary beds**, interconnected networks of vessels (FIGURE 9.6). Liquid and materials in the capillaries are forced out into the tissues through pores due to higher blood pressure near the arterial end of the bed. This process of filtration leaves large proteins behind in the bloodstream, increasing the blood's solute concentration.

At the downstream, venous (vein) end of the capillary bed, this difference in solute concentration between the tissues and the blood causes water to return to the capillaries by osmosis. Wastes in high concentration in the tissues also flow into the capillaries down their concentration gradient.

Muscles called **precapillary sphincters** surrounding the arterial ends of capillary beds can contract to cut off blood flow through the bed. The dynamic regulation of blood flow maintains homeostasis during different physical conditions by directing oxygen and nutrients to immediately essential organs. For instance, during a stressful situation, precapillary sphincters surrounding the digestive organs will contract, increasing blood flow to muscles and putting off digestion until a more restful period.

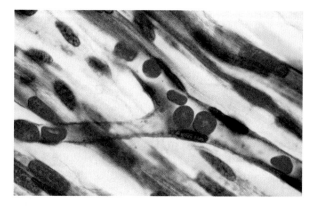

FIGURE 9.5 Tiny blood vessels. The diameter of capillaries is so narrow as to allow only a single file of red blood cells to pass through.

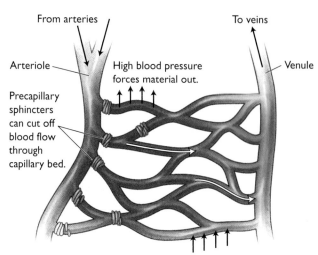

FIGURE 9.6 Capillary bed. Exchange between the body tissues and the bloodstream occurs within capillary beds.

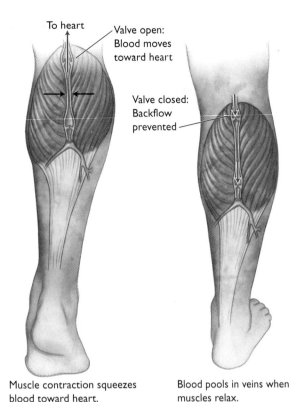

FIGURE 9.7 **Flow of blood in veins.** Blood in the veins is under low pressure and returns to the heart as a result of the contraction of skeletal muscle. Backflow is prevented by one-way valves.

Veins: The Path Back to the Heart

The small vessels that collect blood from the downstream end of a capillary bed are called **venules**. Venules drain into veins, which are somewhat like sewage pipes in a water system, delivering dissolved carbon dioxide to the lungs for disposal.

In contrast to arteries, veins have much thinner, less-elastic walls, and the pressure of the blood is much lower once it reaches these vessels. Vein walls are very stretchy, so blood tends to pool within them, a phenomenon you can easily see by lowering a hand to your side. Blood pooling in the veins on the back of the hand will make these veins become distended and bulge out from the surface of the skin. Because of their thin walls and low pressure, veins serve as a reservoir for the blood—nearly two-thirds of the blood's volume is found in the veins at any given time.

Movement of blood from the veins back to the heart is facilitated by the contraction of skeletal muscles, which compress the veins and squeeze the blood through them. The blood flows in only one direction, toward the heart, due to the presence of one-way valves within the veins (FIGURE 9.7).

Muscle movements associated with breathing also help move blood from the abdomen toward the heart. As we inhale, pressure in the abdominal cavity increases while pressure in the thoracic (chest) cavity decreases. Abdominal veins become compressed and thoracic veins dilate. This change in pressure, called the **respiratory pump**, which is not a structure but a physical process, forces blood toward the heart.

Veins have some capacity for dilation and contraction and thus can control the volume of blood returning to the heart. When a greater degree of blood circulation is required, as when an individual is exercising, the veins constrict, reducing the amount of blood pooled in these vessels. When the need for oxygen delivery to body tissues declines, the veins dilate, reducing overall blood flow.

Stop and Stretch 9.1

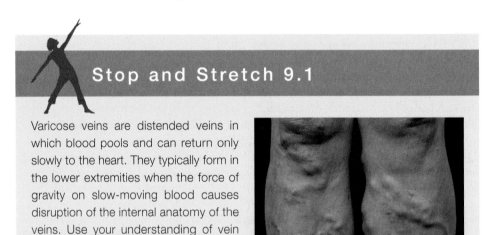

Varicose veins are distended veins in which blood pools and can return only slowly to the heart. They typically form in the lower extremities when the force of gravity on slow-moving blood causes disruption of the internal anatomy of the veins. Use your understanding of vein anatomy to explain why disruption of vein structure can lead to varicose veins.

The Lymphatic System: Draining the Tissues

The amount of liquid returned by osmosis to the venous end of a capillary bed is only about 85% of the amount forced out by blood pressure at the arterial end. In other words, every time blood passes through capillaries, about 15% of the liquid forced from it by blood pressure remains in the surrounding tissue. The tissues would swell and organ function would be compromised if this excess liquid was not collected and returned to the vascular system. Collection and return of excess fluid is performed by a parallel set of channels known as the **lymphatic system** (FIGURE 9.8).

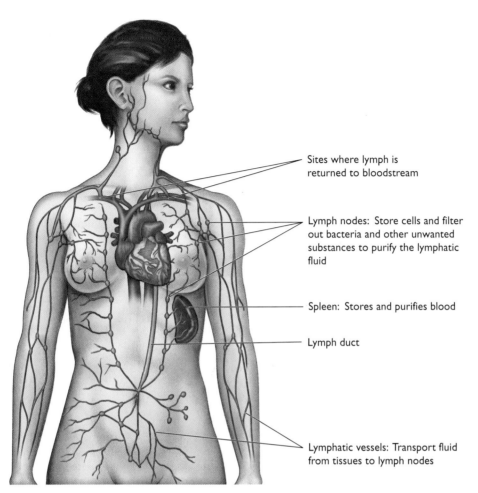

FIGURE 9.8 The lymphatic system. The lymphatic vessels serve as a parallel to the vascular system, returning fluid from the tissues into the bloodstream at venous connections in the chest cavity. The lymph nodes and spleen are among the organs of the lymphatic system that are crucial to immune system function.

VISUALIZE THIS How would a blockage in a lymph vessel affect the body?

Fluid from the tissues along with patrolling white blood cells, together called **lymph**, drain into open-ended lymphatic capillaries, which in turn empty into larger vessels. These larger vessels are similar in structure to veins, in that they contain valves to prevent the backward flow of lymph. Lymphatic vessels merge into **lymph ducts**, which return the lymph to veins in the chest.

The lymphatic system's role as a collector of all fluids flowing across body tissues makes it an essential component of the immune system. The immune function of the lymphatic system will be discussed in detail in Chapter 12.

Control of Blood Pressure

The blood vessels can experience the same types of problems as any pipe: ruptures and clogs. In addition, the arterioles' ability to change their diameter can either reduce the risk of damage or make it more likely, depending on influences from other organ systems.

Blood pressure, the force of the blood against arterial walls, is created in part by the pulse of blood ejected from the contracting heart and in part by the diameter of the arterioles. The elasticity of artery walls helps to dampen the pulse and keeps a relatively steady blood pressure despite the force with which blood is ejected from the heart. The diameter of the arterioles also adjusts to keep blood pressure within narrow bounds, even over a wide range of physical conditions.

When blood pressure rises due to increased heartrate—for instance, to meet the oxygen needs of hard-working muscles—artery walls stretch. The stretch is registered by nervous system receptors in the walls of the carotid arteries serving the head. The stretch receptors trigger a response in the kidneys to decrease water retention, which reduces blood volume. The receptors also send a signal to the cardiovascular center in the brain.

GENES & HOMEOSTASIS

Race-Based Medicine?

In 2005, the U.S. Food and Drug Administration announced that it had approved the drug combination of isosorbide dinitrate and hydralazine hydrochloride (BiDil) as treatment for heart failure in African-American patients. This announcement appeared to be the leading edge of a new field of "race-based medicine," a more targeted approach to medical treatment that recognizes that individuals vary in their disease susceptibility and in their response to drugs. But most genetic research seems to indicate that races are more alike than they are different. Why does BiDil have different effects on individuals who identify as members of different races?

BiDil is marketed as an "African-American" drug as a result of a reanalysis of a clinical study. The purpose of the original study was to investigate the usefulness of the drug combination at reducing death and hospitalization rates due to heart failure. The results of the study indicated that the drug was not effective when examining the entire population; however, researchers noticed some effect on African-American participants. Subsequent research performed only on a black population indicated that BiDil reduced death rates by a dramatic 43% compared to placebo in patients already receiving the standard treatments for heart failure.

BiDil appears to be more effective in African-Americans because it causes the production of nitric oxide, the gas produced by endothelial cells in the arterioles and stimulates vasodilation. On average, African-Americans produce less nitric oxide than white Americans, so the effect of BiDil on their survival is greater. But why do African-Americans produce less nitric oxide? The answer may be related to a completely different genetic factor.

LTA4H is a gene that promotes inflammation in the presence of infection. About 30% of whites carry a variant of this gene that causes a stronger inflammation response. The more active LTA4H variant appears to have evolved in humans that left Africa 50,000 years ago and moved into Europe and Asia. While the mutation may protect against infectious disease, it also increases the risk of heart disease—whites with the more active variant have a 16% higher risk of heart disease than those without. Remarkably, however, the 6% of African-Americans who carry the more active LTA4H variant have a 250% greater risk of heart disease.

Scientists studying this gene hypothesize that the different effect of the LTA4H variant on different populations arises from the length of time the variant has been present in each population. Europeans have experienced this trait for hundreds of generations. Individuals with characteristics that adjusted the homeostatic response to inflammation in the face of highly active LTA4H were less likely to experience heart disease. As a result, these individuals were more likely to survive and produce the next generation. Therefore traits that counteract the damage caused by increased inflammation—traits like increased nitric oxide production—could have become more common in European populations.

African-Americans have only been exposed to this trait in the past four hundred years, when the slave trade in America brought large numbers of Europeans and Africans together and resulted in reproduction between the groups. According to the hypothesis, the current population of African-Americans carries a genetic imprint of that mixing—and some of that imprint includes an increased susceptibility to inflammation without the heart-protective effects of nitric oxide.

The cardiovascular center triggers vasodilation, as well as a reduction in heart rate and the strength of the contraction, reducing the volume of each blood pulse.

High blood pressure also causes a small amount of damage to the endothelial cells lining arterioles. As a result, these cells release *nitric oxide (NO)*, a gas that travels to the arteriole's smooth muscle cells and induces them to relax, dilating the vessel. (The Genes & Homeostasis feature explores a racial difference in NO production that has implications for the treatment of heart disease.) The net effect of all of these changes is a reduction in blood pressure.

If blood pressure drops too low, an opposing series of events occurs—nitric oxide is not produced and the stretch receptors signal the cardiovascular center to increase cardiac output and decrease arteriole diameter to increase blood pressure. The kidneys actively retain more water in order to increase blood volume as well.

Fixing the Pipes

The feedback mechanism described above that keeps blood pressure within normal ranges can fail, and the result may be chronically high blood pressure, or **hypertension**. In the absence of underlying disease, the factors that may cause hypertension include psy-

(a) Normal artery

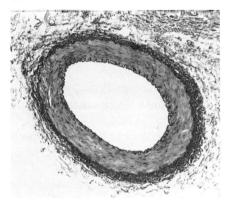

(b) Atherosclerotic artery

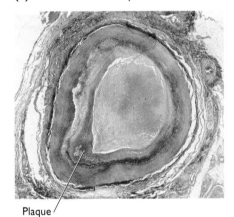

Plaque

FIGURE 9.9 Atherosclerosis. Fat and cholesterol accumulate in the walls of arteries, reducing their diameter and their ability to carry blood.

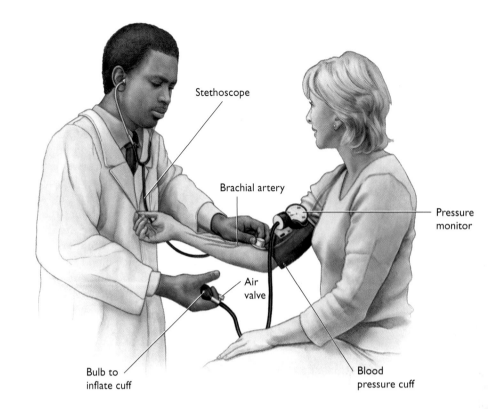

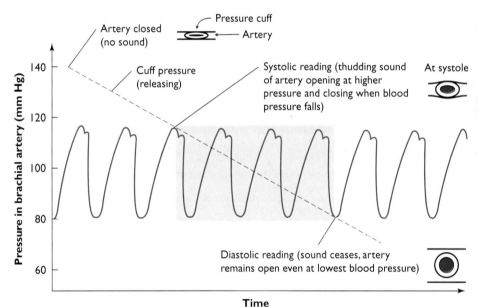

FIGURE 9.10 Monitoring blood pressure. A sphygmomanometer measures the pressure of blood in the brachial artery in millimeters of mercury (Hg) displaced by the force of the blood flow.

chological stress; a high salt intake, which increases blood volume; genetic susceptibility; and obesity. Hypertension can also result from an accumulation of fatty material called *plaques* within the walls of the arteries, a condition called **atherosclerosis** (FIGURE 9.9). Atherosclerosis has a genetic component, but it occurs to some degree in nearly all adults as a result of aging. A high-fat diet and exposure to tobacco smoke contribute to a higher risk of this condition.

Hypertension has few symptoms and is typically detected in patients by blood pressure screening, measured by a **sphygmomanometer** (FIGURE 9.10). A blood pressure reading is established by first cutting off blood flow to the brachial artery by inflating the sphygmomanometer cuff, which encircles the upper arm. As a nurse or other medical professional listens to the artery through a stethoscope, air is slowly released from the cuff. When the cuff is fully inflated, no sounds are heard because no blood is flowing through the artery. As the pressure of the cuff is relieved, the first sounds heard indicate that the artery is opening and closing—opening when a high-pressure pulse of blood is passing and closing when the pulse has passed. This is the highest pressure in the artery, called **systolic pressure**. The cessation of sounds indicates that the artery is

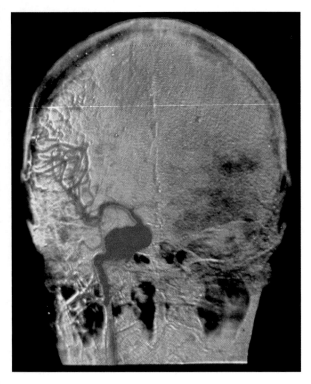

FIGURE 9.11 **Aneurysm.** This scan of a cerebral aneurysm clearly shows the bulge (in bright red) in the blood vessel caused by weakness in the blood vessel walls.

remaining open, and the reading on the cuff at this point indicates the lowest blood pressure in the artery, called **diastolic pressure**.

The sphygmomanometer measures blood pressure on a scale similar to that for air pressure, as millimeters of mercury (Hg) displaced. Normal blood pressure is 120 mm Hg systolic and 80 mm Hg diastolic, typically stated as "120 over 80" or 120/80. High blood pressure is defined as 140/90 or greater.

Hypertension increases the risk of **aneurysm**, a bulge in a blood vessel that can become a rupture (FIGURE 9.11). Aneurysms typically occur in arteries that experience the highest blood pressures, such as the aorta leading from the heart or large vessels in the brain and lower extremities. If blood vessel walls are malformed because of a problem during development, or have been weakened by infection or injury, high blood pressure can exacerbate the damage and cause the vessel to rupture. The resulting loss of blood from the circulatory system can rapidly cause death or lifetime disability. Developing aneurysms are often symptomless and an individual may not know he or she is susceptible until one occurs. Treatment almost always includes surgery, in which the weakened blood vessel is bypassed or reinforced.

Blood pressure can be moderated by changes in lifestyle to reduce psychological stress, by reductions in salt intake to reduce blood volume, and by a variety of drugs, including the beta blockers and ACE inhibitors described in Table 9.1.

Hypertension is not only caused by atherosclerosis, it also accelerates the development of this condition. The high-pressure flow of blood damages the endothelium of blood vessels, providing jagged sites for atherosclerotic plaques to accumulate. A large accumulation of material inside a blood vessel is called a *thrombus*. Continued high blood pressure can force a thrombus of plaque to shear off, producing a floating mass, an *embolus*, that can lodge in another blood vessel. Clogged blood vessels can reduce blood flow to organs, which decreases the oxygen supply and thus causes pain and the death of tissue.

TABLE 9.1 | Drugs for Treating Heart Disease

Drug Type	Examples*	Target	Effect
Beta blockers	Atenolol (Tenormin), metoprolol (Toprol XL)	Blocks receptors on nerve cells that trigger increased heart rate	Inhibits epinephrine effect on heart rate, reduces blood pressure
Angiotensin-converting enzyme (ACE) inhibitors	Lisinopril (Prinivil)	Enzyme that converts inactive chemical into active, blood-pressure-increasing hormone	Inhibits contraction of blood vessels, reduces blood pressure
Statins	Atorvastatin (Lipitor), simvastatin (Zocor)	Blocks liver enzyme required for cholesterol production	Inhibits production of "bad" LDL cholesterol, keeping blood vessels clear
Blood thinners	Aspirin, warfarin (Coumadin)	Platelets	Reduces clot formation, keeping blood vessels clear
Sodium-channel blockers	Quinidine (Quinidex)	Interferes with entry of sodium into heart muscle cells, allowing faster recovery after heart contraction	Stabilizes heart rhythm
Calcium-channel blockers	Amlodipine (Norvasc), diltiazem (Cardizem)	Blocks entry of calcium into heart muscle cells, decreasing the force of contraction and slowing heart rate	Stabilizes heart rhythm, lowers blood pressure, reduces chest pain
Diuretics	Hydrochlorothia-zide (Dyazide), furosemide (Lasix)	Nephrons in kidney	Increases water loss via urine, reducing blood pressure
Nitroglycerin		Increases levels of nitric oxide in blood	Increases dilation of blood vessels, reduces blood pressure

*Drug names in parentheses are brand names; other drug names are generic.

When blockages occur in blood vessels serving the brain, the result can be **stroke**, in which part of the brain tissue dies. If the clog is in vessels serving the heart, the result is heart attack, death of the heart muscle. The formation of blockages in vessels supplying other organs, most commonly the leg muscles, is generically referred to as *peripheral vascular disease.* We discuss strategies for treating clogs in section 9.4, because most of these procedures are performed on the arteries serving the heart muscle itself.

9.2 The Mechanical Heart

The primary function of the **heart** is to pump blood around the body. The structure of this pump is relatively uncomplicated.

Structure of the Heart

In its appearance, the fist-sized heart has only faint similarity to the Valentine's heart. It has a pointed end, called the *apex,* which projects behind the left lung. The heart also has a basic bilateral symmetry, as it consists of two side-by-side muscular pumps. The pumps are completely separated from each other by a muscular wall, or *septum,* and are somewhat independent but also coordinated. One pump, on the right side of the heart, receives oxygen-poor blood from the body and sends it to the lungs, whereas the left pump receives the oxygen-rich blood from the lungs and sends it into general circulation within the body.

The two pumps are each divided into two chambers, a relatively thin-walled but elastic **atrium** (plural: atria) and a thick-walled **ventricle** (FIGURE 9.12). The heart is completely surrounded by the **pericardium**, a membranous sac that holds a small quantity of lubricating liquid. Damage to the heart, including surgery and heart attack, can cause the amount of fluid in the pericardium to increase. As the fluid fills the stiff pericardium, the heart becomes compressed, causing sharp pain.

The chambers of the heart are sealed shut at different stages in its pumping cycle to prevent the backflow, or regurgitation, of blood. The seals take the form of valves that open for blood flow and then close to prevent its passage.

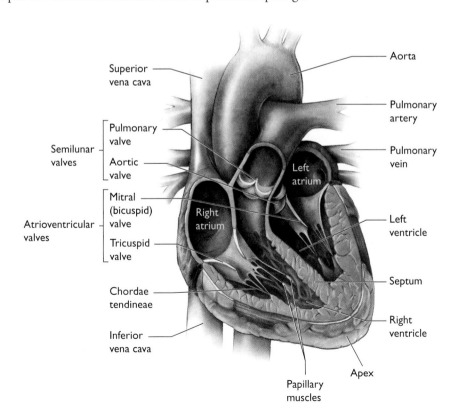

FIGURE 9.12 **The human heart.** The heart consists of four chambers making up two mostly independent pumps. Note that this drawing illustrates the orientation of a person's heart when he is facing you. The left and right sides are relative to that person, not to you.

The atria and ventricles on both sides of the heart are connected via **atrioventricular (AV) valves**. On the right side of the heart, this structure is called the *tricuspid valve,* named for the three flaps that make up its structure, while on the left side the two-flapped *bicuspid* or *mitral valve* (named for its resemblance to a miter, a bishop's cap) connects the left atrium and ventricle. The AV valves are supported by strands of connective tissue called *chordae tendineae* (tendon cords) connected to *papillary muscles* that project from the ventricle walls. The activity of the papillary muscles keeps the chordae tendineae taut, preventing *valve prolapse,* in which flaps of the AV valves collapse back into the atria when the ventricles contract.

Blood pumped out of the heart passes through **semilunar valves**, named for the half-moon-shaped flaps they contain. The valve between the right ventricle and the pulmonary circulation is called the *pulmonary valve,* while the one from the left ventricle into general circulation is the *aortic valve.* These valves close after each heart contraction, preventing the regurgitation of blood into the ventricles.

The Cardiovascular Pathway

Blood flows through the cardiovascular system primarily in two distinct but related circuits, often referred to as a double circulation system. The **pulmonary circuit** moves the blood into the lungs and returns this blood to the heart, where it enters the second circuit. This **systemic circuit** pumps blood to the rest of the body. The path of circulation between the heart, lungs, and body is detailed in FIGURE 9.13.

In Figure 9.13 you can see that oxygen picked up by the bloodstream at the lungs returns to the heart via a pair of large **pulmonary veins**. These veins empty into the left atrium and allow the left side of the heart to fill with blood. At the beginning of a heartbeat, the left atrium contracts to force more blood into the left ventricle. The left ventricle then contracts, sending blood into the **aorta** and from there into the major arteries. These arteries repeatedly branch out into arterioles and lead into capillary beds where oxygen and other components in high concentration within the blood diffuse out and carbon dioxide and wastes diffuse in.

The deoxygenated blood then travels through venules to the systemic veins, which gradually converge into the **superior vena cava** (carrying blood from the head, neck, and arms) and **inferior vena cava** (carrying blood from the abdomen and legs). The two venae cavae both empty into the right atrium and fill the right side of the heart. Contraction of the right atrium forces additional blood into the right ventricle. Contraction of the right ventricle then sends blood into the **pulmonary artery** and from there to the capillary beds of the pulmonary circuit. As the blood flows through the pulmonary capillaries, carbon dioxide diffuses out of it and oxygen diffuses in. A single red blood cell can travel the entire cardiovascular pathway in approximately 1 minute.

Stop and Stretch 9.2

The two sides of the heart eject the same volume of blood with every contraction, but the left ventricle is thicker and thus a proportionally more forceful pump than the right ventricle. Consider the function of each side of the heart to explain why this is the case.

There are two notable diversions from the major cardiovascular circuits: the coronary circulation and the hepatic portal system. The coronary circulation, which provides blood to the heart muscle itself, consists of a short loop from the aorta, over the surface of the heart, and into the right atrium. The **hepatic portal system** is a direct

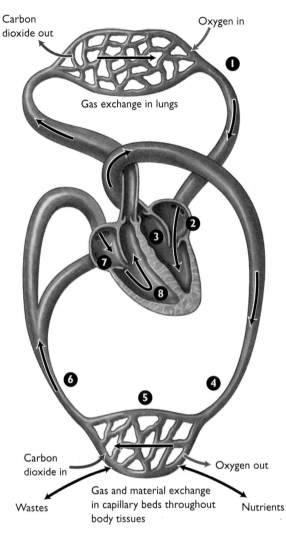

FIGURE 9.13 **The cardiovascular system.** Blood picks up oxygen in the lungs and carries it to the heart, where it is pumped to the rest of the body. The blood then returns to the heart, where it is pumped back to the lungs to dump a load of waste carbon dioxide and pick up more oxygen.

1. Oxygenated blood from the lungs travels through veins to the left side of the heart.
2. Contraction of the heart muscle forces blood in the left atrium to the left ventricle.
3. The slightly delayed contraction of the left ventricle forces blood into the arteries at high pressure.
4. Blood flows through the arteries and arterioles into capillaries.
5. In the capillaries, oxygen and other components in high concentration in the blood diffuse out and carbon dioxide and wastes diffuse in.
6. Capillaries empty into the venules, which carry blood to the veins and eventually to the right side of the heart.
7. Blood is forced into the right ventricle by contraction of the right atrium.
8. Contraction of the right ventricle sends blood into the arteries and capillary beds in the lungs.

connection between most of the organs of the digestive system and the liver, which processes and detoxifies many of the components that enter the body through our diets. A portal system both starts and ends in a capillary bed. In the hepatic portal system, capillaries surrounding the stomach, spleen, pancreas, and intestines drain into the hepatic portal vein, which carries the blood to capillaries in the liver (FIGURE 9.14). Once the blood has passed through the liver, it flows into the inferior vena cava and thus returns to the general circulation.

All of the organs of the body rely on the pumping actions of the heart. When it is ineffective, the outcome is severe impairment, often death. *Cardiologists,* medical doctors who specialize in treating heart disorders, have devised a number of strategies for repairing or replacing a damaged pump structure.

Repairing the Pump

The structure of the heart may be malformed as a result of problems during early development, diseases during childhood or adolescence, or damage due to heart attacks.

The most common developmental malformation is a "hole in the heart," which results from the incomplete formation of the septum between the left and right ventricles or atria. This connection between the two sides of the heart allows oxygenated blood to mix with deoxygenated blood. In severe cases, a newborn with this condition will appear blue because deoxygenated blood flows through the capillaries, including ones in the skin.

FIGURE 9.14 The hepatic portal system. A portal system is a network of blood vessels connecting two capillary beds. The most prominent portal system in the circulatory system connects the organs of the digestive system, with blood flowing from the other organs into the liver.

VISUALIZE THIS If the hepatic portal system did not exist, where would blood from capillaries that surround the intestines flow to? What problems are avoided by diversion of this blood to the liver?

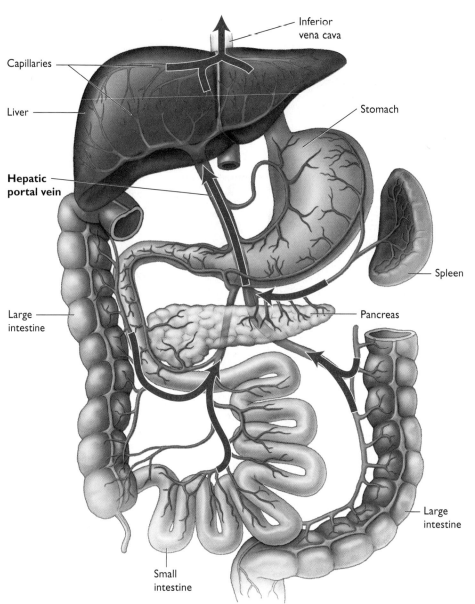

FIGURE 9.15 A stethoscope. The classic stethoscope consists of a diaphragm that is placed on a patient's body surface. Any vibrations received at the surface of the diaphragm are transmitted by sound waves through hollow tubes to the earpieces. Many medical personnel now use electronic stethoscopes, which can amplify sounds as well as vibrations.

A hole in the heart can be repaired by open-heart surgery, in which the chest and ribs are opened so that a cardiologist may work on the heart directly, or by a procedure that threads a plug through the veins and into the heart for placement. As many as 20% of adults have at least a small hole in the septum that does not appear to drastically impair their function.

Another common problem that occurs during development is malformation of one or more heart valves. Valves may also be damaged as a result of an immune reaction to infection by the strep throat bacteria, a condition called *rheumatic fever*. Improper valve function can lead to backflow of blood, the formation of blood clots in the heart, and poor circulation throughout the body. The symptoms of valve malfunction typically include fatigue and shortness of breath. Some valve malfunctions are apparent to a doctor listening to the heart through a stethoscope (FIGURE 9.15) as a swishing or whistling sound, and are described as **heart murmur**.

Cardiologists have several options for helping individuals with valve defects. In the most severe cases, the valve can be replaced with a synthetic heart valve made of plastic or a biological valve from a pig, cow, or deceased human donor (FIGURE 9.16).

Heart attacks and problems with coronary circulation can lead to damage to the heart muscle, impairing its ability to pump blood throughout the body. The development of this situation is referred to as **heart failure**. A large variety of drugs have been developed that can treat the symptoms and causes of heart failure (see Table 9.1), but the condition is often progressive.

A failing heart may become enlarged to compensate for the reduced volume of blood it can move per beat. As the heart continues to weaken, fluid backs up into the lungs, a situation called *congestive heart failure*. In this case, the pumping action of the heart may eventually become too poor to sustain life.

In severe cases, heart failure can be cured with a heart transplant. However, the need for hearts greatly exceeds the supply of donated organs, and hundreds of individuals on the transplant waiting list die every year waiting for a donor heart. Cardiologists have continued to seek new technologies for replacing the heart's pumping function.

One strategy for replacing failing hearts is the production of mechanical pumps. Heart surgeons over the last several decades have used *heart-lung machines* to replace the function of patients' hearts during surgery (FIGURE 9.17). More recently, **artificial hearts** have been produced that replace a diseased heart entirely and run on a battery pack that the patient carries in a shoulder bag or hip pouch. These hearts have extended the lives of some patients, but have only been implanted in a small number of very ill individuals. A more commonly used tool is the *left ventricular assist device (LVAD)*, an additional pump that adds power to a weakened heart but does not replace it.

A nonmechanical alternative for the lack of heart transplants is biological heart replacements. One option is *xenotransplantation*, in which hearts from nonhuman animals replace a patient's diseased heart. Although xenotransplantation may hold some promise, concerns about the transmission of animal viruses to humans and about the ethics of raising animals as "organ factories" have slowed progress.

More recently, scientists have become interested in using stem cells, which have the ability to develop into a number of different types of cells, to repair or replace damaged organs. A study begun in 2006 in the United Kingdom is investigating whether blood stem cells from the bone marrow, when injected directly into a heart, can mature into heart muscle and replace cells killed by heart attack. Some scientists imagine a future in which whole organs may be grown in the lab from embryonic stem cells. However, as discussed in Chapter 21, the reality is that such an achievement is many years of research away from realization.

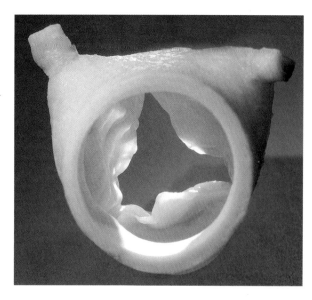

FIGURE 9.16 **Replacement valves.** Damaged or malformed heart valves may be replaced with synthetic or biological alternatives. The biological heart valve pictured here is from a pig heart.

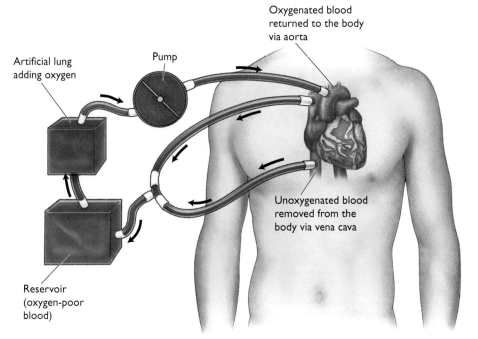

FIGURE 9.17 **A heart-lung machine.** These machines consist of a pump that takes over for the activity of the heart and an oxygenator, which is a synthetic lung made up of a silicone membrane bathed in oxygen. Technicians who operate heart-lung machines are called perfusionists.

9.3 The Electrical Heart

The heart muscle, called the **myocardium**, contracts with a rhythm determined by electrical signals from both within and outside the heart.

The Cardiac Cycle

The myocardium is distinguishable from other muscle by the presence of *intercalated discs,* areas where the membranes of adjacent muscle cells are closely interconnected by multiple gap junctions. Intercalated discs permit direct physical and electrical communication between cells, improving the coordination and strength of each contraction of this powerful muscle.

A small patch of myocardium in the wall of the right atrium, called the **sinoatrial (SA) node**, acts as the heart's pacemaker by sending out electrical signals that first cause both atria to contract. As the electrical signal speeds through the intercalated discs of the heart muscle, it reaches a site between the right atrium and ventricle called the **atrioventricular (AV) node**. This node distributes the electrical signal down the thicker walls of the ventricle via the electricity-conducting neurons in the *atrioventricular bundles* and *Purkinje fibers,* triggering contraction of these chambers (FIGURE 9.18). The signal travels via these fibers in order to stimulate contraction of the ventricles beginning from the heart's apex, effectively forcing blood out of the semilunar valves at the top of the heart.

The AV node acts as a sort of gate, delaying ventricle contraction by about one-tenth of a second after atrial contraction. This delay allows the ventricles to fill completely before they contract, thus forcing the majority of the blood out of the heart. After the heart contracts, it relaxes for a fraction of a second, allowing the pump chambers to refill with blood.

The complete sequence within the heart of filling with blood and then pumping it out is called the **cardiac cycle**. The relaxed period that allows blood to flow into the heart is called **diastole** and the contraction phase is called **systole**—hence the diastolic and systolic blood pressure readings. Systole can be divided into two parts, *atrial systole* when the atria contract, and *ventricular systole,* which occurs a fraction of a second later. During ventricular systole, the AV valves are forced closed by the contraction (making a low sound often described as "lub"). The semilunar valves snap into a closed position as the ventricle relaxes (with a sharp "dupp") at the beginning of diastole. The opening and closing of these valves in response to contraction of the heart muscle thus creates the distinctive heartbeat sound "lub-dupp, lub-dupp" (FIGURE 9.19).

The speed of the cardiac cycle is known as the *heart rate* and is influenced not only by the intrinsic rhythm of the SA node but also by signals from the brain and spinal cord. Heart rate and *stroke volume,* the amount of blood pumped per beat, are finely tuned by the negative feedback system regulating blood pressure. A healthy heart can

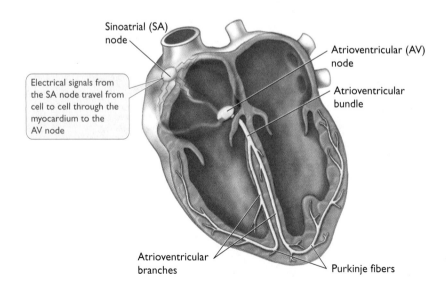

FIGURE 9.18 The heart's electrical circuit. The trigger for a heartbeat begins in the sinoatrial node in the right atrium, travels to the atrioventricular node, and continues along the Purkinje fibers before dissipating.

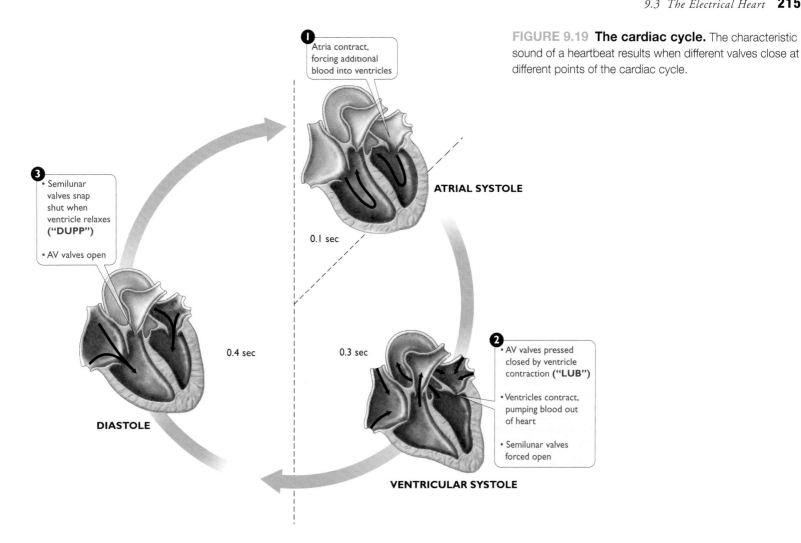

FIGURE 9.19 **The cardiac cycle.** The characteristic sound of a heartbeat results when different valves close at different points of the cardiac cycle.

adjust the *cardiac output,* the total amount of blood ejected into the aorta per minute, by modifying both heart rate and blood volume in order to meet a wide range of demands placed upon it by the body.

Input from other parts of the brain can affect the heart's rhythm, speeding up the heart rate in response to increased carbon dioxide in the blood, intense emotion, or as a result of certain drugs such as caffeine and nicotine. The hormones *epinephrine* and *norepinephrine* (also known as adrenaline and noradrenaline), which are released in times of stress or fear, also trigger an increase in heart rate.

Stop and Stretch 9.3

Caffeine, nicotine, and ephedra, a plant source of epinephrine, have long been used for weight control. Consider cellular respiration, the process that provides energy to cells in the body, as described in Chapter 3. Given that mechanism, explain why a drug that increases heart rate might cause a decrease in weight. Why might these drugs increase the risk of irregular heartbeat?

Because the SA and AV nodes generate electrical signals, electricity-sensing instruments can be used to evaluate an individual's heart health. These instruments generate

FIGURE 9.20 ECG. The trace here illustrates a typical healthy heartbeat. The P wave records the electrical signal produced at the beginning of systole when the atria contract, the QRS wave records ventricular contraction, and the T wave occurs when the ventricle relaxes at the beginning of diastole.

VISUALIZE THIS Go back to Figure 9.19 and label the appropriate portions of the cycle with P, QRS, and T, indicating the electrical signals generated during the particular physical state.

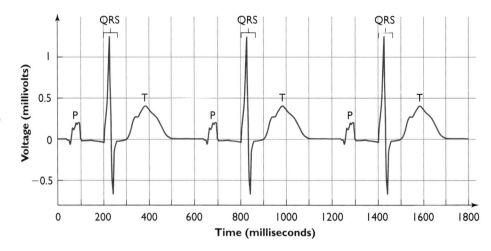

an **electrocardiogram** (**ECG** or **EKG**, FIGURE 9.20). Doctors can use ECGs to detect heartbeat irregularities, which may signify damage to the myocardium, injuries to the electrical conduction pathways, or an enlarged heart.

Steadying the Heartbeat

Disruption of the heart's electrical system causes about half of all deaths due to heart disease—nearly 335,000 per year in the United States alone. The cause of these deaths is often termed **cardiac arrest**, meaning that the heart suddenly stops beating. Cardiac arrest is most often caused when electrical impulses in the heart become too fast, too slow, or disorganized. In many instances, a victim of cardiac arrest appears outwardly healthy. However, most have an undiagnosed underlying heart condition, including an unusual heart rhythm, called **arrhythmia**.

For individuals with known irregularities in the cardiac cycle, treatment may include overriding the SA node with an **artificial pacemaker**. These instruments consist of small battery packs inserted under the skin that send a pulse of electricity via thin wires to one or more regions of the myocardium (FIGURE 9.21). The batteries in a pacemaker last about seven years, after which the pack must be replaced. Most modern pacemakers contain a microprocessor that senses heart rate and sends an electrical pulse only if the rate is too slow or somewhat irregular.

Disruptions of the heart's rhythm may take the form of **fibrillation**, in which the heart muscle contracts chaotically. These contractions are too weak and disorganized to move blood effectively. Fibrillation of the atria can lead to blood pooling in the heart and thus the formation of clots, which can travel through the blood vessels to the brain, causing stroke.

The risk of stroke associated with atrial fibrillation can be reduced if clot formation is suppressed. The therapy of choice for these patients is a drug called warfarin, a blood thinner that reduces the production of clotting factors (see Table 9.1). Blood thinners can have serious side effects, however, including the risk of severe hemorrhage. Most stroke-causing clots form in a small indentation in the left atrium, sometimes called the "heart's belly button" because it is a vestige of fetal development with no adult function. To reduce the risk of clots forming, implantable devices that fill this indent have been developed. Preliminary studies indicate that these implants may reduce stroke incidence by 90% in patients with atrial fibrillation.

Ventricular fibrillation results in little or no blood leaving the heart, causing rapid loss of consciousness and death unless it is reversed. **Cardiopulmonary resuscitation** (**CPR**), in which the chest is compressed to force blood through the vascular system, can help provide limited replacement for a nonpumping heart, but it rarely triggers the heart back into action. CPR is at best a stopgap measure that may prevent damage to oxygen-starved organs until medical help arrives.

Defibrillators, metal paddles that deliver a strong electric current to the chest, are the primary device used to reverse fibrillation. The current produced by a defibrillator

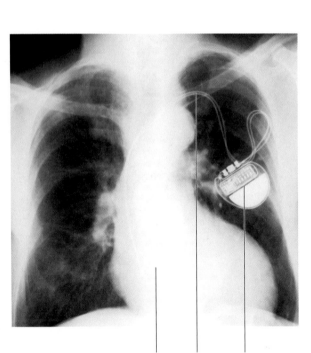

FIGURE 9.21 A pacemaker. The wires from a battery pack inserted in the chest wall are attached to portions of the heart muscle to stabilize the cardiac cycle in a diseased heart.

may shock the SA node back into a regular rhythm. Schools and other public buildings have invested in devices called *automated external defibrillators (AEDs)* that provide feedback to less skilled operators to help them treat an individual stricken by sudden cardiac arrest. Individuals who have had heart damage that predisposes them to fibrillation may receive *implantable automated defibrillators.* These instruments are placed in the chest cavity and automatically deliver a shock whenever an abnormal heart rhythm is sensed.

Damage to the electrical system of the heart often results from heart attack. Heart attacks occur when the power supply to the heart, oxygenated blood, is restricted or cut off.

9.4 Power for the Heart

Compared to muscles elsewhere in the body, the heart is much more dependent on oxygen to generate energy for its activities. Anything that restricts blood flow to the heart muscle can cause portions of the muscle to die because of oxygen starvation. The likelihood of these blockages depends on the condition of the blood vessels serving the heart.

Coronary Blood Vessels

Although the heart is regularly filled with oxygenated blood, the cells of the heart are not nourished by this supply. Instead, all oxygen and nutrients provided to myocardial cells come from blood vessels on the outer surface of the heart muscle.

The coronary circulation consists of **coronary arteries** that receive blood from the aorta; arterioles and capillaries that distribute that blood to the heart muscle; and venules that drain into the **cardiac veins**, which empty into the right atrium (FIGURE 9.22).

Because the circulatory loop is so short, blood flowing through it is under extremely high pressure. The heart's blood vessels are thus more prone to damage than others elsewhere in the body. Damage to the interior surfaces of these vessels provides sites for cholesterol to accumulate, forming thick atherosclerotic plaques.

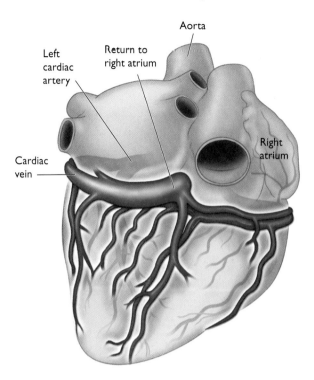

FIGURE 9.22 **The coronary circulation.** The blood vessels supplying the heart with oxygenated blood emerge from the aorta and empty into the right atrium. This illustration shows the dorsal (backside) view of the heart to best illustrate the connections between the coronary blood vessels and the heart.

Maintaining the Heart's Energy Supply

A **heart attack**, more specifically called a **myocardial infarction (MI)**, occurs when blockage in a coronary artery restricts or stops blood flow. As a result, the energy-needy myocardium quickly becomes oxygen starved, resulting in damage or death to parts of the muscle.

The risk of heart attack increases as the thickness of plaques within coronary arteries increases. The gradual reduction of blood flow through coronary arteries is first noticeable as **angina**, or chest discomfort, during exertion. Angina results as the heart muscle experiences oxygen deprivation, and it is typically described as a feeling of chest compression, rather than actual pain. When patients complain of angina, cardiologists will usually subject them to a *stress test,* in which an ECG reading is taken while the patient is running on a treadmill.

Stop and Stretch 9.4

The majority of heart attacks are caused by clots formed when an atherosclerotic plaque in an artery wall ruptures, releasing debris into the blood and causing the rapid formation of a blood clot. Before rupture, most of these vulnerable plaques only minimally narrow the artery. Given this, is a positive outcome on a stress test a clean bill of health?

218 CHAPTER 9 The Cardiovascular System

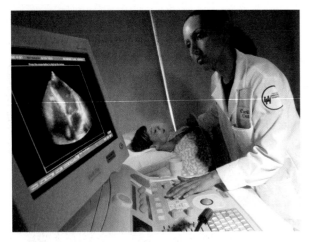

FIGURE 9.23 **An echocardiogram.** This technology allows observation of the heart as it is beating, which can provide more information than a static X-ray image.

If a stress test indicates some abnormality in heart rhythm during exertion, the next step is often either an X-ray *angiogram* or an *echocardiogram,* an ultrasound imaging of the heart and its blood vessels (FIGURE 9.23). If one of these tests indicates a blockage in the coronary arteries—a condition called **coronary heart disease (CHD)**—there are several options for restoring their ability to deliver blood.

If atherosclerotic plaques in the arteries are still relatively small, doctors typically prescribe one of a variety of medications to reduce cholesterol and improve blood flow (see Table 9.1). If blockages are severe, surgeons can increase the internal diameter of coronary blood vessels by compressing plaques against the vessel walls in a technique called **angioplasty** (FIGURE 9.24). A wire scaffold, called a *stent,* may also be implanted in arteries during this procedure to help prop open the vessel walls.

If coronary arteries are severely clogged, another option is **coronary bypass surgery**, in which new blood vessels are attached to the heart to carry blood past the blocked arteries. The number of arteries bypassed figures in the descriptive name of the surgery. For example, a "triple bypass" means that three new blood vessels were attached to bypass three blocked arteries. The new vessels may come from the patient's own vascular system—typically a leg vein—or may be synthetic or a hybrid vessel created from a synthetic scaffold and a patient's own stem cells.

A Healthy Heart

It is clear even from anecdotes that the risk of heart disease is not evenly distributed in the population. Differences in environment, including differences in diet, levels of activity and stress, and exposure to tobacco, play a role in the variation in risk, but genetic traits also matter. By studying families in which heart disease is common, scientists have identified at least 250 genes with variants that increase susceptibility to heart disease. Eventually, testing for gene variants associated with heart disease may enable doctors to target interventions to patients whose risk is highest.

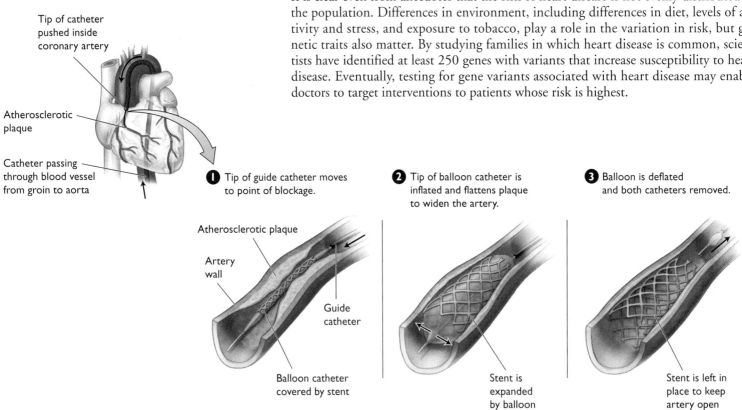

FIGURE 9.24 **Angioplasty.** Guided by ultrasound, a catheter is threaded through veins in the leg to the coronary arteries. Once in place in a narrowed region of the artery, the balloon on the end is inflated, flattening the plaque. At this point, a stent may be inserted to prop the artery open.

VISUALIZE THIS Why is a vein used as a path to the heart rather than an artery? How might the process of threading a catheter through a vein affect the function of the vein even after it is removed?

While cardiologists have made remarkable advances in treating heart disease and improving the quality of life of heart patients, the best "cure" for heart disease is prevention. The Mayo Clinic has summarized the five best strategies everyone can employ to reduce the risk of heart disease:

1. Do not smoke or use tobacco products. Most people believe that lung cancer and other lung diseases are the primary risk associated with exposure to tobacco smoke. In reality, most of the deaths due to smoking result from heart and blood vessel damage. A smoker has twice or three times the chance of dying from heart disease that a nonsmoker has. Much of the damage related to tobacco use stems from the effects of nicotine. Nicotine increases blood clotting, blood pressure, and LDL (low-density lipoprotein, or "bad") cholesterol. Because nicotine is delivered by all forms of tobacco, no tobacco consumption is safe.

Fortunately, the heart-damaging effects of nicotine appear to be reversible. According to the World Health Organization (WHO), the risk of heart disease decreases by 50% in smokers within 1 year of quitting, and the risk to smokers by 15 years after quitting is the same as the risk to people who have never smoked.

2. Exercise. Regular participation in moderate exercise—30 to 60 minutes daily—reduces the risk of heart disease by 25%. Exercise increases blood flow through the coronary arteries and strengthens the heart muscle, making each cardiac cycle more efficient and putting the heart under less strain (FIGURE 9.25).

3. Eat a healthy diet. A diet high in fruits, vegetables, low-fat dairy products, and whole grains can help protect the heart. Replacing saturated and trans fats with unsaturated fats also makes a big difference in blood cholesterol levels and decreases the risk of atherosclerosis.

4. Maintain a healthy weight. Moderate exercise and a heart-healthy diet should also help you keep your weight in a range that is associated with lower risk of heart disease. Even small reductions in weight help reduce blood pressure and cholesterol level.

5. Regular health screening. Regular checkups allow monitoring of blood pressure and blood cholesterol level. High readings of either may signal a problem that can be treated before serious heart damage occurs.

As it turns out, following this advice protects individuals from not only heart disease but a host of other conditions, including diabetes and cancer. The heart-healthy habits you establish now can help protect your health for decades to come.

FIGURE 9.25 **Working out the heart muscle.** Aerobic exercise that raises heart rate increases the strength of the muscle and contributes to its overall health. The exercise does not have to be elaborate. A brisk 30-minute walk that increases your heart and breathing rate but still allows you to carry on a conversation is an excellent option as part of a heart-healthy routine.

Chapter REVIEW

ROOTS TO REMEMBER

The following roots of words come mainly from Latin and Greek and will help you to decipher terms:

angio- refers to a vessel.

card- and **cardio-** relate to the heart.

hyper- means above or over.

myo- means muscle.

-plast- means forming or molding.

pulmon- is from the word for lung.

vas-, **vascul-**, and **vaso-** refer to vessels.

KEY TERMS

9.1 Blood and Lymphatic Vessels: The Circulation Pipes

aneurysm *p. 208*
artery *p. 200*
arteriole *p. 202*
atherosclerosis *p. 207*
blood pressure *p. 205*
blood vessel *p. 200*
capillary *p. 200*
capillary bed *p. 203*
cardiovascular system *p. 200*
diastolic pressure *p. 208*
hypertension *p. 206*
lymph *p. 205*
lymph ducts *p. 205*
lymphatic system *p. 204*
precapillary sphincters *p. 203*
pulse *p. 200*
respiratory pump *p. 204*
sphygmomanometer *p. 207*
stroke *p. 209*
systolic pressure *p. 207*
veins *p. 200*
venules *p. 204*

9.2 The Mechanical Heart

aorta *p. 210*
artificial hearts *p. 213*
atrioventricular (AV) valves *p. 210*
atrium *p. 209*
heart *p. 209*
heart failure *p. 213*
heart murmur *p. 212*
hepatic portal system *p. 210*
inferior vena cava *p. 210*
pericardium *p. 209*
pulmonary artery *p. 210*
pulmonary circuit *p. 210*
pulmonary veins *p. 210*
semilunar valves *p. 210*
superior vena cava *p. 210*
systemic circuit *p. 210*
ventricle *p. 209*

9.3 The Electrical Heart

arrhythmia *p. 216*
artificial pacemaker *p. 216*
atrioventricular (AV) node *p. 214*
cardiac arrest *p. 216*
cardiac cycle *p. 214*
cardiopulmonary resuscitation (CPR) *p. 216*
defibrillators *p. 216*
diastole *p. 214*
electrocardiogram (ECG) *p. 216*
fibrillation *p. 216*
myocardium *p. 214*
sinoatrial (SA) node *p. 214*
systole *p. 214*

9.4 Power for the Heart

angina *p. 217*
angioplasty *p. 218*
cardiac veins *p. 217*
coronary arteries *p. 217*
coronary bypass surgery *p. 218*
coronary heart disease (CHD) *p. 218*
heart attack *p. 217*
myocardial infarction (MI) *p. 217*

SUMMARY

9.1 Blood and Lymphatic Vessels: The Circulation Pipes

- The circulatory system consists of the blood vessels, the heart, and the lymphatic system. Its function is to transport materials and wastes around the body (p. 200).

- Arteries and arterioles carry blood from the heart. These vessels can absorb the pressure produced by a pulse and can also change their diameter to adjust blood pressure (pp. 200–203).

- Capillaries are thin walled to allow the rapid exchange of materials between the body and the blood. Blood flow into capillary beds is dynamically controlled (p. 203; Figure 9.6).

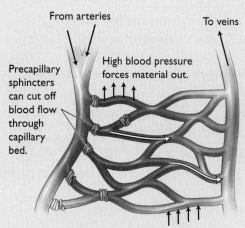

- Veins are the vessels that return blood to the heart. Blood flow in veins is promoted by skeletal muscle contractions and by respiratory movements, and backflow of blood is prevented by the presence of one-way valves (p. 204).

- The lymphatic system drains excess fluid from body tissues, returning it to veins in the vascular system (pp. 204–205).

- Blood pressure is regulated by blood volume and arteriole diameter and is typically maintained at a level that permits effective delivery of materials but prevents injury to blood vessels. Hypertension is chronic high blood pressure, a risk factor for heart disease (pp. 205–206).

- Atherosclerosis occurs when cholesterol in the blood accumulates on the walls of blood vessels, narrowing their diameter and increasing the risk of clots causing heart attack and stroke (pp. 207–209).

9.2 The Mechanical Heart

- The heart is a muscular pump consisting of four chambers: two atria and two ventricles. The chambers of the heart are connected to each other and to the circulatory system via valves that prevent the backflow of blood (pp. 209–210).

- Blood pumped by the heart flows in a double circulation system. Blood ejected from the right ventricle enters the pulmonary circuit to the lungs, where it picks up oxygen. It returns to the left atrium and is forced into the left ventricle, which ejects it through the aorta into the systemic circulation in the body. Blood returns to the right atrium via the veins (pp. 210–211; Figure 9.13).

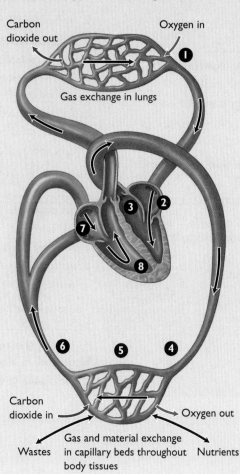

- Damaged heart valves can be replaced by mechanical or biological substitutes. Severe damage to the heart's pumping action, called heart failure, can be repaired by implanting an assistant pump or a donor heart. Artificial replacement hearts are still experimental (pp. 212–214).

9.3 The Electrical Heart

- Contraction of the heart muscle is triggered by electrical signals from the sinoatrial node in the right atrium. These signals are transmitted to the AV node, which delays their transmission slightly so that the ventricles contract a fraction of second after the atria contract (pp. 214–215; Figure 9.19).

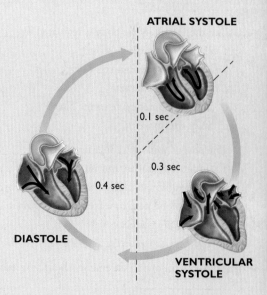

- The basic heart rhythm is established by the negative feedback loop regulating blood pressure. The heart rate can be modified by higher-level brain functions including intense emotions, by certain hormones, and by drugs (pp. 214–215).

- Disruption of the heart's electrical system can lead to stroke or cardiac arrest. Mild disruptions can be modified by signals from an electronic pacemaker, but more severe fibrillations are often fatal. CPR and defibrillators are tools for increasing survival in cases of cardiac arrest (p. 216).

9.4 Power for the Heart

- Coronary arteries supply blood to the heart muscle to power its energy-intensive activities (p. 217).
- If atherosclerotic blockage in coronary arteries becomes too severe, heart attack can result. Treatments for coronary artery disease include angioplasty and coronary artery bypass operations (pp. 217–218).
- Quitting smoking, reducing fat intake, increasing physical activity, maintaining a healthy weight, and regular screening of blood pressure and cholesterol can help reduce the risk of heart disease (pp. 218–219).

LEARNING THE BASICS

1. The blood vessels that carry blood from the heart are known as _____.
   a. veins
   b. arteries
   c. the vascular system
   d. lymph nodes
   e. capillary beds

2. Blood pressure in the body is regulated by _____.
   a. changes in cardiac output
   b. changes in the diameter of arterioles
   c. changes in heart rate
   d. levels of epinephrine in the blood
   e. all of the above

3. The porous walls of capillaries allow _____.
   a. liquid to be forced out into tissues as a result of blood pressure
   b. water to flow back into tissues down an osmotic gradient
   c. white blood cells to leave the bloodstream and patrol the tissues
   d. a and b are correct
   e. a, b, and c are correct

4. All of the following statements about atherosclerosis are true *except* _____.
   a. it refers to the accumulation of plaques within arteries
   b. it may lead to heart attack
   c. there is little one can do to reduce the risk of atherosclerosis
   d. it can be treated by angioplasty
   e. it can be diagnosed and evaluated by angiogram

5. The function of the lymphatic system is to _____.
   a. carry oxygen from the lungs to the body tissues.
   b. recycle dead and dying erythrocytes
   c. drain excess liquid from body tissues
   d. provide a host site for blood stem cells
   e. remove and excrete excess water and waste from the blood

6. Deoxygenated blood from the body first enters the _____ of the heart, and oxygenated blood from the lungs is pumped to the body by the _____.
   a. right atrium, right ventricle
   b. right atrium, left atrium
   c. left atrium, right ventricle
   d. right atrium, left ventricle
   e. right ventricle, left atrium

7. The hepatic portal system _____.
   a. connects most organs of the digestive system to the liver
   b. allows for the detoxification of materials that enter the body through the diet
   c. is made up of a single artery connected to a vein
   d. a and b are correct
   e. a, b, and c are correct

8. A dysfunctional heart valve _____.
   a. can be replaced with a mechanical valve
   b. may be heard as a "murmur" by a physician listening with a stethoscope
   c. can be replaced with a biological valve from a pig
   d. a and b are correct
   e. a, b, and c are correct

9. Electrical signals triggering heart contraction originate at the _____.
   a. SA node
   b. AV valve
   c. Purkinje fibers
   d. ventricle
   e. diastole

10. The sound of a heartbeat as heard through a stethoscope is produced by _____.
    a. the electrical signals sent from the pacemaker
    b. the rush of blood into and out of the heart
    c. the closing of valves between heart chambers
    d. the aorta and pulmonary veins collapsing and refilling
    e. ribs expanding as the heart muscle contacts them

11. A fibrillation _____.
    a. occurs when the heart muscle contracts chaotically
    b. cannot be reversed and inevitably ends in death
    c. regularly occurs during diastole of the cardiac cycle
    d. a and b are correct
    e. a, b, and c are correct

12. The energy to power the heart is supplied by _____.
    a. pulmonary circulation
    b. systemic circulation
    c. coronary circulation
    d. electrical signals from the brain
    e. defibrillation

13. Myocardial infarction (heart attack) occurs when _____.
    a. the myocardium receives an inadequate oxygen supply
    b. the electrical system of the heart becomes disrupted
    c. the atrioventricular valves fail to open
    d. blood pressure drops precipitously
    e. coronary veins drain into the right atrium

14. What is the most important behavior that one can adopt to reduce the risk of heart disease?
    a. quit or do not begin using tobacco products
    b. eat more dark chocolate
    c. reduce levels of strenuous exercise
    d. replace unsaturated fat with trans fat
    e. maintain a high percentage of body fat

15. Match the heart structures to their function.
    _____ right atrium
    _____ semilunar valve
    _____ aorta
    _____ SA node
    _____ myocardium

    a. heart muscle
    b. major pathway of oxygenated blood out of heart
    c. heart's pacemaker
    d. receives deoxygenated blood from body
    e. connects ventricle to outgoing blood vessel

16. Match the instrument with the heart or cardiovascular function it monitors.
    _____ angiogram
    _____ electrocardiogram
    _____ stethoscope
    _____ sphygmomanometer

    a. blood pressure
    b. cardiac cycle regularity
    c. condition of the coronary
    d. heart rate and valve function

17. Match the treatment strategy with the disease or dysfunction it attempts to fix
    _____ angioplasty
    _____ coronary bypass surgery
    _____ defibrillator
    _____ antihypertension drugs

    a. control high blood pressure
    b. reduce the size of atherosclerotic plaques
    c. restore normal heartbeat using electrical shock
    d. replace clogged coronary arteries with other blood vessels

18. The contraction phase of the cardiac cycle is termed _____, and the relaxation phase is called _____.

19. _____ fibrillation can lead to the formation of blood clots, whereas _____ fibrillation can lead quickly to death.

20. **True or False?** Most risk of heart disease is genetic, and we have little opportunity to influence that risk.

ANALYZING AND APPLYING THE BASICS

Use the graph in FIGURE 9.26 to answer questions 1–3.

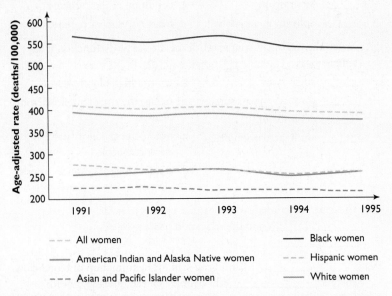

FIGURE 9.26 **Death rates in women due to heart disease, by year and racial category.**

1. Which group of women has the highest death rate due to heart disease? Which has the lowest rate?

2. List five factors that might explain why different groups of women have such different rates of heart disease deaths.

3. Choose one of these factors and describe how you would investigate whether it is an important source of difference between or among groups. What is the hypothesis? What is the prediction based on the hypothesis? How would you eliminate alternative hypotheses in your study?

4. Heart disease strikes women at older ages than men. However, women are almost twice as likely to die as a result of heart attack as men are. Given these facts, what information would you need in order to determine if women are receiving inadequate treatment for heart disease, relative to men's treatment?

5. Why is the pulse felt in an artery but not in a vein?

6. The circulatory system in a frog is similar to that of humans, with one circuit that brings blood to the lungs and back and another that pumps blood to the body tissues. However, a frog's heart has only a single ventricle pumping blood into both circuits. Use your understanding of the human heart to explain how the efficiency of the frog's cardiovascular system differs.

7. During development, a fetus's lungs are filled with fluid, and oxygen is supplied to it by the mother via the vessels in the umbilical cord connected to the placenta. Blood flow in the fetus's body thus bypasses the lungs via a blood vessel called the *ductus arteriosus*. Based on your understanding of cardiovascular pathways, which two blood vessels serving the heart do you think the ductus arteriosus ties together? Explain your reasoning.

8. The Genes & Homeostasis feature explores a hypothesis regarding why blacks may produce lower levels of nitric oxide than whites. Describe a possible test for this hypothesis.

Questions 9 and 10 refer to the following paragraph.

Cardiologists are interested in determining what factors, both genetic and environmental, contribute to risk of cardiac arrest. A research team at Johns Hopkins moved this search forward by focusing on individuals who had a particular variation in their otherwise normal heart rhythm. The variation, an unusually short or long recovery from ventricular contraction, is associated with higher risk of fibrillation and is readily identified on an ECG. The research team examined the DNA sequences of over 2,000 women for specific sequences that showed up more frequently in individuals with this cardiac cycle variation.

9. What information does this research approach provide to the researchers?

10. Explain why the scientists might have limited their research to women.

CONNECTING THE SCIENCE

1. Although heart disease has a genetic component, it is mainly a lifestyle disease that leads to premature death in individuals who do not follow the guidelines discussed in the chapter. Have you heard the guidelines for reducing the risk of heart disease before? If not, why do you think this is the case? If you have, are you following them? Why or why not?

2. Heart disease causes more deaths than any other factor. However, if we examine the top causes of death for each age group, we can see that heart disease is the number-one cause of death only among individuals over age 65. Accidents, homicide, suicide, AIDS, and cancer, while they cause many fewer total deaths than heart disease, are the top causes of death among individuals younger than 65. How does this information influence your view of heart disease? Does it raise any questions?

Chapter 10

The Respiratory System
Secondhand Smoke

LEARNING GOALS

1. Describe the structure and function of the respiratory system.

2. Describe the path of air into the body.

3. Describe the muscles involved in breathing and explain how their movements facilitate air movement into and out of the lungs.

4. Explain the principle of partial pressure and detail how differences in partial pressure cause gas exchange across the respiratory membrane in the lungs.

5. Explain the role of hemoglobin in gas exchange.

6. List the effects of smoking on the various structures and functions of the respiratory tract.

7. Briefly explain the reasons why smoking is powerfully addictive.

The ashtray at a building entrance is often at the center of the "smokers' circle."

A cluster of young adults shuffles and fidgets near the door of the student union. The temperature is well below freezing. The wind whips through the academic quad, ruffling their hair and turning their exposed cheeks ruddy and raw. A steady stream of students and faculty hustles by, mostly ignoring the small gathering. Some passersby even change their path toward the building entrance to avoid walking close to them. One young man in the group turns his back to the wind and cups his hand over the end of his lighter, straining to give the tip of his cigarette enough flame to set it alight.

The smokers' circle. It is a common sight outside of buildings in many states throughout the country. With more public spaces becoming off-limits to cigarettes, pipes, and cigars, smokers are forced to gather in drab alleyways and slushy courtyards outside the range of their nonsmoking friends and colleagues. Despite this enforced segregation, more than 48 million Americans, about 21% of the adult population, smoke at least one cigarette a day.

And the cost of smoking is not limited to social isolation. A pack-a-day habit can cost a smoker $2,500 a year just for the cigarettes, causes an increase in susceptibility to colds, and results in 40% more lost workdays compared to nonsmokers. At a growing number of companies, smokers are charged higher health insurance premiums. The effects of tobacco smoke result in 440,000 premature deaths each year, and the illness caused by smoking costs U.S. employers, insurers, and taxpayers $157 billion annually in lost productivity and health care expense.

In addition to its effects on smokers themselves, cigarette smoke can harm nonsmokers. For example, spouses of smokers have a greatly increased risk of lung cancer compared to spouses of nonsmokers. Children are especially vulnerable to the effects of smoking. Children raised in smokers' households have higher rates of asthma and ear, nose, and throat infections—and may even have impaired brain development.

Information about the risks and costs of smoking is well known to nonsmokers and smokers alike. Billions of dollars have been put into advertising and outreach campaigns to disseminate this information and to prevent

Fewer and fewer public spaces allow smoking.

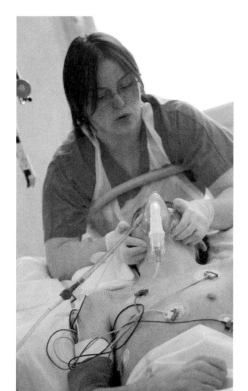

The health risks of smoking are well known.

Why do almost a quarter of American adults still smoke, despite the cost?

children and teens from picking up their first cigarette. Although these campaigns have met with some success, the rate of smoking is still stubbornly high—and among some groups, such as teen girls, it continues to rise.

Why do people continue to be drawn to smoking, and why do they continue to smoke despite all the negative repercussions of the habit? In this chapter, we explore the makeup of tobacco smoke and its effects on the body, with a special focus on the respiratory system, and we will investigate what can be done to reduce the social and economic costs of tobacco use.

10.1 Respiratory System Anatomy: The Path of Smoke into the Lungs

The primary function of the **respiratory system** is to permit gas exchange between the body and the environment, although another important function in humans is its role in producing the voice. The respiratory system is typically divided into two segments, the **upper respiratory tract** containing organs in the head, and the **lower respiratory tract**, made up of organs in the neck and thoracic cavity (FIGURE 10.1).

Upper Respiratory Tract

Smoke first enters the upper respiratory tract (FIGURE 10.2) through the **mouth** or the paired *nostrils* of the **nose**. Smoke entering through the nose passes into the **nasal cavity**,

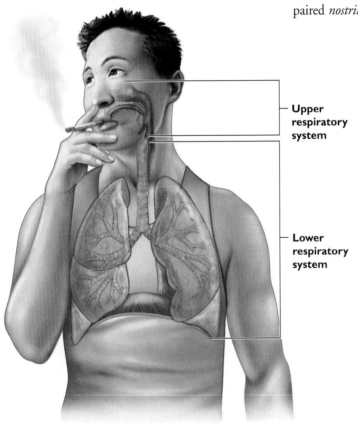

FIGURE 10.1 **The respiratory system.** The respiratory system can be subdivided into two sections: the upper respiratory tract and the lower respiratory tract.

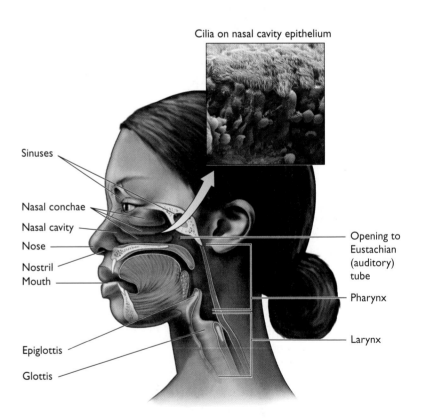

FIGURE 10.2 **The upper respiratory tract.**

a chamber divided into two halves by a **septum** consisting of bone and cartilage, and lined with mucous membrane. The *nasal conchae*, the shelflike bones that create the shell of the nasal cavity, increase the cavity's surface area and create turbulence within the incoming air. As outside air swirls through the nasal cavity, it is filtered, warmed or cooled to body temperature, and humidified. In fact, a primary function of the upper respiratory tract is to protect the lower tract from debris, infectious organisms, and environmental extremes.

Stop and Stretch 10.1

Given the function of the upper respiratory tract, explain why the nostrils are lined with a coat of stiff hairs.

The nasal cavity connects to the ears by the **Eustachian tubes**, or **auditory tubes**, thin passages that allow fluid in the ear to drain and equalize air pressure between the inside and outside of the skull. The nasal cavity is also connected to the **sinuses**, mucous membrane–lined cavities in the skull bones that reduce the weight of the skull, help to filter air, and increase the resonance of the voice.

Epithelial cells on the mucous membranes are covered with hairlike projections called *cilia*, which trap small particles in mucus and actively move the mucus either to the front of the nose, where it can be blown out, or into the throat, where it is expectorated (spat) or swallowed.

Smoke inhaled through the nose or mouth passes through the **pharynx**, or throat, the common passageway for both air and food into the body. Food is prevented from entering the lower respiratory tract by a flap of cartilage called the *epiglottis*. The epiglottis closes during swallowing to cover the *glottis*, the opening that leads to the lungs. A rigid tube of cartilage called the **larynx** surrounds and protects the glottis, allowing air to flow freely into the lower respiratory tract.

Lower Respiratory Tract

The larynx divides the upper respiratory structures from the lower respiratory tract, where the exchange of gases between the body and the environment takes place (FIGURE 10.3).

The larynx is also known as the *voice box*, because it contains the **vocal cords** (FIGURE 10.4). These complex structures, made up of elastic ligaments and folds of epithelial tissue, are responsible for creating sounds. Air passing through the glottis vibrates the vocal cords, producing sound waves.

We can vary the pitch of a spoken word by contracting muscles in the throat that adjust the thickness of the vocal cords. Voice loudness is determined by how fast vocal cords vibrate, which is driven by the rate and volume of airflow passing the cords. High air volumes and speeds produce loud sounds, while lower volumes and speeds produce quiet speech. Voluntary movements of the lips, tongue, and cheek are responsible for the distinct phonics of language.

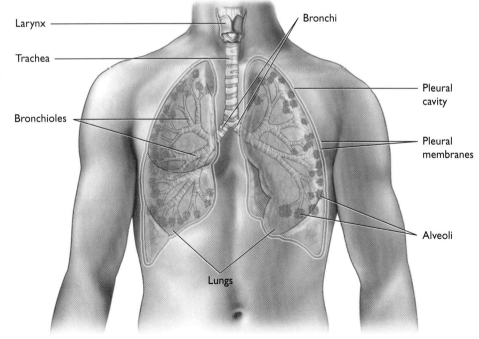

FIGURE 10.3 **The lower respiratory tract.**

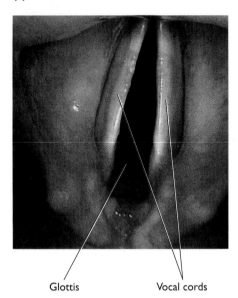
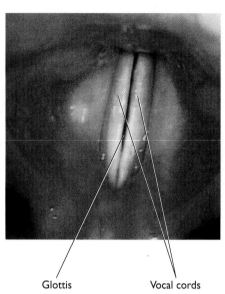

FIGURE 10.4 **Function of the vocal cords.** The vocal cords are made up of cartilage that produces sound when air passing over them causes them to vibrate. Muscular movements that shorten and thicken the cords produce low sounds, while tightly stretched cords produce high sounds.

The size and shape of the vocal cords when speaking, as well as the shape of the mouth and nasal cavity, determine the unique characteristics of each individual's voice. Individuals with higher voices have short vocal cords, while longer vocal cords produce deeper tones.

Below the larynx is the **trachea** (windpipe), which consists of a series of partial rings of cartilage that hold their round shape thanks to a strip of smooth muscle. This muscle can contract quickly during a cough to increase the speed of flow out of the airway (FIGURE 10.5).

Coughing occurs as a result of irritation in the trachea and **bronchi** (singular: **bronchus**), the tubes that distribute air to the lungs. Particles too small to trigger a cough can become trapped in mucus produced by the epithelial tissue lining these structures. Just as in the nasal cavity, these cells are covered with cilia that systematically move irritants and mucus toward the glottis, a movement known as the *mucus escalator*. Once the mucus reaches the glottis, it is shunted into the digestive system. The Genes & Homeostasis feature on page 232 details a relatively common genetic condition, cystic fibrosis, in which the mucus escalator is not as effective at clearing particulates, and mucus, from the lungs.

Healthy **lungs** are pink and cone-shaped (FIGURE 10.6). Within the lungs, the bronchi divide many times, forming a branching structure tipped by **bronchioles** that end in small sacs called **alveoli**. About 300 million air-filled alveoli make up much of the lungs' structure and give them a light spongy texture. Each grapelike cluster of alveoli is surrounded by a net of capillaries, which transfer the gases exchanged in these structures to the entire body (FIGURE 10.7).

Each lung is encased in a two-layered **pleural membrane** that connects the surface of the lung with the chest wall. A small amount of fluid in the *pleural cavity* between the layers keeps the membranes attached to each other by *cohesion*, much as water between two layers of plastic wrap causes the plastic to bind tightly together.

The tiniest particulates in smoke can be drawn deeply into the lungs through the pathway from the bronchus to the bronchioles, and even into the alveoli. Because alveoli lack cilia, the movement of foreign materials out of these structures is much more limited. White blood cells called macrophages patrol the deep areas of the lungs, engulfing and removing small particles, but they cannot keep up with the large amount of debris deposited by chronic exposure to smoke.

10.2 Tobacco Smoke and the Respiratory Tract

Tobacco smoke comes in two distinct forms: the *mainstream smoke*, which is inhaled from the unlit end of a cigarette, cigar, or pipe, and secondhand or *sidestream smoke*. Sidestream smoke emitted by the lit end of a cigarette, combined with the mainstream smoke exhaled by active smokers, is also referred to as **environmental tobacco smoke**. This smoke affects not only the *active smoker*, who is holding and inhaling from the cigarette, but nonsmoking individuals in the area as well. In a smoky environment, nonsmokers are considered *passive smokers* (FIGURE 10.8).

FIGURE 10.5 **Anatomy of a cough.** A cough is produced when the cartilage bands around the trachea are pulled shut, reducing its diameter. When a large volume of air is quickly forced through this narrow tube, the high pressure that results can dislodge foreign objects.

The Composition of Tobacco Smoke

According to the National Cancer Institute, more than 4,500 chemical compounds have been identified in mainstream smoke. Although many of the components in this airborne stew are benign, the list includes dozens of known *carcinogens*—cancer-causing

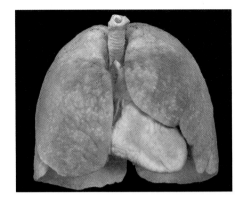

FIGURE 10.6 Healthy lungs. In a healthy individual, the lungs are pink and spongy in texture.

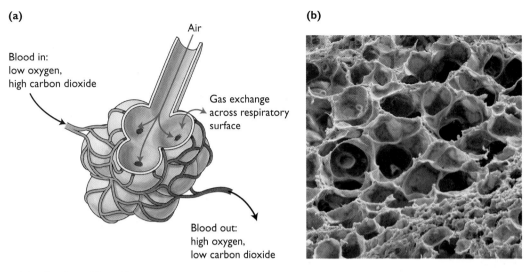

FIGURE 10.7 Alveoli. (a) Alveoli occur in clusters at the end of bronchioles. The clusters are enmeshed in a dense network of thin-walled capillaries, which allows for external respiration to occur. (b) An electron micrograph of alveoli in the lungs.

agents. These compounds can cause changes to the structure or sequence of DNA molecules; that is, they can cause mutations. Cancer results from an accumulation of mutations in the genes controlling cell division. Other compounds in the smoky mixture are known to be toxic to humans, such as ammonia and hydrogen cyanide.

One of the most common gases in environmental tobacco smoke is *carbon monoxide (CO)*, a colorless, odorless gas that is a by-product of the incomplete combustion of any carbon-containing compound. Carbon monoxide makes up about 4% of the total secondhand smoke in a room and forms when tobacco smolders, as at the end of a lit cigarette. You may know of carbon monoxide as a potential source of poisonous air pollution inside homes, resulting from faulty furnaces or other appliances that burn natural gas.

Much of the visible smoke produced by a cigarette consists of tiny particles of partially burned tobacco. These airborne **particulates** have a diameter less than half the

FIGURE 10.8 Exposure to tobacco smoke. The environmental tobacco smoke in a room is a mixture of smoke from the lit end of a cigarette and smoke exhaled by active smokers.

GENES & HOMEOSTASIS

Did an Old Killer Spawn a New One?

Tuberculosis (TB) is a lung disease caused by infection with the bacteria *Mycobacterium tuberculosis*. This disease is characterized by fatigue, fever, a cough that brings up bloody mucus, and pain during breathing. A serious *M. tuberculosis* infection results in extensive scarring of lung tissue, reducing gas exchange with often-fatal results. Between the seventeenth and twentieth centuries, tuberculosis caused a remarkable 20% of all deaths in Europe. With the advent of antibiotics, by the 1950s TB had seemingly become a disease of the past in developed countries, although it still infects and kills millions in poorer regions throughout the world.

A disease that has had such a sustained and severe effect on human survival is almost guaranteed to have caused evolutionary changes in the population. Any traits that were likely to increase an individual's survival in the face of *M. tuberculosis* infection should have become more common, as the individuals who carried these traits survived and reproduced. Some scientists hypothesize that one such trait is a change in the mucous membranes in the lungs—a change that can also be deadly.

Cystic fibrosis is the most common genetic disease in European-Americans, affecting one in every 3,000 babies. In this disease, a mutation in a chloride transporter gene prevents the transporter from embedding in the cell membranes of mucus-producing cells. The chloride transporter normally maintains homeostasis of solutes between the inside and outside of the cell: As the transporter pumps chloride (Cl^-) out of the cell, water follows down its osmotic gradient. In the absence of the transporter, water does not readily leave the cell, making the mucus excreted by the cells thick and sticky (see figure).

Individuals with cystic fibrosis have two copies of the nonfunctional gene, and as a result have large amounts of mucus trapped in their lungs, providing a rich source of food for bacterial invaders. (Thick mucus also interferes with the proper functioning of the digestive system, which requires slippery mucus for movement of materials through organs and to permit the release of various digestive enzymes.) People with cystic fibrosis are prone to life-threatening lung infections, and even with extensive medical intervention, they have much reduced life expectancies.

A single copy of the cystic fibrosis allele is carried by about 4% of people with European heritage. Although it is not clear why the reduced rates of chloride transport found in these individuals may protect them from TB, some research has indicated that carriers do experience lower mortality rates in response to *M. tuberculosis* infection.

Unfortunately, a real-life test of this hypothesis may be unfolding. Ineffective or incomplete treatment of TB infections has led to a dramatic and unsettling increase in antibiotic-resistant *M. tuberculosis* infections that cannot be treated by available drugs. Current evidence suggests that as many as 20% of new TB infections are essentially untreatable by current means. If drug-resistant TB continues to spread in European populations, it may become increasingly clear whether or not the cystic fibrosis allele provides resistance to this old killer.

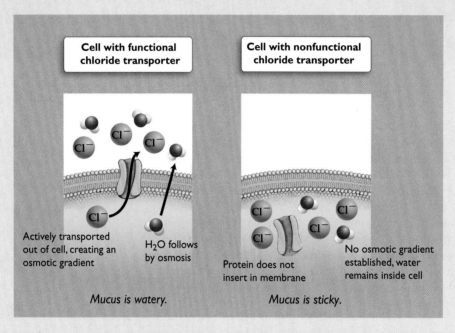

width of a human hair and are commonly known as tar. Because most active smokers inhale through a cigarette's filter tip, the amount of particles acquired via mainstream smoke is reduced. However, particulates present in environmental tobacco smoke enter the respiratory system of anyone who inhales in a smoky room.

Smoke Damages the Respiratory System

In the upper respiratory tract, the mucous membranes lining the nasal passages and sinuses are most susceptible to smoke damage. These membranes increase mucus production in response to smoke, and their cilia are damaged or destroyed by both the solid and gaseous components of smoke. As a result, mucus containing trapped particles remains in the upper respiratory tract longer, promoting inflammation and infection. Controlled studies support the hypothesis that active and passive smokers are more susceptible to common colds than nonsmokers. Some studies suggest that smoke exposure leads to higher rates of **sinusitis** (sinus infection), but the evidence is not yet conclusive.

Stop and Stretch 10.2

Why do you think it is easier to study the relationship between smoking and the common cold than the relationship between smoking and sinus infection?

Smoke-induced inflammation of the Eustachian tubes is especially a problem for small children, since these channels are more horizontal and drain less effectively in their smaller, rounder skulls. Fluid trapped in the ear is a rich source of nutrients for bacteria, so poorly draining Eustachian tubes lead to **otitis media**, an infection of the middle part of the ear. According to the American Academy of Pediatrics, children living in households where they are exposed daily to cigarette smoke have 38% more ear infections than children in smoke-free homes. Chronic otitis media is the leading cause of hearing loss in children. The incidence of ear infections declines as a child grows and his or her Eustachian tubes become more vertical.

In the lower respiratory tract, smoke irritates the vocal cords, causing them to swell and resulting in **laryngitis**. Chronic laryngitis can result in the classic "smoker's voice"—low and husky. A study by researchers at the University of Cincinnati found that nonsmoking women regularly exposed only to environmental tobacco smoke had undergone changes to their vocal cords relative to other nonsmokers, although the changes appeared to have little effect on their voices. Children living in smoky environments also experience higher rates of **tonsillitis**, inflammation of the tonsils, which are the lymphatic tissue at the sides of the throat.

Exposure to tobacco smoke increases the risk of cancer in organs of the upper respiratory tract, larynx, and trachea. Eighty-five percent of the 55,000 Americans who develop head and neck cancer each year are tobacco users. For some, the only treatment is removal of the cancerous organ, often the larynx.

The same factors that cause inflammation of the upper respiratory tract—increased mucus production and damaged cilia—also cause inflammation of the bronchial tubes, a condition called **bronchitis**. This inflammation also increases the risk of a secondary bacterial infection. If a bronchial infection travels to the lung tissue, it may cause **pneumonia**, a serious condition in which fluid begins to fill the alveoli. In longtime smokers, bronchitis may become permanent, or *chronic*, bronchitis. "Smoker's cough," an early morning cough that brings up abundant mucus, is a symptom of chronic bronchitis.

Particulates in smoke are known to exacerbate **asthma**, an abnormally intense immune response that results in the constriction of bronchial walls and an overproduction of mucus. According to the U.S. Environmental Protection Agency (EPA), the number of cases of asthma in American children caused by exposure to tobacco smoke is approximately 26,000 per year (FIGURE 10.9).

We can see the accumulation of particulates in the lungs of a longtime smoker, which are often black with tar trapped inside the alveoli (FIGURE 10.10). When particulates are drawn deeply into the lungs, they stay for months or even years without being cleared. The DNA of lung cells thus remains susceptible to mutation by the carcinogens in tobacco smoke long after smoke inhalation has stopped. Lung cancer is the leading cause of cancer deaths in the United States, causing more mortality than the next three most common causes of cancer (colon, breast, and prostate) combined. An estimated 90% of lung cancers are caused by exposure to tobacco smoke.

FIGURE 10.9 **Asthma treatment.** This child is using an appliance called an inhaler, which provides a fine spray of a bronchodilator, to treat an asthma attack. Bronchodilators are drugs that cause the smooth muscles surrounding the bronchi and bronchioles to relax, allowing air to flow freely again into and out of the lungs.

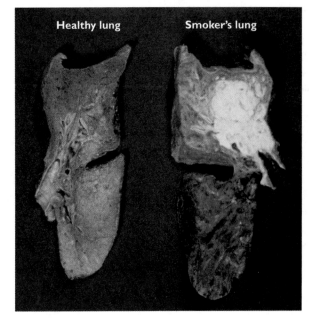

FIGURE 10.10 **Tobacco tar trapped in lungs.** The smoker's lung shows the dark staining caused by the accumulation of tiny airborne particles in tobacco smoke. The white mass in this lung is a cancerous tumor.

Particulates trapped in the lung also promote inflammation and destruction of the alveoli. The effect of this damage can be quite serious, because these delicate air sacs provide the crucial homeostatic function of oxygen uptake and carbon dioxide removal.

10.3 Inhaling and Exhaling

The exchange of gases between the environment and an organism is referred to as **external respiration**. This process is distinct from *cellular respiration*, the process that generates ATP to power cell processes through the consumption of oxygen and production of carbon dioxide as a waste product.

Ventilation, or *breathing*, by physically moving air into and out of the lungs, facilitates external respiration. The connection between the lungs and the circulatory system permits **internal respiration**, the exchange of oxygen and carbon dioxide between circulatory system fluids and the respiring body tissues (FIGURE 10.11).

The Mechanics of Breathing

Two major muscle groups are involved in the control of breathing: the diaphragm and the intercostal muscles between the bones of the rib cage.

In each normal breath, you take in about 500 milliliters—approximately 2 cupfuls—of air and release the same amount. This amount is called the **tidal volume**, because it represents the relatively small volume of air that fluctuates in the "ocean" of air that our lungs can hold. The maximum amount of air you can exhale after a deep breath is called the **vital capacity** and equals about 4,800 milliliters in a healthy person. Our lungs contain an additional 1,200 milliliters of **residual volume** that cannot be exhaled (FIGURE 10.12).

The uptake of air, called either **inhalation** or **inspiration**, is triggered by contraction of a strong, dome-shaped muscle called the **diaphragm**. When the diaphragm contracts, it flattens out, increasing the height of the thoracic (chest) cavity (FIGURE 10.13). At the same time that the diaphragm is contracting, the rib cage is lifted up and outward by **intercostal muscles**, which increase the chest's diameter. The lungs increase in volume along with the chest cavity because they are stuck to the chest wall by the pleural membranes.

As the lungs increase in volume during inhalation, the air pressure within them drops. This drop in pressure creates a partial vacuum in the lungs and is similar to the decline in the air pressure inside a syringe when the plunger is pulled back. If the syringe is in a liquid solution, we can see the liquid flow toward the lower pressure inside the syringe, just as air flows to the lower-pressure environment inside the lungs.

When the lungs are inflating, nerve cells called *stretch receptors* near the alveoli respond to their increasing size. When a critical lung volume is reached, signals from these receptors cause the respiratory center to shut off its signals to the diaphragm and intercostal muscles, which then relax. As a result, the volume of the chest cavity decreases. The air that rushed into the lungs is now squeezed into a smaller volume, increasing the pressure and causing air to flow back out of the body. Thus, when we are resting, the release of air, called **exhalation** or **expiration**, requires no muscle contraction and is called *passive*

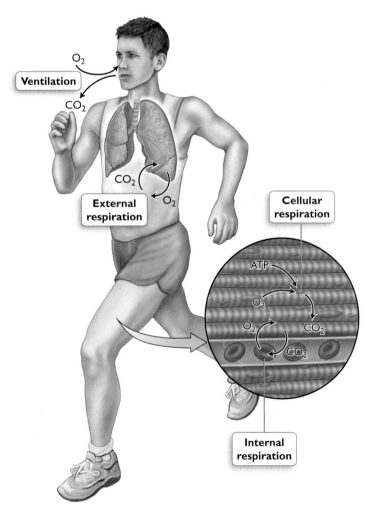

FIGURE 10.11 **Breathing and respiration.** Ventilation brings oxygen into the body and vents carbon dioxide. These gases are exchanged in the lungs in the process of external respiration. Gas exchange between the tissues and the bloodstream is called internal respiration. All of this is required because cellular respiration consumes oxygen and produces carbon dioxide while producing ATP for cellular activity.

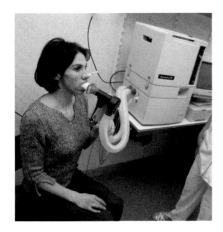
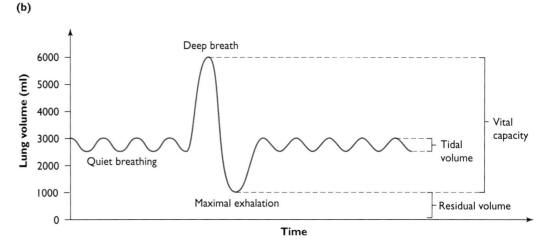

exhalation. Passive exhalation is similar to the release of air from an expanded balloon. Because the elastic walls of the balloon are exerting pressure on the contents, air rushes out when we release the valve, even without our actively squeezing the balloon.

Exhalation becomes an active process during intense aerobic activity. When blood pH levels decline rapidly, indicating a surplus of carbon dioxide, nerve cells in the respiratory center trigger the contraction of muscles in the chest wall and abdomen. Contraction of these muscles squeezes the chest cavity, forcing air out of the lungs in an *active exhalation*. This process is similar to forcing liquid out of a syringe by depressing the plunger and reducing the volume of the chamber.

FIGURE 10.12 Lung capacity. (a) Lung capacity is measured by a spirometer, which calculates the volume of air inhaled and exhaled. (b) This graph illustrates the airflow into and out of the lung of a healthy volunteer as measured by a spirometer. The tidal volume can be observed during quiet breathing. The vital capacity is measured when a large breath is inhaled, followed by a maximal exhalation.

Stop and Stretch 10.3

The Heimlich maneuver is a strategy for forcing an obstruction out of the airway of a person who is choking by forcibly squeezing the abdomen directly under the rib cage. Use your understanding of inhalation and exhalation to explain why this maneuver is effective.

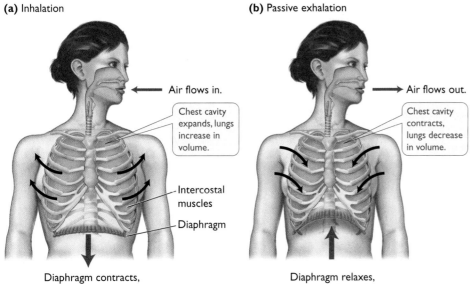

FIGURE 10.13 Mechanics of taking a breath. (a) Contraction of the muscular diaphragm flattens it out, and contraction of the intercostals pulls the rib cage up. As a result, the volume of the chest cavity increases, and air flows into the lungs. (b) During passive exhalation, these muscles relax, allowing the chest cavity to compress and forcing air out of the lungs.

CHAPTER 10 The Respiratory System

The Control of Breathing

You take a breath approximately 12 times per minute, or over 6 million times per year, without thinking much about it. In fact, breathing for the most part is automatically regulated—although we can exercise a surprising degree of control over the process compared to most other automatic functions. The factors that control breathing are summarized in FIGURE 10.14.

The most primitive portion of the brain, the *medulla oblongata*, contains the **respiratory center**, a group of nerve cells that generate a repeated pattern of impulses every few seconds. These impulses trigger contraction of the muscles that permit inhalation. At rest and in a stable environment, a steady breathing rate is maintained by these regular impulses.

During activity, the rate and depth of breathing are modified by sensors that gauge carbon dioxide levels. Carbon dioxide levels are actually measured indirectly by nerve cells sensing the hydrogen ion concentration, or pH, of the fluid surrounding the brain. Carbon dioxide dissolves in water to form *carbonic acid*, causing pH to drop. Declining pH usually signals an increase in cellular respiration and, therefore, a need to take in oxygen and get rid of carbon dioxide to maintain homeostasis. Low blood pH thus causes an increase in the respiratory center's cycle rate, leading to faster and/or deeper breathing. As carbon dioxide is expelled from the lungs, the levels of carbonic acid in the blood decline, pH returns to normal, and breathing rate slows (FIGURE 10.15).

The cycle of the respiratory center can also be modified by oxygen levels in the blood, which are sensed by receptors in arteries leading from the heart. Large declines in oxygen level trigger an increase in the speed of the respiratory cycle. These sensors are much less sensitive than the pH sensors in the brain, meaning that most automatic increases in breathing rate are controlled by the blood's carbon dioxide level, not oxygen level. Oxygen sensors do provide crucial information in certain environments. For example, at high altitudes, where oxygen is less abundant, the sensors stimulate increased respiration rates despite low carbon dioxide production.

FIGURE 10.14 Control of breathing. The respiratory center in the medulla oblongata responds to carbon dioxide levels, oxygen levels, stretch receptors in the lungs, and conscious control by the cortex by coordinating the contraction of muscles that expand the chest cavity. Some nerve cells in the respiratory center respond to very high levels of carbon dioxide in the blood by inducing active exhalation.

FIGURE 10.15 Homeostasis of blood gases. Carbon dioxide and oxygen are maintained at a constant level in the blood by changes in breathing rate in response to activity levels. Blood carbon dioxide levels are most directly responsible for ventilation rate, and thus indirectly affect blood oxygen levels.

VISUALIZE THIS Metabolic acidosis occurs when a disease or dysfunction causes the buildup of acid in the blood. The most immediate symptom of metabolic acidosis is rapid breathing. Use this figure to explain this symptom.

Stop and Stretch 10.4

Hyperventilation, which consists of breathing more rapidly or deeply than is warranted by internal conditions, may be triggered by anxiety or stress. The higher breathing rate reduces carbon dioxide levels in blood below normal, resulting in dizziness and lightheadedness. One treatment for the symptoms of hyperventilation is to breathe into a paper bag. Why is this treatment effective?

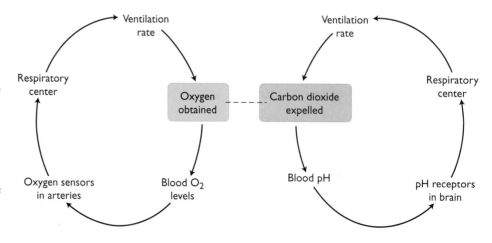

Human beings can maintain conscious control over breathing rates as well. The respiratory cycle can be overridden by signals from the higher levels of the brain, in the cerebral cortex. This voluntary control allows us to speak, hold our breath (at least for a time), or inhale deeply from a cigarette. Once pH levels in the blood drop low enough, however, the involuntary system takes over, forcing us to inhale.

In rare instances, the respiratory cycle is not fully functional, and breathing will cease for 10 to 120 seconds at a time. This condition, called **apnea**, occurs mainly during sleep and can be due to malfunction of the respiratory center or, more commonly, as a result of obstructions in the upper respiratory tract. Sleep apnea can increase blood pressure, cause heart damage, and even result in death. Obesity is a major risk factor for sleep apnea, but smoking also increases the risk of this condition (FIGURE 10.16).

Smoking and Breathing

The act of drawing smoke from a cigarette ("taking a drag") is quite different from a normal inhalation, because the brain cortex is controlling these actions. A typical smoker will take a much deeper breath when drawing on a cigarette, and smokers are more likely to hold their breath for a few seconds before exhaling. The combined effect of these behaviors is to maximize the extent of contact between smoke and the alveoli. As we shall see in the following section, maximizing contact with alveoli increases the efficiency of the transfer of components of smoke to the blood. However, this contact comes with another cost: damage to the structure of the alveoli themselves.

Exposure to particulates causes chronic inflammation in the lungs, leading to scarring and reduced elasticity. As a result, alveoli are placed under higher pressures during active exhalation. During a cough, the pressure can be so great that the walls of some alveoli burst, much like bubbles on the surface of water (FIGURE 10.17). As a result, the

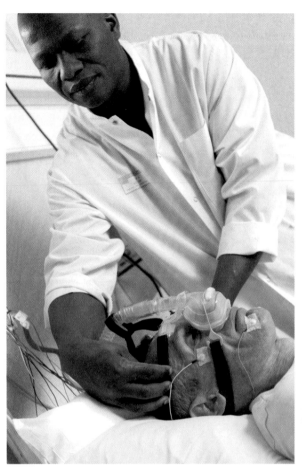

FIGURE 10.16 **Treatment for sleep apnea.** A continuous positive airway pressure (CPAP) device produces a flow of air at elevated pressure relative to room air pressure. The higher pressures keep airways from collapsing, preventing apnea.

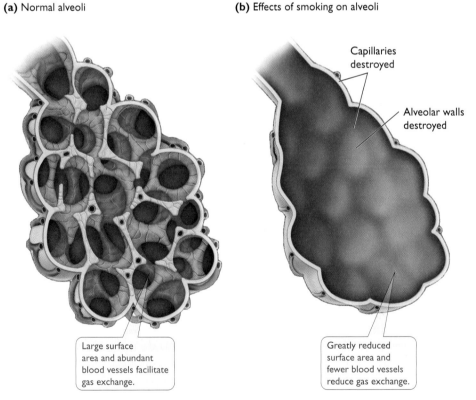

FIGURE 10.17 **Alveoli damage due to smoke exposure.** (a) Normal alveoli look like clusters of grapes. Gas exchange occurs across the "skin" of each of these grapes. (b) Emphysema occurs when lung damage causes alveolar walls to collapse. This reduces the total area for gas exchange, leading to chronic shortness of breath.

lungs cannot respond as well to changes in the shape of the chest, and small bronchioles may collapse during exhalation, trapping air in the lungs. Individuals with this condition, called **emphysema**, often hyperventilate to compensate for the reduced air and gas exchange with each breath. As more and more residual volume is trapped within the lungs, the chest cavity may permanently expand, resulting in a barrel-shaped chest. Emphysema combined with chronic bronchitis occurs in 15% of chronic smokers; the conditions together are now known as **chronic obstructive pulmonary disease (COPD)**.

Individuals with COPD are unable to participate in vigorous activity without becoming breathless and weak. The most severely affected individuals require supplemental oxygen simply to engage in the activities of daily living. Because damaged alveoli cannot be regenerated, COPD is a permanent, irreversible condition. The only "cure" is lung transplant.

10.4 Gas Exchange in the Lungs

The lining of the alveoli represent the **respiratory surface** of our bodies, the epithelial tissue across which gases from inside the body are exchanged with gases in the air. The area of this respiratory surface in humans is approximately 160 square meters (1,725 square feet), about the size of a tennis court. Over the course of a single minute, ventilation causes 4 liters of air to come into direct contact with this surface.

The exchange of carbon dioxide and oxygen at the respiratory surface in the lung occurs by simple diffusion between blood vessels and alveoli. Carbon dioxide in high concentration within the capillaries passes into the alveoli, where exhalation keeps it in low concentration. Oxygen, kept in high concentration in the alveoli by inhalation, passes to the deoxygenated blood in the capillaries.

A Closer Look at Gas Exchange

Although air seems weightless, it is in fact made up of molecules that have mass. The pressure of a gas is the force that the gas exerts on an object by virtue of the mass of the gas. Think of the pushback you feel when you stick your hand out the window of a fast-moving car. This push is created simply by your movement through a mass of gas molecules—in other words, gas pressure. **Gas pressure** was traditionally measured by the displacement of mercury within a barometer, and pressure measurements are still reported as millimeters of mercury (mm Hg).

The concentration of any one gas in a mixture of gases is measured as a partial pressure (or P_x, where x is the chemical symbol for the gas of interest). The **partial pressure** of a gas is equal to its concentration in a mixture of gases multiplied by the total pressure of the combined gases. For instance, at sea level, the total pressure of the atmosphere is 760 mm Hg. Oxygen (O_2) makes up 21% of the gas in the atmosphere, so $P_{O_2} = 760 \times 0.21 = 160$ mm Hg. Gases flow from regions of high partial pressure to regions of low partial pressure.

The pressure difference of oxygen and carbon dioxide between the lungs and the bloodstream is relatively large. On average, P_{O_2} in the alveoli is about 100 mm Hg, and P_{O_2} in the lung capillaries is 40 mm Hg. The P_{CO_2} of blood in the lung capillaries is 46 mm Hg, while in the alveoli, the pressure of carbon dioxide is 40 mm Hg. These differences result in the rapid diffusion of carbon dioxide from the blood into the lungs and oxygen from the lungs into the blood. Diffusion between the alveoli and the capillaries is facilitated by the extremely thin epithelial membrane of the respiratory surface—only 0.2 micrometers, or 1/200 of the thickness of this page.

Carbon dioxide and oxygen cross these membranes only after dissolving in the fluid that coats the respiratory surface. Because the size of the alveoli increases and decreases over the course of a single breath, this layer of fluid must be somewhat slippery to maintain constant coverage of the surface and to prevent the alveolar walls from sticking together by cohesion of the water molecules. A soaplike substance called **surfactant** provides this slippery consistency. The composition of surfactant can be negatively affected by tobacco smoke, and its degradation can reduce levels of gas exchange and make breathing more laborious as alveolar walls tend to stick together rather than expanding with the lungs.

Once external respiration supplies oxygen to the bloodstream in the lungs, the oxygen must be carried to the body tissues to supply the needs of internal respiration.

Recall from Chapter 8 that the oxygen-carrying molecule in the bloodstream is hemoglobin (FIGURE 10.18). The *affinity* of hemoglobin for oxygen, that is, its tendency to pick up oxygen, is affected by the partial pressure of oxygen as well as the temperature and pH of the environment. At a P_{O_2} of 100 mm Hg, the pressure in the alveoli, nearly all hemoglobin molecules will have bound with oxygen. At a P_{O_2} of 40 mm Hg, a typical pressure in the body tissues, hemoglobin's affinity is less strong and only about 75% of the original oxygen load will be retained.

Oxygen uptake is promoted by low temperature and relatively high pH, conditions found in the lungs, which bring in cool outside air and drive the loss of carbonic acid by exhaling carbon dioxide. The opposite effect occurs in tissue where high levels of cellular respiration are occurring—producing excess carbonic acid and higher temperatures through waste heat. Hemoglobin is, therefore, well adapted to facilitate internal respiration, delivering the maximum loads of oxygen to highly active tissue (FIGURE 10.19).

The CO_2 produced by cellular respiration in body cells accumulates in body tissue, raising the P_{CO_2} there relative to the blood flowing from the lungs. Carbon dioxide molecules thus diffuse into the bloodstream. About 10% of the carbon dioxide dissolves directly in the plasma, while the rest enters erythrocytes. There, about 20% of the CO_2 molecules bind with the globin portion of hemoglobin molecules, forming *carbaminohemoglobin*.

The remainder of the carbon dioxide entering the red blood cells combines with water with the assistance of the enzyme *carbonic anhydrase*. This reaction forms carbonic acid, which quickly dissociates into positively charged hydrogen ions and negatively charged *bicarbonate* ions. The bicarbonate exits the erythrocytes and accumulates in the plasma. As will be discussed in Chapter 11, bicarbonate is essential to maintaining homeostasis of blood pH between 7.35 and 7.45.

The excess hydrogen ions from carbonic acid dissociation bind with the protein component of hemoglobin, producing *reduced hemoglobin* and giving blood returning to the heart its dark maroon color. In the lungs, a relatively low P_{CO_2} in the alveoli

FIGURE 10.18 **Hemoglobin.** This protein is responsible for shuttling oxygen from capillaries surrounding the alveoli to the rest of the body tissues.

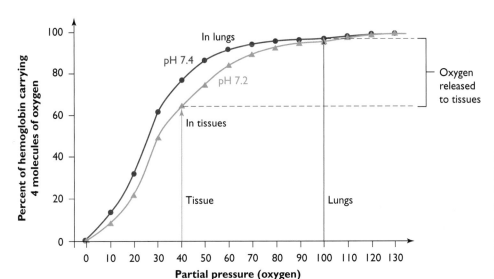

FIGURE 10.19 **Hemoglobin binding.** Hemoglobin picks up oxygen molecules where pH is relatively high, and releases oxygen where pH is low.

VISUALIZE THIS A fetus produces hemoglobin with a higher affinity for oxygen at all partial pressures compared to adult hemoglobin. How would the affinity curve for fetal hemoglobin at pH 7.4 appear, approximately, on this graph?

drives the acid-forming chemical reaction in the opposite direction, causing CO_2 to diffuse from the blood into the air (FIGURE 10.20).

Smoking and Gas Exchange

Hemoglobin is an effective shuttle for oxygen because it will release the molecule under certain conditions. However, the heme group in hemoglobin has an affinity for other, typically more rare molecules that are less likely to be released. Of interest to our discussion is that hemoglobin has a strong affinity for the smoke by-product carbon monoxide. In fact, CO binds to hemoglobin at 240 times the strength with which oxygen binds.

Carbon monoxide inhaled by breathing tobacco smoke is loaded into red blood cells at the respiratory surface. Because the binding of carbon monoxide is so strong, hemoglobin is slow to release it. Even small amounts of carbon monoxide can tie up large amounts of hemoglobin in the body, starving body tissues of oxygen. Although the amounts of carbon monoxide in environmental tobacco smoke are not high enough to cause death from lack of oxygen, chronic low levels of oxygen deprivation caused by carbon monoxide can damage tissues and organs.

Stop and Stretch 10.5

Hyperbaric oxygen therapy is used to treat carbon monoxide poisoning. It consists of placing the patient in a chamber filled with 100% oxygen at a total pressure 2.5 times greater than atmospheric pressure. Use your understanding of the effect of gas pressure on diffusion and hemoglobin binding to explain why this treatment is effective.

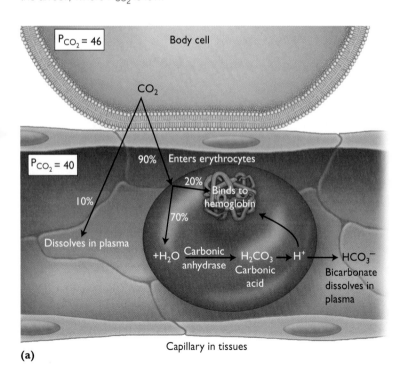

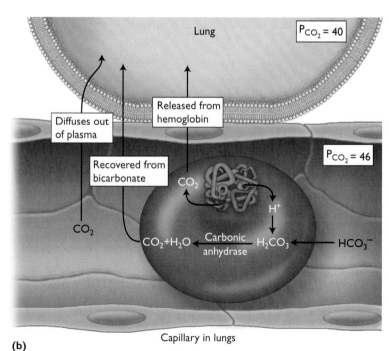

FIGURE 10.20 Carbon dioxide in the blood. (a) About 10% of carbon dioxide molecules entering the bloodstream from body tissues dissolve directly in plasma, while another 20% enter the red blood cells and bind to hemoglobin. The remaining 70% first enter red blood cells and are converted to carbonic acid. When the acid dissociates into bicarbonate and hydrogen, the hydrogen binds to hemoglobin and the bicarbonate returns to the plasma. (b) Carbon dioxide is released from the blood into the air at the alveoli, where P_{CO_2} is low.

Carbon monoxide is especially damaging to developing embryos and fetuses, because they must acquire the oxygen they need through exchange with their mother's blood supply. The lower-than-average birth weights of babies born to mothers who smoke may be due to their relative oxygen deprivation. Some evidence from animal studies indicates that long-term exposure to the carbon monoxide levels found in environmental tobacco smoke may contribute to diminished brain function in infants and children for the same reason—oxygen deprivation.

Nicotine: Why Tobacco Is Habit-Forming

Carbon monoxide is not the only material that can cross the respiratory surface along with oxygen and carbon dioxide. Many of the cancer-causing compounds present in tobacco smoke have been found in the urine of smokers, indicating that the compounds are absorbed into the bloodstream. One of the most potent molecules to cross the respiratory membrane is nicotine, a drug that turns out to be the main reason smokers are willing to accept the health risks of their tobacco use.

Nicotine belongs to a class of naturally produced organic compounds called alkaloids. Plants produce amino acid–derived alkaloids as defenses against their predators. Alkaloids deter predators by their bitter taste and also by being similar in structure to compounds active in the nervous system of mammals, including humans. As a result of this similarity, alkaloids that get into the body can block or enhance nerve signals, causing both physiological and psychological effects. Other familiar alkaloids with effects on the human nervous system include caffeine, ephedrine, codeine, and cocaine.

FIGURE 10.21 Nicotine. Nicotine is an organic compound produced by plants to deter predators.

Nicotine (FIGURE 10.21) is similar in structure to the nerve-activating chemical acetylcholine, which has effects on a variety of bodily functions including breathing, heart rate, appetite, and mood. Because nicotine stimulates the same nerve cells as does acetylcholine, it influences the same systems. Nicotine consumption increases breathing and heart rates, decreases appetite, and promotes a feeling of well-being.

Feelings of well-being are typically induced by daily pleasures, such as a good meal or positive social interactions. These activities make us feel good because they stimulate nerve cells in the "pleasure center" of the brain. This brain structure appears to be a reward system, an adaptation that encourages us to continue to seek out experiences and behaviors that improve our chances of survival and reproduction.

Because nicotine independently stimulates the brain's pleasure center, it is highly addictive, driving users to continually seek the drug. When the initial effect of nicotine wears off, activity in the pleasure center turns down, and users feel depressed and fatigued, which gives them a craving for additional nicotine. In a sense, nicotine and other addictive drugs hijack the brain's reward system. Normal activities no longer feel as pleasurable, and the addictive substance becomes its own reward (FIGURE 10.22).

When a smoker inhales, nicotine crosses the alveolar walls and is carried to the brain in only 10 seconds. The speed of nicotine delivery provides a powerful subconscious connection between the pleasurable sensation and the cigarette. The act of holding the cigarette, lighting up, and drawing smoke into the lungs thus all become part of the addiction. The association of cigarettes with nicotine effects is powerful. At least 40% of people who have ever smoked a cigarette believe they are physically dependent on them, while only 7% of alcohol users feel dependent on that slower-acting drug.

FIGURE 10.22 Addiction to nicotine. Cigarettes are addictive because they deliver nicotine. Nicotine stimulates the pleasure center of the brain, and the smoker associates the activity of smoking with that pleasurable feeling. As the brain adjusts to the stimulation provided by nicotine, the smoker becomes dependent on cigarettes.

VISUALIZE THIS At what point in this positive feedback system do nicotine patches or nicotine gum help to break a cigarette addiction? At what point do drugs that cause nausea when the user is exposed to nicotine have their effect?

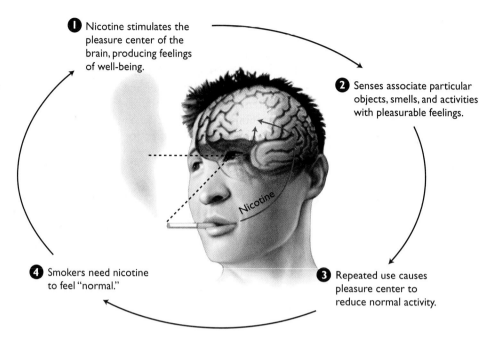

❶ Nicotine stimulates the pleasure center of the brain, producing feelings of well-being.

❷ Senses associate particular objects, smells, and activities with pleasurable feelings.

❸ Repeated use causes pleasure center to reduce normal activity.

❹ Smokers need nicotine to feel "normal."

10.5 Beyond the Lungs

The effect of tobacco smoke on health is established by epidemiological studies, which look for links between environmental or social factors and disease. The mechanics of these studies are described in detail in Chapter 1. Epidemiological studies have indicated some surprising effects of smoking, involving nearly all organ systems (FIGURE 10.23). These organ systems are described in detail in other chapters, but the following provides a brief review.

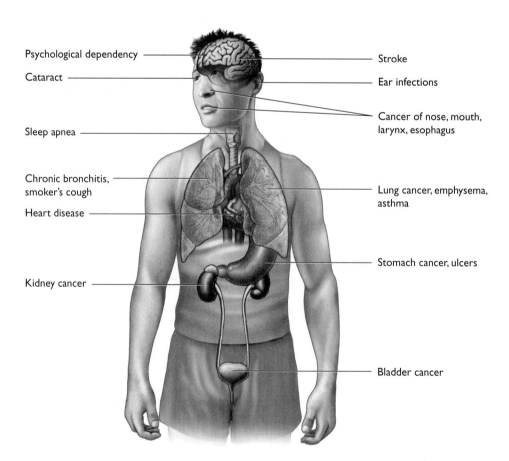

FIGURE 10.23 Tobacco smoke affects nearly all organ systems. A summary of the health risks clearly associated with smoking. Other risks may exist, but links have not been definitively established.

The Effects of Smoke on Other Organ Systems

Cardiovascular System Most people believe that lung cancer and other lung diseases such as COPD are the main risk associated with exposure to tobacco smoke. In reality, most of the deaths due to smoking result from cardiovascular disease.

Most of the effects of smoking on the cardiovascular system are caused by nicotine. The drug appears to increase production of LDL (low-density lipoprotein, the "bad cholesterol"), and reduce production of HDL (high-density lipoprotein, the "good cholesterol"). As a result, individuals exposed to tobacco smoke have higher levels of circulating fats (lipids) and thus higher risk for blockages in blood vessels. Nicotine also stimulates blood clot formation, increasing the risk of both heart attack and stroke.

According to the U.S. Centers for Disease Control and Prevention (CDC), heart disease accounts for about 147,000 deaths a year among active smokers. A smoker has two to three times the chance of dying from heart disease that a nonsmoker has. In addition, the CDC estimates that 35,000 deaths per year from heart disease among passive smokers result from exposure to environmental tobacco smoke.

Digestive and Excretory System The toxic components of tobacco smoke can enter the digestive system when smokers swallow the tar and mucus that is cleared from their airways. This load of toxins results in an increased risk of both stomach and pancreatic cancer. Smoke exposure also exacerbates infections by *Helicobacter pylori*, the leading cause of stomach ulcers.

Many of the toxic components of tobacco smoke are actively excreted from the body in urine. These components thus are concentrated in the kidneys and bladder, the organs that produce and store urine before excretion. As a result, smokers have a 38% greater risk of kidney cancer and a 300% greater risk of bladder cancer than nonsmokers.

Senses and Brain The hydrogen cyanide in tobacco smoke damages or kills sensory receptors in the nose and mouth, diminishing a smoker's senses of taste and smell. Exposure to tobacco smoke also increases the risk of cataracts, an accumulation of dead cells on the lens of the eye, the primary cause of blindness in older adults.

The increased risk of stroke resulting from smoking threatens the brain. Arguably, however, the most important effect of smoking on the brain is nicotine's addictive qualities. Clearly, any attempts to reduce the cost of smoking to both smokers and society needs to address this point.

Stop and Stretch 10.6

Consider the range of health effects of smoke exposure and their causes. Is using smokeless (chewing) tobacco likely to be much less harmful than smoking, based on your understanding of these causes and effects?

Preventing Smoking-Related Illness

Quitting smoking is difficult. Despite the fact that two-thirds of smokers say they wish they could quit, only 1 in 10 succeeds. The strength of cigarette addiction can be made apparent by the following statistics: After surgery for lung cancer, almost half of smokers resume smoking. Among smokers who suffer a heart attack, 38% resume smoking immediately after leaving the hospital. Even when a smoker has his or her larynx removed as a result of cancer treatment, 40% try smoking again.

Smokers who wish to quit need understanding and often assistance. Many benefit from a gradual period of "weaning" from nicotine, replacing the nicotine they obtained via cigarette smoke with nicotine patches, lozenges, or gum. Others benefit from counseling and changing their behaviors to avoid triggers for nicotine craving. Table 10.1 summarizes some of the basic tools of successful quitters.

The addictive power of nicotine is only half of the challenge that we face when trying to reduce the health and economic costs of smoking. The other half is reducing the number of new smokers who pick up the habit. Ninety percent of smokers pick up the habit before they turn 19—that is, an age when many of us are actively rebelling against convention. Because nicotine is an effective appetite suppressant, young women in increasing numbers have started smoking as a strategy for weight loss. Other risk factors for starting smoking include use by other family members and friends, the approval of peers, poor academic achievement, low self-image, and susceptibility to the influence of others.

A review article published in the journal *Pediatrics* in December 2005 reported that the number of PG-13 movies that had characters who smoke outnumbered R-rated movies with smoking in them over the period 2002–2005. When movies and advertisements project smoking as glamorous and "adult" behavior, teenagers may be more willing to model the behavior than if smoking is depicted as negative. In fact, some of the most successful antismoking campaigns do show the ugly side—in ads designed by kids in the age groups most susceptible to starting smoking.

TABLE 10.1 Strategies of Successful Quit-Smoking Programs

Basic Principles	Specific Actions
Prepare.	• Set a quit date. • Get rid of all cigarettes and ashtrays.
Get support and encouragement.	• Tell friends and family of your plans to quit and that you want their support. • Get individual or group counseling—call your health care provider or county health department for referrals.
Learn new behaviors.	• Change your routine to minimize exposure to the triggers that tempt you. • Work on other methods of stress reduction, such as breathing exercises. • Plan something enjoyable to do every day.
Get medication and use it correctly.	• Nicotine replacement products double your chance of quitting—talk to your doctor or pharmacist about options.
Be ready for difficult situations.	• Avoid drinking alcohol. • Avoid other smokers. • Weight gain is common—do not let it distract you from your goal. • Develop strategies for stress relief, such as meditation or exercise programs.

Chapter REVIEW

ROOTS TO REMEMBER

The following roots of words come mainly from Latin and Greek and will help you to decipher terms:

-hale, as in exhale and inhale, comes from a verb meaning to breathe.

-itis is an ending used to describe inflammation (of an organ).

pleur- and **pleura-** come from a word for the sides or flanks.

pneum- and **pulmon-** both mean lung.

-spir- and **-spira-** come from a verb meaning to breathe.

KEY TERMS

10.1 Respiratory System Anatomy: The Path of Smoke into the Lungs
alveoli *p. 230*
auditory tubes *p. 229*
bronchiole *p. 230*
bronchus *p. 230*
Eustachian tubes *p. 229*
larynx *p. 229*
lower respiratory tract *p. 228*
lungs *p. 230*
mouth *p. 228*
nasal cavity *p. 228*
nose *p. 228*
pharynx *p. 229*
pleural membrane *p. 230*
respiratory system *p. 228*
septum *p. 229*
sinus *p. 229*
trachea *p. 230*

upper respiratory tract *p. 228*
vocal cords *p. 229*

10.2 Tobacco Smoke and the Respiratory Tract
asthma *p. 233*
bronchitis *p. 233*
environmental tobacco smoke *p. 230*
laryngitis *p. 233*
otitis media *p. 233*
particulates *p. 231*
pneumonia *p. 233*
sinusitis *p. 233*
tonsillitis *p. 233*

10.3 Inhaling and Exhaling
apnea *p. 237*
chronic obstructive pulmonary disease (COPD) *p. 238*

diaphragm *p. 234*
emphysema *p. 238*
exhalation *p. 234*
expiration *p. 234*
external respiration *p. 234*
inhalation *p. 234*
inspiration *p. 234*
intercostal muscles *p. 234*
internal respiration *p. 234*
residual volume *p. 234*
respiratory center *p. 236*
tidal volume *p. 234*
ventilation *p. 234*
vital capacity *p. 234*

10.4 Gas Exchange in the Lungs
gas pressure *p. 238*
partial pressure *p. 238*
respiratory surface *p. 238*
surfactant *p. 239*

SUMMARY

10.1 Respiratory System Anatomy: The Path of Smoke into the Lungs

- The upper respiratory tract consists of the mouth, nose, sinuses, and pharynx. It functions to clean and condition incoming air (pp. 228–229).

- The lower respiratory tract consists of the larynx, trachea, bronchi, and lungs and delivers inhaled air to the respiratory surfaces in the alveoli (pp. 229–230; Figure 10.7).

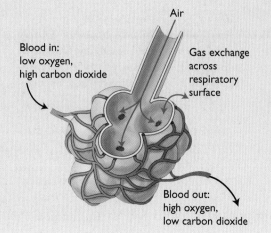

- The human voice is produced by vibrations of the vocal cords in the larynx and by the shape and movements of the mouth (pp. 229–230).

10.2 Tobacco Smoke and the Respiratory Tract

- Mainstream smoke and sidestream smoke consist of hundreds of chemicals, including some that may trigger the development of cancer and others that are potent toxins, as well as tiny solid particulate matter (pp. 230–232).

- Exposure to smoke inhibits the protective functions of the upper respiratory tract,

- leading to higher rates of respiratory infections. Smoke also increases the risk of asthma and of lung cancer (pp. 232–233).

10.3 Inhaling and Exhaling

- Breathing facilitates external respiration, the exchange of gases between the body and the environment. External respiration is necessary to allow internal respiration, the delivery of oxygen to and removal of carbon dioxide from tissues in which cellular respiration is occurring (p. 234; Figure 10.11).

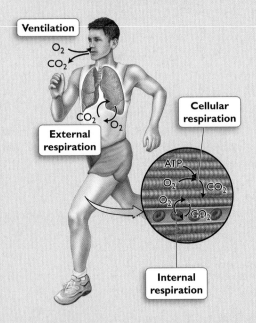

- The air in the lungs consists of the tidal volume, the amount that we breathe in and out in a single breath; the vital capacity, which is the total we can exhale; and the residual volume, the remaining air that we cannot exhale (p. 234).

- Contraction of the diaphragm and the intercostal muscles increases the volume of the lungs. These movements during inhalation draw air in through the upper respiratory tract. When the diaphragm and intercostal muscles relax, the chest cavity shrinks, forcing air out (pp. 234–235).

- The respiratory center of the brain controls the breathing rate under normal conditions, but this rate is modified by homeostatic processes and by the voluntary control of breathing (p. 236).

- A decline in blood pH, caused by increased carbon dioxide production, triggers a more rapid breathing rate (pp. 236–237).

- When activity levels are high and carbon dioxide is more abundant in the blood, signals from the respiratory center cause abdominal and chest muscles to contract, actively forcing greater volumes of air out of the lungs (p. 236).

- Smoke damages the alveoli and makes the lung less elastic, obstructing exhalation and interfering with normal gas exchange (pp. 237–238).

10.4 Gas Exchange in the Lungs

- The respiratory surface in humans is the walls of the alveoli. Oxygen diffuses across this surface into the capillaries, and carbon dioxide diffuses from the capillaries into the alveoli (p. 238).

- The direction of gas movement is from areas of high partial pressure to areas of low partial pressure (p. 238).

- Hemoglobin in blood facilitates the delivery of oxygen to active tissues by binding it strongly at the respiratory surface. Hemoglobin drops its load of oxygen where temperature is higher and pH is lower than in the lungs, in other words, in actively respiring tissues (p. 239; Figure 10.18).

- Carbon dioxide diffusing from the tissues is carried to the respiratory surface primarily in the form of the bicarbonate ion. When it reaches the lung, the bicarbonate is broken down to yield carbon dioxide, which diffuses into the alveoli and is exhaled (p. 239).

- Carbon monoxide in tobacco smoke binds to hemoglobin, depriving tissues of oxygen delivery. The effects of this deprivation are especially severe on developing embryos and fetuses (p. 240).

- Nicotine is the component in smoke that causes addiction by mimicking brain chemicals that induce a feeling of well-being (p. 241).

10.5 Beyond the Lungs

- The components of smoke that cross the respiratory membrane increase the risk of cardiovascular disease, cancer of several organs, ulcer, cataract, and stroke (p. 243).

LEARNING THE BASICS

1. The upper respiratory tract contains all of the following structures except _____.
   a. alveoli
   b. sinuses
   c. pharynx
   d. nose
   e. mucous membranes

2. The function of the upper respiratory tract is _____.
   a. gas exchange
   b. conditioning of air entering the lower respiratory tract
   c. production of vocal sounds
   d. a and c are correct
   e. a, b, and c are correct

3. A high voice is produced by _____.
   a. very flexible vocal cords
   b. a large larynx
   c. individuals who smoke
   d. short vocal cords
   e. the shape of the mouth

4. Cilia in the respiratory tract _____.
   a. sweep mucus and debris out of the respiratory tract
   b. protect the delicate structures of the lungs from contamination
   c. actively direct the movement of materials in the lungs
   d. can be damaged by exposure to tobacco smoke
   e. all of the above are correct

5. Number the following structures in the order in which incoming air flows past them in the respiratory tract:
   _____ bronchioles
   _____ trachea
   _____ nasal conchae
   _____ larynx
   _____ alveoli

6. Alveoli are _____.
   a. small air sacs at the ends of bronchioles in the lungs
   b. the respiratory surface in humans
   c. surrounded by a net of capillaries
   d. subject to damage because of exposure to tobacco smoke
   e. all of the above are correct

7. Compared to mainstream smoke, sidestream smoke _____.
   a. is less concentrated
   b. contains more products of incomplete combustion
   c. contains more particles
   d. a and b are correct
   e. a, b, and c are correct

8. Match each term with its definition.
   _____ cellular respiration   a. inhalation and exhalation
   _____ internal respiration   b. transfer of gases between the blood and body tissues
   _____ external respiration
   _____ ventilation   c. oxygen-requiring process generating ATP inside cells
                                          d. exchange of gases between body and environment

9. Nerve cells in the respiratory center trigger an increased rate of respiration in response to _____.
   a. increased carbon dioxide levels in the blood
   b. decreased pH of the blood
   c. decreased oxygen levels in the blood
   d. a and b are correct
   e. a, b, and c are correct

10. In a normal breath, you take in _____ of the air that the lungs can possibly hold.
    a. all
    b. about 90%
    c. about half
    d. approximately 10%
    e. an undetermined amount

11. At rest, inhalation requires _____.
    a. contraction of the diaphragm
    b. the function of the respiratory pump
    c. creation of a partial vacuum in the capillaries surrounding the alveoli
    d. relaxation of the intercostal muscles
    e. contraction of the abdominal muscles

12. Emphysema leads to enlarged chest size because _____.
    a. affected individuals inhale too much
    b. exposure to tobacco smoke leads to uncontrolled production of additional alveoli
    c. the bronchial tubes are filled with thick mucus, which traps air
    d. buildup of scar tissue makes lungs less elastic, trapping air
    e. lungs are subject to chronic, allergic inflammation

13. All of the following statements about hemoglobin are true, *except* _____.
    a. It is a protein that carries oxygen in the blood.
    b. It binds oxygen when the gas is at low partial pressures and releases it in regions where partial pressure is high.
    c. It strongly binds carbon monoxide.
    d. It contains iron, which is responsible for its red color.
    e. It is found within red blood cells.

14. The partial pressure of a gas in a mixture of gases is equal to _____.
    a. the total number of molecules in the volume of gas
    b. its proportion in the mixture times the total gas pressure
    c. the total gas pressure divided by the number of gases in the mixture
    d. the temperature of the gas relative to the other gases in the mixture
    e. the likelihood that the gas will diffuse from one compartment to another

15. Which of the following conditions promotes the release of oxygen from hemoglobin molecules?
    a. low pH
    b. high temperature
    c. high levels of carbonic acid
    d. low partial pressure of oxygen in nearby tissues
    e. all of the above are correct

16. Carbon dioxide in the blood _____.
    a. can be found in bicarbonate ions
    b. may bind to hemoglobin molecules
    c. does not change blood pH
    d. a and b are correct
    e. a, b, and c are correct

17. Nicotine _____.
    a. activates nerve cells in the brain
    b. is not present in smokeless tobacco
    c. is not found in nature
    d. is not especially addictive
    e. exposure causes depression

18. All of the following conditions make quitting smoking difficult except _____.
    a. the addictiveness of nicotine
    b. the association of smoking with good feelings
    c. associations between smoking and particular habitual behaviors
    d. the health risks of smoking
    e. an association between smoking and personal expression and freedom

19. Which of the following diseases is associated with exposure to tobacco smoke?
    a. heart disease
    b. lung cancer
    c. bladder cancer
    d. asthma
    e. all of the above

ANALYZING AND APPLYING THE BASICS

Use the following information to answer questions 1–3.

Asthma is a respiratory disease whose cause is unclear. It is characterized by attacks that cause chest tightness, wheezing, and shortness of breath. During an asthma attack, cells in the lungs produce thick and sticky mucus. The muscles around bronchial tubes tighten, and the bronchial tubes themselves become inflamed. Asthma attacks can reduce the flow of air into the lungs by 50%. This can decrease the body's ability to obtain oxygen and eliminate carbon dioxide. Asthma attacks can be triggered by a variety of factors, including respiratory infections, cold air, allergens such as mold or pet dander, ozone, cigarette smoke, car exhaust, exercise, dust, perfume, or even emotions.

Asthma is a chronic, but treatable, disease. Drugs are available to prevent or treat asthma attacks. Anti-inflammatory drugs, such as corticosteroids, decrease the production of mucus and reduce inflammation of bronchial tubes. Bronchodilators help to relax the muscles that constrict bronchial tubes during an attack.

1. Researchers studying allergen-triggered asthma found that allergens cause the migration of T cells to the lungs. These T cells release chemicals that cause inflammation. The researchers discovered that T cells lacking a protein called β-arrestin-2 did not migrate to the lungs, so inflammation did not occur.

   Would it be wise for scientists to create a drug that interfered with β-arrestin-2 protein synthesis in T cells in order to treat asthma? Why or why not?

2. The prevalence of asthma has increased significantly in the past few decades. The graphs in FIGURE 10.24 show the percentage of people who belonged to the Kaiser Permanente HMO (health management organization), Northwest Division, between 1967 and 1987 who were treated for asthma. The data are categorized by age group and sex.

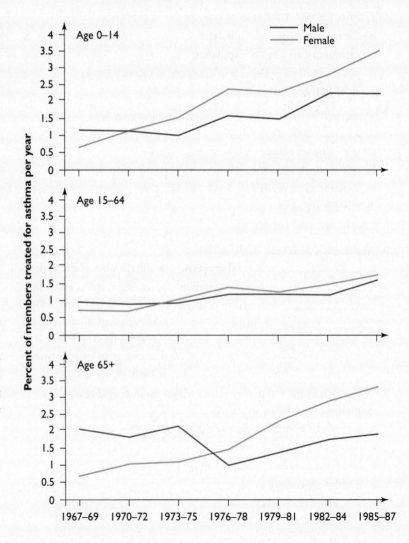

FIGURE 10.24 Percentage of asthma sufferers in Kaiser Permanente HMO, 1967–1987, by age and sex

Source: Vollmer, W. M.; Osborne, M. L.; and Buist, A. S. 1998. 20-Year Trends in the Prevalence of Asthma and Chronic Airflow Obstruction in an HMO. *American Journal of Respiratory and Critical Care Medicine* 157(4): 1079–1084.

What do the graphs indicate about asthma trends during this time period?

3. Some people assume that air pollution is the most likely root cause of asthma, because airborne particulate matter and ozone can trigger asthma attacks. Many scientific studies seem to reject this assumption. One study, comparing asthma in schoolchildren in the German cities of Leipzig (high levels of pollution) and Munich (low levels of pollution) found significantly less asthma in Leipzig. Asthma is most common in children in developed countries, such as Great Britain, Australia, and the Republic of Ireland. Asthma rates in children are lowest in Eastern Europe, China, Uzbekistan, and India.

   Allergens besides air pollution have been shown to potentially protect against the development of asthma. A variety of studies found asthma to be less common in children of animal farmers and children who had pet cats or dogs. Studies also found less prevalence of asthma in youngest children who have many older siblings.

   How would you explain the results of these studies? In light of these studies, what could parents do to help reduce the risk of their young children developing asthma?

4. The 2008 Beijing Summer Olympics had some athletes and coaches concerned, due to the high particulate levels in Beijing's air. To protect human health, the European Commission recommends that particulate levels not exceed 40–50 micrograms per cubic meter of air ($\mu g/m^3$). Particulate levels in Beijing can rise as high as 162 $\mu g/m^3$. Why would high levels of particulates affect athletic performance, or even put an athlete's health at risk?

5. Premature babies lack fully developed lungs and often need mechanical ventilation to help them breathe. Unfortunately, this type of ventilation can damage their lungs. Since the 1980s, scientists have experimented with liquid ventilation to deliver oxygen to preemies. In initial experiments, scientists pumped perfluorinated hydrocarbon containing oxygen into and out of the lungs of 18-week-old lamb fetuses. The hydrocarbon was able to carry more oxygen to the blood of these sheep than forced ventilation could. In 1996, premature infants with breathing difficulties were helped by treatment with oxygenated liquid.

   Describe how liquid ventilation could deliver oxygen to the bloodstream of a premature infant.

6. As tobacco plants grow, radioactive minerals found in soil stick to the plants' leaves. Minerals found in phosphate fertilizer, such as radium, lead-210, and polonium-210, can also accumulate on the tobacco plant. Radioactive substances on tobacco are not removed as the tobacco is processed to make cigarettes. Therefore, each cigarette delivers a dose of radiation along with its dose of nicotine. Someone who smokes about 1.5 packs per day is exposed over the course of a year to a radiation dose equivalent to 300 chest X-rays. How could radioactive minerals in cigarettes produce disease throughout the human body?

CONNECTING THE SCIENCE

1. Some people have argued that the best way to reduce the number of smokers is to make smoking, and the health care costs associated with it, more expensive. This strategy entails placing high taxes on cigarettes and other tobacco products and restricting current and former smokers' access to subsidized medical care. What do you think of these strategies? Do you think they would be successful? Do you think they might be unfair?

2. Although tobacco causes more cases of long-term disability, alcohol is more deadly for young men and women than cigarettes because of automobile and other accidents. Do you think the success of smoking bans means that alcohol will become more strongly restricted? Do you think restricting alcohol advertising and availability is good policy? Why or why not?

Chapter 11

The Urinary System

Surviving the Ironman

LEARNING GOALS

1. Compare and contrast elimination and excretion, and describe the different methods by which wastes are eliminated from the body.

2. List the structures and functions of the urinary system.

3. Summarize steps in the process of urination.

4. List the major components of urine and their sources.

5. List the steps in the process of excretion and briefly describe what happens in each step.

6. Summarize the process of tubular reabsorption, explaining the role of sodium transport in the process.

7. Describe how countercurrent exchange can maximize water retention in the kidneys, and list the effects of hormones on the volume of water lost in urine.

8. Describe how blood pH is regulated in both the short and long term.

9. Summarize the role of the kidney in maintaining salt balance.

10. Explain how dialysis can replace kidney function in individuals with renal disease.

This tropical island paradise hosts one of the most grueling races ever devised—the Ironman triathlon.

The dark streets of Kona, Hawaii, are crowded with people watching racers pass. They cheer wildly as the leading woman racer comes into view. But something is wrong—Julie Moss weaves unsteadily, then collapses to the ground. She gets to her feet quickly, only to have her legs buckle again. With great effort, she stands, jogs another 50 yards, and collapses once more. Waving away the hands helping her up, she staggers and stumbles to within 20 yards of the finish line. Painfully crawling forward, she is passed by a steadily running woman only a few feet from the end. Julie Moss crawls across the line less than a minute later.

It is arguably the most famous second-place finish in sports history. The dramatic end of the 1982 Ironman triathlon was televised around the world. Julie Moss's determination to finish the grueling 2.4-mile swim, 112-mile bike ride, and 26.2-mile footrace inspired millions of everyday athletes to push their bodies to the limit in endurance events.

A swim and bike ride are followed by a marathon.

Why did Julie Moss struggle, while the women's winner steadily gained on her—and while thousands of amateur athletes have succeeded since? How does the human body stand up to the physical challenge of an Ironman?

Our understanding of the body's response to extreme stress has improved athletes' preparation for this challenge. Finishing the triathlon is now the challenging, but accessible, culmination of months of systematic training and careful planning. Nearly 500,000 participants have completed Ironman competitions since the first one in 1978, thanks to application of human physiology research.

What used to be seen as primarily a triumph of willpower . . .

One organ system that is surprisingly crucial to the function of the body under the demanding conditions of an Ironman race is the urinary system.

. . . is now the culmination of systematic training.

11.1 An Overview of the Urinary System

The energy used by Ironman participants during a race, and by everyone for the activities of daily life, ultimately comes from food. The breakdown of food to release energy creates waste. The primary function of the **urinary system**, made up of the kidneys, bladder, and associated organs, is to remove some of that waste.

Homeostasis and the Urinary System

Bulk solid waste is removed from the body, a process called **elimination**, as feces (Chapter 7). For the most part, the waste in feces is made up of materials that never actually left the digestive tract to enter the bloodstream. Inside the body, however, wastes form when the products of digestion are carried to tissues and metabolized to power cellular respiration. These wastes, including ammonia, acids, and salts, accumulate in body tissues and diffuse into the bloodstream. To maintain homeostasis, these poisonous wastes must also be eliminated from the body.

Some metabolic wastes are eliminated by activities of the respiratory and integumentary systems. Carbon dioxide and some of the water produced by cellular respiration are exhaled from the lungs, and additional water and some metabolic wastes are shed in sweat. However, most metabolic wastes produced by body activity are eliminated via the functions of the urinary system. Any process that filters waste from the blood is called **excretion**.

Because the kidneys, the primary organs of the urinary system, filter wastes from the entire bloodstream, the activities of these organs are also essential to maintaining homeostasis of other blood constituents. By regulating the amount of salt excreted, kidneys help maintain water balance between tissues and the blood. By adjusting levels of various ions in the blood, the kidneys also serve to keep blood pH within narrow bounds.

The kidneys also help to maintain levels of other physiologically important ions and release several hormones that have effects on a range of other organ systems (FIGURE 11.1). One of these hormones, erythropoietin, stimulates red blood cell production in response to low oxygen levels in blood and is discussed in detail in Chapter 8.

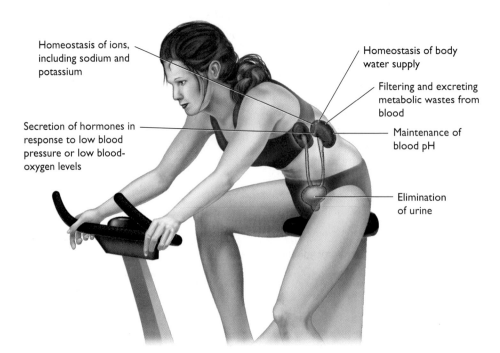

FIGURE 11.1 Functions of the urinary system. The urinary system is responsible for excreting metabolic wastes, balancing body water and blood pH, and maintaining homeostasis of ions and oxygen levels in the bloodstream.

Structure of the Urinary System

The major organs of the urinary system include the **kidneys**, which filter and cleanse circulating blood supplied to them by the *renal arteries*. Waste excreted from the blood is sent through **ureters** to the **urinary bladder**, where it is stored until it is expelled via the **urethra**. Blood that has passed through the kidneys returns to circulation via the *renal veins* (FIGURE 11.2).

The kidneys (FIGURE 11.3) are paired, approximately fist-sized organs that sit behind the liver and stomach in the upper abdominal cavity. Their distinctive beanlike concave shape is due to the large vessels that enter and leave each kidney on its inner-facing, or medial, side. Each kidney is protected by a tough connective tissue membrane called the *renal capsule*.

Inside, the kidney is made up of three distinct layers: the outer **renal cortex**, the inner **renal medulla**, and a hollow center, called the **renal pelvis**. The cortex and medulla are packed with a dense network of microscopic **nephrons**, the functional filters of the kidney. Each kidney contains approximately 1.25 million nephrons, with a combined length in an adult of about 140 kilometers (85 miles). The nephrons are grouped in structures called *renal pyramids* scattered more or less evenly through the medulla.

All of a person's blood passes through the kidneys hundreds of times per day, such that in a resting person, each kidney filters about 1,000 liters of blood every 24 hours. The end product of this filtration is urine.

Once urine is produced by the nephrons (a process described in the next section), it moves via contractions of the muscular ureters to the urinary bladder. The bladder itself is a hollow, elastic organ. Its dimensions vary according to the amount of urine it contains. At its fullest, it can store approximately 1 liter of fluid within its highly expandable walls. The bladder walls are mainly made up of the powerful *detrusor muscle*, which, when it contracts, expels urine through the **internal urethral sphincter**, a small ring of muscle at the top of the urethra.

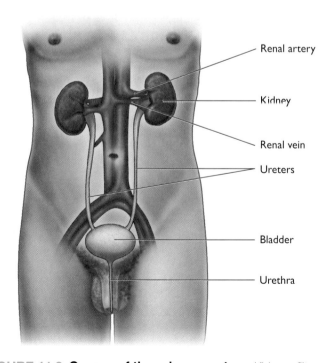

FIGURE 11.2 Organs of the urinary system. Kidneys filter the blood and transfer waste in urine through ureters to the urinary bladder, where it is held until voluntarily released through the urethra.

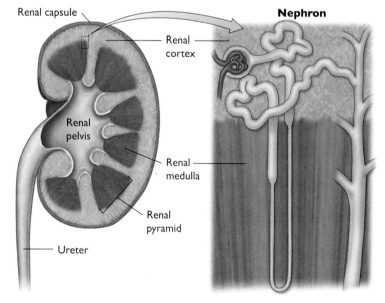

FIGURE 11.3 Kidneys. The kidneys are complex organs made up of functional units called nephrons.

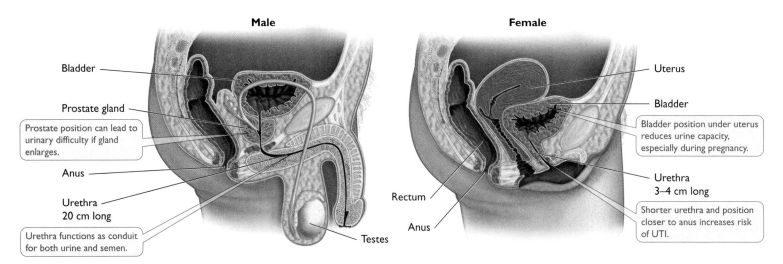

FIGURE 11.4 Sex differences in the urinary system. Although the urinary system is roughly the same in men and women, differences in anatomy lead to some differences in function.

The **external urethral sphincter** controls the flow of urine out of the urethra. The urethra differs in length and function between males and females (FIGURE 11.4). In women, it is 3–4 centimeters (1–1.5 inches) long and functions solely to eliminate urine from the body. The male urethra is much longer, 18–20 centimeters (7–8 inches), and delivers both urine and sperm from the body. The production and release of sperm is covered in Chapter 18.

Urination

As the bladder fills with urine, stretch receptors in its walls trigger contractions of the bladder and also send signals to the brain that we consciously interpret as "needing to urinate" (FIGURE 11.5). Most able-bodied adults can delay the release of urine until an appropriate time and place by keeping the external urethral sphincter contracted and inhibiting the micturition reflex.

The **micturition reflex** consists of the unconscious relaxation of the internal urethral sphincter and simultaneous contraction of the detrusor muscle of the bladder, which occurs when the external sphincter is consciously relaxed. If the bladder continues to fill, eventually this voluntary inhibition of micturition will fail, causing urination.

The bladder empties completely during this process as a result of a positive feedback loop. The flow of urine through the urethra stimulates additional detrusor muscle contractions, which squeeze all of the remaining urine out of the bladder. Positive feedback is effective here because eventually the bladder is empty, urine ceases flowing through the urethra, and detrusor contractions stop.

Stop and Stretch 11.1

Older women who have birthed one or more children are susceptible to urinary incontinence, in which urine leaks from the urethra involuntarily, especially when one laughs, coughs, or lifts heavy objects. What is the probable relationship between giving birth and urinary incontinence? Why might certain activities lead to leakage?

The cycle of bladder filling and emptying in an individual depends on bladder size, sensitivity of the stretch receptors, and water intake. Women generally have a smaller

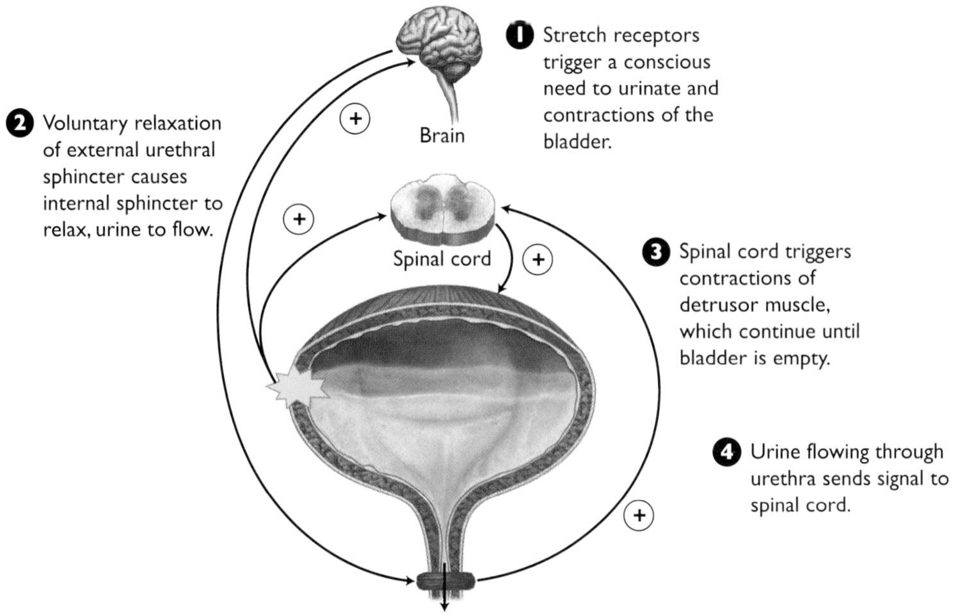

FIGURE 11.5 Positive feedback during urination. The presence of urine in the bladder triggers the micturition reflex. Release of a little urine through the urethra triggers further contractions until the bladder is completely empty.

VISUALIZE THIS What puts an end to this positive feedback process?

bladder capacity than men, mainly because the uterus compresses the bladder in the abdominal cavity. However, an enlarged prostate gland, which squeezes the urethra and bladder, can cause increased frequency of urination in men.

11.2 Excretion

The process of excretion takes place in the kidneys, and urination is its end result. Understanding how excretion occurs in human beings at rest gives Ironman participants information about how their bodies manage the copious metabolic waste they produce during competition.

The Composition of Urine

Urine is a solution containing water and a number of other components with concentrations dependent on diet and activity levels. Prominent among the solutes in urine is **urea**, a modified waste product of protein metabolism. When nitrogen-rich proteins are used for cellular respiration, *ammonia* (NH_3) is produced as a waste product. Ammonia is toxic to cells, and it is immediately processed by the liver into much less dangerous urea. *Uric acid*, a metabolite of DNA metabolism, is found in low levels in human urine as well. If uric acid builds up in the blood, as happens in individuals with a defect in DNA metabolism, it may crystallize in the joints, causing a painful condition called *gout* (FIGURE 11.6).

Urine may also contain ions, including sodium, chloride, potassium, calcium, and hydrogen ions. The kidneys excrete these ions to maintain homeostasis of blood volume and acidity, processes that are described in section 11.3. Trace amounts of other materials that the body does not require, including toxins, drugs, and other metabolic wastes, are also found in urine. Urine's yellow color results from urobilins, waste products of the breakdown of *bilirubin*. As described in Chapter 8, bilirubin is a protein that is produced when hemoglobin, the oxygen-carrying protein in blood, is degraded.

Another important component of urine is *creatinine*, a waste product resulting from intense muscle use. Creatinine forms when the storage molecule creatine phosphate is used in cells to generate ATP. The small store of ATP normally found within a resting cell is enough to power only about three seconds of muscle contraction. In the process, the stored ATP molecules each lose a phosphate group to become ADP. Creatine phosphate can restore ADP to ATP by donating its phosphate group, becoming creatinine (FIGURE 11.7).

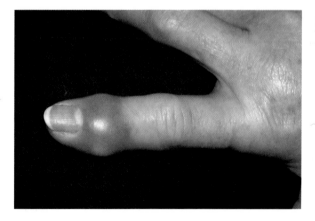

FIGURE 11.6 Gout. A genetic defect in DNA metabolism can lead to the buildup of uric acid in the bloodstream. Even at very low concentrations, the acid will crystallize, forming deposits inside kidneys and on joints and tendons. These deposits become inflamed and may enlarge into prominent nodules. In susceptible individuals, gout is often triggered by excess alcohol consumption (more than two drinks per day) and adult weight gain.

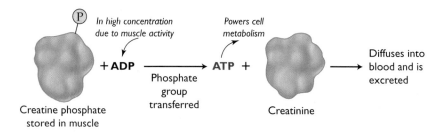

FIGURE 11.7 Creatinine. Creatine phosphate provides a short-term energy burst in hard-working muscles, but it is used up quickly. Creatinine is a waste product of this process, and some is excreted in urine every day.

VISUALIZE THIS Power weight lifters, who compete to lift the largest amount of weight above their heads and hold it there for a few seconds, often take amino acid supplements to increase their store of creatine phosphate. How would an increase in this chemical likely affect their performance?

By using its store of creatine phosphate to regenerate ATP, a muscle cell can support approximately 10 additional seconds of activity. These additional seconds allow time for the process of aerobic respiration to catch up with increased ATP demands. Thus, creatine phosphate is an important source of energy in the first burst of muscular activity, such as at the mass start of the swimming leg of an Ironman competition (FIGURE 11.8). In a resting individual, about 2 grams, 1.6% of the total supply, of creatinine is excreted in urine each day.

Urine Formation

The processing of waste in kidneys has three distinct phases, which we can follow on a diagram of a single, loop-shaped nephron (FIGURE 11.9).

FIGURE 11.8 Ironman mass start. The first leg of a triathlon is a swim in which all competitors enter the water together.

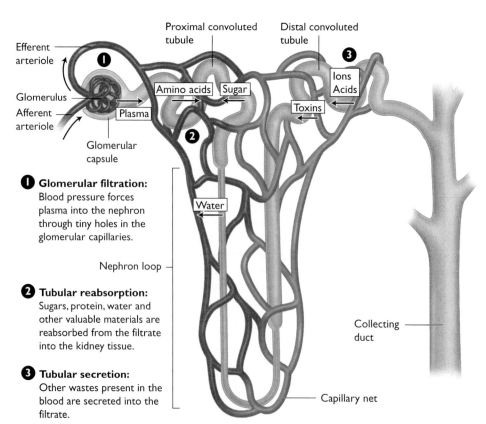

FIGURE 11.9 Nephron structure and function. The nephron first filters plasma from the bloodstream and then allows the reabsorption of essential substances, such as water and salt, from the filtrate. The kidney also actively secretes toxins into the urine, which is then excreted into the bladder.

The receiving end of the nephron is a cup-shaped structure called the **glomerular capsule** (also known as *Bowman's capsule*). The capsule surrounds a mass of capillaries called the **glomerulus**. Fluid filtered from the glomerulus into the glomerular capsule flows through the convoluted looping tubules of the nephron.

The **proximal tubule** of nephrons is embedded in the renal cortex. In about 20% of nephrons, a long **nephron loop**, also called the *loop of Henle* and made up of a descending limb and an ascending limb, brings the tubule into the renal medulla. In the other 80% of nephrons, this loop is small and remains in the cortex.

Each nephron ends in a **distal tubule**, which drains into a common **collecting duct** shared by a number of nephrons. These collecting ducts empty wastes into the renal pelvis.

Step 1: Glomerular Filtration The first step of urine formation occurs when the plasma portion of the blood is forced by blood pressure through pores in the capillary walls of the glomerulus. The plasma flows into the glomerular capsule through cells that extend from the surface of the capsule and encircle the capillaries (FIGURE 11.10).

The blood pressure in the glomerulus, regulated by the diameter of the **afferent arteriole** supplying it, is about twice as high as in any other capillary bed. The higher pressure forces nearly the entire liquid portion of the blood out of the glomerulus and into the nephrons. The cells of the capillary walls and the walls of the glomerular capsule thus act as sieves, retaining cells and larger molecules, such as albumin, in the bloodstream.

The cells and proteins remaining in the capillaries after filtration leave the glomerulus via the **efferent arteriole**. The blood then flows into another capillary bed surrounding the nephrons, where much of the filtrate is returned to the blood. Other wastes that could not exit with the filtrate are actively secreted from these capillaries before the blood returns to the general circulation.

Stop and Stretch 11.2

Chronic high blood pressure damages cells in the glomerulus and glomerular capsule, making the pores between cells much larger. How is the urine of someone with damage to these structures likely to be different from that of someone with an intact filter?

(a)

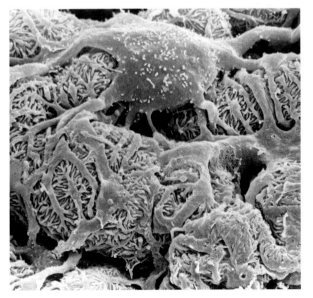

(b)

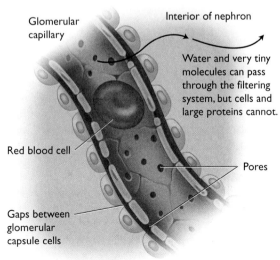

FIGURE 11.10 **Glomerular filtration.** (a) Cells in the glomerular capsule wrap around the capillaries of the glomerulus. (b) Pores in both structures allow passage of water and small molecules from the plasma, but filter out larger molecules and cells.

Step 2: Tubular Reabsorption Because glomerular filtrate contains both wastes and substances required by working muscles, the next step in urine formation is reabsorption of essential nutrients and ions across the walls of each nephron. In fact, nearly all of the glucose, amino acids, and more than 99% of water and sodium are returned to capillaries that surround the tubular portion of the nephron. Although some waste products, such as creatinine, are not reabsorbed, even about 50% of the urea in the filtrate is returned to the blood at this point.

The epithelial cell membranes lining the inside of the proximal tubule are covered with *microvilli*, thin, fingerlike extensions that increase the surface area of the cell and facilitate reabsorption. The complex process of reabsorption across these cells is summarized in FIGURE 11.11. In short, the cells lining the proximal tubule use ATP to actively transport sodium from the nephron epithelial cells into the renal cortex, and this action causes the removal of chloride, water, glucose, and amino acids from the filtrate.

Because sodium is a positively charged ion, its active transport into the renal cortex sets up an electrical gradient between the inside space (the *lumen*) of the nephron and the space outside it. As a result, negatively charged chloride ions diffuse out, equalizing the charge on both sides of the epithelial cell membrane.

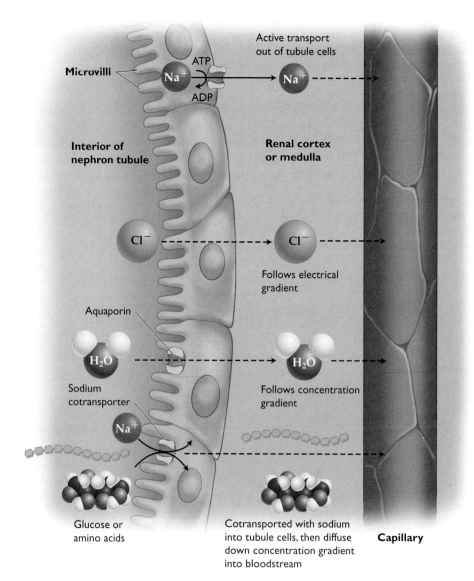

FIGURE 11.11 **Tubular reabsorption.** The active transport of sodium out of the nephron tubule sets up a chain of events that permits recovery of most of the salt, water, glucose, and amino acids from the glomerular filtrate.

VISUALIZE THIS The presence of glucose in urine often signifies high blood-sugar levels resulting from diabetes. Using this diagram, explain why glucose remains in filtrate when its levels are high in plasma.

The accumulation of sodium and chloride ions in the renal cortex creates an osmotic gradient, with the fluid in the nephrons hypotonic to the renal cortex. As a result, water diffuses out of the nephron lumen and into the cortex. Specialized membrane proteins called *aquaporins* facilitate the movement of water through the nephron epithelial cells, allowing rapid reabsorption of two-thirds of the water molecules in the filtrate.

The active transport of sodium out of cells lining the tubule also creates a difference in sodium concentration between the epithelial cells lining the nephron and the filtrate inside the lumen. As a result, sodium from the filtrate diffuses down its concentration gradient into the epithelial cells.

Sodium ions moving from the filtrate into the epithelial cells pass through protein transport molecules in the cell membrane. These transport molecules also carry glucose and amino acids, and thus these substances are removed from the filtrate by a cotransport process, in the opposite direction of their concentration gradient. In this way, the active transport of sodium from the nephron epithelial cells drives the recovery of these valuable substances from the filtrate. Glucose, amino acids, water, sodium, and chloride all eventually diffuse from the nephron epithelial cells into the bloodstream in nearby capillaries.

Step 3: Tubular Secretion Both the proximal and distal tubules of the nephron permit the secretion of certain wastes that are in relatively low concentration in the plasma, such as creatinine (step 3 in Figure 11.9). These wastes move from nearby capillaries into the nephron tubules primarily via active transport. Many drugs, food additives, and inorganic chemicals are removed from the body via tubular secretion. Removal of hydrogen ions to manage acid levels in blood, an especially important function of the kidneys during intense exercise, also occurs via secretion.

In summary, we can think of urine as a fluid that contains materials that have undergone glomerular filtration and have not been reabsorbed, as well as components that have been actively secreted from the blood.

The processing and removal of metabolic wastes is only one function of the kidneys. Another essential function is the management of body water. In an Ironman competition, this function plays an important role in determining whether an athlete successfully completes this grueling race.

11.3 Water, pH, and Salt Balance

All muscular movement produces heat as a by-product, but the intense muscular activity required from a triathlete results in large amounts of waste heat. When the athlete is swimming, the excess heat is lost to the surrounding water, helping to maintain a normal body temperature. However, when the bike segment of an Ironman begins, the athlete's body dissipates heat through evaporative cooling that is the result of constant production of high levels of sweat—as much as 18 to 20 liters (4 to 5 gallons) over the course of a race. This is nine to ten times more water than the approximately 3 liters (0.75 gallon) a resting person loses in a day.

If too much water is lost or if water intake does not keep up with demand, blood volume declines, causing a decrease in blood pressure. Declining blood pressure slows the rate of transfer of materials and wastes around the body, impairing normal metabolic processes. Loss of water also causes the blood to become hypertonic to the tissues, and as a result, body cells lose water to the bloodstream. As discussed in Chapter 2, if too much water is lost, dehydration occurs. The symptoms of **dehydration** include dry mouth, sunken eyes, and lethargy resulting from the shrinkage of brain cells.

Triathletes compensate for sweat loss by consuming vast quantities of liquid, but the fine balance of water in the body is ultimately maintained by the actions of the kidneys. To maintain proper blood volume and dilution, the urinary system can adjust water output from as little as 0.5 liter per day to nearly 1 liter per hour.

Hormones and Water Depletion

Nerve cells in the brain are stimulated when solute concentration in blood crosses a certain threshold, a signal that water levels in the blood may be declining. These neurons trigger the release of **antidiuretic hormone (ADH)** from the pituitary gland in the brain. The site of activity for ADH is the collecting ducts, the shared passageway for multiple nephrons excreting urine into the renal pelvis. ADH increases the permeability of the collecting ducts to water, allowing for increased retention of water in the kidney, where it diffuses back into the blood.

ADH release is also stimulated by low blood pressure, sensed by neurons in the blood vessels. In addition, low blood pressure triggers the release of two hormones—renin and aldosterone—that regulate salt concentration in the body. The protein angiotensin also plays a role in this homeostatic system, which is sometimes called the *RAA (renin-aldosterone-angiotensin) pathway*.

Renin is produced by cells in the *juxtaglomerular apparatus,* a point of contact between the nephron and the afferent arteriole. This enzyme is triggered to release when the pressure of blood flowing through the arteriole declines, in turn reducing pressure on the juxtaglomerular cells and permitting them to expand slightly. Renin activates a large, inert protein molecule (called *angiotensinogen*) that is produced by the liver and circulates in the bloodstream. The eventual result of this modification is the compound **angiotensin II**, a highly active regulatory hormone. Angiotensin II narrows the diameter of blood vessels, immediately increasing blood pressure, and stimulates a structure in the brain to produce the feeling of thirst. Angiotensin II also stimulates the production of the hormone aldosterone by the adrenal glands, endocrine organs that sit on top of each kidney (Chapter 16).

Aldosterone increases overall blood volume by increasing the active transport of sodium from the distal tubule. The enhanced saltiness of the kidney increases its osmotic potential, causing water to flow from the nephrons and collecting ducts into the kidney tissue. Aldosterone, renin, and antidiuretic hormone are part of a negative feedback system, and their release is curtailed when solute concentration in blood declines or blood pressure increases (FIGURE 11.12).

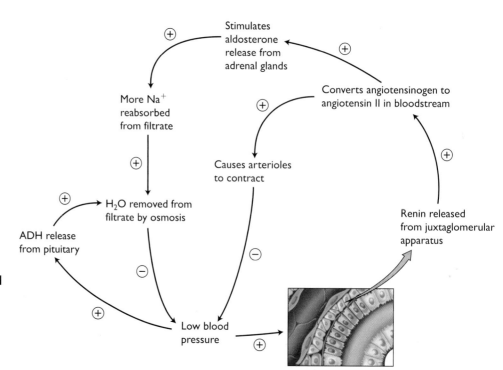

FIGURE 11.12 **Hormones, water loss, and blood pressure.** Water loss from the body causes blood pressure to decline. Nerve cell sensors in brain and kidneys register this low blood pressure and trigger the release of hormones that slow water loss and increase blood pressure. (Stimulation is +, inhibition is −.)

Water retention is enhanced by the anatomy of the 20% of nephrons that extend from the renal cortex to the renal medulla. This structure allows for a process known as countercurrent exchange, which can maximize water retention.

Countercurrent Exchange in the Kidney

Countercurrent exchange occurs when a gas or fluid flows in opposite directions (countercurrent flow) in side-by-side tubes. The close association between the two flows allows for events occurring in each tube to influence what is occurring in the other tube—an exchange. In the kidneys, countercurrent exchange creates a gradient of solute concentration (or osmotic pressure) within the renal medulla. The gradient and exchange allow nephrons crossing into the medulla to return the maximum amount of water to the bloodstream in order to stave off dehydration. The countercurrent exchange of water and solutes in a nephron is illustrated in FIGURE 11.13.

As filtrate flows from the proximal tubule into the descending arm of the nephron loop, the tubule becomes highly permeable to water, but relatively impermeable to salt. Water diffuses out of the nephron lumen passively as the loop passes ever deeper into the solute-rich medulla.

As the nephron makes the turn into the ascending arm, its walls become impermeable to water but allow passage of sodium and chloride. By this time, the filtrate remaining in the tubule is highly concentrated because so much water has escaped. As a result, sodium

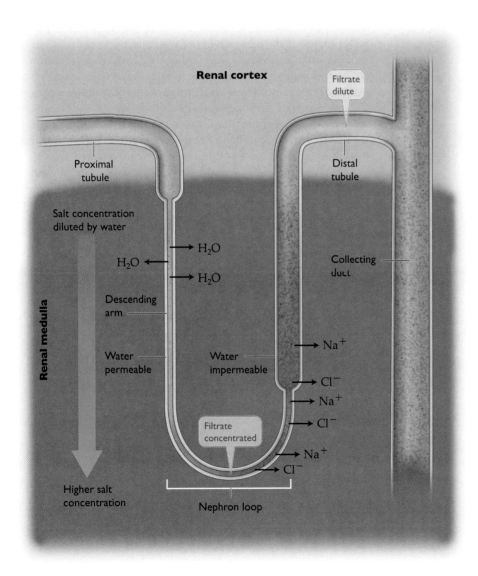

FIGURE 11.13 Water conservation. Countercurrent flow in the nephrons allows for maximum water conservation, in the presence of hormones that permit water flow across the walls of the collecting ducts. If these hormones are not present, water cannot leave the collecting ducts and is excreted in high volumes in the urine.

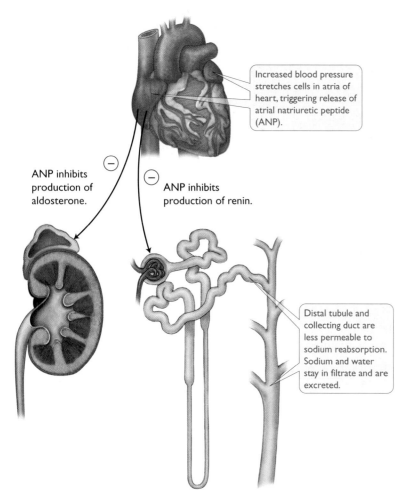

FIGURE 11.14 A diuretic. Atrial natriuretic peptide decreases blood pressure by increasing water excretion. Diuretics such as caffeine have similar effects.

and chloride diffuse out of the tubule, down their concentration gradients. The loss of sodium and chloride from the filtrate at the bottom of the nephron loop helps to create the concentration gradient inside the renal medulla.

The osmotic gradient is also maintained by the counter-current flow of blood through the *vasa recta*, capillaries that parallel the nephron loops. As blood moves slowly through the vasa recta, sodium and chloride diffuse into the bloodstream in the descending capillaries and diffuse out from the ascending capillaries. This mechanism conserves sodium and chloride ions in the deepest part of the medulla. Water leaving the descending nephron loop is quickly absorbed by the solute-rich blood in the ascending arm of the vasa recta and carried away, preventing dilution of the gradient in the medulla.

As the tubule ascends, it widens and becomes impermeable not only to water, but also to most solutes. Because water cannot enter or leave the filtrate along this arm of the nephron, the gradient of salt present in the renal medulla can be maintained. The cell membranes in this wider part of the ascending arm have sodium transporters embedded. As a result of the activity of these transporters, salt is actively excreted from the filtrate into the medulla. At this point, much of the water initially present in the filtrate has been reabsorbed, but the filtrate is relatively dilute as a result of the active removal of salt.

The permeability of the collecting ducts now determines the final concentration of urine excreted into the renal pelvis. Because ADH increases the permeability of the collecting duct to water, the kidneys in an individual who is slightly dehydrated recapture the majority of water remaining in the filtrate. This conservation of water occurs because the collecting duct passes from the dilute exterior to the solute-rich interior of the medulla on the way to the renal pelvis. As the remaining filtrate passes through this increasingly hyperosmotic environment, much of the water that remains in it diffuses out. The result is a small volume of highly concentrated urine.

When solute levels in blood are low as a result of high water intake, ADH production is inhibited. The collecting ducts then are impermeable to water, and water remains in the filtrate even as it passes through the salty medulla of the kidney. In these conditions, large quantities of water are excreted in very dilute urine.

In addition, when blood volume is above normal, certain cells in the atrial chambers of the heart sense the stretching of the heart muscle. These cells release **atrial natriuretic peptide (ANP)**, which inhibits both renin and aldosterone release, and thus increases excretion of both sodium chloride and water (FIGURE 11.14).

ANP is a **diuretic**, a substance that increases the formation of urine. Caffeine is a diuretic that acts in a similar manner to ANP. Alcohol is another diuretic, but one that inhibits ADH release, thus leading to increased water excretion. The lingering effects of alcohol consumption—a hangover—result primarily from the dehydration caused by this effect.

Stop and Stretch 11.3

Diuretic drugs are sometimes used in medical treatment. What sorts of conditions would you expect these drugs to be prescribed for, and why?

Athletes in Ironman competitions or individuals in other situations where sweat production is high know that they must support the water conservation activities of their kidneys by drinking adequate water and avoiding diuretics. Their kidneys will also be called on to rid their blood of the excess acid produced by metabolic activity.

Maintaining Blood pH

Most metabolic activities produce acid—measured as an increase in the concentration of hydrogen ions (H^+) or a decrease in pH. Because blood pH must remain in a range of 7.35 to 7.45, the acid produced by metabolism must be neutralized and removed quickly. If it is not, the low pH causes **acidosis**, increased acidity of blood plasma, affecting nearly all body functions. Acidic conditions cause enzymes to become denatured—to lose their shape. As a result, nearly all enzyme-mediated processes in the body are disrupted by acidosis.

The primary mechanism for pH maintenance in the body consists of buffers in the blood, including the bicarbonate ion (HCO_3^-) and hemoglobin, the oxygen-carrying molecule in red blood cells (FIGURE 11.15). These chemicals absorb excess hydrogen ions when pH is low and release them into the bloodstream when pH is higher. Nephron epithelial cells aid in this process by generating additional bicarbonate from water and carbon dioxide. As described in Chapter 10, most hydrogen ions produced by cellular respiration are lost from the lungs as water during exhalation.

Although the lungs and buffering system can help maintain homeostasis of blood pH in the short term, in the longer term—over the course of a day—the kidneys must excrete the abundance of acid generated by metabolic processes. They do this by secreting excess hydrogen ions into the nephron lumen.

Most of the H^+ ions excreted into the lumen combine with bicarbonate in the filtrate, forming carbonic acid. The enzyme carbonic anhydrase, produced by the nephron cells, degrades the acid into water and carbon dioxide. The water (and thus the excess H^+) is excreted, and the carbon dioxide is returned to the bloodstream and eventually exhaled. When bicarbonate levels in the filtrate are not high enough to deal with excess acidity, H^+ ions in the lumen combine with hydrogen phosphate (HPO_4^{2-}) and are excreted in the urine. If acid levels in the blood are even higher, amino acids inside nephron epithelia can be metabolized into ammonia (NH_3). The ammonia molecules then pick up excess hydrogen ions to form ammonium (NH_4^+) and are excreted in the urine. The amount of ammonium present in urine is thus a measure of acidosis.

The blood buffers and kidneys are so effective at regulating blood pH that athletes, even after an Ironman competition, will have NH_4^+ levels almost indistinguishable from that of a resting individual. Acidosis is not a risk to healthy individuals. It is more common in individuals who are experiencing other, underlying diseases. However, even in healthy individuals, the kidneys may be unable to maintain homeostasis of blood sodium under the stress of high activity levels.

Salt Balance: The Right Amount of Sodium

The Genes & Homeostasis feature on the next page details the effect of a surplus of dietary sodium on blood pressure and fluid balance in certain individuals. In the case of extreme athletic events, the risk is not too much salt, but too little.

In addition to the 20 liters of water evaporated in the athlete's sweat, about 32 grams of sodium are excreted onto the skin over the course of an Ironman race. The amount of sodium lost is about equivalent to that contained in 1/3 cup of table salt. The kidneys play a role in maintaining adequate sodium levels in blood. Salt is retained in the blood via the actions of aldosterone, which increases sodium transport from nephron tubules into the medulla.

The loss of too much sodium can lead to **hyponatremia**, or dilution of the blood. During hyponatremia, water in the now hypotonic blood diffuses into the tissues of the body, including the brain, down its concentration gradient. A low level of sodium

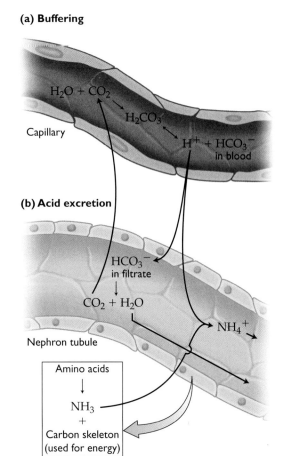

FIGURE 11.15 Buffers in the blood. (a) Blood pH is kept in narrow bounds by buffer systems that pick up hydrogen atoms when they are in excess and release them when they are deficient. (b) In the long term, the excess acid produced by metabolism and obtained from acidic foods are excreted by the kidney.

GENES & HOMEOSTASIS

Please Pass on the Salt

When a doctor diagnoses chronic high blood pressure, or hypertension, one of her first pieces of advice to the patient is to reduce his table salt intake. You might be able to deduce the reasoning behind this advice: More salt in the diet could lead to higher sodium levels in blood, and higher blood-sodium levels draw more water from the tissues, increasing blood volume and blood pressure.

Large studies also support the hypothesis that there is a relationship between dietary sodium and blood pressure. In one study, individuals who reduced their salt intake from 3,800 milligrams (mg) per day to 1,500 mg saw their blood pressure decline an average of 3 to 6 mm Hg (millimeters of mercury) during systole. (Review Chapter 9 for more detail on blood pressure readings.) A reduction of 3 mm Hg in systolic blood pressure is relatively small, but if it occurred in the general U.S. population it would result in an 11% reduction in strokes, a 7% reduction in heart attacks, and 5% fewer deaths.

The suggestion to reduce salt intake to lower blood pressure is not without its critics. It can be quite difficult to maintain a low-sodium diet. Even if you use little or no table salt, your sodium intake can be quite high. This is because most of our sodium intake is in the foods we buy, especially convenience foods, which make up an ever-increasing part of the American diet. Critics say that a difficult-to-follow recommendation that has relatively little effect is likely to be widely ignored. And a perception that changing sodium intake may not affect blood pressure can be dangerous to the approximately 10% of individuals who are extremely salt sensitive.

Scientists have identified many genes that contribute to a susceptibility to hypertension, including those involved in angiotensin production. However, the genetic defects that would lead to sodium-sensitive hypertension must relate to salt balance in the kidneys. In other words, individuals who cannot effectively excrete excess sodium are those who are most likely to experience high blood pressure when salt intake is elevated. The genes involved in sodium-sensitive hypertension are defects in aldosterone production or effect and problems with the sodium channels in the nephron tubules. Researchers are currently attempting to identify the mutations involved in this form of hypertension so they can test individuals to find those who would most benefit from much-reduced sodium intake. In the meantime, the advice to all hypertensive patients remains: Please pass on the salt.

thus leads to a variety of symptoms, including bloating, swelling, nausea, muscle cramps, slurred speech, and disorientation (FIGURE 11.16). If it continues, victims may experience seizures, coma, and even death. Hyponatremia is also known as water intoxication, because it also occurs when pure water intake exceeds water loss.

In most individuals under normal conditions, hyponatremia is not an issue because the kidneys can keep up with salt demand, but in athletes this condition may be surprisingly common. A 1984 analysis of Ironman participants indicated that as many as 30% were hyponatremic. Hyponatremia can also be caused by certain drugs, and it is surprisingly common in elderly patients, affecting nearly 7% in a survey of individuals in a home setting.

11.4 When Kidneys Fail

Kidneys can cease functioning properly for a number of reasons. Some of the causes are correctable—for instance, **kidney stones**, which are masses of calcium, phosphorus, uric acid, and protein crystals that form from urine and collect in the renal pelvis. Although small stones can pass out in urine flow, larger stones can block the ureter, causing great pain. Large kidney stones can be removed surgically or pulverized with ultrasound beams to make them small enough to pass. However, if kidney stones remain trapped in the renal pelvis, urine may back up into the nephrons, destroying them.

Nephrons can also be destroyed by infections that move up the urethra and bladder and by chronic high blood pressure, which damages the glomerulus and glomerular capsule. Nearly 50% of cases of kidney failure are due to diabetes. Untreated diabetes can cause episodes of high blood sugar that increase the solute concentration of the blood, leading to increased blood volume and high blood pressure.

FIGURE 11.16 **The effects of hyponatremia.** Dilution of blood solutes through loss in sweat or overconsumption of water leads to water retention in tissues, resulting in a number of symptoms.

The first sign of kidney failure is the presence of protein that has slipped through the leaky filter of the glomerulus and into the urine. If kidney damage continues, the result is called **end-stage renal disease (ESRD)**. People with ESRD may have less than 10% of normal kidney function, and their kidneys are irreversibly damaged. The only truly effective way to replace the function of kidneys is to transplant a healthy organ from a donor. But the need for kidneys far outpaces their supply. In the meantime, people with ESRD remain alive by undergoing regular dialysis.

The most effective form of **dialysis**, the diffusion of dissolved molecules through a membrane, circulates a patient's blood through an artificial kidney machine. The machine consists of membranous tubes bathed in clean fluid. Metabolic wastes diffuse from the blood, through the semipermeable membranes, and into the fluid (FIGURE 11.17). Similarly, substances added to the fluid can diffuse into the bloodstream. The fluid is continually modified during treatment to maintain concentration gradients that favor the excretion of wastes and accumulation of buffers or other substances. A single dialysis treatment requires three to six hours, but because a single treatment removes a larger proportion of blood urea than functional kidneys can, treatments are needed only three or four times per week.

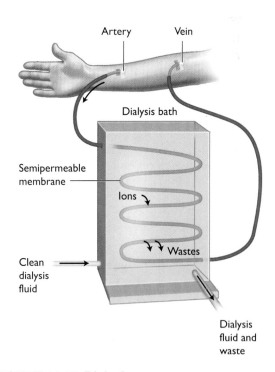

FIGURE 11.17 **Dialysis.** When kidneys are nonfunctional, blood from the body can be pumped through semipermeable tubing that is immersed in a bath of clean fluid. As blood flows through, wastes diffuse from the blood into the fluid.

VISUALIZE THIS Another form of dialysis infuses clean fluid into the abdominal cavity. During four to six hours of "dwell time," a time when the patient can move about and perform normal daily activities, wastes from the blood diffuse into the fluid. At the end of the dwell time, the fluid is drained and replaced with clean fluid. What is the semipermeable membrane in this form of dialysis?

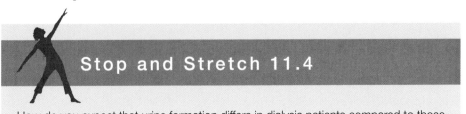

How do you expect that urine formation differs in dialysis patients compared to those with healthy kidneys?

FIGURE 11.18 **Managing water intake.** Aid stations, which provide water and other drinks as well as snacks to passing marathoners, have been moved farther apart in many marathons to reduce the risk of hyponatremia among runners.

Julie Moss's collapse at the end of the 1982 Ironman did not signal underlying kidney disease. Instead, it was a reflection of a then general lack of understanding of how the human body responds to endurance events. According to her symptoms, Julie may have been experiencing hyponatremia—all she had consumed over the 11 hours of the race was water and bananas.

As the popularity of endurance events has soared, so has knowledge of the perils of these events among elite athletes. Most athletic trainers now know to limit pure water intake during endurance events to about 0.5 liter per hour. Organizers of the Ironman and large enrollment marathons also have begun to reduce the number of aid stations to decrease the likelihood of hyponatremia among more casual athletes (FIGURE 11.18).

Ironically, Julie's participation in the 1982 Ironman was, in part, research she was conducting for a graduate degree in exercise physiology. Thanks to her experience, and the research of other scientists about the body's response to exertion, athletes are better prepared for the extremes of an Ironman, and they know what to supply their bodies with during a race in order to maintain homeostasis. Julie Moss returned to the Ironman for its 25th anniversary in 2003—and finished third in her age group, with a smile.

Chapter REVIEW

ROOTS TO REMEMBER

The following roots of words come mainly from Latin and Greek and will help you to decipher terms:

dia- means through (as in diabetes or dialysis).

hypo- is under or below.

-natr- is from a word used to describe sodium and, therefore, salt.

nephr- or **nephro-** relates to kidney.

ren- also describes the kidney.

urin- and **ure-** are roots used to describe the urinary system and its products.

KEY TERMS

11.1 An Overview of the Urinary System
elimination *p. 252*
excretion *p. 252*
external urethral sphincter *p. 254*
internal urethral sphincter *p. 253*
kidney *p. 253*
micturition reflex *p. 254*
nephrons *p. 253*
renal cortex *p. 253*
renal medulla *p. 253*
renal pelvis *p. 253*
ureter *p. 253*
urethra *p. 253*
urinary bladder *p. 253*
urinary system *p. 252*
urination *p. 254*

11.2 Excretion
afferent arteriole *p. 257*
collecting duct *p. 257*
distal tubule *p. 257*
efferent arteriole *p. 257*
glomerular capsule *p. 257*
glomerulus *p. 257*
nephron loop *p. 257*
proximal tubule *p. 257*
urea *p. 255*
urine *p. 255*

11.3 Water, pH, and Salt Balance
acidosis *p. 263*
aldosterone *p. 260*
angiotensin II *p. 260*
antidiuretic hormone (ADH) *p. 260*
atrial natriuretic peptide (ANP) *p. 262*
countercurrent exchange *p. 261*
dehydration *p. 259*
diuretic *p. 262*
hyponatremia *p. 263*
renin *p. 260*

11.4 When Kidneys Fail
dialysis *p. 265*
end-stage renal disease (ESRD) *p. 265*
kidney stone *p. 264*

SUMMARY

11.1 An Overview of the Urinary System

- The urinary system is responsible for excretion, which is the removal of wastes from the bloodstream, as well as maintaining homeostasis of ion concentration, blood pH, and blood pressure (p. 252).

- The urinary system consists of the kidneys, ureters, urinary bladder, and urethra. The kidneys are served by large blood vessels, the renal arteries and renal veins. The functional unit of the kidney is the nephron (p. 253; Figure 11.3).

- When the urinary bladder becomes full, stretch receptors trigger a conscious "need to urinate." Voluntary relaxation of the external urethral sphincter triggers the micturition reflex. A positive feedback loop that triggers bladder contractions while urine flows through the urethra ensures that the bladder empties completely (p. 254).

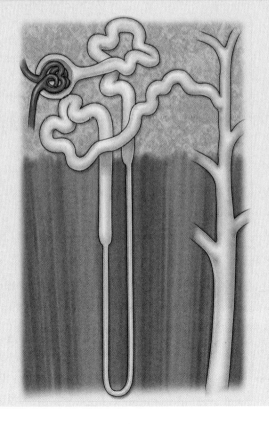

11.2 Excretion

- The urinary system produces urine, which is a solution of metabolic wastes, ions, and toxins (p. 255).

- Plasma is filtered from the capillaries in the glomerulus into the glomerular capsule. Reabsorption along the nephron tubule returns water and other valuable components to the renal cortex or medulla and, eventually, the blood. Toxins are removed from the blood by active transport in the process of secretion (p. 257).

- Reabsorption is a complex process that begins with the active transport of sodium ions out of the epithelial cells of the proximal tubule toward the surrounding capillaries. Chloride ions follow the sodium, causing the renal cortex to become salty. Water then leaves the filtrate for the cortex, following an osmotic gradient. Sodium depletion in the epithelial cells draws sodium from the filtrate, as well as glucose, amino acids, and other small molecules through cotransport proteins (p. 258).

11.3 Water, pH, and Salt Balance

- When blood volume is low because water intake is low or water loss is high, antidiuretic hormone released by the pituitary gland increases the permeability of the kidney's collecting ducts to water. Meanwhile, renin produced by cells in the kidney, and aldosterone produced by the adrenal glands, increase salt reabsorption and cause blood vessel changes to increase blood pressure (p. 260).
- Countercurrent exchange in the nephron loop establishes an osmotic gradient from the outer to the deeper renal medulla, drawing water out of the filtrate along the descending arm of a nephron loop. Along with the increased permeability of the collecting ducts, this system allows for maximum reabsorption of water (p. 261).
- When blood volume is high, cells in the heart produce the hormone atrial natriuretic peptide, which inhibits the reabsorption of sodium and causes the collecting ducts to become impermeable to water, encouraging the excretion of large volumes of dilute urine (p. 262).
- Blood pH is maintained in the short term by a system of buffering chemicals in the plasma. In the long term, kidneys are responsible for excreting the excess acids produced by high levels of metabolism (p. 263).
- The presence of salt keeps the blood isotonic to body tissues. If salt levels drop, blood pressure declines as tissues begin to gain water, a potentially deadly situation. The kidneys help to maintain blood-sodium levels by increasing reabsorption from the urine during times of low blood pressure (p. 263).

11.4 When Kidneys Fail

- Irreversible damage to nephrons caused by kidney stones, infection, or chronic high blood pressure can cause a decline in kidney function. Kidney function can be replaced by dialysis, which requires pumping the blood supply through semipermeable tubes immersed in a bath of clean fluid. Wastes dissolve out of the blood and into the fluid (pp. 264–265; Figure 11.17).

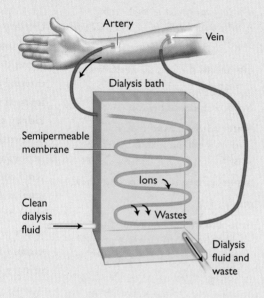

LEARNING THE BASICS

1. All of the following organ systems participate in waste elimination except _____.
   a. the digestive system
   b. the urinary system
   c. the respiratory system
   d. the integumentary system
   e. the musculoskeletal system

2. Which of the following accurately describes the pathway of urine from the body?
   a. kidneys, urethra, bladder, ureters
   b. kidneys, renal artery, bladder, urethra
   c. renal arteries, kidney, bladder, urethra
   d. kidney, renal capsule, ureters, bladder
   e. renal cortex, renal medulla, renal pelvis, bladder, urethra

3. Urination is a positive feedback loop because _____.
   a. drinking water promotes urination, which causes thirst
   b. urine flowing through the urethra stimulates bladder contractions
   c. bladder emptying promotes higher blood pressure in the nephrons
   d. relaxation of the internal urethral sphincter causes immediate relaxation of the external sphincter
   e. bladder contractions cause the internal urethral sphincter to close, cutting off urine flow

4. How do the male and female urinary tracts differ?
   a. Males generally have larger bladders than females.
   b. Females have shorter urethras than males.
   c. In males the urethra has two functions and, in females, only one.
   d. The urethra passes through the prostate in males, but not in females.
   e. All of the above are correct.

5. All of the following are normally components of urine except _____.
   a. large plasma proteins
   b. urea
   c. water
   d. creatinine
   e. excess ions

6. After blood is filtered in the nephrons of the kidneys, _____.
   a. glucose and amino acids are reabsorbed
   b. toxins are actively secreted into the urine
   c. nearly all of the water may be recovered
   d. a and b are correct
   e. a, b, and c are correct

7. During glomerular filtration, _____.
   a. plasma is forced by high blood pressure from the circulatory system into the nephrons
   b. a small amount of water is reabsorbed by the nephron loop
   c. cells and large molecules are secreted into the urine
   d. a and c are correct
   e. a, b, and c are correct

8. The active transport of sodium from the proximal tubule of the nephron _____.
   a. causes chloride to leave the filtrate down an electrical gradient
   b. results in water leaving the filtrate down an osmotic gradient
   c. allows for the cotransport of molecules such as glucose from the filtrate to the renal cortex
   d. drives the reabsorption of water and other valuable materials from the glomerular filtrate
   e. all of the above are correct

9. Dehydration _____.
   a. occurs when water intake exceeds water loss
   b. is caused by an overabundance of salt in the body
   c. causes body cells to shrink as they lose water
   d. cannot be counteracted by the kidneys
   e. is more likely to occur in swimmers than in runners

10. All of the following compounds affect water conservation in the body *except* _____.
    a. erythropoietin
    b. antidiuretic hormone
    c. renin
    d. aldosterone
    e. angiotensin II

11. Which hormone is likely to cause a decrease in blood pressure?
    a. antidiuretic hormone
    b. renin
    c. aldosterone
    d. angiotensin II
    e. atrial natriuretic peptide

12. In a nephron loop that passes into the renal medulla, _____.
    a. water flows as a result of osmotic differences between the filtrate and the medulla on the descending loop
    b. sodium is actively excreted in the upper part of the ascending loop
    c. the ascending loop is impermeable to water
    d. the filtrate becomes more concentrated as it reaches the bottom of the loop
    e. all of the above are correct

13. Countercurrent exchange refers to _____.
    a. processes in which fluid flowing in opposite directions in side-by-side tubes affect each other
    b. loss of sodium from the renal cortex
    c. the secretion of hydrogen ions into urine
    d. the exchange of urea between body tissues and the blood in the kidney
    e. the movement of large proteins from the plasma to the glomerular filtrate

14. Excess hydrogen ions in the blood _____.
    a. raise pH
    b. can be absorbed by bicarbonate ions in the plasma
    c. eventually must be excreted by the kidneys
    d. b and c are correct
    e. a, b, and c are correct

15. Hyponatremia occurs when _____.
    a. salt intake is too high
    b. ADH causes excessive water loss
    c. blood becomes more dilute than body tissues
    d. water intake is limited
    e. aldosterone increases sodium reabsorption

16. Kidney stones _____.
    a. form in the renal pelvis
    b. cannot be eliminated from the body
    c. inevitably lead to end-stage renal disease
    d. result from dialysis
    e. cause gout

17. During dialysis, _____.
    a. water is pumped into the bloodstream to dilute the wastes and encourage urine production
    b. blood flows through the kidney at higher pressures, increasing filtration rate
    c. fluid in the body is controlled by application of aldosterone-blocking drugs
    d. blood flowing through tubes of artificial semipermeable membrane loses wastes via diffusion
    e. individuals must urinate often to purge unwanted toxins produced by kidney disease

18. **True or False?** Metabolic wastes can be excreted by the kidneys, lungs, and skin.

19. **True or False?** The release of urine from the bladder is involuntary in most adults.

20. **True or False?** The yellow color of urine derives from the breakdown of the components of red blood cells.

ANALYZING AND APPLYING THE BASICS

1. Human chorionic gonadotropin (hCG) is a hormone produced by the developing placenta after a fertilized egg implants in the uterine wall. This hormone circulates through the bloodstream, stimulating the ovaries to secrete estrogen and progesterone levels that are high enough to maintain pregnancy. Pregnancy tests measure whether hCG is present in either blood or urine. If hCG is detected, the woman taking the test is pregnant. How does hCG leave the blood, so that it can eventually be detected using a urine test?

Questions 2 and 3 refer to the following information.

Uric acid is a waste product produced when purine molecules (from DNA and RNA) are broken down by the liver, muscles, and intestines. An excess of uric acid can cause health problems, such as gout or kidney stones. However, appropriate levels of uric acid in the blood can have an antioxidant effect, potentially increasing life expectancy in vertebrates such as birds and humans.

In 2002, Enomoto and colleagues identified the transporter molecule that appears to regulate uric-acid reabsorption. The URAT1 transporter moves uric acid from within each proximal tubule of the kidney across the tubule's inner membrane. An unknown transporter then moves the uric acid across the outer membrane and back to the blood.

2. Certain drugs that are used to treat inflammation or high blood pressure can inhibit the URAT1 transporter. What effect might these drugs have on the kidneys?

3. Risk factors for gout include age, body weight, blood pressure, alcohol intake, and consumption of organ meats that are high in purines. Variation in the gene that encodes the URAT1 transporter may also be a risk factor. Why?

Use the following information to answer questions 4 and 5.

In 2002, researchers conducted a study of marathon runners to determine the prevalence of hyponatremia and risk factors associated with it. Researchers collected demographic and training information, information about fluid consumption during the race, and pre- and postrace weights from Boston Marathon participants. At the end of the race, researchers also collected a blood sample from runners. Of the 488 runners who fully participated in the study, 13% had hyponatremia after completing the marathon. Nearly 1% had critical hyponatremia, a condition that can cause death. Analysis of the data found that hyponatremia was associated with weight gain during the race, a racing time of greater than four hours, and extremes of body mass index.

4. Why would weight gain during a race be a signal of hyponatremia?

5. What are three things that marathon race officials could do to guard against hyponatremia?

Use the following information to answer questions 6–9.

Kidney stones occur when salts dissolved in urine precipitate and form solid crystals that adhere to kidney cells. Stone-forming salts include calcium oxalate, calcium phosphate, and uric acid. Kidney stones that break free from the kidney wall may pass out of the urinary system if they are less than 5 millimeters (0.2 inch) in diameter. Larger stones may become lodged in the urinary tract. Doctors use sound waves (extracorporeal shock wave lithotripsy, or ESWL), lasers, or surgery to break up or remove larger stones.

Kidney stones develop due to a variety of physical and/or genetic factors. However, three general conditions are necessary for stone formation. A dissolved salt must be in high enough concentration that it will easily come out of solution (precipitate as a solid). When the salt precipitates, it must be able to adhere to cell surfaces and maintain attachment. To grow, the attached crystallized salt must also be able to promote adhesion of more precipitated crystals.

6. Some disorders of the digestive system can cause incomplete digestion of fatty acids. This leads to an increase in the absorption of oxalate from the digestive system. How could this type of digestive disorder cause the formation of kidney stones?

7. In three randomized, double-blind studies, thiazide diuretics were tested on patients who had previously formed calcium oxalate stones. The results are shown Table 11.1. What conclusions can you draw from the data? Explain the mechanism that could allow a diuretic to prevent kidney stones.

Table 11.1 Effect of Thiazide Diuretics on Formation of Calcium Oxalate Stones

Diuretic	Number of Individuals in Control Group	Number of Individuals in Treatment Group	New Stones, Control Group	New Stones, Treatment Group
Chlorthalidone	26	28	12 (46%)	4 (14%)
Hydrochloro-thiazide	25	23	10 (40%)	4 (17%)
Indapamide	21	43	9 (43%)	6 (15%)

Source: Coe, F. L., Evan, A., and Worcester, E. 2005. Kidney Stone Disease. Journal of Clinical Investigation 115: 2598–2608.

8. Calcium oxalate and calcium phosphate cause most kidney stones (about 80%). Magnesium ammonium phosphate causes 10% of stones; uric acid causes 9%. Cystine, ammonium acid, or drug reactions cause the remainder. What type of kidney stone would most likely be formed by a person who suffers from gout?

9. What is something simple that you can do to avoid kidney stones? Explain your answer.

10. Energy drinks such as Red Bull and Mountain Dew have become popular among young athletes as performance-boosting products. Most of these drinks contain high levels of sugar and caffeine. Why might these drinks actually cause problems for the user?

CONNECTING THE SCIENCE

1. Urinalysis is a common way to screen athletes for the presence of performance-enhancing drugs. In many amateur and professional athletics, competitors are randomly selected for testing for banned substances during training, before major events, and even during competition. Do you think random testing is an effective way to reduce the use of banned substances in sport?

2. Some triathletes think that the attention received by Julie Moss is wrong—that these contests should not be promoted as a triumph of the will. They fear that doing so attracts participants who are not physically prepared for the competition and creates a dangerous situation. Although the number of deaths that result from endurance sporting events is relatively low, most marathons and Ironman qualifying races have several athletes who experience serious injuries or illnesses as a result of the race. Ambulances and rescue crews, as well as dozens of volunteer doctors, nurses, and paramedics, are on standby during these events to help minimize the impact of these injuries. Do you think that potential participants in endurance races should have to be cleared to race by medical professionals before being allowed to register? Why or why not?

Chapter 12

The Immune System

Will Mad Cow Disease Become an Epidemic?

LEARNING GOALS

1. How do genetic and infectious diseases differ?

2. Describe the structure of a typical bacterium and virus.

3. List the means by which infectious diseases are transmitted.

4. Describe the physical and chemical barriers that make up the first line of defense.

5. State the actions of white blood cells, proteins, inflammation, and fever in the second line of defense.

6. Compare and contrast the roles that B cells and T cells play in the immune response.

7. Describe how the immune system differentiates between self and nonself.

8. Why is it necessary to have a flu shot every year but only one inoculation against some other diseases?

9. Why is your immune system usually more effective at fighting off infection after a previous exposure?

10. Outline a genetic reason why one person might die of an infectious disease while another person recovers.

Will scientists stop the spread of mad cow disease?

In the late 1980s, dairy farmers in Great Britain noticed unusual behaviors in a few of their cows. Cows that had previously shown no signs of illness started displaying awkward movements. They would shake and tremble and rub parts of their bodies against walls or fences. Farmers watched with alarm as their cows staggered around fields. Even the cows' dispositions seemed to change. Many of the cows appeared to be irritated or even crazy, leading to the name "mad cow disease."

Several months after the onset of symptoms, the cows inevitably died. Autopsies revealed that the brains of the "mad" cows had holes in them characteristic of a class of diseases known as spongiform encephalopathies. An encephalopathy is a pathology, or disease, of the brain. The diseased cows' brains resembled porous, natural sponges filled with holes. Mad cow disease is a type of spongiform encephalopathy that affects only cows, so it is called bovine spongiform encephalopathy, or BSE.

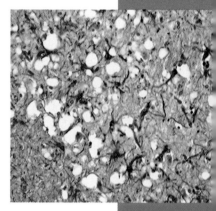

The brains of affected individuals look spongy.

Spongiform encephalopathies had already been diagnosed in sheep, whose skin becomes so itchy that they scrape their bodies against fences. In fact, the disease name, scrapie, describes this behavior. When the spongiform encephalopathy is present in elk and deer, it is called chronic wasting disease, due to the emaciated appearance of affected animals.

Humans can also become ill with spongiform encephalopathies. An obscure disease called kuru has long been known to occur among the peoples of the eastern highlands of New Guinea. Individuals with this disease lose coordination and often become demented. They, too, inevitably die.

The agent that causes this disease is unexpected—and alarming.

A long-recognized but previously very rare condition called Creutzfeldt-Jakob disease (CJD) is another form of spongiform encephalopathy that affects humans. Like the diseased cows, affected humans become dizzy, very agitated, and short-tempered. They experience short-term memory loss, lack of coordination, and slurred speech. Typically, only elderly people display the disease symptoms. However, a recent and alarming trend has been an increase in the rate of diagnosis of CJD in young British patients. As in the case with cows, this disease is lethal in all affected humans.

On average, patients live just over a year after being diagnosed, and when examined at autopsy, their brains look more

This disease can also spread to humans.

similar to those of BSE-infected cattle than CJD-infected humans. The increased number of people diagnosed with the disease, the structure of their brains after death, and the younger age of those infected led scientists to believe that BSE was, in some way, being transmitted from infected cows to these individuals. This new, transmissible form of the disease was named new-variant CJD (nvCJD).

Scientists, doctors, veterinarians, hunters, and patients and their families are concerned about preventing the spread of these diseases. Mad cow disease has occurred in the United States and Canada, bringing concerns about the spread of this disease to U.S. citizens. As a first step toward understanding this disease and preventing its spread in cows and humans, scientists had to determine what type of agent was causing it.

12.1 Infectious Agents

Communicable diseases result from the activity of infectious agents, which gain access to the body and use the body's resources for their own purposes. Such diseases differ from genetic diseases in that they are caused by organisms rather than by malfunctioning genes—although malfunctioning genes can make an organism more susceptible to infection.

Disease-causing organisms are called **pathogens**. When a pathogen can be spread from one organism to another, it is said to be **contagious**. When a pathogen finds a tissue inside the body that will support its growth, it is said to be **infectious**. Organisms that obtain nutrients and shelter required for growth and development from a different organism while contributing nothing to the survival of the host are **parasites**. Most infectious agents are parasites. Note that an organism that infects an individual is contagious only if the infected individual can pass the infection to other individuals.

Organisms that can be seen only when viewed under a microscope are called microscopic organisms, or **microbes**. Microbes cause many sexually transmitted infections, which will be covered in detail in the next chapter (Chapter 13). Here we will concern ourselves with bacteria and viruses.

Bacteria

Bacteria (FIGURE 12.1) are a diverse group of single-celled organisms. They are tiny and numerous. In fact, there are more bacteria in your mouth than there are humans on Earth. Bacteria are commonly rod shaped (bacilli), spherical (cocci), or spiral (spirochetes).

All cells can be placed into one of two categories, prokaryotic or eukaryotic, based on the presence or absence of certain cellular structures. Bacteria are prokaryotic cells. **Prokaryotes** do not have a nucleus, which is the separate, membrane-bounded compartment that contains genetic material in the form of DNA. Nor do prokaryotes contain any membrane-bounded internal structures found in eukaryotes, such as the endoplasmic reticulum or the Golgi apparatus (see Chapter 3). Prokaryotic cells are usually much smaller than eukaryotic cells, and diverse lines of evidence indicate that prokaryotic cells arose over a billion years before eukaryotic cells appeared on Earth.

Instead of a nucleus, prokaryotes have DNA that is coiled up inside a **nucleoid region**. A bacterial chromosome is a double-stranded, circular DNA molecule. In addition to the large DNA chromosome, bacteria may also contain small, circular, DNA molecules, separate from the chromosomes called **plasmids**. In Chapter 4, you learned that scientists use plasmids during genetic engineering. Some of these

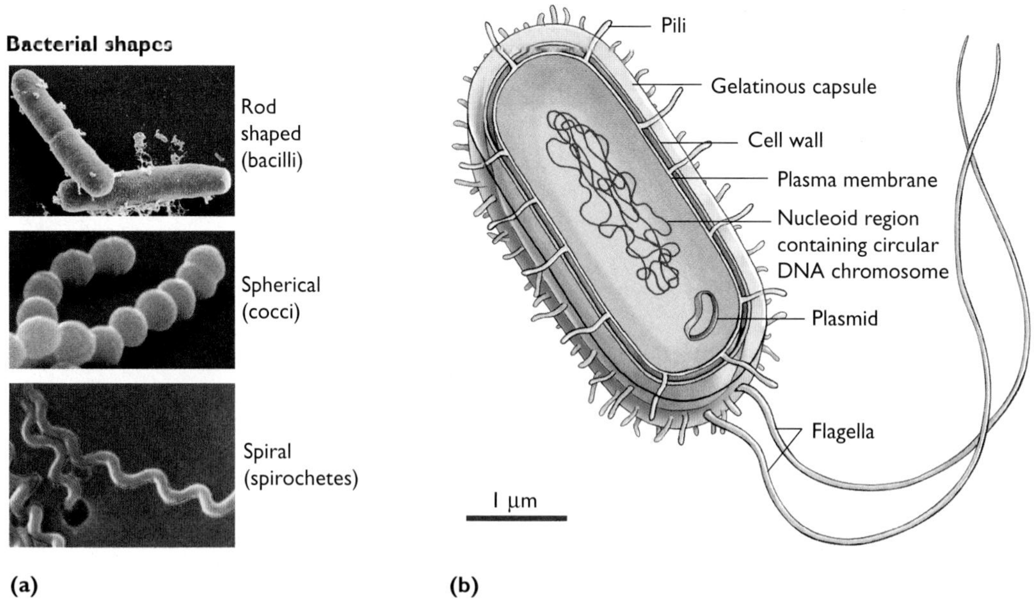

FIGURE 12.1 Bacterial structure. Bacteria (a) as viewed under a microscope and (b) illustrated.

VISUALIZE THIS Note that the bacterial cell does not contain any membrane-bounded organelles such as a Golgi apparatus or an endoplasmic reticulum. Would you expect that this prokaryotic cell could have ribosomes? Why or why not?

plasmids also carry antibiotic-resistant genes that allow bacteria to defend against the drugs that would normally kill them. The Genes & Homeostasis feature addresses the genetic basis of the growing problem of antibiotic resistance in pathogenic bacteria.

Most bacterial cells are surrounded by a **cell wall** that provides rigidity and protection; it is composed of carbohydrate and protein molecules. The cell walls of many bacteria are surrounded by a gelatinous **capsule**, which helps the bacteria attach to cells within tissues they will infect. This capsule also helps some bacteria to escape destruction by cells of the immune system. Bacteria also may have one or more external **flagella** to aid in movement, and **pili**, which help some bacterial cells attach to each other and pass genes.

Bacteria reproduce by a process called **binary fission** (FIGURE 12.2). When a bacterial cell divides by binary fission, the single, circular chromosome that is attached to the plasma membrane inside the cell wall is copied. This copy is then attached to another

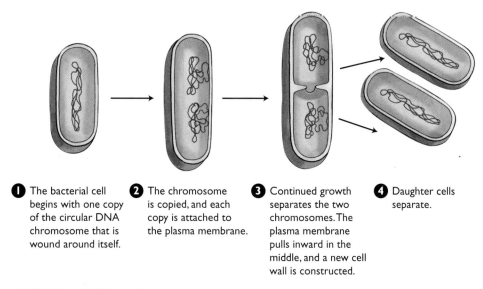

❶ The bacterial cell begins with one copy of the circular DNA chromosome that is wound around itself.

❷ The chromosome is copied, and each copy is attached to the plasma membrane.

❸ Continued growth separates the two chromosomes. The plasma membrane pulls inward in the middle, and a new cell wall is constructed.

❹ Daughter cells separate.

FIGURE 12.2 Binary fission. A bacterial cell reproduces by splitting in two.

GENES & HOMEOSTASIS

Mutations and the Development of Antibiotic Resistance

Antibiotics are medications that doctors prescribe for people with bacterial infections. When you were a child and had an ear infection, your parents probably gave you spoonfuls of a pink medicine. Now when you have a sinus infection, your doctor prescribes other antibiotics. These infections are easily treated and cause you little inconvenience besides a trip to the doctor and the pharmacy. Unfortunately, the days of easily treatable bacterial infections may soon be over because the bacteria that cause many diseases—including tuberculosis, ear infections, and gonorrhea—have become resistant to the antibiotics developed to cure them.

Antibiotic resistance develops as a result of natural selection on a bacterial population. Genetic differences exist within any population of individuals, including a population of bacterial cells. Some bacteria, even before exposure to the antibiotic, carry genes that enable them to resist the antibiotic. Antibiotic-resistant infections can develop when antibiotics kill all but the resistant bacteria, which can then propagate in the absence of competition from other bacteria in your body. Because only a small percentage of bacteria will be resistant, not all people will carry resistant forms. However, antibiotic-resistant infections can be passed from one person to another even if the recipient has never taken antibiotics.

Tetracycline is an antibiotic used to kill the bacteria that cause acne. It works to prevent translation (a step in the synthesis of proteins) in prokaryotes. Because the workbenches of translation, the ribosomes, differ structurally in prokaryotes and eukaryotes, this antibiotic—like all effective antibiotics—selectively kills bacterial cells but not eukaryotic host cells. One bacterial cell in a population of millions may have a different DNA sequence in the region of DNA that encodes a protein that is part of a ribosome. If this changed protein enables the bacterial cell to withstand tetracycline treatment, then the cell will survive and pass on the resistant sequence to its offspring. In this manner, those bacteria that are not killed are selected for, and resistant infections develop.

The problem of resistance in bacteria originated from overuse of antibiotics. Some patients ask their doctors to give them antibiotics for a cold or the flu, both of which are caused by viruses. Viruses are not affected by antibiotics, because viruses do not have the organelles and other structures that antibiotics target. For example, tetracycline would not be able to control a viral infection because the virus is using the patient's own eukaryotic ribosomes for translation and is thus unaffected by the treatment. Meanwhile, the population of antibiotic-resistant bacteria increases in the individual who took the antibiotics when there could be no benefit from doing so.

Adding to the problem of medical overuse is that people often discontinue their prescribed antibiotics when they begin to feel better. A person who does not finish all of the prescribed medication allows any resistant bacteria to proliferate. The infection will recur in a few weeks and be resistant to that antibiotic, and a different course of medication will be required.

Receiving an antibiotic.

site on the plasma membrane, and the membrane between the attachment sites grows and separates the two copies of the original chromosome until it eventually produces two separate daughter cells. Therefore, one bacterial parent cell gives rise to two genetically identical daughter cells.

Bacteria can reproduce rapidly under favorable conditions, doubling their population approximately every 20 minutes or so. In theory, after 8 hours at room temperature, a single *Salmonella* bacterium in a chicken salad sandwich could give rise to millions of bacterial cells. This exponential growth occurs because one cell gives rise to two, and those two yield four, the four divide to become eight, then 16, and so on (FIGURE 12.3). *Salmonella* reproduces quickly in food because the bacteria have access to all the nutrients they need. Fortunately, refrigeration slows the rate at which bacteria divide.

When an infection occurs in your body, the rapidly increasing bacteria support their growth by using your cells' nutrients, effectively preventing your cells from func-

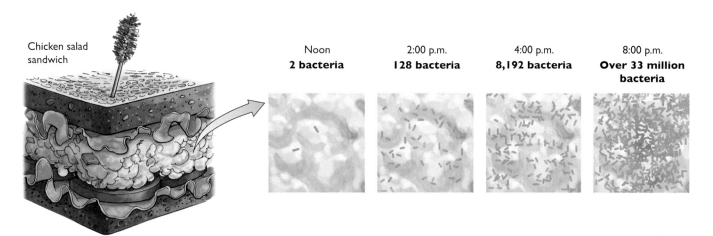

FIGURE 12.3 Exponential growth of bacteria. One kind of bacteria that can cause food poisoning reproduces every 20 minutes. If a sandwich with two bacteria is left out for 20 minutes, each bacterial cell will make a copy of itself, yielding four bacteria. After 40 minutes, there will be eight bacteria. Within 8 hours, there will be 33 million bacteria.

tioning properly. However, most cases of disease-causing bacterial infection result from more than just the large numbers of bacteria in your body. The symptoms of a disease arise due to the effects of molecules called *toxins* that are secreted by the bacterial cells, which can result in cell death. Many examples of bacterially caused diseases are listed in Table 12.1.

While many different types of bacteria cause infections that are troublesome or even fatal to humans, bacteria are more than just a source of disease. They play important roles in human health.

During prenatal development, a human body is free of microbes. However, as soon as a baby makes contact with the outer world, his or her body begins to pick up bacteria. Bacteria are first encountered in the birth canal and more are introduced when the baby is touched or fed. These exposures help the baby to establish the typical mixture of microorganisms that cover its skin, mouth, respiratory, digestive, and reproductive tracts, called the *normal flora*.

Bacteria in the human intestinal tract synthesize and secrete many different vitamins that the human body utilizes. Bacteria of the normal flora also help prevent more pathogenic organisms from colonizing the body by competing with them for nutrients and space. It is only when the host is compromised in some manner that bacteria in the normal flora cause disease.

Viruses, on the other hand, have not been found to function as anything other than parasites.

Viruses

Viruses are not considered to be living organisms for two reasons: They cannot replicate (copy) themselves without the aid of a host cell, and they are not themselves composed of cells. Viruses lack the enzymes for metabolism and contain no ribosomes. Therefore, they cannot make their own proteins. They also lack cytoplasm and membrane-bounded organelles. Nor can they produce toxins, like bacteria can. Viruses are little more than packets of nucleic acid (DNA or RNA) surrounded by a protein coat.

The genetic material, or *genome*, of a virus can be DNA or RNA, either double stranded or single stranded, and either linear or circular. For example, the herpes

TABLE 12.1 | Examples of Infectious Diseases Caused by Bacteria

Disease	Disease Information
Anthrax	Anthrax is an infectious disease caused by the bacterium *Bacillus anthracis*. This microorganism lives mainly in soil and, as a part of its life cycle, can exist as a tough, resistant structure called a *spore*. Anthrax spores can be lethal if inhaled. Anthrax is an example of a disease that is infectious without being contagious because infected individuals do not pass the disease on to other people.
Botulism	The bacterium that causes botulism, *Clostridium botulinum*, produces a powerful toxin that acts on the nervous system, causing paralysis of respiratory and facial muscles. This anaerobic bacterium is commonly found in soil and can make its way into canned goods. When present in canned foods, the bacteria thrive and produce the toxin, which may then be ingested. Because the toxin is the main source of the disease, botulism is really more an example of poisoning than of infection. Cans that have bulges may contain this bacterium, which also produces gas as a by-product of its metabolic activities. Botox injections are injections of the toxin produced by this bacterium, which temporarily paralyze facial muscles to prevent drooping and the appearance of wrinkles.
Escherichia coli (*E. coli* O157:H7) enteritis	*E. coli*, which causes severe diarrhea, is the primary cause of infant mortality in third-world nations due to contamination of water supplies by human sewage. *E. coli* infection can also occur from eating undercooked beef. The meat of infected animals can become contaminated with bacteria from the intestines during slaughter. When meat is ground, bacteria are spread throughout the meat. Bacteria present on the cow's udders or on equipment may get into raw milk, so drinking unpasteurized milk may also allow for infection.
Salmonellosis	Food poisoning is caused by eating contaminated foods. Botulism is one type of food poisoning. Another type can be caused by the *Salmonella enteritidis* bacterium, which is often present in poultry, meat, and eggs.
Tetanus	Toxins produced by the bacteria *Clostridium tetani* affect the nervous and muscular systems, causing the back to arch and spasm. This bacterium can also persist as a resistant spore in the soil. If a person steps on an object containing contaminated soil and a puncture results, bacteria are injected into the bloodstream.
Staphylococcal infections	Bacteria of the genus *Staphylococcus* are common inhabitants of human skin and mucous membranes. Skin infections can be caused when certain species of the bacteria invade and destroy tissue near the site of entry into the body. Toxic shock syndrome is caused by the bacterium *Staphylococcus aureus* when tampons are left in place long enough to allow bacterial growth. Food poisoning can also be caused by a type of *Staphylococcus* which produces a toxin that contaminates creamy dressings, pies, and milk products if they are allowed to remain at room temperature for a few hours. Once the toxin is produced, its damaging effects are not lessened or prevented by heating or boiling.
Streptococcal infections	Streptococcal bacteria occur as chains of cells. Many infections of humans are caused by bacteria of this type. Strep throat is a severe sore throat. If the bacteria that cause strep throat produce a certain toxin, scarlet fever can result. The body's response to this infection includes, in part, the development of a rash on the skin. An even stronger immune response to the toxin can cause rheumatic fever, which can result in very high fever, swollen joints, and heart damage. Streptococci can also cause pneumonia. Flesh-eating disease can be caused by streptococcal bacteria that produce toxins that degrade tissues.
Tuberculosis	Tuberculosis is caused by the bacterium *Mycobacterium tuberculosis*. This infectious agent is spread through infected droplets when a person with the infection coughs or sneezes. Once inside the body, the bacteria take up residence in the lungs. The immune system responds by walling off the bacteria in a manner much like the formation of a scab. The bacteria can live within the walls for years, during which time the infection is not considered to be active nor is it transmissible. If the bacteria escape the immune system's attempts at control, the infection becomes active. Renewed activity is most common in people with compromised immune systems. Approximately one-third of the human population is thought to be infected with TB. The disease is extremely common in areas where impoverished people have minimal health care.

simplex virus has a double-stranded DNA genome, and the polio virus has a single-stranded RNA genome. The genes of a virus code for the production of all the proteins required to produce more viruses inside a host cell.

The protein coat surrounding a virus is called its **capsid**. Many of the viruses that infect humans possess an additional structure outside the capsid called the **viral envelope**. The envelope is derived from the cell membrane of the host cell and may contain additional proteins encoded by the viral genome. Viruses that are surrounded by an envelope are called enveloped viruses. Viral structure is shown in FIGURE 12.4.

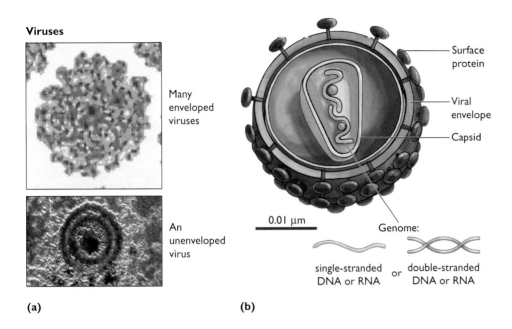

FIGURE 12.4 Viral structure. The structure of a typical virus (a) as viewed under a microscope and (b) illustrated.

VISUALIZE THIS What structures on the virus might help it attach to a human cell?

Infection by an enveloped virus occurs when the virus gains access to the cell by fusing its envelope with the host's cell membrane (FIGURE 12.5). An unenveloped virus uses its capsid proteins to bind to receptor proteins in the plasma membrane and gain access to a host cell. Once inside the host cell, the capsid is removed.

Regardless of how it enters, after the genome gains access to the host cell, the infection continues when the virus makes copies of itself. First the genome is copied; then the virus uses the host cell's ribosomes and amino acids to make viral proteins for building new capsids and synthesizing some of the envelope proteins. Once assembled, the daughter viruses exit the cell, leaving behind some viral proteins in the host's cell membrane. The daughter viruses then move to other cells, spreading the infection.

Some viruses, called **latent viruses**, enter a state of dormancy in the body. During that time, the virus is not replicating. Herpes simplex viruses, which cause outbreaks on the mouth and genitals, for example, can undergo long periods of dormancy during which they are present in the body but not causing symptoms such as painful sores and blisters. Examples of diseases caused by viruses are listed in Table 12.2.

While many infectious diseases are caused by bacteria or viruses, some are also caused by eukaryotic organisms.

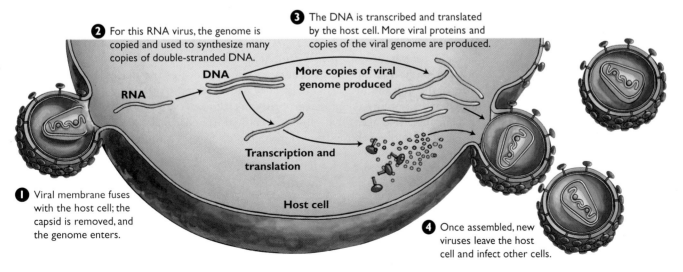

FIGURE 12.5 Replication by an enveloped RNA virus.

TABLE 12.2 | Examples of Infectious Diseases Caused by Viruses

Disease	Disease Information
Common cold	The common cold can be caused by any of close to 200 different viruses, most commonly the rhinovirus. Because so many different viruses cause colds, cold symptoms differ among people and may change from your first cold of the winter to your last. Cold symptoms may involve a runny nose, sore throat, cough, watery eyes, sneezing, and/or congestion. Cold viruses spread easily in droplets when an infected person sneezes or coughs and can be picked up by handling objects such as doorknobs and utensils that an infected person contaminated.
Hepatitis	There are at least six different hepatitis viruses, but the most dangerous is hepatitis C. Exposure to hepatitis C leads to chronic liver diseases such as cirrhosis (irreversible, potentially fatal scarring of the liver), liver cancer, and liver failure. Hepatitis C ranks second to alcoholism as a major cause of liver disease and is the leading reason for liver transplants in the United States. An estimated 3% of the world's population carries this virus, but they are unaware of it because there are often no symptoms until the liver damage is severe. Intravenous drug use and possibly some skin-piercing practices, notably tattooing and body piercing, may also be contributing to the spread of hepatitis C. Hepatitis B is sexually transmitted and causes about half as many deaths as hepatitis C.
Influenza	Flu is caused by the influenza virus, which is prone to mutations that help it evade the immune system response that results from the previous year's exposure. This is why you need a flu shot every year. The flu shot contains parts of three or four flu viruses that scientists at the Centers for Disease Control and Prevention believe may make it to the United States. The viral parts contained in the inoculation stimulate your immune system, giving it a head start before you are exposed to the virus. The virus that causes bird (avian) flu also mutates rapidly. Scientists and medical professionals are concerned that the bird flu virus will spread rapidly and extensively if certain changes to its genetic information occur. When infected, people experience fever, cough, muscle aches, and sore throat, which can progress into pneumonia and respiratory distress and result in death. No one knows whether the avian flu virus will mutate in a fashion that allows it to affect humans, but if it does, the worldwide death toll could be in the millions.
Mononucleosis	Mononucleosis is caused by the Epstein-Barr virus. Most people will have mononucleosis before their mid-30s. Symptoms in children are less severe and may go unnoticed. In older adolescents and young adults the symptoms—fatigue, weakness, fever, headache, and sore throat—can be severe and may last a month or two. The virus can be transmitted in the saliva (kissing) and by coughing and sneezing or by touching an object contaminated by an infected person.
Rabies	Rabies is a viral disease that typically spreads to humans through a bite from an infected animal. The initial signs of infection include flulike symptoms but progress to more serious symptoms including paralysis, convulsions, and hallucinations. Most domesticated animals are vaccinated against rabies, so transmission is most commonly from bites by wild animals.
West Nile virus	The West Nile virus is transmitted to humans via the bite of an infected mosquito. The virus first emerged in the United States in New York and quickly spread west and south throughout the country. In infected persons, the brain and spinal cord can become inflamed, resulting in paralysis and death.

Prions

(a)

Normal prion protein has more helical regions.

(b)

Misfolded prion protein has more pleated regions.

0.001 μm

FIGURE 12.6 Prion structure. (a) Normal and (b) misfolded prions.

Eukaryotic Pathogens

Eukaryotic pathogens such as the single-celled *protozoans*, along with worms and some fungi, cause tremendous human suffering and death worldwide. Examples of diseases caused by eukaryotic pathogens are found in Table 12.3.

Protozoans are often spread by way of water and food contaminated with animal feces, as are parasitic worms. Worms that take up residence in human intestines or other organs can cause extensive damage to internal organs and tissues. Fungi cause diseases of the skin and internal organs. They damage tissue by secreting digestive enzymes into it and absorbing the products of this digestion.

The infectious agent that causes the spongiform encephalopathies is not like any of the agents described above. Instead, this class of diseases is caused by a novel infectious agent called a prion.

Prions

After a cell synthesizes a protein, the protein folds into its characteristic shape. If a protein is folded incorrectly, it can no longer perform its job properly. A **prion** (FIGURE 12.6) is a normally occurring protein produced by brain cells that, when misfolded, causes spongiform encephalopathy.

The word prion is a shortened form of the term "proteinaceous infectious particle." Normal prions are present in the brains of all humans. When highly magnified, the nor-

TABLE 12.3 Examples of Diseases Caused by Eukaryotic Pathogens

Type of Pathogen	Diseases and Disease Information
Protozoans	
	Giardiasis is caused by the water-borne *Giardia lamblia*. During one part of its life cycle, *Giardia* exists as a tough, resistant structure called a cyst. Cysts enter bodies of water or food supplies in contaminated feces. Once the cysts are ingested, *Giardia* completes its life cycle in the intestines of an infected person, causing severe diarrhea and gas.
	African sleeping sickness is caused by a protozoan called *Trypanosoma brucei*. This infection is transmitted by bites of tsetse flies and causes encephalitis or swelling in the brain. This infection can be fatal.
Fungus	
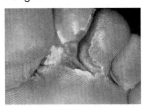	Athlete's foot and jock itch are both types of fungal infections that affect the top layer of skin. Characterized by an itchy, red, circular rash with healthy-looking skin in the center, these infections are caused by the *Tinea* fungus. Because of the circular appearance of the rash, these infections are sometimes called ringworm. Athlete's foot typically grows between the toes.
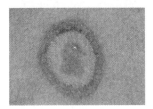	Jock itch affects the skin of the genitals and inner thighs. This fungus spreads easily in public places such as locker rooms and the floors of communal showers. Excessive perspiration can wash away fungus-killing oils of the skin. Due to their routine exposure to locker-room floors and communal showers, and the propensity to sweat during workouts, athletes are more prone to these infections.
Worms	
	Most worms gain access to the body because of poor sanitation or by consumption of contaminated meat or fish. Pinworm infections are caused by poor sanitation. To be infected by pinworms, a person must ingest worm eggs, which hatch inside the body where new worms develop into adult worms. Female pinworms then lay their eggs just outside the anus, causing the infected person to scratch the area, possibly getting eggs on their fingers and touching objects that may allow transmission of the eggs to another person.
	Tapeworm infections are usually caused by ingesting infected raw or undercooked pork or beef. This worm lives in the intestines and can cause abdominal pain and diarrhea.

mal shape of a prion resembles the alpha helical coil you learned about in Chapter 2. The misfolded version of this protein resembles a sheet of paper that has been accordion-folded lengthwise several times (a beta-pleated sheet).

The normal role of the prion in the brain has not yet been determined. Experiments in mice lacking the prion gene, and therefore unable to make the normal version of the protein, indicate that prions may protect mammals against dementia and other degenerative disorders associated with aging. The very rare CJD seen in elderly people is believed either to arise spontaneously when a prion is mistakenly misfolded or to be caused by a mutation in the gene that encodes the precursor protein in humans, leading to the production of a misfolded prion protein.

In contrast, the newly emerging form of the disease that affects younger people, nvCJD, results when an individual is infected by misfolded prion proteins. Remarkably, the misfolded protein refolds proteins that were properly folded into the mutant, disease-causing version (FIGURE 12.7). This behavior is very unusual for a protein. No other known protein has this capability.

Over time, the nerve cells in the brain of an infected individual become clogged with misfolded prions, causing the cells to transmit impulses improperly and, eventually, to stop functioning altogether. Ultimately, the cells burst, freeing their misfolded prions to find and refold normal prions in other cells. Gradually, the brain becomes riddled with empty spaces formerly occupied by normal cells, producing the spongelike structure of the diseased brain.

Prions, unlike viruses, bacteria, and parasitic eukaryotes, have no RNA or DNA, which were once thought to be necessary for any infectious agent to multiply. Furthermore, although most proteins within cells are easily broken down, prions resist degradation, allowing these rogue disease agents to propagate relentlessly.

Stop and Stretch 12.1

Before discovery of prions in the early 1980s, all known infectious agents had their own genomes, and many researchers did not believe that proteins by themselves could be infectious. When scientists attempt to determine if a disease is caused by a bacterial, viral, or protozoal infection, they often try to isolate the disease-causing agent and subject it to harsh treatments such as heat, chemicals, and radiation. Living organisms and viruses are typically rendered nonpathogenic by these types of treatments. How would you expect a disease that was caused by infectious proteins to react to the above treatments? How might the results of these treatments lend credibility to the hypothesis that prions themselves can be infectious?

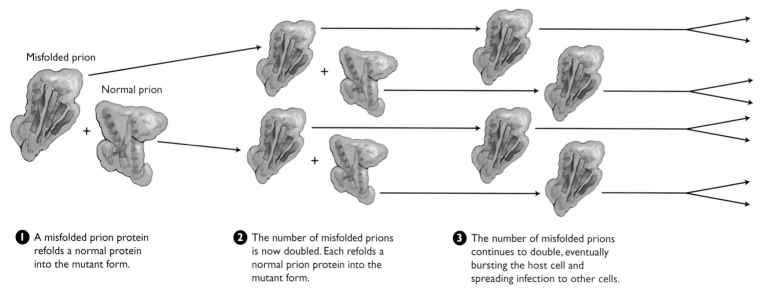

❶ A misfolded prion protein refolds a normal protein into the mutant form.

❷ The number of misfolded prions is now doubled. Each refolds a normal prion protein into the mutant form.

❸ The number of misfolded prions continues to double, eventually bursting the host cell and spreading infection to other cells.

FIGURE 12.7 Prion replication. One misfolded prion can set off a chain reaction of misfoldings.

Understanding the characteristics of prions as infectious agents is only the first step toward preventing spongiform encephalopathies from becoming widespread in humans.

12.2 Transmission of Infectious Agents

Different pathogens are transmitted with varying levels of ease. There are several common methods of transmission, including direct and indirect contact, vector-borne transmission, and ingestion (FIGURE 12.8).

Direct Contact

Coming into contact with a pathogen can often lead to infection. Shaking hands with an infected person can allow transmission of the common cold, for instance. Likewise, exposure to body fluids of an infected person can cause infection. The exchange of semen and vaginal fluids facilitates the spread of sexually transmitted infections (STIs), which are discussed in more detail in Chapter 13. Transmission can also occur when droplets breathed out that contain the pathogen are breathed in by someone else. Transmission of some pathogens is facilitated by coughing, sneezing, or even speaking. This is why people concerned with avoiding or spreading infection wear surgical masks as needed.

Indirect Contact

Contaminated objects such as drinking glasses, paper tissues, telephones, and doorknobs can harbor some pathogens that may be transmitted from the original source to a new host. The flu, for example, may be spread when people touch a surface with flu viruses on it and then touch their nose or mouth.

Vector-Borne Transmission

Transmission of microorganisms can occur through an intermediate organism. The organism that carries disease-causing microorganisms from one host to another is called a **vector**.

Ticks can serve as vectors to transmit the bacterium that causes Lyme disease. Raccoons, bats, and dogs can serve as vectors for the transmission of the virus that causes rabies. Mosquitoes are vectors for the transmission of malaria, as described in Chapter 8.

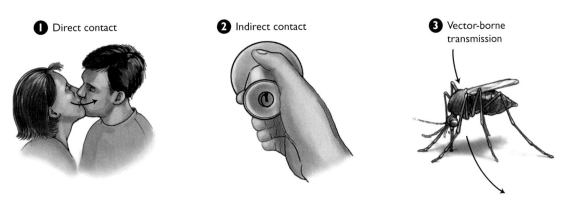

FIGURE 12.8 **Transmission of infectious agents.** Most infectious diseases are transmitted in one or more of these ways.

VISUALIZE THIS Can you think of diseases that might be transmitted by more than one of these mechanisms?

Ingestion

Many ingested pathogens are killed by the acidity of the stomach. However, as you learned earlier, prions can withstand harsh chemical treatments. Therefore, spongiform encephalopathies can be spread via the ingestion of food containing misfolded prions. Kuru results from the tribal custom among some New Guineans of honoring the dead by eating their brains. Scientists think that the cows with mad cow disease ingested misfolded proteins when they ate feed made from the remnants of diseased animals.

After a cow has been slaughtered, what is left of the carcass is cooked and then ground up, producing a type of cattle feed called meat-and-bone meal. Feed mills buy the meal and make it into a toasted, crunchy, cereal-like substance that is fed to cows. When feed mills unwittingly use cows with mad cow disease in making meat-and-bone meal, the infectious prions from those cows are fed to other cows, which become infected.

Humans become infected with misfolded prions when they eat meat from diseased animals. Although the pathogen that causes spongiform encephalopathies in cows and humans infects the brain and spinal cord—parts of the cow that are normally not consumed—meat can be contaminated during the rendering process. For example, hamburger containing parts of the spinal cord that were stripped off the carcass may be infectious.

Stop and Stretch 12.2

Which of the above modes of infection might cause the largest number of people to be infected the most easily?

Although we encounter a multitude of pathogens every day, we do not always become ill. This is because our body is protected by the immune system. For reasons outlined below, spongiform encephalopathies are unique infectious diseases because the immune system is unable to fight an infection by misfolded proteins. Let's first look at how the immune system typically functions, and then at how prions escape recognition.

12.3 The Body's Response to Infection: The Immune System

Humans have evolved three different lines of defenses against pathogens. The skin and mucous membranes attempt to prevent access to the body, and the remaining two lines of defense operate if the pathogen gains access to the body.

First Line of Defense: Skin and Mucous Membranes

The skin, mucous membranes, and their secretions form the first line of defense against pathogens. These external physical and chemical barriers are **nonspecific defenses**; that is, they do not distinguish one pathogen from another. Organisms living on the body's surface are kept out by the skin. In addition to being a physical barrier, the skin also sheds, taking pathogens with it, and has a low pH that can help to repel microorganisms. Likewise, glands in the skin secrete chemicals that slow down the growth of bacteria. For example, tears and saliva contain enzymes that break down bacterial cells, and earwax traps microorganisms.

Digestive secretions, including acids, kill many microorganisms that gain access to the stomach. Mucous membranes lining the respiratory, digestive, urinary, and reproductive tracts also secrete mucus that traps pathogens, which are then coughed or sneezed away from the body, destroyed in the stomach, or excreted in feces. Vomiting can also rid the body of toxins or infectious agents. Pathogens that are able to evade the first line of defense next encounter an internal, second line of defense.

Second Line of Defense: White Blood Cells, Inflammation, Defensive Proteins, and Fever

If a pathogen is able to get past the first line of defense, the internal second line of defense, which is also nonspecific, can often stop the infection. Like the first line of defense, participants in the second line of defense do not target specific pathogens.

White Blood Cells **Phagocytes** are white blood cells that indiscriminately attack invaders by engulfing and digesting them (FIGURE 12.9). **Neutrophils** are phagocytic white blood cells that respond to an infection by destroying bacterial cells and some types of fungi.

Additional phagocytic white blood cells called **macrophages** move throughout the lymphatic fluid, cleaning up dead and damaged cells. To destroy the cells, macrophages extend their long **pseudopodia** (cellular extensions used for eating and moving), grab the invading organism, and engulf it. Enzymes inside the macrophage help break the invader apart. Macrophages can also clean up old blood cells, dead tissue fragments, and other cellular debris. They also release chemicals that stimulate the production of more white blood cells. Much of the destruction of the offending cells occurs in the lymph nodes, which is why those nodes swell when we are ill with an infection.

When an invader is too big to be engulfed by phagocytosis, white blood cells called **eosinophils** cluster around the invader and secrete digestive enzymes that irritate or may even destroy the organism. Invaders that get this treatment include larger organisms such as protozoans and worms.

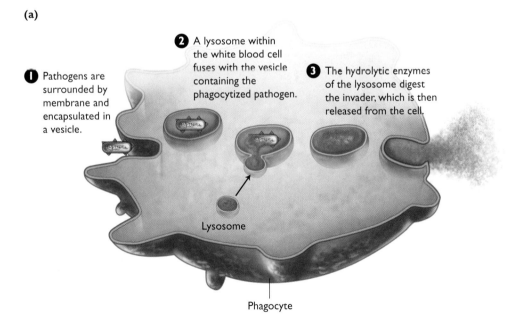

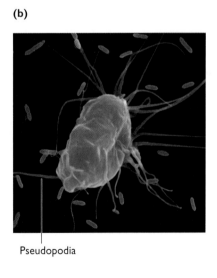

FIGURE 12.9 **Phagocytosis.** (a) A phagocytic white blood cell eats and destroys invaders. (b) Electron micrograph of a macrophage attacking bacterial cells.

Additional white blood cells that circulate through the blood and lymph destroying invaders are **natural killer cells**. These nonspecific cells attack tumor cells and virus-invaded body cells upon first exposure. Natural killer cells release chemicals that break apart plasma membranes of their target cells, causing them to burst.

When white blood cells are actively fighting an infection, the body increases production of white blood cells, so white cell counts can increase dramatically.

Another component of the second line of defense is the inflammatory response.

Inflammation The **inflammatory response** is a reaction producing redness, warmth, swelling, and pain.

Whenever a tissue injury occurs, the inflammatory response begins (FIGURE 12.10). Damaged cells release chemicals that stimulate specialized connective tissue cells called **mast cells** to release **histamine**. White blood cells called **basophils** also secrete histamine. Histamine promotes increased size of blood vessels, or *vasodilation*, near the injury. The histamine-induced dilation of blood vessels allows the endothelial cells of the capillary walls to pull apart, and phagocytes can squeeze between the cells.

As more blood and cells arrive at the site of infection to speed up repair and cleanup, redness, warmth, and swelling occur. Swollen tissues cause pain by pressing against nearby nerves, so the injured person is more likely to rest the affected area. The extra blood flow also brings oxygen and nutrients required for tissue healing.

In the aftermath of the inflammatory response, tissue fluid, dead cells, and microorganisms have accumulated at the site of the infection, producing *pus*. If the pus cannot drain, the body may wall it off with connective tissue, producing an *abscess*.

Defensive Proteins In addition to white blood cells, some proteins act as nonspecific defenders. **Interferons** are proteins produced by virus-infected body cells to help uninfected cells resist infection. When infected cells die, they release interferons that bind to receptors on uninfected cells and stimulate the healthy cells to produce proteins that inhibit viral reproduction.

Another class of defensive proteins, called **complement proteins**, includes at least 20 different proteins that circulate in the blood and help, or complement, other defense mechanisms. These proteins can coat the surfaces of microbes, making them easier for

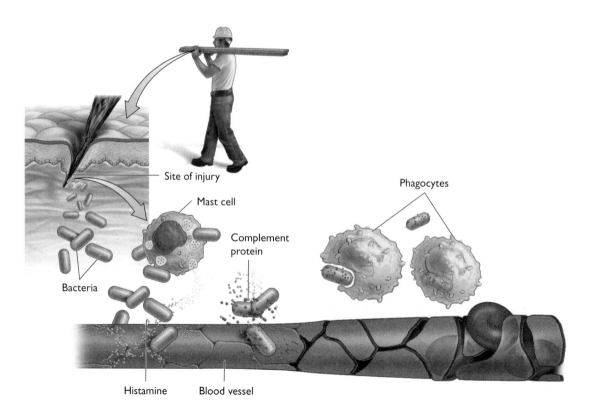

FIGURE 12.10 Inflammation. Injured tissues signal mast cells to release histamine, which dilates blood vessels. Complement proteins can leak out of dilated vessels and bind to bacteria, which marks the invaders for destruction. Phagocytes also squeeze through dilated vessels and engulf and destroy bacteria.

macrophages to engulf. Complement proteins can also poke holes in the membranes surrounding microbes, causing them to break apart (FIGURE 12.11). These proteins also increase the inflammatory response.

Fever A body temperature above the normal range of 36.5–37.5°C (97–99°F) is called a **fever**. Macrophages release chemicals called *pyrogens* as a weapon in their assault on invaders. Pyrogens cause body temperature to increase. A slightly higher than normal temperature decreases bacterial growth and increases the metabolic rate of healthy body cells. Both of these processes help to fight infection by slowing pathogen reproduction and allowing repair of tissue to occur more quickly. When the infection is controlled, macrophages stop releasing pyrogens and body temperature returns to normal.

Third Line of Defense: Lymphocytes

If a pathogen makes it past the first two nonspecific defense systems, the immune system presents a third line of defense. Cells of the immune system identify and attack specific microorganisms that are recognized as foreign. This **specific defense** system consists of millions of **lymphocytes**, one of the groups of white blood cells. Lymphocytes travel throughout the body by moving through spaces between cells and tissues or being transported via the blood and lymphatic systems (FIGURE 12.12).

The specific response generated by the immune system is triggered by proteins and carbohydrates on the surface of pathogens or on cells that have been infected by pathogens. Molecules that are foreign to the host and stimulate the immune system to react are called **antigens**. Examples of antigens include molecules found on invading viruses, bacteria, fungi, protozoans, and worms, as well as molecules on the surface of foreign substances such as dust, pollen, or transplanted tissues.

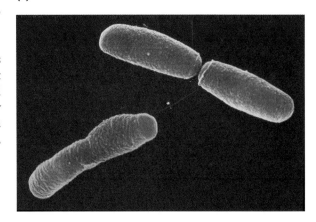
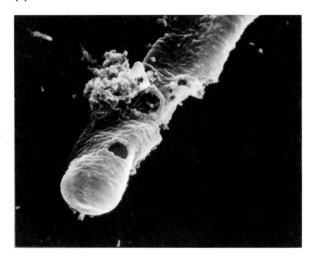

FIGURE 12.11 **Complement proteins.** (a) Bacterial cells before complement proteins attach. (b) Complement proteins cause holes to form in a bacterial cell, penetrating the cell wall and membrane. Holes in the bacterium allow it to fill with fluids until it bursts.

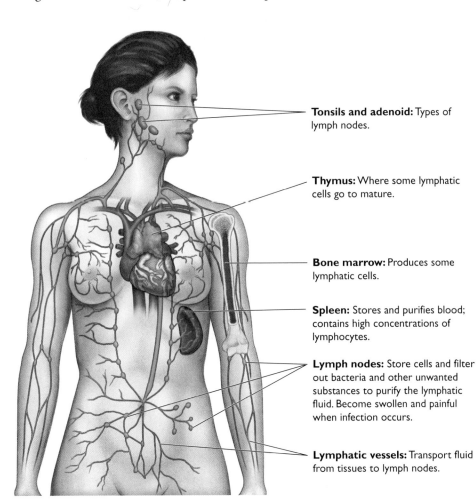

Tonsils and adenoid: Types of lymph nodes.

Thymus: Where some lymphatic cells go to mature.

Bone marrow: Produces some lymphatic cells.

Spleen: Stores and purifies blood; contains high concentrations of lymphocytes.

Lymph nodes: Store cells and filter out bacteria and other unwanted substances to purify the lymphatic fluid. Become swollen and painful when infection occurs.

Lymphatic vessels: Transport fluid from tissues to lymph nodes.

FIGURE 12.12 **The lymphatic system.** Pathogens trigger a response by the lymphatic system. The various organs of this system work to eliminate infections.

FIGURE 12.13 B cells and T cells. (a) The receptors produced by B cells function as antigen receptors on the cell surface or as secreted antibodies that are present in body fluids. (b) T cells have antigen receptors only. These receptors help B and T cells recognize and destroy invaders.

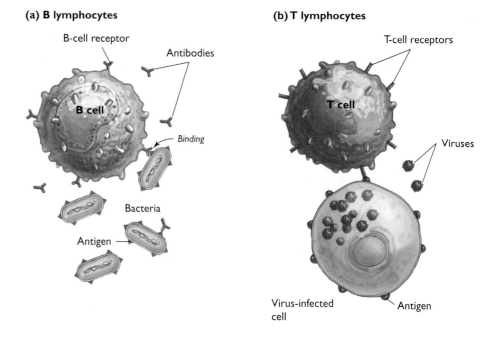

Antigens are present on tissues transplanted from one person to another because every individual's cells have a characteristic set of proteins on their surfaces. These *self markers* are called **major histocompatibility complex (MHC) proteins**. Because MHC proteins are encoded by your genes, each individual has a unique suite of MHC proteins. One person's MHC proteins could serve as an antigen in another person. People receiving organ transplants are given drugs to suppress their immune systems so they do not reject the organ due to its *nonself* markers.

Regardless of the source, when an antigen is present in the body, the production of two types of lymphocytes is enhanced: the **B lymphocytes (B cells)** and **T lymphocytes (T cells)**. Like macrophages, these lymphocytes move throughout the circulatory and lymphatic systems and are concentrated in the spleen and lymph nodes. Lymphocytes display **specificity** because they recognize specific antigens.

This recognition is based on the presence of proteins, called **antigen receptors**, whose shape fits perfectly to a portion of the foreign molecule. The receptor is able to bind to the antigen much the way that a key fits into a lock. These receptors are either attached to the surface of the lymphocyte or secreted by the lymphocyte. B and T cells recognize different types of antigens: B cells recognize and react to small, free-living microorganisms such as bacteria and the toxins they produce (FIGURE 12.13a). T cells recognize and respond to body cells that have gone awry, such as cancer cells or cells that have been invaded by viruses. T cells also respond to transplanted tissues and larger organisms such as fungi and parasitic worms (FIGURE 12.13b).

Antibodies Both B and T cells have receptor molecules on their surfaces, but B cells also secrete special proteins called **antibodies** into the surrounding body fluids to locate and destroy antigens. The structure of an antibody allows it to bind to a specific antigen. All antibodies share the same basic structure (FIGURE 12.14). Each antibody consists of four polypeptide chains arranged in the shape of the letter Y. There are two longer *heavy chains* and two shorter *light chains*. At one end of each of the four chains is a *constant region* where the sequence of amino acids forming the peptide is fixed for a given class of antibodies. At the other end of each chain is the *variable region* where the amino acid sequence varies. It is at the tips of the variable regions that specificity for a particular antigen occurs.

There are five different classes of secreted antibodies, also called immunoglobulins. These are IgG, IgM, IgA, IgD, and IgE. Each immunoglobulin has its own particular functions and locations in the body.

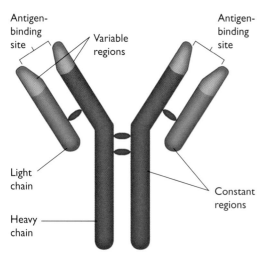

FIGURE 12.14 Antibody structure. Antibodies have a Y-shaped structure. The light and heavy chains are bonded together. All members of one class of antibodies have the same constant regions, but the variable regions differ among antibodies, which allows for specificity in antigen-antibody binding.

VISUALIZE THIS Would you expect that every antibody in your body has a different set of variable regions?

These antibodies are found in the lymph, intestines, and tissue fluids. They are also found in breast milk. When a woman breast-feeds a baby, she passes along some antibodies, conferring short-term **passive immunity** to the child. Because the child did not make the antibodies, this immunity will last only as long as the antibody lasts in his or her blood. An infant's passive immunity is different from the immunity conferred by exposure to an antigen. Exposures to antigens cause the production of antibodies to combat the infection for the individual's lifetime, called **active immunity**.

Allergy An allergy is an immune response that occurs even though no foreign substance is present. The body simply reacts to a nonharmful substance as though a pathogen were present. For example, some people have allergic reactions to peanuts or ragweed pollen. Asthma may also be the result of an allergic reaction.

An allergic reaction can be caused after an initial exposure to an antigen triggers B cells to produce IgE antibodies. IgE antibodies bind to mast cells and basophils. When the same antigen enters the body a second time, it binds to the IgE antibodies that are attached to mast cells and basophils. This binding causes the cells to release histamine, which can result in an allergic reaction (FIGURE 12.15).

The ability of the body's B and T cells to respond to specific antigens begins before humans are born. Thus, we are able to respond to infectious agents the very first time we are exposed. This ability continues into adulthood because these cells are manufactured, at the rate of about 100 million per day, throughout our lives.

The ability to respond to an infection, the **immune response**, actually results from the increased production of B and T cells.

Anticipating Infection

Lymphocytes are produced from special cells, called *stem cells*, that have the ability to become any other cell type. Many parts of the body, including the bone marrow, retain a supply of stem cells that can develop into more specialized cells. Bone marrow stem cells enable the bone marrow to produce blood cells throughout a person's lifetime. Lymphocytes are produced from the stem cells of bone marrow and released into the bloodstream.

Some lymphocytes continue their development in the bone marrow and become B cells. Others take up residence in the thymus gland. The thymus gland, located behind the top of the sternum, stimulates T cells to develop. When immune cells finish their

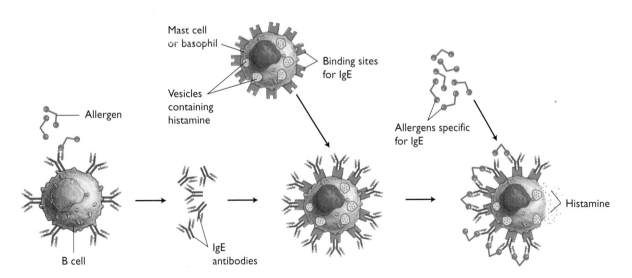

FIGURE 12.15 Allergy. When an allergen enters the body, B cells secrete IgE antibodies. The IgE antibodies bind to histamine-secreting cells which will prepare the cells for future exposures. Subsequent exposure causes histamine release, which causes an inflammatory response.

development in the **b**one marrow, they are called B cells; when they mature in the **t**hymus, they are called T cells.

Lymphocytes can recognize trillions of different antigens. This extraordinary ability results from the trillions of different antigen receptors that are produced by B and T cells. How is it that such an astonishing diversity of receptors can be produced?

Antigen Diversity If every antigen receptor were encoded by its own gene, trillions of genes would have to encode immune system proteins alone. Yet the entire human genome contains approximately 20,000 genes, so this cannot be the case. Instead, B and T cells have evolved a mechanism to generate an enormous variety of receptors from a limited number of genes.

As B and T cells develop, each cell's DNA segments that code for the production of antibodies do something very unusual: They rearrange themselves. Some portions are cut out, and the remaining DNA segments are joined together (FIGURE 12.16). Each unique arrangement of DNA produced by a given cell encodes a different receptor protein. Once synthesized, the proteins move to the surface of the B or T cell and act as antigen receptors. Thus, from a pool of only a few hundreds of genes, a virtually unlimited variety of antigen receptors can be created.

Stop and Stretch 12.3

Using what you know about generating antigenic diversity, explain why a particular individual might become ill from an infection that most other members of the population are able to fight off.

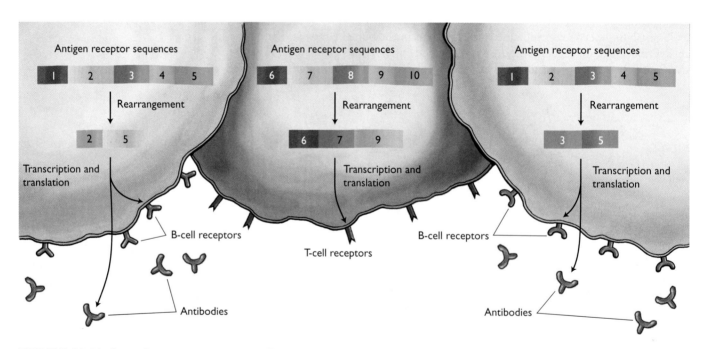

FIGURE 12.16 Genetic rearrangements allow for the production of millions of different antigen receptors. The genetic information encoding antibody structures is stored in bits and pieces. These bits and pieces are arranged in various orders during development. Removal of various DNA segments produces a wide variety of antigen receptors and increases the number of antigens to which each of us can respond.

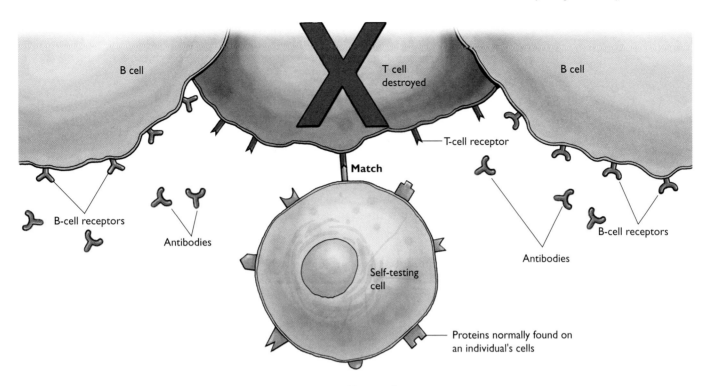

FIGURE 12.17 Testing developing lymphocytes for self-proteins. An individual's cells have characteristic proteins on their surfaces. Developing lymphocytes are tested to determine whether they react to these self-proteins. B cells are tested in bone cells and T cells in the thymus. Lymphocytes that bind to antigen receptors on testing cells are destroyed; those that do not bind are allowed to develop.

Another important facet of the immune system is its ability to determine whether a molecule is part of the host or foreign.

Self Versus Nonself While B and T cells are maturing, their antigen receptors are tested for potential self-reactivity. The cells of a given individual have characteristic proteins on their surfaces, and developing lymphocytes are tested to determine whether or not they will bind to self-proteins. Any developing lymphocyte with antigen receptors that bind to self-proteins is eliminated, making an immune response against one's own body less likely. Lymphocytes with receptors that do not bind are then allowed to develop to maturity (FIGURE 12.17). Thus, the body normally has no mature lymphocytes that react against self-proteins, and the immune system exhibits self-tolerance.

When this testing fails, the immune system is not in homeostasis and the cells of the immune system attack normal body cells. Diseases that result when a person's immune system is attacking the body are called **autoimmune diseases**.

Multiple sclerosis is an autoimmune disease that occurs when T cells specific for a protein on nerve cells attack these cells in the brain. Another is **insulin-dependent diabetes**, in which T and B cells attack cells that produce the hormone insulin in the pancreas. A *systemic* disease—that is, one that can affect many body systems—called **lupus** arises when self-antibodies form that react to the nuclei of *all* cells. These antinuclear antibodies build up in cells, causing inflammation of many tissues in the body. **Rheumatoid arthritis** results from the immune system's attack on synovial membranes that line some joints, including those of the fingers and toes. The resulting inflammation and accumulation of synovial fluid can cause joints to swell and stiffen. When these symptoms are untreated, scar tissue can form in the joints and the bones of the joint can fuse (FIGURE 12.18).

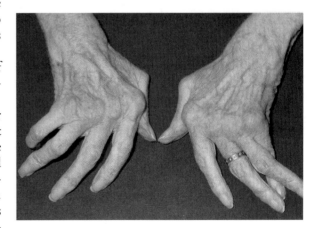

FIGURE 12.18 Autoimmune disease. Rheumatoid arthritis is an autoimmune disease that can result in joint fusion.

Even though we have a single immune system, it is diversified into two subsystems so that we can combat the multitude of infectious agents encountered in our lifetimes. This diversification is a result of the B and T cells' differing approaches to ridding the body of infectious agents once they are found.

Humoral and Cell-Mediated Immunity

Antibodies circulate through the body fluids. The protection afforded by B cells is called **humoral immunity**. T cells provide immunity that depends on the involvement of cells rather than antibodies and is called **cell-mediated immunity**.

Humoral Immunity When a B cell responds to a specific antigen, the B cell immediately makes copies of itself, resulting in a population of identical cells, called **plasma cells**, able to help fight the infection (FIGURE 12.19). This population of cells is called a **clonal population**. The sheer number of cells in the clonal population strengthens the immune system's ability to rid the body of the infectious agent.

The entire clonal population has the same DNA arrangement. Therefore, all the cells in a clonal population carry copies of the same antigen receptor on their membrane and secrete the same antibody. Some cells of the clonal population, called **memory cells**, will help the body respond more quickly if the infectious agent is encountered again. Should subsequent infection occur, the presence of memory cells facilitates a quicker immune response.

Inactivation of the infectious agent occurs when antibodies encounter a pathogen that matches the variable region. The antibody then binds to the antigen, forming the antibody-antigen complex. Formation of this complex marks the pathogen for phagocytosis or for breakdown by complement proteins. Complement proteins cause the foreign and infected cells to burst. The complement proteins circulate in the blood and lymph in an inactive state until they come in contact with antibodies bound to the surface of microorganisms, at which time they are activated. Some antibodies cause the pathogen and attached antibodies to clump together, or *agglutinate*, rendering them unable to infect other cells.

Vaccinations take advantage of the long-term protection provided by antibody-producing memory cells. **Vaccinations** are injections consisting of components of the disease-causing organisms, such as proteins from the plasma membrane of a bacterial cell, parts of a virus, or a whole virus that has been inactivated. The immune system responds to the challenge of the introduced vaccine by producing the clonal population of memory cells that will be prepared for a real infection should it happen. Some vaccines require multiple doses, or boosters, before a sufficient response is generated.

Other vaccines such as those for flu must be given every year because the flu virus rearranges its genetic sequences swiftly. This shifting of genetic sequence results in dif-

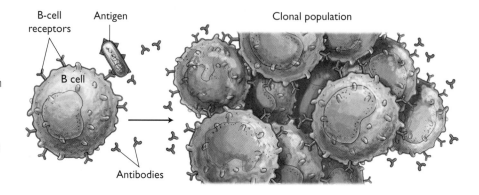

FIGURE 12.19 **Humoral immunity.** B lymphocytes do not directly kill cells bearing antigens. Instead, they make and secrete antibodies specific to antigens. When a B cell binds an antigen, the B cell makes copies of itself, each carrying identical antibodies. A population of such duplicate cells is called a clonal population. Clonal plasma cells secrete antibodies specific to the antigen. After the infection passes, the plasma cells will die but the clonal memory cells will remain and be able to recognize the antigen in the future.

ferent proteins being encoded and placed on the surface of the virus, thereby preventing the body's existing memory cells from recognizing a virus that they have already encountered. Because the proteins on the surface of a flu virus change so quickly, a vaccine that was prepared to protect you from last year's flu virus will not likely work against this year's flu virus. Unfortunately, scientists have not been able to make vaccines for every pathogen—some pathogens are easier to work with than others.

Cell-Mediated Immunity T cells also respond to infection by undergoing rapid cell division to produce memory cells and then becoming specialized cells (FIGURE 12.20). However, unlike B cells, T cells do not secrete antibodies. Instead, they directly attack other cells. Two of these attacking cell types are the cytotoxic T cells and helper T cells. **Cytotoxic T cells** attack and kill body cells that have become infected with a virus. When a virus infects a body cell, viral proteins are placed on the surface of the host cell. Cytotoxic T cells recognize these proteins as foreign, bind to them, and destroy the entire cell. By releasing a chemical that causes the plasma membrane of the target cell to leak, they break down the cell before the virus has had time to replicate.

Helper T cells, also called T4 cells, can be thought of as boosters of the immune response. These cells detect invaders and alert both the B and T cells that infection is occurring. Without helper T cells, there can be almost no immune response. Helper

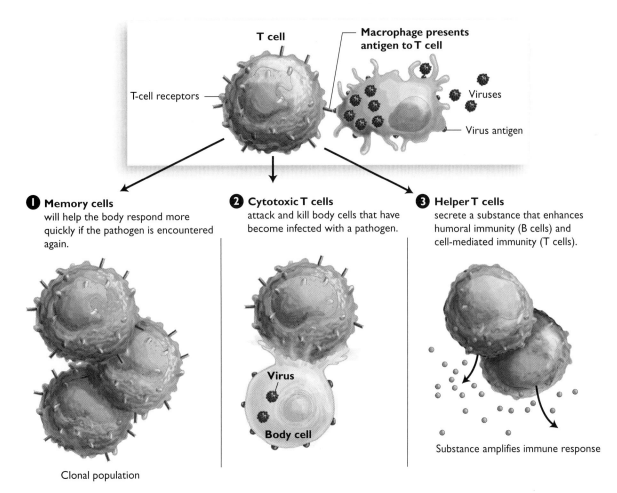

FIGURE 12.20 **Cell-mediated immunity.** T lymphocytes divide to produce different populations of cells: (1) Memory cells carry the specific antigen receptor. (2) Cytotoxic T cells attack and dismantle infected body cells. (3) Helper T cells boost the immune response.

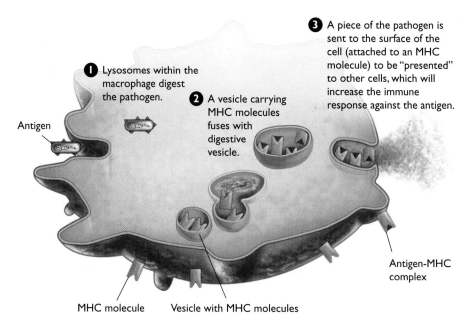

FIGURE 12.21 **Antigen-presenting cell.** An antigen-presenting cell facilitates an increased immune response against a pathogen.

T cells also secrete a substance that greatly increases the level of cytotoxic T-cell response. The AIDS virus (HIV) infects and destroys helper T cells, thus crippling the body's ability to respond to any infection (see Chapter 13).

Macrophages also play a role in cell-mediated immunity by transferring an antigen to their own plasma membranes, thus alerting T cells to a foreign antigen present in the body. A macrophage cell that engulfs foreign particles, partially digests them, and then displays fragments of antigens on its surface is called an **antigen-presenting cell (APC)** (FIGURE 12.21). When the macrophage presents an antigen to a T cell that has the correct receptor, the T cell replicates itself to produce more memory cells, more cytotoxic T cells, and more helper T cells (see Figure 12.20).

As you can see, the various cells (and proteins) of the immune system (Table 12.4) work in concert to strengthen the overall response of the immune system. However, even with all the complexity of the immune system, some substances, such as prions, are able to invade the body without eliciting an immune response.

There Is No Immune Response to Prions

In the case of spongiform encephalopathies, the immune system does not mount a response to misfolded prions at all because they are refolded versions of the normal prion protein. It seems that the refolded version of the protein resembles the normal version enough that the immune system recognizes them both as self. This ability to evade the immune system is quite unusual.

If, after being ingested by humans, the misfolded bovine prions move to the brain and spinal cord, they may cause normal proteins to refold. The destruction of normal proteins damages the brain and eventually causes the patient to die.

There is no effective treatment for prion diseases. Since prions are proteins, not bacteria, viruses, or eukaryotic pathogens, conventional treatments for infectious diseases will not prevent or cure the spongiform encephalopathies. An anti-prion agent would have to be able to determine which prions were properly folded and which were not. Because treating spongiform encephalopathies is not yet possible, it becomes all the more important to prevent their spread.

TABLE 12.4 | Cells of the Immune System

Cell	Function
Antigen-presenting cell	Antigen-presenting cells (APCs) present antigens on their surface so that T cells can recognize them. Macrophages are one type of antigen-presenting cell.
B cells	B cells are types of white blood cells (lymphocytes). Some B cells mature into plasma cells that produce antibodies, while other B cells develop into memory cells. All of the plasma cells that descend from a particular B cell produce the same antibody against the antigen that stimulated it.
Basophil	Basophils are white blood cells that contain and release histamine.
Eosinophil	Eosinophils are white blood cells that increase in abundance due to allergic reactions and parasitic infections.
Macrophage	Macrophages help destroy bacterial, viral, and protozoal invaders. They also release substances that stimulate other immune cells and present antigens to T cells.
Mast cell	Mast cells are connective tissue cells that release chemicals including histamine that cause itching and swelling and may be involved in producing some of the symptoms of asthma.
Natural killer cell	Natural killer cells destroy invading cells even though they have had no previous exposure to the cell.
Neutrophil	Neutrophils are white blood cells containing enzymes that help digest invading microorganisms.
Plasma cell	Plasma cells are white blood cells from B cell lineages that produce and secrete antibodies.
T cell	T cells are any of several different lymphocytes that develop in the thymus, circulate in the blood and lymph, and orchestrate the immune response to infected or cancerous cells. Helper T cells help the immune system by recognizing antigens on the surface of other cells and secreting substances that activate T and B cells. Cytotoxic T cells can find virus-infected cells and induce them to secrete proteins that attract macrophages, which then destroy the infected cells.

12.4 Preventing an Epidemic of Prion Diseases

An **epidemic** is a contagious disease that spreads swiftly and widely among members of a population. An **epidemiologist** is a scientist who attempts to determine who is prone to a particular disease, where risk of the disease is highest, and when the disease is most likely to happen. Epidemiologists try to answer these questions by determining what the victims of the disease have in common. By identifying what factors increase the risk of a disease, epidemiologists help formulate public health policy.

When it is difficult to pinpoint what is causing a disease or how it is spread, deadly epidemics can result. For an epidemic to occur, the infectious agent must cause disease and must be transmissible from one organism to another. Because we know mad cow disease is transmissible, many measures have been enacted by government agencies to try to prevent their spread.

It is now illegal to feed cows, sheep, goats, and deer the meat-and-bone meal produced during the rendering of other mammals. Because certain parts of cattle are at higher risk for harboring prions, materials obtained from cow skulls, brains, eyes, vertebral columns, and spinal cords are prohibited from use in human foods, as is the use of these high-risk materials in cosmetics. These rules become an economic issue for farmers because safe feeding practices are expensive, and the market for these inedible parts of the animal has been restricted.

Despite preventive measures, new cases of mad cow disease still occur in the United States and Canada. Whether cases will continue to appear or even increase in number is difficult for epidemiologists to predict. Whether mad cow disease, or its human counterpart nvCJD, will reach epidemic levels will be decided, in large part, by the level of stringency practiced by the makers of animal feed, by the farmers responsible for feeding cows and observing their behavior, and by the governmental regulatory agencies charged with protecting human health.

Chapter REVIEW

ROOTS TO REMEMBER

The following roots of words come mainly from Latin and Greek and will help you to decipher terms:

auto- means self.

enceph- and **encephalo-** are Greek for the brain (the Latin is **cerebrum**).

epi- means upon, and **-demic** comes for the word for the people (*demos*, as in democratic).

humoral comes from humor, which originally meant a body fluid.

-karyo- is from a Greek word for kernel (and the Latin **nucleus** means kernel, too).

mast (as in mast cell) comes from a word for food, not from the mast on a ship.

-path- or **patho-** relates to disease.

pseudo- means false.

-podia or **-pod** (Greek) and **-ped** (Latin) are from words for the foot.

rheum- or **rheumat-** usually refer to connective tissue.

KEY TERMS

12.1 Infectious Agents
bacterium *p. 274*
binary fission *p. 275*
capsid *p. 278*
capsule *p. 275*
cell wall *p. 275*
contagious *p. 274*
flagellum *p. 275*
infectious *p. 274*
latent virus *p. 279*
microbe *p. 274*
nucleoid region *p. 274*
parasite *p. 274*
pathogen *p. 274*
pilus *p. 275*
plasmid *p. 274*
prion *p. 280*
prokaryote *p. 274*
viral envelope *p. 278*
virus *p. 277*

12.2 Transmission of Infectious Agents
vector *p. 283*

12.3 The Body's Response to Infection: The Immune System
active immunity *p. 289*
antibody *p. 288*
antigen-presenting cell (APC) *p. 294*
antigen receptor *p. 288*
antigen *p. 287*
autoimmune disease *p. 291*
B lymphocyte (B cell) *p. 288*
basophil *p. 286*
cell-mediated immunity *p. 292*
clonal population *p. 292*
complement protein *p. 286*
cytotoxic T cell *p. 293*
eosinophil *p. 285*
fever *p. 287*
helper T cell *p. 293*
histamine *p. 286*
humoral immunity *p. 292*
immune response *p. 289*
inflammatory response *p. 286*
insulin-dependent diabetes *p. 291*
interferon *p. 286*
lupus *p. 291*
lymphocyte *p. 287*
macrophage *p. 285*
major histocompatibility complex (MHC) protein *p. 288*
mast cell *p. 286*
memory cell *p. 292*
multiple sclerosis *p. 291*
natural killer cell *p. 286*
neutrophil *p. 285*
nonspecific defense *p. 284*
passive immunity *p. 289*
phagocyte *p. 285*
plasma cell *p. 292*
pseudopodia *p. 285*
rheumatoid arthritis *p. 291*
specific defense *p. 287*
specificity *p. 288*
T lymphocyte (T cell) *p. 288*
vaccination *p. 292*

12.4 Preventing an Epidemic of Prion Disease
epidemic *p. 295*
epidemiologist *p. 295*

SUMMARY

12.1 Infectious Agents

- Infections result from invasion and multiplication of pathogenic organisms (p. 274).

- Infectious diseases are usually caused by pathogens such as bacteria, viruses, and some eukaryotic organisms (p. 274).

- Bacteria cause disease by using the host's resources to reproduce rapidly and by releasing toxins into the host (pp. 274–277).

- Viruses are composed of nucleic acids encased in a capsid and sometimes an envelope (pp. 277–278).

- Viruses cause disease by using the host cell's resources and by destroying host cells as part of the infectious cycle of the virus (p. 279).

- Viruses replicate their genome inside host cells. The viral genome then directs the synthesis of viral components (p. 279; Figure 12.5).

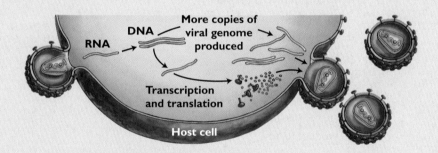

- Eukaryotic pathogens, including protozoans, worms, and fungi, cause many diseases worldwide (pp. 280–281).

- Infectious prions are composed of protein only. They cause disease by refolding the host cell's normally occurring prion proteins (p. 280).

- Misfolded prions can arise spontaneously, as the result of a mutation to the gene that encodes the prion protein. Misfolded prions can also arise as a result of ingesting misfolded proteins from a diseased organism. Once inside the body, the misfolded prions refold normal proteins so that they assume the disease-causing shape (p. 282).

12.2 Transmission of Infectious Agents

The spread of infectious disease results from contact with the disease-causing organism. Contact can occur via direct or indirect exposure to the pathogen, via an intermediate vector, or through ingestion (pp. 283–284; Figure 12.8).

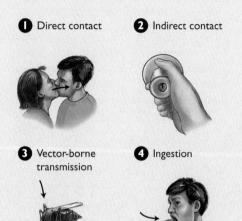

12.3 The Body's Response to Infection: The Immune System

- The skin, mucous membranes, and their secretions are nonspecific defenses that form the first line of defense against infection (p. 284).

- The second line of defense consists of nonspecific internal defenses (p. 285).

- White blood cells such as phagocytic macrophages and neutrophils engulf and digest foreign cells. Eosinophils bombard large invaders with digestive enzymes (p. 285).

- Natural killer cells release chemicals that disintegrate cell membranes of tumor cells and virus-infected cells (p. 286).

- Inflammation attracts phagocytes and promotes tissue healing (p. 286).

- Defensive proteins including interferons, which help protect uninfected cells from becoming infected, and complement proteins, which coat microbes and make them easier for macrophages to ingest, are part of the second-line defenses (pp. 286–287).

- Lymphocytes are the third, specific, line of defense in response to antigens on the surface of pathogens (pp. 287–288).

- Exposure to antigens causes increased production of B and T lymphocytes. B cells secrete antibodies against pathogens, and T cells attack invaders (p. 288; Figure 12.13).

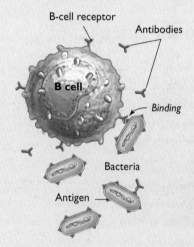

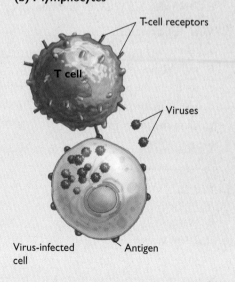

- Antibodies have characteristic Y-shaped structures that vary in one region to allow antibody-antigen binding (p. 289).

- Allergic reactions are reactions to antigenic substances that are not normally pathogens (p. 289).
- Lymphocytes are produced during development and allow the immune system to respond to trillions of different antigens (pp. 289–290).
- Antigen receptors on the surface of immune cells are produced by rearrangements of gene segments (p. 290).
- The antigen receptors of B and T cells are tested for self-reactivity, and those that react against self are eliminated (p. 291).
- The humoral response of the immune system involves B cells, which divide to produce cells that carry and secrete antibodies when exposed to an antigen. Antibody-antigen complexes mark invading cells for attack by phagocytes or by complement proteins, which burst foreign and infected cells (p. 292).
- The cell-mediated response of the immune system involves T cells, which become specialized into different cell types upon exposure to an antigen. These cells speed up the immune response and destroy virus-infected cells (p. 293).
- Prions do not evoke an immune response because the immune system recognizes them as self (p. 294).

12.4 Preventing an Epidemic of Prion Diseases

- Regulating the feeding of animals may decrease the likelihood of a U.S. epidemic (pp. 295–296).

LEARNING THE BASICS

1. Viruses differ from bacteria in that viruses _____.
   a. contain genetic information
   b. cause disease
   c. can be transmitted by a vector
   d. are not cellular
   e. are microbes

2. A virus is composed of _____.
   a. DNA, ribosomes, and a capsid
   b. a DNA or RNA genome and proteins
   c. single cells only
   d. RNA surrounded by a cell wall
   e. DNA, RNA, a cell wall, and ribosomes

3. Viral replication _____.
   a. requires a host cell
   b. utilizes the host cell's ribosomes
   c. results in the production of more viral genomes
   d. results in the production of more viruses
   e. all of the above are true

4. Viruses that are present in cells in an inactive form are called _____.
   a. acute viruses
   b. respiratory viruses
   c. plasmids
   d. latent viruses
   e. immunoviruses

5. Protozoans differ from bacteria in that protozoans _____.
   a. can be infectious
   b. use the host cell's genome
   c. are typically composed of more than one cell
   d. have a nucleus and membrane-bounded organelles

6. Pathogenic worms _____.
   a. transmit viruses to humans
   b. are composed of single cells
   c. are vectors for rabies
   d. produce toxins
   e. can cause disease when they take up residence in human intestines

7. Prions are composed of _____.
   a. nucleic acid only
   b. protein only
   c. cells, proteins, and nucleic acids
   d. nucleic acids and proteins only
   e. cells and nucleic acids only

8. Lyme disease is transmitted by a tick bite. This mode of transmission is _____.
   a. indirect
   b. direct
   c. via a vector
   d. via ingestion

9. Which of the following is not a mechanism of transmission of infectious agents?
   a. contact with infected body fluids
   b. transmission of a defective gene from parent to offspring
   c. exposure to an intermediate host
   d. inhalation of an infectious agent
   e. ingestion of an infectious agent

10. T cells _____.
    a. are lymphocytes
    b. undergo genetic rearrangement to produce an almost infinite variety of receptors
    c. recognize and respond to cancer cells or cells that have been invaded by viruses
    d. differentiate in the thymus
    e. all of the above

11. B cells _____.
    a. develop and mature in the bone
    b. develop and mature in the thymus

12. Autoimmune diseases result when _____.
    a. prions are ingested
    b. liver enzymes malfunction
    c. B cells attack T cells
    d. the immune system fails to distinguish self and nonself cells

13. Which of the following cell types divides to produce cells that make antibodies?
    a. helper T cells
    b. B cells
    c. cytotoxic T cells
    d. all of the above

14. Which of the following is not a lymphocyte?
    a. eosinophil
    b. B cell
    c. basophil
    d. macrophage
    e. interferon

15. Helper T cells secrete substances that _____.
    a. help prevent leukemia
    b. prevent bacteria from entering cells
    c. boost B-cell and cytotoxic T-cell response
    d. inhibit B cells
    e. stimulate the thymus to make more B cells

16. The immune system can recognize a virus you have been exposed to one time because _____.
    a. you harbor the virus for many years
    b. helper T cells kill viruses
    c. a cell that makes receptors to a virus multiplies upon exposure to it and produces memory cells
    d. a copy of the viral genome is inserted into a memory cell

17. Match the bacterial structure to its function.
    _____ nucleoid region
    _____ capsule
    _____ pili
    _____ flagella
    _____ plasmid

    a. helps bacterial cells stick to host tissues
    b. used in the exchange of genetic information
    c. is the location of the bacterial genome
    d. contains extrachromosomal DNA
    e. helps bacterial cells to move

18. Match the immune system component to its function.
    _____ macrophage
    _____ interferons
    _____ natural killer cells
    _____ complement
    _____ antigen

    a. foreign molecules that stimulate immune response
    b. white blood cells that destroy virus-infected cells
    c. white blood cells that move through the lymphatic system cleaning up dead and damaged cells
    d. proteins that coat the surfaces of microbes, making them easier for macrophages to engulf
    e. proteins produced by virus-infected cells that help uninfected cells resist infection

19. **True or False?** Complement proteins bind to the variable region of an antigen to destroy it.

20. **True or False?** The ability of the body's B and T cells to respond to specific antigens begins before human beings are born.

ANALYZING AND APPLYING THE BASICS

Questions 1 and 2 refer to the following information.

Infectious biofilms contain clusters of bacteria that adhere to solid surfaces. Infectious biofilms can develop in the ears and mouth and on medically implanted joints or catheters.

1. How might infectious biofilms lead to difficult-to-treat bacterial ear infections?

2. How might brushing and flossing help to slow the development of dental disease?

Use your understanding of infection, the immune system, and the following information to answer questions 3–5.

The H5N1 strain of avian influenza virus ("bird flu") began spreading through domestic and wild bird populations in Asia in 2003. As of 2007, this strain had moved into the Middle East, Europe, and Africa. Around 240 people have been infected by this virus, and more than 140 of these people have died.

Epidemiologists are concerned that the H5N1 strain could recombine to produce a strain of influenza that would cause a *pandemic*—an epidemic sweeping through human populations around the world. Avian influenza viruses were responsible for strains that caused flu pandemics in 1918, 1957, and 1968. The pandemic in 1918 killed 40–50 million people around the world.

A pandemic-causing virus strain can develop when genetic material from an avian influenza virus mixes with genetic material from a human influenza virus. This mixing can occur within an intermediate host, such as a pig, a species capable of being infected by both avian and human strains of influenza. If the viruses recombine to form a new functional virus strain, that strain can be passed to humans. If the strain can also pass between humans, it may lead to a pandemic.

3. Propose a hypothesis that explains why the H5N1 strain only infected around 240 people, even though it has spread to three continents.

4. Why would the immune system have a more difficult time fighting a virus that had developed partially in bird populations versus one that had developed solely in human populations?

5. Why might it be easier for people to contract an influenza virus from a pig than from a bird?

Questions 6 and 7 refer to the following information.

Wild elk and deer populations in some areas of the United States have been devastated by chronic wasting disease (CWD). CWD is another spongiform encephalopathy. The method of transmission for this disease has been mysterious. Since elk and deer are herbivores, they could not have contracted chronic wasting disease by eating infected tissues from other elk or deer.

Swiss scientists have suggested that CWD may be passed by urine. Their experiments with mice showed that prions from the urine of affected mice were able to cause disease when the urine was injected into healthy mice. They extrapolated that prions from the urine of infected elk or deer could be transmitted when healthy elk or deer eat grass.

6. How could scientists test to see whether or not CWD can be spread among elk or deer by urine?

7. How would the strategies to prevent mad cow disease (BSE) change if BSE could also be transmitted by urine?

8. Malaria is a disease caused by a parasite. After infection, parasites move into the liver where they develop into a mobile form, called the merozoite. Merozoites gain access to liver cells and then kill the cells. Dead liver cells detach and deliver the merozoites to the blood vessels of the liver. Now the parasites can invade red blood cells. Why would it be advantageous for merozoites to use a dead liver cell to move within the body?

9. Antiviral medicines such as oseltamivir (Tamiflu) work by preventing the influenza virus from leaving an infected cell to circulate within the body. How would these antiviral medicines help the immune system to fight influenza?

CONNECTING THE SCIENCE

1. Tuberculosis (TB) is reemerging and becoming a major public health problem across the globe. This reemergence is thought to be due to the development of antibiotic resistance in the bacterium that causes TB and to the prevalance of TB in people with HIV. In fact, people with AIDS are far more likely to develop TB than people who do not carry the virus. Why might this be the case?

2. Some parents refuse to have their children vaccinated. Should these parents be forced to allow their children to be vaccinated to protect society against the reemergence of once-deadly killers such as polio? Defend your answer.

3. In 1796, Edward Jenner intentionally exposed a child to the smallpox virus to test his hypothesis that the child had already acquired immunity via exposure to cowpox virus. Would such treatment be considered ethical today? Why or why not?

Chapter 13

Sexually Transmitted Infections

The Cervical Cancer Vaccine

LEARNING GOALS

1. Define "sexually transmitted infection."
2. List the common sexually transmitted infections, describe their causative agents, and explain how (if at all) they may be cured or prevented.
3. Describe the link between HPV, genital warts, and cervical cancer.
4. Explain how HIV causes the symptoms of AIDS.
5. Describe why HIV eventually wins its battle with the immune system, and how combination drug therapy can prevent this.
6. List the recommendations for the individual behaviors that best prevent HIV and other sexually transmitted infections.

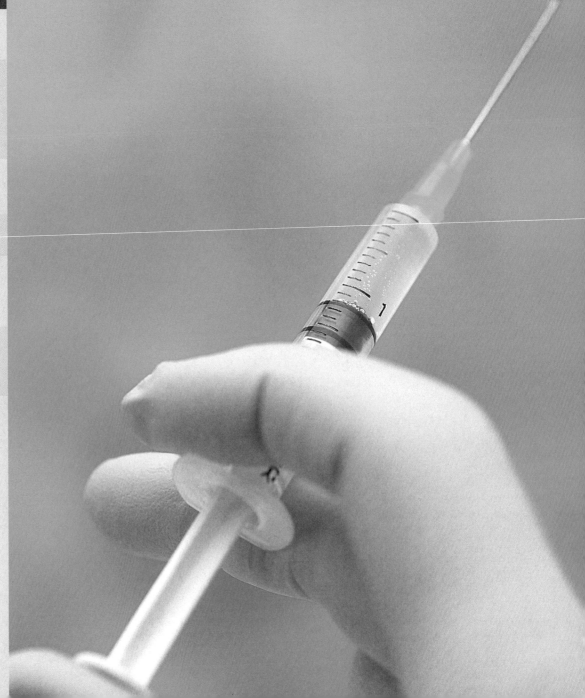

When given to adolescent girls, this simple vaccine . . .

In June 2006, the U.S. Food and Drug Administration (FDA) announced its approval of the vaccine Gardasil, and recommended it for use in girls age 11 to 12. This announcement was met with mixed reaction—hailed by some who see the vaccine as representing a new and promising frontier in medicine and criticized by others who worry about the message that it sends. Gardasil is the first vaccine created expressly to prevent cancer—by preventing infection by a sexually transmitted pathogen.

Vaccination with Gardasil prevents genital warts, the unsightly and sometimes uncomfortable skin growths that can occur on the epithelia of genital organs. The cell changes associated with wart development may also lead to uncontrolled cell division, that is, to cancer. The organ most susceptible to cancer upon infection with genital warts is the cervix, the lower third of the uterus in women.

Ten thousand cases of cervical cancer occur in the United States each year, and infection with genital warts is a contributing factor in nearly all of these cases. Cervical cancer is quite serious. Even with the best medical care, 37% of women with this diagnosis will die from the disease. In poor countries, the toll is even more dramatic—cervical cancer is the leading cause of cancer deaths in the developing world, killing over 300,000 women every year.

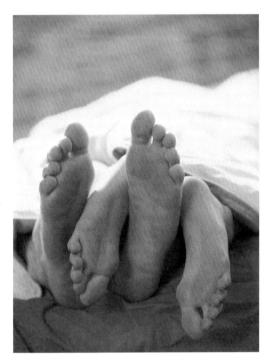

The vaccine also protects against a sexually transmitted infection.

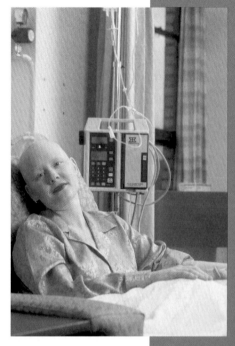

. . . can prevent thousands of cases of cervical cancer.

So why would a vaccine that prevents a deadly cancer be controversial? Because to effectively prevent cervical cancer, girls must receive the vaccine before they become sexually active and thus likely to be exposed to genital warts. According to critics, the vaccination may send young women a message condoning risky sexual activity—a behavior that encourages the spread of other, even more serious sexually transmitted infections. Does the cervical cancer vaccine do more harm than good?

Some people are concerned that 11- and 12-year-old girls are too young to be exposed to this vaccine.

Very high risk
- Vaginal sex without a condom
- Anal sex without a condom

High risk
- Oral sex without a condom or dental dam to protect against exchange of body fluids
- Shared use of sex toys

Moderate risk
- Vaginal sex with a condom
- Anal sex with a condom
- Sexual stimulation using your hand on another person's unprotected genitals

Low risk
- Kissing
- Massage
- Masturbation
- Oral sex with a condom (male) or dental dam (female)
- Shared use of sex toys with condom replaced between partners
- Sexual stimulation using your latex-gloved hand on protected genitals

FIGURE 13.1 Safer sex, but not risk-free. No sexual practice is completely free of the risk of STI transmission, but some practices are less risky than others.

13.1 The Old Epidemics

Genital warts are only one of a number of diseases that are spread by sexual activity. **Sexually transmitted infections (STIs)** are infections that are transferred from one person to another via sexual contact. This contact includes not only vaginal intercourse, but also anal intercourse, genital-genital and oral-genital contact, and the shared use of sex toys such as vibrators (FIGURE 13.1). STIs can lead to illness—symptoms of these infections are known as **sexually transmitted diseases (STDs)**.

Most STDs have been prevalent in the human population for hundreds or even thousands of years. As is often the case with infectious diseases that have a long history in the human population, these older epidemics are often not deadly, although they can cause significant negative consequences and their complications may be fatal. This pattern contrasts with AIDS, the newest epidemic STD. This incurable and nearly always fatal disease is discussed in detail in section 13.2.

The Eukaryotes: Pubic Lice and Trichomoniasis

Most cases of sexually transmitted infection are caused by bacteria and viruses (Table 13.1), but two of the most common diseases are caused by *eukaryotic organisms*: infestation of **pubic lice**, which are insects of the species *Phthirus pubis*, and **trichomoniasis**, caused by a single-celled *protozoan* called *Trichomonas vaginalis* (FIGURE 13.2).

Approximately 3 million new pubic lice infestations are reported in the United States each year. The primary symptom of lice infestation, also called "crabs" because the small insects have large front pincers resembling crab claws, is severe itching in the pubic region, where the lice bite the skin to take a meal of blood. The blood provides energy for reproduction, and female lice will lay eggs, also called nits, at the base of pubic hairs once they accumulate sufficient energy reserves. Treatment for lice infestation includes washing the affected area with soap containing the insecticide permethrin, as well as thoroughly cleaning any clothing or bedding that the infected individual has been in contact with.

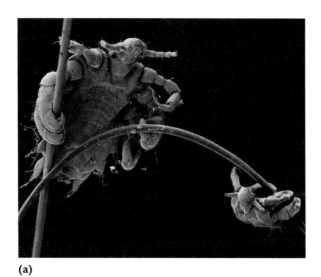
(a)

(b)

FIGURE 13.2 Infectious insects and protozoans. Most sexually transmitted infections are caused by bacteria or viruses. However, common conditions are caused by (a) an insect, *Phthirus pubis*, and (b) a protozoan, *Trichomonas vaginalis*.

TABLE 13.1 | Sexually Transmitted Infections

Disease and Pathogen	Symptoms	Treatment and Prevention	Incidence
Insects, protozoans, and fungi Pubic lice Cause: *Phthirus pubis* (insect)	• Itching of the pubic area caused by an allergic reaction to the bites usually starts about 5 days after initial infection.	• Cured by washing the affected area with a delousing agent.	• Approximately 3 million U.S. cases every year.
Trichomoniasis Cause: *Trichomonas vaginalis* (protozoan)	• In women, vaginal itching with a frothy yellow-green vaginal discharge. • Most men do not have symptoms, but some men experience irritation in urethra after urination or ejaculation.	• Treated with antibiotics. • Condoms decrease the likelihood of transmission.	• An extremely common STD, infecting up to 15% of sexually active women in the United States (over 2 million) per year.
Yeast Cause: *Candida* (fungi)	• Thick whitish discharge from vagina and vaginal itching.	• Antifungal medicines.	• Nearly 75% of all adult U.S. women have had at least one yeast infection.
Bacteria Chlamydia Cause: *Chlamydia trachomatis*	• Pelvic pain, fluid discharge. • If untreated, can lead to pelvic inflammatory disease and infertility. • Many cases are undiagnosed; there may be no symptoms until years after infection.	• Treated with antibiotics. • Condoms prevent transmission.	• Most common bacterial STD. • Approximately 3 million cases each year in the United States.
Gonorrhea Cause: *Neisseria gonorrhoeae*	• Thick discharge from penis or vagina. • 80% of gonorrhea cases are asymptomatic. • Untreated gonorrhea can cause infertility in women if it causes pelvic inflammatory disease.	• Treated with antibiotics. • Condoms prevent transmission.	• An estimated 700,000 cases occur annually in the United States.
Syphilis Cause: *Treponema pallidum*	• Immediate symptoms include a small sore at the site of infection. • Weeks later, a rash, fever, fatigue, and joint soreness. • If untreated, neurological problems, paralysis, and death can occur.	• Antibiotics can cure if caught early.	• 37,000 cases reported annually.
Viruses Herpes simplex Cause: herpes simplex virus type 1 (HSV-1) or type 2 (HSV-2)	• Cold sores or fever blisters on the mouth, face, or genitals. • Can also cause similar symptoms in the genital area, known as genital herpes.	• Antiviral medications lessen the duration and discomfort of herpes outbreaks but do not provide a cure. • Condoms reduce the likelihood of transmission.	• Nearly 20% of the adult population is infected with genital herpes.
Hepatitis B Cause: hepatitis B virus (HBV)	• Inflammation and scarring (cirrhosis) of the liver, which may be fatal.	• Preventable through vaccination. • A few antiviral drugs are effective for treating chronic HBV infection. • Condoms prevent transmission.	• Approximately 60,000 U.S. cases are diagnosed annually. • About 5,000 people die from the infection.

(Continued)

TABLE 13.1 | Sexually Transmitted Infections (Continued)

Disease and Pathogen	Symptoms	Treatment and Prevention	Incidence
Viruses (*continued*) Genital warts Cause: human papillomavirus (HPV)	• Growths or bumps on the pubic area, penis, vulva, or vagina. • Some types of HPV cause genital warts; other types can cause abnormal cell changes on a woman's cervix, leading to cervical cancer.	• HPV vaccine. • Warts can be removed surgically, burned off with lasers, or frozen off with liquid nitrogen. • Condoms reduce the likelihood of infection.	• Some studies estimate that the majority of the sexually active population has been exposed to at least one or more of the over 70 different types.
AIDS Cause: human immunodeficiency virus (HIV)	• Over time, HIV infection weakens the immune system so much that infections that are typically controlled cause severe damage.	• Combination-drug therapies halt progression of the disease. • Cannot be cured. • Condoms reduce the likelihood of transmission.	• Worldwide, about 33 million people are living with AIDS/HIV.

Stop and Stretch 13.1

If you get a lice infestation once, are you likely to be immune from future infestation? Explain your answer using your understanding of immune system function.

There are roughly 5 million new trichomonas infections each year in the United States. Trichomonas causes symptoms mainly in women, as the organism can grow to a large population in the vagina. The symptoms of trichomoniasis include a frothy yellow-green vaginal discharge with a strong odor, as well as itching and discomfort. Trichomoniasis is treated with an orally administered antibiotic. Although her partner may not display any symptoms, a woman can be reinfected by her partner unless he is treated as well.

While not strictly sexually transmitted, fungi in the genus *Candida*, also known as yeast, can cause infections in a woman's reproductive tract. Yeast are normal inhabitants of the vagina and typically coexist in relatively small numbers with other organisms. Under conditions of stress, other illness, or antibiotic treatment, yeast cells may proliferate, resulting in a thick, whitish discharge from the vagina and intense itching. During an active infection, sexual partners can contract the disease as well. More commonly, a yeast infection is a sign of an underlying STI.

The Bacteria: Chlamydia, Gonorrhea, and Syphilis

The two most common STIs are bacterial, and they are often found together in the same individuals. **Chlamydia** is caused by *Chlamydia trachomatis* (FIGURE 13.3), while **gonorrhea**, colloquially known as "the clap," is caused by *Neisseria gonorrhoeae*. These bacteria infect the surfaces of the urethra and anus in both sexes as well as the cervix, endometrium, and oviducts in the reproductive tracts of women. In rare instances, chlamydia and gonorrhea can infect the mucous membranes of the eye, mouth, and throat, especially in infants born to actively infected mothers.

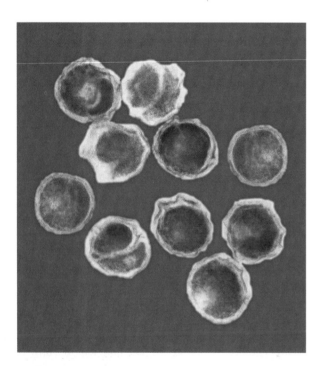

FIGURE 13.3 Bacterial STI. *Chlamydia trachomatis* causes the most common bacterial sexually transmitted infection. Left untreated, *C. trachomatis* will infect the upper reproductive tract in many women, potentially causing sterility.

According to the U.S. Centers for Disease Control and Prevention (CDC), which tracks important communicable diseases, nearly 1 million cases of chlamydia and 330,000 cases of gonorrhea are diagnosed in the United States every year. However, the CDC estimates that many millions more of these infections go unreported. Chlamydia and gonorrhea, along with trichomoniasis, are spread by contact with vaginal discharge or semen; public health authorities thus refer to conditions caused by these types of STIs as discharge diseases.

Infection with either chlamydia or gonorrhea may result in few, if any, outward symptoms. Signs of infection are more common in men and are typically frequent, painful urination and/or discharge from the penis. Women often do not show any signs of infection. The first sign many women have of gonorrhea or chlamydia infection is a positive result from a screening test performed during a routine gynecological exam.

Screening for gonorrhea and chlamydia has become more common in sexually active women under the age of 25. The benefits of screening are enormous. Like most bacterial infections, gonorrhea and chlamydia can usually be cured with inexpensive antibiotics. Left untreated, about 40% of cases result in **pelvic inflammatory disease (PID)**, in which the bacteria infect the internal reproductive organs of women. PID causes debilitating pain and a risk of potentially fatal blood infection. By causing scarring in the reproductive tract, PID also results in infertility in about 20% of cases.

In addition, infection with one or both of these bacteria increases a woman's susceptibility to the virus that causes AIDS—infected women are three to five times more likely than uninfected women to acquire HIV (described below) when exposed to it.

As with most disease-causing bacteria, strains of *C. trachomatis* and *N. gonorrhoeae* that have evolved resistance to multiple antibiotics have begun to appear. Antibiotic-resistant gonorrhea appears to be the largest threat. As many as 13% of cases in Hawaii in 2004 were resistant to most major antibiotics, and in some Asian countries, perhaps 50% of gonorrhea infections are caused by varieties resistant to multiple drugs.

Syphilis is a much less common bacterial STD caused by the organism *Treponema pallidum*. In the United States, syphilis was on the decline during the 1990s, with a record low number of cases (fewer than 6,000) reported in 2000. However, the number of infections has been on the increase since then, especially among homosexual men.

Syphilis has three highly distinct phases. During the primary phase, a painless sore appears on the external genitals or anus (FIGURE 13.4). The sore is teeming with bacteria, and an individual at this stage is highly infectious. Although the sore disappears on its own within three to six weeks, the bacteria may remain in an infected person's body and continue to spread if untreated.

The secondary phase of syphilis may occur months later and includes a characteristic rough skin rash, fever, hair loss, sore throat, or other flulike symptoms. These symptoms also will clear up on their own, but the bacteria may remain in the body. In a small number of infected individuals, *T. pallidum* remains present in the body even after the second stage of the disease. If it does, it continues to do damage.

The third stage of syphilis is characterized by little in the way of overt symptoms. However, the bacteria in this phase may cause damage to the heart, brain, liver, and other internal organs. The final results of syphilis infection may not show up for 5 to 20 years after the initial infection and can include paralysis, blindness, and dementia.

If a woman with untreated syphilis becomes pregnant, the bacteria will infect the fetus, a condition called **congenital syphilis** that results in bone deformations, blindness, or stillbirth. Currently, syphilis is readily treatable and can be cured in the first year of infection with a single injection of penicillin. Syphilis is known or thought to have affected a number of famous individuals, including King Henry VIII of

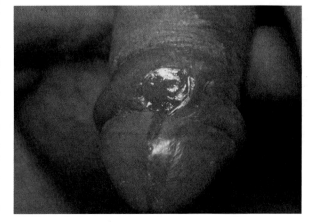

FIGURE 13.4 **Primary phase syphilis.** The first stage of syphilis infection is characterized by a painless sore. Syphilis at this stage is highly infectious.

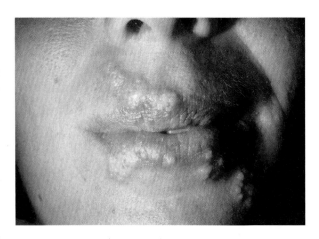

FIGURE 13.5 Herpes. Both strains of the herpes simplex virus can cause cold sores, which appear on the mouth and face.

England, the explorer Meriwether Lewis, the composers Robert Schumann and Franz Schubert, the painters Vincent van Gogh and Henri de Toulouse-Lautrec, and the Soviet leader Vladimir Lenin.

The Viruses: Herpes, Hepatitis, and Genital Warts

Unlike bacterial diseases, viral diseases are rarely completely curable, although some, like warts, may be prevented by vaccination.

Herpes simplex is a common and usually mild disease caused by one of two **herpes simplex viruses (HSV)**, type 1 (HSV-1) or type 2 (HSV-2). HSV-1 most commonly causes oral herpes, resulting in cold sores on the mouth or face (FIGURE 13.5), whereas HSV-2 is more commonly associated with genital herpes, in which blisters may appear on the external genitals. However, both viruses can infect both sites.

After bursting, the blisters leave ulcers on the skin that may last for three weeks. Contact with these blisters helps to spread the virus, although even in the absence of sores, skin-to-skin contact with an infected individual can spread the infection. Herpes and syphilis are known as genital ulcer diseases, because they can be transmitted via contact with the sores (ulcers), even in the absence of vaginal discharge, semen, or blood.

A pregnant woman with an active infection can infect her infant as it passes through the birth canal. HSV infection can be lethal to children or lead to brain damage and blindness. As a result, women with active herpes are typically scheduled for delivery through cesarean section, in which the baby is removed directly from the uterus via an incision in the abdomen, rather than allowed to pass through the birth canal.

Active herpes clears up on its own after several weeks. After the initial infection, however, the herpes virus remains latent in the body. This results in the occasional and seemingly random appearance of blisters throughout an infected individual's lifetime. Medications such as acyclovir and vidarabine disrupt viral reproduction and can lessen the likelihood, duration, and discomfort of herpes outbreaks, but they do not cure individuals of the virus. Nearly 20% of the population in the United States is infected with HSV-1 or HSV-2. There is much interest in developing a herpes vaccine, but none have yet been released.

Hepatitis viruses damage liver cells. With fewer healthy liver cells, the body begins to show symptoms ranging from mild fatigue to more severe symptoms, such as mental confusion. A buildup of bilirubin, the breakdown product of hemoglobin, that results from a failing liver causes the yellowish skin tone of *jaundice*.

Of the six described hepatitis viruses, **hepatitis B virus (HBV)** is most commonly transmitted sexually, although hepatitis C can also be transmitted by the exchange of body fluids. Acute cases of HBV usually run their course in a healthy person's body, but in about 10% of new adult cases, HBV can cause chronic infection. Chronic infection leads to *cirrhosis* (scarring) of the liver and can result in liver cancer, liver failure, and in 15–25% of cases, death. The CDC estimates that about 5,000 people die each year in the United States as a result of chronic HBV. Only a few antiviral drugs are effective for treatment of chronic HBV infection.

During childbirth, hepatitis is readily transmissible from mother to child—a serious event, since the risk of a newborn developing chronic hepatitis from an initial infection is about 90%. Hepatitis B can also be transmitted by blood-to-blood contact; in fact, before 1980 about 20% of cases were in medical personnel who had suffered accidental sticks with contaminated needles. Fortunately, hepatitis B is now preventable through vaccination, and in many states immunization occurs immediately after birth to prevent the risk of chronic disease. As a result of routine vaccination, the number of new HBV cases in the United States has declined from 260,000 in 1980 to about 60,000 today.

The HBV vaccine has several similarities to the cervical cancer vaccine. The main benefit of preventing hepatitis B is the prevention of chronic liver disease, but since one result of chronic liver disease may be cancer, the HBV vaccine is in some sense an anticancer vaccine. And because hepatitis B is sexually transmitted, HBV immunization programs have met with the same sort of resistance.

The Gardasil vaccine is effective against certain types of **human papillomavirus (HPV)**. HPV is a family of viruses that cause warts—different strains cause warts on different areas of the body. You may be most familiar with plantar warts, which form on the soles of the feet, or hand warts that typically form near fingernails. These are caused by HPV-1 and HPV-2, respectively. Gardasil prevents infection by four other strains of HPV, all of which cause genital warts. All strains of HPV are spread by skin-to-skin contact and are thus similar to the genital ulcer diseases.

Genital warts is the most common STD in the United States—the CDC estimates that the number of new HPV infections in the United States exceeds 5 million each year (FIGURE 13.6). While most strains of genital HPV are innocuous, perhaps causing only unsightly growths on the external genitals, a few strains are closely associated with the development of cervical cancer. Inoculation with Gardasil provides immunity against HPV-16 and HPV-18, which are associated with approximately 70% of cervical cancers.

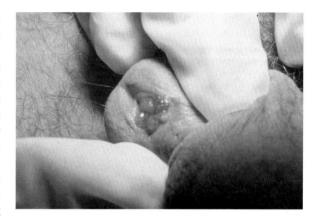

FIGURE 13.6 **Genital warts.** These painless but unsightly genital warts are caused by a variety of the human papillomavirus. Interestingly, these more obvious lesions are not associated with the cancer-causing HPV varieties.

Stop and Stretch 13.2

Which of the STIs described thus far can be prevented by the use of condoms—latex sheaths that cover the penis and prevent the mingling of blood and semen—and which cannot?

Widespread vaccination against HPV would prevent thousands of cases of cancer and hundreds of deaths every year. However, the success of vaccination programs lies not with providing immunity to every susceptible individual, but with reducing the total number of susceptible individuals so that the disease cannot easily spread (FIGURE 13.7).

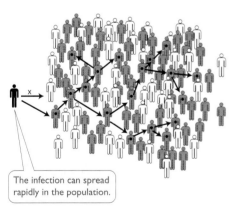

The infection can spread rapidly in the population.

(a) Population where no one has been immunized

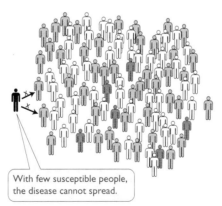

With few susceptible people, the disease cannot spread.

(b) Population where most people have been immunized

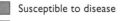

■ Infected with disease ■ Susceptible to disease ☐ Vaccinated
☐ Resistant to disease • Newly infected

FIGURE 13.7 **Disease eradication.** Not all members of a population need to be immunized for a disease to die out. As long as enough individuals are protected, a person with the infection cannot transmit it effectively to someone else, so the disease organism cannot reproduce.

VISUALIZE THIS If large numbers of people begin to refuse vaccination of their children from disease (a phenomenon that is becoming more common, especially in Europe), how does this affect the likelihood that a new epidemic could begin, causing susceptible people to become infected?

Effectively reducing the toll of cervical cancer requires that as many potential carriers of HPV be immunized as possible. In other words, men should receive the vaccine as well to lower the population's infection rate.

HPV-16 and HPV-18 can also cause cancer of the anus or penis in men. However, the low prevalence of these types of cancer (fewer than 3,500 cases per year in the United States) means that many men are unlikely to be motivated to get an anti-HPV vaccine.

To improve its appeal to men, the makers of Gardasil developed it to also serve as a vaccine against the two most common strains of HPV, types 6 and 11. Although these strains are not associated with cervical cancer, they do cause approximately 90% of cases of genital warts. Including these strains in the vaccine will presumably make it more attractive to men interested in reducing the relatively high risk of developing unsightly genital warts.

In effect, the addition of HPV types 6 and 11 transforms Gardasil from a vaccine against cervical cancer to primarily a vaccine against genital warts. If one common and unsightly consequence of incautious sexual activity can be prevented by a vaccine, does this increase the likelihood that people will engage in risky sexual behavior? The answer to this question has serious ramifications. Not only are the old STD epidemics that might spread as a result of increased unsafe sex often painful and debilitating, the new epidemic is a true killer.

13.2 The New Epidemic—AIDS

Acquired immune deficiency syndrome, or **AIDS**, is by far the most troubling of all sexually transmitted diseases. AIDS originated in central Africa, probably in the 1950s, and was only first recognized as a distinct syndrome in 1981. Since that year, more than 1 million people have been diagnosed with AIDS in the United States alone, and more than half of these individuals have died. Worldwide, AIDS has killed 25 million people in the last 25 years, making it one of the most destructive epidemics in recorded history.

A Disease of the Immune System

AIDS was first identified in the United States after hundreds of young gay men in New York City and California were diagnosed with illnesses rarely seen in healthy people. The outbreak of susceptibility to these illnesses appeared to be acquired (that is, caused by exposure to some factor) because it was seen suddenly in large numbers of previously normal-functioning people.

The virus that causes AIDS primarily kills or disables T4 cells (also called helper T cells). Depletion of T4 cells causes immune deficiency—that is, affected individuals experience diseases that are normally controlled by healthy immune systems. These diseases include infections by organisms commonly found on our bodies in low levels, such as *Pneumocystis jirovecii* (formerly known as *Pneumocystis carinii*), a fungus that is found in nearly everyone's lungs by age 30. In healthy people, *P. jirovecii* is held in check by the immune system, but in AIDS patients, this organism causes pneumonia and extensive lung damage. Diseases like *P. jirovecii* pneumonia (PCP) are called **opportunistic infections** because they occur only when the opportunity arises as a result of a weakened immune system.

Because individuals with weakened immune systems may have more than one opportunistic infection, each with its own signs and symptoms, no single condition is always associated with AIDS. Instead, a group of signs and symptoms indicates that an individual has the disease, which is why AIDS is called a syndrome. Primary among those signs is the depletion of T4 cells.

AIDS awareness and activism is a global phenomenon.

The infectious organism that causes AIDS is the **human immunodeficiency virus**, or **HIV**. Worldwide, most HIV transmission occurs via sexual intercourse without a condom. In the United States and Europe, both unprotected sex and the sharing of needles by injection-drug users are primary modes of HIV transmission. The close relationship between HIV infection and AIDS has led many scientists to combine the terms as HIV/AIDS when referring to treatment, research, and prevention.

The reproductive cycle of HIV caused by one HIV particle (a viral particle is an individual virus) is illustrated in FIGURE 13.8. HIV initially binds to the protein receptor CD4 and a co-receptor on the T4 cell membrane and then releases its own RNA and proteins into the cell. Once inside the cell, the viral RNA is converted into viral DNA by the action of one of the viral proteins, the enzyme called reverse transcriptase. Transcription is the process in all cells that rewrites the information in DNA into the language of RNA in preparation for protein synthesis. **Reverse transcription** is simply the converse of that process, in which a strand of RNA is rewritten in the language of DNA. HIV is called a **retrovirus** because of this phenomenon, which seems to run the normal cellular process backwards.

With the help of another viral enzyme, the viral DNA produced by reverse transcription inserts itself into the cell's genome. In many cells, the viral DNA immediately commandeers the proteins and cell organelles required for copying genetic material and producing proteins. These cells now make new copies of the viral RNA, translate its genes into virus proteins, and assemble new viruses. Then the newly made copies of the virus are released from the cell by budding off the cell membrane and go on to infect other cells that possess the CD4 receptor and an appropriate co-receptor.

Infection with HIV usually either disables or kills the host cell. In some cells, however, the HIV DNA is inserted into the genome, but remains dormant, and the cell appears to behave normally. Most of the cells infected with HIV are T4 cells, but other cells that carry the receptor are susceptible to HIV as well.

The Course of HIV Infection

Early symptoms of HIV infection, if they are noticeable at all, resemble the flu. This generalized feeling of fatigue and illness is caused by the initial nonspecific immune response to any viral invader. The immune systems of most people infected with HIV begin to control the virus within 6 to 12 weeks, so they recover from these flulike symptoms.

This seeming recovery from the infection occurs once the immune system develops a specific response to antigens on HIV. Within three months of initial contact with the virus, 95% of infected individuals have high levels of HIV antibodies circulating in their blood. The presence of HIV antibodies designates an individual as **HIV positive**. As a result of high levels of antibodies, the number of HIV particles in the infected person's bloodstream is greatly reduced. However, dormant cells containing HIV genes can remain hidden in organs as diverse as the lymph nodes and the brain.

Once a specific immune response to HIV is fully developed and the number of viruses in the blood has dropped, the levels of T4 cells in an infected individual rebound. At this point, the person is outwardly healthy, or **asymptomatic**, and has a mostly normal immune response. The asymptomatic phase of HIV infection may last for 10 years or more. During this time, the virus and the immune system become locked in an "arms race" that the virus will almost inevitably win.

In nearly all HIV-infected individuals who are not receiving drug treatment, the immune system eventually loses control over the virus. Although researchers are still not certain what triggers the immune system's eventual collapse, one hypothesis is that evolution of the virus within an infected individual eventually allows it to defeat the immune system's defenses. Regularly during an infection, mutation creates new HIV variants with modified antigens that escape immune system detection. Initially, these new variants become more common in the bloodstream. The immune system develops an antibody to each variant's unique antigens, and HIV again begins to be cleared from

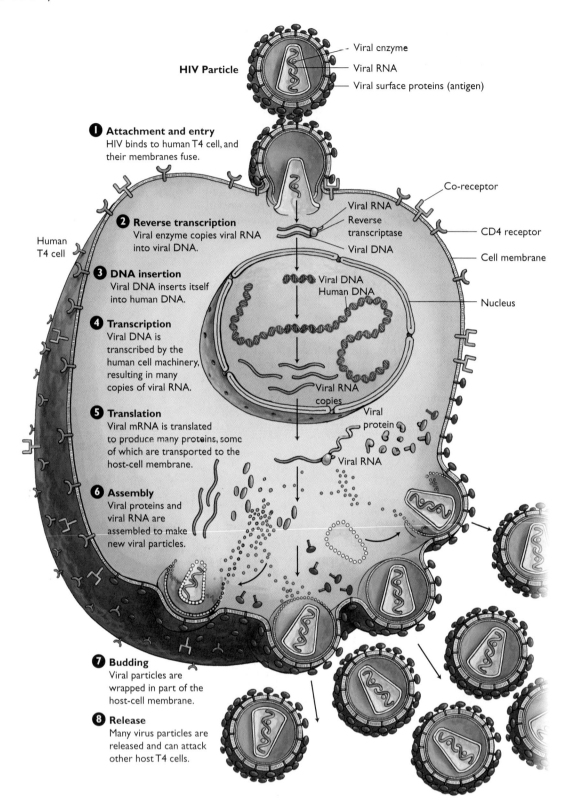

FIGURE 13.8 **The reproductive cycle of HIV.** HIV cannot replicate without infecting a host cell. Once inside, the virus uses the cell's components to make copies. Infection with HIV disables and eventually kills the host cell.

VISUALIZE THIS Imagine that you were designing drugs to attack HIV survival and reproduction. What steps in this process would you concentrate on if you wanted to minimize the number of side effects on human patients?

the bloodstream—that is, until the next new HIV antigen variant arises through mutation. In other words, the population of HIV inside the host is continually evolving and the immune system must continually respond (FIGURE 13.9).

According to the evolutionary hypothesis, the immune system is able to produce antibodies to many different HIV antigen variants, but eventually the sheer number of different HIV variants that it must respond to becomes overwhelming. Finally, one variant arises that escapes immune system control for a long period. Because it is not targeted by antibodies and cleared from the body quickly, it replicates without interference. As a result, large numbers of T4 cells become infected with this variant and are killed or disabled, and the infected individual becomes increasingly immune-deficient. This change initiates the onset of AIDS (FIGURE 13.10).

Treating HIV Infection

Immediately after scientists identified and characterized HIV as the virus that causes AIDS, a search began for drugs that would interfere with HIV's ability to replicate. Early drug therapies had rapid failure rates because HIV quickly became resistant to these drugs.

The current standard of care for HIV infection is the use of **combination drug therapy**, also known as **HAART** or *highly active antiretroviral therapy*, using three or more different drugs to attack the virus. Table 13.2 provides a description of the different types of anti-HIV drugs available. Combination drug therapy has dramatically decreased the number of AIDS cases and deaths due to AIDS in the United States. As an illustration, the average life expectancy of an HIV-positive individual from the time of diagnosis was less than 7 years in 1993; in 2006 it was 24 years.

Stop and Stretch 13.3

Combination drug therapy is more effective than single drug therapy. Even though the likelihood of a virus variant arising that is resistant to a single drug is relatively small, it is still very possible in a patient with 1 billion different HIV variants. However, the likelihood of a virus variant arising with resistance to three drugs in combination is extremely small.

Complete the following analogy relating the evolution of drug resistance in HIV to a lottery. The chance that an HIV particle exists that is resistant to a single drug is analogous to the likelihood that in 1 billion lottery ticket holders, one person will hold the winning combination. The likelihood of a variant being resistant to three different drugs is analogous to a single person winning that lottery _____.

Despite the good news about increased health and prolonged lives of HIV-infected people undergoing combination drug therapy, this approach is far from a solution to the epidemic. Combination drug therapy is expensive, costing over $25,000 per year in 2006; it often results in severe and unpleasant side effects; and it is difficult to follow. Patients may have to take dozens of pills each day, some of which have different requirements. For instance, some pills must be taken on an empty stomach, while others have to be taken with food or significant amounts of water. All of this effort is taken to control an infection that initially may not seem any more severe than a mild flu. Patients who do not follow the drug regimen carefully risk the development of drug-resistant varieties of HIV.

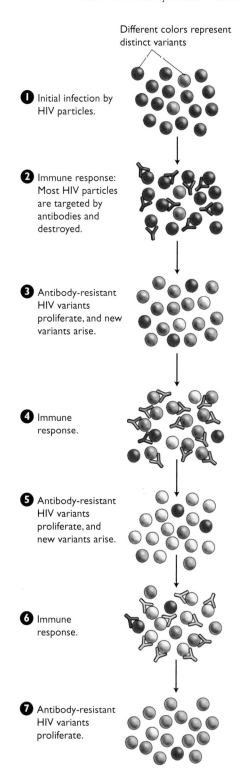

FIGURE 13.9 **The evolution of HIV.** HIV populations evolve in response to changes in the immune system. When the immune system develops a specific response to a strain of HIV, mutants that escape this response proliferate until the immune system develops a response to the new mutant strain.

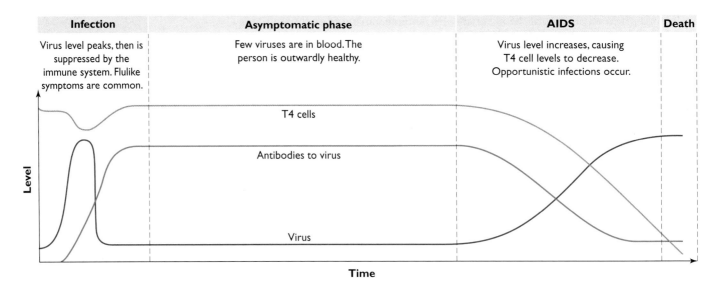

FIGURE 13.10 The typical course of HIV infection. This graph illustrates the change in HIV levels in the blood, the level of anti-HIV antibodies present, and the level of T4 cells over time. After initial infection, most patients produce enough antibodies to control virus levels for months or years. Eventually, however, nearly all HIV-infected individuals develop AIDS.

VISUALIZE THIS How does the long asymptomatic period contribute to the spread of HIV?

Worse yet, there is some evidence that combination drug therapy has made HIV/AIDS appear to be less of a threat, leading to decreased prevention efforts and an upswing in infection rates. The rate of transmission of HIV has not significantly changed in the United States for a decade. About 40,000 new infections are reported every year, and the incidence of unprotected sex is increasing in vulnerable groups—young women and gay men.

Combination drug therapy does not cure HIV infection. Cells containing HIV genes can remain in the body for decades, acting as "Trojan horses" that can release new HIV infectious particles at any time. At best, combination drug therapy is an expensive and long-term commitment that increases an individual's ability to live with this dis-

TABLE 13.2 | Anti-HIV drugs

Class of Drugs	Function	Examples*
Nucleoside analogs	Interfere with reverse transcription by inserting into DNA strand in place of normal nucleotide; insertion stops transcription	AZT, or zidovudine (Retrovir) Lamivudine + Zidovudine (Combivir) Zalcitabine (HIVID)
Protease inhibitors	Interferes with process that converts inactive viral proteins to active enzymes	Nelfinavir (Viracept) Amprenavir (Agenerase)
Non-nucleoside reverse transcriptase inhibitors	Bind to reverse transcriptase, preventing transcription to DNA	Nevirapine (Viramune) Delavirdine (Rescriptor)
Entry inhibitors	Block HIV receptors on cell membranes	Enfuvirtide (Fuzeon)

*Drug names in parentheses are brand names; other drug names are generic.

ease. Increases in transmission, especially of varieties resistant to multiple drugs, may erode the benefits of this powerful therapy over time.

Preventing HIV/AIDS

The Genes & Homeostasis feature on the next page describes a genetic mutation that provides protection from HIV infection, but most of us do not carry that mutation and the human population is unlikely to evolve to become more resistant in the near future. In the absence of natural immunity and an effective vaccine, the best protection from AIDS is avoidance of HIV infection.

HIV is a discharge disease that is transmitted only through direct contact with body fluids—primarily blood, semen, or vaginal fluid, or occasionally breast milk in the case of infants. HIV is frequently spread through needle sharing among injection-drug users, but the main mode of HIV transmission is via unprotected sex with an infected partner. Anal and vaginal sex are the highest-risk modes of transmission, but even unprotected oral sex carries a small risk of infection. Deep kissing carries a very low risk, although a kiss that allowed the commingling of blood (from damaged gums or a cut in the mouth) has been identified as a mode of transmission in a tiny number of cases.

There is no evidence that the virus is spread by tears, sweat, coughing, or sneezing. The virus itself is very fragile and does not survive outside the body. Thus it cannot be spread by contact with an infected person's clothes, a phone, or a toilet seat. HIV is also not known to be transmitted by an insect bite (FIGURE 13.11).

Substantial evidence exists that infection with other STIs increases the likelihood of transmission and acquisition of HIV. Active STDs can increase susceptibility if they result in genital ulcers or sores that allow blood-to-blood transmission. Lesions on the genitals also attract T4 cells, which are the targets of HIV infection, to the site of exposure to the virus. Certain STI infections even increase HIV levels in body fluid in already infected individuals—for example, the concentration of HIV in semen is ten times higher in men infected with both HIV and gonorrhea than in men infected only with HIV.

So, what is the best way to avoid HIV infection—or any STI for that matter? Of course, the best protection from sexually transmitted diseases is practicing

Known modes of HIV transmission:

- Sexual contact with infected person
- Sharing needles or syringes (typically for drug use) with an infected person
- From infected mother to child during birth or breast-feeding
- Contact with infected blood (typically in a health care setting)

No infections are known to have been transmitted by:

- Airborne particles
- Water
- Insect bites
- Social (not passionate) kissing
- Contact with saliva, tears, or sweat of infected person

FIGURE 13.11 **HIV transmission.** HIV is only reliably transmitted via the exchange of body fluids.

GENES & HOMEOSTASIS

Resistance to HIV/AIDS

Scientists have known for many years that a small percentage of people fail to develop an HIV infection despite chronic exposure to the virus. Most of these individuals have a mutation in both of their copies of the CCR5 gene. The CCR5 protein is a co-receptor for HIV on the surface of T4 cells. If CCR5 is absent or nonfunctional, HIV cannot bind to immune system cells, so the infection cannot progress.

About 1% of European-descended whites have a pair of mutant CCR5 genes and are thus HIV resistant. The frequency of this mutation raises two important questions about HIV and the human population. Will humans eventually evolve resistance to HIV? And why is the CCR5 mutation so common in Europeans, 20% of whom carry one copy of the mutant gene?

The evolution of HIV resistance would occur via natural selection, in which individuals with the CCR5 mutation are more likely to survive and reproduce than individuals without it. If this occurs over many generations, eventually the mutation should become more common. However, as the use of combination drug therapy continues and improves, survival and even reproduction of people with functional CCR5 genes may not be reduced by HIV infection. Even if combination drug therapy were unavailable, the evolutionary process would take hundreds of generations because most individuals will never contract the virus. As nonresistant individuals continue to survive and reproduce, the functional CCR5 gene should persist.

The CCR5 mutation may be more common in Europeans because it conferred an advantage to human populations in Europe in the past. Scientists at the U.S. National Cancer Institute analyzed the distribution of the mutation and estimated that it first appeared in Europe approximately 700 years ago. This time period is consistent with the epidemic called the "Black Death," which killed 33% of the European population in the mid-fourteenth century. It could be that the CCR5 mutation also confers resistance to the organism that caused this terrible outbreak, so that individuals who carried it survived the Black Death. Their descendants are now resistant to this new epidemic.

---

abstinence, that is, avoiding sexual activity. Naturally, abstinence is not practiced by a majority of individuals. If you are sexually active, the safest relationship is one that is **monogamous**: Neither partner is sexually active outside of the relationship (and both partners are free of HIV or another STI). If there is any question of your partner's sexual activities outside the relationship or uncertainty about his or her STI history, use a *condom*. No matter what the situation, treat any STI as soon as it appears. Doing so protects you from HIV infection and the negative side effects of untreated infection. And of course, since HIV is also commonly transmitted by needles shared during injection-drug use, don't use illicit injection drugs or share needles.

Perhaps the response to HIV can be used as a barometer to understand how young women will respond to the reduced risk of cervical cancer provided by Gardasil. We know that since the advent of AIDS, the percentage of teens who are sexually active has decreased and the number using condoms when they do engage in sex has increased. Surveys by the CDC have indicated that AIDS education contributes to these behaviors. These results suggest that explaining the risks associated with sexual activity may serve to curb that activity among teens.

Because AIDS is a much more serious disease than genital warts, it is unlikely that reducing the risk of warts in an educated population will increase their rate of risky sexual activity. Ultimately, it appears that a young person's response to HPV vaccination depends not only on the diseases it prevents, but also on other information he or she is receiving about safe sex practices.

Chapter REVIEW

ROOTS TO REMEMBER

The following roots of words come mainly from Latin and Greek and will help you to decipher terms:

a- means without or lacking.

hepa- or **hepat-** is the liver.

-iasis indicates a medical condition.

-itis is a suffix indicating an inflammation.

mono- means one.

retro- is for backward or behind.

KEY TERMS

13.1 The Old Epidemics
chlamydia *p. 306*
congenital syphilis *p. 307*
genital warts *p. 309*
gonorrhea *p. 307*
hepatitis *p. 308*
hepatitis B virus (HBV) *p. 308*
herpes simplex *p. 308*
herpes simplex virus (HSV) *p. 308*
human papillomavirus (HPV) *p. 309*
pelvic inflammatory disease (PID) *p. 307*
pubic lice *p. 304*
sexually transmitted disease (STD) *p. 304*
sexually transmitted infection (STI) *p. 304*
syphilis *p. 307*
trichomoniasis *p. 304*

13.2 The New Epidemic—AIDS
acquired immune deficiency syndrome *p. 310*
AIDS *p. 310*
asymptomatic *p. 311*
combination drug therapy *p. 313*
HAART *p. 313*
HIV *p. 311*
HIV positive *p. 311*
human immunodeficiency virus *p. 311*
monogamous *p. 316*
opportunistic infections *p. 310*
retrovirus *p. 311*
reverse transcription *p. 311*

SUMMARY

13.1 The Old Epidemics

- Pubic lice are a species of insect that can live in pubic hair and survive by ingesting blood from their host. They are transmitted by skin-to-skin contact and controlled by pesticides (p. 304; Figure 13.2).

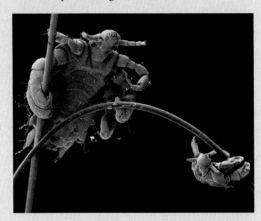

- The parasitic protozoan *Trichomonas vaginalis* is more troublesome to women than to men, causing an uncomfortable and foul vaginal infection (pp. 304–306).

- Yeast is a fungus in the genus *Candida* that can proliferate in the vagina, causing an itchy discharge. Yeast infections often accompany other STIs (p. 306).

- Bacterial STDs include chlamydia, gonorrhea, and syphilis. The first two are much more common, and both can lead to inflammation of the pelvic organs in women and subsequent infertility. All of the bacterial STIs can be cured with antibiotics, although some strains of gonorrhea are becoming resistant to multiple drugs (p. 307).

- Viruses cause herpes, hepatitis B, and genital warts. Herpes is a lifelong disease subject to recurrent outbreaks of painful blisters on the genitals. Infection with certain forms of HPV, the virus that causes genital warts, is associated with an increased risk of cervical cancer. Both hepatitis B and most cases of genital warts can be prevented by vaccination (pp. 308–310; Figure 13.7).

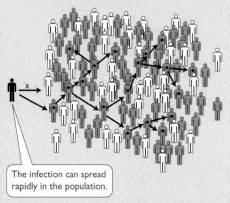

(a) Population where no one has been immunized

- Infected with disease
- Resistant to disease
- Susceptible to disease
- Newly infected

13.2 The New Epidemic—AIDS

- The most dangerous STI is HIV, the virus that causes AIDS. HIV causes immune deficiency by killing helper T cells, resulting in individuals who cannot fight usually innocuous infections (p. 310).

- Although the body initially seems to control HIV infection, the virus continues to survive and evolve in the body. Eventually, in untreated individuals, the immune system loses control and the infection progresses to AIDS (pp. 311–313; Figure 13.9).

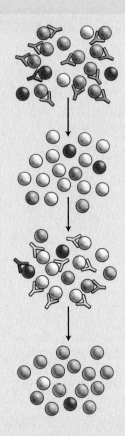

- Progression to AIDS can be halted or delayed by the use of combination drug therapy, in which multiple drugs that prevent viral replication are used at once. Combination drug therapy is not a cure and is expensive, with many side effects (pp. 313–314).

- HIV infection is prevented by practicing abstinence, being engaged in a monogamous relationship with an uninfected partner, and using condoms during intercourse with someone whose HIV status is unknown (pp. 315–316).

LEARNING THE BASICS

For questions 1–4, match the STI listed below with the appropriate statement:

a. chlamydia   b. pubic lice   c. genital warts
d. syphilis   e. herpes

1. _____ May become a lifelong infection
2. _____ A major cause of pelvic inflammatory disease
3. _____ Caused by an insect
4. _____ May be prevented via vaccination

5. Chlamydia and gonorrhea infections may remain untreated in women because _____.
   a. the infections show few outward signs
   b. the infections cannot be prevented by condom use
   c. there is no effective treatment for these infections
   d. monogamous women do not get infected
   e. the infections have few complications, so treatment is unnecessary

6. What are the risks of pelvic inflammatory disease in women?
   a. It is linked with an increased risk of cervical cancer.
   b. It can cause infertility.
   c. It may persist without symptoms for a long time, eventually defeating the immune response.
   d. It cannot be treated with antibiotics.
   e. It is caused by infection with one of the herpes simplex viruses.

7. For which of the following STIs is transmission not reduced by condom use?
   a. chlamydia
   b. AIDS
   c. trichomoniasis
   d. pubic lice
   e. gonorrhea

8. Which of the following STIs can be transmitted from mother to child during birthing?
   a. syphilis
   b. herpes
   c. HIV infection
   d. hepatitis
   e. all of the above

9. Human papillomaviruses are associated with _____.
   a. an increased risk of cervical cancer
   b. higher rates of AIDS infection
   c. genital warts
   d. a and c are correct
   e. a, b, and c are correct

10. HIV causes immune deficiency because _____.
    a. it increases the risk of unsafe sexual activity
    b. it destroys immune system cells
    c. it must escape the immune system to survive
    d. the drugs used to treat HIV kill immune system cells
    e. none of the above are true, because the mechanism of HIV infection is unknown

11. How is reverse transcription different from transcription?
    a. It transcribes RNA into DNA.
    b. It translates proteins into mRNA strands.
    c. It does not require the activity of an enzyme.
    d. It is a process that occurs only in bacteria.
    e. It occurs only on the ribosome of a cell.

12. During the asymptomatic phase of HIV infection, _____.
    a. HIV has been completely cleared from the body by the immune system
    b. HIV is evolving resistance to the immune system response
    c. the immune system is continually responding to new HIV variants
    d. b and c are correct
    e. a, b, and c are correct

13. According to the hypothesis presented in the text, untreated HIV eventually wins its battle over the human immune system because _____.
    a. it remains dormant in skin cells until the immune system is destroyed
    b. infected people are continually reinfected
    c. its high mutation rate allows it to produce more variants than can be controlled
    d. it evolves resistance to all classes of anti-HIV drugs

14. Combination drug therapy prevents HIV infection from progressing to AIDS by _____.
    a. creating an environment in the body that is difficult for the virus to adapt to
    b. killing all HIV variants, regardless of their traits
    c. eliminating all cells that have HIV genes inserted into their genomes
    d. fighting the opportunistic infections caused by immune deficiency
    e. allowing HIV to evolve into a less deadly variety

15. The strategies for prevention of HIV and other STI infection include _____.
    a. abstinence
    b. monogamy
    c. use of a condom
    d. a and c are correct
    e. a, b, and c are correct

ANALYZING AND APPLYING THE BASICS

1. The FDA has licensed only Gardasil for girls and young women between the ages of 9 and 26. Extensive testing on this group has shown the vaccine to be safe and effective. Why did it make sense for the manufacturer to focus on getting Gardasil approved for girls and very young women before testing other groups?

2. In 1932, scientists at the U.S. Public Health Department proposed investigating the course of syphilis in blacks to determine if it was different from that in whites. The scientists identified 399 impoverished black men in Macon County, Alabama, whom they diagnosed with syphilis. The diagnosis was kept from the men, however—they, along with 200 uninfected control subjects, were told they had "bad blood," a local term used to describe a range of conditions. To encourage them to participate, the men were given free medical care, free meals when they visited the Tuskegee Institute for "treatments" such as painful spinal taps, and burial insurance. The study became known as the Tuskegee Syphilis Study.

   What scientific principle were the USPH researchers presumably following when they did not tell the syphilis-infected men about their diagnosis?

3. In the nineteenth century, physician Robert Koch developed a number of postulates to determine the cause of any epidemic disease. If all postulates were true, scientists could be reasonably certain that the suspected infectious agent caused the disease. The postulates are summarized as follows:

   - Association: The suspected infectious agent is found in all individuals suffering from the disease.
   - Isolation: The suspected infectious agent can be grown outside the host in a pure culture (without any other microorganisms).
   - Transmission: Transfer of the suspected agent to an uninfected host produces the disease in the new host.
   - Isolation from new victim: The same agent must be found in the newly infected host.

   Koch's postulates can be used to link the HIV virus to AIDS. Consider the following scenario: Six patients of a Florida dentist contracted the HIV virus after oral surgery. Virus samples from both dentist and patients indicated that the dentist had passed the infection to his patients. The dentist and four of the patients developed AIDS and died.

   Which of Koch's postulates can be used to link HIV to AIDS in the scenario? Explain your answer.

Use the following information to answer questions 4–6.

Charles Nunn and his colleagues at the University of Virginia collected information on the average levels of promiscuity (measured by the number of sexual partners during a fertile period) in various primate species and the concentration of white blood cells in their bloodstreams. The data they collected are summarized in FIGURE 13.12.

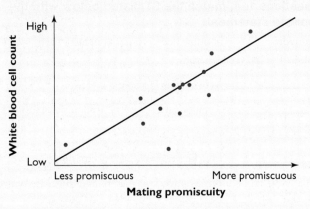

FIGURE 13.12 **Mating promiscuity and white blood cell levels in various primate species**
Source: Adapted from Nunn, C. L., J. L. Gittleman, and J. Antonovics. Promiscuity and the Primate Immune System. *Science* 290(5494): 1168–1170.

Mating promiscuity increases from left to right along the *x*-axis of the graph, from monogamy to extreme polygamy (multiple mates). The relative amount of white blood cells (WBC) increases from bottom to top along the *y*-axis.

4. What type of relationship does this graph illustrate? Describe how white blood cell counts vary according to promiscuity.

5. Why might these two factors be related? Explain your answer in terms of the information from this chapter.

6. Humans have a relatively low WBC count, about a quarter of the way up from the bottom of the graph. If this relationship holds for humans as well, what does that information tell us about promiscuity in human history?

7. In December 2006, the National Institutes of Health announced that it was stopping a study investigating the effect of circumcision on preventing HIV transmission. It had recruited HIV-negative, uncircumcised men in both Uganda and Kenya two years before and randomly assigned them to either remain uncircumcised or to immediately become circumcised, that is, have the foreskin on their penises removed.

   The researchers found that adult male circumcision reduced the risk of acquiring HIV infection by 48% in the study population in Uganda and 53% in the population in Kenya. Given these results, the researchers decided to offer all remaining uncircumcised men in their study the option to become circumcised.

   Why might removal of the foreskin reduce the risk of HIV acquisition? Are there alternative hypotheses that might explain the results the researchers found?

CONNECTING THE SCIENCE

1. Many social conservatives argue that the best method for reducing the risk of pregnancy and STIs among young people is to promote "abstinence-only" sex education. These programs often do not include information about birth control options that may be available to teens. What do you think about the likelihood of success of abstinence-only programs? How could we test whether these programs are effective?

2. The Tuskegee Syphilis Study was described in question 2 of the Analyzing and Applying the Basics section. This study is infamous not only because the men involved were not told of their diagnosis, but also because a cure was withheld from them.

    At the beginning of the study, there was no effective treatment for syphilis, but by 1947, penicillin became the standard cure. This treatment was withheld from the study participants so that the researchers could continue to learn about the progression of the disease. It wasn't until 1972 that the study was exposed and the surviving men given treatment. In the meantime, dozens of men died and many infected their wives, who passed the disease on to their newborns.

    Why do you think treatment was withheld as long as it was? What kinds of safeguards could prevent a Tuskegee Study from happening again?

3. Do you think the availability of Gardasil will increase the rate of risky sexual behaviors? Explain your answer.

Chapter 14

Brain Structure and Function

Attention Deficit Disorder

LEARNING GOALS

1. List the three types of neurons and their functions.
2. Describe the structure of a neuron.
3. Describe the events that occur during an action potential.
4. How is a nerve impulse transmitted from one neuron to the next?
5. Explain what happens during synaptic integration.
6. How would blocking neurotransmitter breakdown in the synapse affect neural function?
7. Identify the major parts of the brain and their functions.
8. Explain the role of the limbic system and identify two of its major structures.
9. List the two types of nerves found in the PNS and explain how they differ from a mixed nerve.
10. Describe the functions of the somatic and autonomic divisions of the peripheral nervous system.

Many schoolchildren take the prescription drug Ritalin.

Lunchtime at a typical U.S. elementary school includes more than crustless peanut butter sandwiches, juice boxes, potato chips, and harried lunchroom aides. For a growing number of children, the lunch period also includes a trip to the nurse's office for a dose of methylphenidate, more commonly known by its trade name, Ritalin. At least 6% of all school-aged boys and 2% of school-aged girls line up to take the drug, which is aimed at treating a syndrome called attention deficit disorder (ADD).

Ritalin is used to control attention deficit disorder (ADD).

In 1980, ADD appeared for the first time in the third edition of psychiatry's most-used reference text, *The Diagnostic and Statistical Manual of Mental Disorders*, or *DSM-III*, published by the American Psychiatric Association. This book guides thousands of mental-health professionals and millions of their patients. The editors of the revised third edition, released in 1987, attempted to increase awareness of the hyperactivity component of ADD by renaming the syndrome as attention-deficit/hyperactivity disorder (ADHD). However, most people still call the syndrome ADD.

Use of this drug has increased over the past few decades.

The ADD diagnosis is typically given to children who consistently display behaviors such as forgetfulness, impulsivity, distractibility, fidgeting, impatience, and restlessness. Virtually all children will display one or a few of these traits. Therefore, the diagnosis is usually given only if a child shows most or all of these behaviors and does so in a way that causes him or her problems in at least two settings, such as at school and at home.

Ritalin belongs to a class of drugs known as stimulants. Stimulants temporarily increase activity in an organism. Giving a stimulant to someone who is having trouble concentrating seems illogical. However, Ritalin has a paradoxical effect in that it actually causes people, especially children, to slow down and pay closer attention. The fact that the drug helps people concentrate has led to its illegal, nonprescription use, sometimes by college students who want to cram for an exam.

Illegal use also fuels some all-night study sessions.

Some people argue that ADD is caused by brain malfunction, so it is best to treat ADD with medicines. Others argue that ADD may be caused, or at least worsened, by our fast-paced society and is therefore better treated by changing the environment of the affected child. To comprehend the debate, an understanding of the biology of the human nervous system is required.

14.1 Nervous System Tissues

Every second, millions of signals make their way to your brain and inform it about what your body is doing and feeling. Your **nervous system** receives and interprets these messages and decides how to respond. For nervous tissue to receive and respond to stimuli, the actions of two different types of cells are necessary. **Neurons** are cells that carry electrical and chemical messages back and forth between your brain and other parts of the body. **Neuroglia** are specialized cells that support and nourish the neuron. Neuroglia do not carry messages. Instead, they supply nutrients to neurons, help to repair the brain after injury, and attack invading bacteria.

Signals are transmitted from one end of a neuron to the other end, between neurons, and from neurons to the cells of muscles and glands. These responsive muscles and glands are called **effectors**. Effectors help the body respond to internal and external environmental changes.

Sensory input, such as changes in the environment, is detected by **sensory receptors**. Receptors are usually neurons or other cells that detect changes in conditions inside or outside the body. The general senses are temperature, pain, touch, pressure, and *proprioception*, or body position. The sensory receptors for the general senses are scattered throughout the body. The special senses are smell, taste, equilibrium, hearing, and vision. The sensory receptors for these five special senses are found in complex sense organs covered in Chapter 15. When receptors are stimulated, signals are generated and carried to the brain or spinal cord.

The nervous system has two main divisions. The **central nervous system (CNS)** is composed of the brain and spinal cord, and the **peripheral nervous system (PNS)** includes the network of nerves that radiates out from the brain and spinal cord extending throughout the body (FIGURE 14.1).

Neuron Structure

A neuron receives signals from sensory receptors or other neurons through its branching **dendrites**. These short extensions radiate from a bulging **cell body**, which houses the nucleus and organelles. Signals received by dendrites are sent to the long, wirelike **axon**, which serves to conduct nerve impulses. The axon terminates in knobby structures called the *axon terminals* or **terminal boutons** (FIGURE 14.2).

Neurons can be grouped into three general categories: (1) **sensory neurons**, which carry sensory input toward the CNS; (2) **motor neurons**, which carry information away from the CNS toward effectors; and (3) **interneurons**, which link sensory and motor neurons within the brain or spinal cord. Most actions within the body require the interaction of all three types of neurons (FIGURE 14.3).

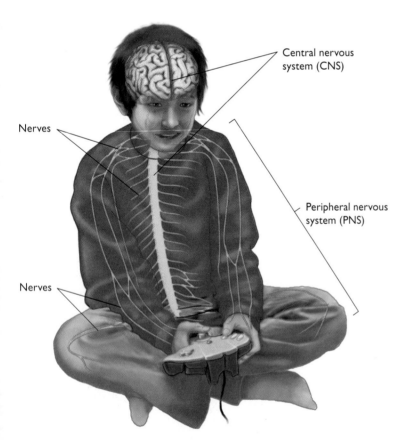

FIGURE 14.1 The central and peripheral nervous systems. The CNS (brain and spinal cord) receives, integrates, processes, and coordinates information from the PNS. The nerves of the peripheral nervous system lie outside the CNS. They communicate sensory and motor information to the rest of the body, and are much more extensive than sketched here.

Myelin Sheath The axons of many neurons are coated with a protective layer called the **myelin sheath** (FIGURE 14.4a). The myelin sheath around axons in the PNS is formed by neuroglial cells called **Schwann cells** (named for the German cell biologist Theodor Schwann) that wrap themselves around the axon. Schwann cells contain the lipid substance myelin in their plasma membrane, which gives them the same glistening white appearance as animal fat. Because of its whiteness, CNS tissue composed of myelinated cells is called **white matter**. The **gray matter** of the CNS contains no myelinated axons. Longer axons tend to have a myelin sheath that acts like insulation on a wire to prevent sideways message transmission, thus increasing the speed at which the electrochemical impulse travels down the axon. Bundles of myelinated axons in the PNS are called **nerves**. When myelinated axons of white matter are bundled together in the CNS, they are called **nerve tracts**.

The myelin sheath is interrupted by spaces between the Schwann cells, leaving tiny, unmyelinated patches called the **nodes of Ranvier**. Nerve impulses "jump" successively from one node of Ranvier to the next (FIGURE 14.4b). This jumping transmission of nerve impulses is up to 100 times faster than signal conduction would be on a completely unmyelinated axon—nerve impulses can travel along myelinated axons at a rate of 100 meters per second (more than 200 miles per hour).

To carry information between parts of the body, sensory neurons bring signals from sensory receptors to interneurons and to motor neurons. Signals are transmitted when neurons pass electrical and chemical signals to each other.

The Creation of Nerve Impulses

Information is typically carried along nerve cells by electrical changes called **nerve impulses**. Neurons transmit impulses from one part of the body to another. Many

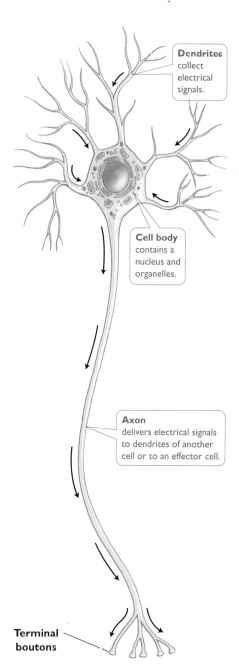

FIGURE 14.2 **The structure of a generalized neuron.** A neuron consists of branching dendrites, a cell body, and an axon with terminal boutons. Nerve impulses are propagated in the direction of the arrows.

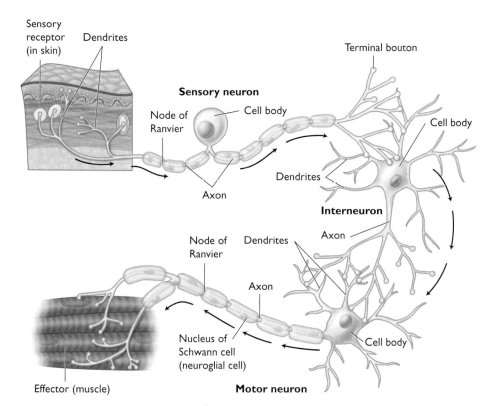

FIGURE 14.3 **Types of neurons.** The nervous system's three main functions are sensory input, integration, and motor output. A nerve signal that begins in a sensory receptor is passed to an interneuron and then to a motor neuron, which stimulates motor output by the effector.

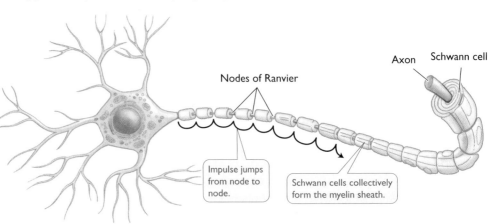

(a) Axons covered in myelin sheaths

(b) Nerve impulses travel more quickly on myelinated than on unmyelinated nerves.

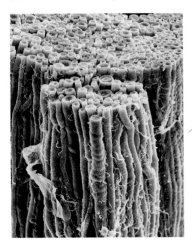

FIGURE 14.4 Myelination. (a) This photo shows a cut bundle of nerve fibers. Each separate fiber consists of a nerve-cell axon (greenish) covered in an insulating layer of myelin (whitish). (b) Nerve impulses travel more rapidly over myelinated neurons because they can move quickly from one node of Ranvier to the next.

types of stimuli, including touch, sound, light, taste, temperature, and smell, have the ability to excite sensory neurons. When sensory neurons are excited, they transmit nerve impulses to your CNS. Your CNS, in turn, sends information through motor nerves, telling your muscles and glands how to respond to the stimulus. To get from point A, the stimulus, to point B, the response, neurons must be able to transmit impulses along their length and from one neuron to the next.

Resting Potential A difference in electric charge or voltage between two regions is called an *electrical potential*.

When an axon is not conducting an impulse, there is a net negative charge inside the cytoplasm of the axon compared to the outside. This is called the **resting potential**. This potential arises from a difference in charge across the plasma membrane of the axon (FIGURE 14.5a) and results from a difference in ionic concentrations. The concentration of sodium ions (Na^+) is greater outside the axon than inside, and the concentration of potassium ions (K^+) is greater inside the axon than outside.

This difference in ions is maintained, in part, by a protein in the membrane called the **sodium-potassium pump**, which actively transports sodium out of the cell and potassium into the cell. The pump transports 3 sodium molecules out of the cell for every 2 potassium molecules it pumps into the cell. Therefore, there are more positive ions outside the membrane than inside. For a nerve impulse to occur, a rapid change in charge difference across the axon's membrane must happen.

Action Potential An **action potential** is a brief reversal of the electrical charge of the axon that transmits a signal along the length of a neuron. The action potential is propagated as a wave of electrical current down the length of the axon. This all-or-none phenomenon takes place only above a certain threshold level. The more intense the stimulus, the more often the axon *fires*, or begins an action potential.

The action potential requires two types of voltage-gated channels in the membrane. One gated channel opens to allow sodium ions to pass, and the other opens to allow potassium ions to pass. Stimulation of a neuron initiates this process by opening some of the gates on these channel proteins in the membrane. Sodium diffuses in first, shifting that particular region inside the cell toward a less negative state. If this local change reaches a critical level, many of the sodium channels open and sodium floods the cell, quickly eliminating the charge difference across the cell, a phenomenon called **depolarization**. This depolarization moves in a wave along the cell, activating sodium channels in adjacent parts of the membrane and thereby moving the impulse along the length of the neuron (FIGURE 14.5b) in much the same manner that toppling one domino causes aligned dominoes to fall.

(a) Resting nerve cell

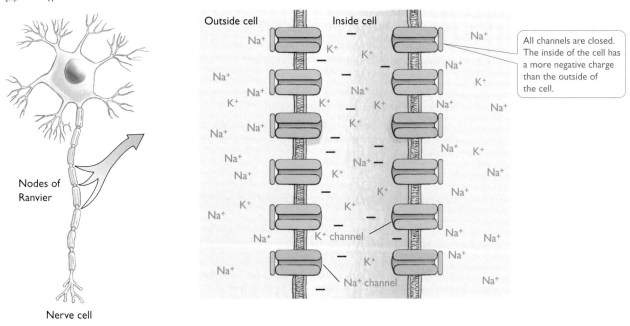

Nodes of Ranvier

Nerve cell

(b) Propagation of an action potential or nerve impulse

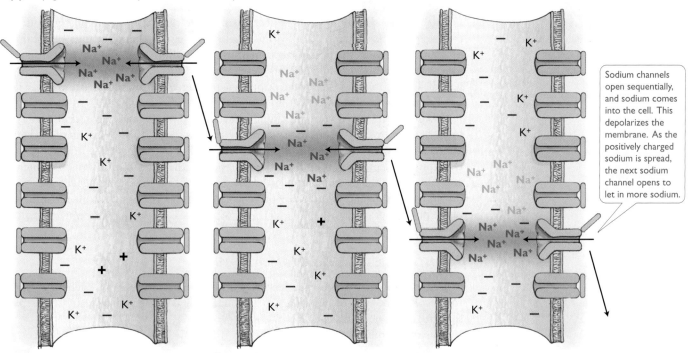

FIGURE 14.5 Generating and propagating a nerve impulse. (a) In a resting nerve cell, the inside of the neuron is more negatively charged than the outside. Closed channels help prevent ions from moving across the membrane down their concentration gradient. (b) A nerve impulse starts with an influx of positively charged sodium ions, which depolarizes a stretch of the nerve cell. A domino effect occurs as the sodium ions spread their charges toward the next voltage-sensitive channel, causing each to open and let in more sodium, so that the depolarization propagates along the length of the neuron.

VISUALIZE THIS If the gene that encodes the sodium-potassium pump is mutated to prevent the pump protein from binding to sodium, how would the ability to generate a nerve impulse be affected?

The entry of sodium into the neuron must be balanced by the exit of another positively charged ion, or the action potential could not decline and return the neuron to its resting state. The action potential declines as sodium channels close and potassium channels open, allowing potassium out of the cell. **Repolarization** of the neuron occurs when potassium ions diffuse out of the cell, and the inside of the cell again becomes more negative than the outside. After an action potential has passed, the sodium-potassium pump restores the resting potential by moving many potassium ions back to the inside and many sodium ions back to the outside.

Once the action potential has moved on, the affected portion of the axon undergoes a **refractory period**, during which time the sodium gates stay closed, preventing another action potential.

Neurotransmitters Carry Signals Between Neurons

Once the signal has traveled along the length of the axon, it must be passed to another neuron or to an effector. Because most neurons are not directly connected to each other, the signal must be transmitted to the next neuron across a region between the two neurons, called the **synapse**. The synapse consists of a terminal bouton of the **presynaptic neuron**, the space between the two adjacent neurons (*the synaptic cleft*), and the plasma membrane of the **postsynaptic neuron**. Saclike structures called vesicles are found in the terminal boutons of the presynaptic neuron.

Each vesicle in a particular neuron is filled with a specific chemical **neurotransmitter**. When an electrical impulse arrives at the terminal bouton of a nerve cell, neurotransmitters are released. Once released by exocytosis, they diffuse across the synapse and bind to specific receptors on the plasma membrane of the postsynaptic neuron (FIGURE 14.6a). Binding of the neurotransmitter to receptors on the postsynaptic cell can excite the postsynaptic cell by once again stimulating a rapid change in the influx of sodium ions into the postsynaptic neuron. The sodium channels open, and sodium ions diffuse inward, causing another depolarization and generating another action potential. In this manner, the nerve impulse is propagated from one neuron to the next until the signal reaches the effector.

Neurotransmitters do not remain in the synapse for long. Some neurotransmitters are broken down by enzymes in the synapse. Others are reabsorbed by the neuron that secreted them via a process called **reuptake** (FIGURE 14.6b). Reuptake occurs when neurotransmitter receptors on the presynaptic cell permit the neurotransmitter to reenter the cell. This allows the neuron that released the neurotransmitter to use it again. Both breakdown by enzymes and reuptake by presynaptic receptors prevent continued stimulation of the postsynaptic cell.

Neurotransmitters and Disease

More than a dozen known chemicals function as neurotransmitters. Each type of neuron secretes one type of neurotransmitter, and each cell responding to it has receptors specific to that neurotransmitter.

Diseases can be caused by defects in neurotransmission. The Genes & Homeostasis feature (see page 330) outlines some genetic changes associated with depression. Another disease affected by changes in neurotransmission is **Alzheimer's disease**, a progressive mental deterioration involving memory loss as well as loss of control of body functions, eventually resulting in death. It is thought that this disease stems from impaired function of the neurotransmitter **acetylcholine**, an important neurotransmitter at nerve-muscle synapses. Drugs that inhibit the enzyme **acetylcholinesterase**, which breaks down acetylcholine, can temporarily improve mental function but cannot stop the progression of this illness.

Parkinson's disease is thought to be caused by the malfunctioning of neurons that produce **dopamine**. Dopamine controls emotions as well as complex movements. The

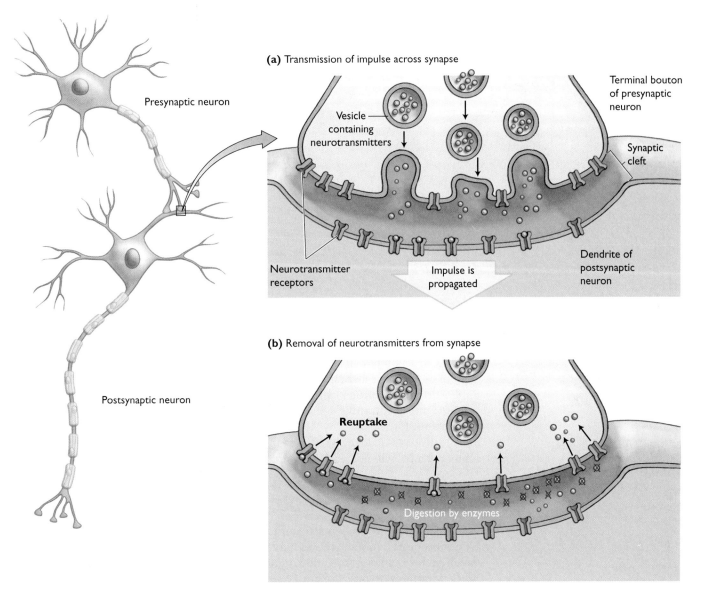

FIGURE 14.6 Transmitting the nerve impulse between neurons. (a) The nerve impulse is transmitted from the terminal bouton of one neuron to the dendrite of the next. Neurotransmitters leave the presynaptic neuron by exocytosis, diffuse across the synapse to the postsynaptic neuron, and bind to receptors on dendrites of the postsynaptic neuron. (b) After the neurotransmitter evokes a response, it is removed from the synapse—either broken apart by enzymes present in the synapse or taken back up by the presynaptic cell.

loss of dopamine causes nerve cells to fire without regulation. As a result, patients have difficulty in controlling their movements and experience tremors, rigidity, and slowed movements. There is no cure for this progressive disease.

Synaptic Integration

Many different axons can synapse with one neuron (FIGURE 14.7a). Therefore, an individual neuron receives many signals, including *excitatory* signals that drive the neuron

GENES & HOMEOSTASIS

A Genetic Link to Depression

Depression is a disease that involves feelings of helplessness and despair. Loss of interest in daily activities, crying spells, and thoughts of suicide may occur as well. Scientists know that depression is associated with disruptions in the levels of several neurotransmitters including serotonin, norepinephrine, and dopamine.

Scientists have isolated genes known to be involved with causing bipolar disorder (also called manic-depressive disorder), which involves recurrent episodes of elation followed by depression. Scientists are trying to find genes associated with the other types of depression as well. It might be the case that mutated genes lead to disruptions in these chemicals and that those with the predisposing genetic imbalance are more likely to become depressed. However, it may also be the case that disruptions in neurotransmitter levels are caused by depression. Not everyone with a family history of depression develops the disease and sometimes people with no family history develop the disease.

Antidepressant medications that block the actions of enzymes that degrade these neurotransmitters or inhibit reuptake help to alleviate many symptoms of this disease in some people. Fluoxetine (Prozac) and sertraline (Zoloft), for instance, fall into the class of specific serotonin reuptake inhibitors (SSRIs) that are prescribed to treat depression. It is also true that for some people, recognizing and changing a situation that is causing them to be depressed can alleviate this problem without medication.

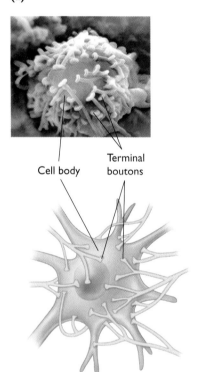

closer to an action potential. Excitatory signals can arrive from many different synapses or from one synapse at a rapid rate. In contrast, *inhibitory* signals sent by neurotransmitters drive the neuron farther from an action potential. An individual neuron must combine these contrasting signals and sum up the overall message via a process called **integration**. If the excitatory signals outweigh the inhibitory signals, the neuron will fire (FIGURE 14.7b). Likewise, if the inhibitory signals swamp the excitatory signals, the action potential will not reach its threshold.

Neurotransmission, ADD, and Ritalin

Abnormal levels of the neurotransmitter dopamine may be involved in producing the symptoms of ADD. Some researchers hypothesize that people with ADD symptoms may have lower than normal dopamine levels. Decreased dopamine levels may cause ADD symptoms because dopamine suppresses the responsiveness of neurons to new inputs or stimuli. Therefore, someone with a low concentration of dopamine may respond impulsively to situations in which pausing to process the input would be more effective.

The cause of decreased dopamine levels in some people with ADD may actually involve an overabundance of dopamine receptors on the presynaptic cell. During reuptake, dopamine receptors on the presynaptic cell remove dopamine from the synapse to prevent continued stimulation of the postsynaptic neuron. Some studies have shown elevated numbers of these receptors in people with ADD. Ritalin, the drug usually prescribed to treat symptoms of ADD, may work to decrease the impact of these extra reuptake receptors.

Ritalin is thought to increase dopamine's ability to stimulate the postsynaptic cell by blocking the actions of the dopamine reuptake receptor (FIGURE 14.8). This blocking leaves more dopamine in the synapse for a longer time, resulting in a decrease in the symptoms of ADD.

FIGURE 14.7 Integration. (a) The photomicrograph and drawing show that many axons synapse with one neuron. (b) Excitatory and inhibitory signals are summed during integration. An action potential is generated when the sum of excitatory and inhibitory signals causes the membrane potential to exceed the threshold needed to open voltage-gated ion channels.

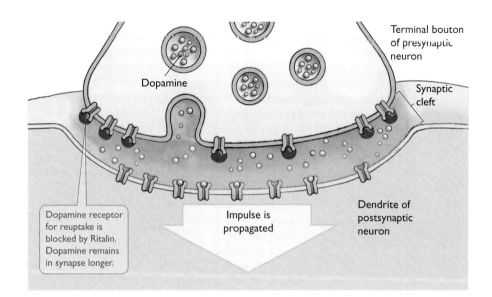

FIGURE 14.8 **The mechanism of Ritalin action.** Ritalin blocks dopamine receptors on the presynaptic neuron. This blocks reuptake, allowing dopamine access to the postsynaptic neuron for a longer time.

VISUALIZE THIS What would be the effect of a drug that broke down dopamine at the synapse?

Other stimulants work through similar mechanisms. In addition to their direct effects on dopamine levels, all stimulants—whether legal like Ritalin and obtained by prescription, or illegal like methamphetamine, cocaine, and amphetamines (collectively called speed)—have similar effects on the body. On their way to the brain, these drugs affect the heart, blood vessels, and lungs by increasing heart rate, elevating blood pressure, and expanding airways in the lungs.

When stimulants are taken in high doses, the user feels euphoric and has more energy and endurance, a sense of power, and a feeling of mental sharpness. As the drug wears off, common effects are heightened fatigue, insomnia, poor concentration, irritability, tearfulness, and depression. Other effects include personality changes, skin rashes, fever, nausea, and headaches. Abusing stimulants can lead to psychotic episodes, delusions, seizures, hallucinations, and sudden death.

Many elementary school students who grow up taking Ritalin to help them with schoolwork are still having their prescriptions filled when they head off to college. Once on campus, these students no longer stand in line for the school nurse to provide the prescribed dose. Instead, they self-administer the stimulant. Some take the prescribed dose, others use the drug only to pull an all-nighter, and still others give pills to their friends.

Due to its availability and status as a prescription drug—not a street drug, such as speed or other amphetamines—students may see Ritalin as a less serious drug than other stimulants. These students are often unaware of the dangers of abusing this drug. Recent studies indicate that about 15% of college students report having used the drug recreationally.

If you have not been prescribed Ritalin but use it to help focus for an exam, there is no evidence that it actually helps you learn. You probably do not have low dopamine levels to begin with. For Ritalin abusers, the euphoria that they experience, coupled with the fact that they are more likely to sit in a chair and focus on their schoolwork, may make them *believe* that Ritalin is helping them learn. As with any drug, the potential for dangerous side effects should be seriously considered before use. Table 14.1 describes many recreational drugs and their effects on the body.

The dopamine reuptake theory is not the only hypothesis about the biological cause of ADD symptoms, nor does it explain all cases of ADD. Not all drugs that alleviate ADD symptoms have an impact on dopamine levels. Many times, drugs must be combined with changes in environmental conditions for the symptoms of ADD to improve. Many scientists who study ADD believe that the brains of those affected differ from the brains of those who are not affected. In the next section, we take a closer look at this remarkable structure.

TABLE 14.1 | Recreational Drugs and the Nervous System

Drug	Mechanisms of Action	Desired Mental Effect	Side Effects
Alcohol A by-product of fermentation. When yeast cells ferment barley, beer is produced. When yeast cells break down the sugars in grapes, wine is produced.	• A depressant that diffuses easily across cell membranes and does not require specific receptors. • Alcohol inhibits neurotransmission in the reticular formation, interfering with the activity of many neurons in the brain.	• Reduced anxiety, a sense of well-being, loss of concern for social constraints.	• Impaired judgment, slurred speech, unsteady gait, slower reaction times, uncontrollable emotions. • Chronic alcohol abuse leads to loss of intellectual ability and liver damage. • Kills nerve cells that cannot be regenerated. The frontal lobes, center for judgment, thought, and reason, are the first to die.
Amphetamines Are used legally to treat obesity, asthma, and narcolepsy. Methamphetamine is an illegal amphetamine. A crystalline form of methamphetamine called *ice* is smoked to produce effects similar to crack cocaine.	• Stimulants that increase activity in the reticular formation. • Mimics the actions of the neurotransmitter norepinephrine, a hormone produced in response to stress. • Blocks reuptake and inhibits the enzyme that breaks down norepinephrine, resulting in prolonged stimulation of the postsynaptic cell.	• Small doses make a person feel more energetic, alert, and confident.	• Effects wear off quickly, causing sudden "crashes" from depleted neurotransmitter stores and leading to depression and fatigue. • Prolonged use results in aggressiveness, delusions, hallucinations, and violent behaviors. • Can cause blood vessels to spasm, clots to form, insufficient blood to enter the heart, fluid to build up in the lungs, and death.
Caffeine A naturally occurring chemical found in plants such as coffee, tea, and cocoa.	• A general stimulant that affects all cells, not just those of the central nervous system. • Gains access to cells, raising the metabolism by increasing glucose production. Higher glucose levels lead the cell to support increased activity.	• Mental alertness, increased energy.	• Insomnia, anxiety, irritability, and increased heart rate.
Cocaine Extracted from the leaves of the coca plant of South America, it can be inhaled or injected. A more potent form, crack, is smoked.	• A stimulant that increases levels of dopamine and norepinephrine by decreasing reuptake.	• A rush of intense pleasure, increased self-confidence, and increased physical vigor.	• Increased heart rate and blood pressure, narrowing of blood vessels, dilation of pupils, a rise in body temperature, and reduction of appetite. • When cocaine wears off, its effects are followed by a period of deep depression, anxiety, and fatigue. • Abuse of this drug can leave a person unable to have positive feelings without the drug.

TABLE 14.1 | Recreational Drugs and the Nervous System (*Continued*)

Drug	Mechanisms of Action	Desired Mental Effect	Side Effects
Ecstasy or MDMA A white crystalline powder that is primarily ingested in pill or capsule form.	• A stimulant and hallucinogenic that acts to prevent serotonin reuptake. Also floods neurons with several other neurotransmitters.	• Euphoria, enhanced emotional and mental clarity, increased energy, heightened sensitivity to touch, and enhanced sexual response.	• Confusion, anxiety, paranoia, depression, and sleeplessness that may last for several weeks. • May permanently damage neurons involved in utilizing serotonin and may also result in permanent memory damage. • When combined with physical exercise, ecstasy use can lead to severe dehydration and the inability to regulate body temperature.
Lysergic acid diethylamide (LSD) and mushrooms A derivative of the fungus *Claviceps purpurea*, which grows on rye, LSD is related to the compound psilocybin, found in certain mushrooms.	• Hallucinogen that binds to serotonin receptors in the brain, increasing the normal response to the neurotransmitter.	• Heightened sensory perception, bizarre changes in thought and emotion, hallucinations.	• Hallucinations can lead users to dangerous actions. • Heavy use leads to permanent brain damage, including losses of memory and attention span, and can lead to psychosis.
Marijuana The leaves, flowers, and stems of the Indian hemp plant *Cannabis sativa*. Also contains a drug called delta-9-tetrahydrocannabinol (THC).	• Receptors for THC are located in the areas of the brain that influence mood, pleasure, memory, pain, and appetite. THC is thought to increase dopamine release.	• Altered sense of time, enhanced feeling of closeness to others, increased sensitivity to stimuli. Large doses can cause hallucinations.	• Marijuana use slows reaction time, reduces coordination, and inhibits one's ability to judge time, speed, and distance. • Impairs short-term memory, slows learning, interferes with one's attention span, and limits the ability to store and acquire information. • With THC receptors on the hypothalamus (the brain area that regulates sex-hormone secretion), long-term use can decrease testosterone production or disrupt menstruation.
Nicotine Found in tobacco plants, nicotine is one of over 1,000 chemicals in cigarette smoke, and one of the most likely to affect the brain.	• A stimulant that triggers neuron receptors in the cerebral cortex to produce acetylcholine, epinephrine, and norepinephrine.	• Increased alertness and awareness, appetite suppression, relaxation.	• Cigarette smoking increases the odds of virtually every type of cancer. Also causes increased heart rate and blood pressure.

(*Continued*)

TABLE 14.1 | Recreational Drugs and the Nervous System (Continued)

Drug	Mechanisms of Action	Desired Mental Effect	Side Effects
Opiates Derived from the opium poppy. Heroin, morphine, and codeine are opiates. Morphine and codeine can be used legally to control pain.	• Bind to opiate receptors in neurons that control feelings of pleasure. • Opiate receptors evolved to bind to opiates produced by the brain in response to exercise; these opiates are called endorphins.	• A quick, intense feeling of pleasure, followed by a sense of well-being and drowsiness.	• Addiction, poor motor coordination, depression. • High doses can cause coma and death. • Thought to change the brain's ability to respond to normal pleasures.

Stop and Stretch 14.1

The decreased dopamine level hypothesis is only one of many current hypotheses about the cause of ADD. Assume it's true that people diagnosed with ADD really do have lowered dopamine levels compared to people who don't have ADD. If you could design a drug to help treat this problem, what neural mechanism (aside from preventing dopamine reuptake as Ritalin does) might you target to increase the effects of the dopamine that is present?

14.2 The Central Nervous System

Neurons in sense organs such as your eyes, ears, or skin transmit information from the world around you to your CNS. There the information is processed and the appropriate response is relayed by other neurons out of the CNS back to your body. In addition to interpreting and acting on information received by the senses, the CNS is the seat of functions such as intelligence, learning, memory, and emotion.

Neurons of the CNS are highly specialized cells that usually do not divide. Therefore, damage to a neuron cannot be repaired by cell division and often results in permanent impairment. Damage to spinal cord neurons, for example, results in lifelong paralysis because messages can no longer be transmitted from the CNS to muscles. Likewise, injury to the brain can result in permanent damage if neurons in the brain are harmed.

The spinal cord and brain are the two components of the CNS. Both of these structures are protected by bone. The spinal cord is protected by vertebrae and the brain by the skull. Also, both of these structures are enclosed in three layers of protective envelopes of connective tissue called **meninges**. In addition, both of these structures sit in a circulating plasma-like fluid, **cerebrospinal fluid**, that fills the cavities of the CNS and protects and cushions the CNS structures.

Spinal Cord

The **spinal cord** serves as a sort of highway that connects the brain to the PNS. The spinal cord extends from the base of the brain to the lowermost rib, between the first and second lumbar vertebrae. It is housed inside the spinal column formed by openings within vertebral bones (FIGURE 14.9a). The spinal cord is composed of nerve fibers that control reflex actions and transmit nerve impulses to and from the brain. Nerve fibers are the axons of neurons. The spinal cord is composed mostly of interneurons.

A cross section of the spinal cord (FIGURE 14.9b) reveals white matter arranged around a butterfly-shaped area of gray matter. White matter is composed of axons that conduct signals, and gray matter is composed primarily of cell bodies, dendrites, and synapses for integration. The outer white matter is divided into dorsal and ventral columns. In the center of the spinal cord is a small central canal, containing cerebrospinal fluid.

Nerve tracts in the white matter carry information between parts of the spinal cord and between spinal cord and brain. Sensory information travels up the ascending tracts, and motor responses are induced by nerve impulses traveling down the descending tracts. Responses can stimulate motor neurons to fire or stimulate glands to secrete their products.

Sensory receptors in the skin send information to the spinal cord through the **spinal nerves**. The cell bodies for these nerve fibers are located in the *dorsal root ganglion*. The dorsal root of each spinal nerve contains sensory fibers that enter the gray matter, and the ventral root contains motor fibers that exit the gray matter. These roots join together to form the spinal nerve that leaves the vertebral column. The nerves themselves are part of the PNS.

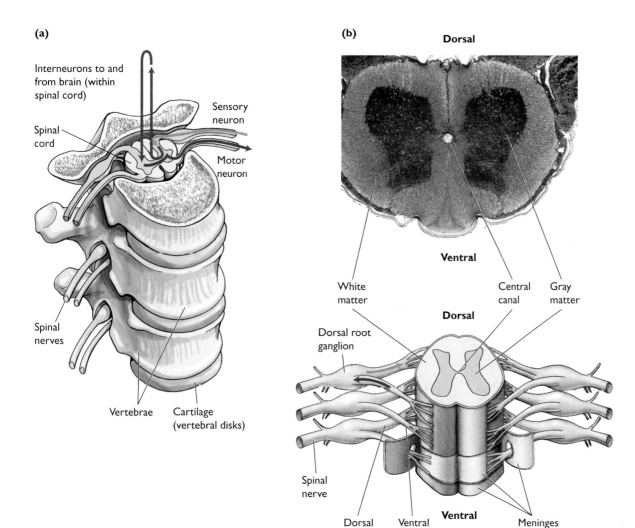

FIGURE 14.9 **The spinal cord and nerves.** (a) Spinal nerves branch out between the vertebrae and go to all parts of the body. (b) The spinal cord in cross section shows white matter surrounding a butterfly-shaped region of gray matter. Spinal nerves contain sensory fibers entering the CNS and motor fibers leaving it.

The Brain

The brain is the region of the body where decisions are reached and where body activities are directed and coordinated. The brain receives information in the form of action potentials from nerves and the spinal cord, integrates these signals, and generates a response. Specialized areas within the brain integrate different types of information or perform specific motor tasks. For example, one area of the brain responds to visual stimuli and another responds to stimuli that help maintain balance.

In addition to housing 100–200 billion neurons, the brain is rich in neuroglial cells. In fact, there are 10 times as many glial cells in the brain as there are neurons. The brain has four fluid-filled **ventricles**, or chambers, which interconnect. Neuroglial cells lining the ventricles produce cerebrospinal fluid stored in the ventricles. Structurally, the brain is subdivided into many important anatomical regions including the cerebrum, thalamus and hypothalamus, cerebellum, and brain stem.

Cerebrum The large **cerebrum** fills the whole upper part of the skull. This part of the brain controls language, memory, sensations, and decision making. It is the last part of the brain to receive sensory input and carry out integration before issuing motor responses. The cerebrum has two hemispheres (FIGURE 14.10a). Each hemisphere is divided into four lobes (FIGURE 14.10b):

1. The **temporal lobe** is involved in processing auditory information and some visual information, as well as memory and emotion.
2. The **occipital lobe** processes visual information from the eyes.
3. The **parietal lobe** processes information about touch and is involved in self-awareness.

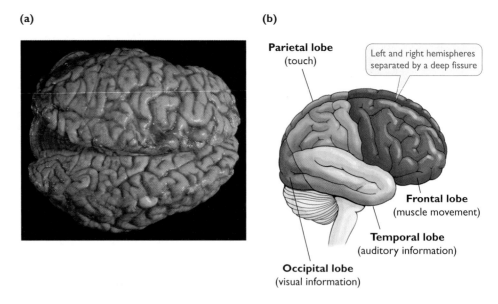
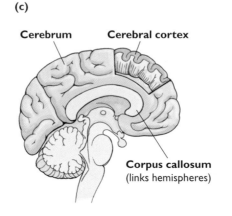
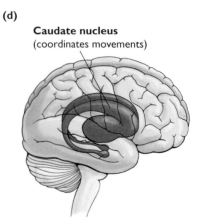

FIGURE 14.10 Structure of the cerebrum. (a) The highly convoluted cerebral cortex is divided down the midline into left and right hemispheres. The cerebrum controls language, memory, sensations, and decision making. (b) Each cerebral hemisphere is divided into four lobes. (c) The hemispheres are connected by the corpus callosum. (d) A caudate nucleus is located within each cerebral hemisphere.

4. The **frontal lobe** processes voluntary muscle movements and is involved in planning and organizing future behavior.

The deeply wrinkled outer surface of the cerebrum is called the **cerebral cortex**. In humans, the cerebral cortex, if it were unfolded, would be the size of a 16-inch pizza. The folding of the cortex increases the surface area and allows this structure to fit inside the skull. The cortex contains areas for understanding and generating speech, areas that receive input from the eyes, and areas that receive other sensory information from the body. It also contains areas that allow planning.

The cerebrum and its cortex are divided into two halves—the right and left cerebral hemispheres—by a deep groove called a **fissure**. At the base of this fissure lies a thick bundle of nerve fibers, the **corpus callosum** (FIGURE 14.10c), which serves as a communication link between the hemispheres. The **caudate nuclei** are paired structures found deep within each cerebral hemisphere (FIGURE 14.10d). These structures are part of the pathway that coordinates movement.

The region of the parietal lobe that receives sensory input from the skin is called the *primary somatosensory* area. The corresponding region of the frontal lobe that initiates motor activity is called the *primary motor area*. Specific regions within each of these areas correspond to individual parts of the body (FIGURE 14.11). Body parts that are very sensitive to touch involve larger portions of the somatosensory area. Likewise, body parts that can perform more complex movements take up more space in the primary motor area.

Thalamus and Hypothalamus   Deep inside the brain, lying between the two cerebral hemispheres, are the **thalamus** and the **hypothalamus** (FIGURE 14.12). The thalamus relays information between the spinal cord and the cerebrum. The thalamus is the first region of the brain to receive messages signaling such sensations as pain, pressure, and temperature. The thalamus suppresses some signals and enhances others, which are then relayed to the cerebrum. The cerebrum processes these messages and sends signals to the spinal cord and to neurons in muscles when action is necessary.

The hypothalamus, located just under the thalamus and about the size of a kidney bean, is the control center for sex drive, pleasure, pain, hunger, thirst, blood pressure, and body temperature. The hypothalamus also releases hormones that regulate the production of sperm and egg cells as well as the menstrual cycle and other physiological processes.

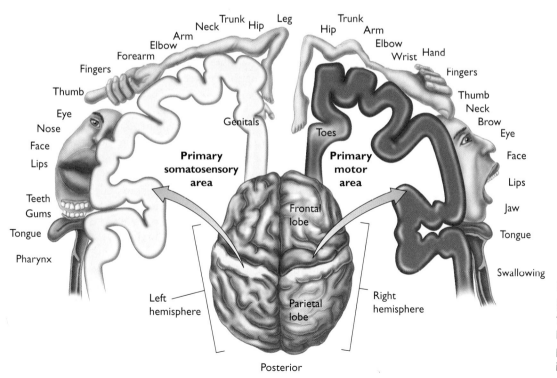

FIGURE 14.11 Cerebral cortex. The primary somatosensory and motor areas of the cerebral cortex. Note that areas of the body that are more sensitive or that perform precise movements take up correspondingly more space on the diagram.

FIGURE 14.12 Anatomy of the brain. This illustration shows the locations of several important structures in a right-facing brain.

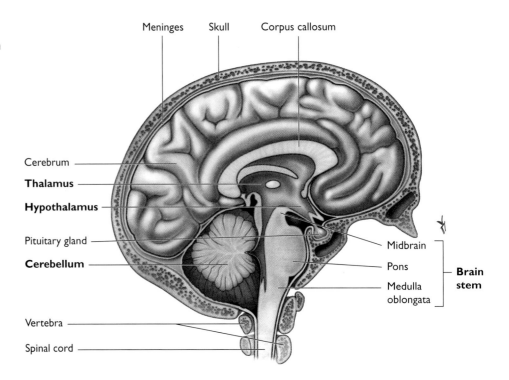

Cerebellum The **cerebellum** controls balance, muscle movement, and coordination. Since this brain region ensures that muscles contract and relax smoothly, damage to the cerebellum can result in rigidity and in severe cases, jerky motions. The cerebellum looks like a smaller version of the cerebrum, and in fact, *cerebellum* means "little brain" in Latin. It is tucked beneath the cerebral hemispheres (Figure 14.12). Like the cerebrum, it has two hemispheres connected to each other by a thick band of neurons. Additional neurons connect the cerebellum to the rest of your brain.

Brain Stem The **brain stem** lies below the thalamus and hypothalamus (Figure 14.12). It governs reflexes and some spontaneous functions such as heartbeat, respiration, swallowing, and coughing. The brain stem is composed of the **midbrain**, **pons**, and **medulla oblongata**. Highest on the brain stem is the midbrain, which adjusts the sensitivity of your eyes to light and of your ears to sound. Below the midbrain is the pons. The pons serves as a bridge, allowing messages to travel between the brain and the spinal cord. The pons is a bundle of axons stretching between the cerebellum and the CNS. The medulla oblongata is a continuation of the spinal cord. It conveys information between the spinal cord and other parts of the brain. The pons and medulla oblongata together help regulate breathing rate. The medulla also controls many other vital functions.

The functions of the brain are divided between the left and right hemispheres. Because many nerve fibers cross over each other to the opposite side, the brain's left hemisphere controls the right half of the body, and vice versa. The areas that control speech, reading, and the ability to solve mathematical problems are located in the left hemisphere. Areas that govern spatial perceptions (the ability to understand shape and form) and the centers of musical and artistic creation reside in the right hemisphere.

The **reticular formation** is an intricate network of neurons that extends the length of the brain stem and radiates toward the cerebral cortex (FIGURE 14.13). The reticular formation acts as a filter for sensory input. It analyzes the constant onslaught of sensory information and filters out stimuli that require no response. This filtering prevents the brain from having to react continuously to repetitive, familiar stimuli such as the sound of automobile traffic outside your bedroom or the sound of your roommate's breathing while you are trying to sleep.

The reticular formation also serves as an activating center by keeping the cerebral cortex alert. Conscious activity originates in the cerebral cortex, but it can do so only if the reticular formation is keeping the cortex alert.

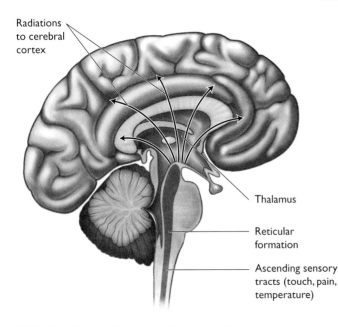

FIGURE 14.13 The reticular formation. The reticular formation lies within the medulla oblongata. It receives and forwards sensory and motor information to the CNS. The reticular activating system arouses the cerebrum and controls alertness.

Brain Activity During Sleep The reticular activating system (RAS) is a group of neurons in the reticular formation. Some of the neurons in the RAS transmit a continuous stream of action potentials to the cerebrum, keeping it awake and alert. The RAS also releases the neurotransmitter serotonin, which induces sleep by inhibiting neurons that arouse the brain. General anesthetics work by suppressing the RAS.

ADD and the Structure and Function of the Brain

Some research suggests that one factor affecting the likelihood of ADD diagnosis is sleep deprivation in young children. This suspected factor accords with the belief of many experts that the increased number of people diagnosed with ADD reflects the increased pressures that children and adults face in our fast-paced, success-driven society.

In the United States, most caretakers of elementary schoolchildren work full time—one parent in single-parent families, and both parents in two-parent families—according to census data. Today's children are involved in more time-consuming after-school activities such as sports, dance, music, theater, and art lessons, leaving less time for sleep. Some opponents of Ritalin point out that, with so much to pay attention to, it is not surprising that some kids have a hard time staying focused.

Structural brain differences may also account for symptoms of ADD. In the same way that X-rays help physicians check for broken bones, scientists have access to brain-scanning technologies that enable them to view the brain. Some *neurobiologists*—biologists who study the nervous system—use these technologies to find physical evidence in the brain structures of people diagnosed with ADD. Some neurobiologists believe they have found subtle differences in the structure and function of the brains of people with ADD.

In one study comparing the brains of people with and without ADD, researchers found that the corpus callosum was slightly smaller in people with ADD. A similar study discovered size differences in the caudate nuclei (involved in coordinating movement). The caudate nucleus in the right hemisphere of people with ADD was slightly smaller than it was in people who did not have the disorder. Other studies have shown that the cerebral cortex and cerebellum of affected persons tend to be smaller than those in unaffected people.

Additional studies have detected functional differences between the brains of people with and without ADD. Some researchers have gathered evidence indicating that one difference may be in how effectively the reticular formation filters signals. It allows too much information to be sent on to the cerebral cortex in people diagnosed as having ADD. One research team observed decreased blood flow through the right caudate nucleus in people with ADD.

Another study showed that the cortex metabolizes less glucose in adults with ADD. Glucose metabolism is an indicator of activity—the more glucose metabolized, the more active an area is. This finding could indicate that the cortex, which is involved in predicting the consequences of various actions, might be less active in people who have ADD.

Brain differences between people with ADD and people who do not have the diagnosis may have arisen because of each person's unique experiences. The brains of people with ADD could reflect inborn biological differences. Or brain differences may arise from some combination of these two factors, in much the same way that people develop different muscles in response to exercise.

Stop and Stretch 14.2

Would it be important to know whether people diagnosed with ADD who participate in brain differences studies had been treated with medicines? Why or why not?

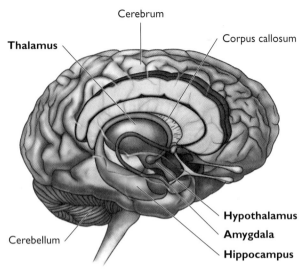

FIGURE 14.14 The limbic system. The neuronal components of the limbic system are located near the base of the cerebrum.

14.3 The Limbic System and Memory

Control of emotional behavior, motivational drive, and the formation of some memories resides in the **limbic system**.

Limbic System Structures

Structurally, the limbic system involves many neuronal areas located near the base of the cerebrum (FIGURE 14.14). The hypothalamus, for instance, exerts control over self-gratifying behaviors such as hunger, thirst, and sexual desire. Of the remaining structures making up the limbic system, the hippocampus and the amygdala are especially important. The **hippocampus**, located inside the temporal lobe, also plays a role in memory by keeping the prefrontal area aware of past experiences. The **amygdala** is an almond-shaped group of neurons located deep in the temporal lobe that is involved in processing memories and emotional reactions. The amygdala can cause experiences to have emotional overtones.

Memory

Memory involves storing information that can later be retrieved. *Short-term* or *working* memory is the ability to retrieve information that was stored within the past few hours. The ability to store and retrieve information from short-term memory involves the limbic system. New sensory information (such as hearing the address of a web site you would like to visit) stimulates a quick burst of action potentials in the limbic system. You can easily recall the address for a short time after hearing it because neurons that have just fired are primed to fire again in the short term.

However, if you don't have a chance to access the web site immediately, you will likely forget the address because the effects of priming the neurons to fire are short-lived. Consequently, this piece of information does not move into *long-term memory*. If the address is very important to you, and you work hard to ensure that you can recall it, for instance, by reciting it several times after first hearing it, or immediately finding a networked computer to gain access to the web site, you will be more likely to store this information in long-term memory. If you use the address to access the web site repeatedly over the next few weeks, the information may be transmitted to your cerebral cortex for storage in long-term memory centers. During this process, neurons create additional synapses to connecting neurons, allowing the pathway toward recall to be activated more quickly in the future.

Stop and Stretch 14.3

Because of the role of the limbic system in helping to control emotional reactions, many studies have looked for differences in the limbic structures of those with and without ADD. One 2006 imaging study published in the *Archives of Psychiatry* showed the hippocampus to be larger in children with ADD. It also showed that children with ADD had a different connection between the amygdala and prefrontal cortex than children without ADD. What might cause the connection between the amygdala and prefrontal cortex to differ between the experimental group and the control group in this study?

14.4 The Peripheral Nervous System

The **peripheral nervous system** serves as the relay center for sensory and motor impulses between the central nervous system and the rest of the body. The PNS is composed of nerves and ganglia. Nerves are bundles of axons within the PNS. The axons found in nerves are called *nerve fibers* (FIGURE 14.15). **Ganglia** are clusters of cell bodies of peripheral neurons that form swellings.

The Nerves of the Peripheral Nervous System

Two kinds of nerves are found in the PNS: cranial nerves and spinal nerves. **Cranial nerves** take impulses to and from the brain. Spinal nerves take impulses to and from the spinal cord.

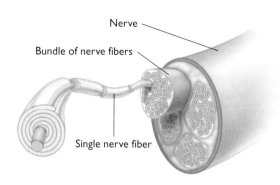

FIGURE 14.15 **Nerves.** Nerves are bundles of axons housed in a connective tissue sleeve. The axons found in nerves are called nerve fibers.

Cranial Nerves The 12 pairs of cranial nerves that attach to the brain are commonly identified by Roman numerals (FIGURE 14.16). Some cranial nerves contain only sensory fibers that bring information to the CNS, whereas others contain only motor fibers that bring information away from the CNS. **Mixed nerves** carry both sensory and motor fibers. Cranial nerves largely control the head, neck, and facial regions of the body.

Spinal Nerves The 31 pairs of spinal nerves exit the spinal cord through openings between vertebrae. You can see in Figure 14.9b that spinal nerves originate from two roots. One, the dorsal root, contains sensory fibers that conduct impulses toward the spinal cord. The second root, the ventral root, contains motor fibers that conduct impulses away from the cord. Therefore, all spinal nerves are mixed nerves. Each spinal nerve serves the region of the body in which it is located.

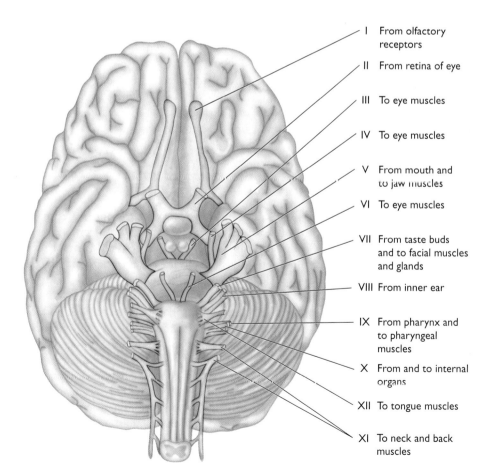

FIGURE 14.16 **Cranial and spinal nerves.** The attachments of the 12 cranial nerves as seen from the base of the brain (an inferior view).

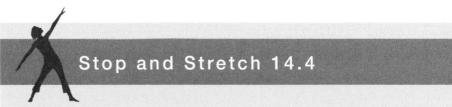

Stop and Stretch 14.4

Paralysis can occur at various levels, including paraplegia (paralyzed legs) or quadriplegia (paralysis of the arms and legs). Why might injuries at different locations on the spinal cord cause different levels of paralysis?

Somatic System

The PNS has two subdivisions, the somatic system and the autonomic system. The **somatic system** is the portion of the nervous system concerned with the control of skeletal muscle and helping the organism to interact with its environment. Nerves in the somatic system carry messages from external sensory receptors to the CNS and take motor commands to the skeletal muscle.

Some of the actions of the somatic system are due to **reflexes**, rapid, automatic responses to stimuli. Reflexes involve involuntary movements of skeletal muscle. Spinal reflexes do not involve the brain, a setup that allows a much faster response because synaptic transmission is the slowest segment of signal conduction. Reflexes are wired in a circuit of neurons called a **reflex arc**, which often consists of a sensory neuron that receives information from a sensory receptor, an interneuron that passes the information along, and a motor neuron that sends a message to the muscle that has to respond.

Reflexes allow a person to react quickly to dangerous stimuli. For instance, the withdrawal reflex occurs when you encounter a dangerous stimulus such as touching something hot. When you touch something hot, sensory neurons from touch receptors send the message to your spinal cord. Within the spinal cord, interneurons send the message through motor neurons to muscles that withdraw your hand from the hot surface (FIGURE 14.17).

While the spinal reflexes are removing your hand from the source of the heat, pain messages are also being sent through your spinal cord to your brain. This message takes a little longer because of the longer distance and the synaptic connections. Therefore, by the time the pain message reaches your brain, your hand has already been removed from the hot surface.

The Autonomic Nervous System

The **autonomic nervous system** is the part of the nervous system that innervates smooth and cardiac muscle and the glands and regulates cardiovascular activity, digestion, metabolism, and thermoregulation. This system works mainly at a subconscious level. It is traditionally partitioned into the **sympathetic division** and the **parasympathetic division**, based on the region of the brain or spinal cord in which the autonomic nerves have their origin (FIGURE 14.18).

The sympathetic system is defined by the autonomic fibers that exit thoracic and lumbar segments of the spinal cord. The parasympathetic system is defined by the autonomic fibers that either exit the brain stem via the cranial nerves or exit the sacral segments of the spinal cord. The two systems work antagonistically with signals from one system counteracting the effects of signals from the other system to maintain homeostasis.

Nerves of the parasympathetic system dominate when the body is not receiving many messages from external receptors. When this system is dominant, energy is used to perform basic housekeeping functions such as digestion and thermoregulation.

The sympathetic division dominates during periods of heightened awareness such as when one is excited or perceives danger. This system is particularly important during

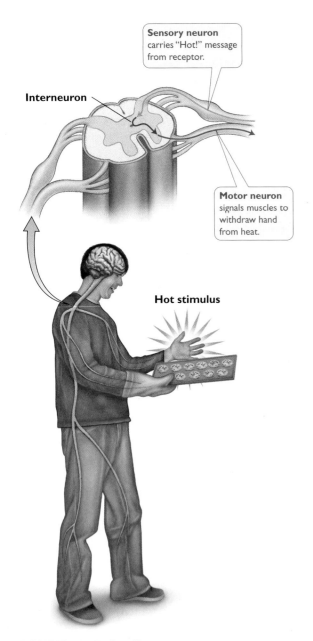

FIGURE 14.17 A reflex arc. A reflex arc can consist of a sensory receptor, a sensory neuron, an interneuron, a motor neuron, and an effector. Touching a hot baking sheet evokes the withdrawal reflex.

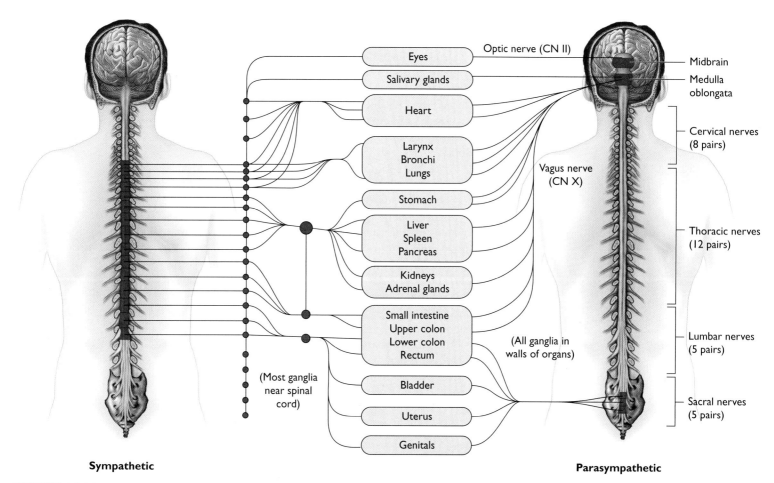

FIGURE 14.18 Autonomic nervous system. The drawing on the left shows sympathetic outflow from the spinal cord. When this system is activated, the heart rate increases, pupils dilate, movement of nutrients through the digestive system slows, and sphincters contract. The drawing on the right shows parasympathetic outflow from the spinal cord and brain. When the parasympathetic system is active, heart rate decreases, pupils constrict, activities of the digestive system increase, and sphincters relax.

emergency situations and facilitates the so-called fight-or-flight response, in which the body readies itself to fight hard or get away fast. In these situations, the housekeeping tasks are put on hold as energy is diverted for other uses.

14.5 What Causes ADD?

The nervous system is a dynamic, changing system, and many different parts of the brain are responsive to environmental influences. There is some evidence of biological differences in the brains of people diagnosed with ADD, but no single brain difference occurs in all people with ADD and, therefore, this disorder has no definitive biological cause.

It may also be the case that environmental factors aggravate ADD. In the last decade, the number of people in the United States diagnosed with ADD has risen from about 1 million to over 6 million. Such a dramatic increase in a very short time is not likely due to biology alone. It would take thousands of years for evolution to effect this magnitude of biological change within a population.

As is often the case in science, the data on increased diagnoses of ADD are difficult to interpret, and there are many different ways to explain the same findings. Doctors may simply be getting better at diagnosing ADD cases, or they may be misdiagnosing overactivity as ADD. It also may be that societal changes are having a negative impact on children.

Determining whether ADD is caused by biology or environment has an influence on treatment options. If this disease is solely one of neurobiology, then using medication alone to treat it makes sense, especially if the medication has few or no side effects. However, like other stimulants, Ritalin use has side effects including insomnia, loss of appetite, and nervousness. It may even be associated with decreased growth rates and may lead to increased stimulant use when these children become adults.

Given the serious side effects and potential for abuse of Ritalin, research on the cause of ADD and investigation of alternative treatments is critical. In some situations, modifying the environment may be the best way of dealing with the problem.

Environmental changes have been shown to help many people with ADD. Behavior modification therapies can improve ADD symptoms in children who are taking Ritalin, and also in children who are not. For instance, parents of ADD children are counseled about effective parenting strategies, such as not overscheduling their ADD child and clearly defining and enforcing rules. Teachers of ADD students often try to minimize distractions by having these students sit close to the front of the room, away from hallways or windows, and by giving them detailed schedules and timelines for assignments.

The success of approaches that combine medical and environmental interventions shows that it may be more useful to think of ADD as an imbalance between the brain's inborn tendencies and the demands of a person's environment.

Stop and Stretch 14.5

Many health problems that are induced, in part, by an individual's lifestyle are treated with medicines. For example, antacids are taken to treat stomach disorders that, in some cases, would be effectively treated through dietary changes. Likewise, heart disease and high cholesterol are medicated when, for many people, dietary changes coupled with exercise would work as well or better. When a health problem can be treated through medicine or via changes in behavior, which approach do you think is preferable? Why?

Chapter REVIEW

ROOTS TO REMEMBER

The following roots of words come mainly from Latin and Greek and will help you to decipher terms:

auto- means self.

cereb- and **cerebr-** are from the Latin word for the brain; the Greek root is **encephalo-**.

cortex means the bark of a tree.

cran- means skull.

hypo- means under or below.

para- means beside, beyond, or aside.

-path- is a root from the Greek word for pain (or disease).

ret- or **reti-** mean a net or network.

soma- and **somat-** are from the Greek word meaning body (the Latin is **corpus**).

sym- means together or joined.

KEY TERMS

14.1 Nervous System Tissues
acetylcholine *p. 328*
acetylcholinesterase *p. 328*
action potential *p. 326*
Alzheimer's disease *p. 328*
axon *p. 324*
cell body *p. 324*
central nervous system (CNS) *p. 324*
dendrite *p. 324*
depolarization *p. 326*
depression *p. 330*
dopamine *p. 328*
effector *p. 324*
gray matter *p. 325*
integration *p. 330*
interneuron *p. 324*
motor neuron *p. 324*
myelin sheath *p. 325*
nerve *p. 325*
nerve impulse *p. 325*
nerve tract *p. 325*
nervous system *p. 324*
neuroglia *p. 324*
neuron *p. 324*
neurotransmitter *p. 328*
node of Ranvier *p. 325*
Parkinson's disease *p. 328*
peripheral nervous system (PNS) *p. 324*
postsynaptic neuron *p. 328*
presynaptic neuron *p. 328*
refractory period *p. 328*
repolarization *p. 328*
resting potential *p. 326*
reuptake *p. 328*
Schwann cell *p. 325*
sensory neuron *p. 324*
sensory receptor *p. 324*
sodium-potassium pump *p. 326*
synapse *p. 328*
terminal bouton *p. 324*
white matter *p. 325*

14.2 The Central Nervous System
brain stem *p. 338*
caudate nuclei *p. 337*
cerebellum *p. 338*
cerebral cortex *p. 337*
cerebrospinal fluid *p. 334*
cerebrum *p. 336*
corpus callosum *p. 337*
fissure *p. 337*
frontal lobe *p. 337*
hypothalamus *p. 337*
medulla oblongata *p. 338*
meninges *p. 334*
midbrain *p. 338*
occipital lobe *p. 336*
parietal lobe *p. 336*
pons *p. 338*
reticular formation *p. 338*
spinal cord *p. 334*
spinal nerves *p. 335*
temporal lobe *p. 336*
thalamus *p. 337*
ventricles *p. 336*

14.3 The Limbic System and Memory
amygdala *p. 340*
hippocampus *p. 340*
limbic system *p. 340*
memory *p. 340*

14.4 The Peripheral Nervous System
autonomic nervous system *p. 342*
cranial nerve *p. 341*
ganglia *p. 341*
mixed nerve *p. 341*
parasympathetic division *p. 342*
peripheral nervous system *p. 341*
reflex *p. 342*
reflex arc *p. 342*
somatic system *p. 342*
sympathetic division *p. 342*

SUMMARY

14.1 Nervous System Tissues

- The nervous system allows an organism to respond to external and internal stimuli (p. 324).

- The nervous system consists of the brain and spinal cord of the central nervous system (CNS) as well as the nerves of the peripheral nervous system (PNS) that carry information to and from the CNS (p. 324).

- Sensory neurons carry information toward the CNS; motor neurons carry information away from the CNS. Interneurons are located between sensory and motor neurons and are within the CNS (p. 324).

- Neurons are specialized cells with a structure that consists of the branching dendrites, a cell body, and an axon with terminal boutons (p. 324; Figure 14.2).

- Axons that are covered in myelin conduct impulses faster and make up the white matter (p. 325).

- Unmyelinated axons conduct impulses more slowly and make up the gray matter (p. 326).

- Nerve impulses, series of action potentials, are generated when depolarization of the plasma membrane occurs. A resting neuron is polarized in the sense that it is negatively charged internally in comparison to the outside of the cell. This polarization is maintained by sodium-potassium pumps in the membrane (pp. 326–330).

- Stimulation of a neuron opens the gates on the sodium-channel proteins in the membrane, allowing sodium to diffuse into the cell and depolarizing the cell. The depolarization moves in a wave down the cell, activating sodium channels in adjacent parts of the membrane and thereby moving the impulse along the length of the neuron (pp. 327–328).

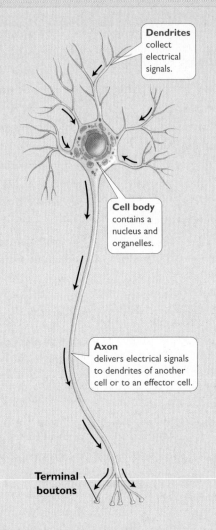

- The electrical impulse is propagated to the ends of the axon, which house chemical neurotransmitters (p. 328).

- Nerve impulses cause neurotransmitters to be released into the synapse, from where they can bind to receptors on a postsynaptic cell (neuron or effector) (p. 328).

- Binding of the neurotransmitter to receptors on the postsynaptic cell may cause depolarization of the postsynaptic cell, generating another impulse (p. 328).

- Neurons receive signals from thousands of other neurons. Each neuron sums up the overall message by integrating excitatory and inhibitory signals (pp. 328–330).

14.2 The Central Nervous System

- The spinal cord and brain are the two components of the CNS. Both structures are protected by bone (vertebrae and skull, respectively), enclosed in meninges, and bathed with and protected by cerebrospinal fluid (p. 334).

- The spinal cord is composed of neurons that control reflex actions and transmit nerve impulses to and from the brain (pp. 334–335).

- The cerebrum of the brain is where most thinking occurs. The two hemispheres of the cerebrum consist of four lobes—temporal, occipital, parietal, and frontal. The outer surface of the cerebrum is the cortex (pp. 336–337).

- The thalamus and hypothalamus are located between and below the two cerebral hemispheres. The thalamus relays information between the spinal cord and the cerebrum, and the hypothalamus regulates many vital body functions such as hunger, thirst, blood pressure, and body temperature (p. 337; Figure 14.12).

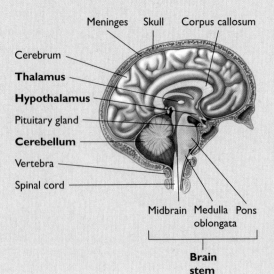

- The cerebellum is located at the base of the brain, beneath the cerebral hemispheres. It regulates balance and coordination (p. 338).

- The brain stem is located below the thalamus and hypothalamus. It controls many unconscious functions such as reflexes, heartbeat, breathing, and swallowing (p. 338).

- The reticular formation filters sensory input and keeps the cerebral cortex alert. The RAS of the reticular formation helps regulate sleep (p. 338).

14.3 The Limbic System and Memory

- Control of emotional behavior, motivational drive, and the formation of some memories resides in the limbic system (p. 340; Figure 14.14).

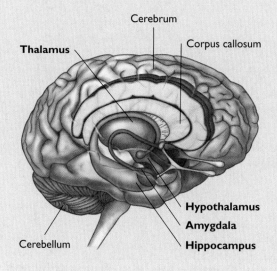

- Structures in the limbic system are located near the base of the cerebrum and include the hippocampus and amygdala. The hippocampus is involved in memory and the amygdala in emotions (p. 340).

- Short-term memories are formed in the limbic system. Long-term memories are formed when neurons create additional synapses, allowing the recall pathway to be activated more quickly in the future (p. 340).

14.4 The Peripheral Nervous System

- The peripheral nervous system relays sensory and motor impulses between the central nervous system and the rest of the body (p. 341).

- Cranial nerves take impulses to and from the brain. Spinal nerves take impulses to and from the spinal cord (p. 341).

- Nerves in the somatic subdivision of the PNS carry messages from external sensory receptors to the CNS, and they take motor commands from the CNS to the skeletal muscle (p. 342).

- Reflexes are fast, involuntary responses to a particular stimulus. Sensory receptors send messages through sensory neurons to the CNS, which in turn relays information through motor neurons to effectors initiating a response (p. 342; Figure 14.17).

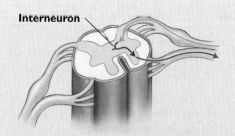

- The autonomic subdivision of the PNS regulates cardiovascular activity, digestion, metabolism, and thermoregulation, and works primarily at a subconscious level. The autonomic system is partitioned into the parasympathetic system and the sympathetic system. Nerves of the parasympathetic system dominate when the body is using energy to perform basic housekeeping functions. The sympathetic division dominates during periods of heightened awareness and facilitates the fight-or-flight response, in which the body readies itself to fight hard or get away fast (pp. 342–343).

14.5 What Causes ADD?

- Members of the scientific community have reached no consensus about the cause or causes of ADD (pp. 343–344).

LEARNING THE BASICS

1. Myelin _____.
   a. is a protective layer that surrounds the axons of some neurons
   b. is formed by neuroglial cells
   c. gives neurons a white appearance
   d. can increase the rate of nerve signal transmission
   e. all of the above

2. How are ions distributed when an axon is not conducting a nerve impulse?
   a. more potassium outside the axon and more sodium inside the axon
   b. more sodium outside the axon and more potassium inside the axon
   c. equal amounts of sodium and potassium inside and outside the axon
   d. sodium and potassium outside the axon only
   e. sodium and potassium inside the axon only

3. An action potential is triggered when diffusion of _____ is accelerated through voltage-gated channels in the plasma membrane.
   a. hydrogen ions
   b. potassium ions
   c. neurotransmitters
   d. sodium ions
   e. cranial nerves

4. Which of the following mechanisms maintains a gradient across a neuron's membrane?
   a. citric acid cycle
   b. electron transport chain
   c. dopamine receptor
   d. sodium-potassium pump

5. An action potential _____.
   a. is a brief reversal of temperature in the neuronal membrane
   b. is propagated from the terminal bouton toward the cell body
   c. begins when sodium channels close in response to stimulation
   d. is propagated as a wave of depolarization moves down the length of the neuron

6. Neurotransmitters _____.
   a. are electrical charges that move down myelinated axons
   b. are released across the nodes of Ranvier to hasten nerve impulse transmission
   c. are released when an electrical impulse arrives at the terminal bouton of the postsynaptic neuron
   d. diffuse across a synapse and bind to receptors on the plasma membrane of the postsynaptic neuron
   e. are steroid hormones released from the thalamus

7. The effects of a neurotransmitter could be increased by _____.
   a. increasing the number of receptors on the postsynaptic cell
   b. preventing reuptake
   c. providing more enzymes involved in synthesizing the neurotransmitter
   d. inhibiting enzymes involved in breakdown of the neurotransmitter from the synapse
   e. all of the above

8. **True or False?** Depolarization is the lessening of charge difference across the membrane of the axon.

9. Use the following list of terms to label the structures in (FIGURE 14.19).

   sensory neuron   dendrite
   interneuron   Schwann cell
   effector   node of Ranvier
   sensory receptor   axon
   cell body   motor neuron

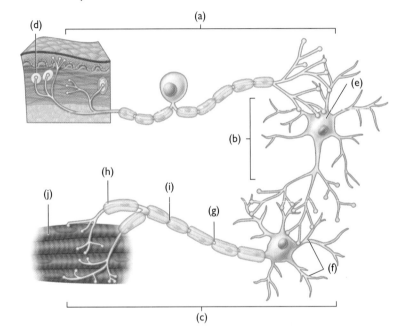

FIGURE 14.19 **Neurons.**

10. **True or False?** Cells found in nervous tissue can be categorized as either neuroglia or neurons.

11. **True or False?** The repolarization phase of an action potential occurs when sodium ions diffuse out of the cell, and the inside of the cell again becomes more negative than the outside.

12. Which brain structure controls balance and coordination?
    a. cerebrum
    b. thalamus
    c. hypothalamus
    d. cerebellum
    e. brain stem

13. Where in the brain is the thalamus located?
    a. in the pons
    b. in the medulla
    c. in the cerebral cortex
    d. between the cerebral hemispheres
    e. in the occipital lobe

14. Use the following list of terms to label the parts of the brain in (FIGURE 14.20).

    thalamus   hypothalamus
    cerebrum   brain stem
    cerebellum

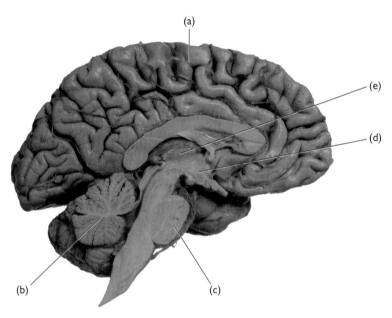

FIGURE 14.20 **Right-facing human brain.**

15. Which of the following is a function of the brain stem?
    a. governs spontaneous functions such as heartbeat and swallowing
    b. coordinates reflexes
    c. connects brain to PNS
    d. monitors internal organs

16. Spinal nerves conduct impulses _____.
    a. to the PNS only
    b. away from the CNS only
    c. both to and from the spinal cord
    d. to the brain stem only

17. Integration _____.
    a. is the summing of excitatory and inhibitory signals
    b. occurs only in the thalamus
    c. involves the release of excess neurotransmitters into the cytoplasm
    d. involves the pons only

18. _____ are clusters of cell bodies of peripheral neurons that form swellings.
    a. Dorsal roots
    b. Tracts
    c. Nerves
    d. Ganglia
    e. Motor nerves

19. **True or False?** Bundles of axons within the CNS are called tracts while those in the PNS are called nerves.

20. **True or False?** The parasympathetic system slows down the body and diverts energy to basic housekeeping tasks.

ANALYZING AND APPLYING THE BASICS

Questions 1 and 2 refer to the following information.

Adrenoleukodystrophy (ALD) is a rare genetic disorder that causes the breakdown of myelin. This disease is due to a mutation of a sex-linked chromosome and mainly affects boys. Once symptoms appear, an affected boy quickly loses his ability to hear, see, speak, or move, and may die within two years.

The normal version of this gene codes for a transport protein that moves fatty acids into peroxisomes, where they are metabolized. Without these transporter proteins, certain fatty acids build up in the cells. This buildup likely causes the inflammation and demyelination of axons that occur in the central nervous system.

Some fatty acids are introduced into the body through diet, and others are synthesized by the body. Therefore, one way to delay progression of this disease is to eliminate consumption of these fatty acids. It is also possible to slow the synthesis of fatty acids. Lorenzo's Oil, a mixture of olive and rapeseed oils, can interfere with the body's ability to synthesize the fatty acids whose buildup causes ALD. A recent study found that 76% of boys with the mutant transport protein associated with ALD but no symptoms still produced normal brain scans after 10 years of taking Lorenzo's Oil. Only 33% of the boys who did not regularly take the oil remained free of the symptoms of ALD after 10 years.

1. Explain how degeneration of the myelin sheath could cause the symptoms of ALD.

2. It is unclear how useful Lorenzo's Oil can be in stopping the destructive effects of ALD once symptoms appear. Lorenzo Odone, the boy whose parents developed the oil, is still alive 17 years after developing symptoms of ALD. However, he can no longer see, hear, or move. Why might Lorenzo's Oil be less effective for boys who already have symptoms of ALD?

Questions 3 and 4 refer to the following information.

Dyslexia is a learning disability that results in difficulty converting spoken words into the letters that stand for sounds. This difficulty can cause people with dyslexia to have trouble reading, writing, and spelling.

Scientists have used brain scan technologies to study the brain activity of dyslexic and nondyslexic adults as they read. These studies showed reduced activity in the region of the dyslexic person's brain that links visual association areas to language areas. The studies also showed increased activity in a region of the dyslexic person's brain called Broca's area—an area normally not activated during reading.

It is also possible to measure metabolic brain activity. One such study showed that dyslexic children used 4.6 times more area in the brain to perform language tasks than did nondyslexic children. Most of the observed activity occurred in the frontal lobe, an area of language function.

3. What do these studies reveal about the connection between dyslexia and the brain?

4. Training programs can help both children and adults with dyslexia to improve their language skills. Children with dyslexia showed more normal brain scans after only a few months of this training. Adults also showed some changes in brain function. What do these findings indicate about the human brain?

5. Alzheimer's disease is the most common form of dementia affecting people older than 60. The hippocampus and the amygdala are regions of the brain that are affected during the early stages of Alzheimer's disease. Use your knowledge of these limbic system structures to predict some of the symptoms of early Alzheimer's disease. Explain your answer.

6. Researchers studying the creation of memory in rats used two study groups: rats with normal brains and rats with damage to their two hippocampi. They presented eight cups of sand to the rats—each cup scented with a different odor. A piece of cereal was buried in three of the cups. They then presented the rats with six cups of sand. Five cups were scented with new odors, but one had the odor of a cup that had previously contained cereal. The researchers found that normal rats and those with hippocampal damage were both able to remember the scent that signaled the reward 75% of the time.

Another group of researchers presented six scented cups of sand (A, B, C, D, E, and F) to rats with normal brains. Each cup contained a piece of cereal. The researchers presented the same rats with six more cups (L, M, N, O, P, Q)—each containing a piece of cereal and each with different odors than the cups in the first group. The researchers then damaged the hippocampi in half of the rats. Each rat was then given one cup (A) from the first group and one (L) from the second. Only the cup from the first group (A) had a cereal reward. Researchers continued to present one cup from each scent group to each rat (B and M, C and N, etc.). Only the cup from the first group ever had a reward. Normal rats were able to understand this pattern 70–95% of the time. Rats with damaged hippocampi were unable to understand the pattern.

Assuming there was no damage to the rat's sense of smell, what do these two studies reveal about the relationship between the hippocampus and memory?

7. Binge drinking and overconsumption of alcohol are serious issues on many college campuses. Alcohol poisoning can be a frightening and sometimes deadly result of overconsumption. When a person's blood alcohol level gets too high, mental confusion and vomiting can result. Breathing may slow down significantly or become irregular. Without medical attention, the drinker may die.

The most dangerous symptoms of alcohol poisoning relate to changes in breathing, blood circulation, and the loss of the gag reflex (which can lead to death by asphyxiation if the person vomits while unconscious). Using this information, discuss how an excess of alcohol might affect the nervous system.

8. Research by Rutter and colleagues (1983) and Biederman and colleagues (1995) found a positive association between a diagnosis of ADD and family-environment risk factors. These risk factors included severe marital problems, large family size, foster care placement, low social class, mental disorder in the mother, and criminal behavior in the father. Choose two of the factors and explain how they might contribute to the development of ADD.

CONNECTING THE SCIENCE

1. Some forms of depression are thought to be caused by decreased amounts of the neurotransmitter serotonin. Does this mean that all depression has a biological cause? Why or why not?

2. Suppose that your friend is an engineering major. This friend finds it very difficult to study for the courses in his major because he finds them uninteresting. He talks to you about his idea that taking Ritalin might help him study. What advice would you have for your friend?

3. Do you think doctors are too quick to prescribe Ritalin to kids? Should parents be forced to try environmental changes before being allowed to give their child this drug? Why or why not?

Chapter 15

The Senses

Do Humans Have a Sixth Sense?

LEARNING GOALS

1. Compare and contrast the five types of sensory receptors, including the type of stimuli to which they respond.

2. Compare and contrast the general and special senses.

3. Describe the steps involved in conscious perception of a stimulus.

4. Describe the two proprioceptors and how they function to maintain body sense and muscle tone.

5. List the stimuli to which touch receptors in the skin respond.

6. Describe the structure of a taste bud and explain how taste is sensed.

7. Compare and contrast the olfactory sense with taste.

8. Explain how sound is sensed in the inner ear.

9. Summarize the structure and function of the vestibular apparatus.

10. Compare and contrast rod and cone cells in the eye and summarize the function of each in vision.

11. Describe how visual stimuli are processed in the brain.

Could these people have predicted the oncoming tsunami?

Did you hear about the young tourist who canceled her trip to the World Trade Center on the morning of September 11, 2001, because she experienced a feeling of dread when she awoke on that beautiful day? Or about the dad who grabbed his young son from the sidewalk just seconds before a speeding car jumped the curb and stopped right where the boy was playing? Have you ever found yourself thinking of an old friend, only to have that friend call you out of the blue? These experiences may suggest to you that some people, some of the time, have a "sixth sense" that allows them to predict the future.

A "sixth sense" is so named to distinguish it from the five senses—touch, taste, hearing, smelling, and sight. These five senses were shaped by millions of years of human evolution to maximize our ability to perceive and respond to the dangers and opportunities experienced by our ancestors. But our senses also have profound limitations, and the challenges we face as individuals today are much different than those faced by our ancestors. Instead of needing to search the landscape for food and having to elude lions and saber-toothed tigers that want to eat us, we focus our attention on computer screens and navigate a world of speeding cars, man-made weapons, and deadly diseases that our ancestors had no experience with. Given these new challenges, it might seem preferable in the modern world to have a sixth sense that can warn us of approaching danger and help us anticipate change in place of one of the other five senses.

In this chapter, we examine the function of our sensory systems and consider how these systems allow us to navigate a world of danger and opportunity. We'll discover that a sixth sense, of a sort, does exist—but its nature may surprise you.

Our senses evolved during the millennia humans were hunter-gatherers.

Or even everyday events, like a call from an old friend?

Do they allow us to sense danger and opportunity today?

15.1 Sensing and Perceiving

The survival of nearly every living thing depends on its ability to sense and respond appropriately to environmental conditions. Plants sense sunlight and orient their leaves and growth patterns to maximize its interception. Bacteria sense a gradient of their particular food source and move toward it. And animals with complex nervous systems can integrate and process many pieces of information to avoid predators and seek out resources. Any information we have about the world comes to us through specialized cells called **sensory receptors**.

Sensory Receptors

The **senses** consist of all the different structures and processes that receive and interpret environmental conditions. The environmental conditions each sensory system responds to are known as **stimuli**.

Although we commonly refer to the "five senses"—hearing, sight, taste, smell, and touch—it is more common in physiology to classify sensory systems by the type of stimuli they typically respond to. Humans have five types of sensory receptors: **mechanoreceptors**, which respond to a change in the shape of the receptor or of nearby cells; **thermoreceptors**, which sense temperature change; **photoreceptors**, which respond to light; **chemoreceptors**, which react to chemical stimuli; and **pain receptors**, also called **nociceptors**, which respond to tissue damage or extreme temperatures.

Sensory receptors also have different focal points. **Externoreceptors** sense conditions outside the body allowing us to respond to the environment, while **internorcceptors** sense internal conditions and contribute to homeostasis. Table 15.1 summarizes the types of sensory receptors and their functions.

Physiologists also classify sensory input by the specialization of receptors. Receptors for touch, pressure, vibration, temperature, body and limb position, and pain are found throughout the body. The information from these receptors is termed the **general senses** or *somatic senses*. In contrast, sight, hearing, taste, smell, and equilibrium (head position), the **special senses**, are sensed via receptors in specialized structures (**sense organs**) in the head.

The general and special senses are the only known pathways for stimuli to enter our consciousness. Unless we discover clear evidence of other methods of sensing the world, the only information an individual can use to sense impending danger is what is gathered by these known systems. Predicting future conditions, a premonition skill that relies on knowledge of events occurring far away from the reach of our sensory receptors, does not appear to be possible for individual humans, given these limitations.

TABLE 15.1	Sensory Receptors	
Receptor Type	**External Stimuli Sensed**	**Internal Stimuli Sensed**
Mechanoreceptor	Air pressure, touch, hearing	Body position, blood pressure
Thermoreceptor	Air, water temperature	Internal body temperature
Photoreceptor	Light	None
Chemoreceptor	Chemicals in air, food, and drink	Carbon dioxide, oxygen, and glucose levels in body fluids
Pain receptor	Chemicals released from damaged cells, increased pressure from swelling	Chemicals released from damaged cells, increased pressure from swelling

The stimuli that our sensory receptors are able to receive reflect a fraction of all possible stimuli in the environment. In general, our senses have evolved to measure aspects of the environment that are crucial to our survival—the colors of light that correspond to our most important food sources, the odors that signify items that are nutritious or dangerous, the frequencies of sound that encompass the range of human speech. Other organisms can detect other sets of stimuli (FIGURE 15.1).

Reading and Understanding the Environment

When a stimulus reaches a sensory receptor, it causes a change in the receptor's plasma membrane. In particular, the electrical charge across the membrane—the membrane potential discussed in Chapter 14—changes. Typically, a stimulus causes the membrane of a sensory receptor to **depolarize**, that is, it causes the inside of the cell to become slightly less negative relative to the outside of the cell. Depolarization then triggers a cascade of events that may eventually result in our perception of the stimulus.

In order for us to perceive the stimulus received by a sensory receptor, the change it triggers in the receptor's plasma membrane must be large enough to stimulate a sensory neuron carrying the information to the central nervous system. The spinal cord, and eventually the brain, receive signals from these neurons and process them. Much of this processing is unconscious, such as the sensory information that allows you to maintain your posture or regulate blood pressure, but some of the information has an impact on our consciousness, producing a **sensation**.

Our interpretation of sensations is called **perception** (FIGURE 15.2). Interestingly, anything that stimulates a sensory neuron will be perceived as typical of the input from that receptor. For example, the chemical menthol, found in over-the-counter pain-relieving creams, stimulates thermoreceptors in the skin to produce a perception of cooling, and a punch in the eye will cause you to "see stars" because the sensory neurons that trigger a perception of light and color have been mechanically disrupted enough to send a signal.

The amount of sensory information that we can perceive would quickly overwhelm our ability to process it if we did not have mechanisms for selectively ignoring input. When a stimulus is recorded continuously over a span of time, perception is turned down, a process called **sensory adaptation**. *Neural adaptation* is a form of sensory adaptation that occurs when a receptor stops sending a signal—for example, the receptors that record the feel of clothing against the skin. A sensory adaptation process called *habituation* occurs when the brain's gatekeeper, the *reticular formation*, stops sending signals from the subconscious brain to the conscious brain. You have experienced the habituation when you have gradually lost awareness of an odor that seemed strong when you initially entered a room—for example, the musty smell of a basement that seems to fade away while you remain there.

(a)

(b)

FIGURE 15.1 Variation in sensory capabilities. Our perception of the world is limited by what we can sense. (a) This is how a flower appears to human sight. (b) Insects can see ultraviolet light, which is invisible to us. The light guides insects to the most desirable flowers and areas of nectar and pollen.

Stop and Stretch 15.1

Animals that have adapted to living in caves often do not have functional eyes, although their closest non-cave-dwelling relatives do. Why would lack of sight be an adaptation in a cave environment?

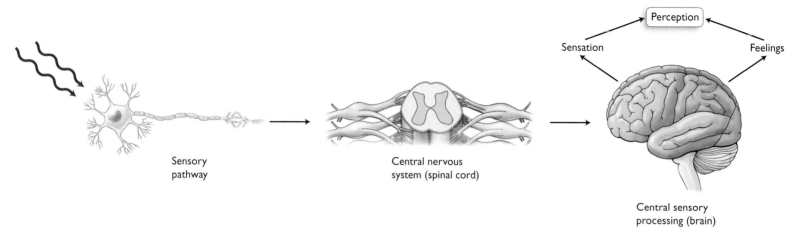

FIGURE 15.2 **The steps of information processing.** A stimulus is detected by a sensory receptor. If it is strong or meaningful enough, the stimulus triggers a cascade of events that may result in perception.

The perception of danger or threat that you might feel in a new situation is influenced not only by the sensations you are experiencing, but also by your expectations and psychological state. Your expectations are determined in part by the capacity of the human brain for drawing patterns from past experiences. For example, you may perceive someone running toward you as a threat or as benign depending on the circumstances of your environment as well as your memories of direct experience and shared stories (FIGURE 15.3). We'll now examine the clues that your senses can gather that may influence your perception of danger and predictions about future events.

15.2 The General Senses

Every square inch of your skin contains dozens of sensory receptors that sense air movement, pressure on the skin, vibration, heat, and pain. Receptors in muscles and tendons monitor body position and posture. The general senses can pick up information about the environment that is not obvious to the eyes, ears, and nose, but which may provide vital information about health and safety.

Proprioception

Crucial to our assessment of external conditions is an understanding of our position in space. Body position is measured by sensory receptors called **proprioceptors**. Proprioceptors are always active, monitoring limb position and helping us to maintain balance and posture. The two main types of proprioceptors are both mechanoreceptors: muscle spindles and Golgi tendon organs.

Muscle spindles are specialized muscle fibers with sensory nerve endings wrapped around them (FIGURE 15.4). These receptors report muscle stretching and help to maintain muscle tone, or tightness. Even when we are at rest, some part of most skeletal muscles is active, contracting to allow us to maintain our position in opposition to gravity. Portions of each muscle are randomly stimulated for short periods, so that constant tension in the muscle is produced without fatiguing individual muscle fibers. Signals transmitted by the muscle spindle fibers ensure that enough of the muscle mass is stimulated to maintain body position.

FIGURE 15.3 **Is this person a threat?** Our perception of danger depends not only on sensory input but also on our beliefs, expectations, and psychological state.

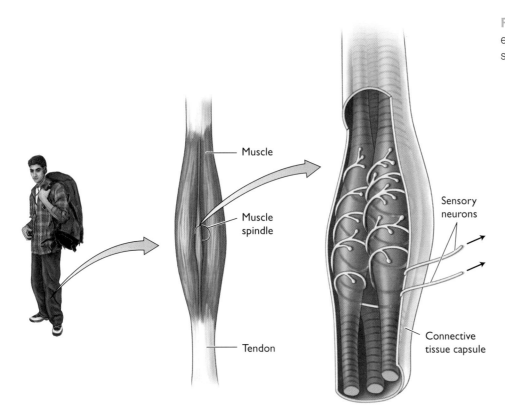

FIGURE 15.4 Muscle spindle. Free sensory nerve endings wrapped around muscle fibers sense muscle stretching and help us to maintain muscle tone.

Stop and Stretch 15.2

Testing the patellar (or "knee jerk") reflex is a typical part of a physical exam. In this test, a small rubber mallet is used to stretch the patellar tendon in the knee. If the reflex is working properly, this deformation causes muscle spindle neurons to fire, stimulating the contraction of the thigh muscle and causing the leg to kick out. In life, this reflex is stimulated as we begin to lose balance, causing us to hop in the air. What is the value of the knee jerk reflex?

Golgi tendon organs measure the degree of tension within tendons attaching muscle to bone. Under typical conditions, these organs provide feedback similar to muscle spindles, permitting the maintenance of balance and posture. However, if these receptors sense very high tension, their firing triggers the rapid reflexive relaxation of its attached muscle. You may have experienced this effect if you have ever tried to suspend yourself by your arms from a pull-up bar (or from monkey bars). Eventually, despite your conscious wish to hang on, your grip will weaken as the Golgi tendon organ triggers muscles to relax. This response protects the muscles, tendons, and bones by dispelling forces that could cause tearing or breakage.

Our understanding of our body's position in the environment relies not only on these internoreceptors but also on our sense of touch, which provides information about air movement, temperature, pressure, and vibration outside the body.

The Sense of Touch

Receptors for touch are mechanoreceptors that transmit signals when they are deformed in some way (FIGURE 15.5). These receptors are unequally distributed, with higher densities in the hands, around the mouth, and on the genitals, all areas where sensitivity is highly useful for survival and reproduction. Touch receptors are less abundant in areas such as the abdomen and back.

The area monitored by a particular sensory neuron is referred to as its **receptive field**. As a result of differences in receptor density, the receptive field for touch ranges in diameter from less than a millimeter (0.04 inch) on the tongue to as large as seven centimeters (3 inches) on the upper arm and back.

The different types of touch receptors consist of the ends of individual nerve cells. Some of these nerve endings are free from any outer covering and are therefore highly sensitive, such as the receptors wrapped around hairs that sense air movement, and **Merkel disks**, which lie at the base of the epidermis and whose expanded cell terminals sense light pressure on the skin.

Encapsulated receptors encase the nerve endings in a sheath of connective tissue that either shields them from the lightest of touches or allows them to adapt to continuous pressure, depending on their function. **Meissner's corpuscles**, which are abundant at the base of the epidermis in hairless skin, detect light pressure and vibration, but adapt to the stimulus very quickly. Both **Ruffini corpuscles** and **Pacinian corpuscles** lie deep in the dermis and respond to harder pressures. Pacinian corpuscles adapt readily, as illustrated in FIGURE 15.6, but Ruffini corpuscles do not adapt and provide information about distortion or stretching of the skin.

Sudden changes in air movement sensed by touch receptors can signal threatening changes in the environment, such as the rapid approach of a car. The ability to sense changes in air pressure is more highly developed in animals other than humans. For example, a cat's whiskers can register very small changes in air pressure, allowing cats to navigate around objects in the dark.

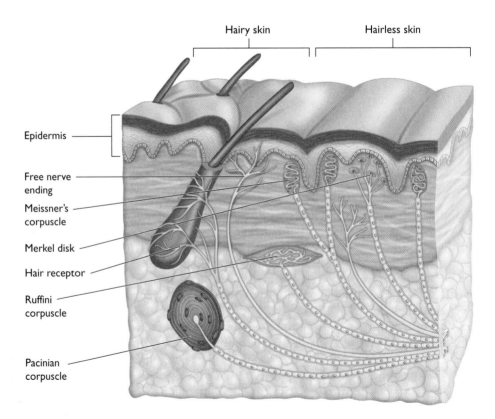

FIGURE 15.5 Touch receptors. A variety of receptors in the skin measure tactile aspects of our environment.

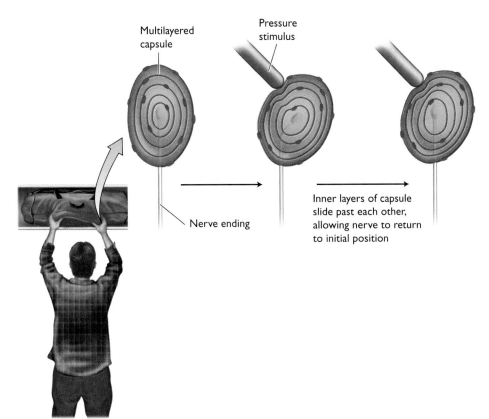

FIGURE 15.6 **Touch receptor adaptation.** The capsule surrounding a Pacinian corpuscle is made up of onion-like layers. When put under pressure, the nerve ending within the capsule deforms, but the layers quickly slide over each other, allowing the nerve ending to straighten out and stopping the signal, even as the pressure stimulus remains.

People in frightening situations may feel the hair on their neck and head "stand up." However, the nerve endings around these hairs are responding to the goose bumps that form as a result of anxiety. Rather than a premonition of danger, the prickling feeling on the back of your neck is a symptom of an already-forming perception of danger, either real or imagined.

After the deadly tsunami that struck nations around the Indian Ocean in December 2004, few dead animals were found among the wreckage. Reports began to filter in that captive animals displayed unusual or frightened behavior before the wave hit. Some scientists have speculated that animals had sensed low-frequency vibrations associated with the earthquake that spawned the tsunami. How could scientists test this hypothesis?

Temperature and Pain

Moderate changes in external temperature are sensed by the free nerve endings of thermoreceptors embedded in the skin. Although all are identical in appearance, certain temperature receptors respond to cool temperatures (10–20°C, or 50–68°F) and a smaller number respond to high temperatures of 25–45°C (77–113°F). The mechanism by which these receptors sense temperature is not well understood.

FIGURE 15.7 A pain chemical. Capsaicin, the active ingredient in hot peppers, binds to pain receptors on the tongue, mouth, and throat, causing the "heat" these spices add to cooked dishes.

Chemicals in certain plants such as peppermint and garlic can also activate thermoreceptors, contributing to a sensation of cold or heat during food consumption. These sensations can be so pronounced that eating spicy food can make you sweat because your brain interprets the signals from the thermoreceptors as an increase in temperature, triggering actions that cool the body. Chemicals that stimulate thermoreceptors likely evolved in plants to discourage certain animals from eating them—although judging by Mexican, Thai, and other spicy cuisines, human beings are not so easily discouraged.

Thermoreceptors in the skin respond to temperature changes but adapt to stable temperatures fairly quickly. After a few minutes in a new temperature environment, you usually feel more comfortable.

Temperature extremes that are dangerous to health and survival are sensed by pain receptors, rather than thermoreceptors. Pain receptors are free nerve endings that are especially abundant in the skin, joints, bones, and blood vessels. As with thermoreceptors, the mechanism by which pain receptors sense temperature extremes is still poorly understood. Pain receptors may be triggered by mechanical disturbance or by chemicals produced by damaged tissue. Chemicals contained in very hot foods, such as habanero chilis (FIGURE 15.7), can also stimulate pain receptors.

In contrast to the skin, the internal organs of the body have few pain receptors. Those organs that do have receptors appear to share nerve pathways with pain receptors in the surface of the body. As a result, damage to internal organs may manifest itself as pain in different areas of the body, a phenomenon known as **referred pain**. Characteristic patterns of referred pain are known to indicate certain conditions—for instance, heart damage often is first noticeable as arm or shoulder pain (FIGURE 15.8).

Deep cuts, burns, or similar injuries trigger sensory neurons that transmit signals extremely rapidly, contributing to a sensation of stabbing or prickling **fast pain**. These sensations trigger immediate withdrawal reflexes—your hand will pull away from a hot surface before you are even conscious of the painful burn. Fast pain is later supplanted by the dull ache of **slow pain**, a perception triggered by the activity of different sensory receptors. Each slow pain receptor has a rather large receptive field, making it difficult to localize the exact origin of certain painful stimuli. Slow pain receptors take longer to adapt, so pain may linger well beyond the point at which we can do anything to change the damage that has occurred.

Chronic pain, which lasts past its usefulness to our survival, is almost always slow pain and may be possible to control by blocking the chemicals that trigger pain receptors. One set of chemicals associated with pain is *prostaglandins*, which cause inflammation in damaged tissue. Aspirin and other anti-inflammatory drugs interfere with the release of prostaglandins, helping to relieve pain. The painkillers morphine and codeine mimic naturally produced pain-relieving chemicals, such as *endorphins*, produced in the brain. These chemicals appear to block pain perception rather than sensory reception.

Although it is unpleasant, pain serves an important function in protecting us from danger. Fast pain provides instant information that the current environmental situation is harmful and triggers an immediate response. Slow pain serves to prevent further damage by discouraging movement during the vulnerable healing process and may serve as a powerful reinforcement to help us avoid the pain-causing behavior in the future.

Fear of pain is an important component of our perception of danger. The perception of danger has two components: *risk*, the likelihood of experiencing pain; and *hazard*, the severity of the pain that could occur. Our perception of risk and hazard strongly influences our sense of the danger inherent in any situation. A low-risk but high-hazard activity may cause a greater perception of danger than a lower-hazard but higher-risk activity.

For example, flying in a plane is a very safe activity, but one in which an accident commonly results in death. To some people it seems much more dangerous than driving a car, where accidents are not as often fatal, but are much more common. In turn, our perception of risk and hazard in different situations colors our perception of stimuli in these environments. As a result, the same conditions may cause us to infer danger in an environment where little risk exists and to miss danger signs where risk is much higher (FIGURE 15.9).

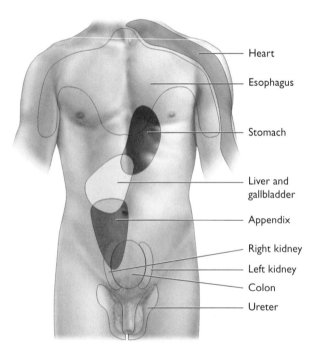

FIGURE 15.8 Referred pain. Pain receptors in internal organs share sensory pathways with skin receptors, so that damage to an internal organ often manifests as pain elsewhere in the body. Damage or disease in internal organs can sometimes be diagnosed by understanding the pattern of referred pain.

FIGURE 15.9 High or low risk? To a passenger in an airplane, a snow storm during approach and landing can be very frightening, despite the fact that few plane crashes occur under these conditions. Automobile accidents in similar conditions are much more common, but travel at these times may not feel as risky to passengers in a car.

15.3 The Chemical Senses

The special senses allow us to make more specific evaluations of aspects of the environment compared to the general senses. Two of these senses, taste and smell, allow us to discern the chemical nature of our environments.

Taste

The **gustatory receptors** (taste receptors) are distributed densely over the top of the tongue and scattered around the mouth and throat in much lower numbers. The rough surface of the tongue is made up of protrusions called **papillae**. Each papilla contains 100–200 taste receptors, or **taste buds**, mainly along its sides. Each of the mouth's 10,000 taste buds is made up of about 25 **taste cells** and an equal number of supporting cells. Each elongated taste cell is crowned with a number of extensions called **taste hairs** (FIGURE 15.10). Cells within a taste bud are completely replaced every 10 days, so that even when cells are damaged by exposure to hot food or drink, overall tasting ability can recover.

The surfaces of taste hairs are covered with plasma membrane receptors that will bind to **tastants**, particular chemicals in food. Binding of a tastant stimulates the release of neurotransmitters from the taste cell, triggering a signal in an associated sensory neuron. Activation of a particular receptor molecule communicates one of five possible qualities of a tastant: sweet, salty, sour, bitter, and savory or meaty (also known as umami). A particular food item may contain several tastants and trigger more than one type of receptor. These sensations, in combination with odor sensations, produce the perception of flavor.

Taste buds and types of taste cells are not evenly distributed over the surface of the tongue—most are concentrated on the margins. Although all four types of taste cells are found in every region of the tongue, there is a higher concentration of sweet receptors on the tip of the tongue and a higher concentration of bitter receptors at the back. When we want to experience the flavor of a food or drink, we move it around our mouths so that it contacts a large number of taste buds. In contrast, when we want to avoid tasting something, we pass it over the center of our tongue and swallow quickly.

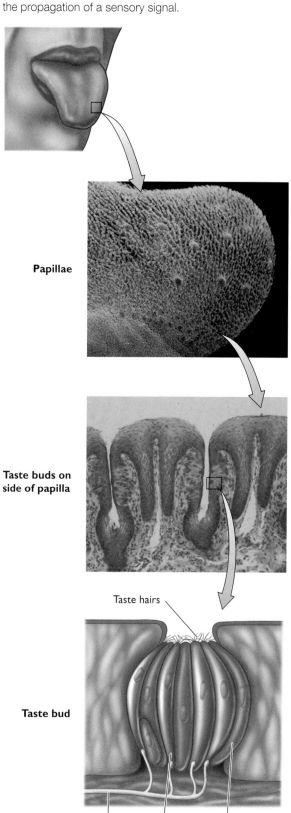

FIGURE 15.10 Taste. Taste buds are found on papillae on the tongue. Hairs on taste cells bind to certain chemicals in food, causing the release of neurotransmitters and the propagation of a sensory signal.

Our reactions to different tastants have been shaped by evolution. Contact with sweet, salty, and savory foods prompts us to swallow, but stimulation of bitter receptors triggers a gag reflex, causing us to spit out the offending item. These are logical responses, since a sweet or salty item is likely to contain essential nutrients, while a bitter food may be toxic. Our bitter receptors are highly tuned to the danger of ingesting poisonous or spoiled food. In fact, these receptors are 100,000 times more sensitive to bitter tastants than sweet receptors are to sweet tastants.

Smell

Olfactory receptors (smell receptors) are concentrated in two mucus-covered patches on the roof of the nasal passage. Unlike taste receptors, these chemoreceptors do not rely on an intermediate neuron to transmit signals to the brain. The receptive surface is a hair-covered knob made up of a dendrite of a sensory neuron. The hairs pro-ject into the nasal cavity and are covered with plasma membrane receptors that bind with chemical **odorants** that become dissolved in the surrounding mucus (FIGURE 15.11). Olfactory receptors are replaced every 60 days and are among the few nerve cells in the human body that can be regenerated.

The cell receptors on olfactory neurons are highly specific—nearly 1,000 different receptors have been identified. Information from numerous olfactory receptors combines

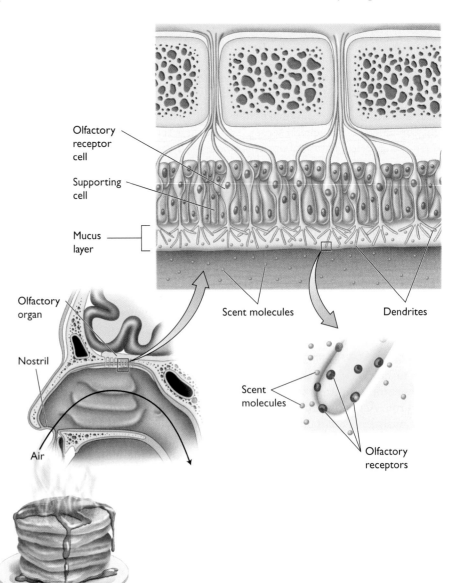

FIGURE 15.11 Smell. The olfactory receptors are found in the upper surface of the nasal passages. These receptors consist of the free ends of sensory neurons containing membrane receptors for specific chemical odorants.

so that we are able to distinguish thousands of distinct smells. The olfactory receptors are also very sensitive—as few as four molecules of an odorant can trigger a neuron to depolarize and send a signal to the central nervous system.

The first stop for information coming from the olfactory receptors is a structure called the **olfactory bulb** in the brain. Here the information from the receptors is partially processed before it is sent to the cerebral cortex. Most olfactory information passes through the limbic system of the cortex, the structures associated with emotion and memory. As a result, odors can trigger profound emotional responses.

There is some evidence that our perception of certain odors is "hard-wired," that is, we have instinctual negative or positive responses to certain smells. Instinctive responses to odors associated with danger are common in other animals—for instance, mice have an immediate withdrawal response when sensing the urine of certain predators. Our pleasure at sweet scents and cooked meat is likely related to the value that food items with these odors had to our ancestors. Similarly, our disgust at odors associated with human waste and spoiled food may be an adaptation to protect us from the dangers of contacting these disease-bearing substances.

Not all potentially dangerous odorants trigger our olfactory receptors. Molecules that are relatively simple and were not especially abundant or dangerous to humans during our early evolution may not have receptors in our noses. Carbon monoxide, for instance, a gas produced by incomplete combustion (especially of fossil fuels, like oil and natural gas), is responsible for hundreds of deaths every year, but it is odorless to us. Carbon monoxide detectors with audible alarms are necessary to protect us from this modern threat. Natural gas piped to many homes for heating and cooking is also odorless, but has a characteristic "rotten egg" odorant added to it to alert people to its presence and the risk of explosion and fire that comes with a gas leak.

Although we have 10 million olfactory receptors in total, we are relatively insensitive to odors compared to many other mammals. For example, a bloodhound has nearly 200 million olfactory receptors and is nearly a hundred million times more sensitive to odors than humans. Dogs are valuable partners to humans in following trails made by game during hunting, retrieving downed birds, tracking other humans on the run from law enforcement, and sniffing out drugs or dangerous chemicals (FIGURE 15.12). Some dog owners insist that their dogs can "smell fear" in people. There is no independent evidence to support this claim, but dogs may indeed be able to pick up chemicals that are produced in human sweat in response to anxiety.

Scents that act as a type of nonverbal signaling system among individuals in the same species are called **pheromones**. Among many nonhuman animals, pheromones can be potent indicators of sexual readiness or methods to communicate information about food sources or dangers. Mice and rats sense pheromones in a patch of sensory neurons removed from the main olfactory receptors, a structure known as the *vomeronasal organ*. Humans do not appear to have a functional vomeronasal organ as adults, and our ability to sense pheromones seems to be extremely limited. It is unclear if humans produce any pheromones and unlikely that humans are able to "sniff out" dangerous people.

FIGURE 15.12 **Smelling cancer?** This dog has been trained to recognize unusual-smelling urine, which may be an indication of bladder cancer. In a recent double-blind trial, several trained dogs were asked to sniff seven urine samples, one of which came from an individual with bladder cancer. The dogs recognized 41% of the diseased samples, significantly better than the 14% that would be expected if their identifications were by chance alone.

Stop and Stretch 15.4

When she was a senior at Wellesley College, Martha K. McClintock asked 135 women in her dorm to keep track of their menstrual cycles. In a summary of her study, published the following year in the journal *Nature*, she showed that the cycles of close friends became more synchronized over the course of the school year than the cycles of more distant acquaintances. McClintock argued that this synchronization illustrates that humans use pheromones to communicate information related to sexuality. Did this research prove human pheromonal communication exists? Why or why not?

15.4 Senses of the Ear

The **ear** contains two sets of mechanoreceptors: One set senses sound waves to produce our sense of hearing and the other senses the position of the head so that we may maintain balance.

Hearing

What we perceive as sound is simply compression waves, most typically waves of compressed air. Consider what happens when you clap your hands together: The air that took up the space between your hands is displaced, creating a pulse of high-pressure air followed by a low-pressure node, resulting in a propagating wave much like the ripples in a pond. All sound waves have this basic structure, created by the vibration of materials.

Individual sound waves do have different characteristics, however, including the *frequency* of the waves (the number of times it cycles from high point to low point and back per second) and the *amplitude* or size of the wave (FIGURE 15.13). Mechanoreceptors in the ear sense these waves, and the nervous system translates them into a perception of sound.

Sound waves make an impact on the **auditory receptors** (sound receptors) in the ear after traveling a complicated path in which the waves are received and amplified. The receiver is the disk-shaped outer ear, also called a **pinna**, which funnels the waves via the **auditory canal** to the eardrum, or **tympanic membrane**. This connective tissue membrane vibrates as sound waves strike it and transfers that vibration to a set of small bones, called **ossicles**, in the middle ear. The ossicles translate the vibration of the tympanic membrane onto a much smaller membrane, called the **oval window**, in the inner ear (FIGURE 15.14). By transferring the energy from a large surface to a small surface, the strength of the vibration is amplified many times.

The inner ear is a complex bony structure that senses both sound and head position. The fluid-filled organ that receives sound, the **cochlea**, is about the size of a pea and shaped like a snail shell. The cochlea consists of two fluid-filled tubes, one inside the other. Vibrations of the oval window are transferred to the fluid in the outer tube of the cochlea. The fluid surge passes over the top of the interior tube, called the **cochlear duct**, around the end of it, and along its bottom, eventually reaching another membrane called the **round window**. The elasticity of the round window allows the energy from the sound wave to dissipate back into the middle ear and eventually out of the *Eustachian* (or *auditory*) *tube*, which empties into the mouth.

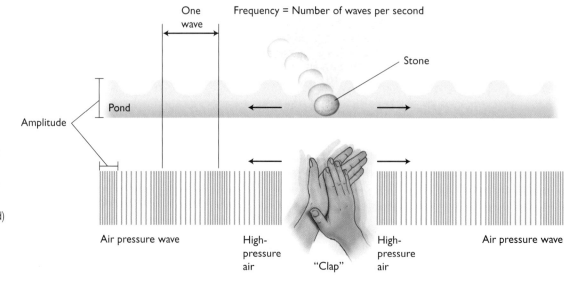

FIGURE 15.13 **Sound waves.** What we perceive as sound is simply pressure waves that cycle in a regular pattern. The frequency of the waves (the number of times that the pressure cycles per second) determines the pitch. The "height" of the waves determines the intensity of the sound.

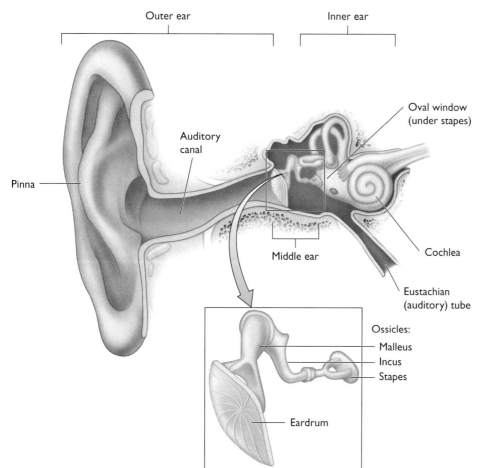

FIGURE 15.14 Pathway of sound to the auditory receptors. Sound waves are funneled by the pinna into the auditory canal, causing the tympanic membrane to begin vibrating. The vibration is amplified as it is transferred through the middle ear by three small ear bones to the inner ear containing the auditory receptors.

The cochlear duct contains the auditory receptors, called **hair cells**, each crowned with around 100 flexible hairlike extensions (FIGURE 15.15). Hair cells are sandwiched between two membranes. The lower surface of the cells sits on an elastic platform called the **basilar membrane**. Resting on the tips of the hair cells is a more gelatinous surface called the **tectorial membrane**. As a fluid wave pulses past the basilar membrane, it causes part of it to vibrate, pushing the hair cells into the tectorial membrane. This contact causes the hairs to bend, triggering the release of a neurotransmitter from the hair cells. The neurotransmitter in turn stimulates a sensory neuron, and a signal is sent via the **auditory nerve** to the brain. The process of sound wave reception is diagrammed in FIGURE 15.16.

Stop and Stretch 15.5

Consider the following question now that you understand the mechanism of sound production and hearing: If a tree falls in a forest and no one is around, does it make any sound?

Sound waves with different properties trigger different neurons as a result of variation in the basilar membrane. The membrane is somewhat like a harp, where the

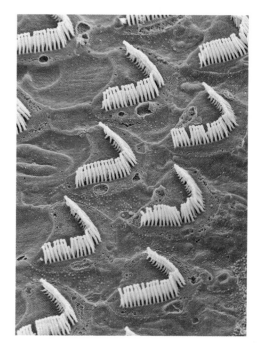

FIGURE 15.15 Auditory receptors. The auditory receptors are hair cells attached to the basilar membrane. When a portion of the membrane vibrates, the hair cells are deformed, sending a signal to a sensory neuron.

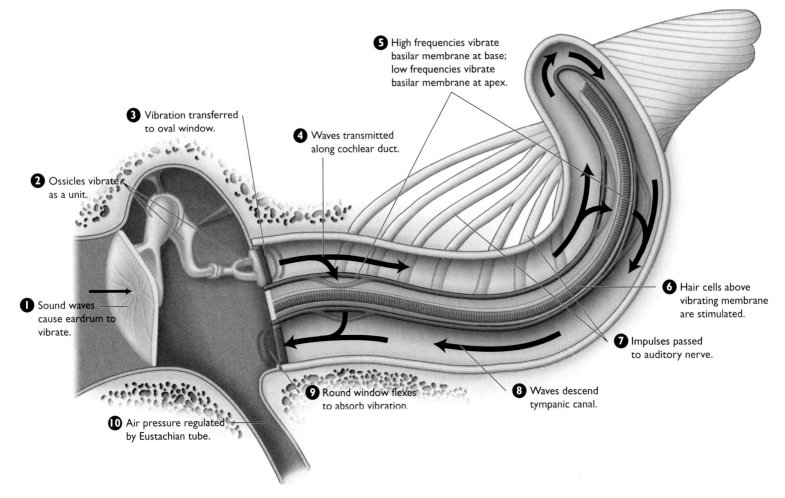

FIGURE 15.16 Function of the cochlea. Vibrations transferred from the middle ear to the oval window of the cochlea produce a pulse of fluid in the organ. The pulse flows around the cochlear duct (here "unrolled" for simplicity), transferring some of its energy to the duct and dissipating at the round window. Waves of different frequency cause different parts of the basilar membrane to vibrate.

"strings" at one end are narrow and stiff and vibrate at a high frequency. The strings graduate in length and decline in tension so that those at the other end vibrate at low frequencies. A high-frequency sound will cause the stiffer part of the basilar membrane to vibrate, while a low-frequency sound will move the looser section of membrane. Only hair cells resting on moving parts of the basilar membrane are forced into the tectorial membrane. Thus, the location on the basilar membrane of stimulated hair cells provides information about the frequency or *pitch* of a sound stimulus. The number of hair cells responding and the strength of their response indicate the amplitude of the sound, which is perceived by us as loudness.

A damaged or dysfunctional cochlea can be replaced by an electronic **cochlear implant** (FIGURE 15.17). These devices transmit sound waves from a receiver in the skull to the cochlea, which has been fitted with electrodes. Different frequency sounds stimulate different electrodes, causing signals to be sent along the auditory nerve to the brain. Although the signals produced by a cochlear implant differ from the signals produced by a functional cochlea, some individuals who have been deaf even since birth can understand speech and hear sounds in the environment as a result of these machines.

Hearing is a key sense for evaluating danger. Sudden loud noises stimulate a rush of the hormone *adrenaline*, which has the effect of preparing us, physiologically, for a

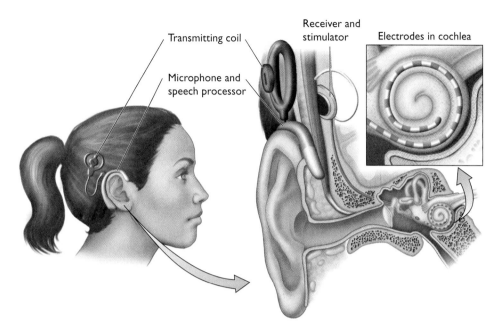

FIGURE 15.17 **A cochlear implant.** This device can restore hearing in even profoundly deaf individuals, as long as their auditory nerve is still functional.

VISUALIZE THIS Return to Figure 15.16 and relate the function of a cochlear implant to the functional cochlea.

dangerous situation. The location of the ears on different sides of the head allows us to better pinpoint the source of a sound by a process of triangulation (FIGURE 15.18).

Our ears are also acutely attuned to the sound of human speech. For example, the pinna is especially good at capturing the sound waves that correspond to the frequencies found in human speech, and the basilar membrane responds best to these frequencies. In a world where survival depended on effective communication among clan members, an auditory system that responded well to the pitch and volume of human speech was an advantage.

However, the ear has its limitations. Sound-wave frequency is measured in a unit called the hertz (Hz), and the basilar membrane responds along its length to frequencies from 20 to 20,000 Hz. Many sounds that might have an impact on our survival are outside this range, including extremely low-frequency sounds, called *infrasounds*, that accompany earthquakes and volcanoes.

Although we do not perceive infrasounds, we may still experience a sensation associated with strong low-frequency compression waves. There is some evidence that experiencing infrasound evokes a feeling of fear and discomfort. When British musicians prepared a concert in which infrasounds were included in some of the pieces and asked audience members to respond to each piece, about 22% of concertgoers described feelings of anxiety, melancholy, or revulsion during the infrasonic events even though they did not report actually hearing these sounds. According to the researchers involved in analyzing these results, their findings support the hypothesis that some allegedly "haunted" sites may actually be places where infrasound is generated by natural forces.

The uncomfortable or sickening feelings some people have when exposed to infrasound may be related to the effect of powerful sound waves on the sensory receptors that monitor head position.

A Sense of Balance

Mechanoreceptors in the inner ear are responsible for sensing the movement of the head and its position in space. These sensory receptors are responsible for our sense of equilibrium or balance. The sensory organ measuring equilibrium is attached to the cochlea

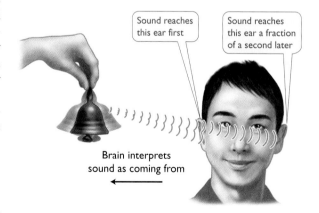

FIGURE 15.18 **Triangulation.** The placement of our ears allows us to locate the source of a sound fairly readily.

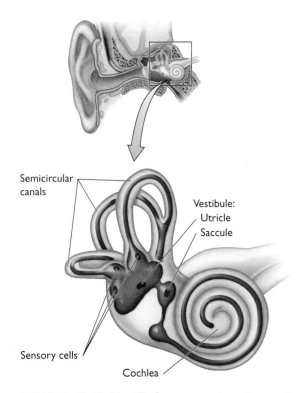

FIGURE 15.19 **Vestibular apparatus.** The equilibrium sense organ in the inner ear consists of a sac with three semicircular canals arising from it. The whole organ is encased in bone and shares space with the cochlea.

and is called the **vestibular apparatus**. This bony structure consists of a basal sac, called the vestibule, with three arching **semicircular canals** attached to it (FIGURE 15.19).

Each semicircular canal projects into one of three dimensions of movement: up-down as in a nod, side-to-side as in a "no," and back-and-forth as when tilting the head. When the head moves in any of these directions, fluid in membranes in the corresponding semicircular canal moves in the opposite direction (much as when a car starts forward, you are pressed back against the seat). The moving fluid impacts a gelatinous structure, called a **cupula**, at the base of the affected canal. Sensory cells that project hairs into the cupula are deformed by the fluid movement, sending signals to the brain for both unconscious and conscious processing (FIGURE 15.20).

The vestibule portion of the vestibular apparatus is divided into two sections, a larger **utricle** and the smaller **saccule**, and measures changes in the head's position relative to gravity, acceleration, and deceleration. Both sections of the sac contain an **otolith organ**, made up of hair cells embedded in a gelatin. In the gelatin are "ear stones," bits of calcium carbonate that move in the gel in response to gravity. As the ear stones move, they distort the hair cells, sending signals about how the head is experiencing changes in position with respect to gravity or to forward progress (FIGURE 15.21).

Signals from the vestibular apparatus are integrated with other sensory signals, especially vision, to help you keep your balance. As a result, any discrepancy between signals from the vestibular system and from other systems can cause motion sickness or a feeling of **vertigo**, an inappropriate sense of motion. When you read in a car or plane, the vestibular system records motion, yet at the same time, your visual system while you look at the page seems to indicate that you are at rest. Often, looking out the window so you can see the moving landscape relieves some of the feelings of illness gener-

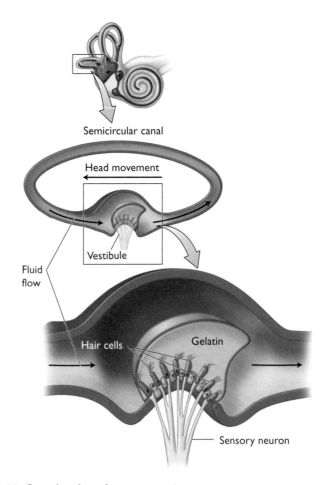

FIGURE 15.20 **Sensing head movement.** Fluid in the semicircular canals begins to flow in response to head movement, triggering mechanoreceptors that send signals to the brain.

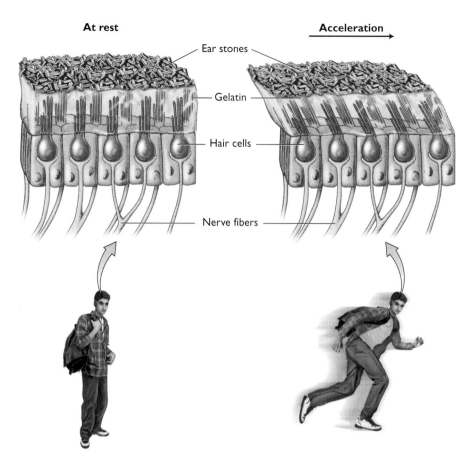

FIGURE 15.21 **Sensing gravity and acceleration.** A stone-filled gelatin rests on the hair cells in the vestibule. When the stones and gelatin move in response to acceleration or vertical movement, the hair cells send signals to a sensory neuron.

VISUALIZE THIS What happens to the ear stones when you stand on your head?

ated by this discrepancy. A feeling of dizziness can also result when cold water or ice applied to the head creates a temperature-related circulation of fluid in the semicircular canals. It may also be that infrasound waves contain enough energy to stir the fluid within the vestibular apparatus, contributing to vertigo in some individuals.

15.5 Vision

By far, our primary sensory system for evaluating danger is sight. The primacy of sight is clear from our own experience, but it is also made obvious by a few facts about our sensory systems. In particular, 70% of all sensory receptors are in the eye, and one-third of the cerebral cortex is involved in visual processing.

Focusing Light

Light is a form of energy that can be thought of in two ways: as particles that are "packets" of energy and as waves that describe the energy level of any given particle. Visual receptors in our eyes respond to light energy between certain wavelengths, a range we call **visible light** (FIGURE 15.22).

Our **eyes** are organs adapted to sensing and focusing visible light. The eyeball itself is a sphere-shaped cavity surrounded by connective tissue (FIGURE 15.23). Much of the surrounding tissue, the **sclera**, is white and opaque, but a small region at the front of the eye, called the **cornea**, is made up of transparent sheets of epithelial tissue that allow light to pass into the eye.

The cornea itself is extremely sensitive to pain—severe damage to this surface is irreversible and could lead to blindness, and the pain receptors help to protect this vulnerable surface. Unfortunately, some forms of damage to the cornea can occur without pain initially, including bacterial and fungal infections that can become established on

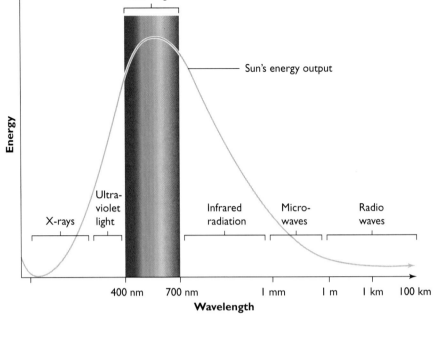

FIGURE 15.22 **The makeup of light.** Light is made up of particles that move in different wavelengths. Shorter wavelengths are violet, while longer wavelengths are red. Only a small part of the spectrum is visible to human beings.

the corneas of contact-lens users. A damaged cornea can be replaced by a transplant of a donated cornea. More recently, artificial corneas have been produced.

Just behind the cornea is a small chamber filled with liquid **aqueous humor**. Aqueous humor is continually recycled, draining through a small channel underneath the eyelid. If this channel becomes blocked, accumulating fluid causes high pressure in the eye, compressing blood vessels and destroying sight. This condition, called **glaucoma**, is a major cause of blindness worldwide. Fortunately, a simple test that measures pressure in the eye can diagnose glaucoma before it causes significant damage. Treatments such as drugs and surgery are available to relieve the pressure.

Light entering the eye is screened by the **iris**, a disk-shaped muscle containing pigment cells that can adjust the **pupil**, the opening that allows light to reach the back of the eye. In bright environments, the iris constricts reflexively to prevent the sensory receptors in the eye from being overwhelmed. In dim light, the iris dilates to permit maximum light gathering. The interior of the eye is so effective at absorbing light from the environment that none is reflected back, and therefore, the pupil appears black. Interestingly, the size of the pupils provides information about an individual's psychological state—they are dilated when we are frightened or interested in something, and constricted when we are bored.

After passing through the pupil, light is focused by the **lens**, another clear structure made of layered proteins, through a cavity filled with liquid **vitreous humor**, onto the **retina** at the back of the eye. The lens is held in place by the **ciliary body**, a structure made up of strands of connective tissue fibers attached to muscles (FIGURE 15.24). The ciliary body can adjust the curvature of the lens to focus images at different distances from the eye onto the retina. To focus on close items, the muscle contracts, causing the fibers to sag. As a result, the lens relaxes and bulges, shortening the focal distance. When the muscle relaxes, the fibers tense, pulling the lens taut and increasing the focal distance.

In about 25% of the population, the relaxed lens focuses the image just short of the retina. This condition is **myopia**, or nearsightedness, and results in blurry images of more distant items. Myopia results when the eyeball itself is too elongated, which can occur during early childhood when children spend a lot of time focusing on close items, tensing the ciliary body and deforming the shape of the eye as it grows. Less common is farsightedness (**hyperopia**), when the eye is shortened such that the focal point for close items falls behind the retina. However, as we age, the lens becomes less flexible and does not relax adequately during close focusing, causing age-related farsightedness and requiring many individuals over 40 to employ reading glasses. Clear focus may also be impaired if the cornea is misshapen, causing light to scatter before it reaches the retina, a condition called **astigmatism**.

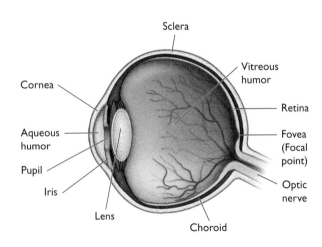

FIGURE 15.23 **The eye.** The external casing of eye is made up of three layers: the outer covering, a middle layer made up of blood vessels and the lens, and the inner layer made up of sensory cells. The center of the eyeball contains fluid.

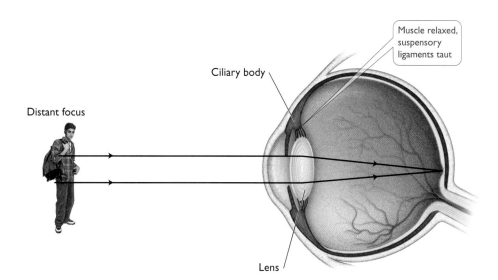

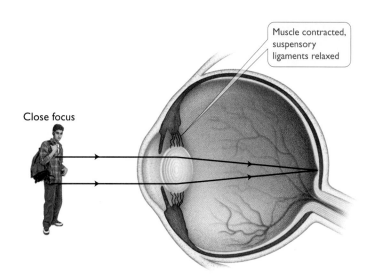

FIGURE 15.24 **Focusing.** The ciliary body contracts and relaxes to modify the shape of the lens. A thinner lens increases the focal distance, and a bulging lens decreases it.

Glasses and contact lenses work by changing the light path from the object to the eye to put the image back on the retina, regardless of the underlying problem (FIGURE 15.25). Procedures that reshape the cornea, such as laser in-situ keratomileusis (better known by the abbreviation LASIK), which uses a laser to remove layers of connective tissue, can also help us regain normal focus.

If the proteins making up the lens become malformed, the lens itself may become cloudy, producing **cataracts**. This condition occurs as a result of diabetes or other diseases, but also happens as a result of aging. Artificial plastic lenses can be surgically inserted to replace lenses clouded by cataracts.

Stop and Stretch 15.6

Exposure to ultraviolet (UV) light, which is common in sunlight, may lead to the development of cataracts. Why might wearing sunglasses that are not treated to protect against UV light transmission actually increase the risk of cataracts as opposed to not using sunglasses at all?

FIGURE 15.25 Improving focus. Irregularly or incorrectly shaped eyes can lead to poor vision. Lenses and surgery can correct these abnormalities.

VISUALIZE THIS LASIK surgery reshapes the cornea to change the way light is focused. How do you expect LASIK changes the shape of an eye with myopia? How about one with astigmatism?

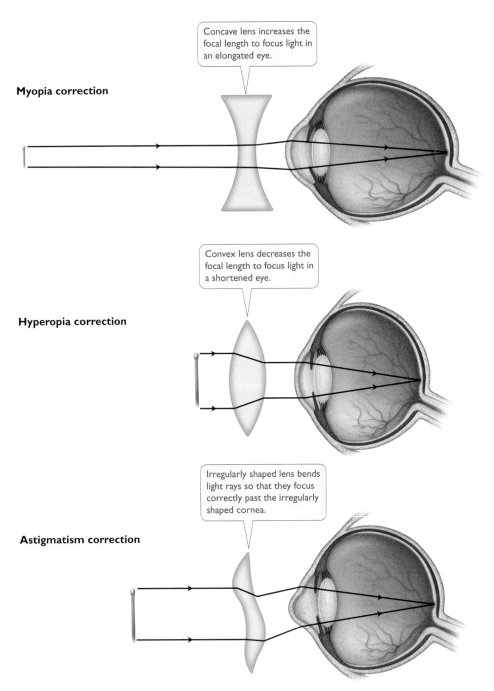

The retina contains the photoreceptors and sits on top of the **choroid**, a layer of tissue containing the capillaries that provide the photoreceptors with nutrients and oxygen. The connection between the choroids and the retina is relatively weak. Any tear or hole in the retina can allow vitreous humor to leak into spaces between the two layers, pulling the retina away in a process called **retinal detachment**. The symptoms of detachment include flashes of light and blurred vision. These danger signs can signal imminent blindness unless the tear is treated immediately, typically with laser treatment that seals the gap. The symptoms of retinal detachment are caused by the impaired and disrupted activities of the photoreceptors.

Photoreceptors

The receptors in the eye respond to light striking a **photochemical**—a compound that changes shape when absorbing light energy. These photochemicals are found in two types of cell: **rods**, which are extremely sensitive and function in night vision, and **cones**, which record color and are used for discerning detail. The eyes contain an astonishing 120 million rods and 6 million cones. Compare that number to the 30,000 hair cells in the ear or even the 10 million olfactory receptors in the nose.

Rods and cones are not replaceable, so their loss results in permanent blindness. In some individuals, the retina begins to break down and is replaced by scar tissue, resulting in "gaps" in the visual field. This condition, called **macular degeneration**, is irreversible and affects up to 30% of people over age 75. Risk factors for macular degeneration include high blood pressure, diabetes, and a family history of aging-related blindness.

Rods and cones are unevenly distributed on the retina. Cones are concentrated at the focal point of the eye, a pinhead-sized dimple called the **fovea**. Rods are much more abundant on the retina as a whole, but they are absent from the fovea and found in maximum concentration in the area surrounding this focal point.

Both rods and cones contain a photopigment called **rhodopsin**, made up of two components, *retinal* (a light-sensitive molecule derived from vitamin A) and a protein, *opsin*. Variation in opsin structure leads to different light absorbances for rhodopsin in different cells. Rhodopsin in rods is very sensitive to light in the middle wavelengths of the visible spectrum, but the forms of rhodopsin in cones are much less sensitive to light energy and are almost completely ineffective in low-light conditions. Variation in rhodopsin forms among cones results in cells that respond mainly to red, green, or blue light wavelengths. These cells are responsible for our ability to see in color (FIGURE 15.26).

Rhodopsin aids in the adaptation of our eyes to different light conditions. Because it becomes disassembled when light strikes it and requires a few minutes to become reassembled, the compound can be "used up" temporarily in bright light. When you first step outside on a sunny day, you will squint to reduce the amount of light entering your eyes. The light that does strike the eye causes the rhodopsin in the photoreceptors to disassemble. As the rhodopsin levels in your eyes decline, the visual system no longer risks being overwhelmed, so your face muscles relax and you cease squinting. On your return indoors, especially to a dim room, it takes a few minutes to see clearly because rhodopsin levels are now low and need to recover before you can adequately sense the lower light levels.

Once a rod or cone responds to light, it sends a signal to an associated sensory neuron called a **bipolar cell** (see FIGURE 15.27 on page 374). Each bipolar cell may serve one or many photoreceptor cells. In general, dozens of rods feed into a single bipolar cell, while each cone has its own dedicated bipolar cell.

The bipolar cells pass their signal to **ganglion cells**, whose long axons make up the **optic nerve** from each eye. More integration of information happens at this step as well. A single ganglion cell will carry information from only a few dozen cones, while another transmits the signals from 240 rods. The differences in integration of rod and cone cells contribute to differences in the acuity of these receptors. Cones provide more clarity than rods because information from many rods is combined before being passed to the brain, producing a fuzzy picture. Rods are thus important for detecting movement in our **peripheral vision**—to our left or right—or at night, but they cannot help us identify the object that is moving peripherally, or help us make out detail in dark conditions.

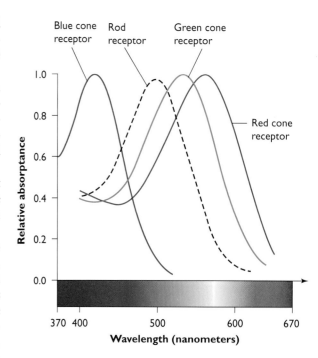

FIGURE 15.26 Color vision. The photoreceptors in the eyes respond to different wavelengths of light. This graph illustrates the absorbance spectra of the four photoreceptors found in a typical eye.

Vision and Perception

Because vision is our primary sense, it is important when evaluating the environment that we are able to trust what we see. Surprisingly, however, what we see is only a subset of surrounding conditions and is strongly influenced by our brain's tendencies to seek patterns. While our visual system is attuned to help us survive and thrive in a variety of environments, it is also easily "tricked."

Consider color vision. Green, red, and blue photoreceptors are not found in equal numbers in each individual. The green photoreceptors nearly always outnumber the other two types. As a result, the color we see is not a perfect representation of the light

FIGURE 15.27 Signal integration. The micrograph shows rods and cones, the sensory receptors in the eye. The light stimuli sensed by rods and cones is integrated into a single signal that is transmitted to the brain. The number of signals integrated together influences the clarity of the signal.

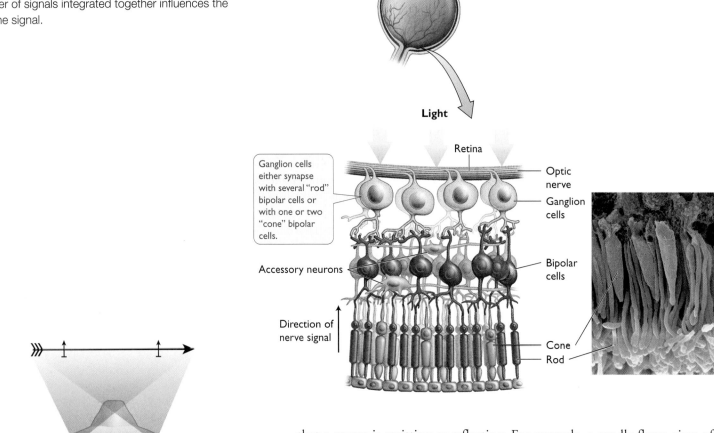

that a source is emitting or reflecting. For example, a candle flame gives off light in the yellow to red spectrum, with more of the energy emitted in the red range, but because we are more sensitive to the yellow light, the flame appears to us to be yellow.

About 15% of the population has some form of color blindness as well, which occurs as a result of mutations in photoreceptor genes. The most common form of color blindness involves the loss of either the red or the green photoreceptor, leading to what is known as red-green color blindness. In this case, red or green appear gray to the observer. The Genes & Homeostasis feature provides additional information about the genetic basis of color blindness.

Finally, as with sounds, there is a range of light wavelengths, including microwaves, infrared light, and X-rays, that our visual systems are not able to perceive. Special cameras, films, and detectors can help us detect these light wavelengths.

The brain performs a complex processing task on incoming visual information (FIGURE 15.28). The image projected on each retina triggers sensory cells that record color, light intensity, or movement. This information is integrated by the ganglion cells and passed via the optic nerves to the **optic chiasm**. At this x-shaped site in the brain, information from each eye is shared so that signals from both eyes are sent to each hemisphere of the brain. Because information is received from both eyes facing the same target, humans are able to perceive the distance between an object and ourselves. In other words, binocular vision (in which the visual field for the eyes overlap) allows us to triangulate objects in our environment and provides us with depth perception.

From the optic chiasm, signals are carried to the **lateral geniculate nucleus** in the brain. Here the information is broken down into its component parts, then passed as a packet of these parts to the visual cortex for reassembly and perception.

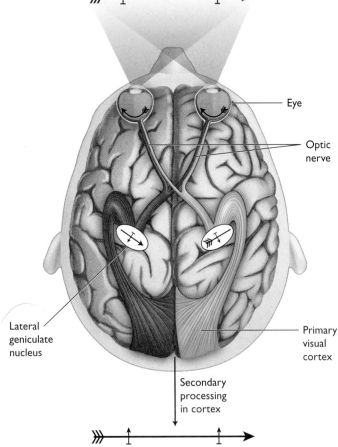

FIGURE 15.28 Visual processing. Perception of visual signals requires a large amount of complex processing, including rotating and reassembling the image formed on the eye.

GENES & HOMEOSTASIS

Color Blindness

John Dalton was one of the early nineteenth century's most famous scientists. He put forth what is now known as the universal atomic theory, which states that all matter is comprised of atoms, that all atoms of a given element are different from all other elements, and that atoms combine with each other to form compounds, which always contain the same relative ratio of types of atoms. For this theory, Dalton is remembered as one of the most influential natural scientists.

Dalton's first paper was on a completely different subject, however; it was entitled "Extraordinary facts relating to the vision of color." In it, he described differences in his color vision relative to others, noting that "orange, yellow, and green seem one color . . . making what I should call different shades of yellow." This 1794 paper is the first scientific description of a relatively uncommon form of color blindness, a deficiency in green cones (see figure).

Charles Darwin noted that color blindness ran in families, but it took the Swiss ophthalmologist Johann Horner, who kept detailed records of his patients' family histories, to deduce the particular pattern of inheritance. In particular, he concluded that "the sons of daughters whose fathers were color-blind, have the greatest chance of being color-blind." This pattern of inheritance is now known to result from the presence of the affected gene on the X chromosome.

Because males carry only one X chromosome (while females have two), a nonfunctional gene on the X chromosome is always apparent in a male. In contrast, a single copy of a nonfunctional gene may be masked in a female who carries a functional gene on her second X chromosome. This type of sex-linked genetic inheritance is discussed in more detail in Chapter 20.

The nonfunctional genes in red-green color blindness appear to be those that code for the opsin proteins of the red- or green-sensitive rhodopsin. These mutations are remarkably common: About 8% of European men are color-blind, and nearly 75% of this group has the deficiency in red rhodopsin, producing nonfunctional red cones.

Why would a mutation that reduces the quality of sensory input be so common? Color carries information about food quality (ripe fruits are rarely green, for instance) and the overall condition of a prospective mate (odd skin color may indicate illness, for example). There is some evidence that individuals with red-green color blindness may be able to distinguish subtle shades of brown and tan that individuals with normal vision cannot. Whether this ability had survival value for our ancestors remains to be seen.

It is also possible that the mutation did not significantly reduce the survival or reproductive rates of color-blind men. As a result, the trait remained in the population. In fact, simple chance could have made it more common. If a few female carriers of the defective genes early in human history had a large number of children either by chance or because they carried another successful trait, the mutation could have become more common in the population. This phenomenon, called genetic drift, also seems to explain other surprisingly common mutations, such as the inability to roll one's tongue into a tube shape, or the presence of a widow's peak (a point in the hairline above the forehead).

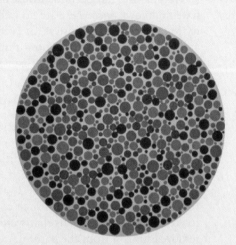

Individuals with normal color vision should be able to distinguish a number from the background pattern in this circle.

All of this visual processing occurs in a matter of nanoseconds—to handle so much information effectively, the brain has evolved certain shortcuts to "fill in" information. You can see a dramatic example of how this works by looking at FIGURE 15.29. The point where the optic nerve exits the retina lacks any photoreceptors—that is, it is a blind spot in the visual field. We do not perceive a blind spot, however, because the visual cortex fills in the missing information to produce a coherent image.

Because our survival depends on our ability to perceive patterns in the environment (to effectively exploit a resource or avoid danger), our visual cortex is well suited to

FIGURE 15.29 **The blind spot.** Close your left eye and stare at the cross mark in the diagram at left. Off to the right you should be able to see the black dot. Now slowly move the book toward your face, while continuing to look at the cross. At about 1 foot from your eye, the dot will seem to disappear because its image falls on the head of the optic nerve on your retina. You perceive a continuous white field there because your brain is "filling in" the gap in your visual field.

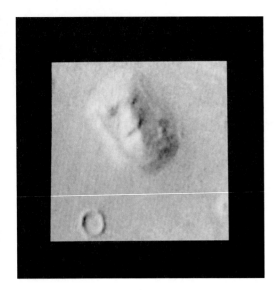

FIGURE 15.30 **Seeking patterns.** Our brains are proficient at perceiving pattern where only a shadow of one exists. Do you see a face in this photograph of the surface of Mars?

create order out of seemingly chaotic visual stimuli (FIGURE 15.30). The downside of this skill is that we will sometimes identify patterns where none exist, which can lead to a heightened sense of danger where no increased risk lies. A familiar example of this is when a child clearly sees a "monster" in a darkened bedroom that in the light is obviously a piece of furniture or pile of clothing.

15.6 Predicting the Future

Upon examining all the known sensory systems in humans, it is clear that no physiological mechanism has been identified that allows us to predict future events for which we have no knowledge. There are no well-supported experimental tests that show any ability to predict future disasters, or even everyday events. With no experimental support and no clear mechanism for how it might happen, it appears that humans do not have a "sixth sense." How, then, do we explain the stories at the beginning of the chapter?

Understanding Premonitions

One factor that can help us understand premonitions is the phenomenon of *recall bias*. Someone who had an experience where their premonition was correct is much more likely to recall that premonition than to recall the hundreds of premonitions they've had that were not fulfilled. The tourist who delayed her visit to the World Trade Center on September 11, 2001, likely has had other vague fears that caused her to change plans, but without such consequence. Another source of recall bias is failing to recall the danger signals that were present before an event occurred. The man who removed his son from the path of a speeding vehicle may not recall that the sounds of squealing tires and a revving engine in the near distance triggered his actions.

Anticipating a call from a friend may also be remembered as a result of recall bias. It could also reflect a shared experience. Learning that the basketball team from your old high school just won the state championship might lead you to memories of school, and to recall some good friends from that time in your life. One of those friends may hear the same news, think of you, too, and follow up with a phone call.

Expanding the Receptive Field

Humans today are born with the same limited sensory capacities as our ancestors evolving in Africa had. However, where biological evolution has left off, social evolution has taken over. Humans have developed a sixth sense of a kind—the ability to make predictions about future conditions—by using our toolmaking skills to expand our receptive fields.

This chapter has described many of these tools already—infrared cameras and receivers for infrasound, for example. Other tools that expand our ability to sense environmental conditions include weather and communications satellites, seismographs for detecting earth movements, and chemical tests for substances in air, water, and even our own blood.

Tools that expand our receptive field alert us to conditions that are fast approaching, for instance, a line of thunderstorms that threatens to spawn tornadoes, or a precipitous change in blood-sugar level that could signal the onset of a diabetic crisis. However, these tools in concert with the scientific method can also give us insight into the more distant future.

The scientific method allows humans to create models of how natural processes function and to test them against reality. If our hypotheses are supported by tests, we can use this information to predict conditions in the future based on conditions today. This is how a physician predicts future heart trouble in a patient with high blood-cholesterol levels, and how a climatologist predicts droughts and coastal flooding in response to rising carbon dioxide levels in the atmosphere. A technological and social sixth sense has allowed humans to use and shape the natural world to fit our needs and desires. Our question is not whether we have the capacity to predict certain future events, but instead how we use this knowledge to chart our course in a challenging world.

Chapter REVIEW

ROOTS TO REMEMBER

The following roots of words come mainly from Latin and Greek and will help you to decipher terms:

Several suffixes in this chapter mean "little," such as **-ula, -uscle,** and **-icle**. For example,

corpus- means "body," so **corpuscle** is "little body."

aqu- is water.

audi- or **audit-** mean hear.

infra- is beneath or below.

cul- and **oculo-** mean the eye.

olfac- comes from a verb for smell.

-opia and **ophthalm-** are words meaning eye.

oss- and **oto-** mean bone.

photo- means light.

proprio- means one's own.

thermo- means heat.

vitr- is glass.

ultra- is a prefix meaning beyond.

KEY TERMS

15.1 Sensing and Perceiving
chemoreceptor *p. 354*
depolarize *p. 355*
externoreceptor *p. 354*
general sense *p. 354*
internoreceptor *p. 354*
mechanoreceptors *p. 354*
nociceptor *p. 354*
pain receptor *p. 354*
perception *p. 355*
photoreceptor *p. 354*
sensation *p. 355*
sense organ *p. 354*
senses *p. 354*
sensory adaptation *p. 355*
sensory receptor *p. 354*
special sense *p. 354*
stimuli *p. 354*
thermoreceptor *p. 354*

15.2 The General Senses
chronic pain *p. 360*
fast pain *p. 360*
Golgi tendon organ *p. 357*
Meissner's corpuscle *p. 358*
Merkel disk *p. 358*
muscle spindle *p. 356*
Pacinian corpuscle *p. 358*
proprioceptor *p. 356*
receptive field *p. 358*
referred pain *p. 360*
Ruffini corpuscle *p. 358*
slow pain *p. 360*

15.3 The Chemical Senses
gustatory receptor *p. 361*
odorant *p. 362*
olfactory bulb *p. 363*
olfactory receptor *p. 362*
papillae *p. 361*
pheromone *p. 363*
tastant *p. 361*
taste bud *p. 361*
taste cell *p. 361*
taste hair *p. 361*

15.4 Senses of the Ear
auditory canal *p. 364*
auditory nerve *p. 365*
auditory receptor *p. 364*
basilar membrane *p. 365*
cochlea *p. 364*
cochlear duct *p. 364*
cochlear implant *p. 366*
cupula *p. 368*
ear *p. 364*
hair cell *p. 365*
ossicle *p. 364*
otolith organ *p. 368*
oval window *p. 364*
pinna *p. 364*
round window *p. 364*
saccule *p. 368*
semicircular canals *p. 368*
tectorial membrane *p. 365*
tympanic membrane *p. 364*
utricle *p. 368*
vertigo *p. 368*
vestibular apparatus *p. 368*

15.5 Vision
aqueous humor *p. 370*
astigmatism *p. 370*
bipolar cell *p. 373*
cataract *p. 371*
choroid *p. 372*
ciliary body *p. 370*
cones *p. 373*
cornea *p. 369*
eyes *p. 369*
fovea *p. 373*
ganglion cell *p. 373*
glaucoma *p. 370*
hyperopia *p. 370*
iris *p. 370*
lateral geniculate nucleus *p. 374*
lens *p. 370*
macular degeneration *p. 373*
myopia *p. 370*
optic chiasm *p. 374*
optic nerve *p. 373*
peripheral vision *p. 373*
photochemical *p. 372*
pupil *p. 370*
retina *p. 370*
retinal detachment *p. 372*
rhodopsin *p. 373*
rods *p. 373*
sclera *p. 369*
visible light *p. 369*
vitreous humor *p. 370*

SUMMARY

15.1 Sensing and Perceiving

- Sensory receptors receive stimuli and send the information gathered via sensory neurons to the brain (p. 354; Figure 15.2).

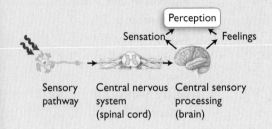

- Sensory receptors can be defined by the type of input they receive—chemical, mechanical, temperature, light, or pain—and by whether they collect data from inside or outside the body (p. 354).

- General senses depend on receptors scattered around the body, while special senses rely on stimuli gathered by specialized sense organs in the head (p. 354).

- Conscious perception of a stimulus can occur only when a stimulus is strong enough or important enough to send a signal all the way to the brain's cortex. Sensory receptors can become adapted to stimuli over time, and their signals can drop below the point of perception (p. 355).

15.2 The General Senses

- Proprioception depends on sensory receptors in the muscles (muscle spindles) and joints (Golgi tendon organs) and helps us maintain balance and posture (pp. 356–357).

- Our sense of touch is mediated by a number of different receptor types in the skin, which measure air movement, light or hard pressure, and vibration (p. 358).

- Temperature is sensed by free nerve endings in the skin and thermoreceptors inside the body (pp. 359–360).

- Pain receptors respond to either mechanical disruption or chemicals released from damaged tissue. The skin is rich in these receptors, but they are less common in internal organs. Damage to internal organs often manifests as referred pain (p. 360).

15.3 The Chemical Senses

- Taste is sensed by taste buds, found mostly on the tongue. These structures consist of chemoreceptors called taste cells that communicate with sensory neurons to produce a perception of flavor as sweet, salty, sour, or bitter when exposed to the chemicals in food (p. 361).

- Our sense of taste serves a protective function, triggering food rejection upon contact with bitter tastants, which are common in poisonous or spoiled foods (p. 361).

- Smell receptors are found on the roof of the nasal passage and are made up of sensory neurons containing specialized chemoreceptors in their plasma (cell) membranes. Humans can recognize thousands of distinct odors (pp. 362–363; Figure 15.11).

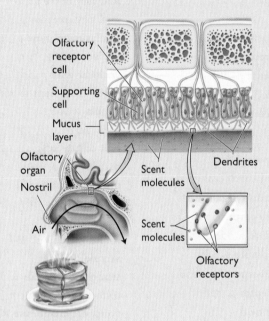

15.4 Senses of the Ear

- Our ears can sense the waves that we call sound (p. 364).

- Sound waves enter the auditory canal and impact the tympanic membrane, causing it to vibrate. This vibration is translated to the cochlea by small bones in the middle ear (p. 364).

- Vibrations transferred to the oval window of the cochlea cause fluid within this organ to pulse. As the fluid passes by the hair cells, it causes part of the membrane holding the hair cells to vibrate. The hair cells deform, triggering an electrical signal that is picked up by sensory neurons and sent to the brain (pp. 365–366).

- Sudden sounds trigger the stress response, but we are unable to perceive very low-frequency sounds that might also signify danger (p. 367).

- The vestibular apparatus senses head position, allowing us to maintain equilibrium (p. 367).

- The semicircular canals measure rotational movement when fluid in the canals moves in response to head movement. When this fluid impacts a structure containing hair cells in the canal, information about movement is transmitted to the brain (p. 368).

- Organs in the vestibular apparatus also measure acceleration and head position in respect to gravity when small pieces of calcium carbonate displaced by forward or vertical movement cause mechanoreceptors to bend (pp. 368–369).

15.5 Vision

- Our primary sense is vision, which is the measurement and interpretation of reflected or transmitted light. Light is sensed in the eye after it passes through the clear cornea, is screened by the iris, and is focused by the lens onto the retina (pp. 369–370; Figure 15.23).

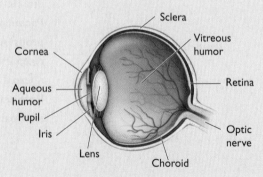

- The lens can change shape to resolve images at different focal distances but may not be able to correct the light path enough so that it falls on the retina. In this case, contact lenses, glasses, or eye surgery can refocus the image on the sensory cells (pp. 370–371).

- The sensory cells of the eye are rods and cones. These cells transmit an electrical signal to a sensory neuron when a photochemical within them changes conformation in response to light absorption. Rods are highly sensitive to light, but the images they produce are black and white, and they are grainy because many rods communicate with a single sensory neuron. Cones provide color vision because the photochemicals of different cones respond to different light wavelengths, and they provide acuity because they have a less diluted connection to the visual center in the brain (pp. 373–374; Figure 15.27).

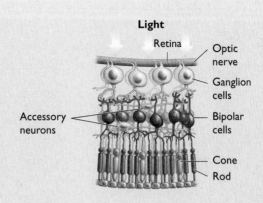

- There are many steps in visual processing, including deconstructing signals from the eye and reassembling them into a coherent image in our brains. Our visual system can be tricked because the brain transforms information into patterns when translating sensory input into an image (pp. 374–375).

LEARNING THE BASICS

1. A stimulus striking the appropriate sensory receptor will _____.
   a. always be detected
   b. always be transmitted via sensory neuron
   c. always reach the central nervous system
   d. always be perceived
   e. none of the above are correct

2. Perception relies on both sensing a stimulus and _____.
   a. responding to the stimulus
   b. processing the stimulus in the cerebral cortex
   c. receiving the stimulus in the brain
   d. adapting to the stimulus
   e. discussing the stimulus with others

3. A continuous signal that was once perceived may fade from perception because _____.
   a. of sensory adaptation
   b. the receptor may stop sending a signal after a time
   c. a gatekeeper in the brain may cease transferring the information to the cortex
   d. a and b are correct
   e. a, b, and c are correct

4. Which type of receptor responds to physical touch or pressure?
   a. mechanoreceptor
   b. thermoreceptor
   c. photoreceptor
   d. chemoreceptor
   e. pain receptor

5. All of the following are general senses except _____.
   a. touch
   b. temperature
   c. pain
   d. taste
   e. vibration

6. Damage to internal organs may be felt as referred pain elsewhere because _____.
   a. damage to one organ causes damage elsewhere
   b. the nervous system is incapable of monitoring internal body conditions
   c. sensory neurons from internal organs share nerve pathways in the spinal cord with body surfaces
   d. the damage releases chemicals that travel through the body, triggering distant receptors
   e. damage to internal organs is not as life threatening as damage to the body surface

7. Proprioceptors in the muscles and tendons _____.
   a. measure muscle stretching and tendon tension
   b. maintain a small amount of activity at rest to maintain muscle tone
   c. help prevent injury by triggering muscle relaxation under intense strain
   d. a and b are correct
   e. a, b, and c are correct

8. Taste cells in the mouth can sense all of the following taste qualities except _____.
   a. sweet
   b. bitter
   c. spicy
   d. salty
   e. sour

9. The number of different types of chemoreceptors in the nose _____ the number of different types in the mouth.
   a. greatly exceeds
   b. is roughly equal to
   c. is equivalent to

10. Which sense is most strongly associated with the limbic system of the brain, the region most closely associated with memory and emotion?
    a. sight
    b. taste
    c. smell
    d. touch
    e. hearing

11. Which of the following represents the path of a sound stimulus in the human body?
    a. outer ear, eardrum, cochlea, ossicles, hair cells
    b. eardrum, ossicles, cochlea, hair cells, auditory nerve
    c. auditory canal, cochlea, eardrum, hair cells
    d. cochlea, eardrum, ossicles, auditory nerve
    e. outer ear, hair cells, cochlea, ossicles

12. How does the ear distinguish between different sound waves?
    a. Different hair cells respond to different amplitudes of sound.
    b. Parts of the membrane supporting the hair cells vibrate at different frequencies.
    c. Louder sounds result in more hair cells being stimulated.
    d. b and c are correct
    e. a, b, and c are correct

13. Receptors in the vestibular apparatus in the inner ear use the movement of fluid or gel to measure _____.
    a. intensity of sound
    b. position of the limbs
    c. muscle tone
    d. change in position of the head
    e. calcium carbonate levels in the body

14. The blind spot in the visual field corresponds to _____.
    a. the site where the optic nerve exits the eye
    b. the focal point, where few rods are found
    c. a space on the lens that cannot focus light rays to the retina
    d. the area that the iris is closing off to light penetration
    e. the region that is most densely packed with rod cells but no cone cells

15. Vision in dim light is less clear (acute) than vision in bright light because _____.
    a. cones are more sensitive to light than rods
    b. the photopigment in rods breaks down rapidly
    c. many rod cells send information to a single ganglion cell
    d. the lens cannot focus the light coming through the widened pupil
    e. colors in the dark are less distinct than colors in the light

16. Match the stimulus with the type of receptor it triggers.
    _____ natural gas odor   a. photoreceptor
    _____ bull's-eye on a target   b. chemoreceptor
    _____ robin's song   c. mechanoreceptor
    _____ heat produced by   d. pain receptor
          sunlight on the skin   e. thermoreceptor
    _____ pinprick

17. Match the following ear structures with their function.
    _____ cochlear hair cells   a. provide information about head movement
    _____ semicircular canals   b. transmit sound to tympanic membrane
    _____ ossicles   c. convert sound waves into electrical signals
    _____ auditory canals   d. amplify sound waves

18. Match the following eye structures with their function.
    _____ rods   a. regulates the amount of light reaching photoreceptors
    _____ cones   b. responsible for color vision
    _____ lens   c. surface where photoreceptors are located
    _____ retina   d. responsible for vision in dim light
    _____ iris   e. focuses light on photoreceptors

ANALYZING AND APPLYING THE BASICS

1. How do your sensory receptors allow you to work in a noisy office or sleep in an apartment near busy traffic?
2. What is likely to happen to the proprioceptors in a young girl as she trains to be a gymnast? Explain your answer.
3. Why would the body be adapted to have more pain receptors on the skin than in the internal organs?
4. People can be divided into two different groups when it comes to the ability to taste the chemical PTC. "Nontasters" taste nothing when exposed to PTC, but "tasters" detect a bitter taste. Some tasters are "supertasters." Supertasters tend to be more sensitive to intense flavors. The bitter compounds in broccoli and other related vegetables, for example, can make those foods unpalatable to supertasters. What are the likely physiological differences between nontasters, tasters, and supertasters?
5. In 2005, a Welsh security company developed the Mosquito—an ultrasonic buzzer that could be played outside of convenience stores and other establishments to discourage loitering by teenagers. The high-pitched frequency of the Mosquito was designed only to be heard by young ears. In 2006, the Mosquito was developed into a ringtone. Teenagers have discovered this tone, and some are using it to their advantage. In the classroom, where cellular phones are banned, the high-pitched ring tone can signal a call or text message without the teacher hearing it. From the information above, what can you deduce about the effects of age on the structures of the ear?
6. High-amplitude sound waves vibrate the basilar membrane to such a degree that the "hairs" on hair cells are sheared off. If scar tissue forms as a result, these hairs cannot regenerate. What does this information tell you about the effect of exposure to loud sounds on lifetime hearing ability?
7. Carrots contain abundant vitamin A. Retinal, a component of rhodopsin, is derived from vitamin A. What aspects of vision would be most responsive to an increase in carrot consumption?

CONNECTING THE SCIENCE

1. You may be aware of the controversy surrounding the use of steroids, chemicals that enhance muscle size and power, in professional and amateur sports. There has been much less controversy about the use of other enhancements. In 1998, golfer Tiger Woods had LASIK surgery, improving his vision to 20/15—meaning that the clarity with which he sees at 20 yards is equivalent to "perfect" vision at 15 yards. Why is Woods's LASIK surgery much less controversial than a baseball slugger's use of steroids? How do you feel about the "fairness" of athletes using the two enhancements?
2. "The Amazing Randi," a famous debunker of supernatural claims, has offered a $1 million prize to the first psychic or medium who demonstrates a power or event that cannot be adequately explained by known phenomena. No one has passed the preliminary tests (a simple demonstration of the phenomenon before a member of Randi's staff). Randi uses this result as evidence that there are no true psychic or extrasensory phenomena. Do you think Randi is correct in his evaluation? Why or why not? What evidence would be necessary to *disprove* the hypothesis of the existence of extrasensory perception?

Chapter 16

The Endocrine System

Worried Sick

LEARNING GOALS

1. Summarize the basic structure and function of the endocrine system.
2. Contrast protein and steroid hormone action.
3. State the difference between an exocrine gland and an endocrine gland.
4. Describe how the hypothalamus and pituitary interact in maintaining homeostasis.
5. List the nine primary endocrine organs and describe where they are found in the body.
6. How does the thymus serve as a junction between the endocrine and immune systems?
7. Provide two examples of locally acting hormones and describe their effects.
8. Describe the endocrine dysfunctions that cause Cushing's syndrome and Graves' disease.
9. How are blood-glucose and calcium homeostasis maintained?
10. How do the actions of prostaglandins differ from those of other hormones?

Modern life is filled with stressors.

The sound of your phone awakens you with a start. You untangle yourself from covers twisted by your tossing and turning during a restless night. You reach for the phone. Your head aches dully as you listen to the caller. Perhaps it is a classmate who is wondering if you can reschedule the group-project meeting for later in the morning. Maybe it's your boss, calling to see if you can come in for a few extra hours. Or maybe it's your child's day-care provider letting you know that she is too sick to care for your kids today. You have to make several phone calls to adjust your plans, but finally you hang up and rush to the bathroom for a quick shower, nearly falling over the pile of towels near the door. You need to get to that laundry later, too.

During your commute, you get stuck behind what seems to be the slowest group of travelers in town. While you are waiting, you try to remember all the items on today's to-do list: There's the rent check that is late and has to be taken to the post office, a meeting with your academic advisor, that big test in human biology to study for, and, oh yes, the laundry.

Some are threats to our physical safety.

And something else is nagging . . . what is it? You rack your brain and distractedly almost cross in front of traffic, a problem that you don't realize until you hear the squealing of brakes. For the second time in a few hours, a loud noise makes you gasp, causes your heart to race. By the time you get to your destination, you feel exhausted, even though the whole day is still in front of you. The headache you had this morning is becoming more insistent, and now you think you notice that scratchy feeling in your throat that signals the start of a cold.

In a prosperous country, though, most stresses are psychological.

Details may differ, but most of us have had days like these, where stressful event follows stressful event. You may have noticed the shakiness you feel after a sudden shock—like a barely avoided car accident. And you may have suspected that there is a

Is chronic stress making us sick?

relationship between the daily stresses in our lives and sleep difficulties or susceptibility to illness. In this chapter, we explore how the body responds to a stressful event and examine the effect of chronic stress on health. Is it possible to make yourself sick by worrying?

16.1 An Overview of the Endocrine System

The endocrine system consists of chemical messengers called hormones and the glandular structures that secrete them. **Endocrine glands** secrete their products into the bloodstream. In contrast, exocrine glands have ducts through which they secrete their products, such as sweat.

Hormones: Chemical Messengers

Hormones are chemicals that travel through the blood and elicit a response from specific cells and tissues. Although hormones are released into general circulation, each only affects a subset of the cells that it comes in contact with. A tissue or cell stimulated by a particular hormone is called that hormone's **target tissue** or **target cell**. The targets contain protein **receptors** for the particular hormone. The relationship between a hormone and its receptor is like that between a lock and a key. Keys are only useful for one type of lock, and once a key is inserted in a lock, it causes something to happen (such as an engine to turn on or a latch to open).

Hormones elicit a specific response from a cell through two general mechanisms: (1) Hormones can bind to receptors on the cell's surface and trigger a change inside the cell, or (2) hormones can diffuse across the cell membrane and bind to receptors inside the cell to trigger the response. How a given hormone elicits a response is a function of its chemistry. Some hormones are not able to cross the membrane and need to act through membrane receptors. Lipid-soluble hormones can cross the membrane to access their intracellular receptors.

Protein hormones consist of amino acids and most cannot cross the membrane. A protein hormone acts on a cell by binding to a receptor on its exterior surface, ultimately triggering a change in the cell.

A chain reaction that relays, or *transduces*, a message from the outside of the cell into an action that occurs inside the cell is called a **signal transduction** pathway (FIGURE 16.1). In the first step of the pathway, a protein hormone, called the *first messenger*, binds to a receptor on the surface of a target cell, and the receptor changes shape. In the next step, this modified receptor signals an intracellular enzyme to convert molecules of ATP within the cell into cyclic AMP (cAMP). Cyclic AMP is a **secondary messenger** that stimulates a cascade of enzyme actions inside the cell.

For example, a signal transduction pathway leads to the conversion of glycogen, an energy-storage molecule, into glucose, as shown in Figure 16.1. Glucose produced by this signal transduction pathway flows into the bloodstream from millions of target body cells, providing an energy boost to muscles.

The other major class of hormones is **steroid hormones**, synthesized from the precursor molecule cholesterol. These nonpolar molecules can diffuse across the membrane. Once inside a target cell, they bind to receptors in the cytoplasm. The hormone-receptor complex then travels to the cell nucleus, initiating transcription of genes that code for enzymes and other proteins that carry out a response to the hormonal signal (FIGURE 16.2).

For example, in the smooth muscle cells ringing blood vessels, a hormone-receptor complex induces the production of angiotensin II receptors. The protein hormone angiotensin II binds to these receptors, causing the vessels to narrow, which increases blood pressure.

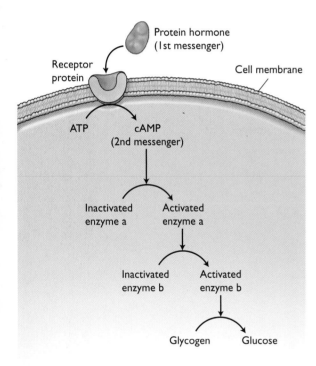

FIGURE 16.1 Protein hormones and signal transduction. Most protein hormones cannot cross cell membranes and must exert their effects via a signal transduction pathway. When a protein hormone binds to a receptor on the surface of a target cell, it can cause a cascade of events to occur inside the cell. One result of hormone binding is the conversion of glycogen to glucose and its release in the bloodstream.

VISUALIZE THIS Can you think of a molecular way to decrease this response?

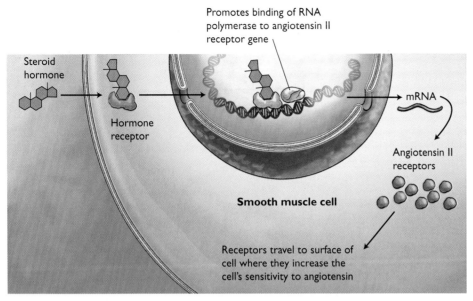

FIGURE 16.2 Steroid hormone function. Steroid hormones such as cortisol can cross cell membranes. If the cell contains a receptor for the steroid, the hormone-receptor complex will move to the nucleus and increase the transcription of certain genes. In smooth muscle cells of blood vessels, the genes expressed code for the production of angiotensin II receptors, making the blood vessels more responsive to hormones that increase blood pressure.

VISUALIZE THIS How would the response to cortisol be different if the cell did not contain a cortisol receptor?

The synthesis of angiotensin II receptors represents a long-term response to stress, tuning the body so that it can respond more efficiently to repeated stressful events. More than 50 hormones have been identified in the human body—this chapter covers only some of the most influential.

Many synthetic hormones injected by unscrupulous athletes mimic the actions of steroid hormones. What types of genes would likely undergo increased expression due to the actions of hormones that increase athletic performance?

Endocrine Glands

The endocrine system was traditionally thought of as consisting of nine glands: the hypothalamus, pituitary, pineal, thyroid, parathyroid, thymus, adrenal, pancreas, and gonads (FIGURE 16.3). As the field of endocrinology has developed, new evidence has led to new thinking about endocrine glands. Scientists now know that endocrine tissue is not confined to these particular organs but can be found throughout the body

Stress and the Endocrine System

A stressful event can be either physical, for example, the threat to life and limb represented by an oncoming car, or psychological, like anxiety over an upcoming exam. Regardless of its source, the biological consequences of stress are remarkably similar: The heart rate, breathing rate, and blood pressure increase, glucose floods into the bloodstream, digestion is halted, pain perception is dulled, and alertness and memory improve. These physiological changes are hallmarks of the **stress response**, which tunes the body for intense muscular activity—a fight (for survival) or rapid flight (that is, retreat from danger).

FIGURE 16.3 The endocrine system. The locations and functions of the major endocrine glands are shown.

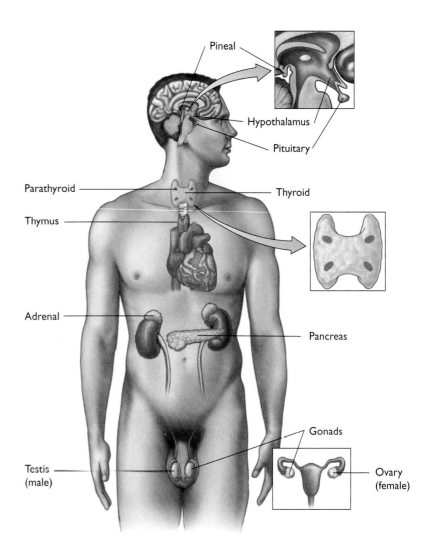

The details of the stress response illustrate the connection between the body's nervous system (Chapter 14), which is made up of neurons and organized to respond quickly to stimuli, and the endocrine system, which triggers a slower, but more prolonged, response to stimuli. In fact, in some aspects of the stress response, the linkage between the two systems is so close that they can properly be called a *neuroendocrine system*.

For example, exposure to a stressor causes neurons of the sympathetic nervous system to activate. Some sympathetic neurons stimulate the heart, blood vessels, and lungs to increase oxygen delivery to muscles right away. Others travel to endocrine glands that release hormones involved in the stress response.

The endocrine glands responsible for the majority of the physiological changes of the stress response are the two **adrenal glands**. Each adrenal gland is a pyramid-shaped structure about the size of a walnut—one atop each kidney. Adrenal glands are made up of two functionally different tissues, the centrally located **medulla**, or inner layer, and the peripherally located **cortex**, or outer layer (FIGURE 16.4).

The adrenal medulla secretes **epinephrine** (also known as *adrenaline*) and **norepinephrine** (*noradrenaline*). The cells of the medulla are actually neurons that represent the end of a nerve pathway extending from the brain. When a stressor causes these neurons to fire, their stores of epinephrine and norepinephrine are secreted into the capillaries inside the adrenal glands.

Epinephrine and norepinephrine cause many of the immediate effects of the stress response: an increase in heart rate, redirection of blood flow from digestive organs to the muscles, and dilation of the pupils of the eyes.

Stressful events also cause the release of two types of steroid hormones from the adrenal cortex. **Mineralocorticoids** are often involved in the retention of minerals. The most common human mineralocorticoid, **aldosterone**, regulates sodium and potassium in the blood. **Glucocorticoids** are steroid hormones that bind to a cortisol receptor. The hormone **cortisone** is produced in response to stress. When converted into its active form, **cortisol**, it gives rise to a longer-term stress response such as increased blood pressure and blood-glucose levels. In fact, glucocorticoids were named for their ability to affect glucose metabolism. Cortisol also has anti-inflammatory effects, and its synthetic form *hydrocortisone* is sold as a topical treatment for inflammation.

16.2 The Endocrine System and Homeostasis

The endocrine system, along with the nervous and immune systems, functions to preserve internal conditions in the face of constantly changing external conditions. For example, when a person moves from a city near sea level (for example, New York) to one at high altitude (say, Denver), the hormone erythropoietin is released in response to lower oxygen levels and increases the concentration of red blood cells. In general, after combating one stressful event, the body becomes prepared to respond even more powerfully to stress in the near future.

The stress response is central to maintaining homeostasis. However, some features of modern life can lead to stress-related disease. Tuning the body to respond to stress effectively was important to survival in our past, when certain stressors were part of a relatively brief event (for instance, during a hunt, when muscles were called repeatedly, and often suddenly, into intense activity). For most of us in today's circumstances, the stressors tend to be psychological—such as an upcoming exam or interactions with a difficult boss—and do not require the same kind of physical response. In addition, because modern stressors often last weeks, months, or even years, a strong stress response can become harmful to health.

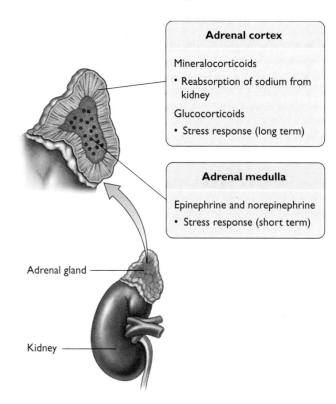

FIGURE 16.4 **The adrenal gland.** Each adrenal gland contains two layers: the outer surface, or cortex, and the inside, called the medulla. Most of the hormones produced by the adrenal glands are associated with the stress response.

The Control Center: The Hypothalamus

When a stressful event happens, nearly all of our reactions to it are controlled by the **hypothalamus**, located deep inside the forebrain. The hypothalamus coordinates a variety of homeostatic activities via the nervous system, including regulating blood pressure, heart rate, and breathing rate, and stimulating enhanced alertness and memory. This complex structure also functions as a neuroendocrine organ, linking the nervous and endocrine systems together.

Much of the regulatory activity performed by the hypothalamus is accomplished by directing the activities of the **pituitary gland**, a grape-sized endocrine organ that sits directly below the hypothalamus and is attached to it by a thin strand of brain tissue. The pituitary gland is composed of two distinct structures, the **anterior pituitary** and the **posterior pituitary** (FIGURE 16.5). The pituitary gland is sometimes called the "master gland," because it produces hormones that affect a wide range of other endocrine and nonendocrine tissues.

The hypothalamus also produces hormones. Two of these, **antidiuretic hormone** (**ADH**) and **oxytocin**, are produced by neuroendocrine cells that extend into the posterior pituitary. These hormones are stored there until an appropriate signal triggers their release.

Unlike ADH and oxytocin, which have effects on a number of different target tissues, the remaining hypothalamic hormones have a single target tissue—the anterior pituitary. These **regulatory hormones**—so called because they control other

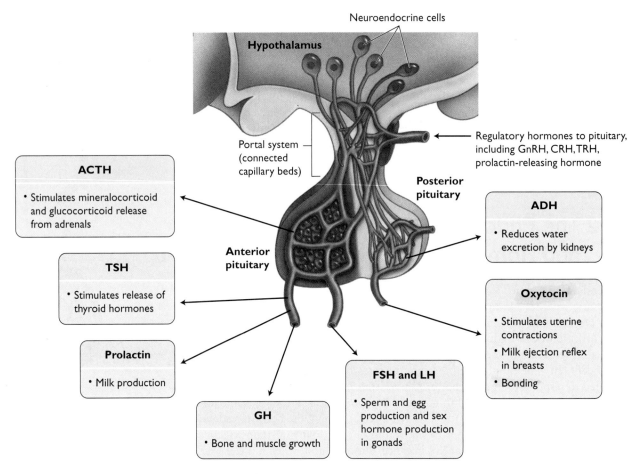

FIGURE 16.5 The hypothalamus and pituitary. The hypothalamus and pituitary are closely connected by neuroendocrine cells and a vascular portal system. Signals from the hypothalamus control the release of hormones from the pituitary. Pituitary hormones influence a large number of other endocrine tissues.

hormones—come in two classes, **releasing hormones** that trigger pituitary hormone production or **inhibiting hormones** that have the opposite effect. Hypothalamic regulatory hormones are produced in tiny amounts and travel through a portal system of two connected capillary beds that provide intimate communication between these organs.

At the onset of a stressor, the hypothalamus sends nerve impulses to the adrenal medulla, triggering the release of epinephrine and norepinephrine. The neurons storing ADH and oxytocin in the posterior pituitary are also triggered, and these hormones flood into the bloodstream. ADH promotes constriction of blood vessels. As a result, blood pressure increases, enhancing the delivery of sugar and oxygen to large muscles.

At first, oxytocin does not seem to be a stress hormone. Its best understood functions are its actions on the uterus and mammary glands in women. During labor, oxytocin is involved in a positive feedback loop that increases the strength of uterine contractions as the pressure of the fetus on the cervix (at the base of the uterus) increases. Oxytocin is also involved in the positive feedback loop that regulates milk release: Suckling of an infant at the breast promotes oxytocin release, triggering the milk ejection reflex, which keeps the infant at the breast, triggering more oxytocin release. Both of the positive feedback loops are self-limiting; eventually the baby is born or ends up with a full stomach and stops suckling.

As a stress hormone, oxytocin also acts on cells in the brain in both men and women, promoting affiliative behaviors such as seeking others' company and feelings of friendship. It appears that under stressful conditions, we are wired to make alliances, a behavior that makes sense in a highly social species such as ours.

Oxytocin is also released after orgasm in both sexes. Why might this be beneficial to both partners, especially in prehistoric times?

Turning Down Hormone Release Through Negative Feedback Loops

Like most homeostatic processes, endocrine responses are usually part of a negative feedback loop in which the release of a hormone triggers a response that inhibits additional hormone release. Endocrine-mediated negative feedback loops have been addressed in Chapter 11 (how renin and angiotensin regulate body water balance), Chapter 7 (how insulin and glucagon released by the pancreas regulate blood sugar), and elsewhere. The hypothalamus and pituitary are linked to a number of different endocrine organs within negative feedback loops that help maintain homeostasis.

During the stress response, the hypothalamus not only directs the release of ADH and oxytocin but also an array of regulatory hormones. The most notable of these is **corticotropin-releasing hormone (CRH)**, which flows into the portal system to the anterior pituitary. CRH promotes the production and release of **adrenocorticotropic hormone (ACTH)**, which travels in the bloodstream to the adrenals, in turn triggering glucocorticoid release. The increase in glucocorticoids causes a decrease in the release of both ACTH and CRH (FIGURE 16.6). As a result of this negative feedback loop along the **hypothalamus-pituitary-adrenal axis**, glucocorticoid levels drop and the individual returns to a stable state.

The hypothalamus-pituitary-adrenal axis effectively regulates stress hormone levels under the conditions experienced by our ancestors. However, in individuals who experience repeated stress responses over the course of a day or week—such as those experiencing more modern psychological stressors—glucocorticoid levels remain high. Because it takes time for these hormones to break down and be flushed from the body, recurring stress leads to their accumulation. The relationship between chronic stress and glucocorticoids is so clear that researchers use the levels of these hormones in the bloodstream of nonhuman animals to measure the amount of stress they have been experiencing.

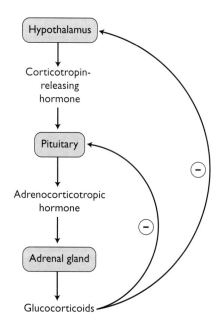

FIGURE 16.6 Negative feedback in the adrenals. The release of adrenal hormones is stimulated by signals from the anterior pituitary and hypothalamus. The presence of these hormones in the blood causes the signals from the controlling glands to turn off in a negative feedback loop.

16.3 Other Endocrine Glands

Thus far, you have learned about the adrenals, hypothalamus, and pituitary gland in terms of their roles in stress situations and in returning the individual to homeostasis. Let's continue our survey of endocrine glands by looking at non-stress-related roles of the endocrine system.

The Pituitary: Regulation of Growth

Growth hormone (GH), released from the pituitary, acts mainly on muscle and bone, where the hormone promotes cell division and growth. In particular, GH stimulates uptake of amino acids into cells, speeding protein synthesis. The majority of growth hormone's effects take place during puberty, before the ends of the long bones solidify and are no longer able to increase in length. In adults, GH primarily serves to regulate metabolism, promoting the use of fats as an energy source.

GENES & HOMEOSTASIS

Size Matters

Eddie Gaedel may be the most famous professional baseball player to record only a single plate appearance. Eddie played in the second game of a double header between the St. Louis Browns and Detroit Tigers on August 19, 1951. Dressed in a Browns jersey with the number 1/8 on the back, the 3-foot, 7-inch Gaedel stepped to the plate and waited while Tigers pitcher Bob Cain tossed four pitches—all high—thus issuing Eddie a free pass to first base. After trotting to first, he was replaced by a pinch runner and left the field to a raucous standing ovation from Browns fans.

In the 1950s, little people who were well below average height, but with limbs proportional to their small trunks—like Eddie Gaedel—were referred to as "midgets." Most of these individuals were deficient in growth hormone as a result of a mutation in genes controlling GH synthesis and release. Beginning in the early 1950s, children whose growth rates were well below average began to be treated with GH harvested from the pituitary glands of cadavers—an expensive proposition, but one that virtually eliminated Eddie Gaedel's body type in the United States.

By the early 1980s, progress in genetic technology allowed scientists to begin to engineer bacteria that carried human genes. The first recombinant human protein that was introduced to market was the hormone insulin; the second was human growth hormone. Recombinant human growth hormone is now widely available and used to treat a variety of little people, even those with the most common type of dwarfism, *achondroplasia*. Individuals with achondroplasia have a dominant mutation in a single gene involved in bone growth. Although achondroplasiacs produce normal amounts of GH, providing excess recombinant growth hormone may increase their rate of growth and their eventual adult height, although this treatment is still experimental.

The success of recombinant GH therapy on little people has inspired parents of some shorter than average (but not necessarily GH-deficient) children. In 2003, the U.S. Food and Drug Administration approved the use of recombinant GH for these children, although the practice was already widespread by that time. GH-replacement therapy requires one or two daily abdominal injections of GH over a period of many months, at a cost of about $1,500 per month. But the result of these injections can be a significant increase in growth rate and final adult height. In a society where below-average height (for men, at least) is associated with lower earning potential and personal happiness, there is significant pressure to be of average height or taller.

Eddie Gaedel, pituitary dwarf

Individuals with a genetic mutation leading to insufficient GH release have a form of dwarfism called *pituitary dwarfism*, discussed in the Genes & Homeostasis feature. A phenomenon called *stress dwarfism* occurs in children who have been severely stressed by abuse or neglect. Stress dwarfism results from suppression of growth hormone and can be reversed by simply removing the child from the stressful situation. Apparently, extreme stress increases the release of a GH inhibitor from the hypothalamus. When the stress is removed, so is the inhibition.

The pituitary also releases **prolactin**. This hormone is most closely associated with the promotion of milk production in pregnant and nursing mothers.

The Gonads: Sex-Specific Characteristics

The activities of the ovaries and testes, or **gonads**, are influenced mainly by the hypothalamus and pituitary gland. In particular, the hypothalamus produces **gonadotropin-releasing hormone (GnRH)**, so named because it causes the release of hormones with the gonads as target organs. GnRH levels regulate the release of **follicle-stimulating hormone (FSH)** and **luteinizing hormone (LH)**. In both sexes, FSH and LH promote the production of sex cells and sex hormones by the gonads. As you will learn in Chapter 18, FSH promotes egg development and LH promotes ovulation in women.

In men, FSH is responsible for triggering sperm development and LH stimulates cells of the testes to produce testosterone. Although the mechanism for the onset of GnRH production is not entirely clear, secretion of this hormone does not begin until puberty in both sexes.

Stressful situations result in impaired fertility by causing a reduction in both release of GnRH and the responsiveness of the pituitary to this hormone. As a result of GnRH suppression, less FSH and LH are produced. In children, the effect of this suppression is delayed puberty. The effects in women are suppression of menstruation and ovulation, and in men, reduction in sperm production and sex drive.

Suppression of puberty, menstruation, and sex drive all derive from the resulting low levels of **sex hormones**, steroid hormones produced by the gonads. The primary sex hormones are **testosterone**, which is more abundant in men, and **estrogen** and **progesterone**, which are more abundant in women. During puberty, these hormones promote the development of *secondary sexual characteristics*, such as the production of pubic hair in both sexes, an increase in pelvis width and fat deposition in girls, and the deepening of the voice and development of muscle in boys. The role of testosterone in muscle development has led to the use of synthetic testosterone derivatives as performance-enhancing drugs in athletes (FIGURE 16.7). Testosterone is also important in maintaining male sex drive.

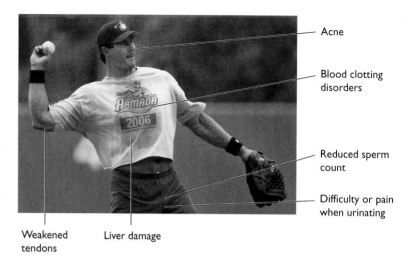

FIGURE 16.7 Steroid abuse. Synthetic male hormones have been used by competitive athletes to increase muscle mass and give them an advantage in competition. These drugs have negative consequences as well, and most professional sporting associations have banned them.

The Pancreas: Regulation of Blood-Glucose Levels

The primary function of the pancreas, as described in Chapter 7, is the synthesis and secretion of digestive enzymes. About 1% of its volume, however, consists of endocrine cells contained in small structures called **pancreatic islets** (formerly known as the islets of Langerhans, named for the scientist who described them). The body of the pancreas contains approximately 2 million islets (FIGURE 16.8). Each islet contains three types of cells: Alpha cells produce **glucagon**, a hormone that raises blood-sugar levels by triggering glycogen breakdown in the liver. Beta cells produce **insulin**, a

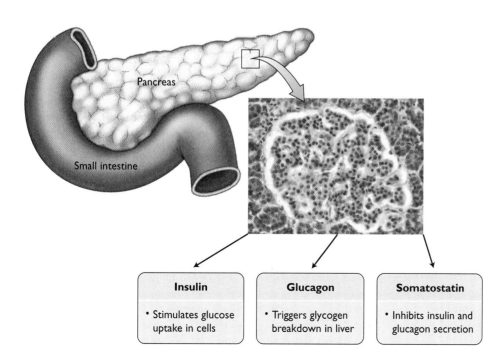

FIGURE 16.8 Pancreatic islets. These small groups of cells scattered throughout the pancreas produce hormones that regulate blood-sugar levels.

hormone that lowers blood-sugar levels by increasing glucose uptake. Delta cells produce the hormone **somatostatin**, which inhibits both glucagon and insulin secretion, among other functions.

In a stressful situation, the sympathetic nervous system triggers the release of glucagon from the alpha cells, contributing to the spike in blood sugar that occurs during the stress response. To maintain these high glucose levels, glucocorticoids act on fat cells to reduce their uptake of nutrients by making them temporarily insulin-resistant.

Glucocorticoids also block the action of **leptin**, a hormone released by fat cells, on its target tissue in the brain where it acts to blunt appetite. The result of leptin suppression is an increase in food intake, leading to higher blood-sugar levels. Beta cells in the pancreas, triggered by the high blood-glucose levels of the stress response, release more insulin, returning blood-sugar levels to normal when glucocorticoids are cleared from the body.

Stop and Stretch 16.3

When leptin is given to genetically obese mice, it causes a reduction in both food intake and body weight. However, it does not seem to have that effect on humans. In fact, many obese individuals with difficulty controlling their appetite have high levels of leptin. What does this tell you about how the appetite suppression system may be malfunctioning in these individuals?

An individual whose stress response is triggered many times in a day will expend significant metabolic energy manufacturing and releasing pancreatic hormones and moving nutrients around to control blood-sugar levels. This is one reason why stress causes fatigue. In addition, both the insulin resistance promoted by glucocorticoids and the overactivity, and eventual burnout, of insulin-producing beta cells can lead to the development of type 2 diabetes.

The relationship between glucocorticoids, fat storage, and diabetes can be seen in individuals with **Cushing's syndrome**, abnormally high glucocorticoid levels caused by diseases of the adrenal or pituitary glands or by excess intake of glucocorticoid drugs. Individuals with Cushing's syndrome store fat in cells that only respond to high levels of circulating insulin—such as those around the midsection and face—and have a high risk of type 2 diabetes (FIGURE 16.9).

The most worrisome effect of stress on the pancreas is that it results in an increase in LDL cholesterol—the "bad cholesterol" discussed in Chapter 9. Because glucocorticoids disrupt the control of nutrients in the blood, high levels of this lipid remain in circulation. Over time, the arteries become clogged and heart disease results.

The Thyroid and Parathyroid: Metabolism and Development

The **thyroid gland** is a bowtie-shaped structure located just below the larynx on the anterior surface of the trachea. Closely associated with it are four **parathyroid glands**, pea-sized structures on the back surface of the thyroid (FIGURE 16.10).

The thyroid and parathyroid glands work together to maintain calcium homeostasis. **Calcitonin** produced by the thyroid decreases blood-calcium levels by promoting deposition of the mineral in bones. **Parathyroid hormone** increases blood-calcium levels by promoting its release from bones as well as increased absorption of calcium from the digestive tract and less excretion of calcium in the urine (FIGURE 16.11).

FIGURE 16.9 Cushing's syndrome. Overproduction of glucocorticoids leads to unusual patterns of fat storage, among other effects. Treatment with substances that block the actions of cortisol reduce symptoms.

Two additional hormones produced by the thyroid are termed T_4 (also called thyroxine) and T_3 (also called triiodothyronine). The numbers refer to the number of iodine atoms each molecule contains.

Although T_4 is typically produced in greater quantity, T_3 has the greater effect on target cells. Unlike other protein hormones, T_3 and T_4 are able to cross the cell membrane of target cells. Once inside target cells, that is, nearly all cells of the body, the hormone-receptor complex activates genes that increase ATP production. As a result, the body's basal metabolic rate (Chapter 3) increases.

Any stressor that increases ATP demand (low body temperature, high altitude, pregnancy) leads to an increase in T_3/T_4 production via the hypothalamus-pituitary-thyroid axis. The stressor stimulates the hypothalamus to secrete TRH (*TSH-releasing hormone*), which causes the anterior pituitary to release TSH (*thyroid-stimulating hormone*). TSH travels through the bloodstream to the thyroid gland, where it stimulates the secretion of T_3 and T_4. The subsequent uptake of thyroid hormone by most cells of the body accelerates ATP production. When ATP demand returns to normal (or when T_3/T_4 levels are restored), release of TSH (and to a lesser extent TRH) is inhibited by negative feedback.

Conditions that lead to low thyroid hormone levels, **hypothyroidism**, can produce weight gain and lethargy. Both of these factors may contribute to the development of depression in some people.

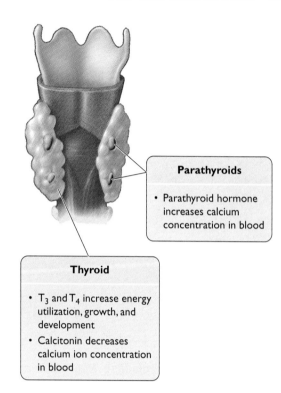

FIGURE 16.10 The thyroid and parathyroids. The thyroid gland and the four associated parathyroid glands produce hormones that regulate blood-calcium levels as well as overall metabolic rate.

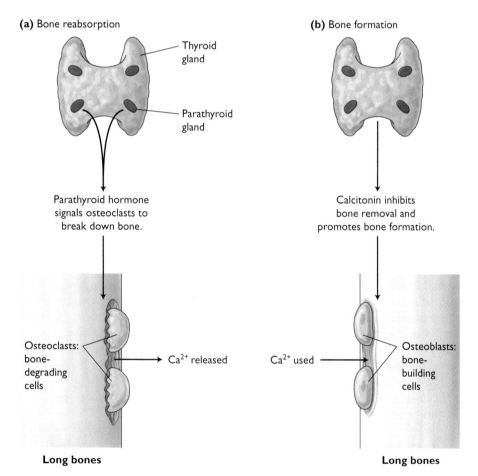

FIGURE 16.11 Calcium regulation. Parathyroid hormone and calcitonin from the thyroid interact in a feedback loop to maintain levels of blood calcium from stores in the bones.

VISUALIZE THIS How would low levels of dietary calcium affect the balance between bone building and bone breakdown?

Hyperthyroidism, oversecretion of thyroid hormones, can be caused by a condition called Graves' disease that is associated with hyperactivity and insomnia. Graves' disease can cause the eyes to protrude (FIGURE 16.12) and a goiter—an enlarged thyroid gland—to form. Removal of part of the thyroid can help cure this condition. The onset of Graves' disease may be associated with severe stress—for example, President George H. W. Bush was diagnosed with this condition not long after the beginning of the first Persian Gulf War. Interestingly, his wife Barbara Bush had earlier been diagnosed with the same condition.

Stop and Stretch 16.4

A goiter, as pictured here, occurs in individuals who do not ingest enough iodine. Because of the lack of iodine, T_3 levels are very low. How do low T_3 levels affect the hypothalamus-pituitary-thyroid axis, and how might this lead to an enlarged thyroid gland in the absence of iodine?

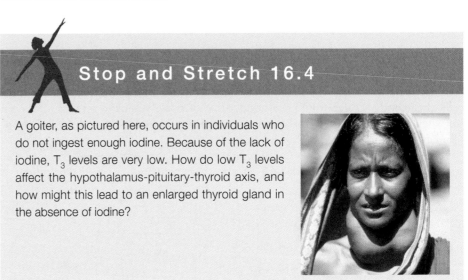

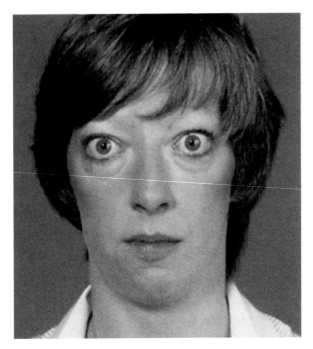

FIGURE 16.12 Graves' disease. Bulging eyes are characteristic of thyroid hormone overproduction. This symptom occurs when tissues around the eye sockets swell, pushing the eyeballs forward.

The Pineal Gland: Hormonal Effects of Light and Darkness

The hormone most associated with sleep is **melatonin**, produced by the pinecone-shaped **pineal gland** located deep within the brain. The pineal gland is attached to the optic nerve and responds to light cues—in fact, melatonin is often called the "hormone of darkness" because its production is associated with lowered light levels. Melatonin production ebbs and flows on a daily cycle synchronized to light levels. Production begins rising during the evening, reaching its peak of 10 times the daytime production during the middle of the night, and then slowly declines until morning. This pattern is nearly the opposite of the daily cycle in glucocorticoid levels (FIGURE 16.13).

Melatonin induces sleep and, as such, tends to reduce glucocorticoid levels. When a disrupted schedule prevents sleep at the point where melatonin levels normally rise, glucocorticoid levels remain high and stress symptoms result. Stress in turn, by keeping the brain alert and aroused, blunts the effect of melatonin and prevents sleep. In other words, poor sleep causes stress and stress causes poor sleep. Because melatonin lays the physical groundwork for sleep, it is sometimes taken as a supplement by individuals suffering from insomnia.

The association between melatonin and light levels may also explain some of the symptoms of seasonal affective disorder (SAD), a condition that may occur in individuals who live in regions where winter days are short and light levels may be low for many months. Individuals in these regions have elevated melatonin levels, and in a percentage of the population, these elevated levels contribute to lethargy and may lead to depression. SAD sufferers can reduce the effects of melatonin by using artificial light to inhibit its production (FIGURE 16.14).

Melatonin is a powerful antioxidant, sponging up oxidizing chemicals that cause DNA mutation. Low melatonin levels, such as those experienced by night workers, are thus associated with higher rates of cancer. If low melatonin levels are accompanied by high stress, the body's immune system may be less able to deal with the mutated cells

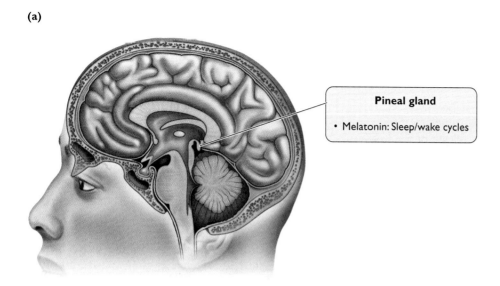

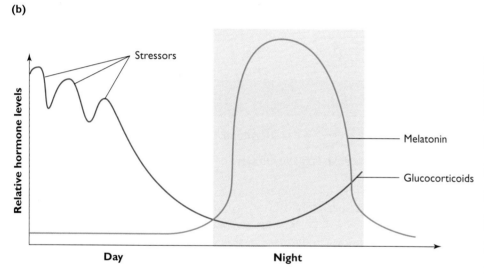

FIGURE 16.13 **Daily hormonal cycles.** (a) Melatonin is produced by the pineal gland. (b) Melatonin levels rise before an individual's typical bedtime, peak in the middle of the night, and fall as morning approaches. Glucocorticoid levels follow the opposite pattern, rising before one awakens and falling by late in the day.

VISUALIZE THIS What is the advantage of a rise in glucocorticoid levels that precedes awakening?

produced. Thus, low melatonin combined with depressed immune system function is a recipe for increased cancer risk.

The Thymus: Junction of the Endocrine and Immune Systems

The **thymus**, an organ directly behind the breastbone, is involved in the development and maturation of the immune system's T cells (FIGURE 16.15). The thymus produces the hormones **thymosin** and **thymopoietin**, which are indispensable to this process.

The activities of the thymus are especially important during childhood, when the immune response is maturing. As we age, the thymus gradually shrinks both in importance and in size. Glucocorticoids greatly accelerate this process. In fact, the effect of glucocorticoids on the thymus is so consistent that before good blood tests for glucocorticoids were devised, the extent of thymus shrinkage was used as an indirect measure.

Does a shrinking thymus cause a decline in immune response? It is clear that stressed individuals have fewer immune cells. Well-controlled studies have established that high levels of stress are associated with higher rates of infection with the common cold virus, but reports examining the relationship of stress and rates of other infectious diseases or cancer are more ambiguous. So while chronic stress clearly affects the immune system, whether or not that impact leads to a greater risk of illness is still not clear.

FIGURE 16.14 **Treatment for SAD.** Melatonin production is higher in darker conditions. During winter in northern regions, short days lead to higher melatonin levels. Sitting in front of a light box can reduce melatonin levels and thus may increase energy levels in individuals with seasonal affective disorder.

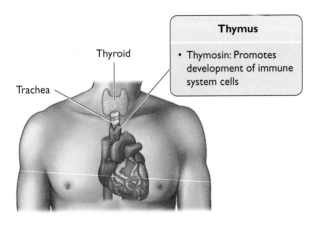

FIGURE 16.15 Thymus. The thymus produces hormones crucial to effective immune system function.

Other Tissues That Produce Hormones

The hormones produced by the nine major endocrine organs are summarized in Table 16.1. Hormones are also produced by a number of organs whose main function is not endocrine in nature (FIGURE 16.16). For example, as discussed in Chapter 8, the kidneys produce erythropoietin, which stimulates production of red blood cells. The kidneys also modify vitamin D, made in the skin, into its active form, *calcitriol*, a hormone that stimulates the absorption of calcium from the intestines. The stress response affects the release of the hormones atrial natriuretic peptide (ANP) from the heart and angiotensin from the kidneys, both of which help to regulate blood pressure.

The digestive tract produces the hormone **gastrin** during food intake. Gastrin stimulates the stomach to release hydrochloric acid for the breakdown of protein. This powerful acid in concert with digestive enzymes can process nearly any protein—including the protein contained in the cells lining the stomach. A thick layer of mucus normally protects the stomach and intestines from damage. Because the stress response diverts blood away from the digestive system, however, the layer of mucus can become thinner. Under conditions of severe physical stress, such as after massive infection, trauma due to accidents or surgery, or severe and extensive burns, stress ulcers may form.

TABLE 16.1 | Endocrine Glands, Their Hormones, Targets, and Actions

Endocrine Gland	Hormones Secreted	Hormone Target and Action
Hypothalamus	• Releasing and inhibiting hormones	• Cause the pituitary to store or secrete hormones
Pituitary	• Antidiuretic hormone (ADH) • Oxytocin • Thyroid-stimulating hormone (TSH) • Adrenocorticotropic hormone (ACTH) • Follicle-stimulating hormone (FSH) • Luteinizing hormone (LH) • Prolactin • Growth hormone (GH)	• Water reabsorption by kidneys • Breast milk release and uterine contractions during delivery • Causes thyroid to secrete T_3 and T_4, which increase metabolic activities of many cells • Causes adrenals to secrete glucocorticoids • Egg and sperm production • Ovulation and testosterone production • Breast milk production • Bone growth, protein synthesis, and cell division
Adrenals	• Glucocorticoids • Mineralocorticoids • Epinephrine and norepinephrine • Small amounts of sex hormones	• Raise blood-glucose levels, promote fat breakdown, and suppress inflammation • Regulate level of sodium and potassium in blood • Act on liver, muscle, and fat to raise blood sugar, increase heart rate, regulate blood vessel diameter, and increase respiration • Promote development and maintenance of sex-specific characteristics
Gonads	• Androgens from testes • Estrogen and progesterone from ovaries	• Promote development of sperm, male reproductive structures, and male-specific sex characteristics • Promote development of egg cells; regulate menstruation, pregnancy, and other female-specific sex characteristics

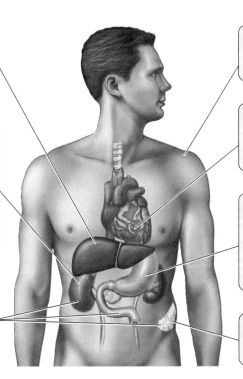

FIGURE 16.16 Other endocrine tissues. Hormones are produced by many tissues and organs not traditionally associated with the endocrine system. A few are illustrated here.

Liver
- Insulin-type growth factor: Promotes repair of damaged tissues

Kidneys
- Erythropoietin: triggers red blood cell production
- Angiotensin: Helps control blood pressure
- Calcitriol: Increases intestinal Ca^{2+} uptake

Many tissues
- Prostaglandins: Promote inflammation locally
- Growth factors: Promote cell division

Skin
- Vitamin D: Calcium absorption

Heart
- Atrial natriuretic peptide: Regulation of blood pressure

Digestive system
- Gastrin: Stimulates stomach acid production
- Secretin: Stimulates release of acid buffer

Fat tissue
- Leptin blunts appetite

TABLE 16.1 | Endocrine Glands, Their Hormones, Targets, and Actions (*Continued*)

Endocrine Gland	Hormones Secreted	Hormone Target and Action
Pancreas	• Insulin • Glucagon	• Lowers blood-glucose levels • Raises blood-glucose levels
Thyroid	• Triiodothyronine (T_3) and thyroxine (T_4) • Calcitonin	• T_3 and T_4 act on most body cells to increase metabolic activities • Acts on the bone to help lower blood-calcium levels
Parathyroids	• Parathyroid hormone (PTH)	• Acts on the bones, digestive tract, and kidneys to raise blood-calcium levels
Pineal	• Melatonin	• Helps regulate sleep
Thymus	• Thymosin and thymopoietin	• Help T cells mature, especially in children

Prostaglandins are an unusual class of hormones in that they are produced by many different endocrine tissues but generally have only a localized effect. For instance, prostaglandin produced by blood vessel cells at the beginning of a stressful event works with epinephrine from the adrenals to regulate blood flow, increasing the flow to "fight-or-flight" muscles and reducing it to less immediately crucial organs.

Prostaglandins also tend to promote inflammation wherever damage occurs, a reaction that increases local blood flow and thus speeds the arrival of immune system cells. For example, cells damaged by stomach acid release prostaglandin to promote rapid healing before an ulcer can form. Because inflammation is generally not beneficial during the stress response, glucocorticoids suppress prostaglandin production. In the case of damaged stomach cells, a decline in prostaglandin may worsen ulcer formation. Some of the most popular pain-relief drugs, including aspirin and ibuprofen, are anti-prostaglandins. As you may know, chronic use of these drugs is associated with ulcer formation as well.

The latest class of hormones identified is the **growth factors**, which stimulate cell growth and division in a wide variety of tissues. Like the prostaglandins, many of these hormones are local and have effects on tissues near their release site. One of the most prominent growth factors is produced by the liver and is called insulin-type growth factor (IGF). IGF promotes the repair of damaged tissues—some scientists refer to it as the anti-aging hormone. High levels of IGF in experimental animals correlate to increased life spans and healthier old age. IGF production is stimulated by growth hormone. Not surprisingly then, chronic stress is associated with low IGF levels.

16.4 Combating Stress

As we have seen, chronic stress can trigger the development of a variety of diseases. By interfering with reproductive hormone production, it impairs fertility in both women and men. By causing high blood pressure and atherosclerosis, it is a major risk factor for heart disease. It interferes with sleep and so can play a role in the development of depression. By disrupting blood-glucose levels, it may increase the risk of diabetes. Stress may cause a decline in immune response that makes us more susceptible to the common cold, perhaps other infectious diseases, and possibly even cancer. It can contribute to the development of ulcers and perhaps cause premature aging. Research on the effect of stress on the brain has even provided evidence that overactivity of the sympathetic nervous system can cause damage to the structures involved in long-term memory.

Even though modern human society differs in many ways from the circumstances of our ancestors, the stress response still has value. By increasing alertness and energy level, stress can motivate us to deal with challenging situations. We also continue to face physical stressors, including infection.

The cost of an inadequate stress response can be seen in individuals with **Addison's disease**, in which glucocorticoid levels are abnormally low. This lack may occur because the adrenal glands are unable to produce these chemicals or the pituitary fails to produce ACTH. If the adrenals are not functioning, the individual with pale skin may exhibit a peculiar bronzing that occurs as excess ACTH stimulates melanin production. President John F. Kennedy had this form of Addison's disease, which gave him the appearance of sun-tanned good health despite severe illness (FIGURE 16.17). With inadequate glucocorticoids, glucose cannot be replenished in the blood once a stressor occurs. The consequences of low glucose are severe weakness, fatigue, and low blood pressure. An Addisonian crisis brought on by stress in affected individuals may be fatal.

How do we manage our stress response so that it benefits us when we need its effects but does not harm us when it is unnecessarily triggered (Table 16.2)? Maintaining a regular sleep schedule reduces the stress-inducing effects of sleep deprivation. Regular exercise, by using up the energy-generating portions of the stress response, can help. Exercise also has the advantage of lessening the risk of heart disease and diabetes as well as decreasing the opportunity for stress to worsen these diseases. Daily meditation, in which you consciously relax and allow your mind to become calmed, also reduces glucocorticoid levels. There are many methods of meditation, from formalized practice, to

FIGURE 16.17 Addison's disease. President John F. Kennedy suffered from an underproduction of glucocorticoids. One symptom of his condition was the noticeable bronze color of his skin, produced by high levels of the glucocorticoid-stimulating hormone, ACTH.

TABLE 16.2 | Techniques for Stress Relief

Technique	Physiological or Psychological Function
Maintain a regular sleep schedule.	Prevents rise in glucocorticoid levels in absence of melatonin. Preserves mental acuity, allowing for more effective psychological responses to stressful situations.
Engage in regular exercise.	Reduces risk of heart disease and diabetes, two diseases that may be worsened by stress. Puts higher blood-glucose levels caused by the stress response to appropriate use in muscle activity.
Meditate.	Reduces glucocorticoid levels. Permits better control over thought processes.
Focus on factors under your control.	Can reduce likelihood of triggering the stress response when faced with a challenge. Puts attention on positive steps that can improve the current situation.
Seek professional help.	Support groups or individual counseling can help develop techniques for dealing with challenging situations. Certain drugs can reduce the overall activity of brain neurons, reducing the stress response but also causing sedation.

prayer routines, to simple deep-breathing exercises—it appears that regardless of the method, the physical result is similar.

Working to modify the emotional response to challenging situations may be even more helpful than avoiding stress entirely. It is probably apparent to you that different individuals faced with the same stressor do not respond the same way. People who are quicker to experience anger or negative emotions are more likely to suffer the ill effects of stress.

Taking stock of what parts of your life you control and trying to minimize the impact of those portions not under your control can reduce the level of anger and frustration in response to challenge. Getting involved in activities that make the world a better place can help you feel that you are having an impact on larger, hard-to-control problems that may cause you worry, such as environmental damage, crime, or poverty. If stress continues to be a problem, seek help. Some drugs can provide relief from anxiety without unwanted side effects, and professional counselors and support groups can help you develop positive methods for dealing with the challenges of life.

Chapter REVIEW

ROOTS TO REMEMBER

The following roots of words come mainly from Latin and Greek and will help you to decipher terms:

hormone comes from the verb *hormon*, meaning "setting in motion."

hyper- means beyond or above.

hypo- means under or below.

-lact- is from the Latin word for milk.

melatonin is based on *melas*, meaning black.

KEY TERMS

16.1 An Overview of the Endocrine System
adrenal gland *p. 386*
aldosterone *p. 387*
cortex *p. 386*
cortisol *p. 387*
cortisone *p. 387*
endocrine gland *p. 384*
epinephrine *p. 386*
glucocorticoid *p. 387*
hormone *p. 384*
medulla *p. 386*
mineralocorticoid *p. 387*
norepinephrine *p. 386*
protein hormone *p. 384*
receptor *p. 384*
secondary messenger *p. 384*
signal transduction *p. 384*
steroid hormone *p. 384*
stress response *p. 385*
target cell *p. 384*
target tissue *p. 384*

16.2 The Endocrine System and Homeostasis
adrenocorticotropic hormone (ACTH) *p. 389*
anterior pituitary *p. 387*
antidiuretic hormone (ADH) *p. 387*
corticotropin-releasing hormone (CRH) *p. 389*
hypothalamus *p. 387*
hypothalamus-pituitary-adrenal axis *p. 389*
inhibiting hormone *p. 388*
oxytocin *p. 387*
pituitary gland *p. 387*
posterior pituitary *p. 387*
regulatory hormone *p. 387*
releasing hormone *p. 388*

16.3 Other Endocrine Glands
calcitonin *p. 392*
Cushing's syndrome *p. 392*
estrogen *p. 391*
follicle-stimulating hormone (FSH) *p. 390*
gastrin *p. 396*
glucagon *p. 391*
gonad *p. 390*
gonadotropin-releasing hormone (GnRH) *p. 390*
growth factor *p. 398*
growth hormone (GH) *p. 389*
hyperthyroidism *p. 394*
hypothyroidism *p. 393*
insulin *p. 391*
leptin *p. 392*
luteinizing hormone (LH) *p. 390*
melatonin *p. 394*
pancreatic islet *p. 391*
parathyroid gland *p. 392*
parathyroid hormone *p. 392*
pineal gland *p. 394*
progesterone *p. 391*
prolactin *p. 390*
prostaglandin *p. 398*
sex hormone *p. 391*
somatostatin *p. 392*
T_3 *p. 393*
T_4 *p. 393*
testosterone *p. 391*
thymopoietin *p. 395*
thymosin *p. 395*
thymus *p. 395*
thyroid gland *p. 392*

16.4 Combating Stress
Addison's disease *p. 398*

SUMMARY

16.1 An Overview of the Endocrine System

- The endocrine system is made up of glands that secrete hormones, which are chemicals that travel through the bloodstream and trigger responses in target tissues distant from the gland itself (pp. 384–385).

- Water-soluble protein hormones cause changes in a target cell by binding to a cell's membrane and triggering a cascade of events inside the cell in a signal transduction pathway (pp. 384–385).

- Fat-soluble steroid hormones can cross cell membranes, bind to receptors inside target cells, and trigger gene expression (pp. 384–385).

- There are nine main endocrine-secreting organs (p. 386; Figure 16.3).

Chapter Review

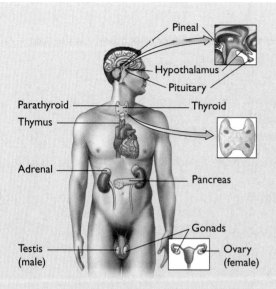

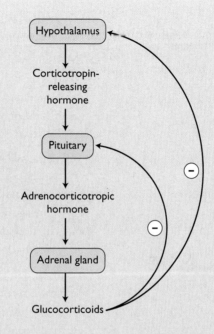

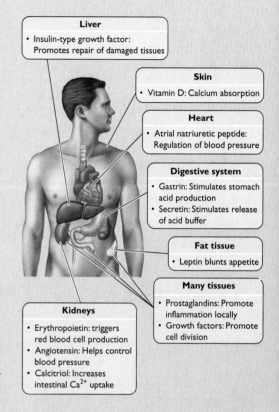

- The adrenal glands are located on top of the kidneys. The adrenal medulla secretes epinephrine and norepinephrine, which increase heart rate, redirect blood flow from digestive organs, and dilate pupils. The adrenal cortex secretes mineralocorticoids, which regulate blood sodium and potassium, and glucocorticoids, which bind to cortisol receptors and trigger longer-term stress responses (p. 387).

16.2 The Endocrine System and Homeostasis

- The hypothalamus is a neuroendocrine organ that regulates many aspects of homeostasis and directs the stress response. It produces and releases hormones that affect the whole body as well as regulatory hormones that control the activities of the pituitary gland (p. 387).

- ADH released by the hypothalamus helps to maintain blood volume and is secreted to increase blood pressure (p. 387).

- The hypothalamus, pituitary, and other endocrine organs are involved in negative feedback loops that help regulate the body's internal conditions. Typically, hormones produced by endocrine glands such as the adrenals affect the release of regulatory hormones from the hypothalamus and releasing hormones from the pituitary (pp. 387–389; Figure 16.6).

16.3 Other Endocrine Glands

- The pituitary gland also releases growth hormone, which promotes the growth of muscle and bone, and prolactin, which triggers milk production (pp. 389–390).

- Gonadotropin-releasing hormone from the hypothalamus controls the production of follicle-stimulating hormone and luteinizing hormone from the pituitary. FSH and LH trigger the production of the steroid sex hormones testosterone, estrogen, and progesterone (pp. 390–391).

- The pancreatic islets secrete glucagon and insulin, hormones that regulate blood-sugar levels. Glucagon secretion is stimulated by stress, and its effects are magnified by glucocorticoids, which block fat uptake into cells and stimulate appetite by blocking the hormone leptin (pp. 391–392).

- The thyroid and parathyroid glands work together to control blood-calcium levels via release of the hormones calcitonin and parathyroid hormone. The thyroid also produces T_4 and T_3 hormones, which regulate metabolic rate (pp. 392–394).

- The pineal gland secretes melatonin, a hormone that prepares the body for sleep (p. 394).

- The thymus produces the hormones thymosin and thymopoietin, required for the maturation of immune system cells. High glucocorticoid levels cause the thymus to shrink, and thus appear to suppress the immune response (p. 395).

- Many tissues play endocrine functions in the body (pp. 396–397; Figure 16.16).

- Gastrin from the digestive system triggers stomach cells to release acid (p. 396).

- Prostaglandins are produced by many different cells and promote inflammation (p. 398).

- Growth factors are a class of hormones produced by many different tissues that promote cell growth and division (p. 398).

16.4 Combating Stress

- While stress causes some disease, it also has value, preparing our bodies and minds for challenges. A number of practices can help us manage the effects of stress, including getting adequate sleep, regular exercise, and engaging in regular meditation. We can also reduce our response to stress by focusing on aspects of our lives that are under our control and reducing the psychological impact of factors outside our control (pp. 398–399).

LEARNING THE BASICS

1. All of the following events are part of the stress response *except* _____.
   a. increased muscular activity
   b. lowered blood pressure
   c. increased heart rate
   d. reduced pain perception
   e. release of glucose into the bloodstream

2. How is the endocrine system like the nervous system?
   a. It relies on networks of cells in close contact to each other.
   b. It relies on chemical and electrical signals.
   c. It helps coordinate the body's response to changing conditions.
   d. It is made up of glands that release hormones.
   e. It provides a prolonged response to changing external conditions.

3. The endocrine organ that sits atop a kidney is _____.
   a. the pituitary gland
   b. the hypothalamus
   c. the ovary
   d. the adrenal gland
   e. the testis

4. The cells that release epinephrine and norepinephrine _____.
   a. are located in the adrenal medulla
   b. are neurons that represent the end of a nerve pathway from the brain
   c. fire in response to stress
   d. all of the above

5. A signal transduction pathway _____.
   a. sends electrical signals down a neuron
   b. is triggered by hormones that attach to receptors on the cell membrane
   c. occurs when a hormone-receptor complex binds to DNA in a cell's nucleus
   d. occurs at equal frequency in target and nontarget cells in response to hormone application
   e. is triggered when a lock on all cell activities is released

6. Steroid hormones act by _____.
   a. causing neurons to fire
   b. increasing muscle contraction
   c. diffusing across membranes and regulating the expression of genes
   d. binding to receptors on the cell surface and transducing a response without entering the cell

7. Why is the pituitary known as the "master gland"?
   a. It controls the activity of the hypothalamus.
   b. It is the only endocrine gland located inside the brain.
   c. It produces hormones that affect the other endocrine organs.
   d. It is the only endocrine gland that responds to negative feedback.
   e. It releases the female hormone estrogen.

8. All of the following are effects of glucocorticoid release on the body *except* _____.
   a. reduced inflammation
   b. shrinkage of the thymus
   c. a decrease in adrenocorticotropic hormone (ACTH) release
   d. reduced uptake of nutrients by fat cells
   e. a decrease in blood-glucose levels

9. Match each endocrine organ with a hormone it produces.
   _____ anterior pituitary   a. testosterone
   _____ pancreatic islets   b. growth hormone
   _____ thyroid   c. T_3
   _____ testes   d. melatonin
   _____ pineal gland   e. insulin

10. Match each hormone with its effect on the body.
    _____ GnRH   a. triggers glycogen breakdown in liver, increasing blood-glucose level
    _____ glucagon   b. stimulates the production and maturation of immune system cells
    _____ T_3 c. triggers the release of FSH and LH from the pituitary
    _____ growth hormone   d. increases synthesis of proteins for ATP synthesis, raising metabolism
    _____ thymosin   e. triggers uptake of amino acids into muscle cells, stimulating growth

11. Chronic stress is associated with increased risk of _____.
    a. heart disease
    b. suppressed immune system
    c. impairment of fertility
    d. disrupted sleep
    e. all of the above

12. **True or False?** The hypothalamus and pituitary are connected by a vascular portal system.

ANALYZING AND APPLYING THE BASICS

Use the following information to answer questions 1–3.

Posttraumatic stress disorder (PTSD) can occur after intense military combat, natural disasters, accidents, or assault. This disorder may become a chronic condition that causes nightmares and flashbacks, insomnia, and social estrangement. Individuals with PTSD often startle easily. Some studies have shown lower than average cortisol levels and higher than average epinephrine and norepinephrine levels in individuals with PTSD. Thyroid hormone levels may also be elevated.

1. Which symptoms of PTSD would most likely be related to elevated levels of epinephrine and norepinephrine? Explain your answer.

2. How could increased thyroid hormone levels affect the symptoms of PTSD?

3. The hippocampus is a part of the brain involved in memory and spatial awareness. People with PTSD often have a smaller than average hippocampus. Some researchers believe that glucocorticoids attack and shrink the hippocampus in these individuals. What might be the long-term effects on a person who does not have PTSD but is exposed to chronic stress?

Questions 4 and 5 refer to the following information.

Repoxygen is a commercial product that induces genes in muscle cells to produce erythropoietin. This substance was created for people with malfunctioning kidneys. However, some officials are concerned that professional athletes might begin using this product when it becomes more widely available.

4. How would an athlete potentially benefit from Repoxygen?

5. How would an athlete get similar benefits without taking Repoxygen?

6. When a person is exposed to a virus or other infectious agent, the body produces substances called endogenous pyrogens. Japanese researchers found that endogenous pyrogens stimulate production of a particular prostaglandin, which acts on the brain to cause fever. What cell type most likely produces this prostaglandin?

7. Jet lag occurs when rapid, long-distance travel disrupts the body's normal daily rhythms. Jet lag can cause grogginess, irregular sleep patterns, headaches, irritability, and nausea. How could a traveler stimulate hormones during a plane ride to reduce the symptoms of jet lag?

8. Women who have been diagnosed with breast cancer often experience changes in the hypothalamus-pituitary-adrenal axis due to increased stress. In these patients, levels of cortisol can become elevated in the bloodstream.

   In a 2000 study, Creuss and coworkers examined the effect of cognitive-behavioral stress management on a group of women with early breast cancer. In the study, 24 women were assigned to a 10-week stress management treatment, while 10 were placed on a waiting list (control group). Stress management treatment included the following: training in assertiveness, anger management, and other stress management techniques; relaxation training; group discussions of personal experiences; role-playing; and homework exercises in stress management and relaxation techniques.

   A statistically significant difference was found between cortisol levels in the treatment versus the control group, with those receiving stress management having lower cortisol levels. What biological benefits might stress management provide for women who are about to undergo surgery for breast cancer?

CONNECTING THE SCIENCE

1. The use of anabolic steroids to build muscle is illegal, but the practice has become widespread in certain professional sports. Do you agree that using steroids is a form of cheating? How is it different from other forms of physical enhancement (for instance, special diets, vision correction, or even drugs that reduce asthma symptoms)?

2. Do you agree that below-average height (in the absence of other health issues) is a disability that should be corrected by injections of recombinant growth hormone if possible? What do you think are the benefits of "height enhancement"? What are the risks?

Chapter 17

DNA Synthesis, Mitosis, and Meiosis

Cancer

LEARNING GOALS

1. Describe the differences between a benign and a malignant tumor.
2. Describe the process of DNA synthesis.
3. Name the enzyme that facilitates DNA synthesis.
4. Define the term *complementary* as it applies to DNA nucleotides.
5. Draw chromosomes in various stages of mitosis.
6. What function do the microtubules of the spindle apparatus play in cell division?
7. List the stages of mitosis, and describe what occurs during each stage.
8. Describe the events of cytokinesis.
9. Compare and contrast the events of meiosis I and meiosis II.
10. What does crossing over accomplish?

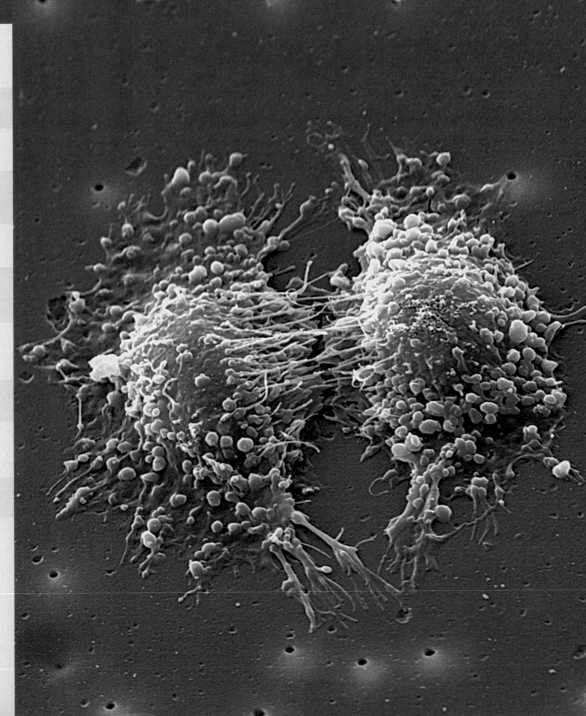

Cancer cells divide when they should not.

Nicole's early college career was similar to that of most students. She enjoyed her independence and the wide variety of courses required for her double major in biology and psychology. She worried about her grades and about finding ways to balance her course work with her social life. She also tried to find time for lifting weights in the school's athletic center and snowboarding at a local ski hill. Some weekends, to take a break from school, she would ride the bus home to see her family.

Managing to get schoolwork done, see friends and family, and still have time left to exercise had been difficult, but possible, for Nicole during her first two years at school. Her ability to maintain this balancing act changed drastically during her third year of school.

Nicole got sick during her junior year of college.

One morning in October of her junior year, Nicole began having severe pains in her abdomen. The first time pain happened, she was just beginning an experiment in cell-biology laboratory. Hunched over and sweating, she barely managed to make it through the two-hour respiration experiment that she and her lab partner were performing. Over the next few days, the pain intensified so much that she was unable to walk from her apartment to her classes without stopping several times to rest.

Later that week, as she was preparing to leave for class, the pain was so severe that she had to lie down in the hallway of her apartment. When her roommate got home a few minutes later, she took Nicole to the student health center for an emergency visit. The physician at the health center first determined that Nicole's appendix had not burst and then made an appointment for Nicole to see a gynecologist the next day.

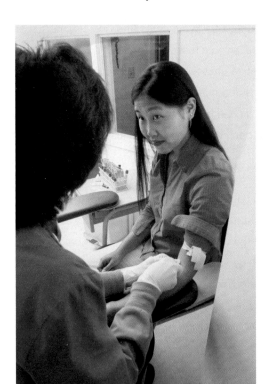

She underwent procedures to see if she had cancer.

After hearing Nicole describe her symptoms, the gynecologist pressed on her abdomen and felt what he thought was a mass on her right ovary. Further tests convinced her gynecologist that she had a large growth on her ovary. He told her that he suspected this growth was a *cyst*, or fluid-filled sac. Her gynecologist told her that cysts often go away without treatment, but this one seemed to be quite large and would have to be surgically removed.

Even though the idea of having an operation was scary for Nicole, she was relieved to know that the pain would stop. Her gynecologist also assured her that she had nothing to

Research prepares her to get more information from her doctor.

worry about because cysts are not cancerous. A week after the abdominal pain began, Nicole's gynecologist removed the cyst and her right ovary, which had been engulfed by the cyst. He then sent the cystic ovary to a physician who specializes in determining if tissues are cancerous. This physician, called a *pathologist*, determined that Nicole's doctor had been right—there was no sign of cancer.

After the operation, Nicole's gynecologist assured her that the remaining ovary would compensate for the missing ovary by ovulating every month. He added that he would have to monitor her remaining ovary carefully to make sure that it did not become cystic, or even worse, cancerous. She could not afford to lose another ovary if she wanted to remain fertile and have children some day.

Monitoring her remaining ovary meant monthly visits to her gynecologist's office, where Nicole had blood drawn and analyzed. The blood was tested for the level of a protein called CA125, which is produced by ovarian cells. Higher-than-normal CA125 levels usually indicate that the ovarian cells have increased in size or number and are associated with the presence of an ovarian tumor.

Nicole went to her scheduled checkups for five months after the original surgery. The day after her March checkup, Nicole received a message from her doctor asking that she come to see him the next day. Because she needed to study for an upcoming exam, Nicole tried to push aside her concerns about the appointment. By the time she arrived at her gynecologist's office, she had convinced herself that nothing serious could be wrong. She thought a mistake had probably been made and that he just wanted to perform another blood test.

The minute her gynecologist entered the exam room, Nicole could tell by his demeanor that something was wrong. As he started speaking to her, she began to feel very anxious—when he said that she might have a tumor on her remaining ovary, she could not believe her ears. When she heard the words *cancer* and *biopsy*, Nicole felt as though she was being pulled underwater. She could see that her doctor was still talking, but she could not hear or understand him. She felt too nauseated to think clearly, so she excused herself from the exam room, took the bus home, and immediately called her parents.

After speaking with her parents, Nicole realized that she had many questions to ask her doctor. She did not understand how it was possible for such a young woman to have lost one ovary to a cyst and then possibly to have a tumor on the other ovary. She wondered how this tumor would be treated and what her prognosis would be. Before seeing her gynecologist again, Nicole decided to do some research to make a list of questions for her doctor.

17.1 What Is Cancer?

Cancer is a disease that begins when a single cell replicates itself although it should not. **Cell division** is the process a cell undergoes to make copies of itself. This process is normally regulated so that a cell divides only when more cells are required and when conditions are favorable for division. A cancerous cell is a cell that divides without being given the go-ahead.

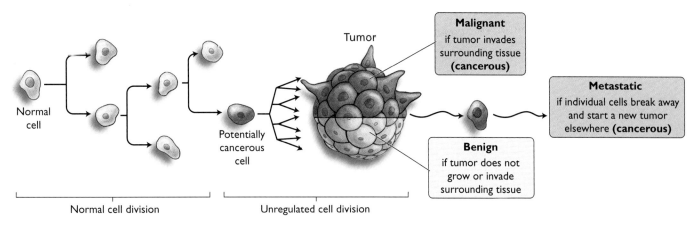

FIGURE 17.1 **What is cancer?** A tumor is a clump of cells with no function. Tumors may remain benign, or they can invade surrounding tissues and become malignant. Tumor cells may move, or metastasize, to other locations in the body. Malignant and metastatic tumors are cancerous.

Unregulated cell division leads to a pileup of cells that form a lump or **tumor**. A tumor is a mass of cells that has no apparent function in the body. Tumors that are slow growing and do not invade surrounding structures are said to be **benign**. Some benign tumors remain harmless; others become cancerous. Tumors that invade surrounding tissues are **malignant** or cancerous. The cells of a malignant tumor can break away and start new cancers at distant locations through a process called **metastasis** (FIGURE 17.1).

Cancer cells can travel throughout the body via the lymphatic and/or circulatory systems. As you learned in Chapter 9, the lymphatic system collects fluids lost from capillaries, and lymph nodes are structures that filter the lost fluids, or lymph.

When a cancer patient is undergoing surgery, the surgeon will often remove a few lymph nodes to see if any cancer cells are in the nodes. If cancer cells appear in the nodes, then some cells have left the original tumor and are moving through the lymphatic vessels. When cancer cells metastasize, they can gain access to the circulatory system. Once inside a blood vessel, cancer cells can drift virtually anywhere in the body (FIGURE 17.2).

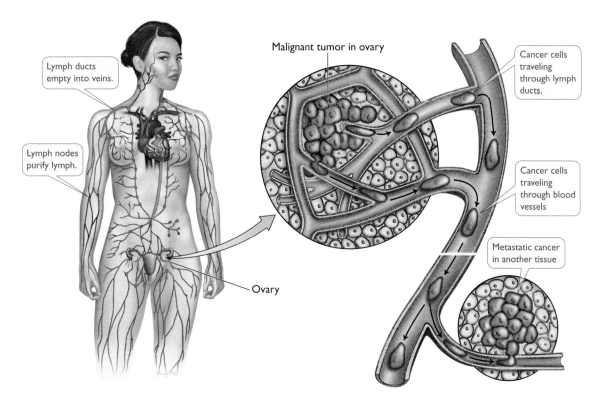

FIGURE 17.2 **Metastasis.** The vessels of the circulatory and lymphatic systems provide a pipeline for cancer cells to move to other locations in the body through a process called metastasis.

Stop and Stretch 17.1

A tumor does not have to become malignant or metastasize to become lethal. Explain why a benign brain tumor could become lethal.

Cancer cells differ from normal cells in three ways: (1) They divide when they should not; (2) they invade surrounding tissues; and (3) they move to other locations in the body. All tissues that undergo cell division are susceptible to becoming cancerous. However, there are ways to increase or decrease the probability of getting cancer.

Risk Factors for Cancer

Certain exposures and behaviors, called **risk factors**, increase a person's risk of acquiring a disease. General risk factors for virtually all cancers include tobacco use, a high-fat and low-fiber diet, lack of exercise, obesity, and increasing age.

Tobacco Use The use of tobacco of any type, whether cigarettes, cigars, pipes, or chewing tobacco, increases your risk of many cancers. While smoking is the cause of 90% of all lung cancers, it is also the cause of about one-third of all cancer deaths. Cigar smokers have increased rates of lung, larynx, esophagus, and mouth cancers. Chewing tobacco increases the risk of cancers of the mouth, gums, and cheeks. People who do not smoke but who are exposed to secondhand smoke have increased lung cancer rates.

Tobacco smoke contains more than 20 known cancer-causing substances called **carcinogens**. For a substance to be considered carcinogenic, exposure to it must be correlated with an increased risk of cancer. Examples of other carcinogens include radiation, ultraviolet light, asbestos, and some viruses.

The carcinogens inhaled during smoking come into contact with cells deep inside the lungs. Chemicals present in cigarettes and cigarette smoke have been shown to increase cell division, inhibit a cell's ability to repair damaged DNA, and prevent cells from dying when they should. Chemicals in cigarette smoke also disrupt the transport of substances across cell membranes and alter many of the enzyme reactions that occur within cells. They have also been shown to increase the generation of *free radicals*, which remove electrons from other molecules. The removal of electrons from DNA or other molecules causes damage to these molecules—damage that, over time, may lead to cancer. Cigarette smoking provides so many different opportunities for DNA damage and cell damage that tumor formation and metastasis are quite likely for smokers. In fact, people who smoke cigarettes increase their odds of developing almost every cancer.

A High-Fat, Low-Fiber Diet Cancer risk may also be influenced by diet. The American Cancer Society recommends eating at least five servings of fruits and vegetables every day as well as six servings of food from other plant sources such as breads, cereals, grains, rice, pasta, or beans. Plant foods are low in fat and high in fiber. A diet high in fat (greater than 15% of all calories obtained from fats) and low in fiber (fewer than 30 grams per day) is associated with increased risk of cancer. Fruits and vegetables are also rich in antioxidants that help to neutralize the electrical charge on free radicals and thereby prevent the free radicals from taking electrons from other molecules, including DNA. There is some evidence that antioxidants may help prevent certain cancers by minimizing the number of free radicals in our cells.

Lack of Exercise Regular exercise decreases the risk of most cancers, partly because exercise keeps the immune system functioning effectively. The immune system helps destroy cancer cells when it can recognize them as foreign to the host body. Unfortunately, since cancer cells are actually your own body's cells run amok, the immune system cannot always differentiate between normal cells and cancer cells.

Obesity Exercise also helps prevent obesity, which is associated with increased risk for many cancers including cancers of the breast, uterus, ovary, colon, gallbladder, and prostate. The abundance of fatty tissue has been hypothesized to increase the odds of hormone-sensitive cancers such as breast, uterine, ovarian, and prostate cancer.

Excess Alcohol Consumption Drinking alcohol is associated with increased risk of some types of cancer. Men should have no more than two alcoholic drinks a day, and women one or none. People who both drink and smoke increase their odds of cancer in a multiplicative rather than additive manner. In other words, if one type of cancer occurs in 10% of smokers and in 2% of drinkers, someone who smokes and drinks multiplies his chances to a rate that is closer to 20% than 12%.

Increasing Age As you age, your immune system weakens, and its ability to distinguish between cancer cells and normal cells decreases. This weakening is part of the reason many cancers are far more likely in elderly people. Additional factors that help explain the higher cancer risk with increasing age include cumulative damage. If we are all exposed to carcinogens during our lifetime, then the longer we are alive, the greater the probability that some of those carcinogens will mutate genes involved in regulating the cell cycle. Also, because multiple mutations are necessary for a cancer to develop, it often takes many years to progress from the initial mutation to a tumor and then to full-blown cancer. Scientists estimate that most cancers large enough to be detected have been growing for at least five years and are composed of close to 1 billion cells.

Nicole's cancer affected ovarian tissue. Why might ovarian cells be more likely to become cancerous than some other types of cells? When an egg cell is released from the ovary during ovulation, the tissue of the ovary becomes perforated. Cells near the perforation site undergo cell division to heal the damaged surface of the ovary. For Nicole, these cell divisions may have become uncontrolled, leading to the growth of a tumor.

17.2 Overview of Cell Division

Cell division produces new cells to heal wounds, replace damaged cells, and help organisms grow and reproduce themselves. Each of us begins life as a single fertilized egg cell that undergoes millions of rounds of cell division to produce all the cells that make up the tissues and organs of our bodies. Organisms whose reproduction requires genetic information from two parents undergo *sexual reproduction*. Humans reproduce sexually when sperm and egg cells combine their genetic information at fertilization.

All dividing cells must first make a copy of their genetic material, or **DNA (deoxyribonucleic acid)**. The DNA is located within the nucleus and carries the genes for building all of the proteins that cells require. The DNA in the nucleus is organized into structures called **chromosomes**. Chromosomes can bear hundreds of genes along their length. Human beings have 46 chromosomes.

DNA molecules are in an uncondensed, stringlike form, called *chromatin*, prior to cell division (FIGURE 17.3a). Just before cell division occurs, the DNA in each chromosome is in a tightly wrapped, condensed form (FIGURE 17.3b). Condensed chromosomes are easier to maneuver during cell division and are less likely to become tangled or broken than the uncondensed and stringlike structures are. When a chromosome is replicated, a copy is produced that carries the same genes. Each chromosome consists

410 CHAPTER 17 DNA Synthesis, Mitosis, and Meiosis

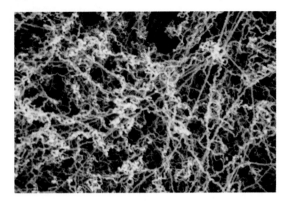

(a) Uncondensed DNA

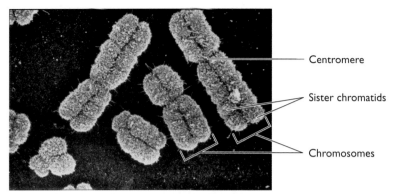

(b) DNA condensed into chromosomes

FIGURE 17.3 DNA condenses during cell division. (a) DNA in its replicated but uncondensed form prior to cell division. (b) During cell division, each copy of DNA is wrapped neatly around many small proteins, forming a condensed structure called a chromosome. After DNA replication, two identical sister chromatids are produced and joined to each other at a region called the centromere.

VISUALIZE THIS In which form, condensed or uncondensed, do you think transcription would be more likely to occur?

of two identical copies of DNA called **sister chromatids**. Each sister chromatid is composed of one DNA molecule. Sister chromatids are attached to each other at a region toward the middle of the replicated chromosome, called the **centromere**.

The structure of the DNA molecule was introduced in Chapter 2 but will be reviewed here. The DNA molecule is double stranded and can be likened to a twisted rope ladder. The backbone, or side, of each strand is composed of alternating sugar and phosphate groups (FIGURE 17.4a). Across the width of the DNA helix (the "rungs" of the ladder) are the nitrogenous bases, paired together via hydrogen bonds such that adenine (A) makes a base pair with thymine (T), and guanine (G) makes a base pair with cytosine (C). The strands of the helix align so that the nucleotides face "up" on one side of the helix and "down" on the other side of the helix. For this reason, the two strands of the helix are said to be antiparallel (FIGURE 17.4b).

FIGURE 17.4 DNA structure. (a) DNA is a double-helical structure composed of sugars, phosphates, and nitrogenous bases. (b) Each strand is composed of nucleotides. The two strands are antiparallel.

(a) DNA double helix is made of two strands.

(b) Each strand is a chain of nucleotides.

DNA Replication

During the process of **DNA replication** that precedes cell division, the double-stranded DNA molecule is copied, first by splitting the molecule in half up the middle of the helix. New nucleotides (molecules composed of a sugar, a phosphate, and a nitrogenous base) are added to each side of the original parent molecule, maintaining the A-to-T and G-to-C base pairings. Because the two strands of a DNA molecule are joined by weak hydrogen bonds, they can be easily unzipped. This process results in two daughter DNA molecules, each composed of one strand of nucleotides from a parent and one newly synthesized strand (FIGURE 17.5a).

To replicate the DNA, an enzyme that assists in DNA synthesis is required. This enzyme, called **DNA polymerase**, moves along the length of the unwound helix and helps bind incoming nucleotides to each other on the newly forming daughter strand (FIGURE 17.5b). When free nucleotides floating in the nucleus have an affinity for each other (A for T and G for C), they bind to each other across the width of the helix. Nucleotides that bind to each other are said to be *complementary* to each other.

The DNA polymerase enzyme triggers, or catalyzes, the formation of the covalent bond between nucleotides along the length of the helix. The paired nitrogenous bases

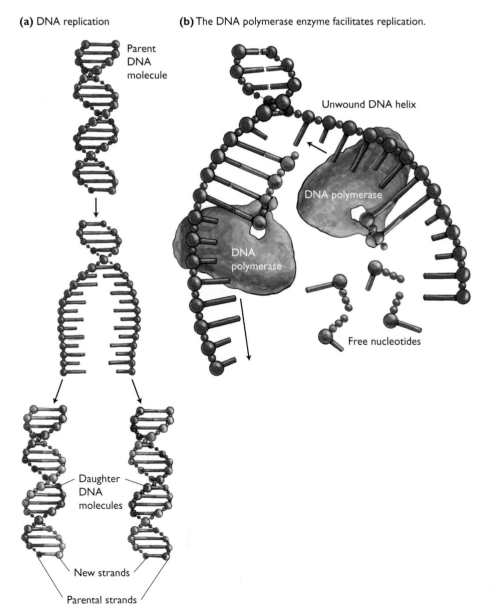

FIGURE 17.5 DNA replication. (a) DNA replication results in the production of two identical daughter DNA molecules from one parent molecule. Each daughter DNA molecule contains half of the parental DNA and half of the newly synthesized DNA. (b) The DNA polymerase enzyme moves along the unwound helix, covalently bonding adjacent nucleotides together on the newly forming daughter DNA strand. Free nucleotides have three phosphate groups, two of which are removed before the nucleotide is added to the growing chain.

FIGURE 17.6 Unreplicated and replicated chromosomes. An unreplicated chromosome is composed of one double-stranded DNA molecule. A replicated chromosome is X shaped and composed of two identical double-stranded DNA molecules. Each DNA molecule of the duplicated chromosome is a copy of the original chromosome and is called a sister chromatid.

VISUALIZE THIS Would you expect to see a duplicated chromosome with different alleles of a gene on each of its two sister chromatids? Why or why not?

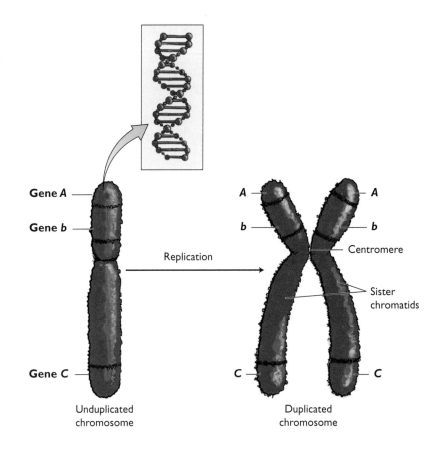

are joined across the width of the backbone by hydrogen bonding, and the DNA polymerase advances along the parental DNA strand to the next unpaired nucleotide. When an entire chromosome has been replicated, the newly synthesized copies are identical to each other. They are attached at the centromere as sister chromatids (FIGURE 17.6). After the DNA is replicated, whether in a normal cell or a cancer cell, the division of the nucleus separates the copied DNA into different daughter cells.

Stop and Stretch 17.2

Two of the scientists credited with determining DNA structure are James Watson and Francis Crick. Watson (on the left in the photo) and Crick reported their hypothesis about the structure of the DNA molecule in a 1953 paper in the journal *Nature*. Although they did not go so far as to propose a detailed model for how the DNA molecule was replicated, they did say, "It has not escaped our notice that the specific pairing we have postulated immediately suggests a copying mechanism for the genetic material." Based on your understanding of the replication of DNA, explain what you think Watson and Crick meant by this statement.

17.3 The Cell Cycle and Mitosis

The cell cycle includes all of the events that occur in going from parent cell to daughter cells during cell division. The cell cycle repeats itself in the sense that parent cells produce daughter cells which can then undergo cell division themselves. The first step of the cell cycle is interphase, the normal functioning of the cell. Interphase is followed by the division of the nucleus and the division of the cytoplasm (FIGURE 17.7a).

Interphase: Normal Functioning and Preparations

A normal cell spends most of its time in **interphase** (FIGURE 17.7b). During this phase of the cell cycle, the cell performs its normal functions and produces the proteins required for the cell to do its particular job. For example, a muscle cell would produce proteins required for muscle contraction, and an epithelial cell of the stomach wall would produce digestive enzymes during interphase. Different cell types spend varying amounts of time in interphase. Cells that frequently divide, like skin cells, spend less time in interphase than do those that seldom divide, such as some nerve cells. A cell that will divide also begins preparations for division during interphase. Interphase can be separated into three phases: G_1, S, and G_2.

During the G_1 (first gap or growth) phase, most of the cell's inner machinery, the organelles, duplicate. Consequently, the cell grows larger during this phase. During the S (synthesis) phase, the DNA in the chromosomes replicates. During the G_2 (second gap) phase of the cell cycle, proteins are synthesized that will help drive mitosis to completion. The cell continues to grow and prepare for the division of chromosomes that will take place during mitosis.

Mitosis: The Nucleus Divides

Mitosis is a division of the cell nucleus that helps produce daughter cells that are exact genetic copies of the parent cell. To achieve this outcome, the sister chromatids of a replicated chromosome are pulled apart, and one copy of each is placed into each newly forming nucleus. Mitosis is accomplished during four stages: *prophase, metaphase, anaphase,* and *telophase.*

During **prophase**, the replicated chromosomes condense, allowing them to move around in the cell without becoming entangled. **Microtubules**, also called *spindle fibers,* form and grow, ultimately radiating out from opposite ends, or poles, of the dividing cell. The growth of microtubules helps the cell to expand. Microtubules also help to move the chromosomes around during cell division. The membrane that surrounds the nucleus, called the **nuclear envelope**, breaks down so that the microtubules can gain access to the replicated chromosomes. At the poles of each dividing cell, structures called **centrioles** anchor one end of each microtubule as it is forming.

During **metaphase**, the replicated chromosomes are aligned across the middle, or *equator,* of each cell. To create the alignment, the microtubules, which are attached to each chromosome at the centromere, line up the chromosomes in single file across the middle of the cell.

During **anaphase**, the centromere splits and the microtubules shorten, pulling each sister chromatid of a chromosome to opposite poles of the cell.

In the last stage of mitosis, **telophase**, the nuclear envelopes re-form around the newly produced daughter nuclei, and the chromosomes revert to their uncondensed form. Cytokinesis divides the cytoplasm, and daughter cells are produced. FIGURE 17.8 summarizes the cell cycle.

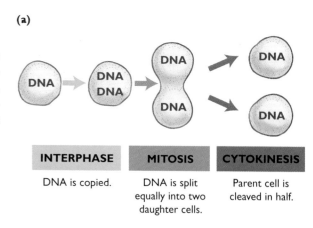

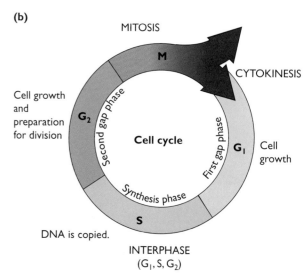

FIGURE 17.7 **The cell cycle.** (a) During interphase, the DNA is copied. Separation of the DNA into two daughter cells occurs during mitosis. Cytokinesis is the division of the cytoplasm, creating two cells. (b) During interphase, there are two stages when the cell grows in preparation for cell division, G_1 and G_2. During the S stage of interphase, the DNA replicates.

414 CHAPTER 17 DNA Synthesis, Mitosis, and Meiosis

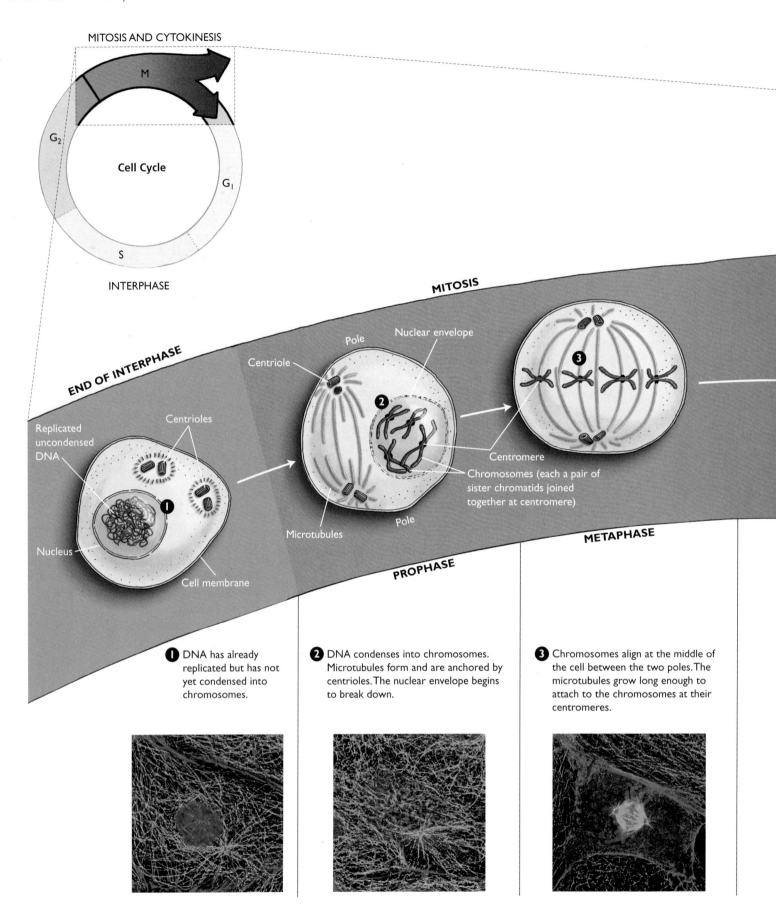

FIGURE 17.8 **Cell division.** This diagram illustrates how cell division proceeds. Photomicrographs show actual animal cells in various stages of cell division.

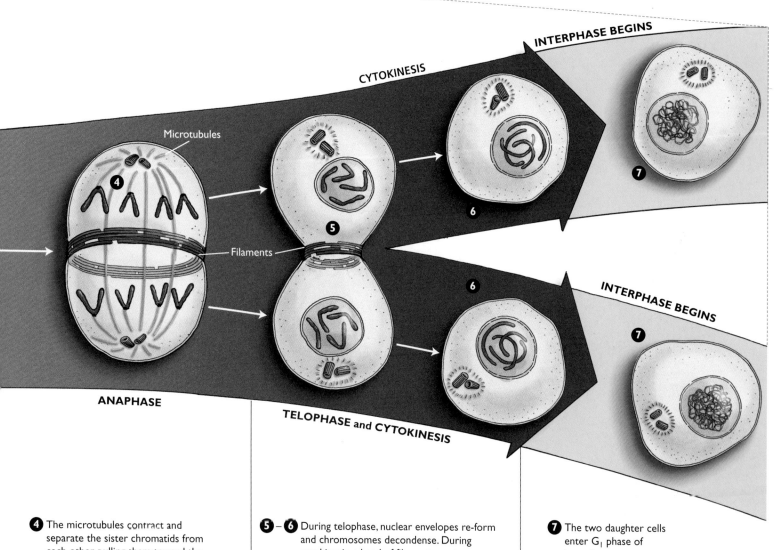

4 The microtubules contract and separate the sister chromatids from each other, pulling them toward the two poles of the cell.

5–**6** During telophase, nuclear envelopes re-form and chromosomes decondense. During cytokinesis, a band of filaments contracts around the equator of the cell, causing two cells to form from the original parent cell.

7 The two daughter cells enter G_1 phase of interphase.

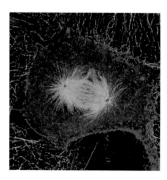
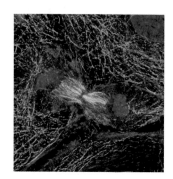

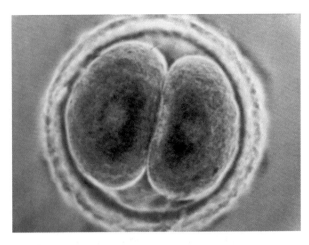

FIGURE 17.9 **Cytokinesis.** During cytokinesis, cells produce a band of filaments that divide the cell in half.

> ### Stop and Stretch 17.3
> Some drugs used to treat cancer attack microtubules and prevent them from growing. Why might this strategy help kill more cancer cells than noncancerous cells?

Cytokinesis: The Cytoplasm Divides

Cytokinesis is the division of the cytoplasm. During cytokinesis a band of proteins encircles the cell at the equator and divides the cytoplasm. This band of proteins contracts to pinch apart the two cells that have formed from the original parent cell. Each daughter cell is genetically identical, having its own nucleus with an exact copy of the parent cell's chromosomes and all the necessary organelles and cytoplasm as well (FIGURE 17.9).

After cytokinesis, the cell reenters interphase, and if the conditions are favorable, the cell cycle may repeat itself. Cells that go through the cell cycle when they should no longer be dividing give rise to tumors.

17.4 Mutations Override Cell-Cycle Controls

When working properly, cell division is a tightly controlled process. Cells are given signals for when and when not to divide. The normal cells in Nicole's ovary and the rest of her body were responding properly to the signals telling them when and how fast to divide. However, the cell that started her tumor was not responding properly to these signals.

Controls in the Cell Cycle

Instead of proceeding through the cell cycle, normal cells halt cell division at a series of checkpoints. During this stoppage, proteins survey the cell to ensure that conditions for a favorable cellular division have been met. Three checkpoints must be passed before cell division can occur. One takes place during G_1, one during G_2, and the last during metaphase (FIGURE 17.10).

Proteins at the G_1 checkpoint determine whether the cell should divide. They survey the cell environment for the presence of other proteins called **growth factors** that stimulate cells to divide. When growth factors are limited in number, cell division does not occur. If enough growth factors are present to trigger cell division, other proteins check to see if the cell is large enough to divide and if all the nutrients required for cell division are available.

At the G_2 checkpoint, other proteins ensure that the DNA has replicated properly and double-check the cell size, again making sure that the cell is large enough to divide. The third and final checkpoint occurs during metaphase. Proteins present at metaphase verify that all the chromosomes have attached themselves to microtubules so that cell division can proceed properly.

If proteins surveying the cell at any of these three checkpoints determine that conditions are not favorable for cell division, the process is halted. When this happens, the cell may undergo a process called *apoptosis*, or programmed cell death.

Proteins that regulate the cell cycle, like all proteins, are encoded by genes. When these proteins are normal, cell division is properly regulated. When these cycle-regulating proteins are unable to perform their jobs, unregulated cell division leads to large masses of cells called tumors.

Mistakes in cell-cycle regulation arise when the genes controlling the cell cycle are altered, or mutated, versions of the normal genes. The Genes & Homeostasis feature outlines some of the mutations that may have led to the development of Nicole's cancer.

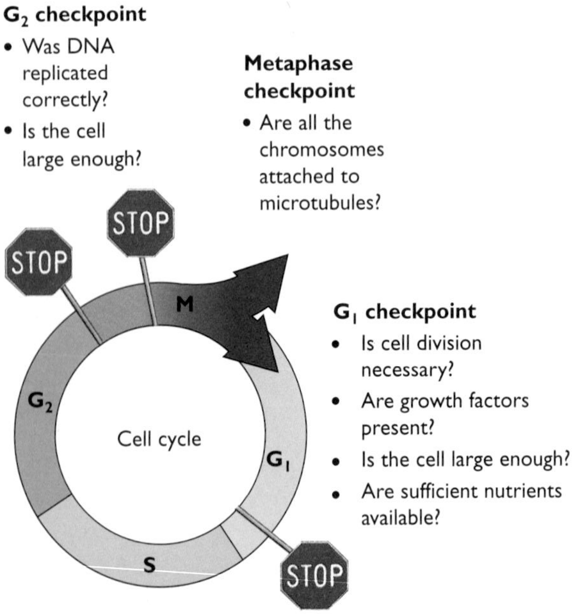

G_2 checkpoint
- Was DNA replicated correctly?
- Is the cell large enough?

Metaphase checkpoint
- Are all the chromosomes attached to microtubules?

G_1 checkpoint
- Is cell division necessary?
- Are growth factors present?
- Is the cell large enough?
- Are sufficient nutrients available?

FIGURE 17.10 **Controls of the cell cycle.** Checkpoints at G_1, G_2, and metaphase determine whether a cell will continue to divide.

GENES & HOMEOSTASIS

Inheritance and Cancer

If mutations occur to genes that encode the proteins regulating the cell cycle, cells can no longer regulate cell division properly.

Genes that encode the proteins regulating the cell cycle are called **proto-oncogenes**. These are normal genes located on many different chromosomes that enable organisms to regulate cell division. When they become mutated, these genes are called **oncogenes**, and are capable of causing cancer.

Many proto-oncogenes provide the cell with instructions for building growth factors. A normal growth factor stimulates cell division only when the cellular environment is favorable and all conditions for division have been met. Oncogenes can overstimulate cell division.

One gene involved in many cases of ovarian cancer is called *HER2*. (Names of genes are italicized, while names of the proteins that they produce are not.) The *HER2* gene carries instructions for building a receptor protein, which binds to a growth factor. When the shape of the receptor on the cell's surface is normal, it signals the inside of the cell to allow division to occur. A mutant receptor protein functions as if many growth factors are present, regardless of actual conditions.

Tumor suppressors are genes that carry the instructions for producing proteins that can detect and repair damage to the DNA. Normal tumor suppressors serve as backups in case the proto-oncogenes undergo mutation. If a growth factor overstimulates cell division, the normal tumor suppressor impedes tumor formation by preventing the mutant cell from moving through a checkpoint.

Researchers believe that a normal *BRCA2* gene encodes a protein that is involved in helping to repair damaged DNA. The misshapen, mutant version of the protein cannot help to repair damaged DNA. This means that damaged DNA will be allowed to undergo mitosis, thus passing new mutations on to their daughter cells. As more and more mutations occur, the probability that a cell will become cancerous increases. The figure summarizing the roles of *HER2* and *BRCA2* in ovarian cancer shows that when ovarian cells have normal versions of these genes, cell division is properly regulated. If a cell in the woman's ovary has a *HER2* mutation, cell division is overstimulated, and a benign tumor can form. If a cell in the woman's ovary has both the *HER2* and *BRCA2* mutations, a cancerous tumor and metastasis are likely.

In Nicole's case, the progression from normal ovarian cells to cancerous cells may have occurred as follows: (1) One single cell in her ovary may have acquired a mutation to its *HER2* gene for a growth factor receptor. (2) The descendants of this cell would have been able to divide faster than neighboring cells, forming a small, benign tumor. (3) Next, a cell within the tumor may have undergone a mutation to its *BRCA2* tumor-suppressor gene, resulting in the inability of the *BRCA2* protein to fix damaged DNA in the cancerous cells. (4) Cells produced by the mitosis of these doubly mutant cells would continue to divide even though their DNA is damaged, thereby enlarging the tumor and producing cells with more mutations. (5) Subsequent mutations could result in angiogenesis, lack of contact inhibition, reactivation of the telomerase enzyme, or overriding of the anchorage dependence. If Nicole were very unlucky, the end result of these mutations could be that cells carrying many mutations would break away from the original ovarian tumor and set up a cancer at one or more new locations in her body.

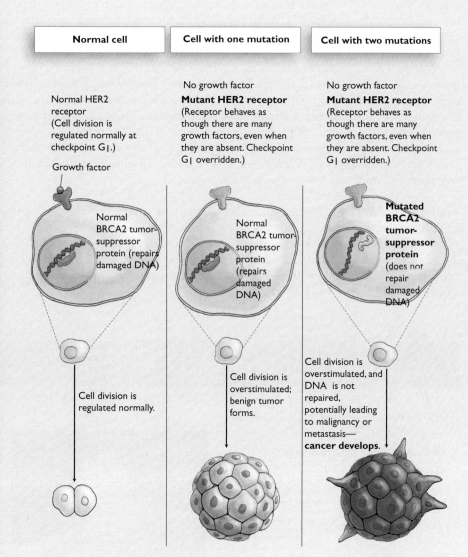

Mutations may be inherited or can arise spontaneously when mistakes in DNA replication occur. Mutations can also be induced by exposure to carcinogens that damage DNA and chromosomes.

From Benign to Malignant Some mutations that occur as a result of damaged DNA being allowed to undergo mitosis are responsible for the progression of a tumor from a benign state, to a malignant state, to metastasis. For example, some cancer cells can stimulate the growth of surrounding blood vessels, through a process called **angiogenesis**. These cancer cells secrete a substance that attracts and reroutes blood vessels so that they supply a developing tumor with oxygen (necessary for cellular respiration) and other nutrients.

Once a tumor has its own blood supply, it can grow at the expense of other, noncancerous cells. Because the growth of rapidly dividing cancer cells occurs more quickly than the growth of normal cells, entire organs can eventually become filled with cancerous cells. When this occurs, an organ can no longer work properly, leading to weakened functioning or organ failure. Damage to organs also explains some of the pain associated with cancer.

Normal cells also display a property called **contact inhibition**, which prevents them from dividing when doing so would require them to pile up on each other. In addition, normal cells need some contact with an underlayer of cells to stay in place. This phenomenon is the result of a process called **anchorage dependence**. Cancer cells override this requirement for some contact with other cells because cancer cells are dividing too quickly and do not expend enough energy to secrete adhesion molecules that glue the cells together. Once a cell loses its anchorage dependence, it may leave the original tumor and move to the blood, lymph, or surrounding tissues.

Most cells are programmed to divide a certain number of times—usually 60 to 70 times—and then they stop dividing. This limits most developing tumors to a small mole, cyst, or lump, all of which are benign. Cancer cells, however, do not obey these life-span limits. Instead, they are **immortal**. They achieve immortality by activating a gene that is usually turned off after early development. This gene produces an enzyme called **telomerase** that helps prevent the degradation of chromosomes. As chromosomes degrade with age, a cell loses its ability to divide. In cancer cells, telomerase is reactivated, allowing the cells to divide without limit.

Multiple Hit Model Because multiple mutations are required for the development and progression of cancer, scientists describe the process of cancer development using the phrase *multiple hit model*. Even though cancer is a disease caused by malfunctioning genes, most cancers are not caused only by inheritance of mutant genes. In fact, scientists estimate that close to 70% of cancers are caused by mutations that occur during a person's lifetime.

Stop and Stretch 17.4

Risk factors for ovarian cancer include smoking and uninterrupted ovulation. Ovarian cancer risk is thought to decrease if ovulation (pictured here) is prevented for periods of time, as it is when a woman is pregnant, breast-feeding, or taking the birth control pill. Why might preventing ovulation help decrease the risk of ovarian cancer?

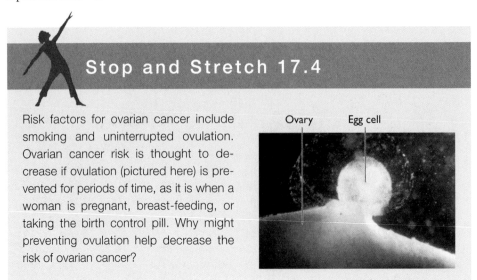

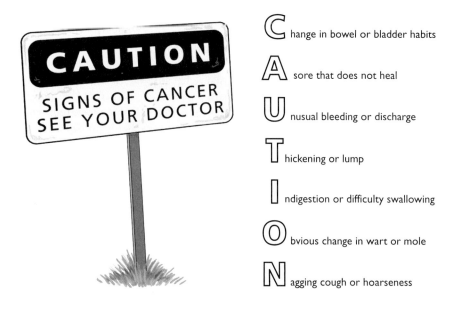

FIGURE 17.11 **Warning signs of cancer.** Self-screening for cancer is possible by looking for these signs. If you experience one or more of these warning signs, see your doctor.

Most of us will inherit few if any mutant genes for cell-cycle control. It is the impact of our environment that will determine whether enough mutations will accumulate during our lifetime to cause cancer.

17.5 Cancer Detection and Treatment

Early detection and treatment of cancer dramatically increase the odds of survival. Being on the lookout for warning signs (FIGURE 17.11) can help alert individuals to developing cancers.

Detecting Cancer

Different types of cancers are detected using different methods. Some tumors, such as those found in the breasts and testicles, can be found by self-examination. Likewise, skin cancer can be detected by visual examination. X-ray images help detect lung cancer. Breast cancer can also be detected by X-ray techniques called mammography (FIGURE 17.12). Some hard-to-reach areas of the body can be examined by a health care worker, for example, the cervix during a pelvic exam and the colon during a colonoscopy. Blood tests can help detect cancers that result in the excess production of proteins normally produced by a particular cell type. Leukemia, cancer of the blood, can be detected by examination of blood cells under a microscope.

Ovarian cancers often show high levels of an ovary-specific protein called CA125 in the blood. Nicole's level of CA125 led her gynecologist to think that a tumor might be forming on her remaining ovary. Once he suspected a tumor, Nicole's physician scheduled a biopsy.

A **biopsy** is the surgical removal of cells, tissue, or fluid that will be analyzed to determine whether they are cancerous. When viewed under a microscope, benign tumors consist of orderly growths of cells that resemble the cells of the tissue from which they were taken. Malignant or cancerous cells do not resemble other cells found in the same tissue. They are dividing so rapidly that they do not have time to produce all the proteins necessary to build normal cells. Their rapid division leads to the often abnormal appearance of cancer cells as seen under a microscope.

A needle biopsy is usually performed if the cancer is located on or close to the surface of the patient's body. For example, breast lumps are often biopsied with a needle to determine if the lump contains fluid and is a noncancerous cyst or if it contains abnormal cells and is a tumor.

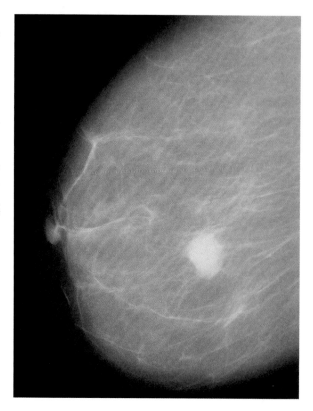

FIGURE 17.12 **Mammography.** A mammogram is an X-ray image of the breast. The dense appearance of the tissue could signal the development of a tumor.

VISUALIZE THIS Why might mammography be recommended mainly for women over 40?

In Nicole's case, getting at the ovary to find tissue for a biopsy required the use of a surgical instrument called a **laparoscope**. For this operation, the surgeon inserted a small light and a scalpel-like instrument through a tiny incision above Nicole's navel.

Nicole's surgeon preferred to use the laparoscope because he knew Nicole would have a much easier recovery from laparoscopic surgery than she had from the surgery to remove her other, cystic ovary. Laparoscopy had not been possible when removing Nicole's other ovary—the cystic ovary had grown so large that her surgeon had to make a large abdominal incision to remove it.

A laparoscope has a small camera that projects images from the ovary onto a monitor that the surgeon views during surgery. These images showed that Nicole's tumor was a different shape, color, and texture from the rest of her ovary. They also showed that the tumor was not confined to the surface of the ovary. In fact, it appeared to have spread deeply into her ovary.

When a cancer is diagnosed, surgery is often performed to remove as much of the cancerous growth as possible without damaging neighboring organs and tissues. Nicole's surgeon decided to shave off only the affected portion of the ovary and leave as much intact as possible, with the hope that the remaining ovarian tissue might still be able to produce egg cells. He then sent the tissue to a laboratory so that the pathologist could examine it. Unfortunately, when the pathologist looked through the microscope this time, she saw the disorderly appearance characteristic of cancer cells. Nicole's ovary was cancerous, and further treatment would be necessary.

Cancer Treatments: Chemotherapy and Radiation

Luckily for Nicole, her ovarian cancer was diagnosed very early. Regrettably, this is not the case for most women with ovarian cancer because the symptoms of ovarian cancer tend to be vague and slow to develop. These symptoms include abdominal swelling, pain, bloating, gas, constipation, indigestion, menstrual disorders, and fatigue. Unfortunately, many women simply overlook these discomforts.

Nicole's cancer was treated early, which increased the likelihood of cure. However, her physician was concerned that some of her cancerous ovarian cells may have spread through blood vessels or lymphatic vessels on or near the ovaries, or into her abdominal cavity, so he started Nicole on chemotherapy after her surgery.

Chemotherapy During **chemotherapy**, chemicals are injected into the bloodstream. These chemicals selectively kill dividing cells. Chemotherapeutic agents act in different ways to interrupt cell division.

Chemotherapy involves many drugs, because most chemotherapeutic agents affect only one type of cellular activity. Cancer cells are rapidly dividing and do not take the time to repair mistakes in replication that lead to mutations. These cells are allowed to proceed through the G_2 checkpoint with many mutations. Therefore, cancer cells can randomly undergo mutations, a few of which might allow them to evade the actions of a particular chemotherapeutic agent. Cells that are resistant to one drug proliferate when the chemotherapeutic agent clears away the other cells that compete for space and nutrients. Cells with a preexisting resistance to the drugs are selected for and produce more daughter cells with the same resistant characteristics, requiring the use of more than one chemotherapeutic agent.

Scientists estimate that cancer cells become resistant at a rate of approximately 1 cell per million. Because tumors contain about 1 billion cells, the average tumor will have close to 1,000 resistant cells. Therefore, treating a cancer patient with a combination of chemotherapeutic agents aimed at different mechanisms increases the chances of destroying all the cancerous cells in a tumor.

Unfortunately, normal cells that divide rapidly are also affected by chemotherapy treatments. Hair follicles, cells that produce red blood cells, white blood cells, and cells that line the intestines and stomach are often damaged or destroyed. The effects of chemotherapy therefore include temporary hair loss, anemia (dizziness and fatigue due to decreased numbers of red blood cells), and lowered protection from infection due to decreases in the number of white blood cells. Also, damage to the cells of the stomach and intestines can lead to nausea, vomiting, and diarrhea.

Several hours after each chemotherapy treatment, Nicole became nauseated. She often had diarrhea and vomited for a day or so after her treatments. Midway through her chemotherapy treatments, Nicole lost most of her hair.

Radiation Therapy Cancer patients often undergo radiation treatments as well as chemotherapy. **Radiation therapy** uses high-energy particles to injure or destroy cells by damaging their DNA, making it impossible for these cells to continue to grow and divide. A typical course of radiation involves a series of 10 to 20 treatments performed after the surgical removal of the tumor, although sometimes radiation is used before surgery to decrease the size of the tumor. Radiation therapy is typically used only when cancers are located close to the surface of the body because it is difficult to focus a beam of radiation on internal organs such as an ovary. Therefore, Nicole's physician recommended chemotherapy only.

Nicole's treatments consisted of many different chemotherapeutic agents, spread over many months. The treatments took place at the local hospital on Wednesdays and Fridays. She usually had a friend drive her to the hospital very early in the morning and return later in the day to pick her up. The drugs were administered through an intravenous (IV) needle into a vein in her arm (FIGURE 17.13). During the hour or so that she was undergoing chemotherapy, Nicole usually studied for her classes.

Nicole did not mind the actual chemotherapy treatments that much. The hospital personnel were kind to her, and she got some studying done. It was the aftermath of these treatments that she hated. During the course of her chemotherapy, Nicole was so exhausted most days that she did not get out of bed until late morning, and on the day after her treatment, she often slept until late afternoon. Then she would get up and try to get some work done or make some phone calls before going back to bed early in the evening.

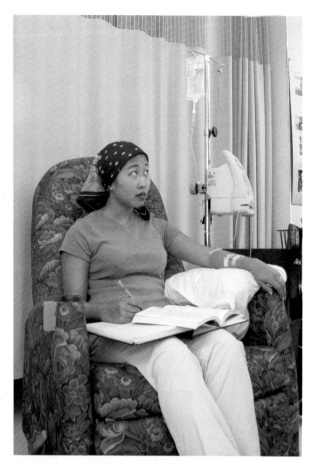

FIGURE 17.13 **Chemotherapy.** Many chemotherapeutic agents are administered through an intravenous (IV) needle.

After six weeks of chemotherapy, Nicole's CA125 levels started to drop. After another two months of chemotherapy, her CA125 levels were back down to their normal, precancerous level. If Nicole has normal CA125 levels for five years, she will be considered to be in **remission**, or no longer suffering negative impacts from cancer. After 10 years of normal CA125 levels, she will be considered cured of her cancer.

Even though her treatments seemed to be going well, Nicole had other worries. She worried that her remaining ovary would not recover from the surgery and chemotherapy, which meant that she would never be able to have children. Nicole had always assumed that she would have children some day. Although she did not currently have a strong desire to have a child, she wondered if her feelings would change. Even though she was not planning to marry any time soon, she also wondered how her future husband would feel if she were not able to become pregnant.

In addition to her concerns about being able to become pregnant, Nicole also became worried that she might pass on mutated, cancer-causing genes to her children. For Nicole, or anyone, to pass on genes to his or her children, reproductive cells must be produced by another type of nuclear division called meiosis.

Stop and Stretch 17.5

Based on what you know about the genetics of cancer, why might a treatment that works for one patient with ovarian cancer not work for another ovarian cancer patient?

17.6 Meiosis: Making Reproductive Cells

Meiosis is a form of cell division that occurs only in specialized cells within the testes of males and the ovaries of females. During meiosis, specialized sex cells called **gametes** are produced. The male gametes are the sperm cells, and the gametes produced by the female are the egg cells.

Meiosis produces gametes that have half as many chromosomes as the parent cell. Both sperm and egg cells contain half as many chromosomes as other non-gamete cells, called **somatic** cells. Somatic cells include cells of the skin, muscle, liver, and stomach tissues. Because human somatic cells have 46 chromosomes and meiosis reduces the chromosome number by one-half, the gametes produced during meiosis contain 23 chromosomes each. When an egg cell and a sperm cell combine their 23 chromosomes at fertilization, the developing embryo will then have the required 46 chromosomes.

The placement of chromosomes into gametes is not random; that is, meiosis does not simply place any 23 of the 46 human chromosomes into a gamete. Instead, meiosis apportions chromosomes very specifically. The 46 chromosomes in human body cells are actually 23 different pairs of chromosomes, and meiosis produces cells that contain one chromosome from every pair.

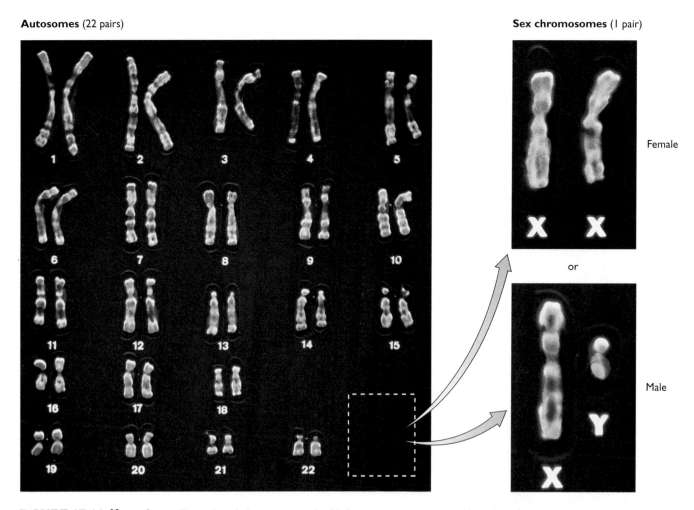

FIGURE 17.14 Karyotype. The pairs of chromosomes in this karyotype are arranged in order of decreasing size and numbered from 1 to 22. The X and Y sex chromosomes are the 23rd pair. The sex chromosomes from a female and a male are shown in the insets.

VISUALIZE THIS Would you consider the X and Y chromosomes to be a homologous pair? Why or why not?

It is possible to visualize chromosome pairs by preparing a **karyotype**, a highly magnified photograph of the chromosomes arranged in pairs. A karyotype is usually prepared from chromosomes that have been removed from the nuclei of white blood cells, which have been treated with chemicals to stop mitosis at metaphase. Because these chromosomes are at metaphase of mitosis, they are composed of replicated sister chromatids and are shaped like the letter X. It is possible to photograph chromosomes and then digitally arrange them in pairs. The 46 human chromosomes can be arranged into 22 pairs of nonsex chromosomes, or **autosomes**, and one pair of **sex chromosomes** (the X and Y chromosomes) to make a total of 23 pairs. Human males have an X and a Y chromosome, whereas females have two X chromosomes. Each chromosome is paired with a mate that is the same size and shape and has its centromere in the same position (FIGURE 17.14).

The pairs of nonsex chromosomes are called **homologous pairs**. Each member of a homologous pair of chromosomes carries the same genes along its length, although not necessarily the same versions of those genes (FIGURE 17.15a). Two chromosomes that are not homologous would be different in size, shape, and position of the centromere and would carry genes that encode different traits (FIGURE 17.15b). Note that there is a difference between the same type of information in the sense that both alleles of this gene code for a cell-cycle control protein, but they happen to code for different versions of the same protein.

Different versions of the same gene are called **alleles** of a gene in the same way that chocolate and vanilla are alternate forms of ice cream. Homologous pairs of chromosomes always carry the same genes at the same locations. Sister chromatids, however, carry the same alleles.

Meiosis separates the members of a homologous pair from each other. Once meiosis is completed, there is one copy of each chromosome (1–23) in every gamete. When only one member of each homologous pair is present in a cell, we say that the cell is **haploid** (*n*)—both egg cells and sperm cells are haploid. All somatic cells in humans

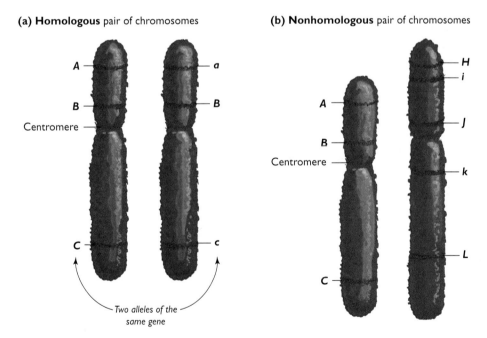

FIGURE 17.15 Homologous and nonhomologous pairs of chromosomes.
(a) Homologous pairs of chromosomes have the same genes (shown here as *A*, *B*, and *C*) but may have different alleles. The dominant allele is represented by an uppercase letter, while the recessive allele is shown with the same letter in lowercase. Note that the chromosomes of this pair each have the same size, shape, and positioning of the centromere. (b) Non-homologous pairs of chromosomes are different sizes and shapes and carry different genes.

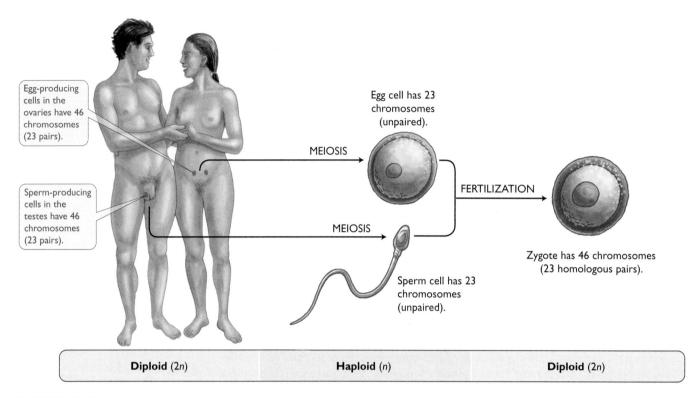

FIGURE 17.16 Gamete production. In humans, the diploid cells of the ovaries and testes undergo meiosis and produce haploid gametes. At fertilization, the diploid condition is restored.

contain homologous pairs of chromosomes and are therefore diploid. For a diploid cell in a person's testis or ovary to become a haploid gamete, it must go through meiosis. After the sperm and egg fuse, the fertilized cell, or **zygote**, will contain two sets of chromosomes and is said to be **diploid (2n)** (FIGURE 17.16).

Like mitosis, meiosis is preceded by an interphase stage that includes G_1, S, and G_2. Interphase is followed by two phases of meiosis, called meiosis I and meiosis II, in which divisions of the nucleus take place (FIGURE 17.17). Meiosis I separates the members of a homologous pair from each other. Meiosis II separates the chromatids from each other. Both meiotic divisions are followed by cytokinesis, during which the cytoplasm is divided between the resulting daughter cells.

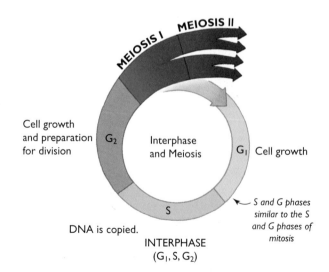

FIGURE 17.17 Interphase and meiosis.
Interphase consists of G_1, S, and G_2, and is followed by two rounds of nuclear division, meiosis I and meiosis II.

Interphase

Like the interphase before mitosis, the interphase that precedes meiosis consists of G_1, S, and G_2. This interphase of meiosis is similar in most respects to the interphase that precedes mitosis. The centrioles from which the microtubules will originate are present. The G phases are times of cell growth and preparation for division. The S phase is when DNA replication occurs. Once the cell's DNA has been replicated, it can enter meiosis I.

Meiosis I

The first meiotic division, meiosis I, consists of prophase I, metaphase I, anaphase I, and telophase I (FIGURE 17.18). During prophase I of meiosis, the nuclear envelope starts to break down, and the microtubules begin to assemble. The previously replicated chromosomes condense so that they can be moved around the cell without becoming entangled. The condensed chromosomes can be seen under a microscope. At this time, the homologous pairs of chromosomes exchange genetic information in a process called *crossing over*, which we will explain in a moment.

At metaphase I, the homologous pairs line up at the equator of the cell. Microtubules of the spindle lengthen and bind to the metaphase chromosomes near the centromere. Homologous pairs are arranged arbitrarily as to which member faces which pole in a process called *random alignment*. At the end of this section, you will find detailed descriptions of crossing over and random alignment along with their impact on genetic diversity.

At anaphase I, the homologous pairs are separated from each other by the shortening of the microtubules. One member of each homologous pair moves to one pole, and the other member moves to the opposite pole of the cell. At telophase I, nuclear envelopes re-form around the chromosomes, partitioning the DNA. The two daughter cells are then separated by cytokinesis. Because each daughter cell contains only one member of each homologous pair, the cells at this point are haploid. Now both of these daughter cells are ready to undergo meiosis II.

Meiosis II

Meiosis II consists of prophase II, metaphase II, anaphase II, and telophase II. This second meiotic division is virtually identical to mitosis and serves to separate the sister chromatids of the replicated chromosome from each other.

At prophase II of meiosis, the cell is readying for another round of division, and the microtubules are lengthening again. At metaphase II, the chromosomes align across the equator in much the same way as they do during mitosis—not as pairs, as was the case with metaphase I. At anaphase II, the sister chromatids separate from each other and move to opposite poles of the cell. At telophase II, the separated chromosomes each become enclosed in their own nucleus. In this fashion, half of a person's genes are physically placed into each gamete; thus, children carry one-half of each parent's genes.

Each parent can produce millions of different types of gametes due to two events that occur during meiosis I—crossing over and random alignment. Both of these processes greatly increase the number of different kinds of gametes that an individual can produce and therefore increase the variation in individuals that can be produced when gametes combine.

Crossing Over and Random Alignment

Crossing over occurs during prophase I of meiosis I. It involves the exchange of portions of chromosomes between members of a homologous pair. Crossing over is believed to occur several times on each homologous pair during each occurrence of meiosis.

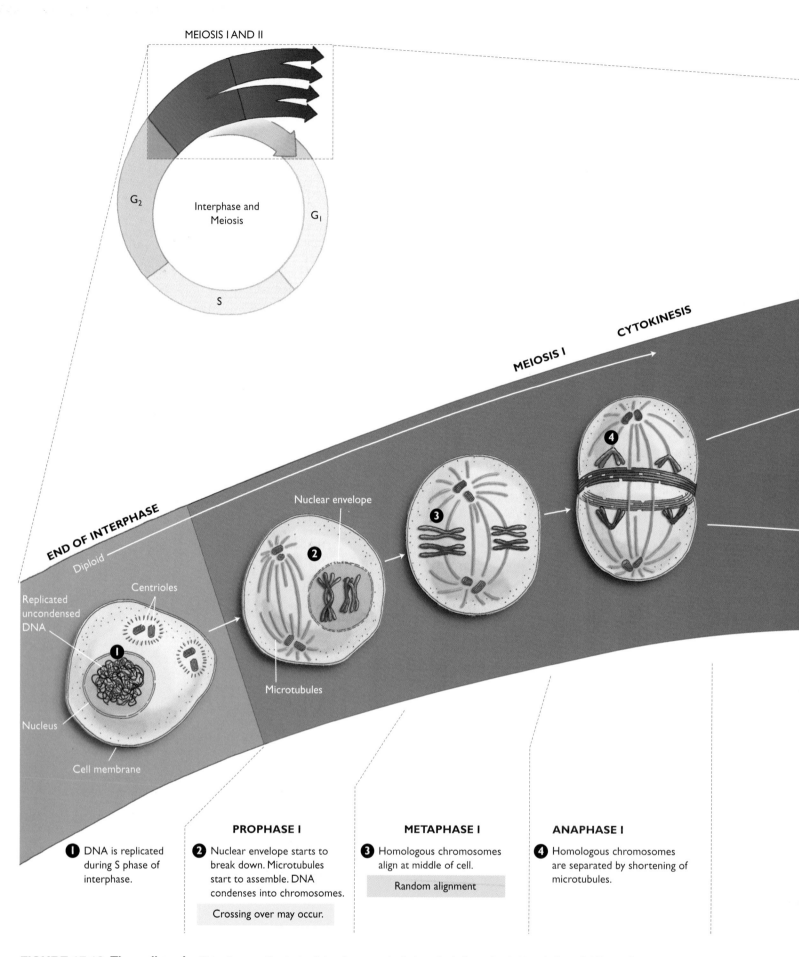

FIGURE 17.18 The cell cycle. This diagram illustrates interphase, meiosis I, meiosis II, and cytokinesis in a dividing cell.

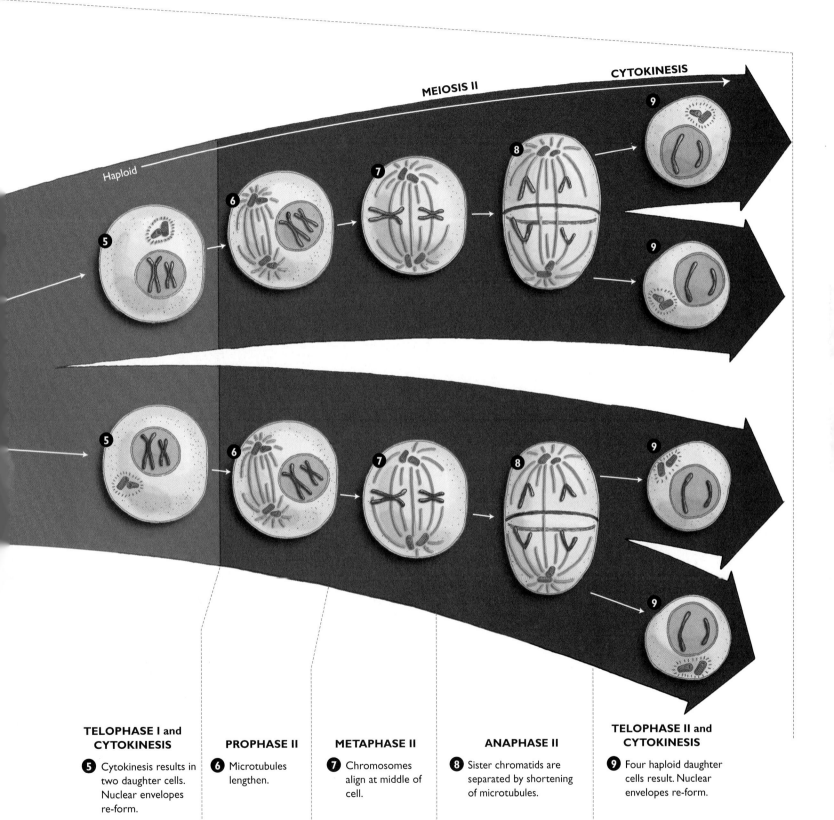

TELOPHASE I and CYTOKINESIS

5 Cytokinesis results in two daughter cells. Nuclear envelopes re-form.

PROPHASE II

6 Microtubules lengthen.

METAPHASE II

7 Chromosomes align at middle of cell.

ANAPHASE II

8 Sister chromatids are separated by shortening of microtubules.

TELOPHASE II and CYTOKINESIS

9 Four haploid daughter cells result. Nuclear envelopes re-form.

To illustrate crossing over, let's consider an example using genes involved in the production of red hair and freckles. These two genes are on the same chromosome and are called **linked genes**. Linked genes move together on the same chromosome to a gamete, and they can undergo crossing over.

If a person has red hair and freckles, the chromosomes may appear as shown in FIGURE 17.19. It is possible for this person to produce four different types of gametes with respect to these two genes. Two types of gametes would result if no crossing over occurred between these genes—the gamete containing the red hair and freckle chromosome, and the gamete containing the nonred hair and nonfreckle chromosome.

Two additional types of gametes could be produced if crossing over did occur—one type containing the red hair and nonfreckle chromosome, and the other containing the reciprocal nonred hair and freckle chromosome. Therefore, crossing over increases *genetic diversity* by increasing the number of distinct combinations of genes that may be present in a gamete.

Random alignment of homologous pairs also increases the number of genetically distinct types of gametes that can be produced. An analogy may help you to visualize

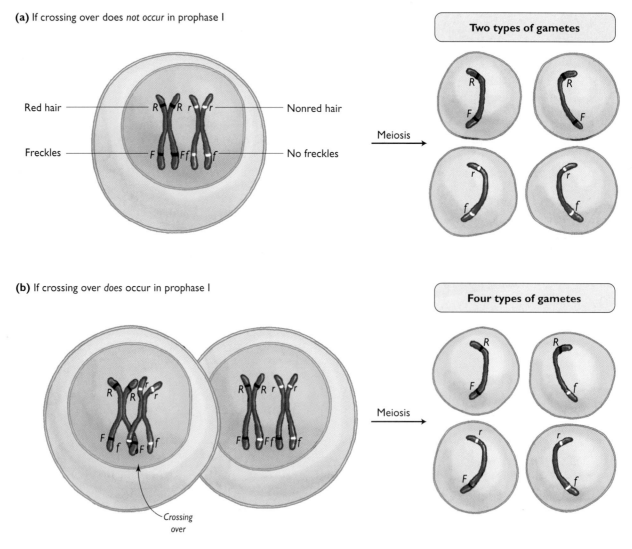

FIGURE 17.19 Crossing over. If an individual with this starting arrangement of alleles undergoes meiosis, he could produce (a) two different types of gametes for these two genes if crossing over does not occur, or (b) four different types of gametes for these two genes if crossing over does occur.

this (FIGURE 17.20). A pair of shoes can be likened to a homologous pair of chromosomes. The two shoes are similar in size, shape, and style but are not exact duplicates, because they fit left and right feet. If you ask 23 students to take off their shoes and place them in a row across the front of the classroom, and they arrange their shoes so that the left shoe is on the left, and the right shoe is on the right, the students could then separate all of the left shoes from the right shoes, just as meiosis separates homologous chromosomes. Separating shoes in this way would produce two different piles—one containing all left shoes, the other containing all right shoes. Each of these piles is equivalent to the set of chromosomes contained within a single human gamete.

Different piles of shoes would result if the very first pair of shoes was reversed so that the left shoe and right shoe exchange places but the other 22 pairs of shoes stayed as they were. When the shoes are separated this time, one pile would have 22 right shoes and 1 left shoe, and the other pile would have 22 left shoes and 1 right shoe. These piles represent two different gametes that could be produced by the same parent. The students could continue making different combinations of left and right shoes for a very long time because there are over 8 million possible ways to line up these 23 pairs of shoes. The same is true for human chromosomes; due to indepen-dent assortment, each individual human can make at least 8 million different types of egg and sperm.

Using Nicole's chromosomes as an example (FIGURE 17.21), let's follow the path of two cell-cycle control genes, *BRCA2* and *HER2*. Mutations to either of these two genes are known to increase the risk of cancer. Let's assume that she did in fact inherit mutant versions of both the *BRCA2* and *HER2* cell-cycle control genes (symbolized *BRCA2*⁻ and *HER2*⁻) and that these genes are located on different chromosomes. The arrangement of homologous pairs of chromosomes at metaphase I determines which chromosomes will end up together in a gamete.

If we consider only these two homologous pairs of chromosomes, then two different alignments are possible, and four different gametes can be produced. For example, when Nicole produces egg cells, the two chromosomes that she inherits from her dad could move together to the gamete, leaving the two chromosomes she inherited from her mom to move to the other gamete. It is equally probable that Nicole could undergo meiosis in which one chromosome from each parent will align randomly together, resulting in two more types of gametes being produced.

From the previous sections, you have learned that cells undergo mitosis for growth and repair, and meiosis to produce gametes. FIGURE 17.22 (see page 431) compares the significant features of mitosis and meiosis.

It is now possible to revisit the question of whether Nicole will pass on cancer-causing genes to any children she may have. Since the surgery on her remaining ovary left as much of the ovary intact as possible, she should be able to have children. Because Nicole developed cancer at such a young age, it seems likely that she may have inherited at least one mutant cell-cycle control gene. She may or may not pass that gene on. If Nicole has both a normal and a mutant version of a cell-cycle control gene, she will be able to make gametes with and without the mutant allele. Therefore, she could pass on the mutant allele if a gamete containing that allele is involved in fertilization.

We have also seen that it takes many "hits" or mutations for a cancer to develop. Therefore, even if Nicole does pass on one or a few mutant versions of cell-cycle control genes to a child, environmental conditions will dictate if enough other mutations will accumulate to allow a cancer to develop.

Mutations caused by environmental exposures are not passed from parents to children, unless the mutation happens to occur in a cell of the gonads that will be used to produce a gamete. Nicole's cancer occurred in the ovary, the site of meiosis, but not all cells in the ovary undergo meiosis. Nicole's cancer originated in the outer covering of the ovary, a tissue that does not undergo meiosis. The cells involved in ovulation are located inside the ovary. A skin cancer that develops from exposure to ultraviolet light will not be passed on, nor will most of the mutations that Nicole obtained from environmental exposures. Therefore, for any children that Nicole (or any of us) might have, the combined effects of inherited mutant alleles and any mutations induced by environmental exposures will determine if cancers will develop.

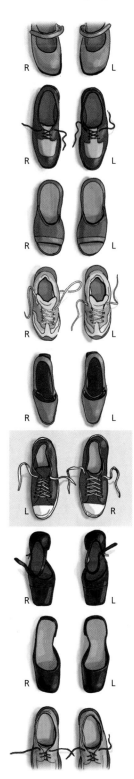

FIGURE 17.20 **Random alignment: an analogy using shoes.** A pair of shoes is comparable to a homologous pair of chromosomes. Meiosis separates the member of one pair independently of other pairs—much like what would happen if students created two piles of shoes by each independently separating and sorting their own pair. Here you can see how two different piles of shoes could be created if one student put her right sneaker in the left shoe pile and vice versa.

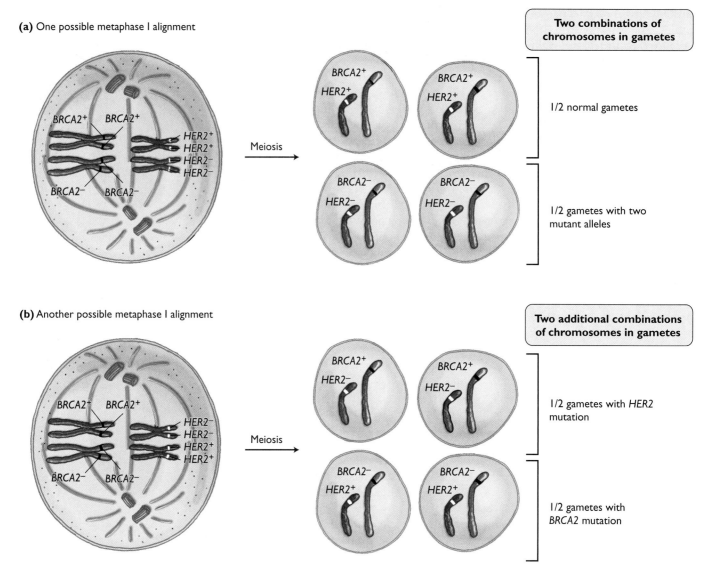

FIGURE 17.21 Random alignment of chromosomes. Two possible alignments, (a) and (b), can occur when there are two homologous pairs of chromosomes. These different alignments can lead to novel combinations of genes in the gametes.

Stop and Stretch 17.6

If Nicole did not inherit a mutated gene for cell-cycle control, can she pass her cancer on to any children she might have? Why or why not?

If Nicole inherited one mutated copy of the *BRCA2* gene for cell-cycle control, what percentage of her children would also inherit that gene from her?

If her children inherit the mutated gene for cell-cycle control, are they destined to get cancer? Why or why not?

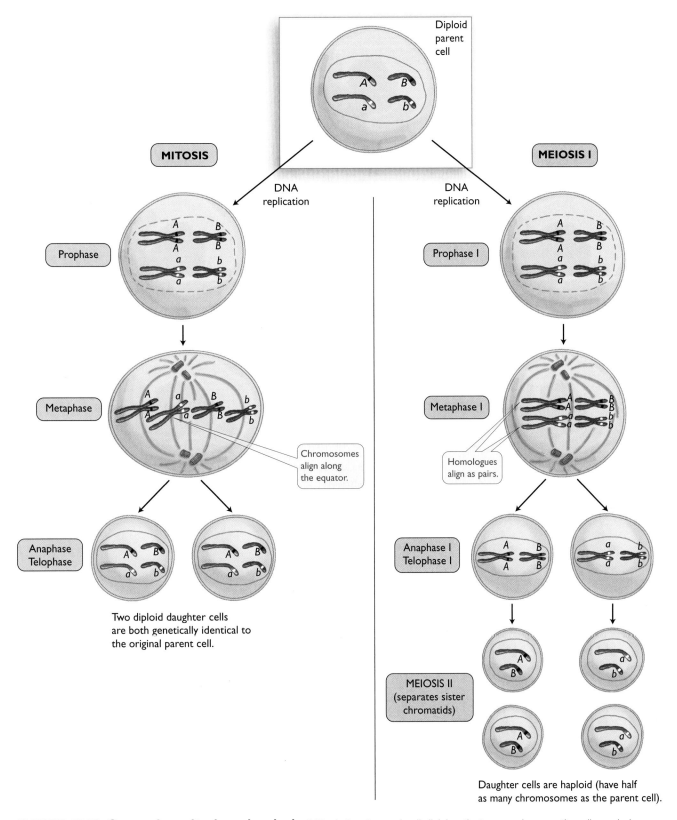

FIGURE 17.22 Comparing mitosis and meiosis. Mitosis is a type of cell division that occurs in somatic cells and gives rise to daughter cells that are exact genetic copies of the parent cell. Meiosis occurs in cells that will give rise to gametes and decreases the chromosome number by one-half. To reduce the chromosome number and still ensure that each gamete receives one member of each homologous pair, the two members of each homologous pair align across the equator at metaphase I of meiosis and are separated from each other during anaphase I.

Chapter REVIEW

ROOTS TO REMEMBER

The following roots of words come mainly from Latin and Greek and will help you to decipher terms:

cyto- and **-cyte** relate to cells.

-kinesis means motion.

mitosis comes from the Greek word for a thread.

onco- means cancer.

proto- means before.

soma-, somato-, and **-some** means body.

telo- means end or completion.

KEY TERMS

17.1 What Is Cancer?
benign *p. 407*
cancer *p. 406*
carcinogen *p. 408*
cell division *p. 406*
malignant *p. 407*
metastasis *p. 407*
risk factor *p. 408*
tumor *p. 407*

17.2 Overview of Cell Division
centromere *p. 410*
chromosome *p. 409*
DNA (deoxyribonucleic acid) *p. 409*
DNA polymerase *p. 411*
DNA replication *p. 411*
sister chromatid *p. 410*

17.3 The Cell Cycle and Mitosis
anaphase *p. 413*
centriole *p. 413*
cytokinesis *p. 416*
interphase *p. 413*
metaphase *p. 413*
microtubule *p. 413*
mitosis *p. 413*
nuclear envelope *p. 413*
prophase *p. 413*
telophase *p. 413*

17.4 Mutations Override Cell-Cycle Controls
anchorage dependence *p. 418*
angiogenesis *p. 418*
contact inhibition *p. 418*
growth factor *p. 416*
immortal *p. 418*
oncogenes *p. 417*
proto-oncogenes *p. 417*
telomerase *p. 418*
tumor suppressors *p. 417*

17.5 Cancer Detection and Treatment
biopsy *p. 419*
chemotherapy *p. 420*
laparoscope *p. 420*
radiation therapy *p. 421*
remission *p. 421*

17.6 Meiosis: Making Reproductive Cells
allele *p. 423*
autosome *p. 423*
crossing over *p. 425*
diploid (*2n*) *p. 424*
gamete *p. 422*
haploid (*n*) *p. 423*
homologous pair *p. 423*
karyotype *p. 423*
linked genes *p. 428*
meiosis *p. 422*
random alignment *p. 428*
sex chromosome *p. 423*
somatic *p. 422*
zygote *p. 424*

SUMMARY

17.1 What Is Cancer?

- Unregulated cell division can lead to the formation of a tumor. Benign tumors are noncancerous tumors that grow slowly and do not invade surrounding structures. Malignant tumors are those that are invasive, or those that metastasize to surrounding tissues. Metastatic tumors move to other locations in the body, starting new cancers (p. 407).

17.2 Overview of Cell Division

- Cell division is a process required for growth and development (p. 409).

- The DNA molecule is composed of antiparallel strands of nucleotides. Nucleotides are composed of a sugar, a phosphate, and a nitrogenous base. Nitrogenous bases include A, C, G, and T. A and T make a base pair with each other, as do C and G (p. 410).

- Once copied, the duplicated chromosome is composed of two sister chromatids that are attached to each other at the centromere (p. 410).

- For cell division to occur, the DNA must be copied and passed on to daughter cell (p. 411).

- During DNA replication or synthesis, each strand of the double-stranded DNA molecule is used as a template for the synthesis of a new daughter strand of DNA. The newly synthesized DNA strand is complementary to the parent strand. The enzyme DNA polymerase ties together the nucleotides on the daughter strand (pp. 411–412).

17.3 The Cell Cycle and Mitosis

- The cell cycle includes all of the events that occur as one cell gives rise to daughter cells (p. 413; Figure 17.7).

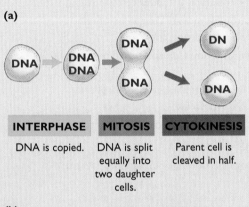

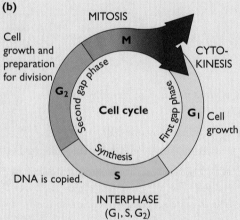

- Interphase includes two gap phases of the cell cycle (G_1 and G_2), during which the cell grows and prepares to enter mitosis or meiosis. During the S (synthesis) phase, the DNA replicates. The S phase of interphase occurs between G_1 and G_2 (p. 413).

- During mitosis, the sister chromatids are separated from each other into daughter cells that are exact genetic copies of the parent cell, for the purpose of tissue growth or repair. In the first stage, prophase, the replicated DNA condenses into linear chromosomes. At meta-phase, these replicated chromosomes align across the middle of the cell. At anaphase, the sister chromatids separate from each other and align at opposite poles of the cells. At telo-phase, a nuclear envelope re-forms around the chromosomes lying at each pole (p. 413).

- Cytokinesis is the last phase of the cell cycle. During cytokinesis, the cytoplasm is divided into two portions, one for each daughter cell (p. 416).

17.4 Mutations Override Cell-Cycle Controls

- When cell division is working properly, it is a tightly controlled process. Normal cells divide only when conditions are favorable (p. 416).

- Proteins survey the cell and its environment at checkpoints as the cell moves through G_1, G_2, and metaphase and can halt cell division if conditions are not favorable (p. 416).

- Mistakes in regulating the cell cycle arise when genes that control the cell cycle are mutated. Mutated genes can be inherited, or mutations may arise spontaneously or be caused by exposure to carcinogens (pp. 416–418).

- As a tumor progresses from benign to malignant, it often undergoes mutations that allow angiogenesis, loss of contact inhibition and anchorage dependence, and immortality. Thus, many changes, or hits, to the cancer cell are required for malignancy (p. 418).

17.5 Cancer Detection and Treatment

- Early cancer detection and treatment increase survival odds (p. 419).

- A biopsy is a common method for detecting cancer. A biopsy involves removing some cells or tissues suspected of being cancerous and analyzing them (p. 419).

- Typical cancer treatments include chemotherapy, which involves injecting chemicals that kill rapidly dividing cells, and radiation, which involves killing tumor cells by exposing them to high-energy particles (pp. 420–421).

17.6 Meiosis: Making Reproductive Cells

- Meiosis is a type of sexual cell division, which gives rise to gametes. Gametes contain half as many chromosomes as somatic cells do. The reduction in the number of chromosomes occurs because diploid cells are used to produce haploid cells (pp. 422–425).

- Meiosis is preceded by an interphase stage, in which the DNA is replicated (p. 425).

- During meiosis I, the members of a homologous pair of chromosomes are separated from each other (p. 425).

- During meiosis II, the sister chromatids are separated from each other (p. 425).

- Homologous pairs of chromosomes exchange genetic information during crossing over at prophase I of meiosis, thereby increasing the number of genetically distinct gametes that an individual can produce (pp. 425–428; Figure 17.19).

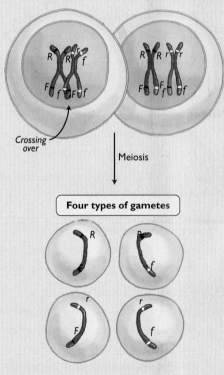

- The alignment of members of a homologous pair at metaphase I is random with regard to which member of a pair faces which pole. This random alignment of homologous chromosomes increases the number of different kinds of gametes an individual can produce (pp. 428–429).

LEARNING THE BASICS

1. Cancer cells differ from normal cells in that _____.
   a. they divide when they should not
   b. they invade surrounding tissues
   c. they can move to other locations in the body
   d. all of the above

2. A cell that begins mitosis with 46 chromosomes produces daughter cells with _____.
   a. 13 chromosomes
   b. 23 chromosomes
   c. 26 chromosomes
   d. 46 chromosomes

3. DNA polymerase _____.
   a. attaches sister chromatids at the centromere
   b. synthesizes daughter DNA molecules from fats and phospholipids
   c. is the enzyme that facilitates DNA synthesis
   d. causes cancer cells to stop dividing

4. The centromere is a region at which _____.
   a. sister chromatids are attached to each other
   b. metaphase chromosomes align
   c. the tips of chromosomes are found
   d. the nucleus is located

5. DNA replication occurs _____.
   a. between G_1 and G_2 of interphase
   b. during G_2
   c. during prophase of mitosis
   d. between metaphase and anaphase
   e. before mitosis, but not meiosis

6. All of the following are characteristics of telophase during mitosis except _____.
   a. cytokinesis begins
   b. the nuclear envelope re-forms
   c. each chromosome is made of two chromatids
   d. the chromosomes are in their uncondensed, stringlike form

7. At metaphase of mitosis, _____.
   a. the chromosomes are condensed and found at the poles
   b. the chromosomes are composed of one sister chromatid
   c. cytokinesis begins
   d. the chromosomes are composed of two sister chromatids and are lined up along the equator of the cell

8. Sister chromatids _____.
   a. are two different chromosomes attached to each other
   b. are exact copies of one chromosome that are attached to each other
   c. arise from the centrioles
   d. are broken down by mitosis
   e. are chromosomes that carry different genes

9. Proto-oncogenes _____.
   a. are mutant genes that a person might inherit
   b. are normal genes that encode cell-cycle control proteins
   c. cause cancer when inherited
   d. are proteins that fail to suppress tumor formation

10. Cytokinesis _____.
    a. is the phase of mitosis that precedes telophase
    b. is the division of the cytoplasm
    c. is the division of the nuclear contents
    d. cleaves one cell into four daughter cells

11. After telophase I of meiosis, each daughter cell is _____.
    a. diploid, and the chromosomes are composed of one double-stranded DNA molecule
    b. diploid, and the chromosomes are composed of two sister chromatids
    c. haploid, and the chromosomes are composed of one double-stranded DNA molecule
    d. haploid, and the chromosomes are composed of two sister chromatids

12. Which of the following cells does *not* contain two sets of chromosomes?
    a. haploid cells
    b. somatic cells
    c. diploid cells
    d. skin cells
    e. liver cells

13. Which of the following is not a function of meiosis?
    a. production of gametes
    b. reduction of chromosome number by one-half
    c. genetic mixing of a person's maternally and paternally inherited chromosomes in the gamete
    d. producing haploid daughter cells
    e. producing daughter cells that are genetically identical to the parent cell

14. Alleles are _____.
    a. alternate forms of a gene
    b. found only at the tips of chromosomes
    c. duplicated chromosomes
    d. formed by random alignment

15. Which of the following is a correct statement about homologous chromosomes?
    a. Their alleles are all identical.
    b. They have the genes for the same traits at the same locations.
    c. They pair up in prophase II.
    d. They are found in haploid cells.

Questions 16–18 refer to the diploid cell in FIGURE 17.23 **with two pairs of chromosomes. The cell at the top is in G_1. For each description in items 16–18, choose the diagram that shows the correct chromosomal condition for that phase.**

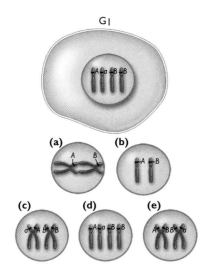

FIGURE 17.23 A cell in G_1 and in other states.

16. A possible daughter cell of mitosis
17. A possible daughter cell at the completion of meiosis I
18. A possible gamete at the completion of meiosis II

19. State whether the chromosomes depicted in each part of FIGURE 17.24 are haploid or diploid.

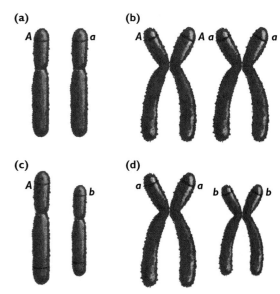

FIGURE 17.24 Haploid or diploid chromosomes?

20. **True or False?** The cells that make up the human body (except sperm and egg cells) contain 46 pairs of chromosomes.

ANALYZING AND APPLYING THE BASICS

1. Why might it be important to know if your family has a history of certain cancers?

2. Assume that you are now a pharmaceutical researcher attempting to treat cancer. If you could design a drug to disrupt any biological process you chose, what would you try to disrupt and why?

3. Would a skin cell mutation that your father obtained using tanning beds make you more likely to get cancer? Why or why not?

4. Cells accumulate mutations in genes over time. Hundreds of genes can be mutated before a cell becomes cancerous. Although the cancerous cell may have hundreds of mutations, as few as one or two oncogenes may actually be responsible for maintaining malignancy. If these few genes are targeted in therapy, the cancer can be controlled. What is likely to happen to a cell that has hundreds of gene mutations but is no longer malignant? Explain your answer.

Questions 5 and 6 refer to the following information.

The Epstein-Barr virus is common in human populations—more than 90% of U.S. adults have been exposed to this virus at some point in their lives. This virus causes infectious mononucleosis ("mono").

This virus has also been associated with several kinds of cancers including breast cancer, Hodgkin's disease, and Burkitt's lymphoma. The virus itself does not appear to be the cause of these cancers. However, the virus can increase the risk that the cancer will metastasize by interfering with the function of a cellular protein that normally prevents the movement of cancer cells.

5. How could knowledge about the connection between the Epstein-Barr virus and breast cancer help doctors to treat patients who have active breast cancer or a family history of breast cancer?

6. How could scientists reduce or eliminate the effects of the Epstein-Barr virus on malignant cancers?

7. In 2006 the findings of a long-term study on the relationships between oral contraceptive use and cancer were published. This study monitored more than 17,000 British women over a period of 30–36 years. Table 17.1 shows some of the results for reproductive cancer incidence in relation to oral contraceptive use. Using the table, summarize the relationships between cancer and oral contraceptive use.

Table 17.1 Oral Contraceptives and Cancer

Cancer Type	Nonuser	Up to 48 mos.	49 to 96 mos.	97 mos. or more
Breast	314	141	182	207
Cervical	6	9	16	28
Uterine	50	12	11	4
Ovarian	58	28	10	10

Duration of Oral Contraceptive Use

Cervical, uterine, and ovarian cancer results were statistically significant—the probability of these results happening by chance is less than 1 in 1,000. Breast cancer results were not significant.

8. Chemotherapy is a common cancer treatment. However, it is not necessarily effective in the treatment of some types of cancer. Its rate of success is only 40%, for example, in patients with the most common type of non-Hodgkin's lymphoma. Scientists now look to genes to try to predict which patients would benefit from chemotherapy. Researchers studied gene expression in the tumor cells of patients who responded well to chemotherapy and those who responded poorly. They found 17 genes that showed a dramatic difference in expression. The scientists then used these differences to create a formula that predicts patient survival. This formula could be used by doctors to decide whether chemotherapy is the best treatment for that particular patient. How could the expression of certain genes in a tumor help to predict the patient's survival rate after chemotherapy?

9. The screening tests in Table 17.2 are recommended for early detection of cancer. Which of these types of cancers are you most at risk for? What screening tests should you or your health care practitioner be performing?

Table 17.2 Recommended Screening Tests for Cancer

Type of Cancer	Test	Frequency
Breast	Self-exam	Monthly after age 18
Breast	Mammogram	Yearly after age 40
Cervical, ovarian, uterine	Pelvic exam and Pap smear	Yearly after age 18
Colon	Rectal exam	Yearly after age 40
Colon	Examination of stool for blood	Yearly after age 50
Prostate	Rectal exam and blood test	Yearly after age 50
Skin	Visual exam	Yearly (more frequent for people at higher risk)
Testicular	Self-exam	Monthly after age 14

10. Bevacizumab (trade name Avastin) is a pharmaceutical agent that acts on blood vessels around a tumor. Avastin inhibits the formation of new blood vessels, causes existing vessels to shrink, and inhibits the growth and branching of existing vessels. When Avastin, in combination with chemotherapy, is used to treat metastatic colorectal cancer, patient survival increases by an average of five months. Why would Avastin be combined with other therapies to more effectively treat this cancer?

CONNECTING THE SCIENCE

1. Should members of society be forced to pay the medical bills of smokers when the cancer risk from smoking is so evident and publicized? Explain your reasoning.
2. Are there changes you could make in your own life to decrease your odds of getting cancer?
3. If you could be tested today to find out whether or not you inherited one or more mutated cell-cycle control genes, would you want that information? Why or why not?

Chapter **18**

Human Reproduction

Fertility and Infertility

LEARNING GOALS

1. Define *sexual reproduction*.

2. Describe the structure and function of the male and female gonads.

3. Diagram the male reproductive system and describe the function of all associated structures.

4. Diagram the female reproductive system and describe the function of all associated structures.

5. Describe gametogenesis in males and females.

6. Compare and contrast spermatogenesis and oogenesis.

7. Describe events of the ovarian cycle.

8. Describe the hormonal regulation of the menstrual cycle.

9. Describe the four-stage model of human sexual response. Explain how the physiological changes associated with these stages promote reproduction.

10. Compare the biological mechanisms of the various methods of birth control.

A newlywed couple wonder when would be the best time to start a family.

A newlywed couple are trying to decide when to have children. They know they want children but have both just finished graduate school and are living in a small apartment in the city. They are in their late 20s and would like to wait a few more years if possible. This will give them both time to settle into their careers and maybe even allow them to save enough money to buy a house with a yard outside of the city. They would both be content to wait a few years if it weren't for the experience that the couple who are their best friends are having with infertility.

Their friends were diagnosed with fertility problems after they were unable to conceive a child after one year of unprotected intercourse. Infertility can be due to problems with the male or female partner of a couple or both. Because doctors have been unable to determine the cause of this couple's infertility, they have been through a battery of tests and expensive assisted reproductive technologies including a procedure where the male's sperm is used to fertilize the female's egg in a petri dish. The infertile couple have spent all their savings trying to have a child, and even took out a second mortgage on their home. Lately they have resorted to charging treatments on their credit card.

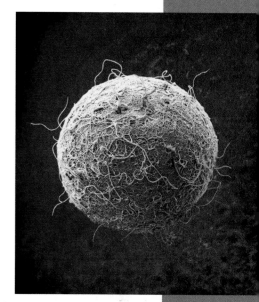

They worry that, like their friends, they might have trouble conceiving.

Multiple births are more likely when reproductive technologies are used, as was the case with the Dilley sextuplets.

Their latest treatment involves the female in the infertile couple taking drugs to increase the number of egg cells she produces. One very real risk of this treatment is a pregnancy that results in multiple births. While this can end well, as was the case with the Dilley sextuplets born in 1993 in Indiana, it usually does not. Multiple-birth pregnancies are at high risk for producing infants who are born prematurely and have a low birth weight. Premature babies require intense medical care before they are able to breathe on their own

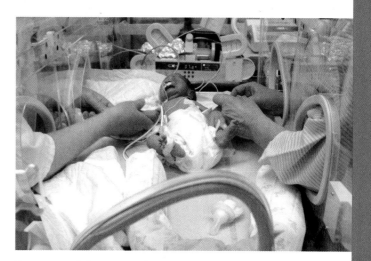

Premature infants may have lifelong health problems.

and to regulate their own body temperature. In addition, they may suffer lifelong health problems and developmental delays.

The newlyweds have followed a more recent case of premature sextuplets with deep sadness. These babies, born weighing between 300 and 550 grams (11 and 19 ounces), did not fare as well as the Dilleys'. Three of the four boys died within a week of their birth and the remaining boy and two girls are at risk for long-term health problems.

Should the newlyweds begin trying to have children now in case they may need extra time and money to conceive, like the roughly 12% of Americans dealing with this problem? Or should they rest assured that they are fertile and wait until they feel more ready to have children? To answer this question, a more thorough understanding of reproduction is required.

18.1 The Human Reproductive Systems

Reproduction is the process whereby organisms produce offspring, allowing for the continuation of their species. During reproduction, cells such as sperm and eggs that are nonessential to the organism's own survival, but indispensable to the perpetuation of the species, are produced and can help give rise to offspring. Humans reproduce by a process called sexual reproduction.

Sexual reproduction involves two genetically distinct parents. Offspring are produced after the sex cells, or **gametes**, from two different individuals combine genetic information at fertilization. The gamete-producing structures are called **gonads**. The male gonads, where gametes called sperm are produced, are called *testes*. The female gonads, where gametes called egg cells are produced, are the *ovaries*.

Because two different individuals contribute genetic information to their gametes, the offspring will be a genetic mixture of both parents. Sexual reproduction produces an almost infinite variety of genetically unique individuals as a result of the mixing of parental genetic information.

Sexual reproduction is distinct from the asexual reproduction favored by some organisms, such as bacteria. During asexual reproduction, one parent organism subdivides itself, producing offspring that are genetic clones of the parent and of each other. This means of reproduction does not allow for genetic variability.

To understand some of the causes of infertility, we must first consider the anatomical structures that make up the human male and female reproductive systems.

The reproductive systems of males and females consist of external and internal structures. These structures are designed to allow the production and maturation of gametes, to synthesize and secrete substances required for reproductive function, and to provide ducts through which gametes are delivered.

The Male Reproductive System

The external and internal components of the male reproductive system are shown in FIGURE 18.1.

The **penis** delivers sperm to the female reproductive tract during sexual intercourse. The penis is composed of spongy erectile tissue. During sexual arousal, this tissue fills with blood. Pressure from the increased volume of blood in the penis seals off the veins that drain the penis of blood. This causes the penis to become engorged with blood, keeping it erect. The erection is necessary for insertion of the penis into the vagina, which facilitates delivery of sperm to the egg.

A tube inside the penis, called the **urethra**, provides passage out of the body for either sperm or urine. A valve prevents sperm and urine from being conducted out of the urethra at the same time. The head, or **glans penis**, consists of highly sensitive skin that is covered by a fold of skin called the foreskin in an uncircumcised male.

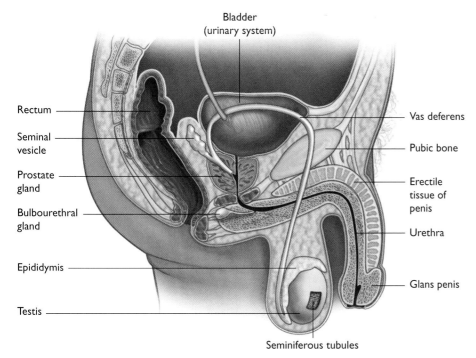

FIGURE 18.1 Male reproductive anatomy. The male external genitalia consist of the penis and scrotum. The scrotum houses the testes. The testicles of a male produce sperm, which are stored in the epididymis and carried by the vas deferens to the urethra, from which they leave the body during ejaculation. Secretions from the seminal vesicles, prostate, and bulbourethral glands help sperm mature and gain motility.

VISUALIZE THIS Can you trace the path a sperm cell would take from its production in a testis to its exit via the urethra at ejaculation?

The **scrotum** is a pouch below the penis that contains the **testes**, or testicles, which produce sperm. The skin of the scrotum is thin and devoid of fatty tissue. It tends to be wrinkled and contain little hair. Below the skin and surrounding each testis is skeletal muscle called the *cremaster* muscle that regulates the position of the testes relative to the body. Sperm production is maximal at a temperature slightly lower than body temperature. Thus, the testes are located outside the body cavity. In colder surroundings, the cremaster contracts to keep the testes closer to the body. This homeostatic temperature-regulating mechanism is disrupted when a man has a varicose vein in the scrotum, called a *varicocele*. The varicocele allows blood to pool in the scrotum, which disrupts normal temperature control and hence prevents sperm from surviving.

In addition to producing sperm, the testes also produce male sex hormones called **androgens**. Each testis consists of many highly coiled tubes, called **seminiferous tubules**, where sperm form, as discussed in section 18.2. The seminiferous tubules are held in place by connective tissues that are rich in androgen-producing cells called **Leydig cells**. Because these cells are located between the seminiferous tubules, they are also called *interstitial cells*.

The process of sperm formation takes about 60 days. From the seminiferous tubules, sperm pass through the coiled **epididymis**, which is an approximately 6-meter (20-foot) tube that coils atop each testis. During the 20 or so days that sperm pass through the epididymis, the sperm become able to move and capable of fertilizing an egg cell.

During ejaculation, sperm are propelled from the epididymis through a sperm-carrying duct called the **vas deferens**. Each duct is covered in smooth muscle and undergoes wavelike peristaltic contractions to move sperm.

Several accessory glands add secretions to the ejaculated sperm as they make their way through the ducts of the reproductive system. The paired **seminal vesicles** secrete a thick fluid containing mucus and fructose (a sugar) as an energy source for sperm. The single **prostate gland** secretes a thin, milky white fluid into the urethra. This secretion contributes to the mobility and viability of sperm. The **bulbourethral glands** are a pair of pea-sized glands that lie below the urethra between the prostate and the penis. Prior to ejaculation, these glands secrete a tiny amount of clear mucus that helps neutralize any acidic urine present in the male urethra. Secretions from the accessory glands combine with sperm to form **semen** with a pH ranging from 7.1 to 8.0. This slightly alkaline fluid is optimal for sperm motility and helps to neutralize acidic conditions in the female reproductive tract.

Semen production may be affected by spinal cord injuries and some diseases, causing infertility. For example, repeated bouts of sexually transmitted infections can cause scarring of sperm-carrying ducts and prevent or decrease sperm passage. If a male gets mumps—a viral infection that usually affects young children—after puberty, the associated inflammation of his testes can impair his sperm production. Likewise, inflammation of the prostate gland, called prostatitis, can also alter the sperm passage.

Stop and Stretch 18.1

In some men, one or both of the testes fail to descend from their location in the abdomen to the scrotum during fetal development. This condition, called cryptorchidism, results in decreased fertility. Why might this be the case?

The Female Reproductive System

FIGURE 18.2a shows the female internal and external reproductive anatomy. The most obvious feature of the female external genitalia is a structure called the **vulva**. The vulva consists of two sets of lips, or labia: the outer **labia majora**, which are fatty and have a hairy external surface, and the inner **labia minora**, which contain neither fat nor hair. At the front of the vulva, the labia minora divide around the **clitoris**, an important organ for female sexual arousal (FIGURE 18.2b).

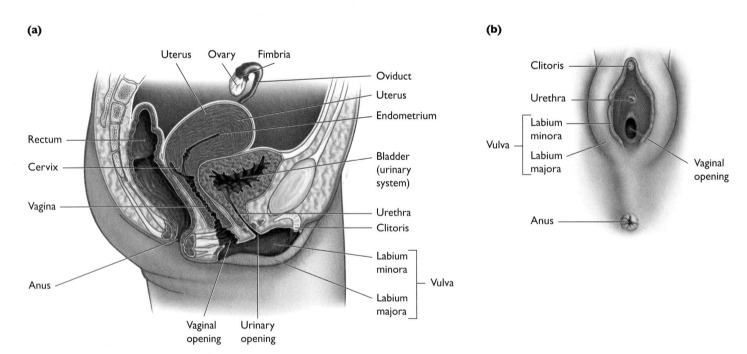

FIGURE 18.2 Female reproductive anatomy. (a) The ovaries of the female produce egg cells and hormones required for reproduction. Egg cells move through the oviducts to the uterus. The uterus can house a developing fetus. The cervix, the bottom portion of the uterus, widens during birthing. The vagina is the passageway into and out of the uterus. (b) The female external genitalia consist of the vulva and the clitoris. Openings to the outside allow for the expulsion of urine (via the urethra) and menstrual tissue or a baby (vagina).

VISUALIZE THIS Can you trace the path an egg cell would take from its production in an ovary to its exit via the vagina if not fertilized?

Between the folds of the labia, there is an opening for the urethra, which serves as a passageway for urine from the bladder. A centimeter or so below the urinary opening is the vaginal opening. Note that the urethra in women is short (4 centimeters, or 1.5 inches, on average) compared to the urethra in men (18 centimeters, or 7 inches, on average). This shorter length of urethra is easier for bacteria and viruses to traverse, making infections of the urinary tract and bladder more common in females than in males.

The organs that make up the internal genitalia are the ovaries, oviducts, uterus, and vagina. The **ovaries** are the female gonads. They produce the gametes and sex hormones that circulate in a woman's body during her reproductive years. Ovaries are about the size of an almond, but at birth, they contain about 2 million ova (eggs) each.

The **oviducts** are actually an extension of the top surface of the uterus. These tubes extend from the body of the uterus toward the ovaries, which are suspended within the abdominal cavity. The oviducts are not attached to the ovaries directly. Instead, an oviduct ends in fingerlike projections collectively called **fimbria** that move over the surface of the ovary. These movements, along with suction created by cilia lining the oviduct, direct an egg released by the ovary into the oviduct.

The **uterus**, where an embryo implants and grows, is about the size of a fist in a woman who is not pregnant. The wall of the uterus is thick (about 1 centimeter, or 1/3 inch) and is composed of some of the most powerful muscles in the human body. These muscular walls contract rhythmically during labor, childbirth, and orgasm. The internal surface of the uterine wall is called the **endometrium**, and it changes in thickness during the course of the menstrual cycle. The lower third of the uterus is narrower than the upper portion and is called the **cervix**. The cervix opens up, or dilates, during childbirth to allow the baby to exit the uterus and be born.

The **vagina** is a muscular organ that acts mainly as a passageway into and out of the uterus. The pathway from the vagina to the uterus to the opening of the oviduct at the fimbria represents the only natural pathway in humans directly from the outside of the body to the inside of the abdominal cavity. This means that women can experience bacterial infections inside their abdominal cavity without experiencing an injury that punctures the cavity wall. As you learned in Chapter 13, infections of the female abdomen can be caused by sexually transmitted microorganisms.

For pregnancy to occur, the oviducts must be open, allowing the sperm to reach the egg and the fertilized egg to travel down the oviduct to the uterus. Infertility can result from physical blockages of the oviducts caused by scarring in response to infection. Infections of reproductive structures are most often caused by exposure to sexually transmitted diseases such as gonorrhea and chlamydia.

Scarred and damaged oviducts also lead to an increased risk of a pregnancy beginning outside the uterus. Such *ectopic* pregnancies can occur if a fertilized egg implants in the oviduct. Ectopic pregnancies do not allow proper fetal development and may kill the mother if left untreated. The risk of ectopic pregnancy increases with each incidence of tubal infection.

Female fertility is also affected by the presence of uterine fibroids. These noncancerous growths in the wall of the uterus are very common in women in their 30s and 40s. If fibroids are large or numerous enough, they can interfere with the ability of the sperm to reach the egg or for the fertilized egg to implant in the uterus.

If the reproductive structures are healthy, the likelihood of infertility decreases. Other sources of infertility are problems with the production and function of sex cells.

18.2 Gametogenesis: Development of Sex Cells

The development of sex cells or gametes, called **gametogenesis**, involves the process of meiosis, the type of cell division that reduces the chromosome number by half. Because human body cells contain 46 chromosomes, meiosis in humans helps to produce gametes that contain 23 chromosomes.

Meiosis alone is not enough to produce a functional gamete. Other changes to sex cells occur as they go through gametogenesis to enable them to mature into gametes capable of being involved in fertilization. For instance, sperm cells lose cytoplasm as they mature, making them smaller and more agile. In contrast, egg production maximizes the amount of cytoplasm present. This increased size and nutrient content allow for the rapid development that takes place as the fertilized egg cell divides to produce the **embryo** that will implant in the uterus and develop into the fetus.

Men produce gametes beginning at puberty and lasting throughout their lifetime, but the number of gametes produced decreases slowly and progressively with age. Women are able to produce gametes only a few days a month from puberty (around age 12) until menopause (around age 50).

Spermatogenesis: Development of Men's Gametes

The production of sperm, called **spermatogenesis** (FIGURE 18.3a) begins to occur in the testes at puberty. Cells that will undergo gametogenesis line the walls of the seminiferous tubules.

During spermatogenesis, each parent cell first duplicates, and then one of the two daughter cells undergoes spermatogenesis. The other daughter cell maintains the function of the parent cell. Because only one of the original two cells undergoes spermatogenesis, a progenitor cell is left behind, and a man will never exhaust the supply of cells that can be used to make sperm cells.

The testes contain around 400 feet of seminiferous tubules, inside of which are the *spermatogonia*. Spermatogonia are stem cells that serve as the starting point for the cell divisions that will produce the actual sperm cells. These cells, located in the testes, continuously divide to provide more spermatogonia to replenish the stem cell pool and to produce *primary spermatocytes*, which become *secondary spermatocytes* after meiosis I. After meiosis II these cells are called *spermatids*. The spermatids undergo further development to produce *spermatozoa*, or mature **sperm**.

Cells that aid the developing sperm, called **Sertoli cells**, are located in the seminiferous tubules. These cells secrete the substances required for further sperm development. Mature sperm are composed of a small head containing the DNA, a midpiece that has mitochondria to provide ATP energy for the journey to the oviduct, and a tail (FIGURE 18.3b). The tail, or flagellum, propels a sperm forward at a pace of around 2.5 centimeters (1 inch) every 15 minutes. The tail arises at the end of spermatogenesis and takes close to two months to form. At the tip of the sperm's head lies the **acrosome**. This structure contains digestive enzymes that help a sperm cell gain access to the egg cell.

Hormonal Control of Spermatogenesis Several hormones regulate the production of sperm. **Testosterone** is an androgen that causes spermatogonia to divide. Testosterone also regulates the growth and development of male reproductive structures, and it stimulates the development of male secondary sex characteristics such as facial hair growth and voice deepening that occur at puberty. Testosterone is secreted by the interstitial Leydig cells located between the seminiferous tubules of the testes.

Recall from Chapter 16 that the hypothalamus, located in the brains of both males and females, secretes **gonadotropin-releasing hormone (GnRH)**, which stimulates another endocrine structure in the brain—the pituitary gland—to release **follicle-stimulating hormone (FSH)** and **luteinizing hormone (LH)** (see FIGURE 18.4 on page 446). These hormones are secreted by both males and females but act on different target organs in each sex. In males, the target organs are the testes, and in females, these hormones act on the ovaries.

When testosterone levels in the blood of a man fall below the level required for maintaining sperm production, the hypothalamus secretes GnRH, which stimulates the release of FSH and LH. Sertoli cells in the testes are the targets of FSH, which helps

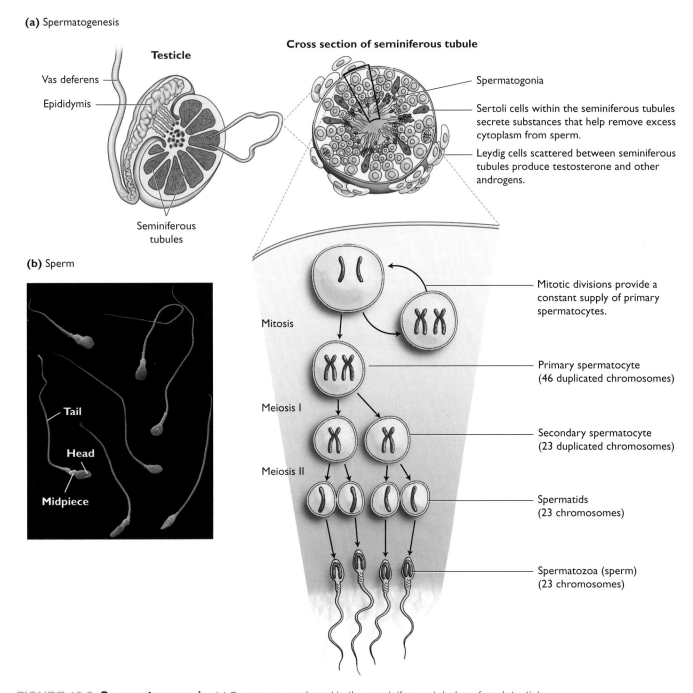

FIGURE 18.3 Spermatogenesis. (a) Sperm are produced in the seminiferous tubules of each testicle. Cells in the testes first undergo a round of mitosis, producing two identical daughter cells with 46 chromosomes. The primary spermatocyte begins meiosis with 46 chromosomes. After meiosis I, the two resultant daughter cells are the secondary spermatocytes. Meiosis II results in the production of four haploid daughter cells containing 23 chromosomes, called spermatids. Spermatids are then modified to become mature spermatozoa. (b) The head of the sperm contains the DNA. The midpiece contains mitochondria to provide ATP energy for the flagellum, or tail.

spermatogenesis commence at puberty. LH stimulates the Leydig cells of the testes to secrete testosterone.

When testosterone in the blood is high, it acts on the hypothalamus to inhibit GnRH release. In addition, when the concentration of sperm is high, Sertoli cells release inhibin, another hornome that prevents the release of GnRH. These negative feedback loops work to maintain homeostatic testosterone levels, thus controlling sperm production and formation.

Problems with Spermatogenesis The vast majority (around 90%) of male infertility cases are due to problems with sperm production and formation. Irregularly shaped sperm don't move properly toward the egg cell, so they impede fertilization. Likewise, sperm that have defects in motility can be too slow to reach the egg cell. Some men have lower than average numbers of sperm in their semen. Normal sperm concentrations are around 20 million sperm per milliliter of semen (about 300 million per ejaculate). When a man's sperm count is below 10 million per milliliter he is considered to be infertile.

Sperm counts begin to decline naturally around age 35. However, various exposures can cause this decrease to occur earlier or less gradually than normal. Exposure to pesticides, herbicides, and other chemicals such as solvents used in paints and glues can result in decreased sperm production. The Genes & Homeostasis feature on page 448 summarizes the current data on how endocrine-disrupting chemicals may act to alter sexual development in humans and other ogranisms. Drug use can also decrease sperm number and quality. Both cocaine and marijuana use are associated with decreases in fertility. Likewise, men who smoke cigarettes have lower sperm counts on average than nonsmokers. Men who use anabolic steroids, which cause the testes to shrink, also have lower than average sperm counts.

Oogenesis: Development of Women's Gametes

The formation and development of female gametes, called **oogenesis**, occurs in the ovaries. A small percentage of these egg cells will be released over a woman's reproductive life span, and an even smaller percentage may be fertilized. Spermatogenesis begins at puberty, but oogenesis actually begins while the female is still in her mother's uterus and then pauses until puberty. At puberty, development of preexisting eggs continues each month until menopause.

When the female embryo is about 8 weeks into its development, mitosis occurs to "seed" the ovaries with cells that can later undergo meiosis. Meiosis begins between weeks 11 and 12 of development; however, the process stops in prophase I. Meiosis I will not be completed until shortly before ovulation. Thus, the primary oocytes remain in prophase I until meiosis resumes at puberty, and some are still there at menopause. A female produces around 2 million of these potential egg cells before her birth. After birth, cells degenerate, causing the total number of cells to decrease progressively until there are about 350,000 left at puberty.

The Ovarian Cycle At puberty, the maturation of egg cells commences. The ovary contains many **follicles**, with each follicle containing an immature egg called an *oocyte*. The **ovarian cycle** includes all the events that occur as a *primary follicle* develops into a *secondary follicle* and then into the mature *Graafian follicle* (FIGURE 18.5a).

The stepwise development of the follicle occurs as the primary follicle (which houses the primary oocyte) secretes estrogen. The secondary follicle includes the secondary oocyte and the cells that surround it. The secondary oocyte is surrounded by pools of fluid and follicle cells that secrete estrogen. The next step in the ovarian cycle includes the development of the mature Graafian follicle from the secondary follicle. The Graafian follicle contains a fluid-filled cavity that increases in volume, causing the ovary wall to balloon out until it bursts. This burst expels the secondary oocyte from the ovary, a process called **ovulation**.

The remnant of the Graafian follicle (minus the oocyte that was ovulated) is called the **corpus luteum**, which secretes reproductive hormones but degenerates after about 10 days if fertilization does not occur.

After ovulation, the secondary oocyte moves into the oviduct where, if sperm are present within 12 hours of ovulation, fertilization is likely.

An ovulated egg cell is large enough to be viewed without a microscope. This is because the egg cell undergoes an off-center meiosis that produces both very small cells that will not be involved in fertilization, called **polar bodies**, and a much larger cell that will give rise to the female gamete (FIGURE 18.5b). This larger cell will contain

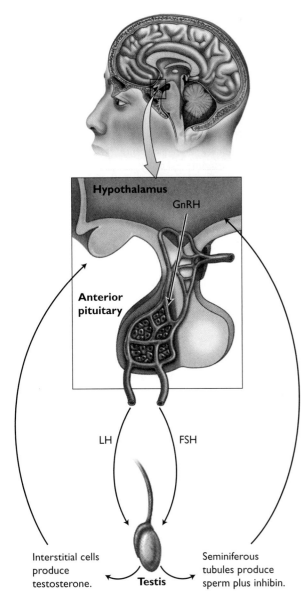

FIGURE 18.4 Hormonal control of the testes. Gonadotropin-releasing hormone (GnRH) stimulates the pituitary to release follicle-stimulating hormone (FSH) and luteinizing hormone (LH). FSH stimulates sperm production and LH causes the testes to release testosterone.

VISUALIZE THIS Testosterone and inhibin exert feedback on the hypothalamus to keep the level of testosterone in the blood within a limited range. Is this feedback negative or positive? Why?

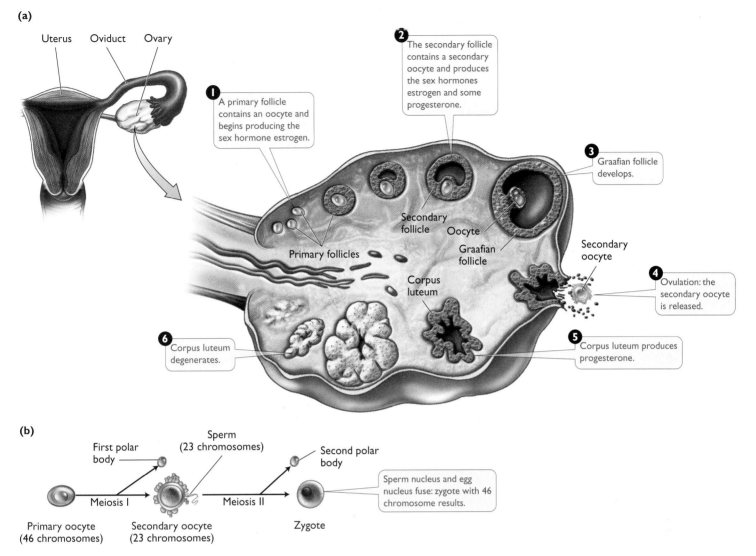

FIGURE 18.5 Oogenesis. (a) One follicle per month goes through steps 1–6 to produce and ovulate an egg cell. (b) Meiosis in human females produces only one fertilizable cell; the others are nutrient-poor polar bodies. If the larger egg cell, with the lion's share of nutrients, is fertilized, it contains all the nutrients and organelles that the fertilized egg (the zygote) will need to serve as the progenitor of all the millions of cells of the human body.

VISUALIZE THIS Would twins produced when ovulation occurs from both ovaries in one month be genetically identical? Why or why not?

enough nutrients to help nourish the zygote and embryo through its very early development.

If not fertilized, the secondary oocyte—an object about the size of a pinhead—takes about three days to move from the ovary through the oviduct, uterus, and cervix and then exit the body through the vagina.

When did the egg cell that helped produce you actually start its development?

GENES & HOMEOSTASIS

Endocrine Disruptors

Are chemical pollutants in the water affecting the expression of sex-specific genes in organisms that live in or drink the water? Because aquatic organisms spend most of their lifetime in the water, they can be used as environmental early warning signals for threats to human health. In 1980, a pesticide spill in Lake Apopka, in central Florida, killed most of the resident alligators. Today, the male alligators in the lake have abnormally small penises. Fish near a waste treatment plant in Colorado's Boulder Creek have both male and female reproductive organs. Populations of Chinook salmon in the Pacific Northwest are almost all female, and most of these female fish have chromosomes typical of males of this species.

Studies of aquatic animals with certain reproductive problems seem to implicate a class of chemicals called *endocrine disruptors*. Some endocrine-disrupting chemicals mimic the actions of the hormone estrogen.

Aside from a very few genes on the Y chromosome, all of the genes that cause development of sex-specific characteristics are present in both sexes. For instance, males have genes that will allow breast development and females have genes that will allow development of the larynx or voice box. Sex-specific genes are turned on only in the presence of sex hormones such as testosterone and estrogen. These hormones cross cell membranes, bind receptors, and, once together, bind to the promoter region of sex-specific genes to initiate gene expression.

When male aquatic animals are exposed to estrogen-mimicking compounds in the water, genes are turned on that would not normally be expressed. For example, when male frogs are exposed to the pesticide atrazine they grow eggs inside their testes.

Are chemical pollutants in our water supply also affecting human reproductive health? One study that compared reproductive health of men from around the United States showed that sperm counts of men from rural Missouri were lower than they were for men from many large cities across the country. Scientists who performed this 2003 study found higher levels of three commonly used agricultural pesticides (including atrazine) in the bodies of men with lower sperm counts than were present in the bodies of men with higher sperm counts. Most of the men with decreased sperm counts did not work directly with these chemicals, leading researchers to question whether these men may have been exposed to chemicals that had found their way into the drinking water.

Another study looked at the effects of exposure to trichloroethylene (TCE), an industrial solvent used to remove grease from metal and found in many household products. Large amounts of this chemical can get into the water supply via air emissions from metal-degreasing plants and via wastewater from factories that produce metals, paints, cleaning products, electronics, and rubber products. A 1996 study at the National University of Singapore showed that males exposed to TCE while working as mechanics or dry cleaners had a larger number of abnormally shaped sperm than did men who had not been exposed to TCE.

Another endocrine disruptor, the chemical di-(2-ethylhexyl)-phthalate (DEHP), is under investigation to determine whether it plays a role in causing endometriosis. DEHP, a type of chemical called a plasticizer, is added to plastics to keep them soft and pliable. Some scientists believe that drinking water from plastic bottles or heating food in plastic containers might increase an individual's exposure to this and other plasticizers that leach out of the plastic and into the water or food. Researchers at the University of Siena in Italy found that women with endometriosis have higher levels of DEHP in their blood than do women who do not have endometriosis.

What's in the water we're drinking?

At first glance, the issue of endocrine-disrupting chemicals in the water supply seems quite disturbing, but how concerned do we really need to be? The correlation between endocrine-disrupting chemicals and reproductive health could result from a shared factor rather than demonstrating a cause-and-effect relationship. For instance, men in agricultural areas may be exposed to more atrazine than urban men are, but they may also have jobs with a greater component of manual labor. It may be that the consistently higher body temperatures experienced by manual laborers interfere with sperm production and that atrazine exposure has little effect.

Likewise, the data on plasticizers show some ambiguity. These studies provide no direct evidence that plasticizers are causing reproductive problems. Women with high levels of DEHP might also have high levels of other, untested for, chemicals that are actually causing the problem. The increased level of environmental chemicals also may be due to some other factor. Research on the hormone-disrupting effects of plasticizers is still in its infancy; before drawing any definitive conclusions, scientists must wait until many studies that provide direct evidence of human reproductive problems caused by plasticizers have been completed.

While there is no conclusive evidence that endocrine disruptors in the water supply are harming human reproductive health, neither is there any reason to disregard the hypothesis. In other words, scientists simply don't yet know whether these chemicals are causing human reproductive problems. Further research is necessary before the role of endocrine disruptors in human reproductive health can be firmly established.

Hormonal Control of the Ovarian Cycle The first half of the ovarian cycle is called the *follicular phase* and the second half is called the *luteal phase*. Estrogen and progesterone are both involved in regulating the ovarian cycle through feedback loops involving the hypothalamus. High and low levels of estrogen provide positive feedback to the hypothalamus. In each case, the hypothalamus secretes GnRH, which acts on the pituitary gland to increase the secretion of FSH and LH. Conversely, high levels of the hormone progesterone have a negative feedback effect on the hypothalamus; GnRH secretion is decreased, and FSH and LH levels decline (FIGURE 18.6).

During the follicular phase of the ovarian cycle, FSH promotes the development of the follicle. The maturing follicle secretes estrogen, and as estrogen levels increase, FSH production declines. After ovulation, the luteal phase commences. During the luteal phase, LH promotes the development of the corpus luteum, which secretes both estrogen and progesterone. If a pregnancy does not occur, the increase in progesterone inhibits the hypothalamus, LH is not secreted, the corpus luteum degenerates, and menstruation occurs.

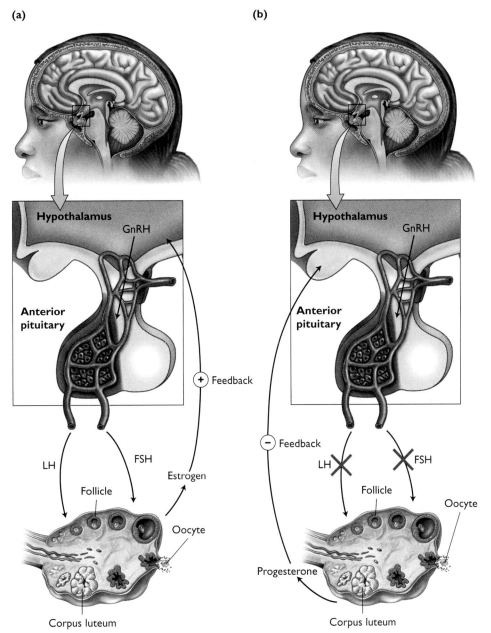

FIGURE 18.6 **Hormonal control of the ovaries.** Gonadotropin-releasing hormone (GnRH) stimulates the pituitary to release follicle-stimulating hormone (FSH) and luteinizing hormone (LH). FSH stimulates the follicle to produce estrogen. LH stimulates the corpus luteum to secrete progesterone. (a) Estrogen exerts positive feedback over the hypothalamus and (b) progesterone exerts negative feedback control over the hypothalamus. The positive feedback loop stimulated by estrogen only occurs a few days before ovulation. At other times in the cycle, estrogen is part of a negative feedback loop.

Problems with Oogenesis When ovulation fails, fertility is disrupted. Sometimes a follicle keeps growing and no egg is ovulated. This can occur if the increase in LH, called the LH surge, does not occur. A follicle that does not rupture can turn into a fluid-filled cyst. Follicular cysts most often are harmless and disappear on their own.

The LH surge might be disrupted by damage to the hypothalamus or pituitary gland, excessive exercise, or anorexia. Anorexia and excessive exercise throw the body out of homeostasis and signals are sent to the hyopthalamus preventing GnRH release.

Another condition that disrupts ovulation, called polycystic ovarian syndrome, is the most common hormonal disorder among women of reproductive age in the U.S., leaving close to 10% of women unable to ovulate. This syndrome is caused by abnormal levels of FSH, LH, and androgens. It is unclear why these hormonal disruptions occur. This condition can also cause irregular menstrual cycles, excess hair growth, and obesity. The name polycystic ovarian syndrome comes from the cystic appearance of the ovaries in women with this disorder.

When ovulation is disrupted, the menstrual cycle is absent or irregular.

18.3 The Menstrual Cycle

The term **menstrual cycle** refers specifically to periodic changes that occur in the uterus. This cycle depends on intricate hormonal relationships between the brain, ovaries, and lining of the uterus. During the course of a single menstrual cycle, a woman's body prepares an egg for potential fertilization and her uterus for a potential pregnancy. If no pregnancy has occurred, the endometrium is shed, and a new cycle begins.

FIGURE 18.7 illustrates the changes in hormone levels, ovarian follicles, and condition of the endometrium that occur throughout the 28-day cycle. Most women's bodies do not adhere precisely to a 28-day cycle—some women have longer cycles, and others have shorter cycles. The first day of a menstrual cycle is considered to be the first day of actual bleeding.

Like many other biological processes, the menstrual cycle is self-regulating, and homeostasis is maintained by responding to feedback. This feedback involves many different hormones.

When the follicle is large enough, it produces enough estrogen to stimulate GnRH release. This leads to a spike in both FSH and LH levels, which lasts for about 24 hours. Ovulation occurs 10 to 12 hours after the LH peak, around 14 days before menstruation in a typical 28-day cycle. This increase in estrogen also serves to stimulate mitosis of endometrial cells and regrowth of blood vessels. This way, if fertilization occurs, the uterine lining will be prepared to give the early embryo a place to implant and begin developing.

After ovulation, the progesterone secreted by the corpus luteum prepares the endometrium for a pregnancy. Progesterone helps maintain the blood flow to the uterine lining, which also supports early fetal development. As you learned earlier, progesterone exerts negative feedback on the hypothalamus by inhibiting continued LH production, and LH levels immediately decline.

If fertilization does not occur, the corpus luteum degenerates 12 to 14 days after ovulation. Because the corpus luteum is no longer secreting the hormones, progesterone and estrogen levels fall, triggering spasms in the arteries that supply the uterine lining. This spasming allows the lining of the uterus to be shed and causes menstrual cramps. Because the endometrial tissue is no longer receiving oxygen and nutrients, endometrial tissues die. Blood escapes from weakened walls of capillaries. When a woman menstruates, the endometrial tissue and blood are shed from the uterus and released through the cervix and vagina.

As the corpus luteum degenerates, decreasing levels of progesterone also serve to release the hypothalamus from inhibitory control. Therefore, LH and FSH levels rise, and the cycle starts over.

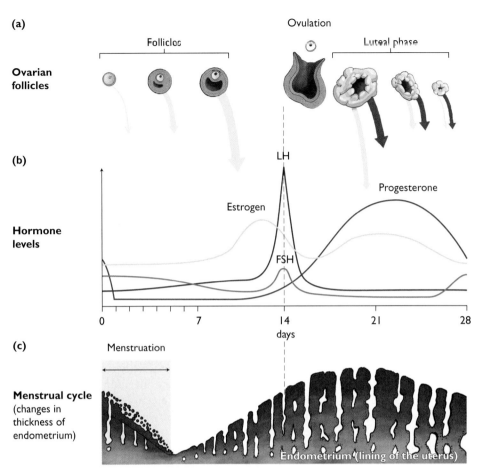

FIGURE 18.7 **The menstrual cycle.** Changes in hormone levels are linked to (a) the state of ovarian follicles, (b) other hormone levels, and (c) uterine condition over the course of the menstrual cycle.

If, however, fertilization has occurred, the process of endometrial breakdown does not occur. The early embryo produces a hormone called *human chorionic gonadotropin* (HCG) that extends the life of the corpus luteum. Women test for the presence of this hormone in most over-the-counter pregnancy tests. With the corpus luteum intact, progesterone and estrogen levels remain high, and the endometrium is maintained. In a pregnant woman, the corpus luteum finally disintegrates after about six or seven weeks of pregnancy, when the *placenta*, an endocrine organ that forms during pregnancy, begins to produce enough progesterone to maintain the uterine lining on its own.

Stop and Stretch 18.3

Women who live together often report that their menstrual cycles are synchronized. In other words, they start menstruating at around the same time each month. Although scientists have tried to determine whether this was actually happening and if so why, results have been inconclusive. Assume that four fertile women move in together. Design an experiment that would measure whether their menstrual cycles were synchronizing over time.

Some cases of female infertility are related to a condition called **endometriosis**, which occurs when tissues that normally line the uterus migrate through the oviducts and implant on other organs such as the bladder, colon, and ovaries. These tissues continue to respond to the hormonal cycle, growing and shedding in sync with the lining of the uterus. Because there is no way for this blood to exit the body, it becomes trapped and can lead to the growth of cysts, scar tissue, and *adhesions*, which are abnormal tissue that binds organs and other structures together. In addition to causing abdominal pain, endometriosis can lead to scarring and inflammation of the oviducts and therefore prevent sperm and egg from uniting.

If a female is ovulating and menstruating normally, infertility is less likely. Sometimes difficulty conceiving a child has less to do with reproductive structures and hormones and more to do with a general sexual problem.

18.4 The Human Sexual Response

In 1966, William H. Masters and Virginia Johnson of Washington University published *The Human Sexual Response*, a book that contained details of research they had performed on human sexuality over the previous several years. In their book, they reported on their pioneering research into the physiology of sexual excitation. Their model of human sexual response still remains the standard for describing phases of sexual activity (FIGURE 18.8).

According to Masters and Johnson's **four-stage model**, the human sexual response begins with *excitement*, in which both male and female bodies prepare for intercourse. Excitement begins with an increase in heart and respiratory rate and rise in blood pressure. In males, the excitement phase is characterized by erection, which occurs when the spongy erectile tissue in the penis becomes engorged with blood. Muscles in the scrotum also tense, drawing the testes closer to the body.

Excitement in women is identified by nipple erection, swelling of the breasts and structures of the external genitals, and the production of lubricating fluid by cells in the walls of the vagina. The woman's uterus also becomes elevated in her body, stretching the length of the vagina by several centimeters.

Excitement is triggered by a number of erotic stimuli, including physical foreplay stimulating the touch receptors as well as an individual's unique perception of visual, auditory, or chemical stimuli. The psychological components of excitement are central to continuing the sexual response cycle.

The *plateau* phase is a period of increasing sexual tension, typically during the act of intercourse. Heart rate and blood pressure continue to increase in both men and women. In men, the bladder sphincter closes tightly during this stage, preventing the mixing of urine and semen. Men may begin secreting seminal fluid from the penis during this stage as well.

In women, the external genitalia and breasts continue to swell, and the opening of the vagina tightens and its lower area swells, creating what Masters and Johnson referred to as the orgasmic platform, which tightly grips the penis and increases pleasurable contact between both organs.

The plateau phase climaxes in **orgasm**, a release of sexual tension. The muscles of the anus, lower pelvis, vagina, and uterus contract rhythmically, and other muscles, including the diaphragm and vocal cords, may contract as well. Males *ejaculate* 2–5 milliliters of semen, with the volume varying according to the time since his last ejaculation and the length of the plateau phase.

The final phase of the sexual response cycle is *resolution*, during which heart rate, blood pressure, and body structures return to baseline conditions. It is possible, especially in women, for the resolution phase to be bypassed by further stimulation and to return directly to the plateau phase.

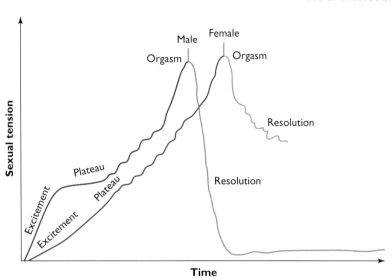

FIGURE 18.8 **The human sexual response cycle.** Masters and Johnson divided human sexual response into four stages, shown here.

Stop and Stretch 18.4

Orgasm decreases the activity of parts of the brain associated with feelings of anxiety or fear. Chemicals are released that contribute to feelings of relaxation and well-being. Why might these positive feelings have been evolutionarily selected for?

Male infertility can be caused by difficulty obtaining and maintaining an erection, which prevents delivery of sperm. Such *erectile dysfunction* occurs when the erectile tissue doesn't expand enough to compress the veins. Pharmaceutical agents such as sildenafil (Viagra) can help males with this condition achieve a full erection. If ejaculation occurs before the penis has entered the vagina, called *premature ejaculation*, sperm and egg will not contact each other. Some males suffer from a condition called retrograde ejaculation. This occurs when semen enters the bladder during orgasm. Many different medical conditions can cause retrograde ejaculation including diabetes, bladder and prostate surgery, and use of some psychiatric drugs.

Until a couple decides to have a child, they have many options for preventing a pregnancy.

18.5 Controlling Fertility

According to a recent survey by the National Center for Health Statistics, nearly 77% of sexually active women in the United States use some form of birth control, or **contraception**, to prevent pregnancy. What follows is a summary of birth control methods currently in use in the United States and a discussion of the future of birth control technology.

Principles of Fertility Control

The complexity of human reproduction offers different options for interfering with this process, from the blocking of sperm transport, to the inhibition of ovulation, to the removal of the fertilized egg or embryo. FIGURE 18.9 summarizes the primary modes of action of the most common methods of birth control.

Because of differences in how they block pregnancy, the methods listed in Figure 18.9 also have different levels of potential effectiveness (see Table 18.1 on page 455). For example, birth control methods that attempt to block the passage of sperm into a woman's uterus must exclude all of the hundreds of millions of sperm in each ejaculation to prevent pregnancy. The chance that one or a few tiny sperm will breach the barrier is nearly impossible to eliminate, so the effectiveness of barrier methods is lower than the effectiveness of methods that attempt to stop the release of (typically) a single egg each month.

Barrier Methods

Birth control methods that physically prohibit sperm from reaching the site of fertilization are known as **barrier methods**. Barrier methods of birth control are popular because they are easy to acquire and use. However, these methods are more likely than others to fail. Table 18.1 lists both the perfect-use effectiveness and the typical-use effectiveness (the effectiveness in the "real world") of different methods of birth control. The gap between perfect and typical effectiveness of barrier contraceptives has to do with

FIGURE 18.9 **Birth control and human reproduction.** Methods of birth control have been developed that interfere with nearly every step in the process of human reproduction.

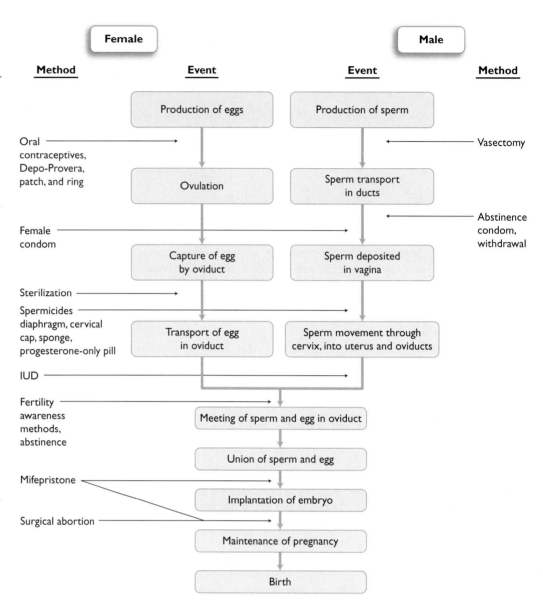

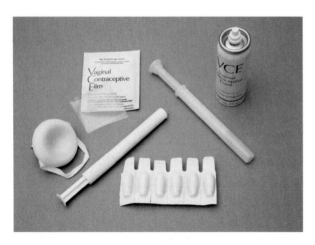

FIGURE 18.10 **Spermicides.** Many forms of birth control utilize sperm-killing chemicals.

how they are used. Many couples find that in the excitement of sexual activity, the proper use of barrier methods can be forgotten.

Withdrawal, in which a man removes his penis from a woman's vagina before ejaculation, may be erroneously thought of as a "barrier" method of birth control, because with this method sperm are supposedly prohibited from entering a woman's reproductive tract. However, as evidenced in Table 18.1, withdrawal is rather ineffective at preventing pregnancy. Small amounts of sperm may be released in the seminal fluid before orgasm, before the penis is withdrawn.

Spermicides inactivate sperm by damaging their cell membranes. Cream, jelly, or foam spermicides can be inserted into the vagina with a plunger-type applicator 15 to 30 minutes before intercourse (FIGURE 18.10). The active chemical in spermicides, nonoxynol-9, is also effective in killing the bacteria that cause the common sexually transmitted infections gonorrhea and chlamydia. However, because nonoxynol-9 can cause open vaginal sores in some women, its use *without condoms* may actually increase the risk of transmission of other STIs.

The **contraceptive sponge** (FIGURE 18.11) is a suppository disk of spermicide-infused polyurethane foam that impedes sperm passage by covering the cervix. The sponge is a relatively ineffective method of birth control on its own, but it has the advantage that it can be inserted many hours before intercourse, and like all

TABLE 18.1 | Effectiveness of Birth Control Methods

Method	% of Women Experiencing an Unintended Pregnancy Within the First Year of Use	
	Typical Use	Perfect Use
No method	85	85
Spermicides	29	18
Withdrawal	27	4
Symptothermal method	N/A	2
Diaphragm	16	6
Contraceptive sponge		
Never given birth	16	9
Previously given birth	32	20
Condom		
Female	21	5
Male	15	2
Combined pill	8	0.3
Contraceptive patch	7	0.3
Vaginal ring	7	0.3
Synthetic progesterone shot	3	0.3
IUD		
ParaGard	0.8	0.6
Mirena	0.1	0.1
Female sterilization	0.5	0.5
Male sterilization	0.15	0.10

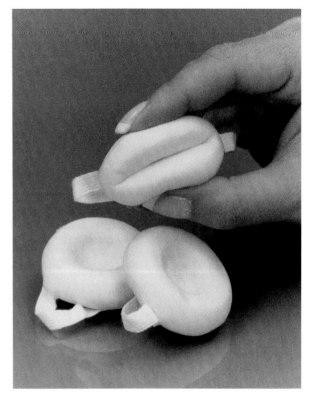

FIGURE 18.11 Contraceptive sponge. The contraceptive sponge has been recently reintroduced as a barrier method of birth control that is most effectively used with a condom.

spermicides, it can increase the effectiveness of other barrier methods in preventing pregnancy.

Condoms for males (FIGURE 18.12) are sheaths made of latex that cover an erect penis and act as a trap for sperm. Since their introduction, condoms have remained a very popular method of barrier birth control. The popularity of this method has grown in recent years—nearly tripling since 1982—as the effectiveness of latex condoms in reducing the spread of STIs has become apparent. Now, according to the Centers for Disease Control and Prevention, more than 20% of sexually active women rely on their partner's condom use for birth control.

A **female latex condom**, designed as a vaginal liner (Figure 18.12), has been introduced in recent years to help women protect themselves from STIs even if their partner will not use a condom. However, few women have adopted the female condom, probably because it is more expensive than the male condom and requires more practice to use effectively.

Diaphragms and **cervical caps** are latex domes with flexible rims (FIGURE 18.13), which cover the cervix to block sperm passage. Both can be inserted up to 6 hours before intercourse and should be left in place for 6 to 24 hours afterwards. When filled with spermicide, diaphragms and caps are effective birth control, but neither provides protection against STIs.

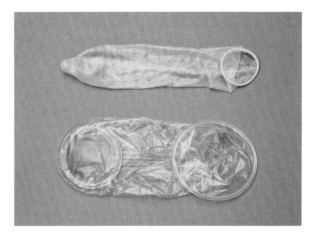

FIGURE 18.12 Condoms. The male condom (top) and female condom (bottom) are made of latex and designed to prevent sperm from entering the female reproductive tract. Aside from abstinence from sexual activity, condoms provide the most effective protection against sexually transmitted infection.

FIGURE 18.13 Barriers to the cervix. The diaphragm (a) and cervical cap (b) both cover the lower end of the uterus, preventing the passage of sperm. They do not prevent the transmission of sexually transmitted infections.

FIGURE 18.14 Hormonal contraceptives. (a) Oral contraceptives, (b) skin patches, and (c) vaginal suppository rings provide synthetic estrogen and progesterone that interfere with egg production and inhibit sperm movement.

Hormonal Birth Control

Most women under age 30 who use birth control use **combined hormone contraceptives**, commonly called "the pill," "the patch," or "the ring" (FIGURE 18.14). The hormones in these methods are synthetic estrogen and synthetic progesterone, which work in concert to inhibit both ovulation and fertilization. The constant levels of estrogen supplied by these methods prevent ovarian follicles from developing and inhibit ovulation. The synthetic progesterone then acts as a backup for the estrogen component in case ovulation does occur, preventing cervical mucus from thinning to the type that facilitates sperm passage and making the endometrium unfavorable for embryo implantation.

Oral contraceptives are pills taken daily as long as contraception is desired. As indicated in Table 18.1, both the theoretical and use effectiveness of birth control pills are quite high. As long as a pill is taken every day, the risk of pregnancy is very low. Missed pills can result in ovulation—in general, missing as few as two pills in a row greatly increases the chance of pregnancy.

The **contraceptive patch** and the contraceptive **vaginal ring** help to reduce the chance of a missed hormonal dose and have a slightly higher effectiveness than pills. The patch is applied to the skin of the stomach, buttocks, or arm once a week, while the ring is inserted in the vagina and removed after three weeks. Estrogen and progesterone diffuse from the patch or the ring and into the bloodstream, having the same effect as the contraceptive pill.

Common side effects of combination hormones are nausea, breast discomfort, weight gain, acne, and headaches. However, most women do not experience any of these symptoms, and many who do so find that these effects diminish after two or three months. Less common but more serious side effects include an increased risk of blood clots (especially in women with cardiovascular disease or who smoke), cervical and liver cancer, and delayed return to fertility once hormone use is stopped. The estrogen in combination hormonal contraceptives does appear to protect women from ovarian and endometrial cancer.

Birth control pills typically come in 28-day packets. Women take pills containing hormones for 21 days, followed by 7 days of placebo, which is a sugar pill that contains no hormones. Patches and vaginal rings are meant to be worn for three out of four weeks in a month, with one week off.

These hormone-free weeks trigger menstruation. However, the menstrual periods induced by these practices do not appear to be necessary. Women who wish not to menstruate often simply skip the hormone-free week. In recognition of this behavior and the desire of many women to eliminate menstruation, oral contraceptive manufacturers have begun producing formulations with fewer, less frequent, or no placebo pills. There are few long-term studies on the effect of having fewer or no periods as a result of these practices, but as yet, there is little evidence that forgoing menstrual periods is dangerous. In 2003, the U.S. Food and Drug Administration (FDA) approved a contraceptive pill that has only four placebo periods per year and in 2007 approved a pill that puts periods off altogether.

Progesterone-only pills are commonly used by breast-feeding women, because they do not interfere with milk production, as estrogen-containing pills will.

The compound medroxyprogesterone acetate mimics the actions of progesterone. Injectable forms of this drug, such as Depo-Provera, are given in the muscles of the arm or buttock. Shots must be administered every three months for continuous contraception.

High doses of birth control pills can be used as **emergency contraception** (EC) in the event of failure of other methods of birth control. These drugs include the progesterone-only "Plan B" pill and must be taken within 72 hours of unprotected sex to prevent pregnancy. Although it is not entirely clear how EC prevents pregnancy, most research indicates that it prevents ovulation or fertilization, depending on where a woman is in the cycle. It may be the case that EC prevents pregnancy by preventing the implantation of a fertilized egg in the uterus.

Other Methods of Birth Control

Methods of birth control that do not fit into one of the above categories include devices inserted into the uterus, fertility awareness, sterilization, and abortion.

Intrauterine Devices (IUDs) As their name implies, **intrauterine devices (IUDs)** function when placed inside a woman's uterus. The two IUDs on the market in the U.S., the ParaGard and the Mirena (FIGURE 18.15), act in slightly different ways to prevent fertilization. Both cause a nonspecific inflammation of the endometrium simply by residing in the uterus. This is sometimes called the *foreign body reaction*, which serves to make the uterus less hospitable to implantation.

Both kinds of IUD rely on a "backup" preventive as well. The ParaGard IUD contains copper, which can be toxic to cells. The slow release of copper ions acts against sperm, eggs, and embryos.

The Mirena IUD contains a small amount of slowly released progesterone. As in birth control pills, progesterone inhibits ovulation, thickens cervical mucus, and thins the endometrium to prevent implantation. Once in place, IUDs act as contraceptives for many years—the Mirena is effective for five years and the ParaGard is effective for ten.

IUDs must be inserted by a clinician using sterile techniques to minimize the risk of infection, which is the major complication of their use. IUDs may be expelled or perforate the uterine wall, which can lead to damage of other internal organs. Although IUDs are highly effective, any pregnancies that occur can be very dangerous to the mother and fetus. A woman who suspects that she is pregnant while using an IUD should see a doctor immediately.

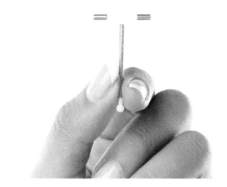

FIGURE 18.15 Intrauterine devices. These small devices fit in the uterus and make it unsuitable for sperm survival and embryo implantation.

Fertility Awareness By using **fertility awareness methods**—becoming familiar with the timing of her cycle and carefully observing the signs of fertility—a woman can identify the days when she should abstain from intercourse if she does not desire pregnancy. These days are typically five days before ovulation through one day after it. The **calendar method** (or "rhythm method") relies only on a woman's awareness of the timing of her typical cycle. If her periods are very regular, she can predict with accuracy the date she will ovulate each month. However, few women's cycles are so regular, making this form of birth control fairly risky.

The other forms of fertility awareness use the calendar as a guide but rely heavily on the physical changes surrounding ovulation to pinpoint its date.

For some women, a reliable indicator of fertility is the presence of thin, stringy cervical mucus that has both the consistency and appearance of raw egg white. Secretions from the cervix change in response to the actions of estrogen and progesterone. During the first half of the cycle, prior to ovulation when estrogen levels are high, the cervical mucus increases in abundance, thins, and becomes slick in texture. The mucus is produced in the "crypts" or folds that line the cervix, and the parallel strands of mucus function as channels for the sperm to move through, preventing them from becoming stuck in the folds of the cervix (FIGURE 18.16). When this type of mucus is present, a woman is fertile. When the level of progesterone is high after ovulation, during the second half of the cycle, cervical mucus is whitish and gummy and can prevent sperm from reaching the egg cell.

A woman's body temperature also changes throughout the menstrual cycle, dropping slowly and irregularly until the day after ovulation, when a rapid rise of at least 0.5°C (0.9°F) occurs (FIGURE 18.17). The temperature rise along with the changes in cervical mucus helps establish when intercourse should be avoided to prevent a pregnancy. Using information from these signs together, called the **symptothermal method**, can be a highly effective means of birth control.

Sterilization One of the most popular modes of birth control in the U.S. is surgical sterilization, which has the advantage of being extremely effective but the disadvantage of being practically irreversible. Sterilization can be achieved either by prohibiting the transport of sperm from a man's testes to his penis, or by prohibiting the passage of an egg from a woman's ovaries to her oviducts.

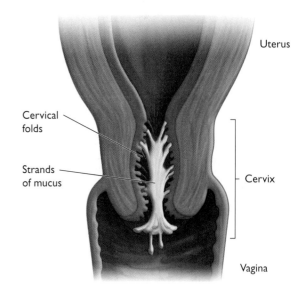

FIGURE 18.16 Fertility and mucus. Before ovulation, mucus produced by the cervix thins and becomes more abundant, providing pathways for sperm to enter the upper reproductive tract without getting stuck in the folds of the cervix.

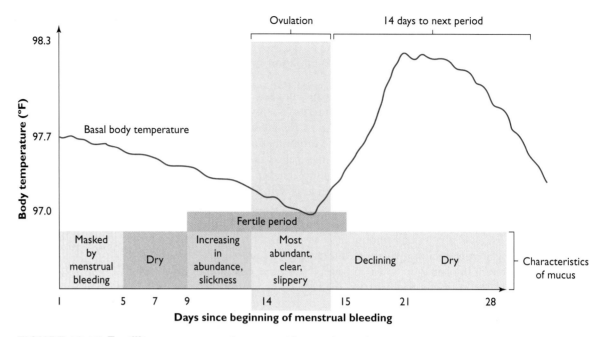

FIGURE 18.17 Fertility awareness. A woman with a regular cycle can observe the characteristics of her cervical mucus and her body temperature to predict her fertile period.

VISUALIZE THIS Can a woman with a cycle length of more or less than 28 days use fertility awareness methods? Explain your answer.

A **vasectomy** requires the severing and tying off of each vas deferens, the ducts that serve as passageways for sperm from the testicles to the urethra (FIGURE 18.18). This operation does not interfere with a man's ability to produce semen or achieve erection or orgasm—the ejaculate he produces is simply missing sperm cells. Vasectomies can take place under local anesthetic and in an outpatient clinic. Male sterilization can sometimes be reversed if needed—about 50% of operations to reconnect one or both vasa deferentia are successful. Aside from the slight risk of infection that accompanies any surgery, there are no apparent short- or long-term side effects of the procedure.

Because the route from ovary to oviduct is inside a woman's abdominal cavity, **tubal ligation** is more complicated than vasectomy and must take place in an operating room.

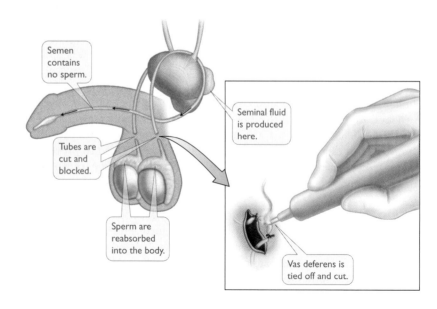

FIGURE 18.18 Vasectomy. Men can be surgically sterilized by cutting the tubes that deliver sperm to the penis. This operation does not interfere with any other aspects of male sexual performance.

The most common operations involve cauterization (FIGURE 18.19a), which melts the tissue with electric current. Tubal ligation is not associated with any health risks outside of those associated with surgery of the abdominal cavity, but recovery time is longer than for vasectomy. Tubal sterilizations are also more difficult to reverse than vasectomies, with a success rate of only about 30% when the oviducts are reconnected. Female sterilization can also occur by the insertion of small springlike tubes into the oviducts to block them (FIGURE 18.19b).

Abortion Nearly half of all pregnancies that occur in the United States every year are unplanned. About half of these unplanned pregnancies (less than a quarter of all pregnancies) are ended by **elective abortion**, the medical termination of pregnancy. Every year, over a million pregnancies are aborted in the United States. Since *Roe v. Wade*, the 1973 Supreme Court ruling that overturned state laws that banned abortion, elective abortion and the way it has been employed have been a hotly debated and divisive issue (FIGURE 18.20).

Pregnancies can be terminated in the first seven weeks in the United States by the ingestion of a chemical called **mifepristone**, commonly referred to by the trade name RU-486. Mifepristone blocks the action of progesterone, the hormone that maintains the endometrium, and thus leads to the shedding of the uterine lining and the loss of the embryo.

The most frequently performed surgical abortion is **vacuum aspiration**. In this method, suction draws the embryo or fetus out of the uterus. Vacuum aspiration of the uterus can only be performed in the first trimester of a pregnancy, as early as 5 weeks but no more than 12 weeks after a woman's last menstrual period.

After 12 weeks, the surgical techniques require more forceful techniques to remove the fetus and placenta. Most involve dilating the cervix, and using forceps, a metal loop called a curette, and vacuum suction to destroy the fetus and extract it.

One form of later-term abortion, called intact dilation and extraction, is sometimes referred to as "partial birth abortion." In this procedure, the lower body of the fetus is pulled through the cervix and then its head is collapsed via pressure from a pair of surgical scissors. Doing so allows the remainder of the body to be removed from the uterus. Federal lawmakers in 2003 banned the use of this particular procedure. Because the federal ban did not allow an exception for the life and health of the mother, it conflicts with the Supreme Court ruling and was overturned by lower courts. A decision to uphold the ban was handed down by the Supreme Court in 2007.

The most common and serious complication of surgical abortion is infection. Some women experience heavy bleeding after an abortion, which may be a sign that some tissue remains in the uterus or that the uterus was perforated during the procedure.

Stop and Stretch 18.5

Some pharmacists have refused to dispense emergency contraception. Why do you suppose there is some opposition to the use of emergency contraception in the United States?

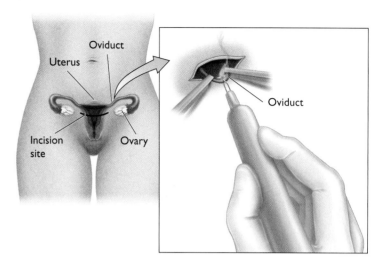

(a) Cautery tool cuts and seals oviduct.

(b)

FIGURE 18.19 Female sterilization. (a) Women can be sterilized by cutting the oviducts, preventing eggs from reaching the site of fertilization. Eggs released into the body cavity degrade and their nutrients are reabsorbed by the bloodstream. (b) The oviducts can be permanently blocked via the insertion of tightly coiled tubes.

FIGURE 18.20 A divisive issue. Abortion remains a hotly debated topic in the United States, with 20–25% of the population favoring unrestricted access, 20–25% favoring a complete ban, and 50–60% of the population falling somewhere in the middle of these two extremes.

The Future of Birth Control Technology

The introduction of hormone patches, contraceptive rings, and emergency contraception and changes in pill formulations and contraceptive sponges have all occurred in the past decade, providing several new birth control options for women. There are few other new options in development. Some researchers have worked on methods of *immunocontraception*, in which women are vaccinated against biological markers of pregnancy so that their immune system will destroy a new embryo. Given the societal debate about abortion, these techniques are extremely controversial, and preliminary testing of the principle in animals is not especially promising.

Development of safe, simple, and reversible methods of birth control for men has lagged considerably behind the progress on female birth control. Most proposed methods of hormonal birth control for men interfere with the process of sperm production. High levels of testosterone, along with drugs that inhibit hormones that stimulate sperm production, and progesterone have all been tested and shown to be effective male contraceptives. However, release of a male birth control to the marketplace is at least several years away. Immunocontraceptive methods that cause a man's immune system to destroy his own sperm are also being investigated.

18.6 Health, Lifestyle, and Fertility

One of the most important things our newlywed couple can do to safeguard their fertility is to take good care of themselves. When the newlyweds, or any couple, decide to have children, they should eat a healthy diet and exercise. They should also try to avoid prolonged or intense emotional stress, which can disrupt the function of some hormones involved in gamete production. Males should also avoid exposing the testes to too much heat. Frequent hot tub or sauna use can impair sperm production. Tobacco, cocaine, and marijuana use reduce the number and quality of sperm and decrease the likelihood of conception. During pregnancy, these drugs will harm the child. Alcohol abuse is associated with general ill health and reduced fertility, and alcohol use during pregnancy can also harm the child.

However, taking good care of oneself is not the only factor. In males and females, age is also a predictor of fertility. A gradual decline in fertility occurs from the mid-30s, which gives our newlyweds plenty of time to make their decision.

Chapter REVIEW

ROOTS TO REMEMBER

The following roots of words come mainly from Latin and Greek and will help you to decipher terms:

cervix means neck, which is why cervical also refers to bones in the neck.

-genesis means generation of or birth of.

glans is the Latin word for acorn and related to the word gland.

gon-, gono-, and **gonado-** are from a Greek word for seed or generation.

luteum is from the Latin word for yellow.

oo- is from the Greek word for egg.

KEY TERMS

18.1 The Human Reproductive Systems
androgen *p. 441*
bulbourethral gland *p. 441*
cervix *p. 443*
clitoris *p. 442*
endometrium *p. 443*
epididymis *p. 441*
fimbria *p. 443*
gamete *p. 440*
glans penis *p. 440*
gonad *p. 440*
labia majora *p. 442*
labia minora *p. 442*
Leydig cell *p. 441*
ovary *p. 443*
oviduct *p. 443*
penis *p. 440*
prostate gland *p. 441*
scrotum *p. 441*
semen *p. 441*
seminal vesicle *p. 441*
seminiferous tubule *p. 441*
sexual reproduction *p. 440*
testis (plural, testes) *p. 441*
urethra *p. 440*
uterus *p. 442*
vagina *p. 443*
vas deferens *p. 441*
vulva *p. 442*

18.2 Gametogenesis: Development of Sex Cells
acrosome *p. 444*
corpus luteum *p. 446*
embryo *p. 444*
follicle *p. 446*
follicle-stimulating hormone (FSH) *p. 444*
gametogenesis *p. 443*
gonadotropin-releasing hormone (GnRH) *p. 444*
luteinizing hormone (LH) *p. 444*
oogenesis *p. 446*
ovarian cycle *p. 446*
ovulation *p. 446*
polar body *p. 446*
Sertoli cell *p. 444*
sperm *p. 444*
spermatogenesis *p. 444*
testosterone *p. 444*

18.3 The Menstrual Cycle
endometriosis *p. 451*
menstrual cycle *p. 450*

18.4 The Human Sexual Response
four-stage model *p. 452*
orgasm *p. 452*

18.5 Controlling Fertility
barrier method *p. 453*
calendar method *p. 457*
cervical cap *p. 455*
combined hormone contraceptive *p. 456*
condom *p. 455*
contraception *p. 453*
contraceptive patch *p. 456*
contraceptive sponge *p. 454*
diaphragm *p. 455*
elective abortion *p. 459*
emergency contraception *p. 456*
female latex condom *p. 455*
fertility awareness method *p. 457*
intrauterine device (IUD) *p. 457*
mifepristone *p. 459*
oral contraceptive *p. 456*
progesterone-only pill *p. 456*
spermicide *p. 454*
symptothermal method *p. 457*
tubal ligation *p. 458*
vacuum aspiration *p. 459*
vaginal ring *p. 456*
vasectomy *p. 458*
withdrawal *p. 454*

SUMMARY

18.1 The Human Reproductive Systems

- Sexual reproduction requires a mating between two parents. Males and females produce gametes in structures called gonads. Gametes unite at fertilization to produce genetically distinct offspring (p. 440).

- The penis of the male reproductive system is involved in sperm delivery. The urethra delivers both sperm and urine, the scrotum houses the testes, and the testes produce sperm and androgens (pp. 440–441).

- The seminal vesicles, prostate, and bulbourethral glands add secretions to sperm that help them develop and provide a source of energy. Semen is composed of the secretions from these glands combined with sperm (p. 441).

- The female reproductive system consists of the external vulva and clitoris; the internal vaginal passageway; the cervix at the base of the uterus, which opens during childbirth; the uterus, which houses a developing baby; oviducts for the passage of gametes; and ovaries, which produce egg cells and hormones (pp. 442–443).

18.2 Gametogenesis: Development of Sex Cells

- Gametogenesis produces gametes and involves meiosis, so that each gamete has half the chromosomes of other body cells (p. 443).

- Sperm are produced when cells in the seminiferous tubules undergo cell division. Sperm are stored in the epididymis, which rests atop the testis. At ejaculation, sperm pass through the vas deferens and out the urethra (p. 444; Figure 18.3).

- Mature sperm are composed of a small head containing the DNA, a midpiece that has mitochondria to provide energy for the journey

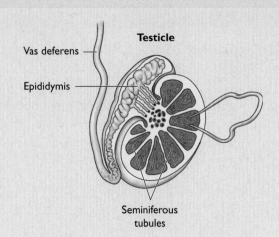

to the oviduct, and a flagellum. An acrosome at the sperm's head contains enzymes that help the sperm gain access to the egg (p. 444).

- Oogenesis occurs in the ovaries and results in the production of egg cells. This process begins while the female is still in her mother's uterus, then pauses until puberty. At puberty, development of preexisting eggs continues each month until menopause. The primary follicle develops into the secondary follicle, and then the Graafian follicle. The secondary oocyte is released from the Graafian follicle at ovulation. The leftover corpus luteum secretes progesterone for a while, and then degenerates if a pregnancy has not occurred (pp. 446–448; Figure 18.6).

- The hypothalamus secretes GnRH, which causes the pituitary gland to secrete FSH and LH. These two hormones regulate gametogenesis in males and females through feedback loops (p. 449).

18.3 The Menstrual Cycle

- High levels of estrogen cause increased FSH and LH production. High levels of progesterone cause decreased FSH and LH secretion (pp. 450–451).

- Estrogen, produced by the follicle, begins to rise during menstruation, eventually stimulating GnRH release and causing a spike in FSH and LH levels. LH causes ovulation to occur (pp. 450–451; Figure 18.7).

- If fertilization does not occur, the corpus luteum degenerates. As a result, progesterone and estrogen levels fall and menstruation ensues. Low levels of estrogen and progesterone stimulate GnRH production. As a result, LH and FSH levels rise, and the cycle is repeated (pp. 450–451).

- If fertilization has occurred, the embryo implants itself within the uterus and secretes HCG, which extends the life of the corpus luteum. Progesterone and estrogen levels remain high. Eventually, the placenta takes over the production of these hormones (p. 451).

18.4 The Human Sexual Response

- The physiology of human sexuality is typically described as a four-stage process: excitement, which prepares the body for intercourse; plateau, where sexual tension builds; orgasm, the release of that tension; and resolution, a return to baseline conditions (p. 452).

18.5 Controlling Fertility

- Unintended pregnancy can be prevented by the use of contraception, methods that interfere with the process of human reproduction (p. 453).

- Barrier methods of contraception inhibit contact between egg and sperm, preventing fertilization. These methods include withdrawal, spermicide, condoms, and diaphragms. Barrier methods can be effective, but present a challenge to couples to use them correctly. Condoms, especially with spermicide, are the only contraceptive that can also prevent many sexually transmitted infections (pp. 453–455).

- Hormonal birth control uses the reproductive hormones estrogen and progesterone to interfere with the process of ovulation and/or make the uterus inhospitable to sperm and fertilized eggs. These methods include pills, the patch, vaginal rings, synthetic progesterone shots, and emergency contraception. If applied correctly, hormonal methods are highly effective at preventing pregnancy, but do not prevent STIs (p. 456).

- IUDs prevent pregnancy by making the uterus inhospitable to sperm and (perhaps) fertilized eggs (p. 457).

- Fertility awareness methods rely on a woman's knowledge of her cycle and the physical changes associated with fertility to time intercourse to avoid pregnancy (p. 457).

- Surgical sterilization prevents sperm from exiting the penis or eggs from reaching the normal site of fertilization (pp. 457–458).

- About one-quarter of all pregnancies in the U.S. end in elective abortion. Abortions can be performed using the chemical mifepristone or via surgical removal of the embryo or fetus (p. 459).

LEARNING THE BASICS

1. Sexual reproduction _____.
   a. involves the production of offspring that are genetically identical to their parents
   b. involves the production of offspring that are genetically identical to their siblings
   c. involves the production of offspring that are a genetic mixture of both parents
   d. allows two parents to produce only a limited number of genetically distinct offspring

2. A sperm cell follows which path?
   a. seminiferous tubules, epididymis, vas deferens, urethra
   b. urethra, vas deferens, seminiferous tubules, epididymis
   c. seminiferous tubules, vas deferens, epididymis, urethra
   d. epididymis, seminiferous tubules, vas deferens, urethra
   e. epididymis, vas deferens, seminiferous tubules, urethra

3. Which of the following pairs is mismatched?
   a. urethra : sperm passage
   b. Leydig cells : male hormone production
   c. vas deferens : semen production
   d. seminiferous tubules : sperm production

4. Luteinizing hormone _____.
   a. stimulates sperm development in males
   b. triggers menstruation in females
   c. stimulates Leydig cells to secrete male hormone
   d. all of the above

5. Follicle-stimulating hormone _____.
   a. stimulates sperm development in males
   b. stimulates egg cell development in females
   c. is released in response to GnRH secretion
   d. all of the above

6. Label the male reproductive structures shown in FIGURE 18.21.

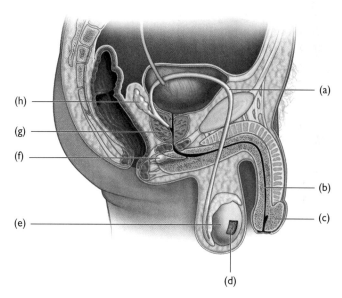

FIGURE 18.21 **Male reproductive structures.**

7. An egg cell that is not fertilized follows which path?
   a. ovary, oviduct, uterus, cervix, vagina
   b. ovary, uterus, oviduct, cervix, vagina
   c. oviduct, ovary, cervix, uterus, vagina
   d. oviduct, ovary, uterus, cervix, vagina
   e. ovary, oviduct, cervix, uterus, vagina

8. The ovary _____.
   a. is attached to the uterus and oviducts
   b. secretes estrogen and progesterone
   c. is ovulated once per menstrual cycle
   d. secretes FSH and LH
   e. all of the above

9. The release of the oocyte from the follicle is caused by _____.
   a. a decrease in estrogen
   b. an increase in FSH
   c. an increase in LH
   d. an increase in progesterone

10. Label the female structures shown in FIGURE 18.22.

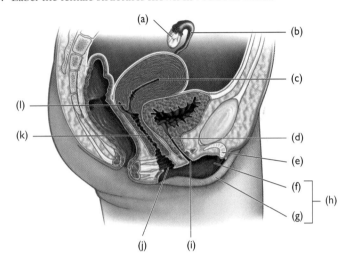

FIGURE 18.22 **Female reproductive structures.**

11. Gametogenesis _____.
    a. begins at puberty in males and females
    b. requires that the Leydig cells of males produce semen
    c. in females, results in the production of four fertilizable egg cells
    d. produces gametes that carry half the number of chromosomes that other body cells have

12. The ovarian cycle is regulated such that _____.
    a. increasing estrogen levels have a positive feedback effect on the secretion of FSH and LH
    b. increasing FSH levels lead to ovulation
    c. as progesterone levels increase, so do FSH and LH levels
    d. ovulation occurs on the fifth day of the cycle
    e. the placenta produces FSH, which stimulates ovulation

13. Match each male reproductive structure with its function.
    ____ urethra
    ____ vas deferens
    ____ epididymis
    ____ seminiferous tubules
    ____ penis
    ____ scrotum
    ____ testis
    ____ seminal vesicle
    ____ bulbourethral gland
    ____ prostate gland

    a. coiled tubes where sperm form
    b. delivers sperm to the female reproductive tract
    c. delivers sperm and urine out of the body
    d. sperm-carrying tube atop a testis
    e. sperm-carrying ducts that move sperm from testes to urethra
    f. sugar-secreting gland
    g. secretes milky white sperm-nourishing fluid into urethra
    h. secretes clear fluid that neutralizes acids in urethra
    i. site of sperm and androgen production
    j. pouch containing the testes

14. Match each female reproductive structure with its description.
    ____ vulva
    ____ clitoris
    ____ corpus luteum
    ____ vagina
    ____ cervix
    ____ uterus
    ____ oviduct
    ____ ovary
    ____ urethra

    a. remnant of the follicle
    b. lower portion of uterus; secretes mucus that affects fertility
    c. most sexually sensitive area for females
    d. consists of two sets of labia
    e. site of egg cell and estrogen production
    f. transports egg cells from ovary to uterus
    g. muscular passageway below the cervix
    h. has opening for urine to exit body
    i. sheds endometrium during menstruation

15. All of the following physiological changes in men and/or women occur during the excitement phase of the sexual response cycle *except* _____.
    a. erection
    b. ejaculation
    c. production of vaginal lubricating fluid
    d. increased heart rate
    e. elevated blood pressure

16. Most of the physiological changes that occur during the sexual response cycle are ultimately meant to _____.
    a. deter sex outside the fertile period
    b. maximize the likelihood of fertilization
    c. improve overall cardiovascular fitness
    d. provide a rich environment for embryo implantation
    e. provide pleasurable feelings

17. Match each method of birth control with the description of how it works.
    ____ combination hormone pills
    ____ IUD
    ____ condom
    ____ vasectomy
    ____ withdrawal

    a. prevents sperm from entering cervical opening
    b. nonspecific uterine inflammation prevents implantation
    c. inhibits ovulation and keeps cervix plugged with sticky mucus
    d. removal of penis from vagina before ejaculation; ineffective
    e. surgical sterilization

18. All of the following are signs that can help a woman diagnose her fertile period *except* _____.
    a. production of copious cervical mucus
    b. changes in body temperature
    c. the timing of her cycle
    d. change in mucus from sticky to slick
    e. swelling of the cervix and vagina

19. **True or False?** The vagina is the outermost portion of the female external genitalia.

20. **True or False?** The epididymis secretes FSH and LH.

ANALYZING AND APPLYING THE BASICS

1. During which stages of the human sexual response could a male impregnate his female partner? Explain your answer.

2. A chemical abortion procedure begins with one oral dose of mifepristone during a clinical visit. This tablet blocks progesterone, triggering the breakdown of the uterine lining. After 36–48 hours, the patient returns to the clinic for an oral or vaginal dose of the hormone misoprostol. Misoprostol is also commonly used to induce labor during a healthy pregnancy. What might this indicate about the mechanism of misoprostol?

3. Which of the forms of birth control covered in the chapter can actually help couples who are having trouble conceiving?

4. Most birth control methods are designed for and marketed to women. Why do you think this is the case? What could be the advantages of creating more birth control methods for men?

Questions 5 and 6 refer to the following information.

Bisphenol-A (BPA) is an industrial compound used in the production of plastic bottles and liners for food and beverage cans. Heat, acidic food, and repeated washing causes leaching of this compound. Because BPA leaches easily and is widely used, nearly all Americans have some measurable level of BPA in their bodies.

A comprehensive literature review of the BPA research (vom Saal and Hughes, 2005) showed 115 published studies that measured the effects of low-dose BPA on laboratory animals. Of these, 94 studies showed significant adverse effects. Thirty-one of those studies showed adverse effects at dose levels equal to or lower than 50 micrograms per kilogram of body weight per day (μg/kg/day)—the dose at which adverse effects would not be expected within a human lifetime, as estimated by the U.S. Environmental Protection Administration. Effects on female rodents whose mothers were exposed to BPA included early sexual maturity and increased mammary gland development. Effects on males from exposed mothers included decreased testosterone and a decrease in sperm production and fertility.

5. Propose a hypothesis that explains the effects of BPA exposure named above.

6. How could you reduce your exposure to bisphenol-A?

7. Researchers in Japan investigated the effect of low doses of bisphenol-A on adult male rats (Sakaue et al., 2001). FIGURE 18.23 shows partial results from the study. Discuss how these results compare to the 50 μg/kg/day dose for BPA estimated by the Environmental Protection Agency to be the dose at which adverse effects would not be expected within a human lifetime.

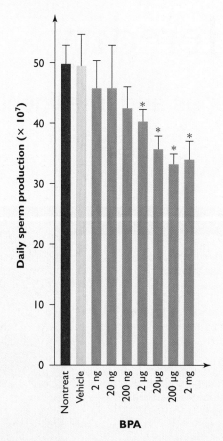

FIGURE 18.23 **Effects of BPA on daily sperm production in adult male rats.** Doses are given in nanograms (ng), micrograms (μg), or milligrams (mg), all per kilogram of body weight.

* = Significantly different from nontreatment or placebo delivered by the same means (vehicle).

Source: Data from Sakaue, M., et al. 2001. Bisphenol A affects spermatogenesis in the adult rat even at a low dose. *Journal of Occupational Health* 43: 185–190.

CONNECTING THE SCIENCE

1. There are many ways to deal with menstrual flow. Most women use so-called sanitary napkins (which actually are not sterile, although menstrual blood is) or tampons. Also available are menstrual cups that collect flow and can be rinsed out and reused. Why do you suppose disposable products are more commonly used than reusable products?

2. What steps can you take to protect your fertility?

3. At what point during development do you think a zygote, embryo, or fetus should have the right to survival despite a woman's right or desire to control her fertility? Explain your reasoning.

Chapter 19

Heredity

Genes and Intelligence

LEARNING GOALS

1. Describe the relationship between genes and chromosomes.

2. Explain how alleles form, and describe the consequences of mutation.

3. Define *independent assortment*, and explain how it contributes to diversity in gametes.

4. What is the relationship between genotype and phenotype?

5. Contrast dominant and recessive mutations.

6. Create and use a Punnett square for a single-gene genetic analysis.

7. Define *quantitative trait*, and describe the factors that generate variability in these traits.

8. What is heritability and how is it calculated in human populations?

9. Describe the difficulties in using the concept of heritability to explain differences between human groups.

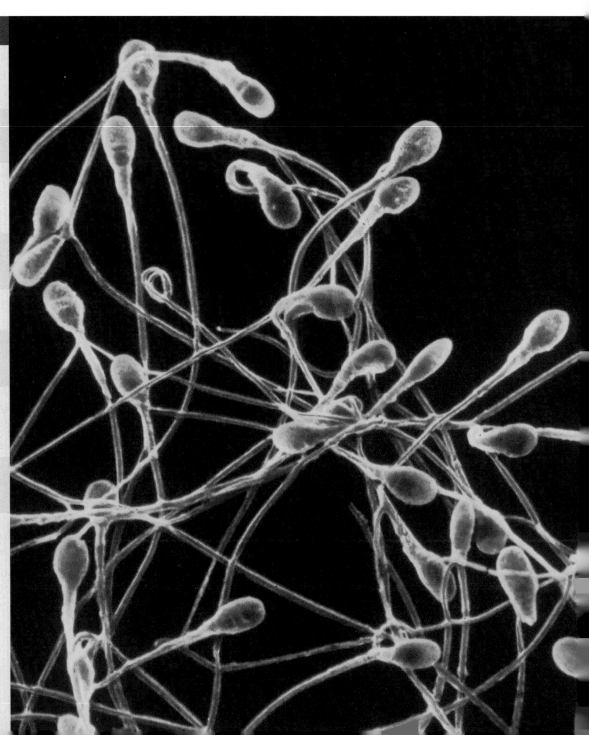

Can a woman choose sperm to create the perfect child?

The Fairfax Cryobank is a nondescript brick building located in a quiet, tree-lined suburb of Washington, D.C. Stored inside this unremarkable edifice are potential answers to the hopes and dreams of thousands of women and their partners. The Fairfax Cryobank is a sperm bank. Inside its many freezers are vials containing sperm collected from hundreds of men. Women can order these sperm for artificial insemination, which may allow them to conceive a child despite the lack of a fertile male partner.

Women who purchase sperm from the Fairfax Cryobank can choose from hundreds of potential donors. The donors are categorized into three basic classes, and their sperm is priced accordingly. Most women who choose artificial insemination want detailed information about the donor before they purchase a sample. While all Fairfax Cryobank donors submit to comprehensive physical exams and disease testing and provide a detailed family health history, not all provide childhood pictures, audio CDs of their voices, or personal essays. Sperm samples from men who did not provide this additional information are sold at a discount because most women seek a donor who seems compatible with their interests and aptitudes.

Will the choice produce a child resembling the mother's partner?

Will the right genes produce a genius?

Most of the donors to Fairfax Cryobank provide this detailed personal information. However, a select number of these donors are placed in a special group called the Doctorate category. These men either are in the process of earning or have completed a doctoral degree. Sperm from this category of donor is 30% more expensive than sperm provided by men whose educational attainment is not as high.

Why would some women be willing to pay significantly more for sperm from a donor who has an advanced degree? Because academic achievement is associated with intelligence. These women want

Or does environment have more influence on a child's intelligence?

intelligent children, and they are willing to pay more to provide their offspring with "extra-smart" genes. But are these women putting their money in the right place? Is intelligence about genes, or does it arise from the conditions in which a baby is raised? In other words, is who we are a result of our "nature" or our "nurture"? As you read this chapter, you will see that the answer to this question is not a simple one—our characteristics come from both our biological inheritance and the environment in which we developed.

19.1 The Inheritance of Traits

Most of us recognize similarities between our birth parents and ourselves. Family members also display resemblances—for instance, all the children of a single set of parents may have dimples. At the same time, though, it is usually quite easy to tell siblings apart. Each child of a set of parents is unique, and none of us is simply the "average" of our parents' traits. Instead, each of us has a combination of their traits—one child may be similar to her mother in eye color and face shape, another similar to mom in height and hair color.

Chapters 4 and 20 deal with the physical nature of inheritance. Here, we will examine the basic principles of how traits are passed from parent to child. We begin by reviewing some useful concepts from previous chapters.

Genes and Chromosomes

Every normal sperm and egg contain information about how to build an organism. The information is in the form of *genes*—which as you know are segments of DNA that code for proteins.

Imagine genes as being roughly equivalent to the words used in an instruction manual. These words are contained on pages, which are analogous to chromosomes. Human cells contain 46 chromosomes, nearly all of which carry thousands of genes. Thus the set of instructions for building a human has 46 pages, each containing thousands of words.

Words can have one meaning when they are alone and another meaning when used in combination with other words (for instance, *saw* versus *see-saw*). Words can also change meaning in different contexts ("I saw the painting" versus "I sharpened the saw"). Some words are repeated often in any set of directions, but other words are not. The presence of certain words and their combination with other words determine the actual instruction given.

Similarly, all cells in a body have the same genes, but the timing of expression and the combination of these genes determine the activities of a particular cell. For instance, eye cells and heart cells in mammals both carry genes that code for the protein rhodopsin, which helps detect light, but rhodopsin is produced only in eye cells, not in heart cells. Rhodopsin requires assistance from another protein to translate light into the actions of the eye cell. This other protein may also be produced in heart cells, but there it is combined with a third protein to help coordinate contractions of the heart muscle.

Thus, a protein may serve two different functions depending on its context. Because genes, like words, can be used in many combinations, the instruction manual for building a living organism is very flexible (FIGURE 19.1).

Producing Diversity in Offspring

During reproduction, each parent contributes instructions to a child. The information must be copied before it is transmitted to the next generation, and it is this process that introduces variation among genes. It is this **genetic variation** that women seeking sperm donors are very interested in—differences among donors' environments shouldn't affect the child that is produced by their sperm, but genetic differences certainly will.

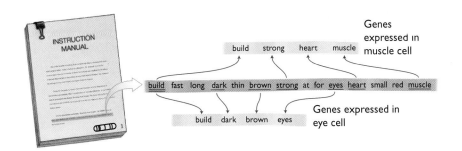

FIGURE 19.1 **Genes as words in an instruction manual.** Different words from the manual are used in cells in different parts of the body. Even when the same words are used in different cells, they are often used in distinctive combinations in each.

Gene Mutation Creates Genetic Diversity You can imagine how a page of instructions might change over many generations if each parent had to type a new copy of the instructions for each offspring. Typographical errors made by a father would be passed on in the manual he gave to his child, and any further errors made by the child during copying would be passed on to the next generation.

Recall the process of DNA replication, covered in Chapter 17. Like a retyped instruction manual page, copies of chromosomes are rewritten rather than "photocopied." As a result, there is a chance of a typographical error, or *mutation*, every time a cell divides. Mutations in genes lead to different versions, or *alleles*, of the gene. Because mutations are random and not expected to occur in the same genes in various individuals, different families should have slightly different alleles for various genes. The various effects of mutation are described in FIGURE 19.2. Repeated mutation creates genetic variation in a population and contributes to differences between families.

Stop and Stretch 19.1

Not all of the DNA in a cell codes for proteins: Some DNA functions as a promoter (where the transcription of a nearby gene begins), other segments may provide structural support for chromosomes, and other parts may be meaningless. What effects might mutations in these non-gene segments of DNA have?

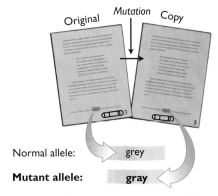

Normal allele: grey
Mutant allele: gray

Mutant allele has the same meaning
(mutant allele functions the same as the original allele).

strong
string

Mutant allele has a different meaning
(mutant allele functions differently than the original allele).

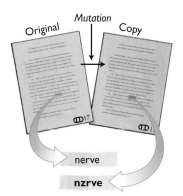

nerve
nzrve

Mutant allele has no meaning
(mutant allele is no longer functional).

FIGURE 19.2 **The formation of different alleles.** Different alleles for a gene, with modified meanings, may form as a result of copying errors.

Segregation and Independent Assortment Create Gamete Diversity

Both parents contribute genetic instructions to each child. But they do not contribute their entire manual. If they did, the genetic instructions carried in human cells would double every generation, making for a pretty crowded cell. Instead, the process of cell division that occurs during production of sperm and eggs (called meiosis and described in Chapter 17), reduces the number of chromosomes in these *gametes* by half.

Although they are only transmitting half of their genetic information in a gamete, each parent actually gives an entire copy of the instruction manual to each child. This can occur because, in effect, our cells have two versions of each page, each containing essentially the same words. The 46 chromosomes each cell contains are actually 23 pairs of chromosomes, with each member of a pair containing essentially the same genes. Each set of two equivalent chromosomes is referred to as a *homologous pair* (FIGURE 19.3). The members of a homologous pair are equivalent, but not identical, because even though both have the same genes, each contains a unique set of mutations inherited from one or the other parent.

The process of meiosis separates homologous pairs of chromosomes and places chromosomes independently into each gamete. These two processes explain why siblings are not identical (with the exception of identical twins). For the most part, it is because parents do not give all of their offspring exactly the same set of alleles.

When homologous pairs are separated, the alleles they carry are separated as well. The separation of pairs of alleles during the production of gametes is called **segregation**. Thus, a parent with two different alleles of a gene will produce gametes with a 50% probability of containing one version of the allele and a 50% probability of containing the other version.

The segregation of chromosomes during meiosis leads to **independent assortment**. Independent assortment arises from the *random alignment* of chromosomes during metaphase I of meiosis (described in Chapter 17). Because the members of each homologous chromosome pair are segregated into daughter cells independently of all the other pairs during the production of gametes, genes that are on different chromosomes are inherited independently of each other. (However, as discussed in Chapter 17, genes located on the same chromosome do not assort independently, but are in fact linked together.)

As a result of independent assortment, the instruction manual contained in a single sperm cell is made up of a unique combination of pages from the manuals a man received from each of his parents. In fact, almost every sperm he makes will contain a unique subset of chromosomes—and thus a unique subset of his alleles. FIGURE 19.4 illustrates this. In the figure, you can see that independent assortment causes an allele for an eye-color gene to end up in a sperm cell independently from an allele for the blood-group gene.

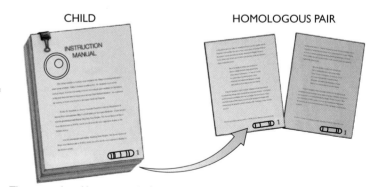

FIGURE 19.3 **Two complete instruction manuals.** A human contains two equivalent sets of instructions, each set inherited from one parent. Equivalent pages of instructions (that is, chromosomes) in the offspring are referred to as homologous pairs.

FIGURE 19.4 Each egg and each sperm is unique. Because each sperm is produced independently, the set of pages in each sperm will be a unique combination of pages. This figure illustrates how one could create very different piles containing one of each instruction manual page when drawing from a pile containing two versions of each instruction manual page. The same process occurs during egg production.

VISUALIZE THIS Four different sperm cells are possible when considering these two alleles on two different chromosomes. Two are pictured here. What combinations of alleles would be found in the other two possibilities?

Because the independent assortment of segregated chromosomes into daughter cells is repeated every time a sperm is produced, the set of alleles that each child receives from a father is different for all of his offspring. The sperm that contributed half of your genetic information might have carried an eye-color allele from your father's mom and a blood-group allele from his dad, while the sperm that produced your sister might have contained both the allele for eye color and the allele for blood group from your paternal grandmother.

Stop and Stretch 19.2

How many different allele combinations in gametes are possible for three genes on three different chromosomes? What is the relationship between chromosome number and the number of possible chromosome combinations that can be produced by independent assortment?

FIGURE 19.5 **Gregor Mendel.** Mendel studied nearly 30,000 pea plants over a 10-year period and publicized the results of his studies in 1865. His paper received little notice before his death in 1884. Only when scientists rediscovered the paper in 1900 did its significance to the new science of genetics become apparent.

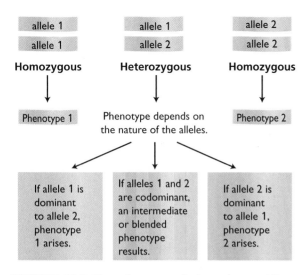

FIGURE 19.6 **Genotypes and phenotypes.** When two alleles for a gene exist in a population, there are three possible genotypes and two or three possible phenotypes for the trait.

VISUALIZE THIS How many genotypes are possible if three alleles for the gene exist in the population?

Random Fertilization Results in a Large Variety of Potential Offspring

As a result of the independent assortment of 23 pairs of chromosomes, each individual human can make at least 8 million different types of egg and sperm. Consider that each of your parents was able to produce such an enormous diversity of gametes. Further, any sperm produced by your father had an equal chance (in theory) of fertilizing any egg produced by your mother.

In other words, gametes combine without regard to the alleles they carry, a process known as **random fertilization**. Hence, the odds of your receiving your particular combination of chromosomes are (1 in 8 million) times (1 in 8 million)—or 1 in 64 trillion. Remarkably, your parents together could have made more than 64 trillion genetically different children, and you are only one of the possibilities.

Mutation creates new alleles, and independent assortment and random fertilization result in unique combinations of alleles in every generation. These processes help to produce the diversity of human beings. And not surprisingly, they make predicting the traits of a child very challenging.

Let us return to the question of predicting the heredity of the genetic traits possessed by a particular sperm donor. Does a smart donor ensure smart offspring?

19.2 Mendelian Genetics: When the Role of a Gene Is Direct

A few human genetic traits have easily identifiable patterns of inheritance. These traits are said to be "Mendelian" because the Austrian monk Gregor Mendel (FIGURE 19.5) was the first person to describe their inheritance accurately.

Mendel developed his understanding of this type of inheritance by studying pea plants, which are easy to grow, produce seeds within only a few months, and can make thousands of offspring. Reproduction in pea plants is also easily controlled by selecting which plants fertilize others.

Although Mendel himself did not understand the physical structure of genes, he was able to determine how traits were inherited by carefully analyzing the appearance of parent pea plants and their offspring. The pattern of inheritance Mendel described occurs mainly in traits that are the result of a single gene with a few distinct alleles.

Although Mendel's work forms the basis for genetics, later observations indicated that not all traits display the straightforward pattern of inheritance he described. Many of the Mendelian traits identified in humans are associated with a readily apparent physical trait or some type of disease or dysfunction. We should note, however, that most alleles in human beings do not cause disease or dysfunction. They are simply alternative versions of genes and thus contribute to human diversity.

Genotype and Phenotype

We call the genetic composition of an individual that person's **genotype** and the physical traits that are displayed the **phenotype**. An individual who carries two different alleles for a gene has a **heterozygous** genotype. An individual who carries two copies of the same allele has a **homozygous** genotype.

The effect of an individual's genotype on her phenotype depends on the alleles she carries (FIGURE 19.6). Some alleles are **recessive**, meaning that their effects can be seen only if a copy of a dominant allele (described in the following paragraph) is not also present. Mutations resulting in alleles that cannot produce functional proteins often are recessive. With these types of mutations, a heterozygous individual carrying two different alleles—one a normal allele and the other a recessive, nonfunctional allele—has a normal phenotype because the instruction produced by the normal allele substitutes for its nonfunctional partner.

Sometimes alleles code for instructions having powerful effects—ones that mask the effects of other alleles for the same gene. These more powerful alleles are termed **dominant** because their effects are seen even when a nondominant allele is present.

For some genes, more than one dominant allele may be produced, and for others, a dominant allele may have a more dramatic effect when the individual carries two copies of the allele. These situations are referred to as *codominance* and *incomplete dominance*, respectively, and are explored in detail in Chapter 20. For the purpose of our discussion, we will focus on simple dominant traits only.

Genetic Diseases in Human Beings

Scientists have identified genetic diseases in human beings that are produced by recessive alleles and dominant alleles. Genetic tests on the sperm and examinations of family health records allow technicians at the sperm bank to screen out donors who are likely to pass on these alleles. Two examples can help illustrate how these genetic conditions work.

Cystic Fibrosis Is a Recessive Condition
Cystic fibrosis is among the most common genetic diseases in European populations, affecting nearly 1 in every 2,500 individuals. It occurs in individuals with two copies of an allele that codes for the production of a nonfunctional protein—specifically, a protein that when functioning normally helps transport the chloride ion into and out of cells lining the lungs, intestines, and other organs. When this ion transporter is not functional, the balance between sodium and chloride in the cell is disrupted. An affected cell produces a thick, sticky mucous layer instead of the thin, slick mucus produced by cells with the normal allele.

Individuals with two copies of the cystic fibrosis allele suffer from progressive deterioration of their lungs and have difficulties absorbing nutrients across the lining of their intestines. Most children born with cystic fibrosis have a dramatically shortened life span (FIGURE 19.7).

People who carry one copy of the nonfunctional allele and one copy of the normal allele, nearly 1 in 25 individuals in European populations, are not affected because the nonmutant protein can still function effectively as a chloride transporter. Cystic fibrosis persists because most carriers have no idea that they could pass on this deadly allele.

Huntington's Disease Is Caused by a Dominant Allele
Huntington's disease is an example of a fatal genetic condition caused by a dominant allele. The Huntington's allele causes production of a protein that clumps up inside the nuclei of cells. Nerve cells in certain areas of the brain are especially likely to contain these protein clumps, and these cells gradually die off over the course of the disease (FIGURE 19.8). As a result, an affected individual gradually loses mental capacity and control of muscle function. Huntington's disease is both progressive and incurable.

Because the mutant Huntington's allele produces protein that builds up in the cell, an individual needs only one copy of the allele to be affected by the disease—that is, even heterozygotes have the disease. Normally, a dominant allele that causes death would not be passed from one generation to the next. However, in the case of Huntington's the symptoms usually do not appear until middle age. As a result, people who carry this allele can unknowingly pass it on to their children. The Genes & Homeostasis feature on page 475 explores the link between the characteristics of the mutant allele and the age of onset of Huntington's disease.

Since the mid-1980s, genetic testing has allowed people with a family history of Huntington's disease to learn whether or not they are affected before they show signs of the disease. Although most sperm banks do not test for the presence of the Huntington's

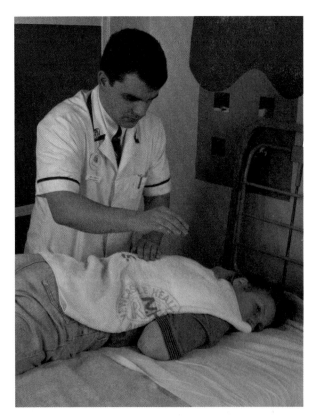

FIGURE 19.7 **Treating cystic fibrosis.** Mucus accumulating in the lungs of cystic fibrosis patients increases the risk of lung-damaging infections. Percussive therapy, seen here, loosens mucus and allows patients to cough up the excess material.

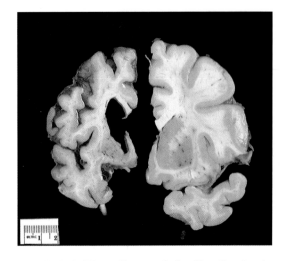

FIGURE 19.8 **The effects of the Huntington's allele.** The difference in brain size between an individual with Huntington's disease (shown at left) and one without is dramatic. It is the loss of brain cells that causes disability and death in affected individuals.

allele, the detailed family medical histories required of sperm bank donors enable Fairfax Cryobank to exclude men whose history indicates a risk of Huntington's disease.

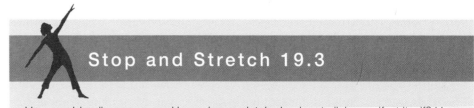

Stop and Stretch 19.3

How would a disease caused by an incompletely dominant allele manifest itself? How would heterozygotes for the allele differ from individuals who carry two copies and from those who carry no copies?

Using Punnett Squares to Predict Genotypes of Offspring

As you have learned in most of the chapters in this text, many diseases caused by a single mutation exist in the human population. However, most genetic diseases lack an easily performed test or obvious family health clue that would allow sperm banks to exclude men who carry these diseases. In other words, sperm from a sperm bank is not guaranteed to be free of all genetic defects, only of those most easily tested for. Where a gene has been identified, the inheritance of these conditions and of other single-gene traits is relatively easy to understand.

We can follow the inheritance of small numbers of genes by using a tool developed by the British geneticist Reginald Punnett. A **Punnett square** is a table that lists the different kinds of sperm or eggs parents can produce relative to the gene or genes in question and then predicts the possible outcomes of a **cross**, or mating, between these parents (FIGURE 19.9).

Imagine a couple in which both members are heterozygotes for cystic fibrosis. Different alleles for a gene are symbolized with letters or number codes that refer to a trait that the gene affects. For instance, the cystic fibrosis gene is symbolized *CFTR* for *cystic fibrosis transmembrane regulator*. The mutant *CFTR* allele is called *CFTR-ΔF508* (Δ is read as delta and means "changed"). However, to make this easier to follow, we will use a simpler key: the letters *F* and *f*, representing the dominant functional allele and recessive nonfunctional allele, respectively. A heterozygote would have the genotype *Ff*. A genetic cross between two heterozygotes could then be symbolized as *Ff × Ff*.

FIGURE 19.9 Calculating the risk of cystic fibrosis. This Punnett square helps determine the likelihood that a woman who carries the cystic fibrosis allele, which is recessive, would have a child with cystic fibrosis if her sperm donor was also a carrier.

VISUALIZE THIS What is the likelihood that a woman will have a child with cystic fibrosis if she is a carrier and the sperm donor does not carry the cystic fibrosis allele?

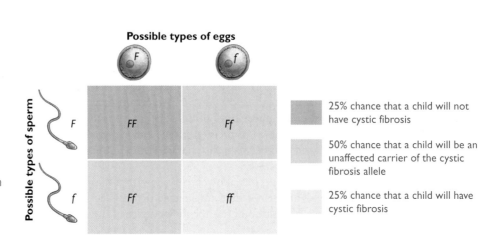

We know that the female in this cross can produce eggs that carry either the *F* or *f* allele, since the process of egg production will segregate the two alleles from each other. We place these two egg types across the horizontal axis of what will become a Punnett square. The male in this cross is also heterozygous, so he too can make sperm containing either the *F* or *f* allele. We place the kinds of sperm that he can produce along the vertical axis. Thus, the letters on the horizontal and vertical axes represent all the possible types of eggs and sperm that the mother and father can produce by meiosis, if we consider only the gene that codes for the chloride transport protein.

Inside the Punnett square are all the genotypes that can be produced from a cross between these two heterozygous individuals. The content of each box is determined by combining the alleles from the egg column and the sperm row.

Note that for a single gene with two alleles, there are three possible types of offspring. The chance of this couple having a child affected by cystic fibrosis is 1 in 4, or 25%, because the *ff* combination of alleles occurs once out of the four possible outcomes. The *FF* genotype is represented once out of four times, meaning that the probability of this couple having a homozygous unaffected child is also 25%. The probability of the couple producing a child who is an unaffected carrier of the cystic fibrosis allele (that is, heterozygous) is 1 in 2, or 50%, because two of the possible outcomes inside the Punnett square are unaffected heterozygotes—one produced by an *F* sperm and an *f* egg, and the other produced by an *f* sperm and an *F* egg.

When parents know which alleles they carry for a single-gene trait, they can easily determine the probability that a child they produce will have the disease phenotype. You should note that this probability is generated independently for each child. In other words, each offspring of two carriers of cystic fibrosis has a 25% chance of being affected. In the case of a dominant allele that causes disease, as for Huntington's, a Punnett square shows a higher probability of the disease phenotype in potential offspring (FIGURE 19.10).

Punnett squares can also be employed to predict the likelihood of a particular genotype when considering multiple genes. Chapter 20 provides an example of a Punnett square using two human traits, hair type and eye color. In fact, as long as each gene of interest is carried on a separate chromosome and the number of alleles for each gene is known, we can predict the likelihood of a particular genotype.

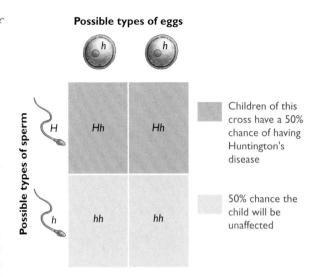

FIGURE 19.10 Calculating the likelihood of a dominant disease. This Punnett square illustrates the outcome of a cross between a man who carries a single copy of the dominant allele for Huntington's disease (*Hh*) and an unaffected woman who carries only the recessive, nondisease alleles (*hh*).

GENES & HOMEOSTASIS

A Genetic Stutter

Huntington's disease is one of 14 known expansion disorders, conditions that result from stretches of DNA that contain the same three-nucleotide sequence repeated many times. Eight of the fourteen diseases, including Huntington's, contain the same repeated sequence of the nucleotides CAG, the codon that codes for glutamine. Interestingly, a common symptom of these disorders is the progressive degeneration of nerve cells, usually affecting people late in life.

Typically in expansion disorders, the more repeats in the mutant allele, the more severe the disease and the younger the age of onset. In Huntington's disease, a total of 40 or fewer glutamine repeats produces a normal protein, called huntingtin. The function of huntingtin in nerve cells is not completely clear, but it appears to assist in producing the scaffolding that determines the shape of a cell.

More than 40 CAG repeats in the gene will result in mutant huntingtin, which builds up in nerve cells, causing them to die off in certain brain regions. The speed of cell degeneration is positively correlated to the number of extra CAG repeats, perhaps because the mutant huntingtin is proportionally larger so fewer copies are needed in each cell to disrupt the cell's function.

Expansion mutations may occur when DNA polymerase slips off the DNA molecule during replication. If the enzyme falls off in a section of repeats, it may reattach to the strand farther upstream, causing a loop of repeated codons to form. Proteins that work to repair DNA then add codons to the template strand, expanding the number of repeats. This process may be responsible for a phenomenon known as anticipation, in which expansion disorders have their effects earlier and earlier in subsequent generations. However, scientists disagree about whether anticipation is common and whether slippage of the DNA polymerase is an important component of this phenomenon.

As you might imagine, as the number of genes in a Punnett square analysis increases, the number of boxes in the square increases, as does the number of possible genotypes. With two genes, each with two alleles, the number of unique gametes a heterozygote can produce is four, the number of boxes in the Punnett square is 16, and the number of unique genotypes that can be produced is 9. With three genes, each with two alleles, the Punnett square has 64 boxes and 22 different possible genotypes. With four genes, the square has 256 boxes, and with five genes, there are over 1,000 boxes! Predicting the outcome of a cross becomes significantly more difficult as the number of genes we are following increases.

Although much of the work in identifying and developing tests for human alleles has so far concentrated on disease alleles, there are many non-disease-causing alleles in the human population that contribute to diversity among us in eye color, hair texture, and blood type, among other features. As scientists identify more of these genes and alleles, the amount of information about the genes of sperm donors or any potential parent will also increase.

Identifying and testing for particular genes in potential parents will allow us to predict the likelihood of numerous genotypes in their offspring. Unfortunately, this increase in genetic testing is not necessarily equaled by an increase in our understanding of how most traits develop, as we shall see in the next section.

19.3 Quantitative Genetics: When Genes and Environment Interact

The single-gene traits discussed in the previous section have a distinct "off-or-on" character. Individuals have either one phenotype (for example, cystic fibrosis) or the other (no cystic fibrosis). Such traits are known as *qualitative* traits. However, many of the traits that interest women who are choosing a sperm donor do not have this off-or-on character.

FIGURE 19.11 **A bell curve.** These people are arranged by height, with the shortest to the right and the tallest to the left. As you can see, although most people are clustered around the average height, the shape the crowd makes is like a bell.

Traits such as height, weight, eye color, musical ability, susceptibility to cancer, and intelligence are called **quantitative traits**. Quantitative traits show **continuous variation**; that is, we can see a large range of phenotypes in a population—for instance, from very short people to very tall people. Wide variation in quantitative traits contributes to the great diversity we see in the human population (FIGURE 19.11).

Why Traits Are Quantitative

One reason that we may see a range of phenotypes in a human population is that many genotypes exist among the individuals in the population. Multiple genotypes can occur when a trait is influenced by more than one gene. Traits influenced by many genes are called **polygenic traits**. As we saw above, when a single gene with two alleles determines a trait, three possible genotypes are present: *FF*, *Ff*, and *ff*, for example. When two genes each with two alleles influence a trait, nine genotypes are possible.

Eye color in humans is a polygenic trait influenced by at least three genes, each with more than one allele. These genes help produce and distribute the pigment melanin to the iris. People with very dark eyes have a lot of melanin in each iris, while blue eyes result when very little melanin is present. The multiple genes and alleles influencing eye color contribute to the wide range of intermediate colors we see in human populations.

Continuous variation may also occur in a quantitative trait because of environmental factors. In this case, each genotype is capable of producing a range of phenotypes depending on outside influences. Thus, even if all individuals have the same genotype, many different phenotypes can result if they are raised in a diversity of environments. FIGURE 19.12 provides a clear example of the effect of the environment on phenotype. These identical twins share 100% of their genes but look quite different. They differ because of variations in their environment—the twin in the right image smoked cigarettes and had much greater exposure to sun than did the twin pictured on the left.

FIGURE 19.12 **The effect of the environment on phenotype.** These identical twins have exactly the same genotype, but they are quite different in appearance due to environmental factors.

Most traits that show continuous variation are influenced by both genes and the effect of differing environmental factors. Skin color in humans is an example of this type of trait. The hue of an individual's skin is dependent on the amount of melanin near the skin's surface. A number of genes have an effect on skin-color phenotype—both those that influence melanin production and those that affect the distribution of melanin in the skin. However, as you know, the environment, particularly the amount of exposure to the sun during a season or lifetime, also influences the skin color of individuals (FIGURE 19.13).

Both genetic factors and environmental factors influence most quantitative traits. Women choosing Doctorate-category sperm donors from Fairfax Cryobank are presumably interested in having smart, successful children, but intelligence has both a genetic and an environmental component. Intelligence depends in part on brain structure and function, and many alleles that interfere with brain structure and function—and thus intelligence—have been identified.

But intelligence also depends on environmental factors. For example, we know that if a developing baby is exposed to high levels of cigarette smoke or alcohol before birth, its brain will develop differently, and it may have delayed or diminished intellectual development. How can the relative influence of genes and environment on intelligence be determined?

(a) Genes  **(b)** Environment

FIGURE 19.13 Skin color is influenced by genes and environment. (a) A difference in skin color due mainly to variations in several alleles that control skin pigment production. (b) A difference in skin color in a single individual entirely due to environmental effects.

The Heritability of Quantitative Traits

If the variation among individuals results from the interaction of genes and environment, then how can we predict if the child of a father with a doctorate will also be capable of earning a doctorate? Scientists most often approach this question by attempting to determine what causes the variation in phenotype among individuals. The genetic component of variation in a given trait is called the trait's **heritability**.

The most direct way to calculate the heritability of traits in humans would be to control reproduction so that only certain individuals were allowed to have children with other selected individuals. This would answer the question of whether greater intelligence can be "bred into" a population. Of course, this is impossible—but scientists have a number of other techniques for estimating the genetic component of variation among humans.

Stop and Stretch 19.4

Scientists can control the reproduction of laboratory animals to test hypotheses about the heritability of quantitative traits. How does the use of animals avoid the problems of doing breeding experiments on humans? What are the limitations of using animal experiments in attempting to understand the heritability of human traits?

A correlation, a technique described in Chapter 1, can tell us how accurately the traits of an individual can be predicted when the traits of a related individual are known. A strong correlation between related individuals indicates that a trait is likely to have a strong genetic component. The correlation between the intelligence of human parents and their children helps us determine how important the intelligence of a donor may be to the mental capacity of his children.

Intelligence is often described by the IQ (intelligence quotient) score, which is generated by performance on an IQ test. These tests were not originally designed to comprehensively measure mental ability. Nonetheless, IQ score remains a convenient

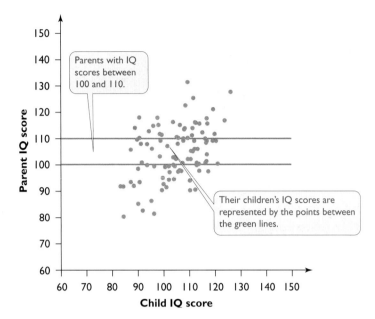

FIGURE 19.14 Correlation between parents and children. Some of the variation among individuals in IQ score can be explained by the IQ scores of their biological parents. The data points rise to the right, meaning that generally, children with higher IQs have parents with higher IQs. But the "scatter" of points indicates that this positive correlation is rather weak.

shorthand often used to describe "natural" or innate intelligence. There is considerable controversy about the use of IQ tests to quantify intelligence. However, no other widely accepted measure has replaced them for this purpose.

Even if IQ tests do not really measure general intelligence, IQ scores are correlated with academic success—meaning that individuals at higher academic levels usually have higher IQs. So, even without having information about their individual IQ scores, we can reasonably expect that donors in the Doctorate category have higher IQs than do other available sperm donors. The question of whether the high IQ of a prospective sperm donor can be passed on to his children still remains.

The average correlation between IQs of parents and their children is 0.42 (**FIGURE 19.14**). Another way of saying this is that 42% of the variation among individuals' IQ scores is explained by the variation among their parents' IQ scores. You might interpret this to mean that 42% of IQ is genetic, and the rest environmental.

However, because most families live together, parents and children share both genes and environment. Therefore, correlations of IQ between the two groups do not distinguish the relative importance of genes from the importance of the environment on influencing IQ score. This inability to distinguish the genes and the environment is the problem found in most arguments about "nature versus nurture." Do children resemble their parents because they are "born that way" or because they are "raised that way"?

The difficulty of separating genetic and environmental influences in most families compels researchers interested in human heritability to use **natural experiments**. These are situations in which unique circumstances allow a hypothesis to be tested without any intervention by researchers.

Human twins are one source of a natural experiment to test hypotheses about the heritability of quantitative traits in humans. Table 19.1 describes

TABLE 19.1 | To What Extent Is IQ Heritable?

A summary of various estimates of IQ heritability and their shortcomings

Method of Measurement	Result	Warnings When Interpreting This Result
Correlation between parents' IQ and children's IQ in a population	0.42	Since parents and children are similar in genes and environment, a correlation cannot be used to indicate the relative importance of genes and environment in determining IQ.
Natural experiment comparing IQ in pairs of identical twins versus nonidentical twins	0.52	People around identical twins treat them as more alike than they do nonidentical twins. Therefore their environment is different from that of nonidentical twins—the heritability value could be an overestimate.
Natural experiment comparing IQ of identical twins raised apart versus nonidentical twins raised apart	0.72	Small sample size may skew results.

natural experiments of IQ heritability using both twins raised together and those raised apart, summarizes the results of these studies, and describes criticisms of the results. Despite concerns about how to interpret the results, it appears that somewhere around 50% of the variation among individuals in IQ score results from genetic differences among individuals.

19.4 Genes, the Environment, and the Individual

We now know that a sperm donor will definitely influence some of his child's traits—eye and skin color, and perhaps even susceptibility to certain diseases. According to the twin studies discussed earlier, the donor will probably also pass on some intellectual traits to the child. In fact, the relatively high value for heritability of IQ appears to indicate that the genes do matter. So it might be a good idea to pay a premium price for Doctorate-category sperm after all.

However, surprisingly, the results of twin studies actually give us very limited information about how closely any individual child will match a sperm donor in intelligence and preferences. To understand why, we need to take a closer look at the practical significance of heritability.

The Use and Misuse of Heritability Calculations

Heritability is a measure of the importance of genes in causing the variation in a quantitative trait in a population. However, the calculated heritability value is unique to the population in which it was measured and to the environment of that population. The fact that heritability measures are specific to a particular environment means that we should be very cautious when using heritability to measure the general importance of genes to the development of a trait.

Differences Between Groups May Be Environmental A "thought experiment" can help illustrate this point. Body weight in laboratory mice has a strong genetic component, with a calculated heritability of about 0.90. In a population of mice where weight is variable, bigger mice have bigger offspring, and smaller mice have smaller offspring.

Imagine that we randomly divide a population of variable mice into two groups—one group is fed a rich diet, and the other group is fed a poor diet. Otherwise, the mice are treated identically. As you might guess, the well-fed mice become fat, while the poorly fed mice become thin. Consider the outcome if we were to keep the mice in the same conditions, allow them to reproduce, and then weighed their adult offspring. The average mouse in the well-fed population could be twice as heavy as the average poorly fed mouse.

Now imagine that another researcher came along and examined these two groups of mice without knowing their diets. He or she would see that body weight is highly heritable and that the two groups of mice are very different. Logically, the researcher could conclude that the groups are genetically different. However, we know this is not the case. Both the heavy and the light mice are offspring of the same original population of parents. It is the environment that the two populations are found in that differs and is the cause of the weight differences (FIGURE 19.15).

Now extend the same thought experiment to human groups. Imagine that we have two groups of humans, and we have determined that IQ had high heritability. In this case, one group of people was raised in an enriched environment, and their average IQ was higher. The other group was raised in a restricted environment, and their average IQ was lower. What conclusions could you draw about the genetic differences between these two populations?

FIGURE 19.15 The environment can have powerful effects on highly heritable traits. If genetically similar populations of mice are raised in radically diverse environments, then differences between the populations are entirely due to environment.

VISUALIZE THIS How would this figure differ if both groups of mice received the same diet?

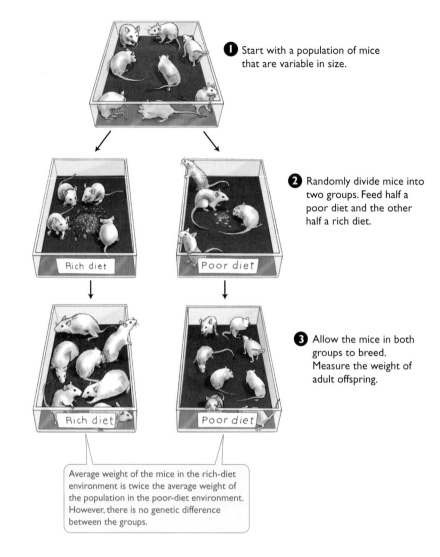

1. Start with a population of mice that are variable in size.

2. Randomly divide mice into two groups. Feed half a poor diet and the other half a rich diet.

3. Allow the mice in both groups to breed. Measure the weight of adult offspring.

Average weight of the mice in the rich-diet environment is twice the average weight of the population in the poor-diet environment. However, there is no genetic difference between the groups.

The answer to this question is none—as with the laboratory mice in the figure, these differences could be entirely due to environment. The high heritability of IQ cannot tell us if two human groups in differing social environments vary in IQ because of variations in genes or because of differences in environment.

A Highly Heritable Trait Can Still Respond to Environmental Change

A high heritability for IQ might seem to imply that IQ is not strongly influenced by environmental conditions. However, intelligence in other animals can be demonstrated to be both highly heritable and strongly influenced by the environment.

Rats can be bred for maze-running ability, and researchers have produced rats that are "maze-bright" and rats that are "maze-dull." Maze-running ability is highly heritable in the laboratory environment; that is, bright rats have bright offspring, and dull rats have dull offspring. The results of an experiment that measured the number of mistakes made by maze-bright and maze-dull rats raised in different environments are presented in FIGURE 19.16.

In the typical lab environment, bright rats were much better at maze running than dull rats. But in both a very boring and a very stimulating environment, the two groups of rats did about the same. In fact, no rats excelled in a boring environment, and all rats did better at maze running in enriched environments, with the duller rats improving more dramatically.

What this example demonstrates is that we cannot predict the response of a trait to a change in the environment, even when that trait is highly heritable. Thus, even if

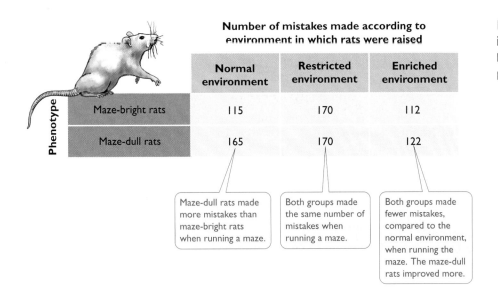

FIGURE 19.16 **A highly heritable trait is not identical in all environments.** The performances of bright and dull rats when completing a maze changed depending on the environment the rats were raised in.

IQ has a strong genetic component, environmental factors affecting IQ can have big effects on an individual's intelligence.

Heritability Does Not Tell Us Why Two Individuals Differ High heritability of a trait is often presumed to mean that the difference between two individuals is mostly due to differences in their genes. However, even if genes explain 90% of the population variability in a particular environment, the reason one individual differs from another may be entirely a function of environment (as an example of this, look back at the twins in Figure 19.12, who were genetically identical but very different in appearance).

How Do Genes Matter?

We know that genes can have a strong influence on eye color, risk of genetic diseases such as cystic fibrosis, susceptibility to a heart attack, and even the structure of the brain. But what really determines who we are—nature or nurture?

Even with single-gene traits, the outcome of a cross between a woman and a sperm donor is not guaranteed. It is only a probability, and probabilities are uncertain. Couple this uncertainty with traits being influenced by more than one gene, and independent assortment greatly increases the types of offspring possible from a single mating. Knowing the phenotype of potential parents gives you relatively little information about the phenotype of their children. So, even if genes have a strong effect on traits, we cannot "program" the traits of children by selecting the traits of their parents.

Our cells carry instructions for all the characteristics of human life, but the process of developing from embryo to adult takes place in a physical and social environment that influences how these genes are expressed. Scientists are still a long way from understanding how all of these complex, interacting circumstances result in who we are.

What is the message for women and couples who are searching for a sperm donor from Fairfax Cryobank? Donors in the Doctorate category may indeed have higher IQs than donors in the cryobank's other categories, but there is no real way to predict if a particular child of one of these donors will be smarter than average. According to the current data on the heritability of IQ, sperm from high-IQ donors will increase the odds of having an offspring with a high IQ, but only if parents provide their children with a stimulating, healthy, and challenging environment in which to mature. Which, of course, would be good for children with any alleles.

Chapter REVIEW

ROOTS TO REMEMBER

The following roots of words come mainly from Latin and Greek and will help you to decipher terms:

cryo- means icy cold.

hetero- means the other, another, or different.

homo- means the same.

pheno- comes from a verb meaning to show.

poly- means many.

-zygous derives from zygote, the cell resulting from the union of an egg and sperm.

KEY TERMS

19.1 The Inheritance of Traits
genetic variation *p. 468*
independent assortment *p. 470*
random fertilization *p. 472*
segregation *p. 470*

19.2 Mendelian Genetics: When the Role of a Gene Is Direct
cross *p. 474*
dominant *p. 473*
genotype *p. 472*
heterozygous *p. 472*
homozygous *p. 472*
phenotype *p. 472*
Punnett square *p. 474*
recessive *p. 472*

19.3 Quantitative Genetics: When Genes and Environment Interact
continuous variation *p. 476*
heritability *p. 477*
natural experiments *p. 478*
polygenic traits *p. 476*
quantitative traits *p. 476*

SUMMARY

19.1 The Inheritance of Traits

- Children resemble their parents in part because they inherit their parents' genes (p. 468).

- Mutations in gene copies can cause slightly different proteins to be produced within cells. Different gene versions are called alleles (p. 469; Figure 19.2).

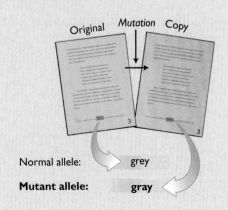

- Due to independent assortment, parents contribute a unique subset of alleles to each of their offspring (pp. 470–471).

19.2 Mendelian Genetics: When the Role of a Gene Is Direct

- The phenotype of a given individual for a particular gene depends on which alleles the individual carries (its genotype), whether it has two different alleles or two copies of the same allele (is heterozygous or homozygous), and whether the alleles are dominant or recessive (p. 472).

- Several human diseases have a genetic basis. These diseases may illustrate the effects of recessive alleles (as for cystic fibrosis) or dominant alleles (as for Huntington's disease) (p. 473).

- A Punnett square helps us determine the probability that two parents of known genotype will produce a child with a particular genotype (pp. 474–475; Figure 19.9).

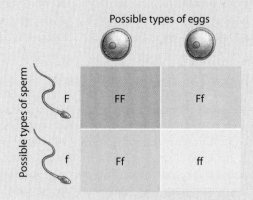

19.3 Quantitative Genetics: When Genes and Environment Interact

- Many traits—such as height, IQ, and musical ability—show continuous variation, which results in a range of values for the trait within a given population (p. 476).

- Quantitative variation in a trait may be generated because the trait is influenced by several genes, because the trait can be influenced by environmental factors, or due to a combination of both factors (pp. 476–477).

- The role of genes in determining the phenotype for a quantitative trait is estimated by calculating the heritability of the trait (pp. 477–478).

19.4 Genes, the Environment, and the Individual

- Calculated heritability values are unique to a particular population in a particular environment. The environment may cause large differences among individuals, even if a trait has high heritability (p. 479; Figure 19.15).

Rich diet

Poor diet

- Knowing the heritability of a trait does not tell us why two individuals differ for that trait (p. 480).

- Our current understanding of the relationship between genes and complex traits does not allow us to predict the phenotype of a particular offspring from the phenotype of its parents (p. 481).

LEARNING THE BASICS

1. Which of the following statements correctly describes the relationship between genes and chromosomes?
    a. Genes are chromosomes.
    b. Chromosomes contain many genes.
    c. Genes are made up of hundreds or thousands of chromosomes.
    d. Genes are assorted independently during meiosis, but chromosomes are not.
    e. More than one of the above is correct.

2. An allele is a _____.
    a. version of a gene
    b. dysfunctional gene
    c. protein
    d. spare copy of a gene
    e. phenotype

3. Sperm or eggs in humans always _____.
    a. have two copies of every gene
    b. have one copy of every gene
    c. contain either all recessive alleles or all dominant alleles
    d. are genetically identical to all other sperm or eggs produced by that person
    e. contain all of the genetic information of the parent

4. A mistake or "misspelling" that occurs during the copying of a gene and results in a change in a gene is called a(n) _____.
    a. dominant allele
    b. dysfunction
    c. mistakes never occur in gene copying
    d. mutation
    e. improvement

5. Which of the following genotypes is heterozygous?
    a. *Aa*
    b. *AA*
    c. *a*
    d. *AA BB*

6. If the effects of an allele are seen only when an individual carries two copies of the allele, the allele is termed _____.
    a. dominant
    b. incompletely dominant
    c. recessive
    d. independently assorted
    e. genotypic

7. A single gene in pea plants has a strong influence on plant height. The gene has two alleles: tall (T), which is dominant, and short (t), which is recessive. What are the genotypes and phenotypes of the offspring of a cross between a TT and a tt plant?

8. What are the genotypes and phenotypes of the offspring of $Tt \times Tt$, referring to the alleles in question 7?

9. Albinism occurs when individuals carry two recessive alleles (*aa*) that interfere with the production of melanin, which is the pigment that colors hair, skin, and eyes. If an albino child is born to two individuals with normal pigment, what is the genotype of each parent?

10. Pfeiffer syndrome is a dominant genetic disease that occurs when certain bones in the skull fuse too early in the development of a child, leading to a misshapen head and face. If a man with one copy of the allele that causes Pfeiffer syndrome marries a woman who is homozygous for the nonmutant allele, what is the chance that their first child will have this syndrome?

11. For the couple in question 10, what is the likelihood that their second child will have this syndrome?

12. A cross between a pea plant that produces yellow peas and a pea plant that produces green peas results in 100% yellow-pea offspring.
    a. Which allele is dominant in this situation?
    b. What are the genotypes of the yellow-pea and green-pea plants in the initial cross?

13. A cross between a pea plant that produces yellow peas and a pea plant that produces green peas results in 50% yellow-pea offspring and 50% green-pea offspring. What are the genotypes of the plants in the initial cross?

14. A woman who is a carrier for the cystic fibrosis allele marries a man who is also a carrier.
    a. What proportion of the woman's eggs will carry the cystic fibrosis allele?
    b. What proportion of the man's sperm will carry the cystic fibrosis allele?
    c. The probability that they will have a child who carries two copies of the cystic fibrosis allele is equal to the proportion of eggs that carry the allele times the proportion of sperm that carry the allele. What is this probability?
    d. Is this the same result you would generate when doing a Punnett square of this cross?

15. The allele *BRCA2* was identified in families with unusually high rates of breast and ovarian cancer. Up to 80% of women with one copy of the *BRCA2* allele develop one of these cancers in their lifetime.
    a. Is *BRCA2* a dominant or a recessive allele?
    b. How is *BRCA2* different from the typical pattern of Mendelian inheritance?

16. A quantitative trait _____.
    a. may be one that is strongly influenced by the environment
    b. varies continuously in a population
    c. may be influenced by many genes
    d. has more than a few values in a population
    e. all of the above are correct

17. When a trait is highly heritable, _____.
    a. it is influenced by genes
    b. it is not influenced by the environment
    c. the variance of the trait in a population can be explained primarily by variance in genotypes
    d. a and c are correct
    e. a, b, and c are correct

ANALYZING AND APPLYING THE BASICS

Questions 1 and 2 refer to the following information.

Currently, scientists know of several genes that control the expression of eye color. On chromosome 15, the *bey2* gene has alleles for brown and blue color. On chromosome 19, the *gey* gene has alleles for blue and green color. Although eye color appears to be controlled by many more genes, scientists have developed a model that uses alleles on the *bey2* and *gey* genes to determine simple inheritance patterns. The following rules create this model:

- The brown allele is always dominant.
- The blue allele is always recessive.
- The green allele is dominant to blue but recessive to brown.

Using these rules, a person who is heterozygous for the *bey2* gene (*Bb*) and heterozygous for the *gey* gene (*Gg*) would have brown eyes. A person who is homozygous recessive for the *bey2* gene (*bb*) and heterozygous for the *gey* gene (*Gg*) would have green eyes.

1. What genotype(s) would produce a phenotype of blue eyes?
2. List the different gametes produced by the following individuals and their frequencies in the individual.

   A: Heterozygous for *bey2* and homozygous recessive for *gey*
   B: Homozygous recessive for *bey2* and heterozygous for *gey*

3. Does a high value of heritability for a trait indicate that the average value of the trait in a population will not change if the environment changes? Explain your answer.

Questions 4 and 5 refer to the following information.

Autism is a disorder characterized by the following behavioral phenotypes:

- Lack of interest in social relationships and failure to develop human attachments
- Use of language for communicating needs but not for developing relationships with other people
- Development of rituals or repetitive behaviors

Researchers have found that parents and siblings of autistic children often share traits that resemble autism. Individuals in these families may have difficulty with communication, lack social skills, and crave ritual and routine. Twin studies found even stronger genetic evidence. Rates of autism in monozygotic versus dizygotic twins produced a heritability estimate of more than 90%.

Interactions between multiple genes have been implicated in the development of autism. However, researchers suspect that different combinations of genes are responsible for the variety and severity of autism phenotypes. Some researchers also suspect that environmental factors contribute to autism.

4. What does a high heritability estimate (>90%) mean in relation to the development of autism? How could this estimate be misinterpreted?

5. How is the autism disorder different from Huntington's disease or cystic fibrosis? What characteristics of autism would make developing a cure more difficult than developing a cure for a disease like cystic fibrosis?

6. The heritability of IQ has been estimated at greater than 50%. If John's IQ is 120 and Jerry's IQ is 90, does John have better "intelligence" genes than Jerry does? Explain your answer.

Questions 7 and 8 refer to the following information.

Scientists have developed different strains of experimental rodents to model human disease. For example, R6/1 HD mice have been bred to model Huntington's disease. The allele that causes Huntington-like symptoms in mice is very similar to the human Huntington's allele. In both cases, the mutation eventually leads to degeneration of brain tissue.

A variety of studies have used rodent models like the R6/1 HD mice to investigate gene-environment interactions for central nervous system disorders. These studies examine the connection between symptoms of each disease and environmental enrichment. The methods for each study are similar. Control groups of rodents are housed in standard laboratory cages. Experimental groups are housed in cages containing a variety of objects that provide environmental enrichment, such as wheels, toys, ladders, and swings. Some of the results are summarized in Table 19.2.

7. Summarize the overall effects of environmental enrichment on the diseases listed in Table 19.2. How does environmental enrichment accomplish these effects?

8. How could the results from the studies listed in Table 19.2 help to strengthen the argument that intelligence depends on environment as well as genes?

Table 19.2 Effects of Environmental Enrichment on Mice with CNS Disorders

Central Nervous System Disorder	Symptoms of Disorder	Notable Effects of Enrichment
Huntington's disease	Degeneration of brain tissue, producing uncontrolled movements, dementia, and depression	Delay in loss of motor skills, delayed degeneration of brain tissue, decreased expression of some proteins
Alzheimer's disease	Degeneration of central nervous system, leading to dementia	Improved spatial memory in females
Parkinson's disease	Tremor, rigid muscles, dementia in late stages	Improved motor function, decreased loss of dopamine-producing neurons
Epilepsy	Unpredictable seizures	Increased resistance to seizures, decreased hippocampal cell death, increased expression of some growth factors involved in synaptic connections, neurotransmitters, and receptors
Fragile X syndrome	Mental retardation	Increased exploratory behavior, increased branching of nerves
Down syndrome	Mental retardation	Increased exploratory behavior, improved spatial learning in females

CONNECTING THE SCIENCE

1. If scientists find a gene that is associated with a trait that is considered a disability (for instance, a genetic predisposition to deafness), will it mean greater or lesser tolerance toward people with that trait? Will it lead to proposals that those affected by the "disorder" should undergo treatment to be "cured," and that measures should be taken to prevent the birth of other individuals who are also afflicted?

2. Does a genetic basis for differences in IQ between people with Down syndrome and people without this condition mean that we should put fewer resources into education for people with Down syndrome? How does your answer to this question relate to questions about how we should treat individuals with other genetic conditions?

Chapter 20

Complex Patterns of Inheritance

DNA Detective

LEARNING GOALS

1. Compare and contrast codominance and incomplete dominance.
2. Use the ABO blood system to explain the concept of multiple allelism.
3. Compare and contrast pleiotropy and polygenic inheritance.
4. Predict the outcome of a dihybrid cross.
5. How do autosomes and sex chromosomes differ?
6. How is sex determined in humans?
7. What are sex-linked genes, and how is their expression different from that of autosomal genes?
8. Analyze genetic pedigrees for various patterns of inheritance.
9. Describe the technique of DNA fingerprinting.
10. Draw a DNA fingerprint that might be generated by two siblings and their parents.

The Romanov family ruled Russia for 300 years, until the Romanovs shown here were overthrown and executed in 1918.

On the night of July 16, 1918, the tsar of Russia, Nicholas II, his wife Alexandra Romanov, their five children, and four family servants were executed in a small room in the basement of the house to which they had been exiled. These murders ended three centuries of rule by the Romanov family over the Russian Empire.

In February 1917, in the wake of protests throughout Russia, Nicholas II had relinquished his power by abdicating for both himself and on behalf of his only son Alexis, then 13 years old. The tsar hoped that these abdications would protect his son, the heir to the throne, as well as the rest of the family from harm.

A toppled statue of Lenin symbolizes the 1991 fall of the communist Soviet Union.

The political climate in Russia at that time was explosive. During the summer of 1914, Russia and other European countries became embroiled in World War I. This war proved to be a disaster for the imperial government. Russia faced severe food shortages, and the poverty of the common people contrasted starkly with the luxurious lives of their leaders. The Russian people felt deep resentment toward the tsar's family. This sentiment sparked the first Russian Revolution in February 1917. Following Nicholas's abdication, the imperial family was kept under guard at one of their palaces outside St. Petersburg.

Were bones found in a grave in Ekaterinburg those of the slain Romanovs?

In November 1917, the Bolshevik Revolution brought the communist regime, led by Vladimir Lenin, to power. Ridding the country of the last vestige of Romanov rule became a priority for Lenin and his political party. Lenin believed that doing so would solidify his regime as well as garner support among people who felt that the exiled Romanovs and their opulent lifestyle had come to represent all that was wrong with Imperial Russia. Fearing any attempt by pro-Romanov forces to save the family, Lenin ordered them to the town of Ekaterinburg in Siberia.

Shortly after midnight on July 16, the family was awakened and asked to dress. Nicholas, Alexandra, and their children—Olga, Tatiana, Maria, Anastasia, and Alexis—along with the family physician, cook, maid, and valet, were escorted to a room in the basement of the house in which they

DNA evidence confirmed that this shallow grave held royal bones.

had been kept. Believing they were to be moved, the family waited. A soldier entered the room and read a short statement indicating they were to be killed. Armed men stormed into the room, and after a hail of bullets, the royal family and their entourage lay dead.

After the murders, the men loaded the bodies of the Romanovs and their servants into a truck and drove to a remote, wooded area in Ekaterinburg. Historical accounts differ as to whether the bodies were dumped down a mine shaft, later to be removed, or were immediately buried. There is also some disagreement regarding the burial of two of the people who were executed. Some reports indicate that all 11 people were buried together, and two of them either were badly decomposed by acid placed on the ground of the burial site or were burned to ash. Other reports indicate that two members of the family were buried separately. Some people even believe that two victims escaped the execution. In any case, the bodies of at least nine people were buried in a shallow grave, where they lay undisturbed until 1991.

The bodies were not all that remained buried. For decades, details of the family's murder were hidden in the Communist Party archives in Moscow. However, after the dissolution of the Soviet Union, postcommunist leaders allowed the bones to be exhumed so that they could be given a proper burial. This exhumation took on intense political meaning because the people of Russia hoped to do more than just give the family a proper burial. The event took on the symbolic significance of laying to rest the brutality of the communist regime that took power after the murders of the Romanov family.

All that remained of these bodies when they were exhumed was a pile of bones, so it was difficult to know if these were the remains of the royal family. A great deal of circumstantial evidence pointed to that conclusion. The bones seemed to indicate that they belonged to six adults and three children. Investigators electronically superimposed the photographs of the skulls on archived photographs of the family. They compared the skeletons' measurements with clothing known to have belonged to the family. Five of the bodies had gold, porcelain, and platinum dental work, which had been available only to aristocrats. These and other data were consistent with the hypothesis that the bodies could be those of the tsar, the tsarina, three of their five children, and the four servants.

However, at this point, the scientists had shown only that these skeletons *might* be the Romanovs. They had not yet shown with any degree of certainty that these bodies *did* belong to the slain royals. The new Russian leaders did not want to make a mistake when symbolically burying a former regime. Unassailable proof was necessary because so much was at stake politically. To begin to answer these questions, scientists turned to the field of genetics.

20.1 Extensions of Mendelism

Solving this puzzle would ultimately require that scientists be able to show relatedness between the tsar and tsarina's skeletons and their children's skeletons. In Chapter 19, you were introduced to the idea that patterns of inheritance can be predicted when genes are inherited in a straightforward manner—for example, a trait that is controlled by one gene with two versions or alleles, one of them dominant and the other recessive. Patterns of inheritance that are a little more complex are called *extensions* of Mendelian genetics. You also learned that traits controlled by more than one gene are said to be inherited in a *polygenic* manner and that traits inherited in a polygenic manner produce a wider array of phenotypes than traits controlled by a single gene.

An additional extension of Mendelism occurs when the offspring of two different parents has a phenotype that is intermediate to that of either parent. The trait is said to display **incomplete dominance**. Hair texture in humans displays incomplete dominance. When a parent with straight hair (genotype *cc*) mates with a parent with curly hair (genotype *CC*), the children they produce will have wavy hair (*Cc*) (FIGURE 20.1). Genetic crosses such as this one, involving one gene, are called *monohybrid crosses*.

Because all that was left of the Romanovs was a pile of bones, the scientists could study only a few genetic traits to show the relatedness of the adult skeletons to two of the four children's skeletons. Genetic traits that were obvious, such as bone size and structure, are controlled by many genes and affected by environmental components like nutrition and physical activity level. Therefore, using bone size and structure to predict which of the adult's skeletons were related to the children's skeletons would have been a matter of guesswork. The scientists had to use more sophisticated analyses.

Forensic scientists include medical examiners, crime scene detectives, and laboratory technicians. One analysis that forensic scientists often use to help determine relatedness

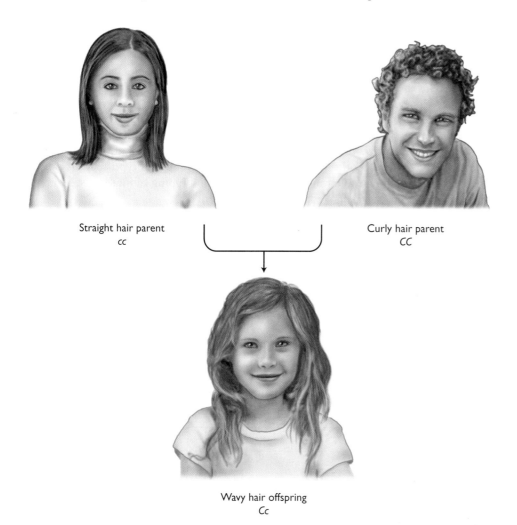

FIGURE 20.1 Incomplete dominance. When a genetic trait is inherited in an incompletely dominant manner, the heterozygote has a phenotype that is intermediate to either homozygote.

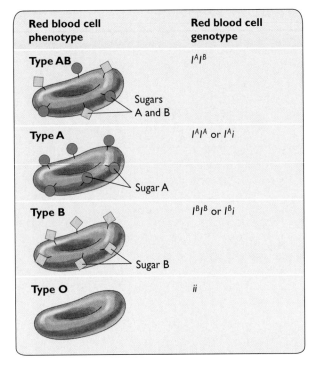

FIGURE 20.2 Red blood cell phenotypes and genotypes. Sugars on the surface of red blood cells determine the phenotype and genotype relative to the ABO blood system.

of people is blood typing, which involves determining if certain carbohydrates are located on the surface of red blood cells. You learned about the importance of matching blood donors and recipients for blood transfusions in Chapter 8. This requirement is due to the genetics of the ABO blood system, which displays two extensions of Mendelism—**codominance**, whereby two different alleles of a gene are both expressed in an individual, and **multiple allelism**, which occurs when there are more than two alleles of a gene in the population. In fact, three distinct alleles of one blood-group gene code for the enzymes that synthesize the sugars found on the surface of red blood cells. Two of the three alleles display codominance with each other, and one allele is recessive to the other two.

FIGURE 20.2 summarizes the possible genotypes and phenotypes for the ABO blood system. The three alleles of this blood-group gene are I^A, I^B, and i. A given individual will carry only two alleles, even though three alleles are being passed on in the entire population. In other words, one person may carry the I^A and I^B alleles, and another might carry the I^A and i alleles. There are three different alleles, but each individual can carry only two alleles.

The symbols used to represent these alternate forms of the blood-group gene tell us something about their effects. The lowercase i allele is recessive to both the I^A and I^B alleles. Therefore, a person with the genotype $I^A i$ has type A blood, and a person with the genotype $I^B i$ has type B blood. A person with both recessive alleles, genotype ii, has type O blood. The uppercase I^A and I^B alleles display codominance in that neither one masks the expression of the other. Both of these alleles are expressed. Thus, a person with the genotype $I^A I^B$, has type AB blood.

Blood typing is often used to help establish whether or not a given set of parents could have produced a particular child. For example, a child with type AB blood and parents who are type A and type B could be related, but a child with type O blood could not have a parent with type AB blood. Likewise, if a child has blood type B and the known mother has blood type AB, then the father of that child could have type AB, A, B, or O blood, which does not help to establish parentage.

If a child has a blood type consistent with alleles that he or she may have inherited from a man who might be his or her father, this finding does not mean that the man is the father. Instead, it is only an indication that he could be. In fact, many other men would also have that blood type (Table 20.1). Therefore, blood type analysis can be used only to eliminate people from consideration. Blood typing cannot be used to positively identify someone as the father of a particular child.

TABLE 20.1 | Percentages of Blood Types in the U.S. Population

Blood Type*	U.S. Percentage
AB⁻	0.5
B⁻	1.5
AB⁺	3
A⁻	5
O⁻	7
B⁺	11
A⁺	32
O⁺	40

*The negative and positive superscripts refer to the absence or presence of the Rh factor you learned about in Chapter 8.

Stop and Stretch 20.1

In an attempt to establish who is the father of a child, blood typing is performed. The mother has type A blood and the child has type AB blood. There are two potential fathers. One has type O blood and one has type B blood. Can it be conclusively determined which of these men is the father?

Blood-typing analysis could have provided scientists with some information about potential relatedness of the skeletons. However, it was not an option in the case of the Romanovs. The very old remains contained no blood, so blood typing was not possible.

A blood-related trait that scientists knew Alexis had is a clotting disorder called **hemophilia**. A person with the most common form of hemophilia cannot produce a protein called clotting factor VIII. When this protein is absent, blood does not form clots to stop bleeding from a cut or internal blood vessel damage. Affected individuals bleed excessively, even from small cuts.

Having this defect in a single gene can lead to multiple effects on an individual's phenotype, by a phenomenon called **pleiotropy**. Due to the direct effects of excessive bleeding, hemophilia can lead to excessive bruising, pain and swelling in the joints, vision loss from bleeding into the eye, and anemia, resulting in fatigue. In addition, neurological problems may occur if bleeding or blood loss occurs in the brain.

Historical records indicate that Alexis, heir to the throne, was so ill with hemophilia that his father actually had to carry him to the basement room where he was executed.

20.2 Dihybrid Crosses

Dihybrid crosses are genetic crosses involving two traits. To analyze the outcome of a cross involving two different genes, each with two different alleles, one can still use a Punnett square. For example, in Mendel's peas, seed color and seed shape are each determined by a single gene, and each are carried on different chromosomes. The two seed-color gene alleles are Y, which is dominant and codes for yellow color, and y, the recessive allele, which results in green seeds when homozygous. The two seed-shape alleles are R, the dominant allele, which codes for a smooth, round shape; and r, which is recessive and codes for a wrinkled shape.

Because the genes for seed color and seed shape are on different chromosomes, they are placed in eggs and sperm independently of each other. In other words, a pea plant that is heterozygous for both genes (genotype $Yy\ Rr$) can make four different types of eggs: one carrying dominant alleles for both genes ($Y R$), one carrying recessive alleles for both genes ($y r$), one carrying the dominant allele for seed color and the recessive allele for seed shape ($Y r$), and one carrying the recessive allele for color and the dominant allele for shape ($y R$).

A Punnett square for a cross between two individuals who are heterozygous for both seed-color and seed-shape genes would contain 16 boxes with four different phenotypes (FIGURE 20.3). The phenotypes produced result in a 9:3:3:1 **phenotypic ratio** where 9/16 include both dominant alleles ($Y\_R\_$); 3/16 include dominant alleles of one gene only (Y_rr); 3/16 include dominant alleles of only the other gene ($yyR\_$); and 1/16 carry recessive alleles only.

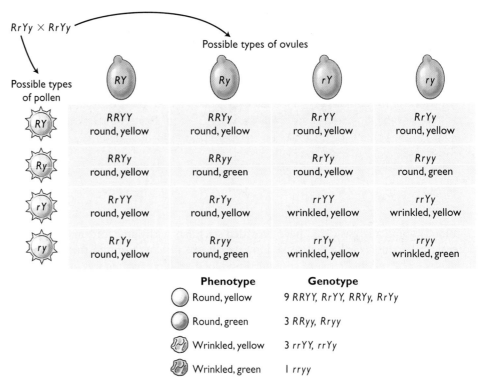

FIGURE 20.3 A dihybrid cross. Punnett squares can be generated to predict the outcome of a cross between individuals when we know their genotypes for more than one gene, as long as those genes are on separate chromosomes. This Punnett square shows a cross between two pea plants that are heterozygous for both the seed-color and the seed-shape genes.

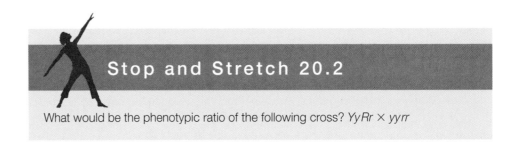

What would be the phenotypic ratio of the following cross? *YyRr* × *yyrr*

The reason that so many different kinds of offspring could be produced from just two genes is the random alignment of homologues, leading to independent assortment of genes, which you learned about in the chapters on cell division (17) and inheritance (19), respectively.

Recall from these earlier chapters that members of a homologous pair of chromosomes align at the cell's equator during meiosis and that their alignment is random with respect to which member of a homologous pair faces which pole. FIGURE 20.4 is a review of random alignment illustrating the process for two chromosomes that carry some of the genes for hair texture and eye color. Alleles for curly hair (*CC*) and straight hair (*cc*) are incompletely dominant, so together they produce wavy hair (*Cc*). Alleles for darkly pigmented eyes (*DD* or *Dd*) are dominant over those for blue eyes (*dd*). Eye color is actually determined by three different genes; but for simplicity, we will follow the inheritance of only one of these genes.

Color photos of the Romanovs allow us to determine their eye color and hair texture, so we will use these traits to illustrate how random alignment leads to inde-

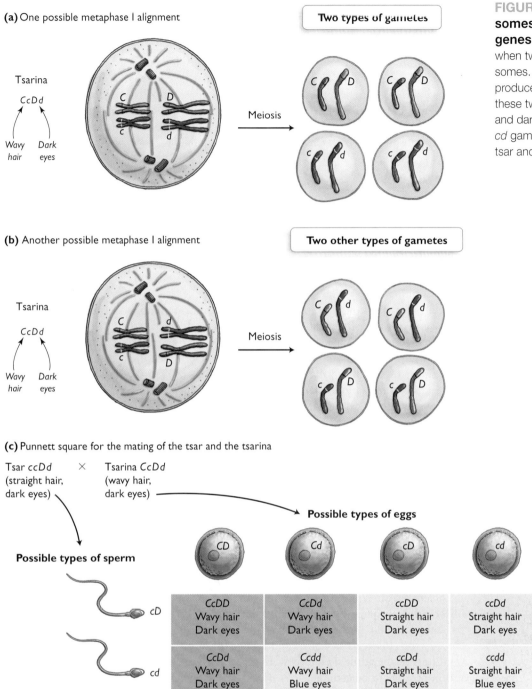

FIGURE 20.4 Random alignment of chromosomes leads to independent assortment of genes. (a, b) Two possible alignments can occur when two genes are located on different chromosomes. (c) Due to random alignment the tsarina could produce four different types of gametes relative to these two genes. Because the tsar had straight hair and dark eyes (*ccDd*), he could produce *cD* and *cd* gametes. Possible offspring from the mating of the tsar and tsarina are shown inside the Punnett square.

pendent assortment of these genes. The tsar had straight hair and dark eyes (*ccDd*), while the tsarina had wavy hair and dark eyes (*CcDd*). We know that the tsar and tsarina were heterozygous for the eye color gene since they had children with blue eyes. Likewise, each member of the royal couple must also have had at least one recessive hair-texture allele, since they had children with straight hair. Because these genes are located on different chromosomes, together the royal couple could produce children with wavy hair and brown eyes (Tatiana and Maria), wavy hair and blue eyes (Anastasia and Olga), straight hair and blue eyes (Alexis), or straight hair and brown eyes.

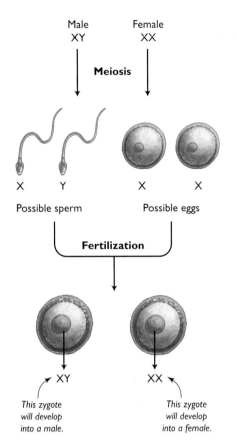

FIGURE 20.5 **Sex determination in humans.** In humans, sex is determined by the male because males produce sperm that carry either an X or a Y chromosome (in addition to 22 autosomes). The egg cell always carries an X chromosome along with the 22 autosomes. When an X-bearing sperm fertilizes the egg cell, a female (XX) results. When a Y-bearing sperm fertilizes an egg cell, an XY male results.

VISUALIZE THIS What is the probability that a couple will have a boy? What is the probability that a couple with four boys will have a girl for their fifth child?

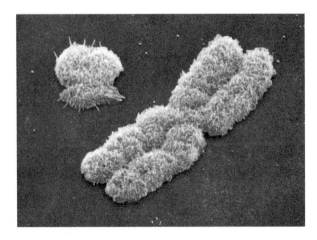

FIGURE 20.6 **The X and Y chromosomes.** The Y chromosome (left) is smaller than the X chromosome.

20.3 Sex Determination and Sex Linkage

It appears that Alexis inherited the hemophilia allele from his mother. We can deduce this pattern of inheritance because we now know that the clotting factor gene is inherited in a sex-specific manner. The clotting factor VIII gene (the gene that, when mutated, causes hemophilia) is located in the X chromosome. Of the 23 pairs of chromosomes present in the cells of human males, 22 pairs are **autosomes**, or nonsex chromosomes, and one pair, X and Y, are the **sex chromosomes**. Males have 22 pairs of autosomes and X and Y sex chromosomes. Females also have 22 pairs of autosomes, but their sex chromosomes are two X chromosomes.

Chromosomes and Sex Determination

The X and Y chromosomes are involved in determining the sex of an individual through a process called **sex determination**. When men produce sperm and the chromosome number is divided in half through meiosis, their sperm cells contain one member of each autosome pair and either an X or a Y chromosome. Females produce gametes with 22 unpaired autosomes and one of their two X chromosomes. Therefore, human egg cells normally contain one copy of an X chromosome, but sperm cells can contain either an X or a Y chromosome. The sperm cell determines the sex of the offspring resulting from a particular fertilization. If an X-bearing sperm unites with an egg cell, the resulting offspring will be female (XX). If a sperm bearing a Y chromosome unites with an egg cell, the resulting offspring will be male (XY). FIGURE 20.5 summarizes the process of sex determination.

Karyotyping is a kind of chromosomal analysis that allows scientists to view an individual's chromosomes. Chapter 17 outlines this process and gives an example of a finished karyotype (Figure 17.14). Karyotyping can allow scientists to diagnose chromosomal anomalies such as broken, duplicated, or missing chromosomes. The Genes & Homeostasis feature explores chromosomal anomalies and their effects on homeostasis more fully.

Karyotype analysis requires that cells are able to divide. This is because chromosomes are best visualized when they are duplicated and condensed, as they would be at metaphase of mitosis. It was not possible to perform karyotyping on the cells of the bones recovered from the Ekaterinburg grave because these cells were no longer dividing. If the scientists had been able to perform karyotyping analysis, they could have determined the sex of the individuals buried in the grave.

Scientists can sometimes determine the sex of an individual based on the structure of the pelvis. Females have evolved to have wider pelvic openings to accommodate the passage of a child through the birth canal. Russian scientists had determined that all three of the children's skeletons and two of the adult's skeletons were probably female (and four of the adult skeletons were male). However, the pelvises had decayed, so it was impossible to be certain.

Sex Linkage

Genes located on the X or Y chromosome are called **sex-linked genes** because biological sex is inherited along with, or "linked to," the X or Y chromosome. Sex-linked genes found on the X chromosomes are said to be X-linked, while those on the Y chromosome are Y-linked. The X chromosome is much larger than the Y chromosome, which carries very little genetic information (FIGURE 20.6).

X-Linked Genes **X-linked genes** are located on the X chromosome. The fact that males have only one X chromosome leads to some peculiarities in inheritance of sex-linked genes. Males always inherit their X chromosome from their mother, because they must inherit the Y chromosome from their father to be male. Thus, males will

GENES & HOMEOSTASIS

Changes to Chromosome Structure and Number

Mutations can occur that alter the structure of an entire chromosome, not simply one gene. For instance, exposing chromosomes to radiation and certain chemicals can cause them to break. If a chromosome break occurs in the gametes of an unaffected person, the broken chromosome can be passed on to a child. A child who inherits one normal chromosome from one parent and a chromosome with missing genetic information from the other parent carries a *deletion*, as the drawing illustrates. This child has only one allele of each gene because of the deleted region.

Having too little genetic information can cause a host of problems. For instance, Williams syndrome is the result of a loss of a portion of chromosome 7. Children with this disorder have a characteristic appearance including upturned noses, wide mouths, and large ears (see the photo). These children typically struggle academically and age prematurely. One copy of the gene that governs the production of the protein elastin is missing, which disrupts the homeostasis of skin cells, and the skin of these children ages much more rapidly than it does in unaffected persons. The decrease in elastin also impairs the cardiovascular system, and these children tend to have shortened life spans.

Segments of chromosomes can also be duplicated. Just as too little genetic information can cause genetic problems, so can *duplications*, which result in too much genetic information.

Sometimes mistakes occur during meiosis that result in the production of offspring with too many or too few chromosomes. This can occur when the homologues (or sister chromatids) fail to separate during meiosis. This failure of chromosomes to separate is called *nondisjunction*. The presence of an extra chromosome is known as *trisomy*. Persons with Down syndrome can have an extra copy of chromosome 21. The absence of one chromosome of a homologous pair is called *monosomy*. Nondisjunction can occur on autosomes and sex chromosomes.

Typically, embryos with too many or too few chromosomes will die because they have way too much or too little genetic information. However, in some situations, such as when the extra or missing chromosome is very small (as in chromosomes 21, 13, and 18), and/or contains very little genetic information (as in the Y chromosome), the embryo can survive.

Having too many X chromosomes is also often compatible with life because females actually inactivate one of their two X chromosomes in every cell during development. The inactivated X chromosome is called a *Barr body* after the discoverer of this phenomenon, Murray Barr. When more than two X chromosomes are inherited, all additional X chromosomes are also inactivated. The table lists some chromosomal anomalies in humans and their effects.

(a) A deletion occurs when part of a chromosome is lost. (b) A child with Williams syndrome.

Autosomal and Sex-Linked Chromosomal Anomalies

Conditions Caused by Nondisjunction of Autosomes	Comments
Trisomy 21, Down syndrome	Affected individuals are mentally retarded and have abnormal skeletal development and heart defects. The frequency of Down syndrome children increases with maternal age. 1 in 1,000 children of mothers under 35 are affected, and 4 in 1,000 children of mothers over 45 are affected.
Trisomy 13, Patau syndrome	Affected individuals are mentally retarded, deaf, and have a cleft lip and palate. Around 1 in 5,000 newborns are affected.

(Continued)

Autosomal and Sex-Linked Chromosomal Anomalies (Continued)

Conditions Caused by Nondisjunction of Autosomes	Comments
Trisomy 18, Edwards syndrome	Affected individuals have malformed organs, ears, mouth, and nose, leading to an elfin appearance. These babies usually die within six months of birth. Around 1 in 6,000 newborns are affected.

Conditions Caused by Nondisjunction of Sex Chromosomes	Comments
XO, Turner syndrome 	Females with one X chromosome can be sterile if their ovaries fail to develop. Webbing of the neck, shorter stature, and hearing impairment are also common. Around 1 in 5,000 female newborns are born with only one X chromosome.
Trisomy X, Meta female	Females with three X chromosomes tend to develop normally. Two of the X chromosomes will condense to become Barr bodies. Approximately 1 in 1,000 females are born with an extra X chromosome.
XXY, Kleinfelter syndrome 	Males with the XXY genotype are less fertile than XY males, have small testes, sparse body hair, some breast enlargement, and may have mental retardation. Testosterone injections can reverse some of the anatomical abnormalities in the approximately 1 in 1,000 males with this condition.
XYY condition	Males with two Y chromosomes tend to be taller than average but have an otherwise normal male phenotype. Around 1 in 1,000 newborn males has an extra Y chromosome.

inherit X-linked genes only from their mothers. Males are more likely to suffer from diseases caused by recessive alleles on the X chromosome because they have only one copy of any X-linked gene. Females are less likely to suffer from these diseases, because they carry two copies of the X chromosome and thus have a greater likelihood of carrying at least one functional version of each X-linked gene.

A **carrier** of a recessively inherited trait has one copy of the recessive allele and one copy of the normal allele and will not exhibit symptoms of the disease. Only females can be carriers of X-linked recessive traits because males with a copy of the recessive allele will have the trait. Both males and females can be carriers of non-sex-linked, autosomal traits.

Even though female carriers of an X-linked recessive trait will not display the recessive trait, they can pass the trait on to their offspring. For this reason, most women carrying the hemophilia allele will not even realize that they are a carrier until their son becomes ill. FIGURE 20.7a illustrates that a cross between a male who does not have hemophilia and a female carrier can produce unaffected females, carrier females, unaffected males, and affected males. FIGURE 20.7b illustrates that no male children produced by a cross between an affected male and a noncarrier female would have hemophilia. All daughters produced by this cross would be carriers of the trait. In the United States, there are over 20,000 hemophiliacs, most of whom are male.

Red-green color blindness is another X-linked trait. This trait affects approximately 4% of males. Red blindness is the inability to see red as a distinct color. Green blindness is the inability to see green as a distinct color. When the genes involved are normal (in this case the dominant alleles are normal), they code for the production of proteins called opsins that help absorb different wavelengths of light. A lack of opsins causes insensitivity to light of red and green wavelengths.

Duchenne muscular dystrophy is a progressive, fatal X-linked disease of muscle wasting that affects approximately 1 in 3,500 males. The onset of muscle wasting occurs between 1 and 12 years of age, and by age 12, affected boys are often confined to a wheelchair. The affected gene is one that normally codes for the dystrophin protein. When at least one allele is normal, dystrophin stabilizes cell membranes during muscle contraction. It is thought that the absence of normal dystrophin protein causes muscle cells to break down and muscle tissue to die.

Y-Linked Genes Y-linked genes are located on the Y chromosome and are passed from fathers to sons. Although this distinctive pattern of inheritance should make Y-linked genes easy to identify, very few genes have been localized to the Y chromosomes. One gene known to be located exclusively on the Y chromosome is called the *SRY* gene (for sex-determining region of the Y chromosome). The expression of this gene triggers a series of events leading to development of the testes and some of the specialized cells required for male sexual characteristics. Genes other than *SRY*, on chromosomes other than the Y, code for proteins that are unique to males but are not expressed unless testes develop.

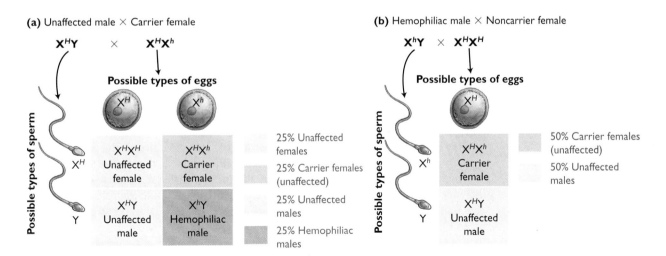

FIGURE 20.7 **Genetic crosses involving the X-linked hemophilia trait.** Cross (a) shows the possible outcomes and associated probabilities of a mating between a nonhemophiliac male and a female carrier of hemophilia. Cross (b) shows the possible outcomes and associated probabilities of a cross between a hemophiliac male and a noncarrier female.

VISUALIZE THIS What genetic cross would result in the highest frequency of affected males?

Stop and Stretch 20.3

A man with a Y-linked gene is planning to have children. What percentage of his male and female offspring are likely to carry that gene?

Since karyotype and pelvic bone analyses were difficult to perform on the decayed bones, scientists analyzed DNA from the bones for sequences known to be present only on the Y chromosome. When DNA was isolated from the children's remains, it became clear that the children's bones all belonged to girls. If these bones did belong to the Romanovs, one of the two missing children was Alexis, the Romanovs' only son.

Another line of evidence was provided by the extensive family trees of the Romanovs and their relatives.

20.4 Pedigrees

Because the hemophilia gene is X-linked, Alexis inherited the disease from his mother, who must have been a carrier of the disease. We can trace the lineage of this disease through the Romanov family by using a chart called a pedigree. A **pedigree** is a family tree that follows the inheritance of a genetic trait for many generations of relatives.

Pedigrees are often used in studying human genetics because it is impossible and unethical to set up controlled matings between humans the way one can with fruit flies or plants. Pedigrees allow scientists to study inheritance by analyzing matings that have already occurred. FIGURE 20.8 gives some of the symbols used in pedigrees, and FIGURE 20.9 shows how scientists can use pedigrees to determine if a trait is inherited as an autosomal dominant or recessive trait, or as a sex-linked recessive trait.

Information is available about the Romanovs' ancestors because they were royalty and because scientists interested in hemophilia had kept very good records of the inheritance of that trait. Hemophilia was common among European royal families in the nineteenth century but rare among the rest of the population. Hemophilia was a result of members of the royal families intermarrying to preserve royal bloodlines.

The pedigree in FIGURE 20.10 (on page 500) shows that the tsarina must have been a carrier of the hemophilia allele because her son had the trait. Her mother, Alice, must also have been a carrier because the tsarina's brother Fred had the disease, as did two of her sister Irene's sons, Waldemar and Henry.

The tsarina's grandmother, Queen Victoria of Great Britain, seems to be the first carrier of this allele in this family because there is no evidence of this disease before her eighth child, Leopold, is affected. Queen Victoria's mother, Princess Victoria, most likely incurred a mutation to the clotting factor VIII gene while the cells of her ovary were undergoing DNA synthesis to produce egg cells. The fertilized egg cell that produced Queen Victoria carried this mutation. When the cell divided by mitosis to produce her body cells, the mutation was passed on to each of her cells. When her sex cells underwent meiosis to produce gametes, she passed the mutant version on to three of her nine children. The extensive pedigree available to the scientists working on the Romanov case, in concert with a powerful technique called DNA fingerprinting, would provide the key data in solving the mystery of the buried bones.

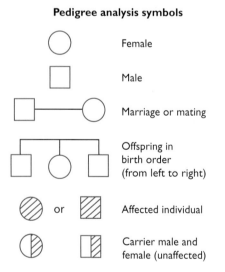

FIGURE 20.8 Pedigree analysis. Symbols used in pedigrees.

VISUALIZE THIS Draw a pedigree showing a female who has a boy with her first husband and a girl with her second husband.

20.4 Pedigrees   499

(a) Dominant trait: Polydactyly

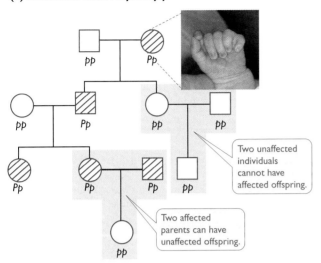

(b) Recessive trait: Attached earlobes

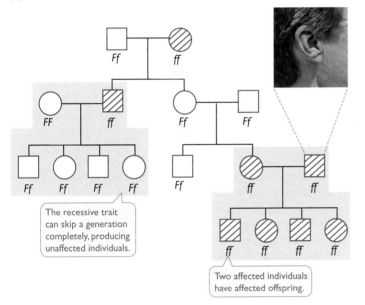

(c) Sex-linked trait: Muscular dystrophy

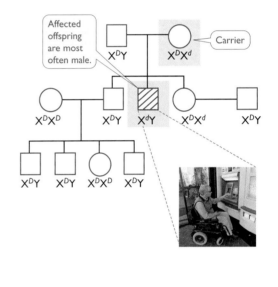

FIGURE 20.9 Pedigrees showing different modes of inheritance. (a) Polydactyly is a dominantly inherited trait. People with this condition have extra fingers and/or toes. (b) Having attached earlobes is a recessively inherited trait. (c) Muscular dystrophy is inherited as an X-linked recessive trait.

Stop and Stretch 20.4

Pedigree analysis can help determine if a family has undergone inbreeding. Inbreeding occurs when relatives mate with each other. Consider the case of a family in which one grandparent carries a rare recessive allele. Diagram a pedigree showing a mating between first cousins and explain why this mating would be more likely to result in a child with two copies of this rare recessive allele than a mating between unrelated individuals.

FIGURE 20.10 **Origin and inheritance of the hemophilia allele.** This abbreviated pedigree shows the origin of the hemophilia allele and its inheritance among the tsarina's family. It appears that the tsarina's great-grandmother underwent a mutation that she passed on to her daughter, who then passed it on to three of her nine children—one of whom was the tsarina's mother, Alice.

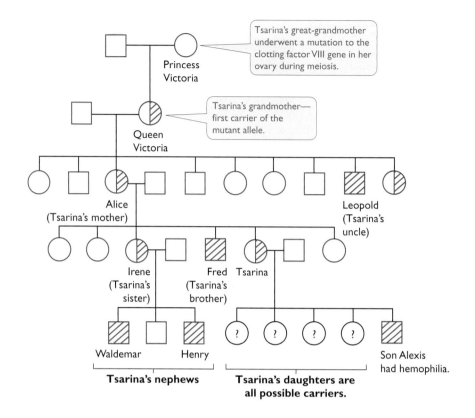

20.5 DNA Fingerprinting

Limits on the power of conventional genetic techniques, such as blood typing and karyotyping, to identify the bones found in the Ekaterinburg grave necessitated the use of more sophisticated techniques. To do so, scientists took advantage of the fact that any two individuals who are not identical twins have small differences in the sequences of nucleotides that make up their DNA. To test the hypothesis that the bones buried in the Ekaterinburg grave belonged to the Romanov family, the scientists had to answer the following questions:

1. Which of the bones from the pile are actually different bones from the same individuals?
2. Which of the adult bones could have been from the Romanovs, and which bones could have belonged to their servants?
3. Are these bones actually from the Romanovs, not some other related set of individuals?

All of these questions were answered using **DNA fingerprinting**. This technique allows unambiguous identification of people in the same manner that traditional fingerprinting has been used in the past. In addition to establishing whether two individuals are related, DNA fingerprinting can also help convict criminals when, for example, DNA evidence is left behind at the scene of a crime. Courts around the world accept DNA evidence in murder and rape cases.

DNA evidence can also be used to exonerate the innocent. The Innocence Project is a nonprofit legal clinic at the Benjamin Cardozo School of Law in New York City that attempts to help prisoners whose claims of innocence can be verified by DNA testing because biological evidence from their cases still exists. Close to 200 inmates have been exonerated since the project started in 1992. Thousands of inmates await the opportunity for testing. Because the Innocence Project has shown that the criminal justice system convicts innocent people, the project has also organized the Innocence Network,

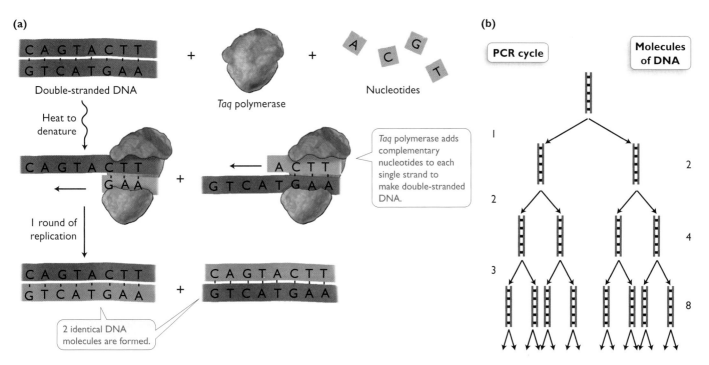

FIGURE 20.11 The polymerase chain reaction (PCR). (a) PCR is used to make copies of DNA. During a PCR reaction, the DNA to be amplified is added to a small test tube, along with the building-block nucleotides required for synthesizing more DNA and the enzyme *Taq* polymerase. The DNA is heated to separate, or denature, the two strands. The *Taq* polymerase uses the single strands as a template for the synthesis of the complementary strand, producing more double-stranded DNA. The new daughter DNA molecules are then heated again, and the cycle repeats itself. (b) Each round of PCR doubles the number of DNA molecules. This type of exponential growth can yield millions of copies of DNA for scientists to work with.

a group of law schools, journalism schools, and lawyers across the country to try to help inmates whose cases do not have biological evidence.

The United States Army collects DNA samples from all enlisted personnel to ease identification of those who are killed during war. American soldiers now provide samples of DNA as a backup for the metal dog tags worn in combat.

To begin the DNA fingerprinting process, it is necessary to isolate the DNA. Scientists can isolate DNA from blood, semen, vaginal fluids, a hair root, skin, and even (as was the case in Ekaterinburg) degraded skeletal remains. When very small amounts of DNA are available, scientists can make many copies of the DNA by first performing a DNA synthesizing reaction.

Copying DNA Through Polymerase Chain Reaction

The **polymerase chain reaction (PCR)** is used to amplify the amount of DNA (FIGURE 20.11a). To perform PCR, scientists place the double-stranded DNA to be amplified, along with the individual building-block subunits of DNA—the nitrogenous bases adenine (A), cytosine (C), guanine (G), and thymine (T)—in a small test tube. Next, an enzyme called ***Taq* polymerase** is added to the tube containing the DNA. This enzyme was given the first part of its name (*Taq*) because it was first isolated from the single-celled organism *Thermus aquaticus*, which lives in hydrothermal vents and can withstand very high temperatures. The second part of the enzyme's name (polymerase) describes its synthesizing activity—it acts as a DNA polymerase. DNA polymerases use one strand of DNA as a template for the synthesis of a daughter strand that carries complementary nitrogenous base (A:T base pairs are complementary, as are G:C base pairs).

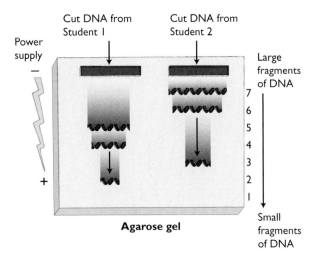

FIGURE 20.12 Gel electrophoresis. DNA from two different individuals is cut with restriction enzymes and loaded on an agarose gel. When these fragments are subjected to an electric current, shorter fragments will migrate through the gel faster than larger fragments.

The main difference between human DNA polymerase and *Taq* polymerase is that the *Taq* polymerase resists extremely high temperatures, temperatures at which human DNA polymerase would be inactivated. The heat resistance of *Taq* polymerase thus allows PCR reactions to be run at very high temperatures. High temperatures are necessary because the DNA molecule being amplified must first be **denatured**, or split up the middle of the double helix, to produce single strands. After heating, the DNA solution is allowed to cool, and the *Taq* polymerase adds complementary nucleotides to the single strands of the DNA molecule, producing double-stranded DNA molecules. This cycle of heating and cooling is repeated many times, with each round of PCR doubling the amount of double-stranded DNA present in the test tube (FIGURE 20.11b).

Once scientists have produced enough DNA by PCR, they can treat the DNA with enzymes that cleave, or cut, the DNA at specific nucleotide sequences. These enzymes are the *restriction enzymes* you learned about in Chapter 4. They act like highly specific molecular scissors and cut DNA only at specific nucleotide sequences. Because each individual has distinct nucleotide sequences, cutting different people's DNA with the same enzymes produces fragments of different sizes.

It is impossible to look at a test tube with digested DNA and determine the size of fragments, so techniques that allow for the separation and visualization of the DNA are required.

Size-Based Separation Through Gel Electrophoresis

The fragments of DNA generated by restriction enzyme cleavage can be separated from each other by allowing the fragments to migrate through a solid support called an **agarose gel**, which resembles a thin slab of gelatin. When an electric current is applied, the gel impedes the progress of the larger DNA fragments more than it does the smaller ones. This size-based separation of molecules when an electric current is applied to a gel is a technique called **gel electrophoresis** (FIGURE 20.12). Once separated by size, the DNA fragments undergo some further treatments and staining so they can be visualized.

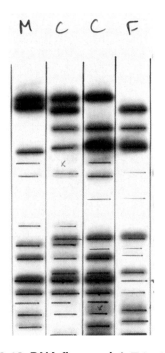

FIGURE 20.13 DNA fingerprint. This photograph of a DNA fingerprint shows a mother (M), a father (F), and their children (C). Note that every band present in a child must also be present in one of the parents.

Stop and Stretch 20.5

When technicians perform DNA fingerprinting analysis, they must wear gloves and be very careful not to contaminate samples with their own DNA or to cross-contaminate samples from two different suspects. How might the legal system ensure that no contaminants or errors have tainted the DNA fingerprinting process?

In 1992, a team of Russian and English scientists used DNA fingerprinting to determine which of the bones discovered at Ekaterinburg belonged to the same skeleton. Their fingerprinting analysis confirmed that the pile of decomposed bones in the Ekaterinburg grave belonged to nine different individuals.

Once scientists had established that the bones from nine different people were buried in the grave, they tried to determine which bones might belong to the adult Romanovs and which belonged to the servants. For the answer to these questions, scientists took advantage of the fact that Romanov family members would have more DNA sequences in common with each other than they would with the servants. This is because the tsar and tsarina each passed half of their DNA on to each of their children via the process of meiosis. FIGURE 20.13 shows an actual DNA fingerprint of two parents and their two children. Note that each band that is present in a child must also be present in one of the parents.

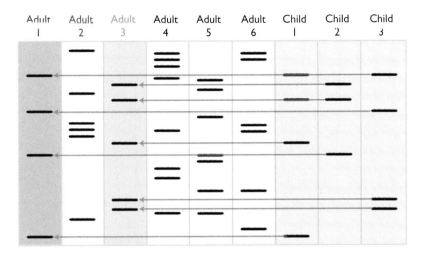

FIGURE 20.14 Hypothetical fingerprint of adult and child skeletons. Shown is a hypothetical DNA fingerprint made from the bone cells of individuals found in the Ekaterinburg grave. From the results of this fingerprint, it is evident that children 1, 2, and 3 are the offspring of adults 1 and 3. Note that each band from each child has a corresponding band in either adult 1 or adult 3. The remaining DNA from the other adults does not match any of the children, so these adults are not the parents of any of these children.

By comparing DNA fingerprints made from the smaller skeletons, scientists were able to determine which of the six adult skeletons could have been the tsar and tsarina. FIGURE 20.14 shows a hypothetical DNA fingerprint that illustrates how the banding patterns produced can be used to determine which of the bones belonged to the parents of the smaller skeletons and which bones may have belonged to the unrelated servants.

The DNA fingerprint evidence helped scientists determine that the two missing skeletons could have belonged to Alexis and one daughter. Thus, DNA evidence put to rest claims made by many pretenders to the throne. Many people from all over the world had alleged that they were either a Romanov who had escaped execution or a descendant of an escapee.

The most compelling of these claims was made by a young woman who was rescued from a canal in Berlin, Germany, two years after the murders. This young woman suffered from amnesia and was cared for in a mental hospital, where the staff named her Anna Anderson (FIGURE 20.15). She later came to believe that she was Anastasia Romanov, a claim she made until her death in 1984. The 1956 Hollywood film *Anastasia*, starring Ingrid Bergman, made Anna Anderson's claim seem plausible. A more recent animated version of the story of an escaped princess, also titled *Anastasia*, convinced many young viewers that Anna Anderson was indeed the Romanov heiress.

Because the sex-typing analysis showed only that one daughter was missing from the grave, but not which daughter, scientists again looked to the fingerprinting data. DNA fingerprinting had been done in the early 1990s on intestinal tissue removed during a surgery performed before Anna Anderson's death. The analysis showed that Anna was not related to anyone buried in the Ekaterinburg grave. She could not be Anastasia.

Thus far, scientists had answered two of the questions posed at the beginning of section 20.5. They had determined (1) that nine different individuals were buried in the Ekaterinburg grave; (2) that two of the adult skeletons were the parents of the three children. The last question was still unanswered. How would the scientists show that these were bones from the Romanov family and not some other set of related individuals?

To answer this question, the scientists turned to a living relative of the Romanovs. DNA testing was performed on England's Prince Philip, who is a grandnephew of

FIGURE 20.15 Anna Anderson. This woman claimed she was Anastasia Romanov.

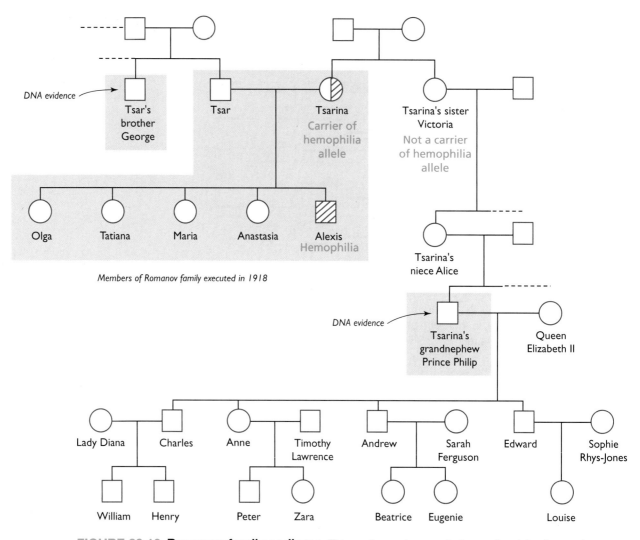

FIGURE 20.16 **Romanov family pedigree.** This pedigree shows only the pertinent family members. DNA from the tsar's brother George showed that he was related to the tsar. Note that Prince Philip is the tsarina's grandnephew. Prince Philip is married to Queen Elizabeth II. Together they had four children, Charles, Anne, Andrew, and Edward, the current British royal family. The tsarina's sister Victoria does not appear to have been a carrier of the hemophilia allele because none of her descendants have been affected by the disease.

Tsarina Alexandra. In addition, Nicholas II's dead brother George was exhumed, and his DNA was tested. FIGURE 20.16 shows the Romanov family pedigree, so that you can see how these individuals are related to each other. The DNA testing performed on these individuals showed that George was genetically related to the adult male skeleton that was related to the children's skeletons, and that Prince Philip was genetically related to the adult female skeleton shown to be related to the children's skeletons. This evidence strongly supported the hypothesis that the adult skeletons were indeed those of the tsar and tsarina. The process of elimination suggests that the remaining four skeletons had to be the servants.

Having shown that the two adult skeletons belonged to the parents of the three smaller skeletons and that they were genetically related to known Romanov relatives, scientists were convinced that the bones found in the Ekaterinburg grave were those of the tsar, tsarina, three of their five children, and their servants. Table 20.2 summarizes how scientists used the scientific method to test the hypothesis that the remains were indeed those of the Romanovs.

TABLE 20.2 | The Scientific Method

A summary of tests and the conclusions that were drawn from them:

Hypothesis: The bones found in the Ekaterinburg grave belonged to the Romanov family and their servants.

Test	Description of Results
Analyze teeth	Expensive dental work was typically seen only in royalty.
Measure skeletons	The skeletons are those of 6 adults and 3 children.
Sex typing	The children in the grave are all female.
DNA fingerprinting	Children in grave are related to two adults in grave.
DNA fingerprinting	Claims to be one of the missing Romanov children or their descendants are disproved.
DNA fingerprinting	The two adults related to the children are related to known Romanov relatives.

Conclusion: When you look at each result individually, the evidence is less compelling than when you look at all the evidence together. As a whole, the evidence strongly supports the hypothesis that it was indeed the Romanovs who were buried in the Ekaterinburg grave.

FIGURE 20.17 A funeral for the Romanovs. In 1998, the remains of the Romanov family found in Ekaterinburg were laid to rest with state honors in the St. Catherine Chapel in the St. Peter and Paul Fortress in St. Petersburg, where all other Russian emperors since Peter the Great lie.

In 1998, 80 years after their execution, the Romanov family were finally put to rest (FIGURE 20.17). The people of postcommunist Russia have now symbolically laid to rest this part of their country's political history.

Chapter REVIEW

ROOTS TO REMEMBER

The following roots of words come mainly from Latin and Greek and will help you to decipher terms:

forensic comes from a Latin word, *forum*, which was the place where legal disputes and lawsuits were heard.

pleio- and **poly-** both mean many.

KEY TERMS

20.1 Extensions of Mendelism
codominance *p. 490*
hemophilia *p. 491*
incomplete dominance *p. 489*
multiple allelism *p. 490*
pleiotropy *p. 491*

20.2 Dihybrid Crosses
phenotypic ratio *p. 491*

20.3 Sex Determination and Sex Linkage
autosome *p. 494*
carrier *p. 496*
sex chromosome *p. 494*
sex determination *p. 494*
sex-linked gene *p. 494*
X-linked gene *p. 494*
Y-linked gene *p. 497*

20.4 Pedigrees
pedigree *p. 498*

20.5 DNA Fingerprinting
agarose gel *p. 502*
denatured *p. 502*
DNA fingerprinting *p. 500*
gel electrophoresis *p. 502*
polymerase chain reaction (PCR) *p. 501*
Taq polymerase *p. 501*

SUMMARY

20.1 Extensions of Mendelism

- Some traits are not inherited in the straightforward manner described by Mendel (p. 489).

- Polygenic inheritance occurs when many genes control one trait (p. 489).

- Incomplete dominance is an extension of Mendelian genetics whereby the phenotype of the progeny is intermediate to that of both parents (p. 489; Figure 20.1).

- Codominance occurs when both alleles of a given gene are expressed (p. 490).

- Genes that have more than two alleles segregating in a population are said to have multiple alleles (p. 490).

- The ABO blood-group system displays multiple allelism (alleles I^A, I^B, and i), and also codominance since both I^A and I^B are expressed in the heterozygote (p. 490).

- Hemophilia is a genetic blood-clotting disorder that illustrates another extension of Mendelian genetics: pleiotropy, in which a single gene leads to multiple effects (p. 491).

20.2 Dihybrid Crosses

- Dihybrids are heterozygous for two traits. It is possible to predict the outcome of a dihybrid cross by using a Punnett square (pp. 491–492).

20.3 Sex Determination and Sex Linkage

- Males have an X and a Y chromosome and can produce gametes containing either sex chromosome. Females have two X chromosomes and always produce gametes containing an X chromosome. When an X-bearing sperm fertilizes an egg cell, a female baby will result. When a Y-bearing sperm fertilizes an egg cell, a male baby will result (p. 494; Figure 20.5).

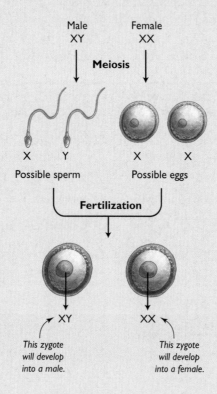

- Meiosis separates both the autosomes and the sex chromosomes from each other. Sperm and egg cells contain one member of each autosome pair and one sex chromosome (p. 494).

- Genes linked to the X and Y chromosomes show characteristic patterns of inheritance. Males need only one recessive X-linked allele to display the associated phenotype. Females can be carriers of an X-linked recessive allele and may pass an X-linked disease on to their sons (pp. 494–496).

- Y-linked genes are passed from fathers to sons (p. 497).

20.4 Pedigrees

- Pedigrees are charts that scientists use to study the transmission of genetic traits among related individuals (pp. 498–499; Figure 20.8).

20.5 DNA Fingerprinting

- DNA fingerprinting is a technique used to connect individuals to DNA evidence, and to show the relatedness of individuals based on similarities in their DNA sequences (p. 500).

- The polymerase chain reaction (PCR) utilizes a special temperature-resistant polymerase called *Taq* polymerase to make millions of copies of a DNA sequence (p. 501).

- Length differences in DNA fragments, generated by restriction digestion of DNA, are characteristic of a given individual (p. 502).

- DNA samples in an agarose gel subjected to an electric current will separate according to their size (p. 502; Figure 20.13).

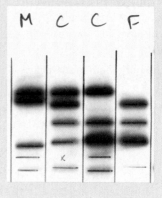

LEARNING THE BASICS

1. Flower color in snapdragons is inherited such that crosses between homozygous red flowers and homozygous white flowers produce pink-flowered offspring. This pattern of inheritance is called _____.
   a. simple dominance
   b. simple recessivity
   c. codominance
   d. sex linkage
   e. incomplete dominance

2. Hair texture is inherited in an incompletely dominant manner, and heterozygotes have wavy hair. If a man and a woman both with wavy hair have children, what types of hair textures would be expected in their children, and in what proportions?
   a. all wavy haired
   b. all curly haired
   c. 1/2 curly haired and 1/2 straight haired
   d. 3/4 curly haired and 1/4 straight haired
   e. 1/4 curly haired, 1/2 wavy haired, and 1/4 straight haired

3. If a man with blood type A and a woman with blood type B have a child with type O blood, what are the genotypes of each parent?
   a. The man is $I^A I^A$ and the woman is $I^B I^B$.
   b. The man is $I^A i$ and the woman is $I^B I^B$.
   c. The man is $I^A i$ and the woman is $I^B i$.
   d. The man is $I^A I^B$ and the woman is $I^B I^B$.
   e. The man is $I^A I^B$ and the woman is $I^B i$.

4. A man with type A blood whose father had type O blood and a woman with type AB blood could produce children with what blood types?
   a. type A only
   b. types A and B only
   c. types A and AB
   d. types A, AB, and B only
   e. types A, AB, B, and O

5. A man who is A⁻, whose father was type O⁺, mates with a woman who is AB⁺. The woman's mother was type B⁻. What proportion of their children would be type A⁻?
   a. 0   b. 1/8   c. 2/8   d. 3/8   e. 5/8

6. What proportion of their children would be AB⁺?
   a. 0   b. 1/8   c. 2/8   d. 3/8   e. 5/8

7. In Mendel's peas, round is dominant to wrinkled, and yellow is dominant to green. What phenotypic ratio would be produced by the cross $RrYy \times RrYy$?
   a. 9:3:3:1   b. 12:3:1   c. 12:4   d. 1:1:1:1

8. In Mendel's peas, round is dominant to wrinkled and yellow is dominant to green. What phenotypic ratio would be produced by the cross $RrYy \times rryy$?
   a. 9:3:3:1   b. 12:3:1   c. 12:4   d. 1:1:1:1

9. Which of the following is a true statement about sex determination in humans?
   a. The ratio of X to Y chromosomes determines the sex of the offspring.
   b. The number of X chromosomes determines the sex of the offspring.
   c. The presence or absence of a Y chromosome determines the sex of the offspring.
   d. The environment in the uterus during development determines the sex of the offspring.

10. Color blindness is an X-linked recessive trait. A color-blind female mates with a non-color-blind male. What percentage of their male offspring will be color blind?
    a. 0%   b. 25%   c. 50%   d. 75%   e. 100%

11. A woman is a carrier of the X-linked recessive color-blindness gene. She mates with a man with normal color vision. Which of the following is a true statement about any offspring they may produce together?
    a. All males will be normal.
    b. All females will be normal and all males will be color blind.
    c. All females will be color blind and all males will be normal.
    d. All offspring will be color blind.
    e. Half of males will be color blind.

12. Which of the following statements would be true of a Y-linked allele?
    a. All children of an affected father would be affected.
    b. All sons of an affected father would be affected.
    c. Daughters of an unaffected father could be affected.
    d. Sons of an unaffected father could be affected.
    e. All of the above are true.

13. A person with Down syndrome (trisomy 21) would most likely have how many chromosomes in his or her somatic cells?
    a. 23   b. 24   c. 25   d. 46   e. 47

14. A gamete affected by nondisjunction could _____.
    a. have a change in the number of chromosomes
    b. have an extra chromosome
    c. have a chromosome missing
    d. give rise to an individual at risk for genetic disorders if it is involved in fertilization
    e. all of the above

15. The pedigree in FIGURE 20.18 illustrates the inheritance of hemophilia (a sex-linked recessive trait) in the royal family. What is the genotype of individual II-5 (Alexis)?
    a. $X^H X^H$
    b. $X^H X^h$
    c. $X^h X^h$
    d. $X^H Y$
    e. $X^h Y$

Pedigree

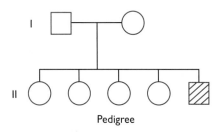

FIGURE 20.18 Pedigree.

16. Which of the following statements is consistent with the DNA fingerprint shown in FIGURE 20.19?
    a. B is the child of A and C.
    b. C is the child of A and B.
    c. D is the child of B and C.
    d. A is the child of B and C.
    e. A is the child of C and D.

DNA fingerprint

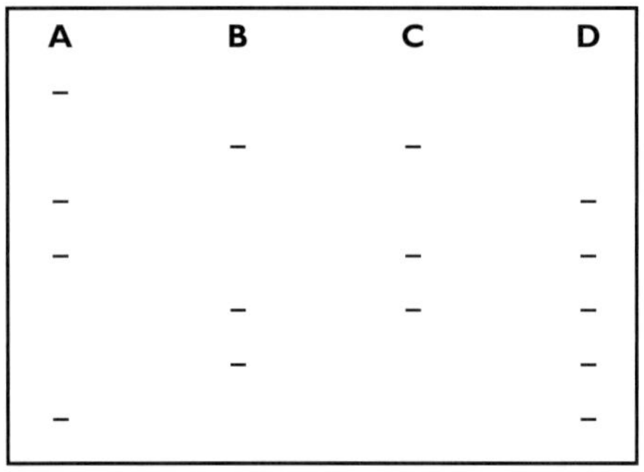

FIGURE 20.19 DNA fingerprinting analysis.

17. **True or False?** Two siblings should have more similar DNA fingerprints than two people who are unrelated.

18. **True or False?** A karyotype consists of unduplicated chromosomes arranged in homologous pairs.

19. **True or False?** The sex chromosomes are only found in the cells that make up the gonads.

20. **True or False?** All the cells in a female's body (except her egg cells) have one active and one inactive X chromosome.

ANALYZING AND APPLYING THE BASICS

Questions 1 and 2 refer to the following information.

The probability of any two individuals (except for identical twins) sharing the same DNA fingerprint approaches zero. This idea is especially important to understand when DNA fingerprinting is used in court cases. Juries must recognize that the chance of another person coincidentally having the same DNA fingerprint as the accused is basically zero.

If DNA evidence has been found at the scene of the crime, the probability of another person having the same fingerprint can be calculated by multiplying together the frequency with which each visible band occurs in the general population.

Assume that bands produced in a DNA fingerprint occur in the general population with the following frequencies: band 1 = 2%; band 2 = 1%; band 3 = 5%; band 4 = 1.5%; band 5 = 1.2%; and band 6 = 3%.

1. What is the probability that this combination of bands would be found in two different individuals in the general population?

2. Compare the probability you calculated with the population of the world (around 6 billion). What does that tell us about the likelihood of two different people having the same DNA fingerprint?

Use the following paragraph to answer questions 3–5.

In 1997, Dr. Eugene Foster began investigating the rumor that Thomas Jefferson had been involved with one of his slaves, a woman named Sally Hemings, and that he had fathered her children. Dr. Foster analyzed blood samples from five male descendants of Jefferson's paternal uncle, to determine the DNA fingerprint of the Jefferson Y chromosome. He also analyzed blood samples from descendants of Thomas Jefferson's nephews, Peter and Samuel Carr. One of these men had been reputed to be the father of Sally Hemings's children. Dr. Foster compared the Jefferson and Carr DNA fingerprints to the Y chromosomes of male descendants of Sally Hemings. A match was found between the Jefferson Y chromosome and the Y chromosome from a descendant of her son, Eston Hemings. The Carr Y chromosome was not a match. Dr. Foster concluded that Jefferson was indeed the father of Eston Hemings.

3. Why would the DNA from the Y chromosome be useful for determining paternity over many generations?

4. What can you conclude about the descendants of Eston Hemings from these findings?

5. Why was the Y chromosome used in these analyses?

Questions 6 and 7 refer to the following information.

Börjeson-Forssman-Lehman syndrome (BFLS), first described in 1962, has been linked to a mutation on the *PHF6* gene. BFLS leads to developmental delay and mild to moderate mental handicap. Physical symptoms include large earlobes, shortened toes, coarsened facial features, small genitalia, obesity, and breast enlargement. This syndrome mainly affects males. Affected females may have some learning problems and reduced physical symptoms.

6. How is BFLS an example of pleiotropy?

7. Construct a pedigree that shows how BFLS could affect a boy who is four generations removed from the original mutation.

CONNECTING THE SCIENCE

1. People who are convicted on the basis of false identification and later released due to DNA evidence have sued court systems and won awards in the millions. Do you think people who spend time in jail for crimes they did not commit are entitled to financial rewards from the criminal justice system?

2. Forensic scientists work in the criminal justice system. A person with a criminal record will almost never be hired for such a position. Do you think that a mistake that a person makes while they are young (driving under the influence, for instance) should prevent them from getting a job later in life?

3. The state of Illinois acknowledges, on the basis of work done at the Center for Wrongful Convictions, that it has put at least 11 innocent men to death. For that reason, former Illinois Governor George Ryan decided to suspend executions. Should states be allowed to execute people even though investigations are uncovering such false convictions?

Chapter 21

Development and Aging

The Promise and Perils of Stem Cells

LEARNING GOALS

1. Describe the differences between adult, fetal, and embryonic stem cells.
2. Describe the process of fertilization in detail.
3. Summarize the development of a preembryo from zygote through blastocyst.
4. Illustrate the structure of the gastrula, and compare and contrast the eventual fate of its three layers.
5. Describe the early stages of organogenesis of the central nervous system.
6. Explain the role of the SRY gene in the development of sex-specific organs.
7. Describe the function of the placenta.
8. Compare and contrast fetal circulation and adult circulation.
9. Describe the three stages of labor.
10. Describe the changes in males and females during puberty.
11. Describe two hypotheses for the causes of aging, and provide evidence for each.
12. Describe some of the effects of aging on organ systems.

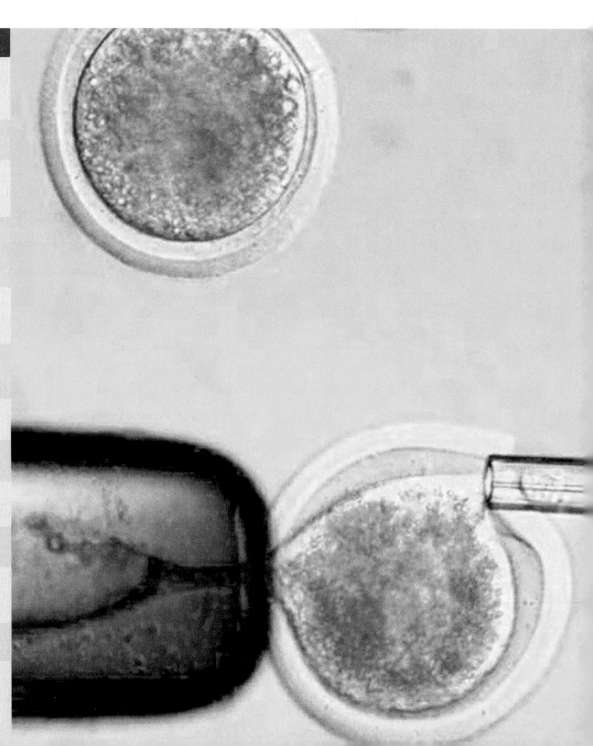

Embryonic stem cells hold the promise of cures for a number of diseases.

In June 2006, President George W. Bush broke one of the longest streaks in modern presidental history—the most number of days into a presidency before issuing a veto of a bill. The bill did not concern taxes, Social Security, defense spending, or terrorism. Instead, it addressed science policy. The Stem Cell Research Enhancement Act that he rejected had been passed by Congress a few days before. This law directed the federal government to provide funding for research that utilizes embryonic stem cells collected anytime after August 9, 2001. On that day, the President had signed an executive order allowing federal funding for embryonic stem cell research, but only using those cell lines that had been established by then. Since he signed this order, it has become clear that his directive has greatly hampered research on embryonic stem cells in federally funded laboratories, reducing the rate at which an understanding of these cells is progressing.

President Bush opposed research on these cells for moral reasons.

The president's veto set off a storm of criticism from Democrats and Republicans alike, who see in stem cells the possiblity of cures for a number of serious conditions. Stem cell research has also attracted a number of famous advocates, including the late Christopher Reeve, the actor paralyzed in a horse-riding accident; Nancy Reagan, whose husband, former President Ronald Reagan, died of Alzheimer's disease; and the actor Michael J. Fox, who has Parkinson's disease. Opinion polls taken around the time of the veto indicated that 70% of Americans support funding embryonic stem cell research.

Michael J. Fox and others think it is immoral to prevent this research.

Why would the president choose a popular bill funding possibly lifesaving research as his first veto? In large part, he vetoed the bill because he believes that embryonic stem cell research is morally wrong. "If this bill were to become law, American taxpayers for the first time in our history would be compelled to fund the deliberate destruction of human embryos," he wrote in his veto statement. "Crossing this

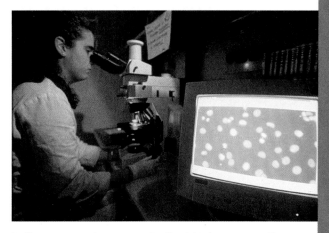

Is there a way to answer both sides' concerns?

line... would be a grave mistake and would needlessly encourage a conflict between science and ethics that can only do damage to both and harm our nation as a whole." The president went on to assert that research on embryonic stem cells is unnecessary and new scientific advances will soon render their research advantages obsolete.

Advocates of stem cell research, such as Michael J. Fox, make a moral argument as well. Fox was diagnosed at age 31 with Parkinson's disease, a degenerative disease of the brain that causes uncontrollable tremors, eventually leading to severe disability. As Fox wrote in a letter to the president before his veto, "Embryonic stem cell research transforms embryos already marked for destruction into potentially life-saving research. I can think of no better affirmation of the culture of life. I am asking that you, with all compassion, stand up for what is right not only for America's continued leadership in health, science, and medicine, but for what is right for the future of the [millions of] American families touched by debilitating diseases."

In the middle of this debate are hundreds of American scientists who wish to explore the potential of stem cells but are limited by the funding restrictions currently in place. What can science tell us about the potential of stem cell research? Is a solution to this moral dilemma to be found in the laboratory?

21.1 The Production of Embryonic Stem Cells

Unlike most cells in the body, a **stem cell** is able to give rise to many different kinds of cells and tissues. Because every cell in an individual begins life with the same genetic information, one might expect that all cells are **totipotent**, that is, having the capability to become any other type of cell. However, during the process of **cell differentiation**, in which each cell obtains its specific function, most become fixed as one type of cell or another.

Embryonic stem cells retain their flexibility and, with the correct prodding, can develop into any of the more than 200 cell types in the human body. These cells are called **pluripotent**, to distinguish them from cells that can also become part of the embryonic membranes described below. As a result of their pluripotency, they are of great interest to researchers in the field of **regenerative medicine**, which seeks to replace damaged organs and tissues with fresh, new material. Imagine that you are remodeling an old home, and you have a type of material that you can mold into anything you might need for the remodeling job—brick, tile, pipe, plaster, and so forth. Having a supply of this material would help you fix many different kinds of damage. Scientists believe that pluripotent stem cells may serve as this type of all-purpose repair material in the body.

There are three major types of stem cells: adult, fetal, and embryonic. Individuals produce adult stem cells after birth and into adulthood. These types of cells are **multipotent**, meaning that they are typically partially differentiated and able to produce daughter cells of only one or a few types. For example, stem cells in the red bone marrow can produce a range of red and white blood cells. Bone marrow transplants in leukemia patients whose own marrow has been destroyed by chemotherapy take advantage of these stem cells to help patients reestablish their red and white blood cells.

Adult stem cells hold some promise in regenerative medicine, but they have other important limitations in addition to being relatively inflexible. Adult stem cells are uncommon and difficult to find in the body, and they can produce only a limited number of daughter cells, making it hard to grow them in large batches.

Fetal and embryonic stem cells are produced during the stages of development before birth. In humans, the term **embryo** is used to describe the stage of development from the second week after fertilization to about the ninth week. (The stage between fertilization and embryo formation is referred to as the **preembryo** stage.) From the end of the ninth week until birth, the developing human being is referred to as a **fetus**. **Embryonic stem cells** are produced during the first weeks of development, and **fetal stem cells** are produced during the fetal stage. Both of these stem cells can divide indefinitely, are easy to obtain, and give rise to a larger number of cell types.

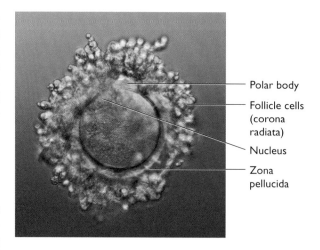

FIGURE 21.1 **A human egg cell before fertilization.** This egg cell displays all the characteristics of a healthy egg that can be fertilized successfully. The presence of the polar body indicates that the egg has completed meiosis I.

Fertilization: Forming the Ultimate Stem Cell

The single cell that is the product of fusion of sperm and egg represents the ultimate stem cell—every cell in the human body (as well as cells that make up some of the supporting membranes during development) derives from this single cell. The embryonic stem cells most interesting to researchers are usually isolated from preembryos several days after fertilization.

The source of embryonic stem cells used by researchers is preembryos that are left over after fertility treatments. **In vitro fertilization (IVF)** is a common fertility treatment that generates embryonic stem cells. The first successful IVF procedure took place in 1977. Since then, millions of "test-tube babies" have been born around the world.

IVF begins with the "harvesting" of eggs from a woman's ovaries after drug therapy to stimulate excess follicle development and the ovulation of multiple eggs. Eggs are harvested from the ovaries by drawing them into a needle inserted through the wall of the vagina. An embryologist evaluates these eggs and removes those that have not reached metaphase II of meiosis, since these eggs are too immature to complete fertilization (FIGURE 21.1).

In the next step of in vitro fertilization, eggs at metaphase II are incubated with prepared sperm in a glass dish for 48 hours. Sperm preparation primarily consists of the process of **capacitation**, which weakens the plasma membrane surrounding the acrosome at the tip of the sperm cell. In natural human reproduction, capacitation of sperm occurs in the female reproductive tract. In IVF, sperm are capacitated by incubation in a medium that chemically mimics this environment.

Fertilization, the fusion of sperm and egg, is nearly assured in vitro, unlike the situation in natural reproduction. While roughly 200 million sperm are released in each ejaculation, only a few thousand reach the site of fertilization in the oviduct. The number of healthy, ready sperm in a petri dish is much greater, because the sperm in this environment do not face the perils that sperm in the female reproductive tract do— high acidity in the vagina (which works to help prevent bacterial infections in women but also can kill sperm) and detours that trap sperm in the folds of the cervix and vagina, the uterine wall, and in the oviduct that does not contain an egg.

In natural reproduction, prostaglandin released in semen triggers muscular contractions of the uterus, drawing the sperm toward the upper portion of the oviduct. If an egg is present there, fertilization can take place approximately 6 hours after intercourse, once capacitation is complete. Eggs are only capable of being fertilized within 24 hours of their release from the ovary, and fertilization is most likely within the first 12 hours. However, since most sperm live for 24 hours and some can live for much longer, fertilization in natural reproduction is possible if intercourse takes place up to five days before ovulation and until one day after.

Once sperm and egg come in contact, the process of fertilization is the same whether it occurs within the oviduct or in a glass dish (FIGURE 21.2). First, the beating of their flagella force sperm cells through the mass of follicle cells that surround and nourish the egg. The sperm must then cross the **zona pellucida**, a translucent covering on the egg cell. The zona pellucida shields the egg from mechanical damage and also acts as a species-specific barrier. Getting through this barrier requires binding of a specific protein receptor on the head of the sperm to a receptor on the zona pellucida. Only sperm produced by a male of the same species as the female that produced the egg can traverse the zona pellucida.

FIGURE 21.2 Fertilization. The steps involved in fertilization are illustrated in this figure.

VISUALIZE THIS At what stage would the process be halted if the sperm and egg were from different species?

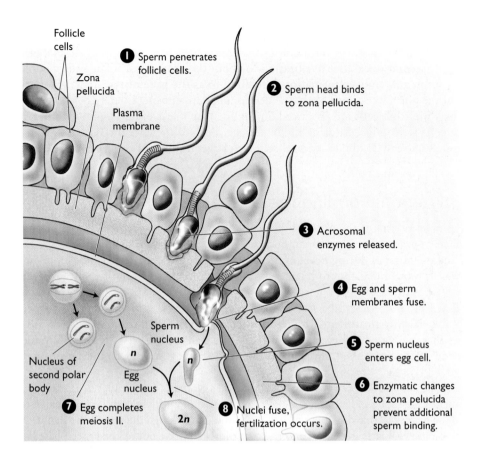

The binding of the sperm head to the zona pellucida triggers the release of enzymes from the acrosome. The enzymes digest a tunnel through the covering toward the egg cell's plasma membrane. Once a sperm cell breaks through the zona pellucida, its plasma membrane and the egg's plasma membrane fuse, causing a cascade of events.

First, the egg's plasma membrane depolarizes, resulting in a much-reduced electric gradient across it. This occurs by much the same process as in neurons during signal transmission (Chapter 14). Depolarization of the egg causes a release of enzymes that detach the zona pellucida from the egg's surface and destroy its sperm receptors, inhibiting sperm attachment. The zona pellucida does remain in place, as a sort of shell protecting the developing egg from damage, for several more days.

Second, the egg completes meiosis II, producing a single **ovum**, or mature egg, that contains the majority of the cytoplasm, and a tiny *polar body* that quickly degrades.

Finally, the sperm nucleus and egg nucleus fuse (step 8 in Figure 21.2, where $2n$ indicates the presence of pairs of chromosomes). The resulting cell is the **zygote**. The follicle cells are then shed, and the zygote undergoes its first round of cell division.

In natural human reproduction, the number of eggs available for fertilization during any ovarian cycle is typically one. In roughly 3 to 5 of every 1,000 births, though, a woman has released two or more eggs for fertilization. If two eggs are fertilized at the same time, the result is **dizygotic twins** (or "fraternal" twins). During IVF, many eggs may be fertilized—up to 10 or more—resulting in multiple zygotes.

After three to five days of test-tube development, one to four of the healthiest-looking preembryos that result from IVF are implanted into the woman's uterus. The remaining preembryos are stored so that more attempts can be made if pregnancy does not happen or if the couple desires more children. When the couple is finished with the fertility treatments, the remaining preembryos may be discarded, donated to other infertile couples, or used for stem cell research.

In both natural and assisted reproduction, the remainder of **development**, the series of events that take place after fertilization, occurs in a woman's uterus and continues after birth. Development in the uterus leads to the formation of a new multicellular organism.

Preembryonic Development

Preembryonic development consists of a series of changes that produce a multicellular embryo capable of producing adult tissues. There is no growth in size during the preembryonic period, only cell division and changes in form, or **morphogenesis**.

The first step is a series of rapid mitotic cell divisions, called **cleavage**, that begin while the zygote is still in the oviduct. The first cell division occurs about 24 hours after fertilization. The second and third happen the following day. At about this stage, connections called *gap junctions* form between adjacent cells, causing the cell mass to contract into a solid ball.

Before the cells are compacted, a preembyro may break apart, resulting in two cell masses containing identical cells. If both of the resulting preembryos develop, the result is **monozygotic twins** ("identical" twins). If a preembryo fragments but does not split completely, the result is **conjoined twins**.

Stop and Stretch 21.1

For parents known to carry a harmful genetic mutation, IVF clinicians will take a cell from an eight-cell preembryo to test it for the presence of the defective gene. Taking a single cell causes no apparent harm to the developing structures. How does this compare to the process of monozygotic twinning, and what does this tell you about the capabilities of preembryonic cells?

Another round of division results in a 16-cell mass called the *morula*. Further division of the morula over the next two days results in the production of a fluid-filled sphere called a **blastocyst**, made up of about 250 tiny cells (FIGURE 21.3).

At this point of development, the cells of the preembryo are already somewhat distinct from each other. Inside the blastocyst, a cluster of 30 to 34 cells called the **inner cell mass** has formed. Some of these cells will give rise to the body of the embryo. The 200 or so outer cells of the blastocyst are called the **trophoblast**. The trophoblast includes the early *chorion*, one of the **extraembryonic membranes**, structures formed from the preembryo but not part of the developing individual (Table 21.1). The chorion together with tissues from the mother will eventually make up the placenta, which is described in section 21.4.

Most IVF embryos are introduced into the uterus at the blastocyst stage. This stage of development is also when cells are harvested from the inner cell mass to create stem cell lines. These *cell lines* are colonies of cells that can replicate themselves, having been given the proper nutrients and hormones for growth. Cells can be removed from the colony and used for experimentation without destroying the colony itself. Harvesting embryonic stem cells from a blastocyst to begin these cell lines, however, destroys the preembryo. This is the source of opposition to stem cell research—that establishment of new stem cell lines require the destruction of potential human life.

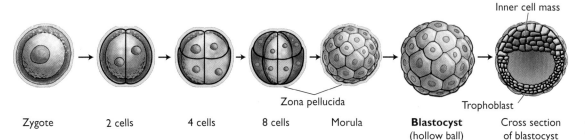

FIGURE 21.3 **Development of the early embryo.** The human zygote undergoes cleavage divisions to produce the blastocyst.

TABLE 21.1 | Extraembryonic Membranes

Membrane	Function in Human
Yolk sac	Source of early blood cells and germ cells that will become eggs or sperm. Functions only briefly, very early in development.
Amnion	Contains the amniotic fluid, which cushions the developing embryo and fetus.
Allantois	Becomes blood vessels in the umbilical cord.
Chorion	Becomes the fetal side of the placenta, the organ for exchange with the mother.

21.2 Early Embryonic Development

The blastocyst enters the uterus from the oviduct approximately one week after fertilization and begins the process of **implantation**. Scientists estimate that one-third to one-half of embryos fail to implant, most probably due to genetic abnormalities. These embryos typically pass from the body unnoticed. Failure to implant is distinct from **miscarriage**, spontaneous abortion of an embryo. While numbers are difficult to estimate, about 20% of pregnancies end in miscarriage, most resulting from genetic or developmental abnormalities of the embryo.

The rate of implantation of IVF embryos is even lower—around 15%—probably because the timing of development and the suitability of the uterus must be ideal for this process to happen. This low success rate is one reason for transferring more than one embryo. Although implantation is relatively rare for any single IVF embryo, the pregnancy rate is raised to around 30% by transferring two or more.

During implantation, the embryo breaks out of the zona pellucida and the trophoblast begins to secrete enzymes that digest part of the *endometrium*, the lining of the uterus. Over the course of several days, the embryo sinks deeper into this rich tissue, becoming almost completely engulfed (FIGURE 21.4). Occasionally, an embryo will implant instead in the walls of an oviduct, producing an **ectopic pregnancy**. Ecotopic pregnancies cannot be successful and must be surgically removed to prevent serious injury to the mother.

Implantation marks the beginning of **clinical pregnancy**, that is, the time when pregnancy can be detected, as it also coincides with the release of **human chorionic gonadotropin (HCG)** from the trophoblast. Pregnancy tests detect the presence of HCG in urine to confirm pregnancy. HCG inhibits the degradation of the corpus luteum in the ovary so that it continues to produce progesterone and estrogen. These two hormones maintain the endometrium and inhibit menstruation.

Endometrium   Embryo

FIGURE 21.4 Implantation. The embryo at implantation is almost completely engulfed by the nutrient-rich uterine lining.

VISUALIZE THIS The "abortion pill" mifepristone can be provided to women only during very early pregnancy. The pill stimulates menstruation. Why does this lead to loss of the pregnancy?

Stop and Stretch 21.2

The length of a pregnancy can be measured in various ways. The average number of weeks between the last menstrual period (LMP) and birth is 40, the average number of weeks between conception and birth is 38, and the average number of weeks between the onset of clinical pregnancy and birth is 37 weeks. Why do so many methods of "counting the weeks" exist?

Levels of progesterone and estrogen produced by the corpus luteum continue to increase through the first five weeks of pregnancy, possibly contributing to the nausea experienced by many women during this time. Although commonly called "morning sickness," this nausea can strike at any time of the day.

After implantation, two additional extraembryonic membranes form from the trophoblast: the **yolk sac**, which will eventually produce blood cells, and the **amnion**, which surrounds the amniotic cavity. The amniotic cavity is filled with **amniotic fluid**, a liquid that serves as an insulator against extreme temperatures and absorbs any shock generated by the mother's movements.

The inner cell mass detaches from the trophoblast and is now termed the **embryonic disk**. This mass of cells makes a dramatic change in conformation to produce the **gastrula**, a structure containing the three primary germ layers (FIGURE 21.5). The formation of these three germ layers is the first clear differentiation among body cells, eventually leading to the production of adult cells with specific functions.

The outer primary germ layer, or **ectoderm**, gives rise to the skin, nervous system, and sensory organs. The middle layer, or **mesoderm**, produces muscles, kidneys, heart and blood vessels, sex organs, and the skeleton. The innermost layer, or **endoderm**, lines the gastrointestinal tract and becomes the organs associated with digestion and breathing. Embryonic stem cells grown in the lab are considered pluripotent because they can be triggered to develop into each of these primary germ layer.

Cell differentiation remains one of the most mysterious processes in biology. How do individual cells "know" what to specialize in? What are the factors that cause one undifferentiated cell to become a neuron and another to become a muscle cell? Studies of embryonic development in other organisms have provided preliminary answers to these questions (FIGURE 21.6). For example, in other vertebrate animals, a cell's differentiation into ectoderm, mesoderm, or endoderm depends partly on the chemicals present in the cell's cytoplasm. These chemical signals are unequally distributed in a zygote—the particular chemical makeup of a cell in the preembryo depends on the particular portion of the zygote's cytoplasm that it received during cell division.

Once the embryonic disk forms, differentiation seems to happen in response to chemical signals that pass between different cell layers and among adjacent cells. In both early and later differentiation, the chemical signals turn on some genes in a particular cell and leave others off, leading to development of a cell's unique structural and chemical makeup. Research on human embryonic stem cells can help to shed light on the steps required to produce particular cell types in humans.

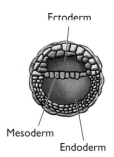

FIGURE 21.5 Gastrulation. The process of gastrulation is the first visible differentiation of the cells that will become the fetus.

21.3 Organ Formation

Cell differentiation is the first step on the path toward the formation of organs and organ systems, the process of **organogenesis**. By the end of the embryonic period, organogenesis is nearly complete, and the embryo has a distinct human appearance (Table 21.2). The embryonic period thus represents the most vulnerable period of development, when environmental exposure to toxic compounds can disrupt the structure of organs essential for normal functioning.

FIGURE 21.7 illustrates the periods of organ system development in the embryo and fetus. As you can see, one of the first systems to develop is also one that is sensitive to environmental damage for the longest period: the central nervous system.

Cell Migration and Death

Organogenesis results from not only cell differentiation, but also **cell migration**, the movement of differentiating cells from one region of the embryo or fetus to another, and **apoptosis**, programmed cell death.

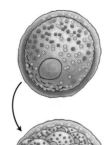

Egg
Unequal distribution of proteins, mRNAs, and other molecules in the cytoplasm determines the polarity of zygote (top versus bottom).

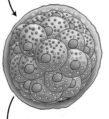

Morula
Cells in blastula do not have identical cytoplasm. Different constituents trigger transcription of different genes.

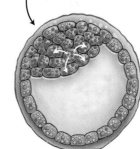

Blastocyst
Protein synthesis in each cell produces signals affecting the developmental pathway of self and neighboring cells. The concentration of various signals also affects transcription.

FIGURE 21.6 The process of cell differentiation. Although scientists still have much to learn about how stem cells become differentiated, some basics are known and are summarized here.

TABLE 21.2 | Embryonic Development

Weeks since fertilization: 2
Implantation and gastrulation (formation of three primary germ layers) occur.

Weeks since fertilization: 3
Neural tube forms.

Weeks since fertilization: 4

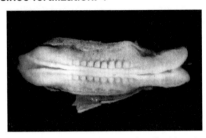

Heart begins development as a tube that begins pumping blood. Pharyngeal arches form; they will contribute to structures of face, neck, and mouth.

Weeks since fertilization: 5
Stalk connects tail of embryo with chorion. Limb buds form, and head and sense organs are more apparent.

Weeks since fertilization: 6

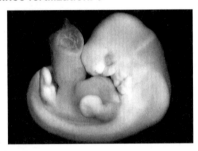

Hand plates form.

Weeks since fertilization: 7
Fingers begin to appear when cells between finger rays undergo apoptosis.

Weeks since fertilization: 8

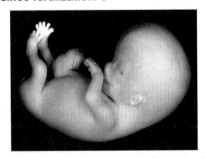

Torso begins to straighten out, bone formation is more obvious, nose takes shape, toes begin to form. Embryo is 1.5 cm (0.6 in) long and has a distinct human form.

Note: Top image (CS10, specimen 20297), middle image (CS17, specimen 4655), and bottom image (CS23, specimen 10074) courtesy of Congenital Anamoly Research Center, Kyoto University.

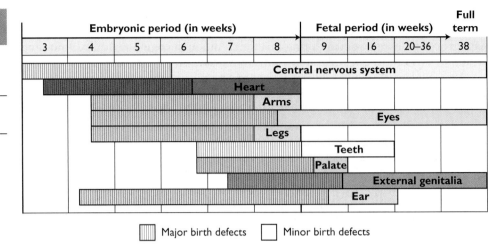

FIGURE 21.7 Organ system development. This timeline depicts when each major organ system develops and, therefore, when each is most susceptible to environmental damage resulting in major birth defects. Minor birth defects are possible outside of organogenesis, as indicated by the lighter bars.

VISUALIZE THIS During which months of development is the developing embryo or fetus at risk of birth defects in the largest number of systems?

Cell migration is facilitated by chemical signals produced by other cells along the route to, and at, the final destination of the migrating cells. The role of chemical signals in migration is apparent because certain environmental toxins can disrupt the process. For example, nicotine interferes with the migration of bone marrow stem cells from the fetal liver—likely causing the cardiovascular and immunological problems seen in the babies of heavy smokers.

Like cell migration, apoptosis is common during development but poorly understood. One example of programmed cell death is the shaping of fingers and toes from the paddle-like embryonic hands and feet, which must lose cells in a particular pattern in order to form properly. An understanding of the triggers for apoptosis may be gained by embryonic stem cell research, allowing scientists to develop therapies that selectively kill tumor cells in cancer patients.

Early Organogenesis: Development of the Nervous System

The central nervous system begins forming during the gastrula stage, when a faint line called the *primitive streak* appears on the midline of the embryonic disk. The streak develops by the third week into a thickened ridge, called the *neural plate*, which runs along the entire backside of the embryo. Below this plate is the notochord, a tube of mesoderm that marks the eventual location of the backbone. The plate then folds partially, causing a depression, and the ridges on either side of the depression fuse to form a hollow **neural tube** covered with a layer of ectoderm.

The neural tube will become the brain and spinal cord. Cells that initially connect the tube to the ectoderm either undergo apoptosis or migrate to other areas of the body, forming structures such as the peripheral nerves, and thus the neural tube separates from the other tissues of the embryo. The mesoderm on either side of the neural tube undergoes cell division and migration to form lumps called *somites*, which will eventually make up some muscles, skin, and parts of the skeleton (FIGURE 21.8).

Environmental factors can disrupt the process of organogenesis even this early in development. **Spina bifida**, a birth defect found in about 2 of every 1,000 newborns, occurs when the neural tube fails to close and separate from the ectoderm completely. As a result, part of the spinal cord may be exposed in a cyst that is outside the spine and that may emerge from the skin (see FIGURE 21.9 on the next page). The effects of spina bifida range from mild numbness to severe paralysis. The development of spina bifida and other neural tube defects is made more likely by a deficiency of the vitamin folic acid in the mother, and its incidence has declined as folic acid supplementation of flour and processed foods has increased.

As development proceeds, differential gene expression plays a larger role in organogenesis than environmental factors. The production of reproductive organs illustrates the role of genes in later organogenesis—but also illustrates how environmental conditions can still alter the course of development.

Later Organogenesis: The Reproductive Organs

Before the seventh week of development, male and female embryos are indistinguishable to the eye—both contain embryonic *gonads*, which will produce reproductive cells, and two sets of ducts that have the potential to develop into either male or female reproductive structures. Differentiation of these embryonic structures into male or female reproductive organs is determined by the expression of sex-specific genes, some of which are found on the sex chromosomes.

In both sexes, development of the gonads begins as cells migrate to a region in the abdomen of the embryo called the *gonadal ridge*. Once there, these cells swiftly make copies of themselves, yielding two **indifferent gonads**. These gonads are considered indifferent in the sense that they could become either the ovaries or testicles.

Inside the indifferent gonads are germ cells that could become either sperm or egg. There are also supporting and steroid-producing cells that become either the structural and estrogen-secreting cells of the ovary or the structural and testosterone-secreting cells of the testes.

Both sexes also contain two different sets of ducts: the **Wolffian duct**, which may develop into male structures, and the **Mullerian duct**, which may develop into female structures. Which of these structures remains and develops past seven weeks depends on the sex hormones the embryo is exposed to.

For the indifferent gonads to develop into testes and produce testosterone, the ***SRY* gene** on the Y chromosome must be present. The *SRY* gene produces a protein that stimulates the transcription of other genes. Within the gonads, the SRY protein turns on the genes required for testicular development.

Once indifferent gonads become embryonic testes, they begin to produce the male hormone testosterone. Testosterone in turn directs the development of the vas deferens, epididymis, and urethra from the Wolffian duct. The testes also produce *anti-Mullerian hormone* beginning in the eighth week of development, so that the Mullerian duct regresses, preventing development of female reproductive organs.

In the absence of the *SRY* gene, the indifferent gonads develop into ovaries. And without the apoptosis triggered by testosterone, the Mullerian duct persists and develops into the oviducts, uterus, cervix, and vagina. Lacking the stimulation of testosterone, cells of the Wolffian duct die and male reproductive structures are not formed.

As with the development of internal reproductive structures, the formation of external genitalia in embryos depends on the presence or absence of testosterone. Externally, male and female embryos are indistinguishable until about 12 weeks after fertilization, both possessing an undifferentiated **genital tubercle** and **labioscrotal swellings**. The tubercle and swellings differentiate into the penis and scrotum, respectively, in the presence of dihydroxytestosterone (DHT), a testosterone derivative. Late in development, the testes descend into the scrotum in males, where cooler temperatures facilitate sperm production. In females, the absence of testosterone triggers development of the clitoris from the genital tubercle and vulva from the labioscrotal swellings. FIGURE 21.10 shows the embryonic structures before they start to differentiate by sex, and how they develop in females and males, respectively.

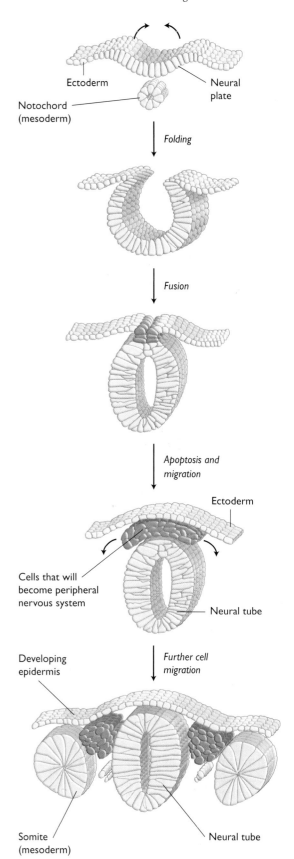

FIGURE 21.8 **Formation of the neural tube.** The neural tube, which is where the brain and spinal cord will develop, forms when a ridge on the very young embryo folds and closes.

520 CHAPTER 21 Development and Aging

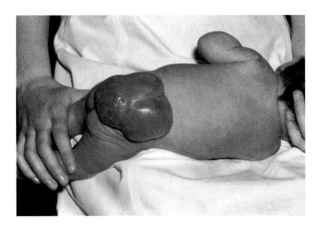

FIGURE 21.9 Neural tube defects. The cyst on this infant's back contains part of the spinal cord, which emerges from an incompletely formed neural tube.

Stop and Stretch 21.3

The X and Y chromosomes pair during metaphase I of meiosis. Recall that at this stage, crossing over may occur. What would be the ultimate result of a crossing-over event between an X and a Y chromosome that caused a copy of the *SRY* gene to be deleted from the Y chromosome and added to the X?

About 2 in every 1,000 children born in the U.S. have genitalia that are neither clearly female nor clearly male. These ambiguous genitalia vary in form and usually result from exposure to abnormal levels of sex hormone during development. Abnormal exposures typically arise as a result of a genetic condition, but increasingly, environmental exposures may be influencing development.

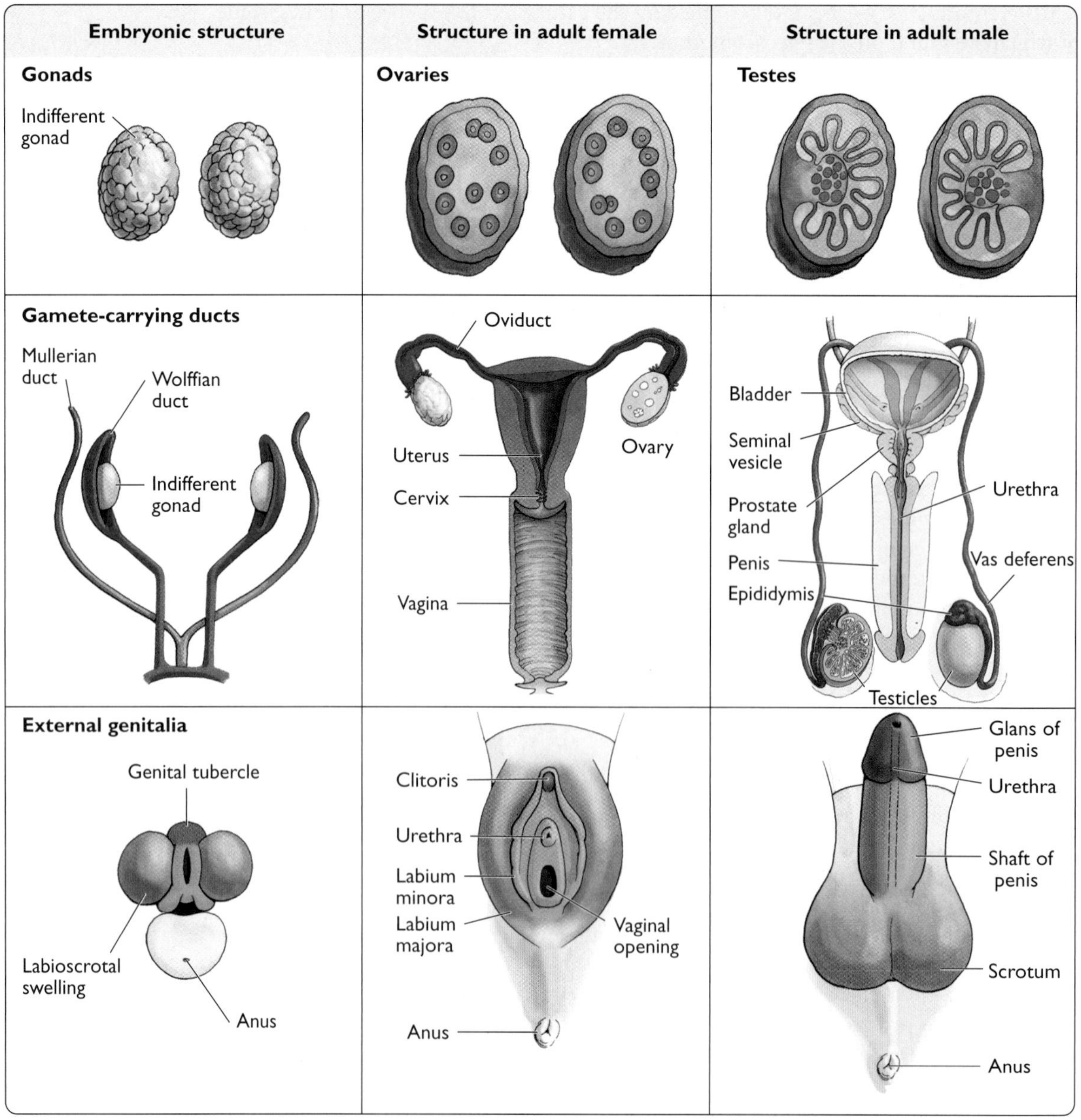

FIGURE 21.10
Determination of sex of reproductive organs. The embryo's indifferent gonads, dual duct system, and external genitalia each have two possible final forms, which depend on the embryo's chromosomal makeup. Colors identify which embryonic structures develop into the various adult structures of the ducts and external genitalia.

A mutation in the enzyme that converts testosterone into DHT can result in a mismatch between chromosomal sex and physical appearance. As a result of this mutation, DHT is not produced and male external genital structures do not fully develop, so a baby who is genetically male will appear to be a girl. However, because the Wolffian ducts respond to testosterone and not DHT, a male internal duct system develops. At puberty, testosterone causes the typical male secondary sex characteristics to arise.

Exposure to hormones or hormone-disrupting chemicals in the environment can also affect the development of reproductive structures. DES, a synthetic estrogen given to pregnant women to combat miscarriage and morning sickness from the 1940s into the 1970s, caused malformation in the uteruses and ovaries of their daughters. Lower levels of environmental hormones present in plastics, pesticides, and other chemicals also correlate to abnormalities in male reproductive organs, although a definitive link between these chemicals and developmental problems is unproven.

Major birth defects, those that occur during the embryonic period and result in malformed organ systems, appear in about 2–3% of live births. Embryonic stem cells hold some promise for treating these defects, including spina bifida and its negative effects on nervous system development. Perhaps more promising is the improved understanding of embryonic development that can come from studying embryonic stem cells. By observing how different environmental exposures or genetic mutations affect organogenesis in a petri dish, scientists may be able to develop strategies for reducing the incidence of often-devastating birth defects.

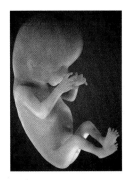

12 weeks
- Most cartilage has been replaced by bone.
- Heartbeat can be heard via stethoscope.
- Size at end of four months: 15 centimeters (6 inches) long, 170 g (6 ounces).

21.4 Fetal Development and Birth

The fetal period, from week 10 through birth at about week 38, is characterized by organ maturation and the growth of the fetus. FIGURE 21.11 illustrates the changes that occur during these weeks.

By the beginning of the fetal period, the organ that provides the physical connection between mother and offspring is fully developed—and it is exerting its influence on both.

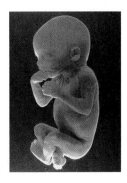

24 weeks
- Mother can feel movement.
- Skin is covered with downy hair called lanugo, and coated with greasy vernix, which protects fetus from amniotic fluid.
- Eyelids are formed, and eyes can open.
- Size at end of six months: 30 cm (12 in) long, 1.4 kg (3 lb).

The Purpose of the Placenta in Pregnancy

The **placenta** is a unique organ because it consists of both fetal and maternal tissue. Its development begins at implantation, when the trophoblast digests part of the uterine lining. Blood vessels in the endometrium are damaged by this process and leak blood, forming a pool in which the trophoblast sits. Oxygen, nutrients, and wastes are exchanged between embryo and mother via diffusion within this pool of blood. As the embryo develops blood vessels, the chorion grows fingerlike projections called **chorionic villi** into the surrounding uterine tissue, increasing the surface area for exchange between the blood supplies of the mother and embryo (FIGURE 21.12).

The placenta produces progesterone and estrogen and, by the beginning of the fetal period, has taken over the functions of the corpus luteum of maintaining the endometrium and preventing follicle formation. Progesterone produced by the placenta relaxes the smooth muscles in both the uterus and the gastrointestinal tract, allowing the uterus to expand in size, but also leading to acid reflux (heartburn) and constipation.

Because of its effects on smooth muscles, progesterone functions to relax the arteries, causing blood pressure to drop. As described in Chapter 11, a decrease in blood pressure triggers the kidneys to release hormones that increase blood volume. In addition, increasing estrogen levels promote water retention, and these two effects together result in a 40% increase in blood volume in a pregnant woman. Blood volume peaks at 7 liters about 30 weeks after fertilization.

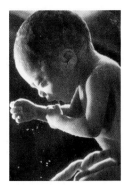

38 weeks
- Weight increases as a result of fat deposition under skin.
- Fetus rotates so head is near cervix.
- Size at term: 53 cm (20.5 in) long, 3.4 kg (7.5 lb).

FIGURE 21.11 **Fetal development.** Although organs have substantially formed by the beginning of the fetal period, many changes are still in store for the developing fetus.

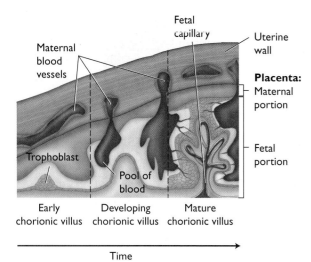

FIGURE 21.12 Chorionic villi. The chorionic villi greatly increase the surface area for exchange between the embryo or fetus and the mother.

The placenta produces additional hormones with various functions. One hormone makes the mother's cells resistant to insulin. This change may encourage higher blood-sugar levels to supply the rapidly growing fetus. However, this hormone may also induce pregnancy-related diabetes. The placenta also synthesizes the hormone **somatomammotropin**, which works with estrogen, progesterone, and the pituitary hormone **prolactin** to stimulate development of the mother's mammary glands and breasts (FIGURE 21.13).

Fetal Circulation

The placenta is required for the survival of the developing fetus because it provides all its fuel and other materials, and disposes of its wastes. The exchange of materials across this organ requires some differences from the adult circulatory system, both at the placenta surface and within the developing fetus.

Dissolved substances such as nutrients, oxygen, carbon dioxide, and wastes in the blood flow across the placental membrane in response to a concentration gradient. For materials to enter or leave the fetus efficiently, the concentration gradient must be in the proper direction compared to the mother's blood supply in the uterus. Glucose in the mother's bloodstream is elevated as a result of the actions of placental hormones, and the sugar's high rate of use in the fetus ensures steady delivery to the fetus's bloodstream.

Oxygen preferentially flows to the fetal blood supply across the placenta because the fetus produces hemoglobin that has a higher affinity for oxygen than adult hemoglobin. Production of the fetal hemoglobin molecule begins a few weeks before the beginning of the fetal period and continues through birth. Fetal hemoglobin levels in the newborn's bloodstream decline steadily until adult hemoglobin completely replaces it by approximately one year of age.

Carbon dioxide preferentially flows in the opposite direction—from fetus to mother. As the uterus grows it widens the thoracic cavity, leading to a 40% increase in the mother's lung capacity and a resulting 20% decline in maternal blood CO_2 levels. The fetus's high heart rate also ensures the rapid transfer of dissolved gases to and from the placenta, helping to maintain the concentration difference across its surface.

The placenta, however, is an imperfect gatekeeper for substances crossing from the maternal blood supply into the developing embryo and fetus. It contains specialized receptors for a number of different molecules, including amino acids, but it also allows certain materials to cross unimpeded. Among these freely passing chemicals are those that can cause great harm to the developing embryo and fetus, such as alcohol, which can lead to skull deformities and reduced mental capacity (FIGURE 21.14). Once any substance crosses the placenta, it travels rapidly throughout the body of the fetus via the developing circulatory system.

The bloodstream of the fetus is connected to the placenta via the **umbilical cord**, which contains the umbilical blood vessels. Blood pumped from the fetal heart travels through various body arteries to the *umbilical arteries* carrying blood to the placenta. Blood returns to the fetus from the placenta via the *umbilical vein*, which eventually merges with the inferior vena cava (FIGURE 21.15). Because the nutrients passing into the fetal bloodstream have already been processed by the maternal liver, most of the returning blood bypasses the fetal liver by passing through a temporary shunt called the **ductus venosus**.

Fetal lungs are nonfunctional until birth, so a gap called the **foramen ovale** (oval window) between the two atria in the heart allows some of the blood entering the right atrium to bypass the lungs and flow directly to the left atrium. The left ventricle then pumps this blood into the systemic circulation. Blood that reaches the right ventricle bypasses the lungs via a shunt (the **ductus arteriosus**) that empties directly into the aorta. After birth, the foramen ovale and the two ducti close, and the remains of the shunts and the umbilical blood vessels gradually convert to connective tissue.

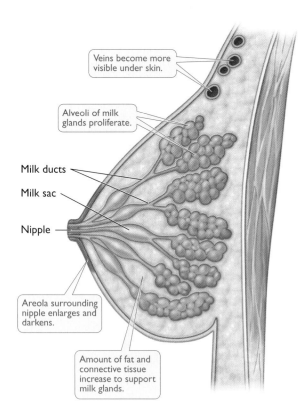

FIGURE 21.13 Breast development during pregnancy. Placental hormones trigger an increase in the number and size of milk ducts in the breast in preparation for producing milk.

Stop and Stretch 21.4

In a small percentage of newborns, the ductus arteriosus remains open (or patent) after its function is no longer needed. Use your understanding of normal circulation to explain what the consequences of this defect likely are.

Stem Cells and the Fetal Period

Although most birth defects occur during the embryonic period, nearly all screening for the presence of these defects occurs during the fetal or *prenatal* (before birth) period. These tests include **amniocentesis**, in which amniotic fluid containing fetal cells is removed for genetic testing, and **ultrasound**, a procedure that allows direct observation of organs. A number of other screening tests can also aid in diagnosing birth defects during the early fetal period (Table 21.3).

Problems identified by prenatal screening tests can sometimes be repaired by fetal surgery, in which a surgeon operates on the fetus while it is still in the womb. One example of such a surgery is repair of a spina bifida lesion that may cause serious disability if the spinal cord remains exposed to amniotic fluid.

Fetal stem cells collected from amniotic fluid may be employed in efforts to repair minor, but potentially deadly, birth defects as well. Researchers at Boston's Children's

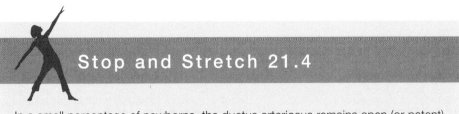

FIGURE 21.14 Fetal alcohol syndrome. Exposure to high levels of alcohol during organogenesis can have profound effects on a child.

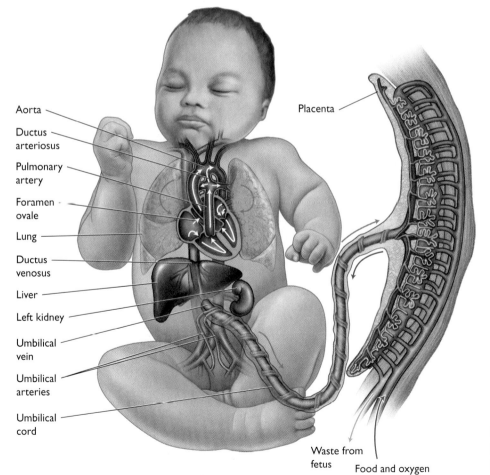

FIGURE 21.15 Fetal circulation. Because the lungs and liver are bypassed and all material exchange occurs across the placenta, circulation in the fetus differs from adult circulation.

TABLE 21.3 | Screening Tests Used During Pregnancy

Test	When and How Performed	Diagnoses
Chorionic villi sampling (CVS)	8–10 weeks after fertilization. Sample of chorionic villi obtained by suction either through the cervix or directly through the abdominal wall.	Genetic diseases for which tests are available (e.g., cystic fibrosis, Huntington's disease). The sampling increases risk of miscarriage.
Alpha-fetoprotein (AFP)	14–16 weeks after fertilization. Maternal blood test.	An elevated level indicates a possible neural tube defect.
Amniocentesis	14–16 weeks after fertilization. Needle inserted through abdominal wall removes a small amount of amniotic fluid.	Genetic diseases, more accurate evaluation of AFP level.
Ultrasound	16–18 weeks after fertilization. Image formed by sound waves bounced off the fetus and back to a receiver.	Malformations of body and internal organs, position of the placenta.

Hospital have coaxed stem cells drawn from amniotic fluid in sheep to develop into cartilage that can be used to repair holes in the sheep's diaphragm or trachea. These tissue patches can even be implanted during fetal surgery. Scientists at the University of California at San Francisco and other centers are investigating the possibility of using healthy fetal bone marrow stem cells to repair genetic diseases that produce abnormal blood or immune system cells. Introducing healthy replacement cells before birth can minimize or eliminate some of the lifelong negative effects caused by the presence during the fetal period of defective bone marrow stem cells.

Fetal stem cells may be useful in therapies for individuals after birth, as well. Fetal stem cells have already been used to repair hearts damaged by chronic heart disease and show some promise in restoring nerve function in a damaged spinal cord and brain. Fetal stem cells have wider multipotency than adult stem cells and have the potential to develop into the wide range of cell types found in the organ they were initially drawn from.

The use of fetal stem cells also avoids the ethical issues posed by the use of embryonic stem cells. They can be collected from amniotic fluid as early as 10 weeks after conception, and do not cause the death of the embryo. In fact, because they are already partially differentiated, fetal cells may be more helpful to medicine in the short term than embryonic stem cells—until scientists have a better understanding of the steps required to differentiate embryonic cells into particular tissues. However, most researchers feel that undifferentiated embryonic stem cells will prove to be more effective than partially differentiated fetal cells in many applications.

The fetal period ends when the placenta and fetus begin to produce hormones that stimulate labor and birth.

The Process of Childbirth

The exact triggers of **parturition** (childbirth) are not entirely clear, and as yet it is impossible for a woman to predict when labor will begin. What is known is that as the end of its intrauterine development approaches, the fetus releases a flood of hormones from its hypothalamus, pituitary, and adrenal glands, perhaps in response to hormones released by the placenta. The fetal hormones stimulate the placenta to release more estrogen, and the estrogen surge triggers an increase in oxytocin and prostaglandin. Oxytocin stimulates the smooth mucles in the uterus to begin contracting.

The first contractions of childbirth often are not noticed by the mother, but they begin the process of dilating the cervix to allow the fetus to pass through. As the cervix dilates, the mucous plug that has blocked its entrance since the beginning of pregnancy is expelled, a symptom referred to as "the bloody show," indicating that labor is approaching. As the child's head presses against the cervix, the amniotic sac may burst, releasing the amniotic fluid. This is referred to as a woman's "water breaking" and is a sign that labor will commence within 24 hours.

Labor involves strong, rhythmic contractions of the uterus. Labor results from one of the few positive feedback processes in human biology. Oxytocin causes uterine contractions, which force the child's head against the cervix. Stretching of the cervix stimulates the release of more oxytocin, increasing the strength of the contractions and putting even more pressure on the cervix. At first, contractions last for 40 seconds or more and happen every 15 to 20 minutes. As labor progresses, contractions may last longer and be separated by only a few minutes. Labor is divided into three stages, defined by the major events during each stage (FIGURE 21.16).

During the first stage of labor, the cervical opening dilates from a tiny, pencil-lead-sized opening to 10 centimeters (4 inches) in diameter. This process is often referred to as effacement or thinning of the cervix because the uterus is actually being pulled up toward the baby's head. The first stage of labor takes several hours or even days to complete.

The second stage of labor occurs as the baby moves into the birth canal. As the baby's head descends into the vagina, the mother's urge to push by actively contracting her abdominal muscles becomes greater. Contractions during this stage are frequent and long, and this stage is often the most painful part of childbirth. Women today have many options for dealing with the pain of childbirth.

A small incision called an *episiotomy* is commonly made at the mouth of the birth canal near the end of the second stage. While the intent of this incision is to ease the passage of the baby's head, there is no scientific evidence that this prevents further damage to the mother's tissues or helps the baby in any way. Despite recommendations that episiotomies should only be used in certain cases, this small operation is still surprisingly common.

Once the head and shoulders of the baby are delivered, the rest of its body slides out rapidly. The third stage of labor occurs about 15 minutes later, the delivery of the placenta (or "afterbirth"), triggered by continued contractions.

The human pelvis, adapted for an upright stance, is narrow and twisted, requiring the infant to pass through a very small opening and change positions several times. When a woman has a small pelvis, is carrying an exceptionally large baby, or when the infant is **breech** (that is, with its head facing away from the cervix), a **Caesarean section** may be used to deliver the baby. In this operation, an incision is made through the lower abdomen and the infant is removed directly from the opened uterus. C-sections

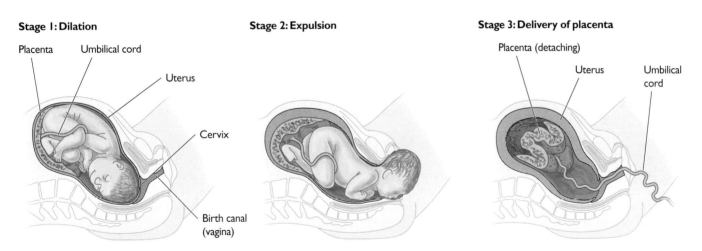

FIGURE 21.16 **The three stages of labor.** During labor, (1) the cervix dilates, (2) the baby is expelled, and (3) the placenta is delivered.

were once reserved for emergency situations, but have become more common over time, especially among women and physicians who wish to schedule the timing of a child's birth.

Even before the placenta is delivered, the infant's connection with it is severed when the umbilical cord is cut. As a result, carbon dioxide begins to build up in the baby's bloodstream, triggering the breathing reflex in order to restore homeostasis of blood gases. The first forceful inhalation inflates the lungs and initiates gas exchange. In babies born prematurely, underdeveloped lungs may fail to inflate or to remain inflated, requiring the use of a respirator.

Some of the blood cells within the umbilical cord are stem cells, and rather than simply disposing of the cord, some parents have begun banking these cells. Research on umbilical cord stem cells indicates that they may be more flexible than adult stem cells, but it is not yet clear whether they can be induced to produce cells other than blood or bone.

Stop and Stretch 21.5

Opponents of stem cell research note that while no diseases have been cured using embryonic stem cells, cells from cord blood have provided treatments for dozens of blood- and bone-related diseases. Why haven't embryonic stem cells cured any diseases?

21.5 Development After Birth

Development is a continuous process. Change and maturation continue to occur after birth and into adulthood.

Growth and Maturation

During the first month of life, the newborn or **neonatal** period, the child's digestive system begins to function. He or she then begins to deposit additional fat underneath the skin. This fat allows the child to begin controlling body temperature more effectively. The brain also develops rapidly during this time, and the neonate becomes gradually more able to adjust to loud stimuli (FIGURE 21.17) and to express basic emotions and hold short-term memories. The skull plates, which are bony but not connected, begin to knit together, although the "soft spots," or *fontanels*, on the head where the bone has not fused remain apparent until about two years of age.

The health of a neonate depends on the infant's nutrition, which was exclusively provided by breast milk until infant formula became available. The first few days after birth, the mammary glands produce *colostrum*, a yellow fluid low in fat but rich in proteins, antibodies, minerals, and vitamin A. Prolactin secreted by the mother's pituitary after childbirth stimulates **lactation**, the production of mature milk. Suckling by the infant triggers the release of oxytocin, which stimulates letdown, the release of milk into the milk ducts (FIGURE 21.18).

The remaining periods of childhood development are termed **infancy** (from 1 to 15 months), **childhood** (through ages 12 to 15), **pubescence** (puberty, which occurs between 10 and 16 years of age in girls and 13 to 16 in boys), and **adolescence** (through

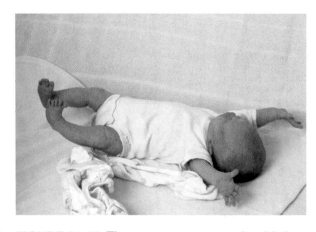

FIGURE 21.17 **The nervous system after birth.** This infant is demonstrating the startle reflex—throwing out her legs and arms and arching her back—in response to a loud noise. This reflex disappears a few weeks after birth as the brain matures.

ages 19 to 21). Some development occurs even during **adulthood**, when maturation is substantially finished—primarily bone growth and brain maturation, both of which are complete around age 25.

Growth from infancy through childhood is relatively steady, with the overall growth rate declining until the child approaches puberty. The onset of puberty is marked by a growth spurt, after which little additional growth occurs.

The pattern of growth after birth is irregular, with some body parts growing fast and others more slowly. This differential pattern is termed **allometric growth** and is illustrated in FIGURE 21.19. You may have noticed how short an infant's legs are relative to his or her body—this relationship changes as the limbs grow much faster than the trunk throughout childhood and puberty. The beginning of puberty is marked by more than a growth spurt. Puberty also commences the process of sexual maturation.

Puberty

Development of mature sexual organs and **secondary sex characteristics** (traits found in sexually mature adults) occurs as a result of a series of hormone triggers. The first step is poorly understood—somehow, the hypothalamus is triggered to begin secreting *gonadotropin-releasing hormone (GnRH)*. This hormone in turn triggers the release of *luteinizing hormone (LH)* and *follicle-stimulating hormone (FSH)* from the pituitary. In both sexes, LH and FSH stimulate the gonads to produce sex hormones. Although estrogen is more abundant in girls and is considered the female hormone and testosterone the male, both hormones are produced by both sexes.

In girls, the beginning of puberty becomes noticeable as the breasts and pubic hair begin to develop. Breast development begins around age 10 to 11 and occurs over approximately two years. Pubic hair growth usually lags behind breast development slightly. The first menstruation (**menarche**) usually occurs about two years after breast development begins, and in about another two years, periods are fairly regular. The Genes & Homeostasis feature explores the trend toward earlier menarche that has occurred in the last 100 years.

Body hair, especially under the arms, on the upper lip, and on the arms and legs becomes thicker and darker in girls during puberty as well. An increase in the size of the lower pelvis and increased fat storage leads to larger hips and contributes to the adult woman's curvy silhouette. Growth in female height is complete in most women by about two years after menarche.

The first sign of puberty in boys is enlargement of the testes and penis, which have remained the same size since early childhood. Libido increases dramatically, and erections become more common. Boys are able to ejaculate quite early in puberty, but typically few or no sperm are present in the ejaculate for several years. Pubic hair

FIGURE 21.18 Lactation. Suckling at the breast stimulates letdown, release of milk from the milk ducts. Milk production by the mother consists of a feedback loop in which production increases in response to increased suckling and decreases when the baby is less hungry.

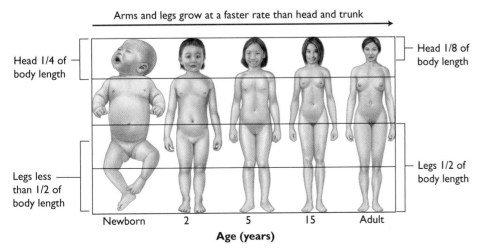

FIGURE 21.19 Allometric growth. Growth throughout development is not distributed equally across all parts of the skeleton. In general, the limbs grow much more than the head and torso, resulting in a changing relationship between head and body size.

VISUALIZE THIS Girls complete their growth earlier than boys. If both sexes grow allometrically in the same way, what differences in relative limb length should you see between boys and girls?

GENES & HOMEOSTASIS

Are Girls Becoming Women Too Young?

In 1962, British pediatrician James Turner shocked scientists with this bit of information: from the 1830s to the 1960s, the average age of first menstruation (menarche) dropped from 17 to 12.8. The rate of decline was an amazing four months per decade. Since then, there have been reports that the age of menarche has continued to drop dramatically: One study saw a decline of between two and nine months during the period 1973 to 1994. Many parents already believe that their kids grow up too fast. Is this study proof there is reason for concern? And why is it happening?

As it turns out, age of menarche declined over the twentieth century—but not nearly as much as Turner estimated. His estimate of menarche for girls in the 1830s was based on data from a small, unrepresentative group of impoverished British girls. Additional data collected from a variety of sources seem to indicate that average age of menarche in the nineteenth century was about 14 to 15 years, just two years older than today's average of 12 to 13.

The decline in age of menarche over the past 150 years appears to be due to better nutrition. Estrogen levels rise throughout adolescence, eventually crossing a threshold that triggers menarche. While most estrogen is produced by the ovaries, some is produced by fat cells. Thus, girls with higher fat stores reach the estrogen threshold earlier than those with less body fat. As the population as a whole became better fed and thus fatter in the time since the Industrial Revolution of the early 1800s, the average age of menarche dropped.

Even with the population-level decline in age of menarche, there is still significant variability among girls in this stage of development. This variability appears to have a genetic component. For instance, the age of menarche is closely correlated in mothers and daughters, siblings, and identical twins. Some evidence shows that earlier menarche is associated with particular alleles for genes controlling estrogen synthesis and estrogen metabolism. Gene variants that contributed to higher estrogen levels are associated with earlier onset of menarche.

Interestingly, some of the same genes associated with early menarche are also correlated with breast cancer. This link may explain why early menarche is considered a risk factor for breast cancer later in life. The elevated estrogen levels that trigger early menstruation may also stimulate cancerous cells in the breast to form tumors faster and earlier.

---

begins to appear. Rapid growth of the larynx (voice box) results in the deepening of the voice by a full octave. The voice change is followed by characteristic facial and body hair changes. The growth spurt in boys starts later than in girls and lasts longer, resulting in a 10-centimeter (4-inch) height difference, on average, between adult men and women.

Stop and Stretch 21.6

The brain completes most of its development by age 25, but throughout life neurons continue to make new connections and remodel themselves to adapt to new situations. What does this tell you about thought processes and learning in a child, teenager, young adult, and older adult?

Adults produce multipotent stem cells in at least the bone marrow (for blood and various types of connective tissue), skin (for dermal cell types), and brain (for neurons and supporting cells). Some of these stem cells can be triggered to become pluripotent under certain growth conditions. Unfortunately, these cells are low in number in these tissues and difficult to grow in the lab, two factors that have limited their use in research.

Little additional growth occurs after age 25, when both men and women are at their peak of physical potential. We can remain vigorous for decades, but eventually the effects of aging appear.

21.6 Aging

Senescence, or aging, consists of the progressive changes throughout life that contribute to an increased risk of disease, disability, and death. Although habits and behaviors from childhood and early adulthood may contribute to senescence, the signs of aging begin in earnest for most individuals around age 40. One application of embryonic stem cells may be to halt or reverse the effects of aging on the body.

Why Do We Age?

The average life span in the United States is about 72 years for men and 79 years for women, and the maximum life span, as indicated by the oldest surviving individual, is 125 years. Several hypotheses seek to explain what limits human life span. The most commonly cited are described here.

Some scientists suspect that our maximum life span is coded in our genes. This hypothesis is supported by the observation that adult cells can divide only a certain number of times before dying. This limit is enforced by the activity of **telomeres**, long repeated DNA sequences found at the ends of chromosomes.

Every time a cell divides, a bit of the telomere is lost, and when only a short segment of the telomere remains, the cell dies. Certain cells, including sperm-generating cells, cancer cells, and fetal and embryonic stem cells, produce an enzyme called telomerase that restores the telomere after division, resulting in the continued ability to divide.

Researchers studying other organisms have also identified genes associated with life span—mutations in these genes led to unusually shortened or lengthened lives. Genes with similar DNA sequences have been identified in humans, athough it is as yet unclear if these have the same function in us.

Other scientists hypothesize that the cumulative assaults our cells are exposed to gradually cause cells to lose function and die, contributing to the loss of function of various organ systems. This hypothesis is supported by the observation that individuals with mutations that cause DNA damage age prematurely (FIGURE 21.20).

DNA damage is often caused by free radicals, molecules that scavenge electrons from other chemicals. Free radicals are produced by aerobic respiration in all cells, and suppression of free radical production as a result of caloric restriction may explain why animals fed a very low-calorie diet live longer than their well-fed siblings.

Because the body is an integrated system, damage to one system affects others. In many ways, aging is a positive feedback process. For example, once the endocrine system begins to weaken, the immune system, which relies on hormonal signals, also loses function. As a result, the elderly are more susceptible to illness, cancer, and autoimmune disorders such as rheumatoid arthritis.

Effects of Aging

Exactly how an individual ages depends on genetic factors as well as environmental ones. Many of the effects appear to be related to the formation of chemical cross-links between proteins, interfering with the proteins' normal function. The rapidity with which these cross-links form appears to have a genetic basis.

However, health habits throughout our lifetimes also have an influence. For example, high levels of sun exposure cause skin collagen to break down earlier, leading to premature wrinkling. Cigarette smoking, alcohol intake, and inadequate calcium consumption reduce bone mass and contribute to early bone weakness.

Some of the most common effects of aging on the organ system are:

- **Skin:** Less elasticity and reduced fat reserves cause sagging and wrinkling. Reductions in fat, hair follicles, and sweat glands impair the ability to thermoregulate. Pigment cells decline in number but increase in size, so the skin becomes pale and blotchy, and hair grays. Fewer sebaceous glands and reduced

FIGURE 21.20 **Premature aging.** This 15-year-old child has progeria syndrome, a rare disease in which the nucleus is unstable and DNA is subject to high levels of damage. As a result, the effects of aging occur very early in life.

cell division cause skin to become dry and heal more slowly. The cornea of the eye may become cloudy, reducing visual acuity.
- **Cardiovascular:** Cardiac cell size declines, reducing heart size and cardiac output. Arteries become less elastic and contain more atherosclerotic plaques, increasing blood pressure and the risks of heart attack and stroke. Blood flow to the liver declines, reducing the rate of metabolism of drugs.
- **Digestive:** The digestive organs lose muscle tone, increasing susceptibility to heartburn and constipation. Fewer calories are needed as metabolic rate declines (for example, at age 50, the metabolic rate is 80–85% of the childhood rate). If calorie intake remains stable, weight rises. Long-term neglect of dental health may result in tooth loss.
- **Excretory:** Kidneys become less efficient at maintaining salt/water balance, making dehydration more likely. Poor muscle tone reduces the ability to empty the bladder fully, increasing the risk of bladder infection. In women, pelvic muscles weakened by childbirth may lead to incontinence; in men, an enlarged prostate may cause similar urinary problems.
- **Musculoskeletal:** The number of muscle cells declines as a result of lack of use. Average muscle mass at age 80 is 50% of the mass at age 30. Bones begin losing calcium, and thus mass. Women are especially susceptible and may lose half of their bone mass by age 70. Intervertebral disks deteriorate, causing height to decrease by about 1 centimeter per decade. Joint cartilage degrades, resulting in arthritis in about 15% of older adults. Cell division and repair decline, so damaged muscles and joints are slower to recover.
- **Reproductive:** Sex hormone levels decline. In women this leads to cessation of ovulation and menstruation (menopause), hot flashes, and reduced vaginal lubrication. Males take longer to achieve erection, but remain fertile throughout life.
- **Nervous:** Brain mass declines about 10% by age 80, which has little effect on its own. Reduction of blood flow to brain causes nerve cell death and results in reduced learning capacity, delayed reaction time, and poorer short-term memory. The function of the sensory system declines: Destruction of auditory hair cells over time reduces hearing acuity, and loss of taste cells reduces sensitivity to flavors. The lens of the eye becomes less flexible, interfering with near vision.

The field of **gerontology** seeks to identify ways to prolong the healthy life span. One of the main recommendations of gerontologists is for individuals to employ a sensible exercise program and healthy diet to stave off the negative effects of aging for as long as possible.

Stop and Stretch 21.7

Consider the hypotheses about why we age addressed in the previous section. Does the advice of gerontologists support one or both of these hypotheses?

In this chapter, you have already learned of examples where stem cells have been used to reduce the effects of aging—for example, to repair hearts and reestablish a healthy blood supply. Research on stem cells may also allow scientists in the future to grow organs to replace those that are damaged or have failed, as well as to restore the function of various tissues. One potential use of stem cell therapy is to provide cures for the age-associated diseases that many individuals fear the most—loss of central nervous system function.

Restoring the Brain

Alzheimer's disease (AD) begins when tangled clumps of protein fibers form inside brain neurons. Affected neurons die, and whole regions of the brain lose function. The result is progressive memory loss, dementia, and, eventually, death. AD is not a normal symptom of aging. Yet it affects nearly 3% of people over age 65 and nearly half of those 85 and older. In 2004, nearly 4.5 million Americans had AD. With the aging of the baby boom generation, estimates are that 11 to 14 million will be afflicted by 2050. AD results in significant disability, and as of now, we have few treatments for it.

Because the memory function of the brain is associated with the connections between neurons, replacing damaged neurons with stem cells without reestablishing the previous connections is unlikely to restore memory. Despite Nancy Reagan's advocacy, research on embryonic stem cells is unlikely to result in a cure for AD, although their study may provide insight into the molecular basis of the disease.

Stem cells offer more promise in treating **Parkinson's disease (PD)**, which afflicted Michael J. Fox at age 31 but is typically associated with aging. Most cases of PD are diagnosed after age 60. Parkinson's is slightly less common than AD. Estimates of affected individuals range from 1 to 3 million people, and this number is also expected to rise as the elderly population increases.

PD occurs when cells in the brain that produce the neurotransmitter dopamine cease functioning for an unknown reason. Because the cause of the disease is limited to a relatively small number of cells with a particular function, it may be possible to use stem cells as replacements. In fact, mouse embryonic stem cells have been used to cure a PD-like disease in rats.

The experiment on rats underscored the hope that stem cell therapy offers to Parkinson's sufferers. Yet it also illustrated one of the negative side effects of stem cell use. In 20% of the rats injected with mice stem cells, the cells did not differentiate in the brain but instead grew into an undifferentiated tumor. Until researchers learn how to control the proliferation of undifferentiated stem cells, they will not be useful as therapy.

So much is unknown about stem cells and the processes of cell differentiation and organogenesis that it is impossible to say how embryonic stem cells may be used to treat currently incurable diseases. President Bush's decision to prohibit federal funding for research using new stem cell lines has demonstrably impeded progress on understanding the true promise of these cells. Whether the moral issue of destroying human embryos to obtain stem cells is worth the still unknown benefits of their study remains unclear.

However, states such as California, Missouri, and New York, and countries in the European Union, have stepped into the funding gap left by the federal policy and are stimulating research on embryonic stem cells. Most of these jurisdictions restrict the research to lines created from unwanted or donated IVF embryos, sidestepping some (but not all) of the moral objections.

There is also progress in understanding how to reprogram already differentiated cells to become multi- or pluripotent. In November 2007, scientists at both Kyoto University in Japan and the University of Wisconsin announced that they had "tricked" ordinary human skin cells into behaving as if they were pluripotent embryonic stem cells by inserting four new genes into the cells. This announcement delighted both opponents and supporters of embryonic stem cell research. By reprogramming adult cells, almost all of the ethical challenges posed by harvesting cells from days-old embryos can be avoided.

Currently, it is not clear whether reprogrammed adult cells will be able to differentiate like normal embryonic cells or if the genes inserted by the researchers will cause problems in tissues generated from these cells. But if history is any guide, what we learn from research on any of these intriguing cells will surprise us.

Chapter REVIEW

ROOTS TO REMEMBER

The following roots of words come mainly from Latin and Greek and will help you to decipher terms:

ec- is from a word meaning outside (ectopic means out of place).

lact- is from a word for milk.

meso- means middle.

-morph and **morpho-** mean form.

multi- (many, much), **pluri-** (several, many), and **toti-** (all, wholly) come from words for the quantity of things.

neo- comes from the word for new.

KEY TERMS

21.1 The Production of Embryonic Stem Cells
blastocyst *p. 515*
capacitation *p. 513*
cell differentiation *p. 512*
cleavage *p. 515*
conjoined twins *p. 515*
development *p. 514*
dizygotic twins *p. 514*
embryo *p. 513*
embryonic stem cells *p. 513*
extraembryonic membranes *p. 515*
fertilization *p. 513*
fetal stem cells *p. 513*
fetus *p. 513*
in vitro fertilization (IVF) *p. 513*
inner cell mass *p. 515*
monozygotic twins *p. 515*
morphogenesis *p. 515*
multipotent *p. 512*
ovum *p. 514*
pluripotent *p. 512*
preembryo *p. 513*
regenerative medicine *p. 512*
stem cell *p. 512*
totipotent *p. 512*
trophoblast *p. 515*
zona pellucida *p. 513*
zygote *p. 514*

21.2 Early Embryonic Development
amnion *p. 517*
amniotic fluid *p. 517*
clinical pregnancy *p. 516*
ectoderm *p. 517*
ectopic pregnancy *p. 516*
embryonic disk *p. 517*
endoderm *p. 517*
gastrula *p. 517*
human chorionic gonadotropin (HCG) *p. 516*
implantation *p. 516*
mesoderm *p. 517*
miscarriage *p. 516*
yolk sac *p. 517*

21.3 Organ Formation
apoptosis *p. 517*
cell migration *p. 517*
genital tubercle *p. 519*
indifferent gonads *p. 519*
labioscrotal swellings *p. 519*
Mullerian duct *p. 519*
neural tube *p. 518*
organogenesis *p. 517*
spina bifida *p. 518*
SRY gene *p. 519*
Wolffian duct *p. 519*

21.4 Fetal Development and Birth
amniocentesis *p. 523*
breech *p. 525*
Caesarean section *p. 525*
chorionic villi *p. 521*
ductus arteriosus *p. 522*
ductus venosus *p. 522*
foramen ovale *p. 522*
labor *p. 525*
parturition *p. 524*
placenta *p. 521*
prolactin *p. 522*
somatomammotropin *p. 522*
ultrasound *p. 523*
umbilical cord *p. 522*

21.5 Development After Birth
adolescence *p. 526*
adulthood *p. 527*
allometric growth *p. 527*
childhood *p. 526*
infancy *p. 526*
lactation *p. 526*
menarche *p. 527*
neonatal *p. 526*
pubescence *p. 526*
secondary sex characteristics *p. 527*

21.6 Aging
gerontology *p. 530*
Parkinson's disease (PD) *p. 531*
senescence *p. 529*
telomeres *p. 529*

SUMMARY

21.1 The Production of Embryonic Stem Cells

- Adult and fetal stem cells are partially differentiated and multipotent. Embryonic stem cells are undifferentiated and pluripotent, having the potential to become any cell in the human body. Embryonic stem cells' pluripotency makes them likely candidates for the production of replacement tissues and organs (p. 512).

- To fertilize an egg cell, capacitated sperm slide through the follicle cells that surround the egg to reach the zona pellucida. Binding of the sperm head to the zona pellucida triggers the release of enzymes in the acrosome, which break down the zona pellucida and allow the sperm to reach the egg cell's plasma membrane (p. 513; Figure 21.2).

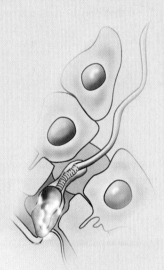

- After fusion of the egg and sperm plasma membranes, the egg becomes impermeable to other sperm. The egg completes meiosis II, producing an ovum. When the sperm and ovum nuclei fuse, the cell becomes a zygote (p. 514).

- The zygote undergoes rapid cleavage divisions to produce the preembryo. Further division produces the blastocyst. The inner cell mass of the blastocyst will become the embryo and eventually the fetus, while the outer cells develop into some of the extraembryonic membranes. The inner cell mass is the source of embryonic stem cells for research (p. 515).

21.2 Early Embryonic Development

- The blastocyst implants in the uterine wall and establishes a nutritional connection to the mother. At implantation, the blastocyst also begins to secrete human chorionic gonadotropin, the first signal of pregnancy (p. 516).

- The inner cell mass of the blastocyst then changes conformation to produce a three-layered gastrula. Each layer of the gastrula (ectoderm, mesoderm, and endoderm) differentiates into specific tissues (p. 517).

21.3 Organ Formation

- Along with differentiation, other key components of organogenesis are cell migration, in which partially or fully differentiated cells move to the proper final position in the body, and apoptosis, programmed cell death (pp. 517–518).

- The brain and spinal cord are one of the first organ systems to develop, when the neural plate along the midline of the embryonic disk folds and forms a neural tube covered with ectoderm. Environmental factors can interfere with neural tube development, resulting in birth defects (pp. 518–519).

- The expression of sex-specific genes and hormones determines whether a particular embryo will become male or female. Embryonic gonads become either testes or ovaries. Ductal structures exist side by side in male and female embryos. In each sex, one structure regresses. Male and female external genitalia are fashioned from the same embryonic material (pp. 519–520; Figure 21.10).

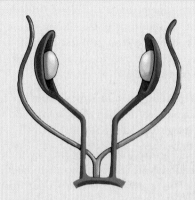

21.4 Fetal Development and Birth

- The placenta forms in the uterus from the trophoblast of the blastocyst and uterine tissues. Chorionic villi on the placenta ensure material exchange between the fetus and mother. Hormones produced by the placenta help maintain the uterine lining, prevent ovulation, facilitate material exchange, prepare the breasts for lactation, and also lead to many of the side effects of pregnancy (p. 522; Figure 21.13).

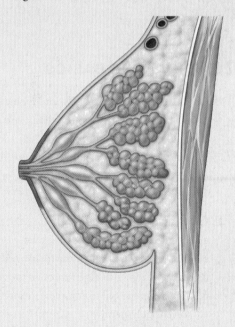

- Because all nutrient and gas exchange occurs across the placenta and blood flows through the umbilical cord, circulation in the fetus can bypass the lungs. Fetal circulatory patterns change over to the adult pattern after birth (p. 522).
- Stem cells, including those from the fetus, can be used to repair birth defects and other damage (pp. 523–524).
- The signals that trigger the beginning of childbirth are still unknown, but the process itself is a positive feedback loop that increases contractions as a result of pressure on the cervix. Parturition involves three stages of labor—dilation of the cervix, expulsion of the baby from the uterus, and delivery of the placenta. Cells from blood remaining in the umbilical cord after birth may prove to be useful stem cells (pp. 524–525).

21.5 Development After Birth

- The neonatal period is marked by rapid gain of fat tissue and some maturation of the brain and digestive system. Infants at this stage are dependent on liquid food, often provided by breast milk. Breast milk production is another positive feedback process, where milk release is triggered by suckling (p. 526).
- The pattern of growth until adulthood is allometric, with some parts of the body growing more rapidly than other parts (p. 527; Figure 21.19).

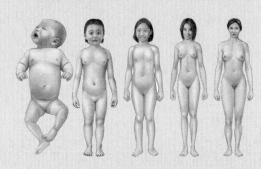

- Puberty occurs when male and female hormone production rises, leading to the development of secondary sex characteristics. These characteristics include breast and genital development, production of pubic hair, and the onset of menstruation and ejaculation (p. 527).

21.6 Aging

- Senescence begins in earnest around age 40, perhaps in response to a limitation on cell division or as a side effect of cumulative damage done to cell DNA (p. 529).
- The effects of aging include superficial markers such as skin wrinkling, loss of hair pigmentation, and more serious effects such as loss of muscle and bone mass and cardiovascular disease (pp. 529–530).
- Two diseases associated with aging have different prospects for treatment by stem cell therapy. Alzheimer's disease results in the loss of connections between neurons, which stem cells are unlikely to be able to replace. Parkinson's disease is caused by the loss of neurons that produce dopamine, a situation that may be repaired by stem cells (p. 531).

LEARNING THE BASICS

1. Compared to adult stem cells, embryonic stem cells _____.
   a. can become more types of cells
   b. can divide many more times
   c. have more genetic material
   d. a and b are correct
   e. a, b, and c are correct

2. Capacitation allows sperm to _____.
   a. enter the female reproductive tract
   b. work their way through the follicle cells surrounding the egg
   c. release enzymes to digest through the zona pellucida
   d. trigger the egg cell to complete meiosis II
   e. prevent polyspermy

3. Match the type of twins with the description of how they form.
   _____ dizygotic twins
   _____ monozygotic twins
   _____ conjoined twins

   a. two eggs fertilized by two sperm during the same cycle
   b. incomplete separation of the cells of a preembryo
   c. splitting of a preembryo

4. The first distinct structure of the preembryo that will develop into the fetus (and the part with cells that can be harvested for embryonic stem cell research) is called the _____.
   a. chorion
   b. inner cell mass
   c. trophoblast
   d. blastomere
   e. mesoderm

5. At or around implantation, _____.
   a. the embryo begins releasing human chorionic gonadotropin
   b. women begin noticing the first symptoms of pregnancy
   c. the trophoblast secretes digestive enzymes that dissolve endometrial tissue
   d. the blastocyst becomes embedded in the lining of the uterus
   e. all of the above

6. Gastrulation _____.
   a. consists of rapid cell divisions that form the blastocyst
   b. allows a zygote to implant in the lining of the uterus
   c. produces the placenta
   d. results in the formation of three layers of cells that will give rise to specific adult tissues
   e. causes sperm and egg release

7. Match the germ layer with the tissues it develops into.
   _____ ectoderm   a. lining of gastrointestinal tract, lungs
   _____ mesoderm   b. skin, nervous system
   _____ endoderm   c. muscles, blood vessels, skeleton

8. Toxins are most damaging to embryos during the first nine weeks of development because _____.
   a. the embryo consists of few cells, all of which can develop into a new embryo
   b. the placenta is not fully formed, so it cannot insulate the embryo from harm
   c. this is the period of organ formation, when errors can have profound effects
   d. it is before the brain forms, so the nervous system cannot respond to toxic effects
   e. the umbilical cord brings toxins directly to the bloodstream at this point

9. The neural tube is _____.
   a. an early stage in development of the brain and spinal cord
   b. an early stage in development of the vertebrae of the spine
   c. a structure composed of endoderm
   d. a defect that forms as a result of spina bifida
   e. more than one of the above is correct

10. The movement of partially differentiated cells from one part of an embryo to another _____.
    a. is called cell migration
    b. occurs in response to chemical signals produced along the route and at the cell's final destination
    c. is still not well understood
    d. a and c are correct
    e. a, b, and c are correct

11. The indifferent gonads can become _____.
    a. ovaries or testes
    b. sperm or eggs
    c. oviducts or vas deferens
    d. testosterone or estrogen
    e. Mullerian or Wolffian ducts

12. In the developing embryo, the presence of testosterone _____.
    a. causes growth of facial hair
    b. allows the Mullerian duct to persist in spite of anti-Mullerian hormone
    c. causes the Wolffian duct to develop into the epididymis and other male structures
    d. if converted to dihydrotestosterone (DHT), causes the formation of the vulva and clitoris
    e. suppresses secondary sexual characteristics from developing

13. All of the following statements about the structures that increase the surface area of the placenta are true except _____.
    a. they are called the chorionic villi
    b. they are engulfed in a pool of blood
    c. they are sites for oxygen and carbon dioxide exchange
    d. they are no longer necessary once the umbilical cord forms
    e. they allow the passage of certain toxins from the mother's bloodstream into the developing embryo and fetus

14. How is oxygen flow between the mother's blood and the fetal blood maintained?
    a. The fetus is small and has low oxygen needs, so maintenance is unnecessary.
    b. The fetal heart rate is slow, allowing blood to pool in the placenta, permitting maximum oxygen saturation.
    c. Fetal hemoglobin has a higher affinity for oxygen than adult hemoglobin.
    d. The increased size of the mother's thoracic cavity promotes higher oxygen levels in her bloodstream.
    e. Blood flow in the fetus bypasses the lungs, preventing oxygen accumulation.

15. Labor consists of a positive feedback loop, which is briefly described as _____.
    a. the fetus releases a hormone that causes the mother to release a hormone, triggering the onset of labor
    b. dilation of the cervix causes the mucous plug sealing it to fall out, allowing the amniotic fluid to leak out
    c. oxytocin stimulates the uterus to contract, forcing the fetus against the cervix, triggering the release of additional oxytocin
    d. delivery of the newborn is followed by delivery of the placenta
    e. buildup of carbon dioxide in the newborn's bloodstream stimulates its first inhalation

16. All of the following statements about puberty are true except _____.
    a. it coincides with a growth spurt
    b. it begins earlier in boys than in girls
    c. it results in sexual maturity in both sexes
    d. it is accompanied by increased growth of body hair in both sexes
    e. it is the last major developmental change before adulthood

17. Telomeres _____.
    a. limit the number of times a cell can divide
    b. reduce the risk of damage to DNA
    c. are made up of repeating sequences of DNA
    d. a and c are correct
    e. a, b, and c are correct

18. Alzheimer's disease is probably not a good candidate for stem cell therapy because _____.
    a. the loss of connections between neurons cannot be replaced by new cells
    b. people with AD are already elderly and likely to die sooner rather than later
    c. the cause of AD is unknown and therefore we cannot treat it
    d. only some cells are affected by AD, so stem cells are too unspecialized to treat it
    e. there are relatively few individuals with AD, so the cost of developing the therapy outweighs the benefits

19. All of the following are effects of aging except _____.
    a. sagging, wrinkling skin
    b. reduced heart volume and cardiac output
    c. increased risk of cardiovascular disease
    d. loss of pigment cells in skin and hair
    e. increasing bone mass

20. Which of the following behaviors can reduce or delay the effects of aging?
    a. eating a diet rich in antioxidants
    b. engaging in regular exercise
    c. avoiding cigarette smoke
    d. avoiding excessive exposure to the sun
    e. all of the above

ANALYZING AND APPLYING THE BASICS

1. In June 2006, researchers at Johns Hopkins University reported that they were able to restore partial movement to 11 paralyzed adult rats using mouse embryonic stem cells. The rats had suffered spinal cord damage due to infection by the Sindbis virus—a virus that kills motor neurons. After embryonic stem cells were injected, over 4,100 new motor neurons formed in the spinal cords of the rats. Some of these motor neurons formed connections between nerves and muscles. This allowed the rats to walk on hind legs that had been paralyzed.

   Discuss the implications of this research for human diseases.

2. The umbilical cord contains undifferentiated stem cells that can develop into any other type of blood cell. Parents may save the umbilical cord from their newborn baby in an umbilical cord bank. If that child, or a close family member, develops a blood or immune system disorder, cord blood can be used to produce healthy blood cells for transplant.

   Cord blood transplant is similar to bone marrow transplant, in that both transplants use stem cells. However, cord blood transplant has the following advantages:

   - No long wait for a suitable donor
   - No donor surgery
   - Less chance of rejection
   - Less risk of transmitting infectious diseases
   - More blood-producing cells present

   Why aren't cord blood cells a major focus of stem cell research?

3. Drinking alcohol during pregnancy can lead to fetal alcohol syndrome (FAS), a series of alcohol-induced birth defects in the fetus. Alcohol can damage the developing central nervous system, leading to such problems as learning disability, problems with memory or attention span, poor reasoning skills, developmental disability, hyperactive behavior, or mental retardation. Fetal alcohol syndrome can also cause abnormal growth and facial development.

   Discuss why drinking alcohol at *any point* in a pregnancy could lead to central nervous system damage in the developing fetus.

4. Studies by D.J.P. Barker and others have shown a link between fetal malnutrition and the later development of heart disease or type 2 diabetes.

   Heart disease: A study examining cause of death for more than 15,000 British men and women found a positive correlation between low birth weight (less than 5.5 lb) and eventual death from heart disease. This relationship was independent of other risk factors, such as smoking or poor diet.

   Type 2 diabetes: Rat studies showed that a diet deficient in protein altered the fetal liver. Cells that synthesize enzymes that produce glucose were increased, while cells that synthesize enzymes that break down glucose were decreased. These changes could eventually lead to the development of type 2 diabetes in adulthood.

   What kind of additional scientific information is needed to show a strong link between fetal malnutrition and these two diseases?

5. Breast milk contains the perfect proportions of fatty acids, amino acids, lactose, and water for a human baby's growth and development. In addition, breast milk provides the baby with antibodies from the mother. This provision of antibodies enhances development of the baby's immune system. A breast-fed baby gains protection from diseases such as pneumonia, bronchitis, influenza, staph infection, meningitis, ear infection, and German measles. The baby may also be protected from the later development of allergies, type 1 diabetes, and certain intestinal disorders.

   Why might it actually be a good idea for a mother to breast-feed a baby when she is fighting a cold?

6. The developing fetus in the mother's womb is completely free of microbes. First exposure to bacteria occurs as the baby travels down the birth canal. During its first two years of life, the baby establishes a microbial community that may remain fairly stable over his or her lifetime.

   Microbes in the digestive system, called gut microflora, help the body to obtain nutrients and energy from food. Gut microflora may also play a role in the development of human obesity. In a study of germ-free mice, Bäckhed and Gordon found that, although the mice ate more food than normal mice, they had 60% less fat. Much of the food passed through their system without being metabolized. Once gut microflora were transplanted into the germ-free mice, they became as fat as normal mice. In a related study, Samuel and Gordon (2004) found that mice with two particular types of bacteria in their gut microflora, *Bacteroides thetaiotaomicron* and *Methanobrevibacter smithii*, were 13% fatter than mice that had only one of these species of microbes.

What would be the developmental advantages of having a large community of gut microflora? What would be the disadvantages?

7. In the mid-nineteenth century, the average American man was 3 inches shorter and weighed 50 pounds less than an American man today. He was also much less healthy than a modern American. A study of Civil War records found that 1/6 of all teenagers who tried to join the army were rejected due to a disability. Around 80% of former Union Army soldiers had heart disease by age 60. In fact, it was common for nineteenth-century Americans in their 40s and 50s to develop chronic diseases, such as heart, lung, and liver disease. These people often died in their 50s or 60s, after a lifetime of arthritis, back pain, hernia, or other debilitating diseases.

Describe some of the factors that have helped to improve the health and longevity of modern people.

CONNECTING THE SCIENCE

1. Mercury and lead are toxins that can have severe negative effects on the developing fetus and that easily cross the placenta. Although the release of these chemicals into the environment is now regulated, they are still common. For example, mercury released when coal is burned eventually accumulates in fish, and lead used in paints before 1977 is abundant in homes built before that time. Pregnant women are warned to limit fish intake and avoid exposure to lead paint during pregnancy.

   Whose responsibility do you think it is to protect developing embryos from these (and other) exposures? Should it be primarily the mother's responsibility, the government's responsibility, or industry's responsibility, and why? How could you enforce this "responsibility taking"?

2. Do you think that President Bush made the right decision when he decided to prohibit federal funding for research on newly derived stem cell lines? Why or why not?

Chapter 22

Evolution

Where Did We Come From?

LEARNING GOALS

1. Define *biological evolution*.

2. Describe the theory of common descent.

3. Describe how Linnaeus's classification system supports the theory of common descent.

4. Define *vestigial structures*.

5. Describe the evidence for evolution provided by biogeography.

6. Explain how fossils form, and illustrate how they support the hypothesis of common descent.

7. Summarize the process of speciation and the biological species concept.

8. Define *fitness* as used in the context of evolution and natural selection.

9. Explain why scientists think there is a single common ancestor for all living organisms.

10. List the steps required for life to form from nonliving precursors.

11. Describe how evolution can occur via genetic drift.

Should public school students be taught about alternative hypotheses to evolution?

When most folks think of Kansas, they probably imagine picturesque small towns surrounded by miles of gently rolling yellow wheat fields. Quiet, resilient folks, accustomed to the vagaries of weather and crop prices and who value their independence. A relaxed, even slow, pace of life. What most people probably don't picture when they think of Kansas is its place in the center of a contentious debate that has been brewing for more than a century.

The subject of the debate: high school biology.

The issue: what students should know about the subject of evolution.

The brouhaha began in August 1999, when the Kansas State Board of Education, an elected body that sets the educational standards for schools around the state, released its science guidelines. Conspicuously missing from the list of topics young Kansans were required to be tested on was a portion of the subject of biological evolution. In particular, the standards excluded the theory that describes the descent of modern species from extinct ancestors.

One idea about the origin of humans: special creation.

Another idea: evolution.

"We are only being honest with our students," said board member Steve Abrams at the time. "Evolution is only a theory." Voters seemed to disagree. In 2000, public backlash against the board resulted in the defeat of several antievolution members and the reinstatement of these topics in the science standards.

The pendulum swung the other way in 2004, when many of the ousted members were reelected, and the board took a new approach—mandating the teaching of alternatives to evolution, including an idea called "intelligent design."

The balance shifted again in 2006, when pro-evolution candidates again won a majority. They reinstated standards emphasizing evolution as the best explanation for the origin of life and removing the requirement that any other explanations be presented.

What's the matter with Kansas? Actually, the furor over teaching evolution is not isolated. In the last decade, legislators in states around the nation, including Michigan, Minnesota, Washington, Hawaii, and Louisiana, have considered

Why do biologists insist that only evolution be taught?

requiring science curricula that include discussion of the "challenges" to evolution or even outright rejection of the topic. At the same time, school boards in local districts from Wisconsin to Delaware to Texas have faced the same issue. In fact, the debate over whether or not children should learn about evolution is nearly as old as the theory itself.

The subject of evolution probably would not cause so much controversy, in Kansas and elsewhere, if it did not address fundamental questions of human existence: Who are we, and where do we come from? Many religious traditions teach that humans were designed by an intelligent supernatural being. In contrast to these beliefs, evolutionary theory argues that humans are the descendants of ancient apes. On the surface, these ideas appear to be competing scientific hypotheses. However, U.S. federal courts have consistently ruled that the Constitution prohibits instruction of religious beliefs about human origins in American public-school science classrooms.

Is the Kansas State Board of Education on the cusp of a revolution of science teaching—one that recognizes that supernatural explanations deserve equal time with those that rely solely on the laws of nature? Or does it represent a dangerous trend in U.S. education—one that sets religious belief on equal footing with scientific understanding? In this chapter, we examine these questions by exploring the theory of evolution and the origin of humans as a matter of science.

22.1 Evidence of Evolution

Biologists claim that evolution is one of the most well-supported ideas in all of science. Here we examine the evidence supporting it.

What Is Evolution?

Evolution has two different meanings to biologists. The term can refer to either a process or a theory, which is a general principle that explains or predicts observations.

Generally, the word *evolution* means "change," and the process of biological evolution is related to this definition. **Biological evolution** is a change in the characteristics of a population of organisms that occurs over the course of generations. More specifically, an evolutionary change is one that results in a change in the frequency of particular alleles in a population.

The **theory of evolution** is a principle for understanding how **species**, that is, kinds of organisms, originate and why they have the characteristics that they exhibit. In its most simple form, the theory is stated as follows:

> All species present on Earth today are descendants of a single common ancestor, and all species represent the product of millions of years of evolutionary change.

In other words, modern plants, fungi, microbes, and animals, including human beings, are related to each other and have been diverging from their common ancestor by various processes since the origin of life on this planet.

The origin of modern species from a common ancestor is what most nonscientists think of when they hear the word *evolution*. This part of the theory of evolution, called the **theory of common descent**, is illustrated in FIGURE 22.1.

Charles Darwin's Revolution

Charles Darwin is credited with bringing the theory of evolution into the mainstream of modern science (FIGURE 22.2). In 1831, at age 22, Darwin set out on what would become his life-defining journey—the five-year voyage of the HMS *Beagle*.

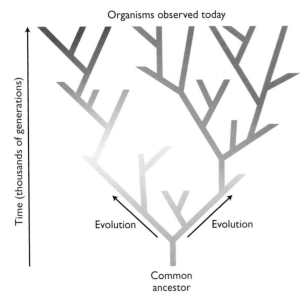

FIGURE 22.1 **The theory of common descent.** This theory states that all modern organisms descended from a single common ancestor. Each branching point on the tree represents the origin of new species from an ancestral form.

The *Beagle*'s primary mission was to chart the coasts and ports of South America. Charles Darwin was added as an unpaid member of the crew to act as an assistant naturalist, and to collect, catalogue, and describe the natural resources of the regions the ship visited.

The stop that had the most influence on Darwin as he made his journey was at a small archipelago of volcanic islands off the coast of Ecuador, the Galápagos Islands. During the month that the *Beagle* spent sailing the islands, Darwin collected an astonishing variety of organisms. Many of the birds and reptiles that he observed appeared to be unique to each island. While all the islands had populations of tortoises, the type of tortoise found on one island was different from the types found on other islands (FIGURE 22.3).

As he reflected on his observations and specimens, Darwin began to realize that they fit a pattern. He saw that they supported the then-radical hypothesis of common descent. Fear of being rejected by his scientific colleagues caused Darwin to carefully document and research additional support for the hypothesis over more than 20 years. In 1859, when he published his most influential book—*On the Origin of Species by Means of Natural Selection, or the Preservation of Favoured Races in the Struggle for Life*—the evidence he had accumulated was overwhelming.

Alternative Hypotheses: Scientific and Religious

Most scientists no longer question the theory that modern species evolved from extinct ancestors. Indeed, most biologists would agree that common descent is a scientific fact. Let's explore that statement, "common descent (or evolution) is a fact," more closely.

When *The Origin of Species* was published, most Europeans believed that *special creation* explained how organisms came into being. According to this belief, God created organisms during the six days of creation described in the first book of the Bible, Genesis. This belief also states that organisms, including humans, have not changed significantly since this beginning. According to some biblical scholars, the Genesis story indicates that creation also occurred fairly recently, within the last 10,000 years.

Consider special creation as an alternative to the theory of common descent for explaining how modern organisms came to be. Recall the definition of science presented in Chapter 1. Special creation is not itself a scientific hypothesis. Because a supernatural creator is not observable or measurable, there is no way to determine the existence or predict the actions of such an entity via the scientific method.

The belief of special creation does provide some testable hypotheses—for instance, the assertion that organisms came into being within the last 10,000 years and that they have not changed substantially since their creation. We can call this hypothesis the *static model*, indicating that organisms are recently derived and unchanging.

The hypothesis of *intelligent design*, the "alternative" to evolution supported by members of the Kansas State Board of Education, has the same problem as special creation. Supporters of this hypothesis assert that certain aspects of the natural world are so complex that they could not have arisen without a designer—a supernatural creator—to guide them. Thus stated, intelligent design is not science. Supporters of intelligent design generally favor alternatives to the static model that fall between this hypothesis and common descent.

To evaluate these different hypotheses, we must use observations of the natural world. As we make our observations, we will address one of the most controversial

FIGURE 22.2 **Charles Darwin.** As a young naturalist, Darwin conceived of and developed the theory of evolution. He spent nearly 25 years collecting data and building evidence for his hypothesis before publishing it.

(a) Tortoise from Santa Cruz Island, an island with abundant vegetation.

(b) Tortoise from Española Island, an island with sparse vegetation.

FIGURE 22.3 **Giant tortoises of the Galápagos.** Tortoises with dome-shaped shells (a) are common on islands with abundant vegetation, while those with flatter shells (b) are found on islands where vegetation is less abundant. The flatter shell may allow the tortoise to reach leaves on low shrubs as well as on the ground.

questions underlying the debate about teaching evolution: Are humans really related to apes?

Stop and Stretch 22.1

Faith can be defined as the acceptance of ideals or beliefs that are not necessarily demonstrable through experimentation or direct observation of nature. How does faith differ from science? Can statements of faith be tested scientifically? Explain.

Evidence from Biological Classification

As modern science was developing in the sixteenth and seventeenth centuries, various methods for organizing biological diversity were proposed. Many of these **classification systems** grouped organisms by similarities in habitat, diet, or behavior. Some of these classifications placed humans with the great apes, and others did not.

Into the classification debate stepped Carolus Linnaeus, a Swedish physician and botanist. Linnaeus developed a new way to organize living organisms according to shared physical similarities. His classification system was arranged hierarchically. Organisms that shared many traits were placed in the same narrow classification, while those that shared fewer, broader traits were placed in more comprehensive categories. Linnaeus's hierarchy today takes the following form, from broadest to narrowest groupings:

Domain
  Kingdom
    Phylum (or Division)
      Class
        Order
          Family
            Genus (the "type" described in the binomial system)
              Species

Other scientists quickly adopted the logical and orderly Linnaean system of classification, and it became the standard practice for organizing biological diversity. Modern biologists, to more finely classify groups of organisms, have added sub- and/or superlevels to each step of this hierarchy (for instance, subfamily as a category between family and genus). The major domains and kingdoms in which all of life is organized are described in Table 22.1.

Linnaeus's references to "God's plan" make it clear that he believed in special creation. However, Darwin argued that Linnaeus's system was so effective because it reflected the true evolutionary relationships among living organisms. Darwin noted that the levels in Linnaean classification could be interpreted as different degrees of relationship. In other words, all species in the same family share a relatively recent common ancestor, while all families in the same class share a more distant common ancestor.

Using his classification system, Linnaeus placed humans, monkeys, and apes in the same order, which he called Primates, because humans have forward-facing eyes and coordinated hands like other primates. The modern classification of humans reflects only refinements of Linnaeus's ideas. Among living primates, humans are most like apes. Humans and apes share a number of characteristics, including relatively large brains, erect posture, lack of a tail, and increased flexibility of the thumb. Scientists now place humans and apes in the same family, Hominidae.

Humans and the African great apes (gorillas, chimpanzees, and bonobos) share even more characteristics, including elongated skulls, short canine teeth, and reduced hairiness. They are placed together in the same subfamily, Homininae.

TABLE 22.1 | The Classification of Life

Kingdom Name	Kingdom Characteristics	Examples	Approximate Number of Known Species	Domain Name and Characteristics
Plantae	Eukaryotic, multicellular, make own food, largely stationary	Pines, wheat, moss, ferns	300,000	**Eukarya** All organisms contain eukaryotic cells, in which the genetic material is contained in a nucleus. (In prokaryotes, genetic material is not separated from the rest of the cell by a membrane.)
Animalia	Eukaryotic, multicellular, rely on other organisms for food, mobile for at least part of life cycle	Mammals, birds, fish, insects, spiders, sponges	1,000,000	
Fungi	Eukaryotic, multicellular, rely on other organisms for food, reproduce by spores, body made up of thin filaments called hyphae	Mildew, mushrooms, yeast, *Penicillium*, rusts	100,000	
Protista	Eukaryotic, mostly single-celled forms, wide diversity of lifestyles, including plantlike, funguslike, and animal-like types	Green algae, *Amoeba*, *Paramecium*, diatoms, chytrids	15,000	
Monera	Prokaryotic, mostly single-celled forms, although some form permanent aggregates of cells	*Escherichia coli*, *Salmonella*, *Bacillus anthracis*, *Anabena*, sulfur bacteria	4,000	**Bacteria** Prokaryotes with cell wall containing peptidoglycan. Wide diversity of lifestyles, including many that can make their own food.
		Thermus aquaticus, *Halobacteria halobium*, methanogens	1,000	**Archaea** Prokaryotes without peptidoglycan and with similarities to Eukarya in genome organization and control. Many known species live in extreme environments.

The orange boxes indicate the six categories currently used to classify the diversity of life.

Five-Kingdom System

Three-Domain System

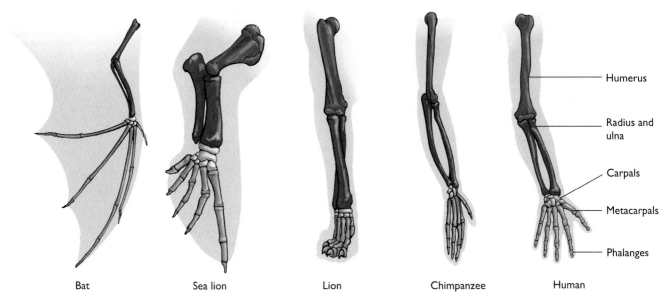

FIGURE 22.4 Homology of mammal forelimbs. The bones in the forelimbs of these mammals are very similar, although they are used for different functions. Equivalent bones in each organism are shaded the same color. The similarity in basic bone structure is evidence of shared origin.

Evidence from Homology: Related Species Are Similar

The tree of relationship implied by Linnaeus's classification forms a hypothesis that can be tested. If modern species represent the descendants of ancestors that also gave rise to other species, we should be able to observe other, less obvious similarities between related modern species in anatomy and genetic material.

FIGURE 22.4 illustrates the concept of **homology**, which is the similarity in characteristics that has resulted from common ancestry. Each of the mammal forelimbs pictured in the figure has a different function. However, each of these limbs shares a common set of bones that are in the same relationship to each other. The most likely explanation for the similarity in the underlying structure of these limbs is that each species inherited the basic structure from the same common ancestor. In this case, the homology is in anatomy, that is, physical structure. However, homology can be observed in other features of organisms from their behavior to their DNA.

While similarities in the structure of functional limbs provide support for the hypothesis of common ancestry, even more compelling evidence comes from similarities between functional traits in one organism and nonfunctional features, or **vestigial traits**, in another. FIGURE 22.5 provides one example of a vestigal trait in humans. Great apes and humans have a tailbone like other primates, yet neither great apes nor humans have a tail. Another familiar vestigial trait is goose bumps. All mammals possess tiny muscles called arrector pili at the base of each hair. When the arrector pili contract under conditions of emotional stress or cold temperatures, the hair is elevated to increase the perceived size of the animal and improve the insulating value of the hair coat. In humans, the same conditions produce only small temporary pimples, which provide neither benefit. In other words, the function of the arrector pili is vestigial in humans.

Darwin maintained that the hypothesis of common descent provided a better explanation for vestigial structures than did the hypothesis of special creation represented by the static model. A useless trait such as goose bumps is better explained as the result of inheritance from our biological ancestors than as a feature that appeared independently in our species.

Another striking similarity among organisms that Darwin and his contemporaries observed was the resemblance of different chordates early in their development. All chordates—animals that have a backbone or closely related structure—produce structures early in development called pharyngeal slits, and most have tails as early embryos (FIGURE 22.6).

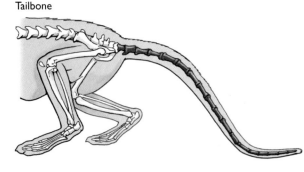

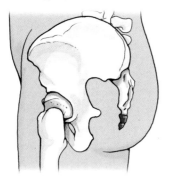

FIGURE 22.5 Vestigial traits reflect our evolutionary heritage. Humans and the great apes do not have tails, but they do have a vestigial tailbone, which corresponds to the functional tailbone of a monkey.

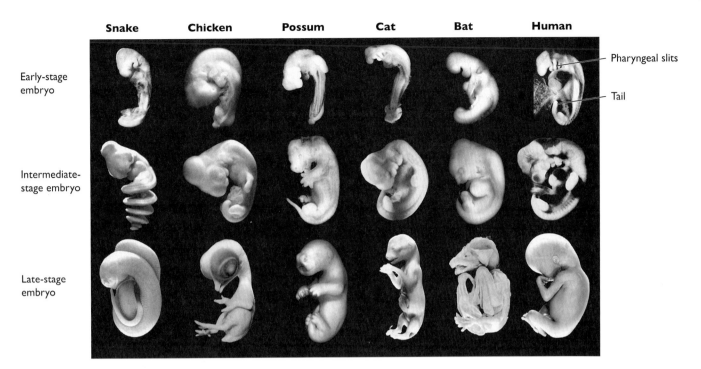

FIGURE 22.6 **Similarity among chordate embryos.** Vertebrate embryos are much alike in the first stage of their development, shown here in the top row. This observation provides evidence that these diverse organisms share a common ancestor that developed along the same pathway.

These similarities in early development support the hypothesis that humans, bats, chickens, and snakes derived from an ancestor that developed along a similar pathway and that these species thus share an evolutionary relationship with all other chordates.

Scientists now understand that differences among individuals arise largely from differences in their genes. It stands to reason that differences among species must also derive from differences in their genes. If the hypothesis of common descent is correct, then species that appear to be closely related must have more similar genes than do species that are more distantly related.

Certain genes are found in nearly all living organisms. For instance, genes that code for proteins used in electron transport chains are found in algae, fungi, fruit flies, humans, and all other organisms that use aerobic respiration. Among organisms that share many aspects of structure and function, such as primates, many more genes are shared. However, the sequences of these genes are not identical.

If we compare the sequence of DNA bases for the same gene found in two different primates, we find that the more similar their classification is, the more similar their genes are. In other words, if classification indicates that two species share a recent common ancestor, then their DNA sequences are more similar than those of two species that share a more ancient common ancestor.

A comparison of the sequences of dozens of genes that are found in humans and other primates is pictured in FIGURE 22.7. The DNA sequences of these genes in humans and chimpanzees are 99.01% similar, whereas the DNA sequences of humans and gorillas are identical over 98.9% of their length. More distantly related primates are less similar to humans in DNA sequence. This pattern of similarity in DNA sequence exactly matches the biological relationships implied by physical similarity.

Evidence from Biogeography

Even without the DNA evidence that modern biologists have, Darwin was able to draw on many examples of anatomical and behavioral homology to support the hypothesis that humans are closely related to chimpanzees and other great apes. He predicted that

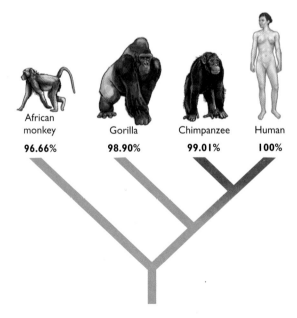

FIGURE 22.7 **Similar organisms have similar DNA sequences, which mirrors their overall similarity.**

additional evidence for this hypothesis would be the discovery of fossils of human ancestors in Africa. Darwin based this prediction on similar patterns he had seen of the distribution of species on Earth, that is, of **biogeography**, the geography of life.

Earlier in this chapter, we described one biogeographical pattern observed by Darwin that supported the hypothesis of common descent—the relationships between different populations of giant tortoises on different islands in the Galápagos. Tortoises on each island are distinct in appearance. These varieties of tortoises all belong to the same species, however, and thus must have the same ancestor. This observation convinced Darwin that all of the populations had changed over time independently on each islands.

Even more dramatic to Darwin was the appearance of unique species of mockingbirds on each island, all similar in appearance to a different species found on mainland Ecuador. Darwin noted that although the unique bird species on the Galápagos appeared to be related to bird species on the nearby mainland, the groups of species on these islands were very different from species found on other, more distant island.

These kinds of observations support the hypothesis that species in a geographic location are generally descended from ancestors in that geographic location. If the alternative hypotheses that species appeared independently were true, we would expect to see a set of "tropical-island-adapted" species that are found on all tropical islands.

We can now see why Darwin predicted, based on biogeographical patterns, that evidence of human ancestors would be found in Africa, the home of chimpanzees and other great apes. If humans and apes share a common ancestor, then highly mobile humans must have first appeared where their less-mobile relatives can still be found.

Evidence from the Fossil Record

As with nearly all evidence in science, the observations described above allow us to infer the accuracy of a hypothesis but do not prove the hypothesis correct. This type of evidence is similar to the "circumstantial evidence" presented in a murder trial, such as finding the murder weapon in a car belonging to a suspect.

But as in a murder trial, direct evidence is always preferred to establish the truth—for instance, the testimony of an eyewitness or a recording of the crime by a security camera. Of course, there are no human eyewitnesses to the evolution of humans, but we have a type of "recording," in the form of the fossil record. The evidence from the fossil record provides even more convincing support for the theory of common descent.

Fossils are the remains left in soil or rock of living organisms. Most fossils of large animals are rocks that have formed as the organic material in bone decomposed and minerals filled the spaces left behind (FIGURE 22.8).

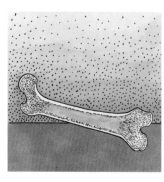 ❶ An organism is rapidly buried in water, mud, sand, or volcanic ash. The tissues begin to decompose very slowly.

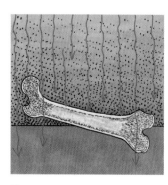 ❷ Water seeping through the sediment picks up minerals from the soil and deposits them in the spaces left by the decaying tissue.

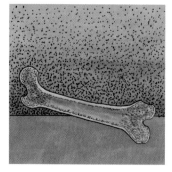 ❸ After thousands of years, most or all of the original tissue is replaced by very hard minerals, resulting in a rock model of the original bone.

 ❹ When erosion or human disturbance removes the overlying sediment, the fossil is exposed (as shown here looking from above).

FIGURE 22.8 **Fossilization.** When bones fossilize, the material that makes up much of their structure slowly decays and is replaced by minerals from water seeping through the surrounding sediments. Eventually what remains is a rock "model" of the original bone.

The process of fossilization requires special conditions because most organisms either quickly decompose after death or are scavenged by other organisms. Therefore, the fossil record is not a complete "recording" of the history of life—it is more like a security video that captures only a small portion of the action, with many blank segments. Just as the blank segments in a security video do not make the video an incorrect record of events, "gaps" in the fossil record do not diminish its value.

One key difference between humans and other apes is our mode of locomotion. While chimpanzees and gorillas use all four limbs to move, human beings are bipedal; that is, we walk upright on only two limbs.

Bipedalism evolved through several anatomical changes. The foramen magnum, the hole in the skull through which the spinal cord passes, is found on the back of the skull in other apes but at the base of the skull in human beings. In addition, the structures of the human pelvis and knee are modified for an upright stance. The foot is changed from being grasping to weight-bearing. The lower limbs are also elongated relative to the front limbs (FIGURE 22.9). For a fossil to be classified as a relative of modern humans—a **hominin**—its skeleton must show evidence of bipedalism.

While many hominin fossils have been found in Europe and Asia, the oldest fossils have been discovered in southern and eastern Africa, including the famous "Lucy." Lucy is a remarkably complete, 3.2 million-year-old skeleton of the species *Australopithecus afarensis* that was discovered in Ethiopia in 1974. Lucy's skeleton included a large section of her pelvis, which clearly indicated that she walked upright.

Scientists determined Lucy's age and that of other hominin remains by estimating the age of the rock that surrounded the fossil. **Radiometric dating** relies on a natural process of change in particular chemical elements. This process, called radioactive decay, involves the breakdown of radioactive elements in the rock into different, unique elements known as *daughter products*. Each radioactive element decays at its own unique rate.

The rate of decay is measured by an element's half-life—the amount of time required for one-half of the amount of the element to decay into the daughter product. By determining the ratio of radioactive element to daughter product in a rock sample, scientists can use the half-life of the radioactive element to estimate the number of years since the rock formed.

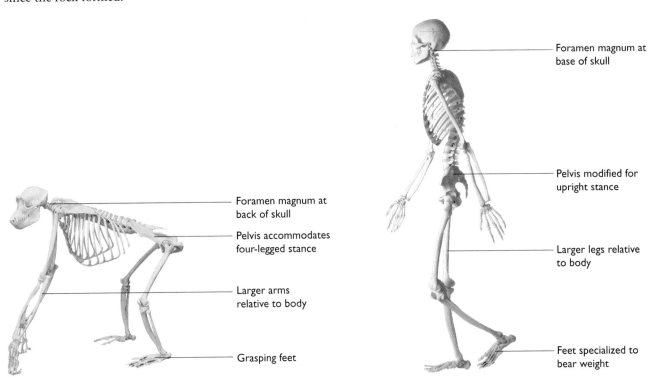

FIGURE 22.9 **Anatomical differences between humans and chimpanzees.** The evolution of bipedalism required several anatomical changes of four-footed ancestors. If any of these features are present in a fossil primate, the fossil is classified as a hominin.

Using radiometric dating, scientists have also determined that the most ancient hominin fossil, the species *Ardepithecus ramidus*, is 5.2 to 5.8 million years old. These very early fossils probably represent hominins that are quite similar to the common ancestor of humans and chimpanzees. As Darwin predicted, this species was discovered in Ethiopia, in equatorial Africa.

As the number of described hominin fossils has increased, a tentative genealogy of human beings has emerged. The species can be arranged in a pedigree from most ancient to most modern species by determining the ages and anatomical similarities among organisms (FIGURE 22.10). The pedigree indicates that modern humans are the last remaining branch of a once diverse group of hominins. But does the pedigree provide convincing evidence that modern humans evolved from a common ancestor of other apes?

Besides being bipedal, human beings differ from other apes in having a relatively large brain, a flatter face, and a more extensive culture. The oldest hominins are bipedal but are otherwise similar to other apes in skull shape, brain size, and probable way of life. More modern hominins show greater similarity to modern humans, with flattened faces and increased brain size (FIGURE 22.11). Even younger fossil finds indicate the existence of art and other abstract symbols and extensive tool use, trademarks of modern humans. The fossil record of hominins showing a progression from more apelike to more humanlike features provides compelling evidence of our evolution from ancient apes.

The physical evidence we have discussed thus far allows us to clearly reject the static model. Radiometric dating has established that Earth is far older than 10,000 years, and

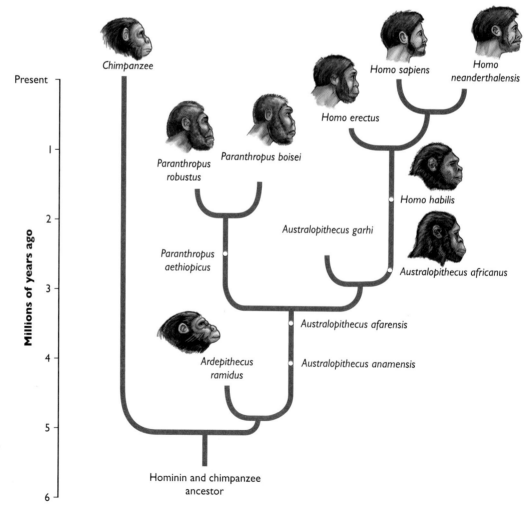

FIGURE 22.10 The evolutionary relationships among hominin species. This tree represents the current consensus among scientists who are attempting to uncover human evolutionary history. Modern humans are the last remaining species of a group that was once highly diverse.

VISUALIZE THIS According to this tree diagram, were there time periods during which more than one hominin species lived on Earth?

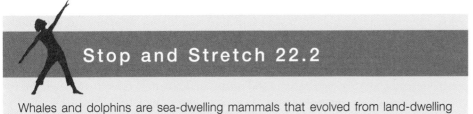

FIGURE 22.11 The ape-to-human transition. Ancient hominins display numerous apelike characteristics, including a large jaw, small braincase, and receding forehead. More recent hominins look more like modern humans.

the fossil record provides evidence beyond a reasonable doubt that the species that have inhabited this planet have changed over time. Homologies provide evidence of shared ancestry. Evolution from common ancestors is also supported by evidence about the emergence of new species.

Stop and Stretch 22.2

Whales and dolphins are sea-dwelling mammals that evolved from land-dwelling ancestors. Describe two pieces of evidence you can imagine that would support this hypothesis.

22.2 The Origin of Species

According to an idea called the **biological species concept**, a species is defined as a group of individuals that can interbreed and produce fertile offspring but that cannot reproduce with members of other species. The inability of pairs of individuals from different species to produce fertile offspring is known as **reproductive isolation** (see FIGURE 22.12 on the next page). How does a single species give rise to two different, reproductively isolated species?

Speciation: How One Becomes Two

This evolution of one or more species from an ancestral form is called **speciation**. For one species to give rise to a new species, most biologists agree that three steps are necessary:

1. Subgroups, or **populations**, of the species become isolated.
2. Evolutionary changes happen in one or both of the isolated populations, causing them to **diverge** (see FIGURE 22.13 on the next page).
3. When the divergence is great enough, reproductive isolation between these populations evolves, preventing any future breeding.

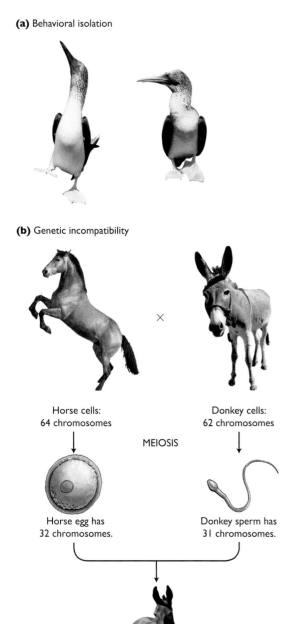

FIGURE 22.12 **Reproductive isolation.** (a) These blue-footed boobies are isolated from other birds by their behavior. Female blue-footed boobies will not mate with males who do not perform a very specific mating dance. (b) Horses and donkeys are isolated by genetic incompatibility. Even though they can produce a hybrid mule offspring, the hybrid is sterile because it cannot produce normal sperm or egg cells.

Populations can become isolated from each other for several reasons. Often a small population becomes isolated when it migrates to a distant location or by the intrusion of a geologic barrier.

Separation between two populations may also occur even if the populations are living physically near each other, if certain individuals specialize in particular food items or mate at slightly different times.

When populations have diverged in reproductive behavior or become genetically incompatible, individuals from different populations can no longer reproduce together. Each population can now be called a different species. Why do isolated populations diverge from each other? There are several mechanisms by which divergence happens. The most prominent, and the mechanism that Darwin elucidated in *The Origin of Species*, is natural selection.

The Theory of Natural Selection

Within 20 years of publishing *The Origin of Species*, Darwin could see that the theory of common descent had been accepted by most scientists. However, it was another 60 years before the scientific community accepted Darwin's ideas about how the great variety of living organisms had come about—the process he called natural selection.

The **theory of natural selection** is elegantly simple. It is an inference based on four general observations.

1. Individuals within populations vary. Observations of groups of humans support this statement—people do come in an enormous variety of shapes, sizes, colors, and facial features. We can add all kinds of less obvious differences to this visible variation. For example, people differ in the percentage of dietary fat they convert to cholesterol. Each different type of individual in a population is termed a *variant*.

2. Some of the variation among individuals can be passed to their offspring. Although Darwin did not understand how it occurred, he observed many examples of the general resemblance between parents and offspring. Darwin noticed that pigeon breeders took advantage of the inheritance of variation to produce fancy birds. Darwin hypothesized that offspring tend to have the same characteristics as their parents in natural populations as well.

When Gregor Mendel's work on inheritance in pea plants (discussed in Chapter 19) was rediscovered in the 1900s, the mechanism for this observation became clear—natural selection operates on genetic variation that can be passed from one generation to the next.

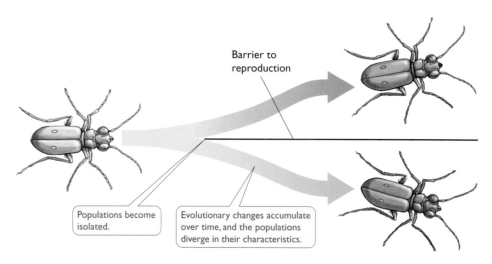

FIGURE 22.13 **Isolation of populations leads to divergence of traits.** In this model situation, populations of beetles diverge as each adapts to its own particular environmental conditions.

3. Populations of organisms produce more offspring than will survive. This observation is clear to most of us—the trees in the local park make literally millions of seeds every summer, but only a few of the much smaller number that sprout will live for more than a few years.

In *The Origin of Species*, Darwin gave a graphic example of the difference between offspring production and survival. He used elephants, animals that live long lives and are very slow breeders. A female elephant does not begin breeding until age 30, and she produces about one calf every 10 years until around age 90. Darwin calculated that even at this very low rate of reproduction, if all the descendants of a single pair of African elephants survived and lived full, fertile lives, after about 500 years their family would have more than 15 million members—many more than can be supported by all the available food in African savannas! The capacity for reproduction far outstrips the resources of the environment, so many individuals do not survive to maturity.

4. Survival and reproduction are not random. In other words, the subset of individuals that survive long enough to reproduce is not an arbitrary group. Some variants in a population have a higher likelihood of survival and reproduction than other variants do.

The relative survival and reproduction of one variant compared to others in the same population is referred to as its **fitness**. Traits that increase an individual's fitness in a particular environment are called **adaptations**. Individuals with adaptations to a particular environment are more likely to survive and reproduce than are individuals lacking these adaptations.

Darwin referred to the results of differential survival and reproduction as natural selection and noted that it could lead to evolution. For example, among the birds called Darwin's finches, scientists have observed that when rainfall is scarce, a large bill is an adaptation. Birds with larger bills are able to crack open large, tough seeds—the only food available during severe droughts. Thus individuals with larger bills were more likely to survive the drought than other, smaller-billed members of the population; that is, they were selected for. As shown in FIGURE 22.14, the 300 survivors of a 1977 drought had an average bill depth that was 6% greater than the average bill depth of the original population of 1,300 birds.

Scientists now understand that the random process of gene mutation generates the raw material—variations—for evolution. Natural selection acts as a filter that selects for or against new alleles produced by mutation.

Darwin's inference: Natural selection causes evolution. The result of natural selection is that an adaptation becomes more common in a population as the individuals with this adaptation contribute greater numbers of their offspring to the next generation. Natural selection results in a change in the traits of individuals in a population over the course of generations—voilà, evolution.

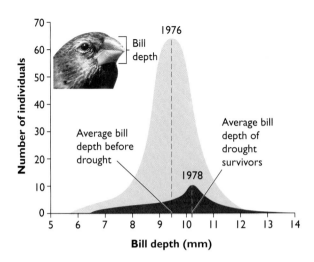

FIGURE 22.14 **Survival and reproduction are not random.** The pale purple curve summarizes bill depth in Darwin's finches in the Galápagos Islands before a drought. The same population after the drought of 1976 and 1977 (dark purple curve) had an average bill depth approximately 0.5 mm greater. During these years, finches with larger bills that could crack larger seeds had higher fitness than did smaller-billed birds.

Critical Thinking About Natural Selection

Natural selection is a fairly simple idea, but it is surprisingly easy to misunderstand. The following list details some common misconceptions about this process.

1. Fitness is a relative trait, not an absolute one. One misunderstanding stems from a common phrase used to describe natural selection—"survival of the fittest." It is important to recognize that natural selection results in the survival and reproduction of those individuals that are best adapted to the current environment given the current variants in the population. The survivors are not "fittest" in an absolute sense, only relatively, meaning better adapted than others who lack those traits.

2. Natural selection does not result in "ideal" organisms. Natural selection causes populations to become adapted to their environment, but the result of that process is not necessarily "better" organisms—simply ones that are better survivors in the current situation. Changes in traits that increase survival and reproduction in one environment may be liabilities in another environment. Nearly all adaptations have a trade-off—increased success in one environment or in one aspect of survival leads to decreased success in another environment or aspect of survival.

FIGURE 22.15 The panda's thumb. The thumb is not made up of hand bones but of a wrist bone adapted to help these animals strip leaves from bamboo shoots, their main food.

3. Natural selection does not result in the "progress" of a population toward a predetermined goal. Instead, natural selection is situational. This fact helps explain why nonhuman bipedal apes do not exist today. These apes were only successful in an environment that humans did not inhabit. The fitness advantage of a new bipedal ape would be very small, because another, very successful biped already exists. Although human beings may seem to be superior animals, chimpanzees are well adapted to their habitats and way of life.

4. Natural selection does not cause the emergence of new variants. The process of natural selection acts only on the variants that are available within a population. These variants are produced by the random process of mutation and are typically modifications of the organism's underlying structure rather than new traits. The constraints on adaptations of organisms are apparent throughout nature in what evolutionary biologists call *jury-rigged design*, meaning "made using whatever is available."

One of the most famous examples of jury-rigged design is the "thumb" found on a giant panda's front paws (FIGURE 22.15). These animals apparently have six digits: five fingers composed of the same bones as our fingers, and a thumb constructed from an enlarged bone equivalent to one found in our wrist. This structure in pandas is an adaptation that increases their ability to strip leaves from bamboo shoots, their main food source. A more effective design for an opposable thumb is our own, adapted from one of the basic five digits. However, in the panda population, this variation did not exist. Individuals with enlarged wrist bones did exist, so what evolved in giant pandas was a jury-rigged thumb that does its job but is not as flexible as our own thumb.

The intelligent design textbook *Of Pandas and People* has in its title an explicit reference to the panda's thumb, and it uses a long-settled debate over its origin as one way to attack the supposed uncertainty of evolutionary theory.

5. Natural selection cannot change the characteristics of individuals, only populations. This misconception about natural selection can be illustrated by the following erroneous statement: "The dodo could not adapt when human hunters arrived, so it went extinct." This assertion seems to presume that individuals must change within their lifetime to survive environmental changes. However, evolution can occur only when traits that influence survival are present in a population and have a genetic basis. Natural selection cannot cause an individual to change the alleles it carries.

Stop and Stretch 22.3

Explain the misunderstandings about evolution that are apparent in the following statement: If global warming causes polar bears to go extinct, it is the bear's fault. All organisms should adapt to the warming climate. Polar bears could adapt by switching from hunting seals to eating fish, insects, and berries the way grizzly bears do now.

Evolution: A Robust Theory

The most compelling evidence for the single origin of all life is the universality of both DNA and of the relationship between DNA and proteins. As noted in Chapter 4, genes from cows can be transferred to bacteria, and the bacteria will make a functional cow protein. This is possible only because both bacteria and cows translate genetic material into functional proteins in a similar manner. If bacteria and cows arose separately, we could not expect them to translate genetic information similarly.

Other evidence also supports the hypothesis of a single common ancestor. For instance, Chapter 3 discussed the structures of human cells—structures that are also found in all organisms containing eukaryotic cells (cells with a nucleus enclosing the genetic material). While it may be difficult to imagine that organisms as different as humans, pine trees, and ladybugs all share a single common ancestor, our shared cellular processes point to an ancient relationship.

Scientists favor the theory of common descent because it is the best scientific explanation for how modern organisms came about. The theory of evolution—including the theory of common descent—is robust. In other words, evolution explains a wide variety of observations and is well supported by a large body of evidence from anatomy, geology, molecular biology, and genetics.

Proponents of teaching "alternatives to evolution" in science courses should explain how their ideas provide an explanation of the natural world that is as effective as the theory of evolution. These proponents have not done so. The hypothesis of intelligent design, currently the most popular alternative, does not have any experimental support. Instead, it is based primarily on a willful misreading of data, manufactured or overblown controversies about current research, a reliance on false statistical arguments about probability, and the idea, known as *irreducible complexity*, that certain structures or processes are too complex to have arisen without a conscious designer.

Irreducible complexity has been around since well before Darwin. In fact, Darwin tackled this argument in *The Origin of Species* by explaining how the very complex eye found in humans could have evolved from much simpler eyes in a series of steps. As evidence for this process, Darwin described a series of eye types found in other organisms that, when put together, could illustrate the steps in this process (FIGURE 22.16).

Evolution provides a framework for understanding how even very complex organs and organisms could have arisen on Earth by purely natural processes. The alternative hypotheses to evolution either have been convincingly falsified or lack observational and experimental support. School boards and legislatures do not serve their children well by mandating the teaching of these alternatives as equally likely explanations for the diversity of life.

Understanding evolution also helps us piece together the history of human life—and gives us the tools for thinking about our prospects on this planet.

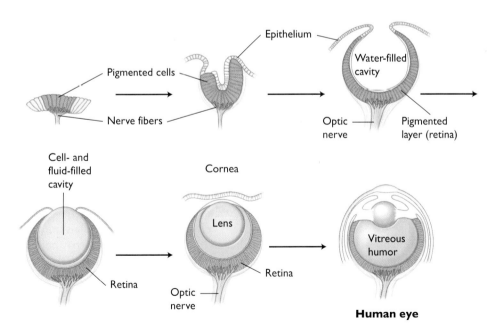

FIGURE 22.16 Eye evolution. The major steps in eye evolution are illustrated here. Darwin identified modern organisms with each of these light- or image-sensing organs, supporting the principle that complexity could have arisen in steps from simpler functional units.

22.3 Human Evolution

Linnaeus coined the binomial name *Homo sapiens* (*Homo* meaning "human being," and *sapiens* meaning "knowing" or "wise") to describe the human species. Modern biologists have added a subspecies name, *Homo sapiens sapiens*, to distinguish modern humans from earlier humans.

Humans are one of the most successful species on Earth. We are abundant, inhabit a huge range of habitats, and have modified large areas of the globe to meet our needs. People come in a variety of shapes, colors, and sizes—all variants that have evolved since we shared our most recent common ancestor. Traditionally, scientists have divided humans into racial groups based on physical differences. Linnaeus famously did so, even assigning stereotypical behavioral traits to each race—and egocentrically noting that his own race was the superior type (FIGURE 22.17).

How did the diversity among human populations arise? Is this diversity meaningful, as Linnaeus's classification seems to suggest? Or are human differences truly only "skin deep"?

According to fossil evidence, all modern human populations descended from an African population beginning about 200,000 years ago. In other words, the physical differences we see among human populations must have arisen in about 10,000 human generations. In evolutionary terms, this is not much time. This recent shared ancestry of human groups supports the hypothesis that human races are not very different from each other.

Analysis of DNA sequence differences and genetic data indicates that no clear boundaries within the human species correspond to classifications made on skin color. Because human populations have always interbred, there are no genes unique to a single race and no set of alleles that can clearly identify someone as belonging to one race or another. The differences among Linnaeus's races appear to be little more than skin deep.

In fact, most biologists reject the idea that humans can be grouped into races at all. Although human populations are remarkably similar, it is clear that certain physical traits are more common in some groups than in others. These physical differences have arisen as a result of evolutionary change within partially isolated human populations.

Why Human Groups Differ: Selection and Genetic Drift

Differences among partially isolated human groups evolved via selection and chance and are reinforced by human mating patterns. In this section, we use the term "race" to refer to the still commonly used population groupings of Africans (blacks), Europeans (whites), Asians, American Indians, and Pacific Islanders. Although these groups have relatively few biological differences, they are still used by Western scientists to group smaller human populations into larger related groups.

Natural Selection Causes Some Differences Between Groups When scientists compare the average skin color in a native human population to the level of ultraviolet (UV) light to which that population is exposed, they see a nearly perfect correlation—the lower the UV light level, the lighter the skin. As a result, different populations within a race can have very different skin colors. In other words, the skin color of a population appears to be a function of the environmental conditions that population inhabits rather than a sign of genetic similarity to same-colored populations.

UV light is light energy in a range that is not visible to the human eye. UV light interferes with the body's ability to retain the vitamin folic acid, which is required for proper development in fetuses and for adequate sperm production in males. Men with low folic acid levels have low fertility, and women with low folic acid levels are more likely to have children with severe birth defects.

Darker-skinned individuals absorb less UV light and thus have higher folic acid levels in high-UV environments than light-skinned individuals do. In other words,

---

I. HOMO.

Sapiens. Diurnal; varying by education and situation.

- Copper-coloured, choleric, erect. *American.*
  Hair black, straight, thick; *nostrils* wide, *face* harsh; *beard* scanty; *obstinate*, content free. *Paints* himself with fine red lines. *Regulated* by customs.

- Fair, sanguine, brawny. *European.*
  Hair yellow, brown, flowing; eyes blue; gentle, acute, inventive. Covered with close vestments. *Governed* by laws.

- Sooty, melancholy, rigid. *Asiatic.*
  Hair black; eyes dark; severe, haughty, covetous. Covered with loose garments. *Governed* by opinions.

- Black, phlegmatic, relaxed. *African.*
  Hair black, frizzled; *skin* silky; *nose* flat; *lips* tumid; *crafty,* indolent, negligent. *Anoints* himself with grease. *Governed* by caprice.

FIGURE 22.17 Linnaean classification of human variety. Linnaeus published this classification of the varieties of humans in the tenth edition of *Systema Naturae* in 1758. The behavioral characteristics he attributed to each variety reflect the common biases among Western scientists of the time.

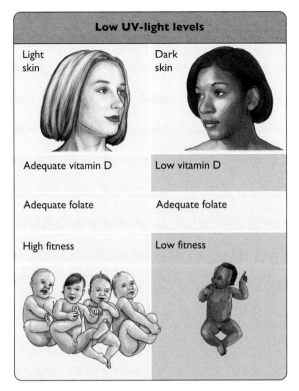
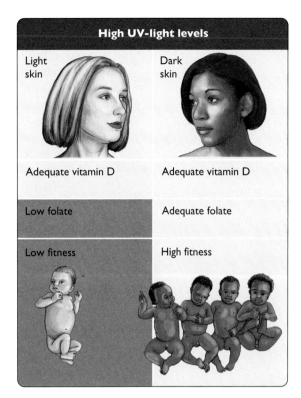

FIGURE 22.18 The relationship between UV light levels, folate, vitamin D, and skin color. Populations in regions where UV light levels are high experience selection for darker, UV-resistant skin. Populations in regions where UV light levels are low experience selection for lighter, UV-transparent skin.

in environments where UV light levels are high, dark skin is favored by natural selection—it is an adaptation to maintain proper levels of folic acid.

On the other hand, absorption of UV light is required for the synthesis of vitamin D, which is crucial for the proper development of bones. In areas where levels of UV light are low, individuals with lighter skin are able to maximize their absorption of what light is available and thus have higher levels of vitamin D. In these environments, light skin is an adaptation (FIGURE 22.18).

Sexual Selection Also Creates Differences Men and women within a population may have preferences for particular physical features in their mates. These preferences can cause populations to differ in appearance. When a trait influences the likelihood of mating, that trait is under the influence of **sexual selection**.

Darwin proposed sexual selection in 1871 as an explanation for many of the differences between male and female animals. For instance, the enormous tail on a male peacock exists because female peahens choose mates with showier tails. Because large tails require a lot of energy to display, and males with big tails are more visible to predators, peacocks with the largest tails must be both physically strong and smart to survive. Peahens can use the size of the tail, therefore, as a measure of the "quality" of the male (FIGURE 22.19).

In humans, there is some evidence that the difference in overall body size between men and women is a result of sexual selection. That is, females may prefer larger males, perhaps because size may indicate overall fitness. Populations vary in this trait—in some populations, men and women are approximately the same size, while in others the differences are great. A hypothesis that explains this difference is that the process of sexual selection differed among isolated human groups.

FIGURE 22.19 The effects of sexual selection. Sexual selection is responsible for many unique and fantastic characteristics of organisms, such as the peacock's tail.

Differences Can Arise by Chance Human populations are prone to travel and to send small groups to colonize new areas. As these movements occur, evolutionary changes can occur by chance, through a process called **genetic drift**.

A common cause of genetic drift occurs when a small sample of a larger population moves away. The new population founded by these migrants is rarely an exact copy of its source, leading to a divergence called the **founder effect**. Genetic diseases that are unusually common in certain populations often result from the founder effect. For example, Ellis–van Creveld syndrome, a recessive disease that causes dwarfism, is 5,000 times more common in the Pennsylvania Amish population than in other German-American groups. This difference comes from a single founder in the original population of Pennsylvania Amish who carried this very rare allele.

Genetic drift may also occur as the result of a **population bottleneck**, a dramatic reduction in population size followed by a rapid increase in population. A sixteenth-century bottleneck on the island of Pukapuka in the South Pacific resulted in the evolution of the population there: The 17 survivors of a tsunami on Pukapuka were all petite, and their modern descendants are significantly shorter compared to nearby island populations.

Even without a bottleneck, evolutionary change can occur by chance. When an allele is low in frequency within a small population, only a few individuals carry a copy. If one of these individuals fails to reproduce, the frequency of the rare allele may drop in the next generation. The smaller a population is, the more likely alleles will be lost in this way. For example, the Hutterites are a religious sect with small, isolated communities in South Dakota and Canada. Genetic drift in this population over the last century has resulted in a near absence of type B blood among the Hutterites, compared to a frequency of 15% to 30% in other European-Americans.

Humans are and have been a highly mobile species. Most early human populations were also probably quite small. These factors make human populations susceptible to the founder effect, genetic bottlenecks, and genetic drift, and have contributed to the differences among modern human groups.

Stop and Stretch 22.4

The disease phenylketonuria (PKU) results from the inability to metabolize the amino acid phenylalanine. In homozygous individuals, it results in severe mental retardation if untreated. The allele that causes PKU has a frequency in Irish populations of 1 in every 7,000 births, while the frequency is 1 in 18,000 in urban British populations and only 1 in 36,000 in Scandinavian populations. Give two reasons that this allele may be found in different frequencies in these populations.

Differences Are Maintained by Assortative Mating Some differences between human populations may be reinforced by how people choose their mates. Individuals usually prefer to marry someone who is like themselves, a process called positive **assortative mating**. For example, there is a tendency for people to mate assortatively by height—that is, tall women tend to marry tall men—and by skin color.

When two populations differ in obvious physical features, the number of matings between them may be small if the traits of one population are unattractive to members of the other population. Positive assortative mating maintains and even exaggerates physical differences between populations. In highly social human beings, assortative mating may reinforce differences between groups.

Although human populations may show superficial differences due to natural selection, genetic drift, sexual selection, and assortative mating, many of these differences are literally no more than skin deep. Beneath a veneer of physical differences, humans are basically the same.

Evolution in the Classroom

Earlier, we referred to evolution as a robust theory with ample scientific support. In this section, we have seen how the theory of evolution can help us understand the source of minor differences among human groups. But does evolution need to be a part of the high school curriculum?

Evolutionary theory helps us understand the functions of human genes, comprehend the interactions among species, and predict the consequences of a changing global environment for modern species. Students who do not have a grasp of this fundamental biological principle may lack an appreciation of the basic unity and diversity of life. They also may fail to understand the effects of evolutionary history and change on the natural world and ourselves. Evolution is not only a foundation of biological science, it is one of the most powerful ideas in human history.

Chapter REVIEW

ROOTS TO REMEMBER

The following roots of words come mainly from Latin and Greek and will help you to decipher terms:

homini- means humanlike.

homolog- means similar or shared origin. From a word meaning "in agreement."

KEY TERMS

22.1 Evidence of Evolution
biogeography *p. 546*
biological evolution *p. 540*
classification systems *p. 542*
fossils *p. 546*
hominin *p. 547*
homology *p. 544*
radiometric dating *p. 547*
species *p. 540*
theory of common descent *p. 540*
theory of evolution *p. 540*
vestigial traits *p. 544*

22.2 The Origin of Species
adaptation *p. 551*
biological species concept *p. 549*
diverge *p. 549*
fitness *p. 551*
populations *p. 549*
reproductive isolation *p. 549*
speciation *p. 549*
theory of natural selection *p. 550*

22.3 Human Evolution
assortative mating *p. 556*
founder effect *p. 556*
genetic drift *p. 556*
population bottleneck *p. 556*
sexual selection *p. 555*

SUMMARY

22.1 Evidence of Evolution

- The process of evolution is the change that occurs in the characteristics of organisms in a population over time (p. 540).
- The theory of evolution, as described by Charles Darwin, is that all modern organisms are related to each other and arose from a single common ancestor (p. 540).
- Linnaeus classified humans in the same order with apes and monkeys based on his observations of physical similarities between these organisms. Darwin argued that the pattern of biological relationships illustrated by Linnaeus's classification provided strong support for the theory of common descent (p. 542; Figure 22.7).

- Similarities in the underlying structures of a variety of organisms and the existence of vestigial structures are difficult to explain except through the theory of common descent (p. 544).
- Similarities in embryonic development among diverse organisms are best explained as a result of their common ancestry (p. 545).
- Modern data on similarities of DNA sequences among organisms provide an independent line of evidence supporting the hypothesis of common descent (p. 545).
- Biogeographical patterns support the hypothesis of common descent because species that appear related physically are also often close to each other geographically (p. 546).
- As predicted by the theory of common descent, fossils of ancient hominins have more apelike characteristics than do more modern hominins (pp. 546–547).

22.2 The Origin of Species

- All humans belong to the same biological species, *Homo sapiens sapiens*. A biological species is defined as a group of individuals that can interbreed and produce fertile offspring (p. 549).
- Speciation occurs when populations of a species become isolated from each other. These populations diverge from each other, and reproductive isolation between the populations evolves (pp. 549–550).
- The theory of natural selection consists of four observations and an inference: (1) Individuals in a population vary, and (2) some of this variation can be passed on to offspring. (3) Not all individuals born in a population survive to adulthood, and not all adults produce the maximum number of offspring possible. (4) Advantageous traits, called adaptations, increase an individual's fitness: his or her chance of survival and/or reproduction. The increased fitness of individuals with particular adaptations causes the adaptation to become more prevalent in a population over generations. Inference: Natural selection results in evolution (pp. 550–551; Figure 22.14).

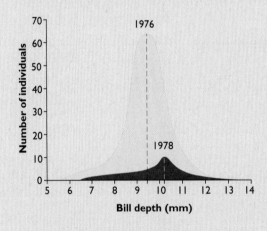

- Natural selection can act only on the variants currently available in the population, does not push a population in the direction of a predetermined "goal," and results in a population that is better adapted to its environment but usually not perfectly adapted, as a result of trade-offs (p. 551).
- Selection can cause the traits in a population to change in a particular direction but in some environments may cause certain traits to resist change and in other environments cause multiple variants to evolve (p. 552).
- Shared characteristics of all life, especially the universality of DNA and the relationship between DNA and proteins, provide evidence that all organisms on Earth descended from a single common ancestor rather than from multiple ancestors (pp. 552–553).
- Life could have evolved from nonliving precursors in a multistep process (p. 553).

22.3 Human Evolution

- Most evidence indicates that all modern humans share a single common ancestor that lived less than 200,000 years ago (p. 554).
- Genetic evidence indicates that human groups have been mixing for thousands of years. Modern human groups do not show evidence that they have been isolated from each other and formed distinct races (p. 554).
- Similarities among human populations may evolve as a result of natural selection. Light skin is more common in areas where the UV light level is low, an adaptation that is a result of natural selection in these environments (pp. 554–555).
- Sexual selection, whereby individuals—typically females—choose mates that display some "attractive" quality, may also be responsible for creating differences among human populations (p. 555).
- Human populations may show differences due to genetic drift, which is defined as changes in allele frequency due to chance events such as founder effects or population bottlenecks (p. 556).
- Positive assortative mating, in which individuals choose mates who are like themselves, can reinforce differences between human populations (pp. 556–557).

LEARNING THE BASICS

1. What observations did Charles Darwin make on the Galápakgos Islands that helped convince him that evolution occurs?
    a. the existence of animals with eyes that are simpler than human eyes
    b. the presence of species he had seen on other tropical islands far from the Galápagos
    c. the appearance of pigeons with elaborate fan-shaped tails
    d. fossils of human ancestors

2. The process of biological evolution _____.
    a. is not supported by scientific evidence
    b. results in a change in the features of individuals in a population
    c. takes place over the course of generations
    d. b and c are correct
    e. a, b, and c are correct

3. The theory of common descent states that all modern organisms _____.
    a. can change in response to environmental change
    b. descended from a single common ancestor
    c. descended from one of many ancestors that originally arose on Earth
    d. have not evolved
    e. can be arranged in a hierarchy from "least evolved" to "most evolved"

4. Which of the following lists shows the classification levels in order from broadest grouping to narrowest grouping?
    a. family, phylum, genus, order
    b. phylum, family, genus, class
    c. order, genus, species, phylum
    d. kingdom, order, genus, species
    e. class, phylum, family, order

5. All of the following structures are homologous to each other *except* _____.
    a. a sea lion flipper
    b. a bat wing
    c. fish gills
    d. a human forearm
    e. a cat leg

6. What characteristics of a fossil can paleontologists use to determine whether the fossil is a part of the human evolutionary lineage?
    a. the position of the foramen magnum
    b. the structure of the pelvis
    c. the structure of the foot
    d. a and c are correct
    e. a, b, and c are correct

7. Match each observation to the type of evidence it provides for the theory of evolution.
    _____ Fossils of the oldest hominins are found in Africa near modern great apes.
    _____ Both humans and chimpanzees have a vestigial appendix.
    _____ The brain size of ancient hominins was less than that of modern humans.

    a. fossil record
    b. homology
    c. biogeography

8. The fossil record of hominins _____.
    a. does not indicate a relationship between humans and apes because a missing link has not been found
    b. dates back at least 5 million years
    c. indicates that bipedal apes first evolved in Africa
    d. b and c are correct
    e. a, b, and c are correct

9. Of the following possibilities, which is the best way to measure an organism's evolutionary fitness?
    a. Measure its heart rate.
    b. Determine how long it takes to tire on a treadmill or other similar device.
    c. Count the number of offspring it produces that survive to adulthood.
    d. Expose it to cold temperatures and record how long it survives.
    e. Calculate the number of sexual partners it had in its lifetime.

10. For two populations of organisms to be considered separate biological species, they must be _____.
    a. reproductively isolated from each other
    b. unable to interbreed to produce living, fertile offspring
    c. physically very different from each other
    d. a and b are correct
    e. a, b, and c are correct

11. According to the most accepted scientific hypothesis about the origin of two new species from a single common ancestor, most new species arise when _____.
    a. many mutations occur
    b. populations of the ancestral species are isolated from each other
    c. there is no natural selection
    d. a Creator decides that two new species would be preferable to the old one
    e. the ancestral species decides to evolve

12. Which of the following observations is *not* part of the theory of natural selection?
    a. Modern organisms descended from a single common ancestor.
    b. There is variation among individuals in a population.
    c. Populations of organisms have more offspring than will survive.
    d. Traits can be passed on from parent to offspring.
    e. Some variants in a population have a higher probability of survival and reproduction than other variants do.

13. The best definition of evolutionary fitness is _____.
    a. physical health
    b. the ability to attract members of the opposite sex
    c. the ability to adapt to the environment
    d. survival and reproduction relative to other members of the population
    e. overall strength

14. An adaptation is a trait of an organism that increases _____.
    a. its fitness
    b. its ability to survive and replicate
    c. in frequency in a population over many generations
    d. a and b are correct
    e. a, b, and c are correct

15. The heritable differences among organisms are a result of _____.
    a. differences in their DNA
    b. mutation
    c. differences in alleles
    d. a and b are correct
    e. a, b, and c are correct

16. For life to form from nonliving building blocks, _____.
    a. simple molecules common in nature would have to self-assemble into more complex molecules
    b. a mechanism for storing and transmitting information would have to spontaneously appear
    c. materials would have to be segregated in enclosed compartments
    d. cell-like structures would have to self-replicate
    e. all of the above

17. Similarity in skin color among different human populations appears to be primarily the result of _____.
    a. natural selection
    b. chance
    c. their racial category
    d. a and b are correct
    e. a, b, and c are correct

18. All of the following are examples of genetic drift *except* _____.
    a. the founder effect
    b. the loss of alleles that occurs when a population suddenly loses a large number of its members
    c. the loss of rare alleles from small populations
    d. the change in frequency in a trait when a small group migrates away from its source population to found a new colony
    e. sexual selection

19. The tendency for individuals to choose mates who are like themselves is called _____.
    a. natural selection
    b. sexual selection
    c. positive assortative mating
    d. the founder effect
    e. random mating

20. **True or False?** The theory of intelligent design is nearly as well supported as the theory of evolution and is equally robust as a scientific explanation for life on Earth.

ANALYZING AND APPLYING THE BASICS

1. The classification system devised by Linnaeus can be "rewritten" in the form of an evolutionary tree. Draw a tree that illustrates the relationship among these flowering plant species, given their classification (note that "subclass" is a grouping between class and order):

    Pasture rose (*Rosa carolina*, family Rosaceae, order Rosales, subclass Rosidae)
    Live forever (*Sedum purpureum*, family Crassulaceae, order Rosales, subclass Rosidae)
    Spring avens (*Geum vernum*, family Rosaceae, order Rosales, subclass Rosidae)
    Spring vetch (*Vicia lathyroides*, family Fabaceae, order Fabales, subclass Rosidae)
    Multiflora rose (*Rosa multiflora*, family Rosaceae, order Rosales, subclass Rosidae)

2. In late 2006, paleontologists announced the discovery of the fossilized remains of a 3.3-million-year-old child, found in Ethiopia. Dubbed "Lucy's Baby" by some, the three-year-old *Australopithecus afarensis* female is the best-preserved early human ever found. Because her skeleton is so complete, researchers could verify that *A. afarensis* was adapted for bipedal movement. However, the skeleton also shows arms and shoulders that appear adapted for tree-climbing. The skeleton has a braincase about the same size as a chimpanzee's. Her one intact finger is also curved like a chimpanzee's. A rare intact hyoid bone—a bone found in the voice box—appears apelike.

    How does the *Australopithecus afarensis* fossil support the hypothesis that humans evolved from ancient apes?

Use Table 22.2 to answer questions 3 and 4.
Table 22.2 provides some characteristics of early hominins. Hominins are listed in chronological order, from ancient to modern.

3. Suppose you are a paleontologist who has just discovered a new fossil skeleton that has a mix of apelike and human features. What characteristics would you look for to help place the fossil correctly within Table 22.2?

Table 22.2 Early Hominin Characteristics

Hominin	Characteristics
Ardipithecus ramidus	Teeth intermediate between early apes and A. afarensis, possibly bipedal
Australopithecus afarensis	Apelike face, flat nose, bony ridge over the eyes, brain size 375–550 cubic centimeters (cc), large back teeth, small but pointed canines, jaw intermediate between humans and apes, pelvis and leg bones humanlike, long fingers and toes, able to walk upright
Homo habilis	Brain size 500–800 cc, teeth reduced but larger than modern human, primitive face, protruding jaw, able to walk upright
Homo erectus	Brain size 750–1,225 cc, primitive face, protruding jaw, large molars, thick brow ridge, no chin, walks upright
Homo sapiens	Brain size around 1,350 cc, small eyebrow ridges, prominent chin, small teeth, walks upright

4. What characteristics of the fossil would make it difficult to accurately place it in Table 22.2?

Use the following information to answer questions 5–7.
Consider a hypothetical situation: A species of aster normally blooms between mid-August and late September. The aster's range is located in a climate that usually produces freezing temperatures in late October. The aster's pollinators (bees, ants, and other insects) are active between late April and early October.

Mutations can occur in the gene that controls flowering time for some individuals of this aster, causing them to flower earlier, later, or longer than normal. The following table shows possible flowering times.

Aster Type	Flowering Time
Normal	August 15–September 30
Early mutation	July 1–August 15
Late mutation	October 15–November 30
Expanded mutation	July 15–September 30

5. Describe how a population of these asters could gradually change to expand its flowering time into July.

6. Which mutation is unlikely to be selected for? Explain your answer.

7. Which mutation could eventually produce a new species of aster? Explain your answer.

Use the following information to answer questions 8 and 9.
A relatively small (16,000-base-pair) loop of DNA is found within each eukaryotic cell's mitochondrion. This mitochondrial DNA, abbreviated mtDNA, is different from the much longer (about 3-billion-base-pair) strands of DNA found in the cell nucleus. Mitochondrial DNA is also inherited in a different manner than nuclear DNA is. During fertilization, nuclear DNA in both the sperm and the egg combine. Therefore, offspring inherit nuclear DNA from both father and mother. However, when the sperm fertilizes the egg, the sperm's mtDNA is excluded from the egg cell. Therefore, offspring only inherit mtDNA from the mother.

8. How could mtDNA be useful in tracing the history of early humans?

9. What other genetic material is also inherited from only one parent and thus could be used to trace the history of early humans? Explain your answer.

CONNECTING THE SCIENCE

1. Humans and chimpanzees are more alike genetically than many very similar-looking species of fruit fly are to each other. What does this similarity imply regarding the usefulness of chimpanzees as stand-ins for humans during scientific research? What do you think it implies regarding our moral obligations to these animals?

2. The theory of natural selection has been applied to human culture in many different realms. For instance, there is a general belief in the United States that "survival of the fittest" determines which businesses are successful and which go bankrupt. How is the selection of "winning" and "losing" companies in our economic system similar to the way natural selection works in biological systems? How is it different?

3. Creationists have argued that if students learn that humans descended from animals and are, in fact, a type of animal, these impressionable youngsters will take this fact as permission to act on their "animal instincts." What do you think of this claim?

Chapter 23

Ecology

Is Earth Experiencing a Mass Extinction?

LEARNING GOALS

1. What does the study of ecology entail?

2. Describe exponential growth.

3. Define *carrying capacity* and its effect on a population.

4. Why are species at higher trophic levels more vulnerable to extinction?

5. Describe the four causes of extinction that human beings contribute to.

6. Compare and contrast the four ecological interactions within a community—predation, parasitism, mutualism, and competition—and give an example of each.

7. Describe how energy flows in an ecosystem and why there is less energy available at higher trophic levels than at lower levels.

8. Give two reasons why large populations are needed to remove species from threatened status.

Farmers and their supporters broke open irrigation gates to allow water to flow to their fields.

They came with chainsaws, pry bars, blowtorches, and American flags to challenge the authorities. On their way to the facility, they passed throngs of cheering supporters holding handmade signs and carrying buckets to fill with purloined resources. Then the farmers from normally sleepy Klamath Falls, Oregon, set upon the fences and gates that were preventing them from working and threatening to destroy their way of life. It was July 2001, and events in the Klamath Basin had come to a head.

The wrath of the people of Klamath Falls and surrounding communities was generated by the federal government's requirement to protect species at risk of extinction. In the case of the Klamath crisis, the species at risk are two fish—the Lost River sucker and the shortnose sucker. In the midst of a multiyear drought and dangerously low water levels in Upper Klamath Lake, home to these endangered fish, the U.S. Fish and Wildlife Service stopped the outflow of water from the lake in April 2001. The irrigation canals that had fed barley, potato, and alfalfa fields in the high desert of the Klamath Basin since the early 1900s suddenly went dry. Thousands of farmers were unexpectedly unable to produce crops and faced the prospect of bankruptcy, foreclosure, and loss of their livelihood.

Their interests are opposed by the interests of this fish.

Do fish or farmers have a greater right to this lake?

Lost River and shortnose suckers are dull-colored fish that feed on the mucky bottoms of lakes and streams in the region. Humans have not harvested these fish in large numbers for several decades. In contrast, the crops produced annually by irrigated fields in the region produce millions of dollars in income.

Ty Kliewer, then a student at Oregon State University and who still farms in the basin, summarized the feelings of many when he spoke at a public hearing about having learned the importance of balancing mathematical and chemical equations. "It appears to me that the people who run the Bureau of Reclamation and the U.S. Fish and Wildlife Service slept through those classes," Kliewer said. "The solution lacks balance, and we've been left out of the equation."

Why should we care about the fate of an endangered species?

Although the intensity of the crisis in Klamath Falls has declined somewhat, the people in the Klamath Basin are still at odds with the federal government over water allocations. A federal judge ruled in March 2006 that irrigation must be cut off to preserve the endangered fish if water levels fall below a certain level. The next drought could bring the issue back to a boiling point.

The problems in the Klamath Basin are not unique. Thousands of people all over the United States have had their jobs threatened or eliminated by the government's attempts to protect endangered species. Why should the survival of one or a few species come before the needs of humans?

Many biologists and environmentalists say that the Klamath Falls bumper sticker "Fish or Farmers?" misstates the dilemma. Instead, they argue, humans depend on the web of life that creates and supports natural ecosystems, and they worry that disruptions to this web may become so severe that our own survival as a species will be threatened. In this view, protecting endangered species is not about pitting fish against farmers; it is about protecting fish to ensure the survival of farmers. In this chapter, we explore the causes and consequences of the loss of biological diversity.

23.1 Limits to Population Growth

To devise effective strategies for meeting the needs of both fish and farmers, scientists must understand the interactions among organisms as well as between organisms and their environment. **Ecology** is the field of biology that focuses on these interactions.

Ecology can be studied at many levels—from the individual, to populations of the same species, to communities of interacting species, and finally to the effects of biological activities on the nonbiological environment, such as the atmosphere.

Principles of Population Ecology

From an ecological perspective, a **population** is defined as all of the individuals of a species within a given area. Populations exhibit a structure, which includes the spacing of individuals (that is, their distribution) and their number (abundance).

A population's distribution and abundance together provide a partial snapshot of its current circumstances. However, to better understand how a population is responding to its environment, we have to determine how it is changing through time. Historians have been able to use archeological evidence and written records to determine the size of the human population on Earth at various times during the past 10,000 years. This record, presented in FIGURE 23.1, dramatically illustrates the pattern of human population growth. As evidenced by the graph, improvements in technology have allowed the human population to explode in numbers since the beginning of the Industrial Revolution.

The graph of human population growth is a striking illustration of **exponential growth**—growth that occurs in proportion to the current total. In other words, the population does not have a fixed number of offspring every year. Instead, the quantity of new offspring is an ever-growing number. Exponential growth results in the J-shaped growth curve seen in Figure 23.1.

We know from studying nonhuman populations that unending exponential growth is impossible. The elk population in Yellowstone National Park suffered enormous mortality throughout the 1970s after it grew so large that it degraded its own rangeland. The massive migrations of Norway lemmings that occur every five to seven years and lead to many deaths result from crowding. Even the yeast used in brewing beer become large populations that eventually use up their food source and die off during the fermenting process.

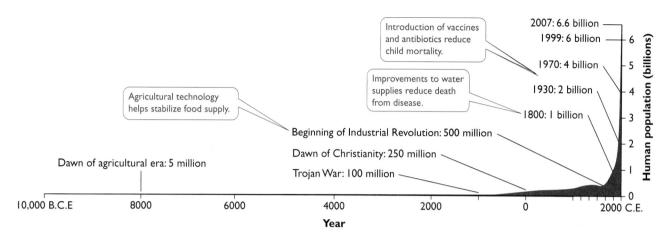

FIGURE 23.1 **Human population growth.** Estimates of human populations indicate that the number of people on Earth grew relatively slowly from the origin of agriculture through the seventeenth century. Beginning around the time of the Industrial Revolution, growth rates and population numbers began to soar.

These examples illustrate a basic biological principle. Although populations have the capacity to grow exponentially, their growth is limited by the resources—food, water, shelter, and space—that individuals need to survive and reproduce. The maximum population that can be supported indefinitely in a given environment is known as the environment's **carrying capacity**.

A simplified graph of population size over time in resource-limited populations is S shaped (FIGURE 23.2). This model shows the growth rate of a population declining to zero as it approaches the carrying capacity. In other words, birth rate and death rate become equal, and the population stabilizes at its maximum size. This pattern is known as **logistic growth**.

The declining growth rate near a population's carrying capacity is caused by **density-dependent factors**, which are population-limiting factors that increase in intensity as the population increases in size. Density-dependent factors include limited food supplies, increased risk of infectious disease in more crowded conditions, and an increase in toxin concentration caused by increased waste levels. Density-dependent factors cause declines in birth rate or increases in death rate.

Density-independent factors that influence population growth rates include events such as severe droughts that increase the death rate in plant populations regardless of their density, or increased temperatures that increase the birth rate in insects that only breed in warm weather. Density-independent factors do not occur in a vacuum. They can have more or less severe effects depending on the size of a population. For example, a density-independent factor such as an unusually cold winter can be deadly to individuals in a white-footed mouse population, but the likelihood of survival is also a function of how much food each individual has stored for the winter. How much food is stored depends on the density of mice competing for food sources during the autumn.

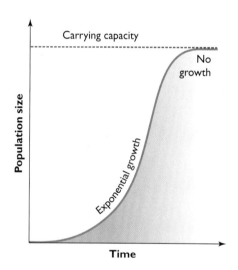

FIGURE 23.2 **The logistic growth curve.** An S-shaped curve describing population growth over time results from a gradual slowing of the population growth rate as it approaches carrying capacity.

VISUALIZE THIS Where on the graph is the population increasing in number most rapidly? Why is this point not at the very left edge of the graph?

Stop and Stretch 23.1

Examine the figure in this box. What is the carrying capacity of the culture bottle for fruit flies? How would you expect the carrying capacity of the population to change if the flies are supplied with a greater amount of food? What other factors might influence the carrying capacity in this environment?

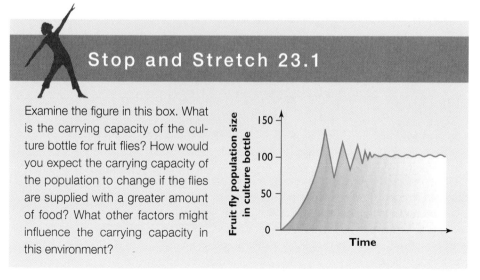

Population Crashes

Some ecologists are concerned that human populations may have a tendency to "overshoot" the Earth's carrying capacity, with disastrous results. Ecologists have long known that when populations have high growth rates, they may continue to add new members even as resources dwindle. The members of this large population are then competing for far too few resources, and the death rate soars while the birth rate plummets. This combination of factors results in a **population crash**, a steep decline in number.

A population overshoot and subsequent crash affected the human population on the Pacific island of Rapa Nui (also known as Easter Island) during the eighteenth century (FIGURE 23.3). This 163-square-kilometer (64-square-mile) island is separated from other landmasses by thousands of kilometers of ocean. Therefore, its people were limited to using only the resources on or near their island. Archaeological evidence suggests that at one time, the human population on Rapa Nui was at least 7,000—apparently a number far greater than the carrying capacity of the island. By 1775, the subsequent overuse and loss of Rapa Nui's formerly lush palm forest had resulted in a rapid decline to fewer than 700 people, a population likely much lower than the initial carrying capacity of the island.

The story of Rapa Nui may serve as a warning for current human populations. It is possible that humanity's use of the stored energy in fossil fuels may be allowing us to overshoot Earth's true carrying capacity—if the energy we use to produce abundant food runs low, the entire globe could perhaps experience a population crash.

Biological populations may also overshoot carrying capacity when there is a time lag between when the population approaches carrying capacity and when it actually responds to that environmental limit. Scientists who study human populations note a lag between the time when humans reduce birth rates and when population numbers respond. While parents may be reducing their family size, their children will begin having children before the parents die, causing the population to continue growing. Even when families have an average of two children, just enough to replace the parents, this momentum causes a human population to grow for another 60 to 70 years before reaching a stable level.

The future growth potential of a population can be estimated by looking at its **population pyramid**, a summary of the numbers and proportions of individuals of each sex and each age group. As FIGURE 23.4 illustrates, when the age structure most

FIGURE 23.3 The crash of a human population. On Rapa Nui, also known as Easter Island, the human inhabitants created these large statues to honor their ancestors. Soon after completely deforesting this small island, possibly in order to move these statues where they are, the large population suffered a severe crash.

FIGURE 23.4 A time lag in human population growth. (a) In a rapidly growing human population like that of South Africa in 2005, most of the population is young, and the population will continue to grow as these children reach child-bearing age. (b) In a stable or declining human population, there are fewer children relative to adults, as in the United States in 2005.

VISUALIZE THIS If child survival is the same in both South Africa and the United States, and growth rates remain the same, how would these graphs look in 2030?

closely resembles a true pyramid, with a large proportion of young people, the population will continue to increase. In more stable populations, the proportion that is young is not significantly larger than the proportion that is middle aged, and the pyramid looks more like a column.

Whether or not our reliance on stored resources and the potential demographic momentum in human populations will result in an overshoot of Earth's carrying capacity—followed by a severe crash, as on the island of Rapa Nui—remains to be seen. Regardless, the current human population and probable future growth in the short term is endangering the survival of thousands of other species.

23.2 The Sixth Extinction

Let's return to the dilemma introduced at the beginning of the chapter. Growing human populations and human activity in the Klamath Basin have led to the decline of certain fish native to that region. The federal government has stepped in, in an attempt to prevent the extinction of the shortnose and Lost River suckers.

The government was acting under the authority of the **Endangered Species Act (ESA)**, a law passed in 1973 with the purpose of protecting and encouraging the population growth of threatened and endangered species. With populations of fewer than 500 and minimal reproduction, these two species of fish are in danger of **extinction**, defined as the complete loss of a species. Critically imperiled species such as the Lost River and shortnose suckers are exactly the type of organisms that legislators had in mind when they enacted the ESA.

The ESA was passed because of the public's concern about the continuing erosion of **biodiversity**, the entire variety of living organisms. Critics of the ESA argue that the goal of saving all species from extinction is unrealistic. After all, extinction is a natural process—the approximately 10 million species living today constitute less than 1% of the species that have ever existed.

Measuring Extinction Rates

If ESA critics are correct, then the extinction rate today should be roughly equal to the rate in previous eras. The rate of extinction in the past can be estimated by examining the fossil record.

Since the rapid evolution of a wide variety of animal groups beginning approximately 580 million years ago, the number of families of organisms has generally increased. The fossil record shows a constant turnover of species, at a rate of about 1 extinction per million (0.0001%) per year. In addition, it indicates that the history of life on Earth has been punctuated by five **mass extinctions**—species losses that are global in scale, affect large numbers of species, and have dramatic impacts.

Many scientists argue that we are now seeing biodiversity's sixth mass extinction, this one caused by the massive global changes resulting from human activity. There is evidence that these scientists are correct; if we examine the past 400 years more closely, we see that the extinction rate has actually increased since the start of this historical record (FIGURE 23.5) to about 0.01% per year, making the current rate 100 times higher than the calculated background rate.

There are also reasons to expect that the current elevated rate of extinction will continue into the future. The World Conservation Union (known by its French abbreviation, IUCN), a highly respected global organization composed of and funded by states, government agencies, and nongovernmental organizations from over 140 countries, collects and coordinates data on threats to biodiversity. According to the IUCN's most recent assessment, 12% of birds, 20% of mammals, and 29% of amphibians are in danger of extinction, and human activities on the planet pose the greatest threat to most of these species.

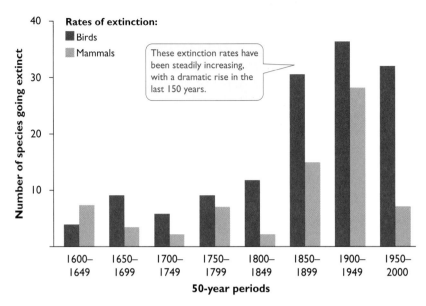

FIGURE 23.5 Rate of extinction. The number of species of mammals and birds known to have become extinct since 1600.

Causes of Extinction

A variety of human activities can put species at risk of extinction. The most severe threats belong to one of four general categories: loss or degradation of habitat, introduction of nonnative species, overharvesting, and effects of pollution. However, these four categories are not equal. The IUCN estimates that 83% of endangered mammals, 89% of endangered birds, and 91% of endangered plants are directly threatened by damage to or destruction of the places where they live.

Habitat Destruction and Fragmentation The dramatic reduction in numbers of shortnose and Lost River suckers in Upper Klamath Lake is almost entirely due to human modification of these species' **habitat**, the place where they live and obtain their food, water, shelter, and space. At one time, 350,000 acres of wetlands regulated the overall quality and amount of water entering the lake. Most of these wetlands have been drained and converted to irrigated agricultural fields now. This disruption of natural water flows has reduced the number of offspring the fish produce by as much as 95%.

The outright loss of habitat experienced by the Lost River and shortnose suckers is commonly called **habitat destruction**, and it is not limited to species in the developed world (FIGURE 23.6). Rates of habitat destruction caused by agricultural, industrial, and residential development accelerated throughout the twentieth century in concert with human population growth. In effect, humans have reduced the carrying capacity of Earth for a large variety of species.

The relationship between the size of a natural area and the number of species that it can support follows a general pattern called a **species-area curve**. A species-area curve for reptiles and amphibians on Caribbean islands is illustrated in FIGURE 23.7a. The pattern of these curves measured for many different groups of species is that the number of species in an area increases rapidly as the size of the area increases, but the rate of increase slows as the area becomes very large. The general approximation derived from the studies is shown in FIGURE 23.7b. From this graph, we can estimate that a 90% decrease in natural landscape area will cut the number of species living in the remaining area by half.

Applying species-area curves to estimate extinction rates requires that we calculate the amount of natural landscape that has been lost in recent decades. Using images from satellites, scientists have calculated that approximately 20,000 square kilometers (about 7,700 square miles, an area the size of Massachusetts) of rain forest are cut each year in South America's Amazon River basin. At this rate of habitat destruction, tropi-

FIGURE 23.6 Lost species. Lemurs are found only on the island of Madagascar, first settled by humans 1,500 years ago. Of the 48 species of lemur present on the island 2,000 years ago, 16 have become extinct, and 15 are at risk of extinction.

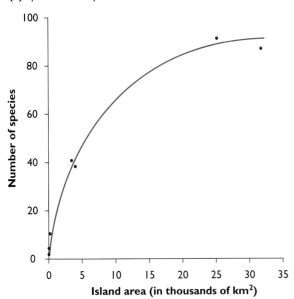
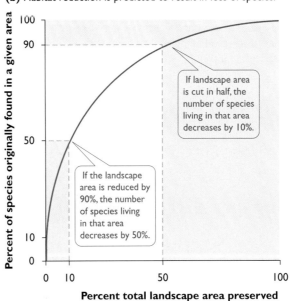

FIGURE 23.7

Predicting extinction caused by habitat destruction.
(a) This curve demonstrates the relationship between the size of a Caribbean island and the number of reptile and amphibian species living there. (b) We use a generalized species-area curve to roughly predict the number of extinctions in an area experiencing habitat loss.

cal rain forests will be reduced to 10% of their original size within roughly 35 years. If we apply the species-area curve, the habitat loss translates into the extinction of about 50% of species living in the Amazonian rain forest.

Of course, habitat destruction is not limited to tropical rain forests. When all of Earth's biomes are evaluated, freshwater lakes and streams, grasslands, and temperate forests are also experiencing high levels of modification. According to the IUCN, if habitat destruction around the world continues at its present rate, nearly one-fourth of all living species will be lost within the next 50 years.

Stop and Stretch 23.2

Review Figure 23.7a. The graph depicts the relationship between island size and the number of amphibian and reptile species found on islands in the Caribbean. How many species of reptiles and amphibians would you expect to find on an island that is 15,000 square kilometers in area? Imagine that humans colonize this island and dramatically modify 10,000 square kilometers of the natural habitat. What percentage of the species that were originally found on the island would you expect to become extinct?

Human activity rarely results in the complete loss of a habitat type. Often what results is **habitat fragmentation**, in which large areas of intact natural habitat are broken up. Habitat fragmentation is especially threatening to large predators, such as grizzly bears and tigers, because of their need for large hunting areas.

Large predators require large, intact hunting areas due to a basic rule of biological systems: Energy flows in one direction within an ecological system along a **food chain**. This energy flows from the sun to **producers** (plants and other photosynthetic organisms) to the primary **consumers** that feed on them, to secondary consumers (predators that feed on the primary consumers), and so on. Along the way, most of the calories taken in at one **trophic level** (that is, a level of the food chain) are used to support the activities of the individuals at that level and therefore are not available for

FIGURE 23.8 A trophic pyramid. The relationship between producers, primary consumers, and secondary consumers in a biological community. Because most of the energy consumed by a trophic level is used within that level for maintenance, biomass decreases as rank in the food chain increases.

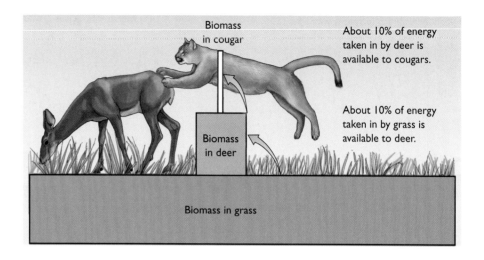

use by organisms at the next level. In other words, a major part of the solar energy initially captured by producers is given off as heat at each level within a food web. You can see this process in your own life. An average adult needs to consume between 1,600 and 2,400 Calories a day simply to maintain his or her current weight.

The flow of energy along a food chain leads to the principle of the *trophic pyramid*, the bottom-heavy relationship between the **biomass** (total weight) of populations at each level of the chain (FIGURE 23.8). Habitat destruction and fragmentation can cause the lower levels of the pyramid to shrink, reducing the carrying capacity for larger species.

Other Human Causes of Extinction According to the IUCN, the remaining threats to biodiversity posed by human activity play a role in about 40% of all cases of endangerment. Native species all over the world are threatened by **introduced species**, moved by humans from their home range to a new environment. Because the native organisms have not evolved in the presence of these exotics, the natives may be vulnerable to predation or competition by the introduced species. For example, domestic cats imported into the United States have devastated ground-nesting birds, and the introduced zebra mussel has crowded out native mussels in the Great Lakes.

When the rate of human use of a species outpaces its reproduction, the species experiences **overexploitation**. Overexploitation often occurs when particular organisms are highly prized by humans. For instance, four of the planet's ten tiger subspecies have become extinct, and the remaining species are gravely endangered. The main causes are the demand for their bones for their purported ability to treat arthritis, and their genitals, which are erroneously believed to reverse male impotence.

The release of poisons, excess nutrients, and other wastes into the environment—a practice otherwise known as *pollution*—poses an additional threat to biodiversity. For example, fertilizer pollution from farms in the Klamath Valley poses a risk to the shortnose and Lost River suckers. Increased levels of nitrogen and phosphorus from fertilizer have increased the growth of algae in Upper Klamath Lake. When these algae explode in numbers, the bacteria that feed on them flourish and rapidly use up the available oxygen in the water. This process of oxygen-depleting *eutrophication* results in large fish kills, not just in Upper Klamath Lake but also in many other bodies of water that receive fertilizer runoff from farms.

The evidence just discussed suggests that ESA critics who describe modern extinction rates as "natural" are incorrect. Over the past 400 years, humans have caused the extinction of species at a rate that appears to far exceed past rates, and it is clear that human activities continue to threaten thousands of additional species around the world. Earth appears to be on the brink of a sixth mass extinction of biodiversity—and the pervasive global change causing this extinction is human activity.

The Consequences of Extinction

Concern over the loss of biodiversity is not only about the survival of nonhuman life. Humans have evolved with and among the variety of species that exist on our planet, and the loss of these species often results in negative consequences for us.

Loss of Resources The Lost River and shortnose suckers were once numerous enough to support fishing and canning industries on the shores of Upper Klamath Lake. Even before the arrival of European settlers, the native people of the area relied upon these fish as a mainstay of their diet. The loss of these species represents a tremendous impoverishment of wild food sources and thus imposes an economic cost. One estimate places the value of wild products in the United States at $87 billion a year, about 4% of the gross domestic product. Clearly, humans have a direct interest in preserving these species.

Wild species can also provide resources for humans in the form of unique biological chemicals. One example of a natural origin for valuable chemicals is Madagascar's rosy periwinkle (*Catharanthus roseus*). Two drugs derived from it, vincristine and vinblastine, have contributed to major gains in the likelihood of survival from leukemia and Hodgkin's disease. If we are unable to screen living species due to their extinction, we will never know which ones might have provided compounds that would improve human lives.

Wild relatives of plants and animals that have been domesticated, such as agricultural crops and cattle, are also important resources for humans. Genes and alleles that have been "bred out" of domesticated species are often still found in their wild relatives. For example, the Mexican teosinte species *Zea diploperennis*, an ancestor of modern corn, is resistant to several viruses that plague cultivated corn. Some genes that confer this resistance have been transferred to domestic varieties to improve disease resistance.

Additionally, there is value in preserving wild relatives of domesticated crops in their natural habitats. Often the wild organisms in these communities provide the key to reducing pest damage and disease in the domestic crop. For example, the wasp *Catolaccus grandis* consumes boll weevils and is used to control infestations of these pests in cotton fields. *C. grandis* was discovered in the tropical forest of southern Mexico, where it parasitizes a similar pest in wild cotton populations.

Disruption of Ecological Communities Although humans receive direct benefits from thousands of species, most threatened and endangered species are probably of little or no use to people. Even the Lost River and shortnose suckers are not especially missed as a food source. No one has starved simply because these fish have become less common. In reality, most species are beneficial to humans because they are connected to other species and natural processes in a biological **community**, consisting of all the organisms living together in a particular habitat area.

The complex linkage among organisms in a community is often referred to as a **food web** (FIGURE 23.9). Because relationships between species may be based on requirements other than food, ecologists often use the phrase "web of life" to illustrate all of these interactions. The web of life consists of both antagonistic and cooperative interactions. The most obvious interaction is **predation**, in which one species, the predator, feeds on another, the prey. Predation includes dramatic interactions such as the one between hawks and jack rabbits, as well as more benign appearing ones, such as between jack rabbits and prarie grasses. When two different predators feed on the same prey species, they both experience the effect of **competition**. Competition can also occur over other resources, including living space. In some cases, two species will interact in a **mutualism**, which benefits both. Table 23.1 (on page 574) provides examples of these types of interactions.

As with a spider's web, any disruption in one strand of the web of life is felt by other portions of the web. Some tugs on the web cause only minor changes to the community, while others can cause the entire web to collapse. A collapse can be caused by the accumulation of indirect effects. Look again at the food web in Figure 23.9.

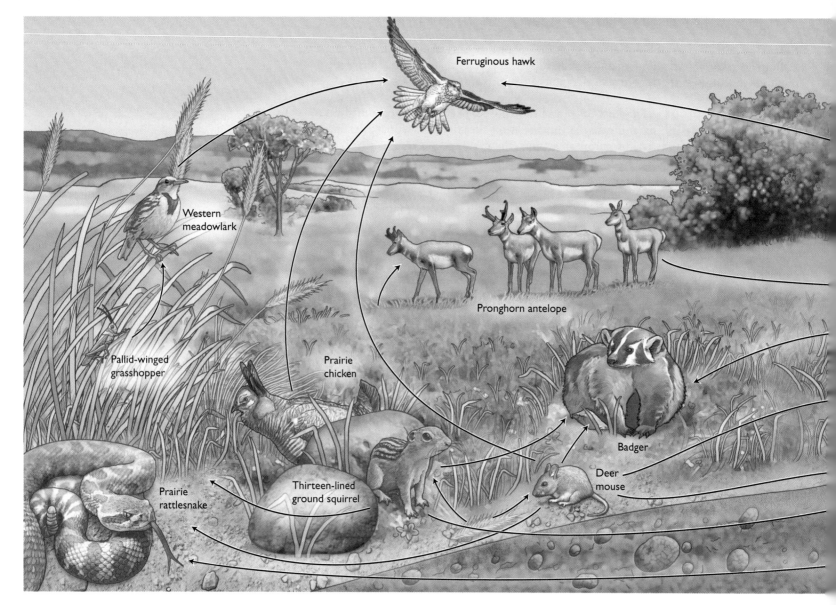

FIGURE 23.9 The web of life. Species on the prairie, as well as in other ecosystems, are connected to each other and to their environment in various, complex ways. Black arrows represent feeding relationships.

Stop and Stretch 23.3

Examine the web of relationships among organisms depicted in Figure 23.9. Which of the following species pairs are likely competitors? In each case, describe what they compete for.

a. badger, jackrabbit; **b.** bison, coyote; **c.** rattlesnake, badger; **d.** ground squirrel, deer mouse; **e.** jackrabbit, prairie dog

How could you test your hypothesis that these animals are in competition with each other?

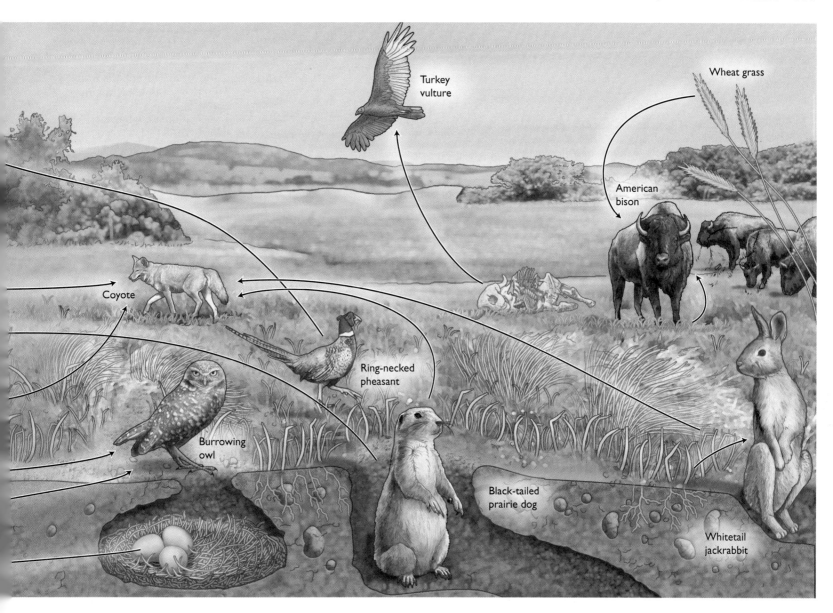

You can imagine that badgers, by preying on deer mice, have a negative effect on rattlesnakes, which they compete with for these mice, and a more indirect positive effect on ground squirrels, which compete with mice for grass seeds.

The existence of indirect effects has led ecologists to hypothesize that in at least some communities, the activities of a single species can play a dramatic role. These organisms are called **keystone species** because their role in a community is analogous to the role of a keystone in an archway—crucial and larger than expected (FIGURE 23.10). Remove the keystone, and an archway collapses; remove the keystone species, and the web of life collapses.

Damaged Ecological Systems Ecologists define an **ecosystem** as all of the organisms in a given area, along with their nonbiological environment. The function of an ecosystem is described in terms of the rate at which energy flows through it and the rate at which nutrients are recycled within it. The loss of some species can dramatically affect both of these ecosystem properties.

In nearly all ecosystems, the primary energy source is the sun. Producers convert sun energy into chemical energy during the process of photosynthesis, and the energy is passed through trophic levels making up a food chain. Energy is then partitioned

FIGURE 23.10 **Keystone species.** Wolves reintroduced into Yellowstone National Park scare elk away from open areas near streams. The dispersal of elk allows aspen trees to grow, in turn permitting beavers to return to the park.

TABLE 23.1 | Types of Species Interactions and Their Direct Effects

Interaction	Example	Effect on Species 1	Effect on Species 2	Interactions that Benefit Humans
Mutualism: Association increases the growth or population size of both species.	1. Ants 2. Acacia tree	**+** The swollen thorns of the acacia provide shelter for the ants. The acacia provides protein and nectar that the ants harvest for food.	**+** Ants kill herbivorous insects and destroy competing vegetation, benefiting the acacia.	Mutualism between bees and flowering plants provides fruit for human consumption.
Predation and parasitism: Consumption of one organism by another.	1. Brown bear 2. Salmon	**+** The brown bear catches the salmon and eats it, obtaining nourishment.	**−** The salmon does not survive.	Predation of songbirds on forest pests increases wood production and decreases fire risk.
Competition: Association causes a decrease or limitation in population size of both species.	1. Dandelion 2. Tomato plant	**−** The dandelion does not grow as well in the presence of the tomato plant. Dandelion produces fewer seeds and fewer offspring.	**−** The tomato plant does not grow optimally in the presence of the weed. Tomato plant produces fewer flowers and fruit.	Competition between benign microorganisms and dangerous ones prevents disease outbreaks.

among trophic levels in a bottom-heavy trophic pyramid (review Figure 23.8). The biodiversity found in an ecosystem can have strong effects on energy flow within it.

For example, by comparing experimental prairie "gardens" planted with the same total number of individual plants but with different numbers of species, scientists at the University of Minnesota and elsewhere have discovered that the overall plant biomass tends to be greater in more diverse gardens. This research indicates that a decline in diversity, even without a decline in habitat area, may lead to less energy being made available to organisms higher on the food chain, including people who depend on wild-caught food.

When essential mineral nutrients for plant growth pass through a food web, they are generally not lost from the environment—hence the term **nutrient cycling**. FIGURE 23.11 illustrates the nitrogen nutrient cycle in a natural prairie.

You should note as you review Figure 23.11 that plants absorb simple molecules from the soil and incorporate them into more complex molecules. These complex molecules move through the food web with relatively minor changes until they return to the soil. Here, complex molecules are broken down into simpler ones by the action of **decomposers**, typically bacteria and fungi.

Changes in the soil community can greatly affect nutrient cycling and thus the survival of certain species in ecosystems. Scientists investigating the effects of introduced earthworms in forests throughout the northeastern United States have observed dramatic reductions in the diversity and abundance of plants on the forest floor. These changes in the plant community may be related to the effects on nutrient cycling resulting from this change in the soil community.

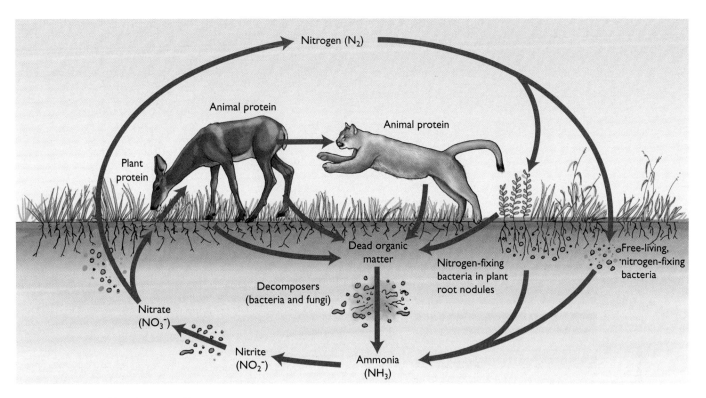

FIGURE 23.11 Nutrient cycling. Nutrients are recycled in an ecosystem, flowing from soil to producers to consumers and then back into the soil, where complex nutrients are decomposed into simpler forms.

VISUALIZE THIS Consider how people obtain nutrition and what happens to our waste and remains after death. How are nutrient flows in human systems different from natural cycles?

23.3 Saving Species

So far in this chapter, we have established the possibility of a modern mass extinction occurring as a result of growing human populations, and we have described the potentially serious costs of this loss of biodiversity to human populations. Because current elevated extinction rates are largely a result of human activity, reversing the trend of species loss requires mostly political and economic, rather than scientific, decisions. But what can science tell us about how to stop the rapid erosion of biodiversity?

Protecting Habitat

Without knowing exactly which species are closest to extinction and where they are located, the most effective way to prevent loss of species is to preserve as many habitats as possible. The same species-area curve used to estimate the future rate of extinction also gives us hope for reducing this number. Recall that according to this curve, species diversity declines rather slowly as habitat area declines. Thus, in theory, we can lose 50% of a habitat but still retain 90% of its species. This estimate is optimistic because habitat destruction is not the only threat to biodiversity. Nevertheless, the species-area curve tells us that if the rate of habitat destruction is slowed or stopped, extinction rates will slow as well.

Protecting the Greatest Number of Species Given the growing human population, it is difficult to imagine a complete halt to habitat destruction. To better focus protection efforts, the environmental group Conservation International has

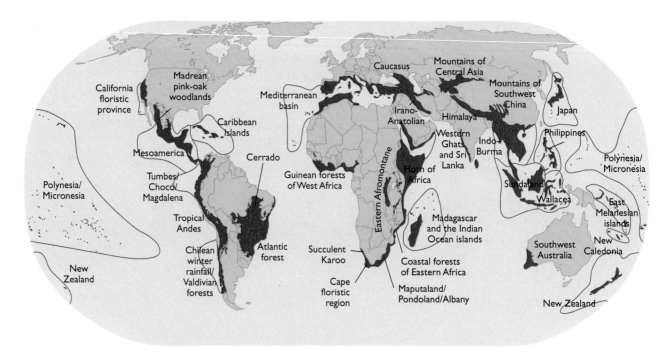

FIGURE 23.12 Diversity "hotspots." This map shows the locations of 34 identified biodiversity hotspots around the world.

VISUALIZE THIS Island areas are overrepresented in the hot spot listings. Why do you think islands tend to be hotspots of diversity?

identified 34 biodiversity "hotspots" (FIGURE 23.12). These areas make up less than 3% of Earth's land surface, yet contain up to 50% of all mammal, bird, reptile, amphibian, and plant species, many unique to those areas. These hotspots are also severely threatened by human activity. Stopping habitat destruction in these areas could greatly reduce the global extinction rate. By focusing conservation efforts on hotspots at the greatest risk, humans can very quickly prevent the loss of a large number of species.

Protecting Habitat for Critically Endangered Species Although preserving a variety of habitats ensures fewer extinctions, species already endangered require a more individualized approach. The ESA requires the U.S. Department of Interior to designate *critical habitats* for endangered species; that is, areas in need of protection for the survival of the species.

The biological part of a critical habitat designation includes conducting a study of habitat requirements for the endangered species and setting a population goal for it. However, federal designation of a critical habitat results in the restriction of human activities that can take place there. The U.S. Department of Interior has the ability to exclude some habitats from protection if there are "sufficient economic benefits" for doing so—a political decision.

Decreasing the Rate of Habitat Destruction Preserving habitat is not simply the job of national governments that set aside lands in protected areas or of private conservation organizations that purchase at-risk habitats. All of us can take actions to reduce habitat destruction and stem the rate of species extinction.

Conversion of land to agricultural production is a major cause of habitat destruction, so reducing your consumption of meat and dairy products from animals that are fed field crops is one of the most effective actions you can take. Reducing your use of wood and paper products, and limiting your consumption of these products to those harvested sustainably (that is, in a manner that preserves the long-term health of the forest), can help slow the loss of forested land.

Other measures to decrease the rate of habitat destruction require group effort. For instance, increased financial support for developing countries may help slow the rate of habitat destruction, by allowing them to invest money in technologies that decrease their use of natural resources. Strategies that slow the rate of human population growth offer more ways to avoid mass extinction. You can participate in group conservation efforts by joining nonprofit organizations focused on these issues, writing to politicians, and educating others.

Ensuring Adequate Population Size

For species on the brink of extinction—like the shortnose and Lost River suckers—preserving habitat is not enough. Populations can become so small that they can disappear, even with adequate living space. Recovery plans for both the Lost River and shortnose suckers set a short-term goal of one stable population made up of at least 500 individuals for each unique stock of suckers.

Five hundred individuals is sufficient to protect the species over the short term. This many are needed because the longer a population remains small, the more it is at risk of experiencing a catastrophic event that could eliminate it entirely. The story of the heath hen is a cautionary tale about dangers facing small populations (FIGURE 23.13).

The heath hen was a small wildfowl that once numbered in the hundreds of thousands on the east coast of the United States. Loss of habitat and heavy hunting caused the hen to become endangered. Despite the establishment of a nature reserve and some success in increasing the population size, a series of unpredictable events caused its extinction in the 1930s.

The final causes of heath hen extinction were natural events—fire, harsh weather, predation, and disease. But it was human-caused habitat loss and human hunting that caused the population to become more endangered by these challenges. A population of 100,000 individuals can weather a disaster that kills 90% of its members but leaves 10,000 survivors. A population of 1,000 individuals will be nearly eliminated by the same circumstances. Even when human-caused losses to the heath hen population were halted, the species' survival was still extremely precarious.

Small populations also lose genetic variability as a result of genetic drift, a process described in detail in Chapter 22. Populations with low levels of genetic variability have an insecure future for two reasons. First, when alleles are lost, individuals are homozygous for more genes. In general, high levels of homozygosity lead to lower reproduction and higher death rates. Second, populations with low genetic variability may be at risk of extinction because they cannot evolve in response to changes in the environment. When few alleles are available for any given gene, it is possible that no individuals in a population will possess an adaptation that allows them to survive an environmental challenge. There are many examples of populations that suffer because of low genetic variability. The Irish potato provides perhaps the most dramatic story.

When the organism that causes potato blight arrived in Ireland in September 1845, nearly all of the planted potatoes became infected and rotted in the fields. The few potatoes that by chance escaped the initial infection were used to plant the following year's crops. However, apparently very few or no Irish potatoes carried alleles that made them resistant to blight, and in 1846, the entire Irish potato crop failed. Because of this failure and another in 1848, nearly 1 million Irish peasants died of starvation and disease, and another 1.5 million peasants emigrated to North America.

Irish potatoes descended from a small group of plants that were missing the allele for blight resistance, so even an enormous population of these plants could not escape the catastrophe caused by this disease. Similarly, because small populations lose genetic variability rapidly through genetic drift, preventing endangered species from declining to very small population levels may be critical for avoiding a similar genetic disaster.

FIGURE 23.13 **A victim of small population size.** Although the heath hen was protected when its population fell to nearly 50 individuals, a series of unexpected disasters, including fire, an introduced poultry disease, and increased predation, caused its extinction.

FIGURE 23.14 A restoration project. Scientists hope to learn enough about ecological systems to be able to restore their essential functions after they have been degraded by human activity. This is a restoration project in the Klamath Basin.

The historical situations of the heath hen and Irish potato support the current need to preserve adequate numbers of Lost River and shortnose suckers, even at the expense of crop production in the Klamath Basin, to save these species from extinction.

Meeting the Needs of Humans and Nature

Saving the Lost River and shortnose suckers from extinction requires protecting all of the remaining fish and restoring the habitat they need for reproduction. These actions cause economic and emotional suffering for humans who make their living in the Klamath Basin. In fact, many actions necessary to save endangered species result in immediate problems for people.

If the debate in the Klamath Basin follows the pattern set by other ESA controversies, a political solution that causes some economic hardship while ensuring the immediate survival of the fish will prevail. Biologists working on the problem agree that the recovery goal of 500 individuals for shortnose and Lost River suckers is high enough to ensure short-term survival (50 years), but it is not high enough to ensure both species' long-term survival (500 years). The biologists' assessment is based on computer models predicting how the population will respond to predicted environmental changes. The recovery population size is large enough to withstand environmental catastrophes in the short term, but in the long term, continued loss of genetic variability results in the extinction of populations of only 500 fish in their models.

The risk to the long-term survival of the fish of a population of 500 helps balance the cost to the farmers of the Klamath Basin. The short-term cost to farmers was somewhat alleviated when they received federal disaster assistance to help them adjust to the loss of lake-derived irrigation water. While recent increases in rainfall have helped provide enough water for both farmers and fish and reduced the level of conflict, federal courts have ruled that in the event of another drought, water deliveries to farmers will be cut off again.

In order to prevent another crisis like the one in 2001, the federal government is employing **restoration ecology** in the Klamath Basin, attempting to return the lake to a state where natural processes are functioning effectively for the fish (FIGURE 23.14). To restore Upper Klamath Lake, the government is purchasing farmland from willing sellers in the Basin to preserve water supply and restoring wetlands that protect the fish. The U.S. Fish and Wildlife Service hopes that this long-term solution will help provide adequate habitat for the survival of both the shortnose and Lost River suckers.

As with any challenge that humans face, the best strategy for preserving biodiversity is to prevent species from becoming endangered. Table 23.2 provides a list of actions that can help preserve biodiversity. Meeting this challenge requires some creativity, but it is often possible to provide for the needs of people while preserving our natural heritage. It will take efforts from all of us to help keep the equation in balance.

TABLE 23.2 | Taking Action to Preserve Biodiversity

Objective	Why do it?	Actions
Reduce fossil fuel use.	• Mining, drilling, and transporting fossil fuels modifies habitat and leads to pollution. • Burning fossil fuels contributes to global climate change, further degrading natural habitats.	• Buy energy-efficient vehicles and appliances. • Walk, bike, carpool, or ride the bus whenever possible. • Choose a home near school, work, or easily accessible public transportation. • Buy "clean energy" from your electric provider, if offered.
Reduce the impact of meat consumption.	• The primary cause of habitat destruction and modification is agriculture. • Modern beef, pork, and chicken production relies on grains produced on farms. One pound of beef requires 4.8 pounds of grains, or about 25 square meters of agricultural land.	• Eat one more meat-free meal per week. • Make meat a "side-dish" instead of the main course. • Purchase grass-fed or free-range meat.
Reduce pollution.	• Pollution kills organisms directly or can reduce their ability to survive and reproduce in an environment.	• Do not use pesticides. • Buy products produced without the use of pesticides. • Replace toxic cleaners with biodegradable, less harmful chemicals. • Consider the materials that make up the goods you purchase, and choose the least-polluting option. • Reuse or recycle materials instead of throwing them out.
Educate yourself and others.	• Change happens most rapidly when many individuals are working for it.	• Ask manufacturers or store owners about the environmental costs of their goods. • Talk to family and friends about the choices you make. • Write to decision makers to urge action on effective measures to reduce human population growth and curb habitat destruction and species extinction.

Chapter REVIEW

ROOTS TO REMEMBER

The following roots of words come mainly from Latin and Greek and will help you to decipher terms:

eco- comes from a Greek word meaning house or habitation.

troph- comes from a Greek word for food.

KEY TERMS

23.1 Limits to Population Growth
carrying capacity *p. 565*
density-dependent factor *p. 565*
density-independent factor *p. 565*
ecology *p. 564*
exponential growth *p. 564*
logistic growth *p. 565*
population *p. 564*
population crash *p. 566*
population pyramid *p. 566*

23.2 The Sixth Extinction
biodiversity *p. 567*

biomass *p. 570*
community *p. 571*
competition *p. 571*
consumer *p. 569*
decomposer *p. 574*
ecosystem *p. 573*
extinction *p. 567*
food chain *p. 569*
food web *p. 571*
habitat *p. 568*
habitat destruction *p. 568*
habitat fragmentation *p. 569*
introduced species *p. 570*
keystone species *p. 573*

mass extinction *p. 567*
mutualism *p. 571*
nutrient cycling *p. 574*
overexploitation *p. 570*
predation *p. 571*
producer *p. 569*
species-area curve *p. 568*
trophic level *p. 569*

23.3 Saving Species
restoration ecology *p. 578*

SUMMARY

23.1 Limits to Population Growth

- A population is defined as all individuals of the same species living in a defined area. The structure of a population can be described by the number of individuals and their dispersion (p. 564).

- The human population has grown very rapidly over the last 250 years and exhibits a pattern of exponential growth, which is an increase in numbers proportional to the current population size (p. 564).

- Nearly all populations eventually reach the carrying capacity of their environment. Near carrying capacity, density-dependent factors cause an increase in death rate, a decrease in birth rate, or both (p. 565; Figure 23.2).

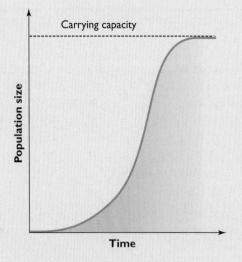

- Fast-growing populations that overshoot their environment's carrying capacity may experience a crash or go through periodic booms and busts (p. 566).

23.2 The Sixth Extinction

- The loss of biodiversity through species extinction is exceeding historical rates by 50 to 100 times. This may be part of a mass extinction, in which species losses are extreme and global. Five mass extinctions have occurred previously in Earth's history (p. 567).

- Species-area curves help us predict how many species will become extinct due to human destruction of natural habitat (pp. 568–569).

- Species at the top of the food chain are more susceptible to extinction because less energy is available for survival at higher trophic levels (p. 569).

- Additional threats of habitat fragmentation, introduced species, overexploitation, and pollution also contribute to species extinction (p. 569).

- Species are important to us as resources, either directly as consumed products or indirectly as organisms used to provide potential medicines or genetic resources (p. 571).

- Species are members of communities. Their loss as mutualists, predators, competitors, and keystone species may change a community, making it less valuable or even harmful to humans (p. 571).

- Species also play a role in ecosystem function, including effects on energy flow and nutrient cycling. Changes to the biological components of an ecosystem may change its nonbiological properties as well (p. 574; Figure 23.11).

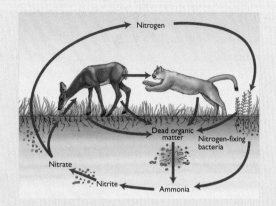

23.3 Saving Species

- If habitat protection is focused on a few well-defined biodiversity hotspots, then the number of organisms becoming extinct can be markedly reduced (p. 575).

- When species are already endangered, restoring larger populations is critical for preventing extinction (p. 576).

- Small populations are at higher risk for extinction due to environmental catastrophes. Genetic variability is lost in small populations because of genetic drift—the loss of alleles from a population due to chance events. Therefore, small populations may be less able to evolve in response to environmental change (p. 577).

- The political process enables people to develop plans for helping endangered species recover from the brink of extinction while minimizing the negative effects of these actions on people (p. 578).

LEARNING THE BASICS

1. The growth of human populations over the past 150 years has increased mainly due to _____.
   a. increases in death rate
   b. increases in birth rate
   c. decreases in death rate
   d. decreases in birth rate
   e. increases in net primary production

2. Carrying capacity refers to _____.
   a. the number of prey animals a predator can eat in one sitting
   b. the capacity of an environment to support a population of a particular organism
   c. the likelihood that density-independent factors will affect a population
   d. the severity of resource limitation in a given environment
   e. the consequences of the unequal distribution of resources in an environment

3. Which of the following is a density-independent factor that can cause a decrease in population?
   a. severe weather
   b. increased competition for resources
   c. because of crowding, individuals must inhabit low-quality den sites
   d. increased rate of disease spread
   e. increased predation by predators attracted to a very large prey population

4. Populations that rely on stored resources are likely to overshoot the carrying capacity of the environment and consequently experience a _____.
   a. clumped distribution
   b. negative carrying capacity
   c. decrease in death rates
   d. population crash
   e. exponential growth

5. All of the following factors are associated with declines in a country's population growth rate except _____.
   a. an increase in per capita income
   b. an increase in female literacy
   c. an increase in women's social status
   d. a preference for daughters over sons

6. Current rates of species extinction appear to be approximately _____ historical rates of extinction.
   a. equal to
   b. 10 times lower than
   c. 10 times higher than
   d. 50 to 100 times higher than
   e. 1,000 to 10,000 times higher than

7. The background rate of extinction _____.
   a. is determined by examining the fossil record
   b. can be estimated by calculating the average life span of a species
   c. is several times lower than the current rate of extinction
   d. may be an underestimate, based on the type of species that leave fossil remains
   e. all of the above are correct

8. The relationship between the size of a natural habitat and the number of species that the habitat supports is described by a(n) _____.
   a. habitat fragmentation measure
   b. inbreeding depression matrix
   c. species-area curve
   d. overexploitation scale
   e. ecosystem services cost

9. According to the generalized species-area curve, when habitat is reduced to 50% of its original size, approximately _____ of the species once present there will be lost.
   a. 10%
   b. 25%
   c. 50%
   d. 90%
   e. it is impossible to estimate

10. Which cause of extinction results from humans' direct use of a species?
    a. overexploitation
    b. habitat fragmentation
    c. pollution
    d. introduction of competitors or predators
    e. global warming

11. A mass extinction _____.
    a. is global in scale
    b. affects many different groups of organisms
    c. is caused only by human activity
    d. a and b are correct
    e. a, b, and c are correct

12. The web of life refers to the _____.
    a. evolutionary relationships among living organisms
    b. connections between species in an ecosystem
    c. complicated nature of genetic variability
    d. flow of information from parent to child
    e. predatory effect of humans on the rest of the natural world

Choose the relationship that represents each ecological interaction in items 13–16.

a. predation   b. competition   c. mutualism

13. _____ Moles catching and eating earthworms from the moles' underground tunnels

14. _____ Cattails and reed canary grass growing together in wetland soils

15. _____ Cleaner fish removing parasites from the teeth of sharks

16. _____ Colorado potato beetles consuming potato plant leaves

17. According to many scientists, the most effective way to reduce the rate of extinction is to _____.
    a. preserve habitat, especially in highly diverse areas
    b. focus on a single species at a time
    c. eliminate the risk of genetic drift
    d. produce less trash by recycling more
    e. encourage people to rely more on agricultural products and less on wild products

18. The risks faced by small populations include _____.
    a. erosion of genetic variability through genetic drift
    b. decreased fitness of individuals as a result of increased homozygosity
    c. increased risk of experiencing natural disasters
    d. a and b are correct
    e. a, b, and c are correct

19. The field of biology that attempts to re-create functioning ecological systems in areas that have been affected by human activity is called _____.
    a. ecology
    b. environmentalism
    c. conservation biology
    d. genetic engineering
    e. restoration ecology

ANALYZING AND APPLYING THE BASICS

1. A researcher captures 50 penguins, marks them with a spot of paint on their bills, and releases them. One month later she returns, again captures 50 penguins, and notes that only 1 has a previous mark. What is the likely size of the total penguin population in the researcher's study area?

2. Imagine two human populations, each one made up of 5 million individuals. In one population, over 50% of the members are in the age group 0–20 years and about 2% are over 65. In the other population, about 20% are 0–20 years old and about 20% are over 65. Which of these populations will probably stabilize at a larger number, and why?

3. The piping plover is a small shorebird that nests on beaches in North America. The plover population in the Great Lakes is endangered and consists of only about 30 breeding pairs. Imagine that you are developing a recovery plan for the piping plover in the Great Lakes. What sort of information about the bird and the risks to its survival would help you to determine the population goal for this species as well as how to reach this goal?

4. By studying the fossilized teeth of Spanish rodents, scientists have identified cycles of extinction that appear linked to the Milankovitch cycles—changes in Earth's climate caused by the changes in Earth's orbit, tilt, and wobble. The scientists identified two cycles of extinction: one cycle every 2.4–2.5 million years and one every million years. During each cycle, around 30% of the rodent species living during that time became extinct. One of these extinctions coincided with the cooler climate produced when Earth's orbit became more circular (and less elliptical). How could climate changes lead to extinctions?

Questions 5 and 6 refer to the following information.

The kakapo is a nocturnal parrot found only in New Zealand. This bird is flightless, with moss-green feathers striped with black bands. Kakapos have a potential life span of 60 years. Due to their long lives, the onset of breeding is delayed until the age of 5 years for males and 9–11 years for females. Even then, kakapos only breed once every three to five years—during mast years when trees produce an overabundance of fruit. Kakapos reproduce by gathering together in leks—groups in which males compete for females by calling and displaying. After mating, female kakapos leave their mates to lay eggs. Nest sites are commonly under plants on the ground or in hollow tree trunks.

Kakapo populations were decimated by black rats, domestic cats, and short-tailed weasels. With populations nearly extinct, scientists moved all surviving kakapos to predator-free islands, where they are now carefully monitored. Scientists provide supplemental food to the kakapos, to encourage more frequent breeding. Currently, the kakapo population numbers around 85 birds.

5. What characteristics of kakapos made their populations particularly vulnerable to predation?

6. Is it likely that kakapos will once again thrive in the wild? Support your answer.

7. Around 1,200 species of woody bamboo grow in forests in Asia, South America, and Africa. These plants provide food and shelter for giant and red pandas, mountain gorillas, lemurs, mountain bongos (a type of antelope), spectacled bears, mountain tapirs, bamboo frogs, lesser bamboo bats, plowshare tortoises, and a variety of bird species. People also use bamboo for food and building materials. A recent United Nations report found that half of these bamboo species are endangered. Describe an indirect effect that could occur if bamboo were to become extinct in a particular area.

CONNECTING THE SCIENCE

1. How are the factors that limit fruit-fly populations in a culture bottle similar to the factors that limit human populations on Earth? How are they different?

2. From your perspective, which of the following reasons for preserving biodiversity is most convincing? (a) Nonhuman species have roles in ecosystems and should be preserved in order to protect the ecosystems that support humans; or (b) nonhuman species have a fundamental right to existence. Explain your choice.

3. If a child asks you the following question 20 or 30 years from now, what will be your answer, and why?

"When it became clear that humans were causing a mass extinction, what did you do about it?"

Chapter 24

Biomes and Natural Resources

Where Do You Live?

LEARNING GOALS

1. Compare and contrast *weather* and *climate*. What are the major components of climate?

2. List the three factors that influence global and local temperatures, and briefly describe their effects.

3. How do global factors and nearness to water and mountains affect precipitation patterns?

4. Compare and contrast tropical forests, temperate forests, boreal forests, and chaparral.

5. Summarize the characteristics of grasslands, desert, and tundra.

6. What are the differences between freshwater and marine habitats?

7. Define *ecological footprint*, and describe how individual choices can change footprints.

8. What is the evidence that increased carbon dioxide in the atmosphere will lead to global climate change?

What is your "biological address"?

How do you answer the question, "Where do you live?" Most of us would respond with a neighborhood or street address to someone from our community, a city or town name to someone from elsewhere in our state, or a state or country name to someone who lives far away. Not too many of us would give a reply such as "the Sonoran Desert" or "the boreal forest." These descriptions of the natural environment in which we live can be thought of as biological addresses. Our biological addresses include the native vegetation and the resident animals, fungi, and microbes that share, or once shared, our living space.

Who are your neighbors?

Unlike previous chapters in which we followed a story, in this chapter, you are the storyline. In particular, we will explore the Earth's natural neighborhoods by guiding you to learn more about your own. Can you answer the following questions?

A human-designed landscape may mask a biological address.

- Is the native vegetation of the place where you live forest, grassland, or desert?
- What are the seasonal weather changes in your area?
- Can you describe the physical characteristics of three plant species that are native to your area?
- What is the largest mammalian predator that lives, or once lived, in the habitat that is now your neighborhood?
- Can you describe three native bird species that breed in your region?

Is it important to know the answers to questions like these? Many Americans would have a difficult time answering at least some of them. Knowing the answers to these questions will lead to bioregional awareness, an understanding of local environmental conditions.

One consequence of a general lack of bioregional awareness that you may have experienced is local summertime water shortages caused by water used to keep alive thirsty suburban lawns. Other costs of lack of bioregional awareness may be more severe, including construction of homes in areas where periodic fires or floods are common or building on sandy coastlines, which are very unstable in their natural condition.

How can you determine where your basic resources come from?

Having a deep awareness of one's own bioregion may allow people to build human settlements that are better both for humans and the natural environment. Taking into account one's bioregion when developing human habitats can occur in many ways. For example, in the southwestern desert regions of the United States, environmentally sensitive housing developers use xeriscaping, a kind of landscaping that relies on native, drought-tolerant plants. Xeriscaping not only prevents the overconsumption of water but also provides a habitat for resident wildlife.

Human populations, like all natural populations, require resources and produce waste. Surprisingly few Americans have an understanding of where our resources come from and where our waste goes. See if you can answer these questions:

- What is the source of your tap water?
- What primary agricultural crops are grown in your area?
- How is your electricity generated?
- Where does your garbage go?
- What happens to the wastewater that is flushed from your toilet or runs down the drain?

Why does knowing the answers to these questions matter? Humans remain dependent on the natural world for our resources and waste disposal. An understanding of the capacity of the natural environment in our bioregion can help us to design ways of living that take advantage of a region's natural gifts and to respect its limits. The consequences of a lack of understanding of how our human communities fit into the surrounding biological community can include air and water pollution and the negative effects of that pollution on ourselves and our biological neighbors.

In this chapter, we explore how the ecology of a bioregion intersects with the biology of human habitats.

24.1 Terrestrial Biomes

A major factor that determines the nature of a particular "biological address" is the region's **climate**, the average weather as measured over many years. Climate should be distinguished from **weather**, which is the current conditions in terms of temperature, cloud cover, and **precipitation** (rain or snowfall). Put another way, the weather in a place will tell you if you have to shovel snow tomorrow morning, while the climate in a place will tell you if you even have to own a snow shovel.

The major components of climate in a geographic area are temperature, precipitation, and the variability of these two factors over the course of a year. In general, temperatures are cooler further from the equator because of the curvature of Earth. While the sun is directly overhead at the equator, it is closer to the horizon at the poles, and its warming rays are diffused over a larger area of atmosphere and land surface. Seasons are more pronounced closer to the poles because of the tilt of Earth's axis. As Earth travels its orbit year, each pole makes a complete circuit from facing toward to facing away from the sun. The differential heating of Earth's surface that results from curvature and tilt drive global patterns of precipitation as well. Table 24.1 summarizes these and other factors that influence global and regional climate.

Plants (and animals) native to a region have adapted, via the process of natural selection, to the water availability and temperatures experienced there. Water availability

TABLE 24.1 | Factors That Influence Local and Regional Climate

	Temperature	Precipitation
Latitude	• Cooler near the poles, hotter near the equator • Greater variation closer to poles	• Wet at equator, dry at 30° north and south latitudes
Proximity of surface water	• Moderates temperature during seasons • Ocean currents may bring heat from tropics to temperate region and vice versa	• More precipitation on side of water body opposite prevailing wind
Mountains	• Cooler at high altitudes • Locally, cooler in valleys where denser cold air drains	• More precipitation on side of mountain facing prevailing winds • Less on sheltered side
Characteristics of land surface	• Ice and snow reflect sunlight, so air does not heat • Dark surfaces in cities absorb sunlight, increase temperature	

is obviously a function of total precipitation, but it is also influenced by temperature. Frozen water cannot be taken up by plants.

There are four basic land **biome** categories, or primary vegetation types: forest, grassland, desert, and tundra. Each of these categories may contain several biome types. For instance, a grassland may be either prairie, steppe, or savanna. The relationship between climate and biome type is illustrated in FIGURE 24.1. Which of these biomes is your home? And how has the human population changed the environment of your biological neighborhood?

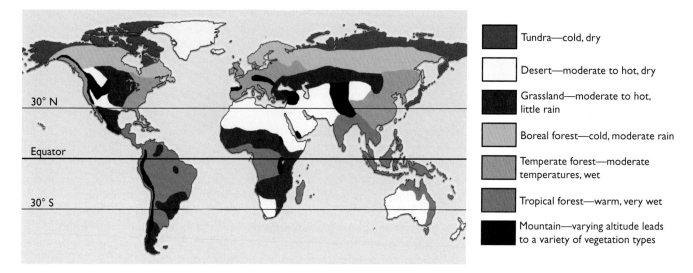

FIGURE 24.1 The distribution of Earth's land biomes. The primary vegetation type in a given area is determined by the region's climate.

Forests and Shrublands

Forests are vegetation communities dominated by trees and other woody plants. They occupy approximately one-third of Earth's land surface and contain about 70% of the biomass found on land (FIGURE 24.2).

Extensive areas of **tropical forest** were once found throughout Earth's equatorial region—in Central and South America, central Africa, India, Southeast Asia, and Indonesia. Tropical forests contain great biological diversity. One hectare (10,000 square meters, about 2.5 acres) may contain as many as 750 different tree species. However, the increasing human population of tropical countries is causing widespread loss of this productive biome.

Some areas of tropical forest, especially those farther from the equator, may display seasonal changes between annual wet and dry periods. But most people associate major seasonal change with forests in temperate areas, where winter temperatures can drop well below freezing. Large areas of **temperate forest** still cover eastern North America, but only remnants of these forests are left in Europe and eastern China.

Because their soils are relatively easy to turn over and are rich in nutrients, nearly all of the forested lands in the eastern United States were converted to farmland within 100 years of the American Revolution. In the late nineteenth and early twentieth centuries, however, farms in the eastern United States were abandoned as production moved south and west. Now, many of these abandoned farms have returned to forest. These second-growth forest sites are once again threatened—this time by expanding urban development and the negative effects of air pollution. The World Wildlife Fund estimates that worldwide, only 5% of temperate deciduous forests remain relatively untouched by humans.

The largest biome on Earth is the **boreal forest**, covering vast expanses of northern North America, northern Europe, and Asia, and high-altitude areas in the western United States. Coniferous plants that produce seed cones instead of flowers and fruits dominate boreal forests. In fact, boreal forests are among the only land areas where flowering plants are not the dominant species. Trees in these landscapes are valuable for both building materials and paper products, and logging in the boreal forest is extensive. There are increasing concerns that the boreal forest in North America is being cut at rates faster than it can be replaced.

One major biome is dominated by woody plants but is not a forest. This landscape is known as **chaparral**, and its vegetation consists mostly of spiny evergreen shrubs. Chaparral is found extensively in areas surrounding the Mediterranean Sea and in smaller patches in southern California, South Africa, and southwestern Australia.

In southern California, the flammability of chaparral vegetation has come directly into conflict with rapid population growth and urbanization. In response to recurrent massive fires, recommendations by a state task force have called for a policy of putting wildfire protection ahead of protection of wildlife and wildlife habitat. Because of this policy, in combination with human modification of chaparral habitats, this unique biome is one of the most threatened on the planet.

Grasslands

Grasslands are regions dominated by nonwoody grasses. They support few or no shrubs or trees. These biomes occupy geographic regions where precipitation is too limited to support woody plants.

Tropical grasslands are known as **savannas** and are characterized by the presence of scattered individual trees. Savanna covers about half of the African continent as well as large areas of India, South America, and Australia. Temperate grassland biomes include tallgrass **prairies** and shortgrass **steppes** (FIGURE 24.3). Generally, the height of the vegetation corresponds to the precipi-tation—greater precipitation can support taller grasses. Prairies and steppes are found in central North America, central Asia, parts of Australia, and southern South America. These landscapes are generally flat to slightly rolling and contain few trees.

(a) Tropical rain forest

(b) Temperate forest

(c) Boreal forest

FIGURE 24.2 Forest biomes. (a) Tropical rain forests are highly diverse and efficient at capturing light before it reaches the ground. (b) Temperate forests contain mostly deciduous trees that drop all of their leaves once a year. (c) Boreal forests are often highly uniform, made up of one or two species of coniferous trees.

In cooler temperate regions, decomposition is relatively slow, and the soil of prairies and steppes is rich with the partially decayed roots of grasses and other plants. These soils provide an excellent base for agriculture, and most native prairies and steppes have been plowed and replanted to crops. In North America, less than 1% of native prairie remains.

Desert

Where rainfall is less than 50 centimeters (20 inches) per year, the landscape is called **desert**. This biome can be found throughout the world, but the world's great deserts include the Sahara in northern Africa, the Gobi of central Asia, and the deserts of the Middle East, central Australia, and the southwestern United States. Most of these deserts are close to 30° north or 30° south of the equator. Most deserts have vegetation, although it can be sparse.

The sunny, warm, and dry climate of the deserts in Arizona, New Mexico, and Nevada is appealing to people. This region has the greatest rate of human population growth in the United States. The increasing population is putting stress on water supplies in these dry states, threatening the ability to support the water needs of cities and agricultural production as well as depleting water sources for native animals.

Tundra

The biome type where temperatures are coldest—close to Earth's poles and at high altitudes—is known as **tundra** (FIGURE 24.4). Here plant growth can be sustained for only 50 to 60 days during the year, when temperatures are high enough to melt ice in the soil.

High temperatures in tundra regions are not sustained long enough to melt all of the soil's ice. Therefore, soils are underlain by **permafrost**, icy blocks of gravel and finer soil material. The permafrost layer impedes water drainage, and soils above permafrost are often boggy and saturated.

Tundra is very lightly settled by humans, but it is threatened by our dependence on fossil fuels—oil, natural gas, and coal that formed from the remains of ancient plants. Some of the largest remaining untapped oil deposits are found in tundra regions. The infrastructure required to extract this oil damages tundra vegetation, produces toxic pollution, and disrupts animal migration.

The use of fossil fuels also appears to be causing Earth to warm, as covered at the end of this chapter. Global warming has been greatest at the poles, where tundra is predominant. Winters in Alaska have warmed by 2–3°C (4–6°F), whereas elsewhere they have warmed by about 1°C. As climate conditions change, areas that were once tundra have begun to support shrub and tree growth and are changing into boreal forest, further threatening the fragile tundra biome.

Stop and Stretch 24.1

Describe the native vegetation of the place where you live. How much native vegetation remains? What ecological factors (climate, fire, and so forth) have influenced the native vegetation type?

FIGURE 24.3 A prairie in late summer. Although the dominant plants on prairies are grasses, some of the less numerous plants produce large, colorful flowers.

VISUALIZE THIS Why are flowers most abundant in prairies in summer, when they are absent from forests at this time?

FIGURE 24.4 Tundra. The short growing season and the year-round likelihood of below-freezing temperatures limit the vegetation of the tundra to ground-hugging plants.

FIGURE 24.5 **Wetlands.** Wetlands provide an important habitat for a wide variety of creatures as well as a natural filtration system for water.

24.2 Aquatic Biomes

Nearly all human beings live on Earth's land surface, but most people also live near a major body of water and are both influenced by, and have an influence on, these **aquatic** systems. Aquatic biomes are typically classified as either freshwater or saltwater. As you read this section of the chapter, consider the water bodies that affect you most and how you affect them.

Freshwater

Freshwater is characterized as having a low concentration of salts—typically less than 0.1% of total volume. Scientists usually describe three types of freshwater biomes: lakes and ponds, rivers and streams, and wetlands.

Bodies of water surrounded by land are known as **lakes** or **ponds**. Fertilizers applied to agricultural lands and residential lawns near lakes and ponds can also increase their algae populations as the nutrients leach into the water. Ironically, too many nutrients added in this way can lead to the death of these bodies of water. Their degradation occurs because of eutrophication, a process described in Chapter 23. During eutrophication, large populations of algae (and the microorganisms that feed on them) lead to low oxygen levels and the death of fish.

Rivers (and their generally smaller counterparts, streams, brooks, and creeks) are flowing water moving in one direction. Rivers are threatened by the same pollutants that damage lakes. But their habitats also face wholesale destruction with the development of dams and channels. Dams are built to provide hydropower or maintain reservoirs for drinking water or for cooling fossil-fuel- or nuclear-powered electricity-generating plants, and channels are changes to the river banks to simplify and expedite boat traffic.

Areas of standing water that support above-water aquatic plants are called **wetlands** (FIGURE 24.5). Wetlands result from the high nutrient levels found at these interfaces between the aquatic and land environments. Besides their importance as biological factories, wetlands provide health and safety benefits by slowing down the flow of water. Slower water flows reduce the likelihood of flooding and allow sediments and pollutants to settle before the water enters lakes or rivers.

Since the European settlement of the continental United States, over 50% of wetlands have been filled, drained, or otherwise degraded. Extensive efforts by environmental organizations over the past 25 years have led to legislation that has greatly slowed the rate of wetland loss in the United States. However, about 58,000 acres of swamp, marsh, and bog are still destroyed every year.

FIGURE 24.6 **Coral reefs.** In this picture, algae mutualists give the coral animals their reddish color.

Saltwater

About 75% of Earth's surface is covered with saltwater, or **marine**, biomes. **Saltwater** forms when the only outflow from a body of water is via evaporation, which removes water molecules but leaves dissolved materials behind. Marine biomes can be loosely grouped into three types: oceans, coral reefs, and estuaries.

The open ocean covers about two-thirds of Earth's surface. About 50% of the oxygen in Earth's atmosphere is generated by single-celled photosynthetic *plankton* in the open ocean. These areas also generate most of Earth's freshwater. The water molecules evaporating from their surface condense and fall on adjacent landmasses as rain and snow. The open ocean is heavily exploited by human fishing fleets—a review published in the journal *Science* in 2006 provided evidence that nearly 1/3 of all species of ocean fish are severely endangered and that global catch and ocean biodiversity are declining rapidly.

Coral reefs are unique biomes in that the structure of the habitat is composed of the skeletons of the dominant organism in the habitat (FIGURE 24.6). Reef-building coral live in large colonies, and each individual coral secretes a limestone skeleton that protects it from other animals and from wave action. Coral reefs are made up of the accumulations of billions of these skeletons.

Coral reefs are found throughout the tropics, in warm and well-lit water, providing ample resources for abundant plankton and algae growth. Reefs are comparable to terrestrial tropical rain forests in number of species per unit area.

The zone where freshwater rivers drain into salty oceans is known as an **estuary**. The mixing of fresh and salty water that occurs in estuaries, combined with water-level fluctuations produced by tides, creates a unique habitat that is extremely productive.

Estuaries provide a habitat for up to 75% of commercial fish populations and 80–90% of recreational fish populations. They are sometimes called the nurseries of the sea. Estuaries are also rich sources of shellfish—crabs, lobsters, and clams (FIGURE 24.7). Vegetation surrounding estuaries provides a buffer zone that stabilizes a shoreline and prevents erosion. Unfortunately, estuaries are threatened by human activity as well, including fertilizer pollution and outright loss as a result of housing and resort development.

FIGURE 24.7 **Estuary.** The estuary in Chesapeake Bay is the largest in North America. It supports an enormous number of organisms, some with significant commercial value, such as the blue crab.

Stop and Stretch 24.2

What are the aquatic habitats nearest to you? Can you name some of the dominant species in these habitats? What threats do these habitats face in your area?

As is clear from our discussion of both aquatic and terrestrial biomes, no habitat on Earth has escaped the effects of humans. Consequently, to truly understand our biological addresses, we must learn how human populations use the environment. Preserving our biological neighborhoods requires being able to meet human needs and at the same time respecting the needs of other species with which we coexist.

24.3 Human Habitats

According to the United Nations Food and Agriculture Organization, humans have modified 50% of Earth's land surface for our own use (FIGURE 24.8). Most of this modification has resulted from agriculture and forestry, but a surprisingly large amount—2–3% of Earth's land surface (larger than the area of India)—has been modified for human settlements.

Energy and Natural Resources

Consider the requirements of a forest. The only energy required for its growth is the solar radiation striking the area, and the only nutrients available to support this growth are present in the soil. In addition, nearly all of the waste produced by the organisms in a forest biome is processed on site; that is, it becomes part of the soil or air and is recycled endlessly.

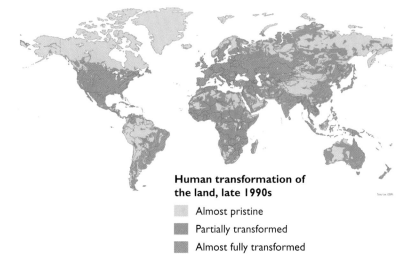

Human transformation of the land, late 1990s
- Almost pristine
- Partially transformed
- Almost fully transformed

FIGURE 24.8 **Human modification of Earth's land surface.** About half of Earth's land surface remains relatively untouched by humans—most of the pristine land lies in the vast boreal forest covering much of Canada and Siberia, the great deserts of Africa and Australia, and the dense Amazonian rain forest in northern and central South America.

VISUALIZE THIS Look at the pattern of human impact across the globe. What factors do you think influence whether a biome is heavily or lightly affected by human activity?

FIGURE 24.9 The cost of fossil fuel extraction. The technique of "mountaintop removal" consists of using explosives to blast off the rock above the coal deposit and dumping the waste rock into nearby valleys. Since 1981, more than 1,300 square kilometers (500 square miles) of West Virginia have been affected by this method of mining.

In contrast, cities rely extensively on energy imported from elsewhere and tend to be the central link in a linear flow of materials—from natural or agricultural landscapes, through the cities, and into waste disposed of elsewhere. Cities differ greatly from most typical biomes.

Energy Use In more developed countries, much of the energy required to power the activities within cities is derived from fossil fuels. In other words, the energy we use is not associated with the bioregion in which we live. The environmental impacts of acquiring and transporting fossil fuel can be substantial, ranging from oil spills that degrade oceans and estuaries to the wholesale dismantling of forested mountains during coal mining (FIGURE 24.9).

In nonindustrial countries, fossil fuel use is still relatively small, and energy sources are more directly tied to the surrounding natural environment. A primary source of energy for heating and cooking in these countries is wood and other plant-based materials (including the dung of plant-eating animals). As urban areas in less developed countries grow, surrounding forests are stripped of trees, sometimes at a faster rate than they can be replaced.

Stop and Stretch 24.3

What is the main source for the electricity you use in your home? Is your bioregion rich in any environmentally friendly energy resources?

What agricultural products are produced in your bioregion? How easy is it to buy local produce? What other natural resources does your bioregion supply?

Natural Resources In addition to energy, human settlements require materials for survival—food from agricultural production and harvesting of natural resources, metals and salts from mines, freshwater for human and industrial consumption, petroleum for asphalt and manufactured goods, and trees for processing into paper and packaging. Developing and extracting all these raw materials changes biomes far from a city's center.

The amount of material needed to support a city can be tremendous. A year 2000 estimate calculated that the **ecological footprint** of the city of London—that is, the amount of land needed to support the human activity there—was 293 times the actual size of the city, equal to twice the entire land surface of the United Kingdom. Clearly, the resources to support London must come from other countries around the world. There is no reason to believe that London is exceptional; most large modern cities likely have similarly sized footprints.

Cities' footprints can be reduced by their citizens, especially if they reduce the amount of energy used for transportation and make sensitive consumer choices, including buying locally produced foods and less meat.

Waste Production

The ecological footprint calculated for London includes not only the land needed to provide resources but also that required to handle the city's waste. Human wastes can be liquid, in the form of wastewater; solid, in the form of garbage; or gaseous, in the form of air pollution.

Wastewater In developed countries, cities have sewage treatment systems to handle the **wastewater** that drains from sinks, tubs, toilets, and industrial plants. These systems typically treat water by removing semisolid wastes through settling, using chemical treatment to kill any disease-causing microorganisms, and eventually discharging the treated water to lakes, streams, or oceans. Unfortunately, older treatment systems can be overwhelmed by storm water, from rain or melting snow, causing the discharge of large volumes of wastewater directly into waterways. Untreated wastewater can cause nutrient levels to spike and algae and bacteria growth to increase in these waterways, leading to the closing of beaches to swimming and other recreation.

Semisolid wastes, called sludge, are often composted (that is, allowed to decompose via the action of bacteria and fungi) and applied to land as fertilizer or trucked to landfills for burial. Land application of sewage sludge has its own problems. This material contains not only human waste—a valuable fertilizer if properly composted—but also industrial wastewater, which can contain a wide variety of toxic chemicals. A major challenge of wastewater treatment in developed countries is the safe and effective disposal of sewage sludge.

In less developed countries, safe disposal of treated sewage from rapidly growing cities may be a distant dream. In many regions, the emigration of people to urban areas has overwhelmed antiquated and inadequate sewer systems. Large numbers of people in these cities live in slums where running water is scarce, and untreated human waste flows in open gutters. The consequences of inadequate disposal of human waste can be severe—intestinal diseases transmitted by contaminated drinking water result in the death of more than 2 million children under five years of age every year.

Garbage and Recycling In addition to dealing with wastewater, people in urban areas also must find ways to control their **solid waste**—that is, their garbage. In more developed countries, most of the solid waste finds its way into sanitary landfills, which are pits lined with resistant material such as plastic. Landfills have systems for collecting liquid that drains through the waste, and exhaust pipes to vent dangerous gases that may build up as a result of decomposition (FIGURE 24.10).

FIGURE 24.10 Disposal of garbage. Solid waste contained in a sanitary landfill is locked within a thick liner and surrounded by a drainage system that collects all water released from or leaching through the waste. Garbage is compacted and "capped" with soil or other material as soon as it is dumped.

To stave off a looming shortage of landfill space, many states and communities mandate recycling of paper, glass, and metal. Several have or are considering community-wide composting programs to reduce the amount of food and yard waste trucked to landfills. Unfortunately, even in cities and states where recycling rates are relatively high, household garbage production continues to increase.

In less developed countries, the problem of solid waste disposal is more severe. Many large cities in these areas have large, open dumps in which unstable, fire-prone piles of garbage provide living space for desperately poor immigrants in the city.

Air Pollution Because of human use of fossil fuels, urban areas produce large amounts of gaseous waste. These air emissions include carbon dioxide, a chief contributor to global warming, as well as nitrogen and sulfur oxides, small airborne particulates, and fuel contaminants such as mercury. Exposure to sunlight and high temperatures can cause some of these by-products to react with oxygen in the air to form ground-level ozone or **smog** (FIGURE 24.11). For individuals with asthma, heart disease, or reduced lung function, increased ozone levels can lead to severe illness, even death.

When the gaseous pollution produced by human settlements enters the upper reaches of Earth's atmosphere, it can be carried on air currents throughout the globe. Air emissions from coal-fired power plants throughout the Midwest cause severe acid rain in the northeast United States, and airborne toxins such as benzene and PCB have been found in high levels in animals around the North Pole.

Air pollution is a problem in both developed and less developed countries. In less developed countries, pollution control is weak or lacking altogether. In more developed countries, the sheer volume of fossil fuel use contributes to poor air quality. In the United States, the number of miles driven by car per household has nearly doubled in the last 25 years, partially due to an increase in the distance that individuals live from their workplaces.

FIGURE 24.11 The effect of waste gas emissions. The brownish haze over downtown Los Angeles, California, in this picture is smog. Smog results from automobile emissions and other air pollutants causing the conversion of oxygen to ozone in sunny, calm conditions.

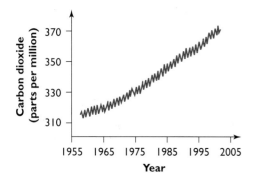

FIGURE 24.12 Carbon dioxide levels in the atmosphere. Carbon dioxide levels measured atop the Hawaiian peak Mauna Loa have steadily risen over the course of several decades. The yearly variation is due to a peak of carbon dioxide emissions in winter, when removal of the gas from the atmosphere by plants is limited.

Stop and Stretch 24.4

Where does your wastewater go? Does your community have any problems handling wastewater and sewage?

Where does your garbage go? What is the rate of recycling in your community, and how can it be improved?

How is the air quality in your community? What are the major air pollutants and their sources?

Climate Change

One air pollutant has potential global effects that may be far more serious than all others. That pollutant, carbon dioxide, has only been widely recognized as a global threat in recent decades. This gas, produced by burning fossil fuels such as oil, coal, and natural gas, has been steadily accumulating in the atmosphere over the past 150 years (FIGURE 24.12). At current rates of production, levels of carbon dioxide in the atmosphere are expected to double by the year 2075.

Scientists who study climate agree that such a large and rapid change in carbon dioxide levels influences weather patterns on Earth. Atmospheric carbon dioxide contributes to the **greenhouse effect**, keeping Earth relatively warm by preventing some of the heat generated by the sun shining on Earth's surface from escaping into space (FIGURE 24.13). Given this effect, an increase in carbon dioxide in the atmosphere should result in an increase in global temperature.

Long-term records indicate that carbon dioxide levels and global temperatures have always fluctuated. However, the concentration of carbon dioxide in the atmosphere is much higher now than at any time in the past 400,000 years.

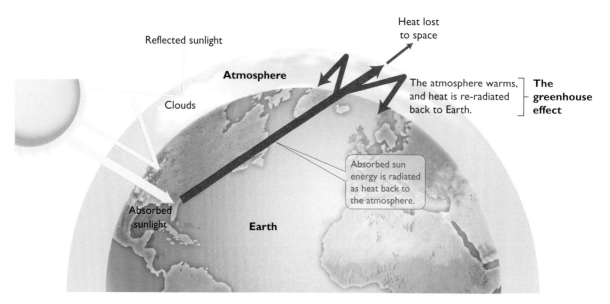

FIGURE 24.13 The greenhouse effect. Carbon dioxide causes the Earth to warm because it retains some heat radiated from the Earth within the atmosphere, rather than allowing it to dissipate.

VISUALIZE THIS An electric blanket generates heat to warm up a sleeper, while a down blanket traps the heat the sleeper produces to keep the bed and sleeper warm. Is increased carbon dioxide in the atmosphere more like an electric blanket or a down blanket?

Most computer models of Earth's climate predict that average global temperatures will increase between 1.5° and 4.5°C (3° and 8°F) by 2075 as a result of modern carbon dioxide emissions. These models also predict that warming will not be uniform, because ocean currents may change and typical patterns of snow and cloud cover will be modified. Certain areas of the globe, such as at the North and South poles, are warming faster and to a greater degree than other areas. Some regions may even cool slightly. Accompanying this temperature shift will be a change in rain and snowfall—again, a change that varies from region to region. The disruptions caused by the greenhouse effect are collectively known as **global climate change**.

Models predict the future long-term effect of warming on Earth, but there is plenty of evidence that our planet is already warming—from the retreat of alpine glaciers to the gradual upward creep in average yearly temperatures over the past several decades. Melting of glaciers and ice caps will cause sea levels to rise, flooding low-lying coastal areas and oceanic islands (FIGURE 24.14). With more heat energy in the atmosphere, storms will become more frequent and extreme. Warmer winters and longer summers will allow some species of plant-eating insects to thrive, potentially causing severe damage to many different ecosystems. Insects that carry disease may also thrive in a warmer climate, threatening the health of human and nonhuman populations alike.

Global warming is already posing a threat to biodiversity. A review published in the journal *Nature* in January 2003 described the "fingerprints" of global warming, and listed various species and ecosystems that have been affected by climate change. Many of these species are temperature sensitive and must move closer to the poles or higher in elevation to find regions with the proper climate. Some of the responses to warming include changes in leafing or blooming times for flowering plants, earlier migration dates for birds and insects, earlier mating seasons for amphibians, and changes in range for all groups of species.

Global warming may work in concert with human-caused habitat destruction to deliver a knockout blow to some species making this last response—the move to find appropriate habitat under warming conditions. If climate change renders the current habitat of an organism unsuitable for its survival, its only hope of persisting is dispersal to a more appropriate habitat. If the species is slow to disperse or if it cannot cross human-modified landscapes, it may not be able to establish a new population in an appropriate habitat before it becomes extinct in its rapidly changing home. Global climate change is another uncertainty created by humans that threatens the survival of thousands of species and even entire biomes.

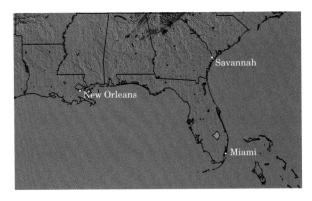

FIGURE 24.14 Projected sea level rise. Melting of glaciers and ice caps, as well as thermal expansion of water, will cause sea levels to rise. The increase in sea level may be as much as 1 meter (3.3 feet) by the end of this century, an amount that would put Miami and New Orleans underwater.

The Future of Our Shared Environment

The effect of human settlements on surrounding natural biomes can be significant and severe, as we have discovered in this chapter. However, many of these impacts can be mitigated with thoughtful planning and the use of improved technology. In the United States, laws such as the Clean Air Act and the Clean Water Act—passed in 1970 and 1972, respectively—have greatly reduced air and water pollution and have contributed to the recovery of once severely impaired habitats.

Cities throughout the more developed world are supporting projects aimed at creating sustainable communities that are both economically vital and environmentally "intelligent." Perhaps by getting to know our biological neighbors and understanding how our choices affect these organisms, we can be inspired to help create communities that are safe and healthy for humans and other species.

Chapter REVIEW

ROOTS TO REMEMBER

The following roots of words come mainly from Latin and Greek and will help you to decipher terms:

boreal is derived from a name for the north wind, Boreas.

decid- comes from the roots **de-** (off) and **cid-** (strike or fall).
terre- is from the word for Earth.

KEY TERMS

24.1 Terrestrial Biomes
biome *p. 587*
boreal forest *p. 588*
chaparral *p. 588*
climate *p. 586*
desert *p. 589*
forest *p. 588*
grassland *p. 588*
permafrost *p. 589*
prairie *p. 588*
precipitation *p. 586*
savanna *p. 588*
steppe *p. 588*
temperate forest *p. 588*
tropical forest *p. 588*
tundra *p. 589*
weather *p. 586*

24.2 Aquatic Biomes
aquatic *p. 590*
coral reef *p. 590*
estuary *p. 591*
freshwater *p. 590*
lake *p. 590*
marine *p. 590*
pond *p. 590*
river *p. 590*
saltwater *p. 590*
wetland *p. 590*

24.3 Human Habitats
ecological footprint *p. 592*
global climate change *p. 595*
greenhouse effect *p. 594*
smog *p. 593*
solid waste *p. 593*
wastewater *p. 592*

SUMMARY

24.1 Terrestrial Biomes

- Climate is the weather of a place as measured over many years. Climate in an area is determined by global temperature patterns, which are driven by solar radiation and local factors such as proximity to large bodies of water (p. 586).

- Categories of primary vegetation types found on Earth's land surfaces are called biomes (p. 587).

- Forests are dominated by trees and categorized by distance from the equator into tropical, temperate, and boreal types (p. 588).

- The chaparral biome is dominated by woody shrubs (p. 588).

- The major vegetation type on grasslands is nonwoody grasses. These biomes are categorized by distance from the equator into tropical and temperate types (p. 588).

- Deserts are found where precipitation is 50 centimeters (20 inches) per year or less and contain mostly drought-resistant plants (p. 589).

- Tundra occurs in areas where the growing season is less than 60 days, both near the poles and at high altitudes (p. 589).

24.2 Aquatic Biomes

- Freshwater biomes include lakes, rivers, and wetlands. These habitats are threatened by pollution and habitat degradation (p. 590).

- Marine biomes include oceans, coral reefs, and estuaries. Oceans are experiencing dramatic changes in their ecosystems as a result of overfishing. Coral reefs are especially threatened by climate change (p. 590).

24.3 Human Habitats

- Humans have modified 50% of Earth's land surface; much of this modification now supports the activities of cities (p. 591).

- Cities must rely on imported energy and other resources to survive; extraction of these resources carries an environmental cost borne by other bioregions. The environmental effects of a city can be expressed by its ecological footprint, the amount of land required to sustain it (p. 592).

- Waste disposal is a significant challenge for large urban areas. Sewage must be treated to avoid contaminating water for drinking; garbage must be disposed of effectively; and air emissions must be controlled (pp. 592–593).

- Global climate change, caused by carbon dioxide emissions that increase the insulative properties of the atmosphere, may pose the largest threat to both human and animal survival of any of our modifications to Earth. These changes include disrupted climate patterns, higher sea levels, and greater storm ferocity, as well as the extinction of some species and the expansion of others (pp. 594–595).

LEARNING THE BASICS

1. What factors determine the kind of terrestrial biome that will develop in a particular geographic area?
   a. climate
   b. average temperature
   c. average precipitation
   d. b and c are correct
   e. a, b, and c are correct

2. A land area's climate is determined by _____.
   a. annual average solar radiation
   b. whether or not it is near a large body of water
   c. the amount of variation in solar radiation over the course of a year
   d. the area's altitude
   e. all of the above

3. Match each biome with its distinguishing characteristic.
   _____ temperate forest   a. evergreen coniferous trees
   _____ boreal forest   b. very low average temperatures
   _____ chaparral   c. fire-adapted shrubs
   _____ desert   d. deciduous, broad-leaved trees
   _____ tundra   e. very low average precipitation

4. Tundra _____.
   a. is found where average temperatures are low and growing seasons are short
   b. is found near the poles
   c. is found at high altitudes
   d. contains only very stunted trees
   e. all of the above are correct

5. Which of the following biomes has a structure made up primarily of the remains of its dominant organisms?
   a. coral reefs
   b. freshwater lakes
   c. rivers
   d. estuaries
   e. oceans

6. All of the following are marine biomes *except* _____.
   a. oceans
   b. coral reefs
   c. rivers
   d. estuaries
   e. intertidal zones

7. Which of the following biomes is most common on Earth's terrestrial surface?
   a. temperate forest
   b. desert
   c. lakes
   d. boreal forest
   e. estuaries

8. In developed countries, wastewater containing human waste _____.
   a. is dumped directly into waterways
   b. is applied directly to farm fields
   c. is piped into sanitary landfills
   d. is treated to remove sludge and disease-causing organisms and released into nearby waterways
   e. is a major source of infectious disease in human populations

9. An ecological footprint _____.
   a. is the position an individual holds in the ecological food chain
   b. estimates the total land area required to support a particular person or human population
   c. is equal to the size of a human population
   d. helps determine the most appropriate wastewater treatment plan for a community
   e. is often smaller than the actual land footprint of residences in a city

10. The greenhouse effect is caused by _____.
    a. greater solar radiation at the equator than the poles
    b. the burning of fossil fuels releasing excess heat into the atmosphere
    c. carbon dioxide trapping heat within the atmosphere
    d. larger and more intense storms
    e. melting of mountain glaciers

11. Which of the following is not considered a likely outcome of global climate change?
    a. increased solar radiation
    b. rising sea levels
    c. increased global average temperatures
    d. extinction of some climate-limited species
    e. increased incidences of coral bleaching

ANALYZING AND APPLYING THE BASICS

1. Consider the following geographic factors, and predict both the climate and biome type found in the location described. Explain the reasoning that you used to determine your answer. This small city is:
   - On the east coast of the Pacific Ocean
   - 20° north of the equator
   - 200 meters (650 feet) above sea level
   - At the base of a mountain range

2. One prediction of global climate change models is that significant amounts of melting ice will change the salt content of the ocean, which may cause the Gulf Stream in the Atlantic Ocean to stop altogether. How will this change likely affect Europe?

3. What can you infer about the geographical relationship among the cities in the following table and the Cascades, the primary mountain range that influences their climate?

City	Approximate Average Annual Rainfall
Bend, OR	12 inches
Eugene, OR	43 inches
Portland, OR	36 inches
Tacoma, WA	39 inches
Yakima, WA	8 inches

Use the following information to answer questions 4 and 5.

Cicadas are large, flying insects that feed on tree branches and roots. Periodical cicadas develop underground for many years and emerge in a synchronized group every 13 or 17 years. When they emerge, the males create a loud, buzzing sound to attract females. Hundreds of thousands of cicadas per acre can cover the landscape when a brood emerges. This can cause deafening noise and damage to young trees. The table below shows the periods and geographical ranges of seven broods of cicadas.

Brood	Period	Years	Region
IX	17 years	1969, 1986, 2003	NC, VA, WV
X	17 years	1970, 1987, 2004	DE, GA, IL, IN, KY, MD, MI, NC, NJ, NY, OH, PA, TN, VA, WV
XIII	17 years	1973, 1990, 2007	IA, IL, IN, MI, WI
XIV	17 years	1974, 1991, 2008	KY, GA, IN, MA, MD, NC, NJ, NY, OH, PA, TN, VA, WV
XIX	13 years	1972, 1985, 1998	AL, AR, GA, IN, IL, KY, LA, MD, MO, MS, NC, OK, SC, TN, TX, VA
XXII	13 years	1975, 1988, 2001	LA, MS
XXIII	13 years	1976, 1989, 2002	AR, IL, IN, KY, LA, MO, MS, TN

4. Which brood in the table spans the widest geographic range and when will that brood emerge next?

5. Brood X emerged from underground in 2004. Its range included New Jersey, New York, Pennsylvania, Maryland, and Virginia. Describe how human development may have affected the size of brood X between 1970 and 2004.

6. Why are most state capitals in the U.S. and many national capitals worldwide located on rivers?

Use the following information to answer questions 7 and 8.

Your personal ecological footprint represents the amount of material that you need to support your standard of living. An ecological footprint can be calculated by considering the following:
- How often do you eat animal products?
- How much of the food that you eat is processed or packaged?
- How much of the food that you eat is grown locally?
- How much waste do you generate?
- How large is your house/apartment and how many people live there?
- Do you conserve electricity?
- How many miles do you travel each week and what type of transportation do you use?
- How often do you bike or walk each week?
- How many miles per gallon does your car get?
- How often do you carpool?

The average ecological footprint in the United States is 24 acres per person. Materials equal to 5.4 Earths would be needed if everyone had the footprint of the average American.

7. To maintain the current human population sustainably, the total ecological footprint of humanity will need to equal the total land area available on Earth. Given that the average footprint of 24 acres per person would require 5.4 Earths, how many acres would make up the sustainable ecological footprint?

8. Describe six steps that you could take to reduce your ecological footprint.

CONNECTING THE SCIENCE

1. How many biomes do you rely on to supply your food? Many grocery stores label the origin of their produce. The next time you go to the grocery store, try to determine the number of different countries from which your produce comes. Could you easily change your diet and shopping habits to rely on locally produced food? Why or why not?

2. Consider the setting of your home (such as placement of windows, direction of prevailing winds, access to sidewalks, accessibility of stores and schools, distance from the center city, landscaping, and so on). What kind of changes to the environment of your home would help it fit better into the bioregion? Are these changes feasible? Are they desirable?

APPENDIX

Metric System Conversions

To Convert Metric Units:	Multiply by:	To Get English Equivalent:
Length		
Centimeters (cm)	0.3937	Inches (in)
Meters (m)	3.2808	Feet (ft)
Meters (m)	1.0936	Yards (yd)
Kilometers (km)	0.6214	Miles (mi)
Area		
Square centimeters (cm^2)	0.155	Square inches (in^2)
Square meters (m^2)	10.7639	Square feet (ft^2)
Square meters (m^2)	1.1960	Square yards (yd^2)
Square kilometers (km^2)	0.3831	Square miles (mi^2)
Hectare (ha) (10,000 m^2)	2.4710	Acres (a)
Volume		
Cubic centimeters (cm^3)	0.06	Cubic inches (in^3)
Cubic meters (m^3)	35.30	Cubic feet (ft^3)
Cubic meters (m^3)	1.3079	Cubic yards (yd^3)
Cubic kilometers (km^3)	0.24	Cubic miles (mi^3)
Liters (L)	1.0567	Quarts (qt), U.S.
Liters (L)	0.26	Gallons (gal), U.S.
Mass		
Grams (g)	0.03527	Ounces (oz)
Kilograms (kg)	2.2046	Pounds (lb)
Metric ton (tonne) (t)	1.10	Ton (tn), U.S.
Speed		
Meters/second (mps)	2.24	Miles/hour (mph)
Kilometers/hour (kmph)	0.62	Miles/hour (mph)

To Convert English Units:	Multiply by:	To Get Metric Equivalent:
Length		
Inches (in)	2.54	Centimeters (cm)
Feet (ft)	0.3048	Meters (m)
Yards (yd)	0.9144	Meters (m)
Miles (mi)	1.6094	Kilometers (km)
Area		
Square inches (in^2)	6.45	Square centimeters (cm^2)
Square feet (ft^2)	0.0929	Square meters (m^2)
Square yards (yd^2)	0.8361	Square meters (m^2)
Square miles (mi^2)	2.5900	Square kilometers (km^2)
Acres (a)	0.4047	Hectare (ha) (10,000 m^2)
Volume		
Cubic inches (in^3)	16.39	Cubic centimeters (cm^3)
Cubic feet (ft^3)	0.028	Cubic meters (m^3)
Cubic yards (yd^3)	0.765	Cubic meters (m^3)
Cubic miles (mi^3)	4.17	Cubic kilometers (km^3)
Quarts (qt), U.S.	0.9463	Liters (L)
Gallons (gal), U.S.	3.8	Liters (L)
Mass		
Ounces (oz)	28.3495	Grams (g)
Pounds (lb)	0.4536	Kilograms (kg)
Ton (tn), U.S.	0.91	Metric ton (tonne) (t)
Speed		
Miles/hour (mph)	0.448	Meters/second (mps)
Miles/hour (mph)	1.6094	Kilometers/hour (kmph)

Metric Prefixes

Prefix		Meaning
giga-	G	$10^9 = 1{,}000{,}000{,}000$
mega-	M	$10^6 = 1{,}000{,}000$
kilo-	k	$10^3 = 1{,}000$
hecto-	h	$10^2 = 100$
deka-	da	$10^1 = 10$
		$10^0 = 1$
deci-	d	$10^{-1} = 0.1$
centi-	c	$10^{-2} = 0.01$
milli-	m	$10^{-3} = 0.001$
micro-	μ	$10^{-6} = 0.000001$

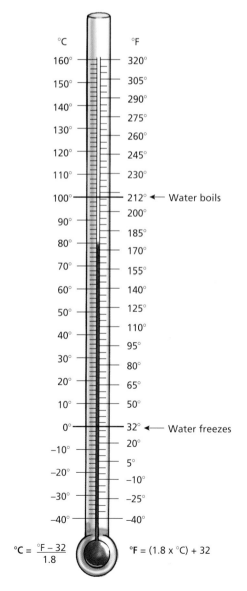

$$°C = \frac{°F - 32}{1.8} \qquad °F = (1.8 \times °C) + 32$$

ANSWERS

Answers are provided for the Stop and Stretch boxes, which appear within the chapters, and for Learning the Basics and Analyzing and Applying the Basics questions, which appear at the end of each chapter.

Chapter 1

Stop and Stretch

1.1. Because the God in this hypothesis is a supernatural entity, it does not have to follow natural laws. In other words, any result of a test of a supernatural entity is a possibility, so there is no way to falsify the existence of such a being.

1.2. No, someone else could have eaten the doughnuts. The truth of the prediction (the doughnuts are gone), does not prove the truth of the hypothesis (you ate them).

1.3. Independent variable is diet, dependent variable is cholesterol level.

1.4. The confidence intervals overlapped. With a standard error of 3%, the actual level of support for candidate B has a 95% probability of being somewhere between 48 and 54%, and the actual level of support for candidate A has a 95% probability of being somewhere between 44 and 50%.

Learning the Basics

1. a **2.** d **3.** b **4.** a **5.** d **6.** d **7.** e **8.** a **9.** c
10. e **11.** b **12.** b **13.** b **14.** a **15.** e **16.** e
17. c **18.** c **19.** d **20.** false

Analyzing and Applying the Basics

1. Because neither the students nor the providers were blinded to the treatment, and because there is a widespread perception that vitamin C is an effective treatment (or preventive) for colds, both ill students in the vitamin C group and their care providers may have tended to underreport cold symptoms. Subject expectation and observer bias can influence the results in this type of study.

2. Only that a correlation exists. One might use this result to hypothesize that exercise increases BDNF levels and run a controlled experiment to test this. Alternatively, higher BDNF levels may increase the desire to run—this is more difficult to run an experiment on, but possible if BDNF can be artificially manipulated.

3. By not knowing the racial group that any skull belonged to before or during measurement.

4. Product 2. It has at least been subject to experimental tests by an independent laboratory—possibly controlled experiments, given the comparative data presented.

5. The nature of the experimental tests, in particular whether the experiment was controlled, whether researchers were double-blinded, and whether the sample size was sufficiently large. On a practical level, it would be useful to know if the claimed effects were actually visible to the naked eye and persistent enough so that the cost of the product justified the benefits gained.

6. A high-fat, high-calorie diet reduces life span in mice relative to a more typical diet. The addition of resveratrol to the high-calorie diet appears to reduce the diet's negative effects on life span.

7. That a high-calorie diet is generally bad for health and may shorten life span. However, a chemical present in red wine may reverse these bad effects, so that a high-fat, high-red-wine diet may be more healthful.

8. The amount of red wine one would have to consume to reduce the negative effects of a high-fat diet appears to be ridiculously high. The claims at the beginning of the article are not supported by the actual research report.

Chapter 2

Stop and Stretch

2.1. Covalent bond because the oxygen molecules each have 6 valence electrons, and therefore need 2 to fill their valence shell. Carbon has 4 valence electrons and can share with each oxygen to fill its shell.

2.2. Different cells use different genes.

Learning the Basics

1. d **2.** a **3.** e **4.** d **5.** b **6.** b **7.** d **8.** e
9. a **10.** d **11.** c **12.** a **13.** b **14.** a **15.** e
16. $C_{12}H_{22}O_{11}$ **17.** polymers; condensation
18. vitamins and minerals **19.** false **20.** false

Analyzing and Applying the Basics

1. Oxygen will have a partial negative charge; carbon will have a partial positive charge.

2. Liquids become gases when pressure is low.

3. Hydrogen ions lead to lower pH. Acidic stomach juices aid in digestion.

4. Sometimes it is necessary to supplement during an endurance event. Otherwise, a person should eat whole foods to get enough fiber, vitamins, and minerals in the diet.

5. The function of a chemical is tied directly to its structure.

6. 3′-TTGCTAGGC-5′

7. Heat-killed bacteria would still be virulent.

8. Was not lipid or protein. Could be nucleic acid.

9. Viral DNA, not protein, was being used to hijack cells.

10. The danger is that energy drinks mask symptoms of drunkenness. The alcohol is still having negative effects; your body just perceives the impairment less.

Chapter 3

Stop and Stretch

3.1. Yes, fats require carbon, hydrogen, and oxygen, and all three are found in amino acids and proteins.

3.2. 10,000

3.3. Cells could swell and burst if placed in a hypotonic solution or shrivel if placed in a hypertonic solution.

3.4. a. isotonic, b. hypotonic, c. hypertonic

3.5. The substrate would be converted to product at a much slower rate.

3.6. If hydrogen ions diffused through the tube, the absence of a gradient would prevent the synthesis of ATP.

3.7. High levels of glucose in blood means water will leave cells, causing dehydration.

ANS-1

Learning the Basics

1. e **2.** c **3.** c **4.** d **5.** e **6.** c **7.** c **8.** b **9.** e **10.** d **11.** b **12.** d **13.** d **14.** d **15.** c, d, b, e, a **16.** true **17.** true

Analyzing and Applying the Basics

1. Genetics, age, sex

2. Because it contains more mitochondria, brown fat can produce more energy for the body. This energy may be most useful as heat energy, to warm the vital organs of young mammals and of adult mammals that live in cold temperatures.

3. When cells run low on NAD^+, they use fermentation to regenerate NAD^+. With ample NAD^+, cells can switch back to respiration (in the presence of oxygen), because NADH is available to transfer electrons to the top of the electron transport chain.

4. Like an enzyme, the CB1 receptor must have properties that only allow certain molecules to interact with it to produce a reaction. The drug must have properties that allow it to "fit" the CB1 receptor, like a substrate fits into the active site on an enzyme. The drug is different from a substrate, however, because it blocks a reaction instead of inducing one.

5. Sample answer: When the stomach is empty, energy may be low, so the stomach releases ghrelin to stimulate appetite. As the person eats, the gastrointestinal tract senses food and releases cholecystokinin to signal a sense of fullness. As material moves through the intestinal tract, the colon senses that there is enough energy in the system. It releases PYY(3-36) to inhibit eating.

6. A decrease in leptin levels, increase in NPY, and decrease in melanocortin would all stimulate appetite, making it harder for the person to continue dieting.

7. Possibly. Because leptin is known to suppress appetite in humans, injections may help to suppress appetite and promote weight loss. However, every person may not react to leptin in the same way. For people that have a faulty leptin-producing gene, leptin injections may help a lot. But others who produce adequate amounts of leptin and still have a weight problem may not benefit from injections.

8. Try to lose weight slowly, so that the body's hormones are able to gradually change and adjust. Losing weight quickly could trigger a sharp drop in leptin, which would trigger a large increase in hormones that stimulate appetite. Eat small meals throughout the day. This would prevent the stomach from becoming empty and stimulating appetite. Having food in the digestive tract would also stimulate the release of hormones that give a sense of fullness and reduce appetite. Eat high-fiber, bulky foods, such as vegetables and whole grains, to obtain a sense of fullness without high caloric intake. Exercise to increase the number of Calories burned.

Chapter 4

Stop and Stretch

4.1. 2

4.2. RNA polymerase can't bind, and transcription decreases.

4.3. Need to know what enzymes to use to cleave the gene from the genome.

4.4. Artificial selection takes many generations to have an effect. Artificially selected organisms have been modified in the sense of selection for desirable traits.

Learning the Basics

1. GCUAAUGAAU **2.** gln-arg-ile-leu **3.** d **4.** d **5.** b **6.** c **7.** b **8.** c **9.** c **10.** b **11.** a **12.** d **13.** d **14.** a **15.** true **16.** true **17.** false **18.** b **19.** c **20.** e

Analyzing and Applying the Basics

1. Mutations to the third position often result in the same amino acid being called for. For example ACU, ACC, ACA, and ACG all call for threonine.

2. Transcription of all genes would be affected.

3. AT**GAATTC**CGTCCG

3. TA**CTTAAG**GCAGGC

4. While most pests would be killed, some with preexisting resistance would survive and reproduce. Because their competitors are killed, the resistant pests would flourish.

5. CAUAUGGAUCAU

6. arg-thr-arg-gly-ala-val

Chapter 5

Stop and Stretch

5.1. Strength training helps keep joints properly aligned and prevents some of the friction associated with joint surfaces rubbing against each other. Exercises that don't jar the joints are best for persons with joint injuries. Examples include swimming, biking, and the use of elliptical training equipment.

5.2. If the brain is still healing from a concussion, blood vessels may not be fully repaired. A second concussion causes even more swelling, which can cause the brain to shut down.

5.3. Positive

Learning the Basics

1. c **2.** e **3.** a **4.** d **5.** d **6.** b **7.** c **8.** d **9.** a **10.** d **11.** a **12.** d **13.** d **14.** a **15.** d **16.** k **17.** a **18.** g **19.** c **20.** b

Analyzing and Applying the Basics

1. Muscle and the connective tissue in bone become stronger with exercise. Nervous tissue releases more endorphins and other neurotransmitters that may help to fight depression and anxiety. Increased circulation due to exercise can help to nourish the epithelial and connective tissues in skin. This removes toxins and helps to reduce the signs of aging. Exercise can also help lubricate connective tissues, such as tendons and ligaments. However, too much exercise can damage connective tissues.

2. Tall, column-shaped cells would have a larger surface area than flatter cells, so they would allow the small intestine to absorb more nutrients.

3. Retinoids increase the production of collagen and/or elastin fibers, in order to improve the strength of skin and its connection with underlying muscle.

4. The heart and the lungs are both located in the thoracic cavity. The lungs take in oxygen that is delivered to cells by the pumping of the heart. As the heart pumps blood through the body, the blood also removes waste carbon dioxide. This carbon dioxide is passed to the lungs and eliminated from the body.

5. The amount of lubricating fluid within the membrane increases and presses on the heart to cause pain.

6. The diameter of the balloon should get larger and larger over the twelve months, because exercise can increase your lung capacity.

7. Your nervous system allows you to track the ball. The nervous system signals the muscular system to move the skeletal system, and the respiratory and circulatory systems deliver oxygen to your muscles, so that you can run to hit the ball. The nervous system also signals the integumentary system to dissipate heat, so that you can maintain a comfortable core body temperature as you play.

Chapter 6

Stop and Stretch

6.1. More complex breaks are more difficult to align properly. More complex breaks cause more damage and increase healing time.

6.2. Marathoners—slow twitch; weight lifters—fast twitch

6.3. Differences within a group are greater than between group differences.

Learning the Basics

1. c **2.** a **3.** b **4.** b **5.** a **6.** e **7.** true **8.** true
9. d, b, c, e, a **10.** c, e, a, d, b **11.** b **12.** b
13. b **14.** e **15.** c **16.** See Figure 6.24 **17.** b
18. c **19.** false **20.** true

Analyzing and Applying the Basics

1. A drop in estrogen allows the absorption of bone by osteoclasts to outpace deposition by osteoblasts, even in the presence of normal or high calcium levels.

2. She could increase weight-bearing exercise and take calcium supplements to help build up her bone mass. She could also consider estrogen-replacement therapies as her menstrual periods become irregular and decline.

3. No. Due to the activity of osteoclasts, the coral graft would gradually be replaced by human bone.

4. They are aligned so that they can work together to move the body more effectively.

5. Hypogonadism would decrease the development of muscle mass.

6. Because men sweat more and begin sweating at a lower core body temperature, they should be able to maintain a lower average core temperature than women can during exercise. A lower temperature would allow the systems in the male body to operate within a more normal range, with reduced stress. This could enhance the athletic performance of men over women. However, because they are losing more water, men may be more susceptible to dehydration. This susceptibility could reduce athletic performance.

7. Cells must continually rise to the surface to replace protective cells that are rubbed off.

Chapter 7

Stop and Stretch

7.1. Transcription and translation of the gene that codes for the lactase enzyme might slow with age.

7.2. The appendix seems not to have any function. However, this does not prove that the appendix is nonfunctional. There may be an effect that scientists have not noticed or correlated with appendix removal. Further study is needed before any real conclusions can be drawn.

7.3. Non-obese patients having abdominal surgery control for any effect the actual surgery might have on ghrelin levels.

7.4. Resistin is named for its purported role in making cells resistant to the effects of insulin.

7.5. Same overall trend but not the same absolute percentages. One study is not really enough information to base this decision on.

Learning the Basics

1. a **2.** a **3.** c **4.** d **5.** b **6.** e **7.** e **8.** e **9.** a
10. c **11.** d **12.** d **13.** b **14.** e **15.** e **16.** c
17. e, c, b, a, d **18.** d, c, e, a, b **19.** clockwise from top: pharynx, esophagus, diaphragm, stomach, small intestine, large intestine, pancreas, gallbladder, liver **20.** true

Analyzing and Applying the Basics

1. Acidity. Stomach acids and acidic foods and beverages seem to lead to erosion.

2. The risk of dental erosion would be around 8 times higher than the risk for a person without those risk factors.

3. A bulimic person's teeth are exposed to stomach acids that erode enamel.

4. Answer: D, B, C, A. (Pepsinogen is activated by acid.)

5. Sample answer: If beta cell destruction is due to misfolded proinsulin, therapies could focus on introducing beta cells from individuals that are not prone to diabetes or fixing the mechanism in beta cells that leads to misfolding.

6. If the cytochrome P-450 system were more active, it might take over the processing of acetaminophen at lower doses. This would produce more toxic NAPQI. With a reduction in glutathione, the increased concentrations of NAPQI would remain toxic and build up in the liver. Therefore, alcohol would allow lower doses of acetaminophen to cause liver damage.

7. Although it is lower, the single dose would probably be more toxic. A dose of 3,000 mg taken at one time would quickly overwhelm the sulfation, glucuronidation, and glutathione systems, to allow NAPQI to build up quickly and cause damage. A larger dose over time might not overwhelm the normal systems as dramatically.

Chapter 8

Stop and Stretch

8.1. Lack of protein in the diet leads to reduced albumin production. As a result, the osmotic potential of the blood is lower than that of the tissues (that is, the blood is more watery). Thus, water diffuses from the bloodstream into the body tissues, causing swelling.

8.2. The bone marrow is the source of the defective stem cells. Removing the marrow may remove the affected cells, and replacement with donor marrow can reestablish a healthy population of white blood cells.

8.3. O^- blood cells do not contain any of the common surface antigens that can cause an agglutination reaction.

8.4 A thrombus or embolus will severely impair or cut off oxygen and nutrient delivery to tissues served by the blood vessel in which they form or lodge.

Learning the Basics

1. c **2.** e **3.** b **4.** a **5.** d **6.** b **7.** a **8.** d **9.** e
10. e **11.** d **12.** b **13.** d **14.** d **15.** a **16.** d
17. c **18.** a **19.** b **20.** false

Analyzing and Applying the Basics

1. Individuals deficient in the RB protein would have difficulty producing an adequate amount of erythrocytes, leading to chronic anemia. In addition, the immune system would likely be compromised, since the number of mature macrophages would be low as well.

2. Not likely. The iron is likely present in individuals with porphyria, it just cannot be attached to the heme group efficiently without the porphyrin enzyme present in the mitochondria.

3. Yes, because this would increase the number of functional erythrocytes, reducing anemia symptoms.

4. They have severe anemia and would be lethargic, easily fatigued, and have difficulty breathing. Kind of like Dracula.

5. Regular menstruation removes blood, and thus iron, from a woman's body.

6. Suppression of prostacyclin allows platelet activity to increase. The increased activity of platelets in turn would lead to an increase in blood clotting. Some of these clots can form in the heart or brain or travel there and cause heart attack or stroke by cutting off blood supply to portions of these organs.

7. Individuals on heparin have a decreased ability to make clots, so a severe injury could result in traumatic blood loss.

8. G6PD deficiency may protect individuals against malaria, by causing the prophylactic destruction of red blood cells in the presence of a stressor like infection with a malaria parasite. Even though anemia would result, the parasite is less likely to establish a large population in an individual if the blood cells it infects are more likely to die.

9. One could look for a correlation between malaria infection rate and G6PD deficiency.

10. The presence of one or two copies of allele *C* and allele *S* appeared to provide resistance to malaria. Among those *AC* and *AS* heterozygotes who did become infected, they were less likely to experience severe malaria than individuals of genotype *AA*.

Chapter 9

Stop and Stretch

9.1. If the anti-backflow valves in veins fail, blood under the force of gravity will tend to drop to lower levels in the veins.

9.2. The right side of the heart distributes blood to the lungs, a relatively short distance away from the heart. However, the left side distributes blood to the whole body, so the pressure created by each compression of the heart must be greater.

9.3. The heart requires energy to beat. If it is beating at a higher rate, more energy is required, thus more calories are burned.

9.4. No, because if these dangerous plaques do not disrupt blood flow, they probably do not cause a worrisome stress test result.

Learning the Basics

1. b **2.** e **3.** e **4.** c **5.** c **6.** d **7.** d **8.** e **9.** a
10. c **11.** a **12.** c **13.** a **14.** a **15.** d, e, b, c, a
16. c, b, d, a **17.** b, d, c, a **18.** systole, diastole
19. Atrial, ventricular **20.** false

Analyzing and Applying the Basics

1. Black women have the highest rates, Asian and Pacific-Islanders have the lowest rates.

2. (1) Genetic differences (2) Economic differences (3) Educational differences (4) Dietary differences (5) Differences in stress level (others are possible as well)

3. The key is to match individuals as much as possible. One could compare racial groups in different economic categories, for instance, to determine the role of economic differences on these outcomes.

4. Whether when comparing groups of individuals of *the same ages*, the outcomes for women were poorer than those for men.

5. Blood pressure is significantly lower in veins, but also, the less rigid walls of veins are less likely to register a "bulge" of blood going by.

6. Oxygenated and deoxygenated blood mix in the single ventricle and both types are pumped to the lungs and body. Thus, the oxygen concentration in the arteries is lower in frogs than in humans, reducing the maximum amount of energy they can expend in any given period.

7. The pulmonary artery and the aorta. This allows blood that would go to the lungs to be diverted into the aorta for circulation through the body.

8. One could compare heart disease rates and outcomes between blacks who are descended from slaves, and thus likely to have some white ancestors, and blacks who are more recent immigrants (or the children of immigrants) from Africa.

9. It provides a list of genetic sequences that correlate to increased risk of atrial fibrillation. Correlated sequences might contain genes that have an effect on cardiac cycle.

10. To control for any differences between men and women that may relate to differences in their predisposition to cardiac cycle disruptions.

Chapter 10

Stop and Stretch

10.1. Most inhaled air comes through the nose. Nose hairs trap some of the particles before they are drawn farther into the respiratory system, where they may cause damage.

10.2. Common colds are much more common than sinus infections, so any correlation between colds and smoking should be easier to measure.

10.3. The Heimlich maneuver forces the diaphragm upward into the chest cavity, dramatically and rapidly reducing lung volume. Air in the lungs is forcefully ejected through the trachea, which may dislodge the obstruction. In many ways, the maneuver is like a manual cough.

10.4. The air in the paper bag will become more rich in carbon dioxide the longer one breathes into it. The carbon dioxide thus inhaled can diffuse into the blood, helping to reverse the loss of carbon dioxide brought about by an increased exhalation rate.

10.5. The higher gas pressure drives oxygen across the respiratory surface and helps to displace the carbon monoxide from hemoglobin.

10.6. Nicotine is responsible for a large number of deaths due to heart disease. So while smokeless tobacco users are unlikely to experience the decline in lung function or increased risk of lung cancer that smokers face, they are more likely to suffer the heart and cardiovascular system damage caused by nicotine. Both types of tobacco use are harmful.

Learning the Basics

1. a **2.** b **3.** d **4.** e **5.** (1) nasal conchae (2) larynx (3) trachea (4) bronchioles (5) alveoli
6. e **7.** e **8.** c, b, d, a **9.** e **10.** d **11.** d **12.** d
13. b **14.** b **15.** e **16.** d **17.** a **18.** d **19.** e

Analyzing and Applying the Basics

1. Reducing migration of an immune system cell to the lungs may have serious negative side effects, including an increased risk of lung infection, that might not outweigh the benefit in reducing asthma attack frequency.

2. Asthma rates have been rising steadily in children, have risen slightly in older individuals and quite a bit in older women, but have dropped in older men.

3. It appears that exposure to dirty air may help individuals avoid the development of asthma. Perhaps regular low-level exposure is the normal condition and lack of exposure can lead to improper development of the lung's immune system. One strategy for reducing the risk of asthma in children may be to own pets and/or encourage play in un-air-conditioned or outdoor environments, where allergen concentrations are higher.

4. Particulates can cause inflammation and infection, interfering with gas exchange and thus effective function.

5. As long as the partial pressure of oxygen in the liquid is higher than that of the blood, oxygen will diffuse across the respiratory surface and into the bloodstream, whether the medium is liquid or gaseous.

6. If the minerals can be absorbed across the respiratory membrane, they can accumulate in the bloodstream.

Chapter 11

Stop and Stretch

11.1. The muscles of the tissues surrounding the vaginal opening can be stretched or damaged during childbirth. They may no longer contract as strongly around the urethral sphincter after this damage. The loosened sphincter may allow drops of urine to escape when it is put under high pressure, as during a cough.

11.2. The urine will contain excess protein and other large molecules, perhaps even blood cells.

11.3. High blood pressure, to reduce blood volume and damage to blood vessels. Also heart disease, to reduce the amount of blood that must be moved through the heart at each contraction.

11.4. Dialysis patients should produce relatively little urine, and it should be more dilute than in individuals with intact kidneys.

Learning the Basics

1. e 2. c 3. b 4. e 5. a 6. e 7. a 8. e 9. c
10. a 11. e 12. e 13. a 14. e 15. c 16. a
17. d 18. true 19. true 20. true

Analyzing and Applying the Basics

1. HCG must be excreted by the kidneys. It is likely part of the filtrate that does not get reabsorbed.

2. It would allow uric acid to remain in the filtrate, potentially crystallizing in the renal pelvis into kidney stones.

3. URAT1 transporters that were highly active and recovered most uric acid from the filtrate would allow uric acid to accumulate in the blood, potentially crystallizing on joints.

4. Tissues begin to accumulate water, increasing overall body weight.

5. Reduce the number of aid stations to reduce water intake. Provide sports drinks containing solutes. Provide salty snacks at aid stations.

6. Excess oxylate from the digestive system could combine with calcium to form calcium oxylate, which is a mineral found in stones.

7. All of the diuretics reduced the production of kidney stones. By increasing urine production, diuretics may help flush the kidney of stone-forming compounds and also keep the fluid in the renal pelvis dilute.

8. A uric acid stone, produced by an overall excess of uric acid in the blood.

9. Drink a lot of water. Frequent urination keeps the kidney flushed of accumulating minerals, and dilute urine will reduce the likelihood that crystallization will occur.

10. Caffeine is a diuretic and high-solute concentrations in a drink can rob tissues of water. In combination, these drinks may be more likely to cause dehydration than rehydration.

Chapter 12

Stop and Stretch

12.1. (1) Prions are not affected by such treatment. (2) Prions' unusual response to treatments that usually kill other pathogens lends support to the prion hypothesis.

12.2. Direct and indirect contact

12.3. A person might not have the genes required to produce a particular antigen no matter how much rearrangement occurs.

Learning the Basics

1. d 2. b 3. e 4. d 5. d 6. e 7. b 8. c 9. b
10. e 11. a 12. d 13. b 14. e 15. c 16. c
17. c, a, b, e, d 18. c, e, b, d, a 19. false 20. true

Analyzing and Applying the Basics

1. The infectious biofilm would adhere to surfaces inside the ear, allowing bacteria to grow and reproduce easily. The biofilms could form a protective barrier preventing antibiotics from accessing bacteria. The structure of the ear itself would make it difficult to remove the infectious biofilm by physical means.

2. Brushing and flossing disrupt biofilms that contain bacteria in the mouth. Rinsing then removes these broken-up biofilms and bacteria. This leads to fewer bacteria in the mouth, making it less likely that dental disease will develop.

3. The H5N1 strain has not yet recombined to form a new strain that is able to both infect humans and be passed effectively between humans.

4. The human immune system would not have encountered much of the genetic material that had developed in bird populations, so it would have a difficult time mounting a specific defense. It may have encountered viruses that were similar to the one that had developed solely within human populations, so it would be better able to recognize and respond to that virus.

5. Pigs and humans are both mammals, so they are more closely related than humans and birds are. Therefore, they are more likely to be affected by similar viruses.

6. Scientists could test the urine of infected elk or deer, to see if prions were present. They could feed grass covered with urine from infected animals to healthy animals, to see if those animals contracted CWD.

7. Farmers would need to keep feed protected from animal wastes. They would also need to maintain hygienic stalls and protect individuals from coming into contact with another animal's waste as much as possible.

8. The liver cell is part of the body, so it would not trigger an immune response. This allows the merozoites to travel within the body without being attacked by lymphocytes.

9. The antiviral would help to slow the rate of infection by preventing viral cells from infecting more cells. This would allow the immune system to produce enough lymphocytes to effectively battle the infection.

Chapter 13

Stop and Stretch

13.1. No, because the immune system does not control the lice, so specific immunity does not develop against this infestation.

13.2. Trichomoniasis, chlamydia, gonorrhea, and hepatitis can be prevented via condom use. Syphilis, herpes, genital warts, and lice infestations cannot.

13.3. Three times.

Learning the Basics

1. e (d or c are also reasonable answers) 2. a
3. b 4. c 5. a 6. b 7. d 8. e 9. d 10. b 11. a
12. d 13. c 14. a 15. e

Analyzing and Applying the Basics

1. Few girls at this age are sexually active, so few are already infected with HPV. Once infection occurs, the vaccine is not expected to reduce the risk of cervical cancer.

2. They were "blinding" the subjects of the study so that the subjects' behavior would not affect the study results.

3. Association and transmission.

4. A positive correlation. White blood cell counts increase as promiscuity increases.

Chapter 14

Stop and Stretch

14.1. Sample answers: Could block receptors on postsynaptic cell or could block exocytosis from presynaptic cell.

14.2. (1) Yes, because the medicine itself could change brain structures or alter functions. (2) Sample answer: Children with more severe ADD may have more severe structural differences or may make less neurotransmitter.

14.3. Sample answers: The enlarged hippocampus in ADD children might arise as the brain attempts to overcome some of the symptoms associated with ADD, or children with ADD might have been born with a larger hippocampus.

14.4. Damage to spinal nerves higher on the cord hinders the ability of all organs and tissues below them. The higher the injury, the more damage occurs.

14.5. Most physicians recommend attempting changes to diet and exercise before turning to prescription medicines, because medicines can have negative side effects while a healthy diet and exercise program provides only benefits.

Learning the Basics

1. e **2.** b **3.** d **4.** d **5.** d **6.** d **7.** e **8.** true
9. a. sensory neuron **b.** interneuron **c.** motor neuron **d.** sensory receptor **e.** cell body **f.** dendrite **g.** axon **h.** Schwann cell **i.** node of Ranvier **j.** effector **10.** true **11.** false **12.** d **13.** d
14. a. cerebrum **b.** cerebellum **c.** brain stem **d.** hypothalamus **e.** thalamus **15.** a. **16.** c
17. a **18.** d **19.** true **20.** true

Analyzing and Applying the Basics

1. The myelin sheath protects the nerves from damage and prevents sideways message transmission. Degeneration of the myelin sheath could interfere with nerve impulses that control movement or sensory perception. Degeneration could also damage nerves so that they can no longer function. This could cause the complete loss of sensory perception, the loss of the ability to move, and the eventual loss of the ability to breathe.

2. Once symptoms have appeared, a large concentration of fatty acids would have already built up in the boy's body. Lorenzo's Oil interferes with biosynthesis, but it does not decrease existing concentrations of fatty acids. If fatty acids are already in the body at high levels, they could continue to cause damage to the myelin sheath, even if the boy was taking Lorenzo's Oil.

3. The studies show that the brain of a person with dyslexia performs differently than the brain of someone without the disability. The dyslexic brain has reduced activity in the region that connects visual association areas with language areas. Because this connection is reduced, other areas of the brain must work harder to perform reading tasks.

4. The brain has some plasticity, or ability to change its function with training. Changes can occur both during the brain's development and after it has matured.

5. Sample answer: Because the hippocampus plays a role in creating memories, forgetfulness would be an early symptom. Because the amygdala causes experiences to have emotional overtones, moodiness would also be an early symptom.

6. Simple memory associations can be created in the absence of a functioning hippocampus. However, more complex associations between new and previous memories cannot be created without the hippocampus.

7. An excess of alcohol depresses nerves in the autonomic nervous system. This leads to changes in respiration and circulation that may lead to death.

8. Sample answer: The pressure to compete for attention in a large family and constant fighting by the parents would produce stress in the child, causing him or her to become more distracted.

Chapter 15

Stop and Stretch

15.1. Nervous tissue supporting vision in a dark cave environment is not useful, so it is wasteful: The energy used to create and support unnecessary structures could be used to improve the supporting cells of another sensory system. Therefore, individuals with mutations that made them sightless may have more effective sensory receptors for sound or touch.

15.2. A hop in the air gives you a split second to recenter your feet under your body, preventing a fall.

15.3. Expose the animals to randomly occurring low-frequency vibrations and look for reactions that differ from their behavior when no vibrations are occurring.

15.4. No, there are other factors that McClintock did not control for, including shared visual cues that might trigger a change in cycle, or convergence in diet or alcohol consumption between roommates, among other shared experiences and environmental conditions.

15.5. Yes, it makes sound in the form of waves of compressed air. Without the presence of an ear and brain, however, that sound is not perceived.

15.6. The iris protects the lens from UV radiation. Sunglasses without UV protection allow the pupil to dilate (that is, the iris to pull back), permitting more UV light to strike the lens than if the individual did not wear sunglasses at all.

Learning the Basics

1. e **2.** b **3.** e **4.** a **5.** d **6.** c **7.** e **8.** c **9.** a
10. c **11.** b **12.** d **13.** d **14.** a **15.** c **16.** b, a, c, e, d **17.** c, a, d, b **18.** d, b, e, c, a

Analyzing and Applying the Basics

1. Sensory adaptation, in which receptors or central processing functions are "turned down" in response to constant stimulation.

2. They are likely to become adapted to the stressful movements she performs, allowing her to complete movements that might damage weaker muscles without causing reflexive muscle relaxation.

3. If any part of the outer surface of the body is damaged, infection can result. Thus any action that can potentially damage this outer surface should be avoided. Pain is a good motivator to prevent those potentially dangerous actions from occurring. Because acute damage to internal organs is much more rare, pain receptors on these organs are not as useful for survival.

4. Nontasters are likely to be missing a particular tastant receptor that responds to PTC. Supertasters may have an abundance of receptors, especially ones that respond to bitter chemicals.

5. The hair cells that respond to the highest-pitch noises must fail or be destroyed over time.

6. Exposure to loud noises permanently degrades hearing. Chronic exposure to loud noises while young may lead to serious hearing impairment in later adulthood.

7. Light sensitivity should improve with increased vitamin A consumption.

Chapter 16

Stop and Stretch

16.1. Genes that code for the synthesis of skeletal muscle proteins.

16.2. Affiliative behaviors may increase the likelihood of parents sticking around to raise the kids.

16.3. Brain is not responding normally to leptin. May be that receptors are absent or functioning at a suboptimal level.

16.4. In response to low T3 levels, the pituitary releases TSH to increase T3 and T4 synthesis, which causes the thyroid to grow by increasing cell division.

Learning the Basics

1. b **2.** c **3.** d **4.** d **5.** b **6.** c **7.** c **8.** b **9.** b, e, c, a, d **10.** c, a, d, e, b **11.** e **12.** true

Analyzing and Applying the Basics

1. Epinephrine and norepinephrine cause an increase in heart rate and blood flow to the muscles. These hormones would be likely to cause hyper-arousal of the nervous system and an elevated startle response. They could also be involved in nightmares and flashbacks, since they cause symptoms that are similar to the fear response. They could be involved in insomnia; during the stress response, the body would not easily relax into sleep.

2. Increased thyroid hormone levels could increase hyper-arousal of the nervous system and increase insomnia.

3. Chronic stress would release glucocorticoids, so the hippocampus could potentially shrink and cause memory loss as the individual ages.

4. *Repoxygen* induces cells to produce erythropoietin, a hormone that increases the concentration of red blood cells in the body. An increase in red blood cells would create a competitive advantage for athletes, as it would increase the supply of oxygen to working muscles.

5. Train at a high altitude. Lower oxygen levels would naturally trigger the release of the hormone erythropoietin, which would increase the concentration of red blood cells in the athlete's body.

6. The prostaglandin is likely produced in the nerve cells of the brain. Prostaglandins generally only have a localized effect, so this type would need to be produced in nerve cells near its receptors in the brain in order to induce fever.

7. Adjust watch to the new time zone. Keep the lights on, or try to remain exposed to sunlight, during the time when the destination is experiencing day. This would inhibit melatonin production and ward off sleepiness. Turn off the lights during the time when the destination is experiencing night. This would increase melatonin production and help the traveler's body to establish a new daily rhythm in sync with the destination.

8. May improve immune function by inhibiting the stress response. This could help in a woman's recovery from surgery. The treatment could also give the woman many techniques for dealing with stress, which would allow her to minimize the development of a chronic stress response that could also interfere with her recovery.

Chapter 17

Stop and Stretch

17.1. Because the skull surrounds the brain, when the tumor forms it will have to press other parts of the brain, potentially altering their function. It is also very difficult or impossible to surgically remove a brain tumor without removing brain tissue that may function in keeping a person alive, breathing, etc.

17.2. That unwinding the DNA and adding complementary bases would be a simple way for DNA to replicate itself.

17.3. Because cancer cells divide rapidly and require microtubules to move chromosomes around, they will be much more affected by this treatment than will normal cells.

17.4. Ovulation requires that a cell be released from the ovary. Releasing this cell damages ovarian tissue. Mitosis occurs to provide new, healthy cells to replace the damaged cells. This increased cell division means mutations can occur, thus increasing the likelihood of cancer.

17.5. Different people with cancer have different mutations. A treatment that affects a specific type of mutation would only help those with that mutation.

17.6. (1) Only if she acquired a mutation to a cell cycle control gene in her ovarian cells that are undergoing meiosis to produce gametes. (2) 50% (3) No, because it takes many mutations to get cancer.

Learning the Basics

1. d **2.** d **3.** c **4.** a **5.** a **6.** c **7.** d **8.** b **9.** b **10.** b **11.** b **12.** a **13.** e **14.** a **15.** b **16.** d **17.** a **18.** b **19.** a and b are diploid. c and d are haploid. **20.** false

Analyzing and Applying the Basics

1. Mutations can be inherited. Diet and lifestyle considerations become even more important if family history is prevalent.

2. Sample answers: Disrupt any part of cell division because cancer cells divide more rapidly than other cells. Prevent blood vessels from accessing tumors. Prevent detachment of cells required for metastasis.

3. No, because a mutation to skin cells is not going to impact the cells of his testes that undergo meiosis to produce gametes.

4. The cell is likely to die. If the cell is no longer malignant, controls in the cell cycle could detect abnormal gene expression in the cell and signal the cell to stop dividing.

5. In patients with active breast cancer, doctors would know to monitor for metastasis more often if the patient had infectious mononucleosis in the past, or had other exposure to the virus. In patients with a family history of breast cancer, doctors would know to test for evidence of cancer more often if the patient has been exposed to the Epstein-Barr virus.

6. Scientists could develop a vaccine, so that fewer people would carry the virus. They could also develop a way to block the virus's interaction with the cellular protein that prevents cancer cell metastasis.

7. There is no significant relationship between breast cancer and oral contraceptive use. There is a positive relationship between oral contraceptive use and cervical cancer, and using oral contraceptives for a longer time seems to increase the risk. There is a negative relationship between uterine or ovarian cancer and oral contraceptives, so oral contraceptives may offer some protection from these cancers.

8. Sample answer: Some genes may produce substances or activate pathways that could allow the tumor to be resistant to chemotherapy in some way. If the tumor is resistant, chemotherapy might weaken the patient without killing the cancer. This could lead to a lower survival rate.

9. Answers will vary based on age and sex.

10. Avastin can inhibit the formation and growth of blood vessels around a tumor, but cannot completely destroy the vascular network that surrounds and feeds the tumor. Other therapies are still needed to work on destroying cancer cells in the tumor.

Chapter 18

Stop and Stretch

18.1. Too warm in the abdomen for efficient sperm production.

18.2. The egg that produced you began developing in your mother's ovary while she was in her mother's (your grandmother's) uterus!

18.3. Sample answer: Track cycles before the women move in together and after many months of living together and see if the start dates cycle more closely over time.

18.4. These positive feelings help to ensure sexual activity is pleasant and people will want to have sex, thereby increasing the likelihood of fertilization and propagation of the species.

18.5. Opposition to emergency birth control is usually from those who believe a fertilized egg has the right to survive. This treatment may result in the inability of a fertilized egg to implant in the uterus.

Learning the Basics

1. c **2.** a **3.** c **4.** c **5.** d **6. a.** vas deferens **b.** urethra **c.** glans penis **d.** seminiferous tubules **e.** testis **f.** bulbourethral gland **g.** prostate gland **h.** seminal vesicle **7.** a **8.** b **9.** c **10. a.** ovary **b.** oviduct **c.** endometrium **d.** urethra **e.** clitoris **f.** labia minora **g.** labia majora **h.** vulva **i.** urinary opening **j.** vaginal opening **k.** vagina **l.** cervix **11.** d **12.** a **13.** c, e, d, a, b, j, i, f, h, g **14.** d, c, a, g, b, i, f, e, h **15.** b **16.** b **17.** c, b, a, e, d **18.** e **19.** false **20.** false

Analyzing and Applying the Basics

1. A male can impregnate a female during the plateau phase because seminal fluid (which does contain sperm) is secreted, and during orgasm when semen is ejaculated.

2. Misoprostol is a hormone that causes uterine contractions and expulsion of the embryo and uterine lining.

3. Fertility awareness

4. May be easier to contain one egg versus millions of sperm. Women are the ones who become pregnant so they may be more invested in using birth control. Shared responsibility for birth control could prevent many unintended pregnancies and prevent STDs.

5. BPA appears to mimic the actions of estrogen.

6. You could use glass or ceramic containers in the microwave, limit your intake of soda and canned foods, especially acidic ones, and use ceramic or glass dishes.

7. Bisphenol-A lowered daily sperm production at doses as low as 20 μg/kg body weight. This dose is lower than the dose the EPA suggests is unharmful for humans.

Chapter 19

Stop and Stretch

19.1. The effect of mutation in a noncoding portion of the DNA obviously depends on the function of the noncoding portion. If it is a "junk" region, the mutation should have no effect; if it is in a promoter region, it may have the effect of eliminating or enhancing the transcription of the associated gene; and if it is in a structural region, it could have devastating effects on chromosome structure and many associated genes.

19.2. With two pairs of chromosomes, 4 types of gametes are possible (e.g., *AB, Ab, aB,* and *ab*). With three pairs, 8 gametes are possible (*ABC, ABc, AbC, Abc, aBC, aBc, abC,* and *abc*). The relationship is that for every additional chromosome, the number of possible gametes goes up by the power of two (the number of gametes equals 2^n where n = the number of chromosomes).

19.3. There would be three phenotypes. Unaffected individuals with no copies of the defective allele, more mildly affected individuals with one copy of the defective allele, and more severely affected individuals with two copies of the defective allele.

19.4. Using animals allows the scientists to control the reproduction of individuals in ways impossible with humans (preventing some individuals from reproducing while encouraging others, and making matches between individuals). Many human traits of interest are not found in other animals, for instance a capacity for language or very flexible learning skills, so certain traits cannot be studied in animals.

Learning the Basics

1. b **2.** a **3.** b **4.** d **5.** a **6.** c **7.** All are genotype *Tt*, tall. **8.** 75% are tall, because 25% of the offspring have genotype *TT* and 50% of the offspring have genotype *Tt*. The remaining 25%, genotype *tt*, are short. **9.** *Aa* **10.** 50% **11.** 50% **12. a.** yellow **b.** *YY* × *yy* **13.** *Yy* × *yy* **14. a.** 50% **b.** 50% **c.** .5 × .5 = .25, or 25% **d.** Yes **15. a.** dominant **b.** Not all individuals carrying the allele get the disease. **16.** e **17.** d

Analyzing and Applying the Basics

1. *bb gg*

2. A: 50% *B g*, 50% *b g*
B: 50% *b G*, 50% *b g*

3. No. For instance, if body weight is extremely heritable, the weights of children can be predicted from the weights of their parents. But if food supply is much greater in the children's generation, they will be on average heavier than their parents.

4. Characteristics of communication and social skills are transmitted socially (that is, through the environment) as well as genetically. Thus a high correlation between parents and children or between twins on these traits may be partially due to environmental similarities. If the environment plays a role and similarly diagnosed individuals share an environment, then the estimate of heritability will be too high.

5. Multigene disorders are more difficult to treat because individual genes may have relatively small effects and it is the combination of these effects that cause the disorder.

6. No, we can't use heritability to explain differences among specific individuals—these boys could be identical twins, just exposed to different environments during early life.

7. The overall effect of environmental enrichment is to delay progress of the diseases and improve overall mental functioning. These effects are probably due to increased connections between nerve cells in the brain triggered by greater diversity of activity and more problem solving.

8. Here we have genetic traits that are clearly linked with a reduction or gradual decline in intelligence that can be reversed with environmental enrichment. Clearly, genes are not destiny even in these extreme cases.

Chapter 20

Stop and Stretch

20.1. The type O man cannot be the father. The type B man might be the father.

20.2. 1:1:1:1

20.3. 100% of males; 0% of females

20.4. First cousins have the same grandparents. If one of the grandparents carries a rare recessive allele it can be passed to each of the cousins. If these cousins mate, their offspring have a 1/4 chance of having this recessively inherited disorder.

20.5. Have separate labs analyze the DNA.

Learning the Basics

1. e **2.** e **3.** c **4.** d **5.** c **6.** b **7.** a **8.** d **9.** c
10. e **11.** e **12.** b **13.** e **14.** e **15.** e **16.** b
17. true **18.** false **19.** false **20.** true

Analyzing and Applying the Basics

1. 5.4×10^{-11} (or one in 540 billion—the product of all six frequencies)

2. Since there are only 6 billion people in the world, it is virtually impossible for two people to have the same DNA fingerprint.

3. DNA from the Y chromosome does not undergo recombination, so its specific fingerprint persists as it is passed from father to son to grandson and beyond. This makes it a useful tool for tracing paternity over generations.

4. There must have been a continuous line of sons linking the living male descendant to Eston Hemings. Otherwise, the Jefferson Y chromosome would have been lost.

5. No. Because all the living descendants trace back to Martha, none of them carries the Jefferson Y chromosome. These descendants would carry some alleles that had originated with Thomas Jefferson in their autosomal chromosomes, but these would be mixed with alleles from other families. Therefore, it would be more complicated to confirm a match between the DNA of autosomal chromosomes from direct descendants versus the DNA from Y chromosomes of descendants within an extended family.

6. Pleiotropy occurs when a single gene mutation causes multiple effects on the individual's phenotype. In the case of BFLS syndrome, a mutation of the *PHF6* gene causes multiple effects on the phenotype, including mental handicap, large earlobes, shortened toes, coarsened facial features, small genitalia, obesity, and breast enlargement.

7. The pedigree should show that this is an X-linked syndrome. The original mutation would have occurred in a female, who would have passed the mutant gene to her sons and daughters. Her daughters would have been carriers who passed the gene along to some of their daughters. A female carrier in the third generation would have passed the syndrome down to her son.

Chapter 21

Stop and Stretch

21.1. Removing a cell from a preembryo without damaging development is like monozygotic twinning because even if a preembryo splits in two, the two halves have the capacity to develop into complete human beings. This tells us that the cells of the preembryo are very similar to zygotes, in that they are totipotent.

21.2. Counting in weeks from LMP is typically easiest for most women, who are more aware of their menstrual cycle (and thus know the last time they menstruated) than they are of their ovarian cycle. Weeks since conception provides a straightforward way to talk about the process of development from the moment the zygote forms. Weeks since pregnancy can be detected provides information on the changes that are occurring in the mother.

21.3. If the *SRY* gene was deleted from a Y chromosome, a child could be chromosomally male but in every other way female. If an X chromosome picked up an *SRY*, the opposite could occur—a chromosomal female that is in every other way male.

21.4. In these cases, the heart pumps a mixture of oxygenated and deoxygenated blood into general circulation. This could result in inadequate oxygen delivery to the tissues and organs of the body.

21.5. Our understanding of development is incomplete, so scientists still do not know all of the triggers that are required to result in development of many different cells from a pluripotent cell. Because cord blood cells are multipotent (can develop only into blood and marrow cells), major steps in the cell's development have already taken place.

21.6. Children and teenagers have a great capacity to learn new skills but do not have all the brain architecture that adults do, and may have difficulty making adult decisions that require understanding other perspectives and long-term planning. Young adults have these structures and are more capable, but more years of life experience can still cause changes in the brain.

21.7. The advice of gerontologists seems to support the second hypothesis that aging is caused by accumulated damage to cells—damage that may be caused by lack of exercise, poor diet, and risky behaviors.

Learning the Basics

1. d **2.** c **3.** a, c, b **4.** b **5.** e **6.** d **7.** b, c, a
8. c **9.** a **10.** e **11.** a **12.** c **13.** d **14.** c **15.** c
16. b **17.** d **18.** a **19.** e **20.** e

Analyzing and Applying the Basics

1. This research lends support to the hypothesis that embryonic stem cells have the ability to cure spinal injuries that result in paralysis—conditions that are currently incurable.

2. Cord blood stem cells are multipotent, but they only have the ability to transform into blood or bone marrow cells. Bone marrow transplants can already treat the conditions treated by cord blood cells, so even though cord blood is an improvement, the situation for individuals suffering from these conditions is less dire.

3. Because the nervous system is developing during the entire pregnancy (in contrast to, say, the heart, which is fully formed very early in development), exposure to a substance that disrupts the nervous system can have effects at any time during pregnancy.

4. For the heart disease study, it is important to control for other factors, besides malnutrition, that might result in low birth weight (for instance, premature labor and birth unrelated to fetal health). For the type 2 diabetes study, the same effect seen in rats would have to be shown in humans. In addition, it must be clear that the precursor conditions seen in the rats deprived of protein during development do actually lead to higher rates of diabetes in these animals.

5. As she is fighting the cold, she is making antibodies to the virus that she is infected with. Because her child would be in close proximity to her, it is likely that the child will be exposed to the virus as well—her antibodies could help protect the child from severe infection by this virus.

6. The advantages are that the gut microflora can help make accessible nutrients that our own digestive processes cannot liberate from food. This in turn provides the developing individual with more energy and the building blocks needed to grow. A large community of microflora may, however, contain representatives that maximize fat accumulation, which in today's environment has negative consequences.

7. Improved nutrition: Today we have access to more consistent food and higher quality food than in many times in the past. Improved health care: We have better access to medicines, especially antibiotics and vaccinations, to prevent serious disease that can damage organs. Easier work lives: Fewer individuals are engaged in demanding physical occupations that could compromise future health. Greater financial security: Reduces stress that can compromise health. Healthier environment: Cleaner air and water and reduced smoking rates lessen damage to the body.

Chapter 22

Stop and Stretch

22.1. The key difference is that the foundations of faith are not supported by direct observation or experience. Statements of faith cannot be tested scientifically because there is no observation that can definitively refute these statements.

22.2. Several possible answers: Fossils of land-dwelling animals that show clear evidence of ancestry to whales and dolphins. The presence of vestigial structures in these animals that can only be explained as useful to ancestors on land. Homologies in DNA sequence that link these animals with modern land animals.

22.3. Natural selection cannot cause the appearance of new variants, such as individual polar bears that seek and can survive on a grizzly bear diet. Natural selection does not work on the characteristics of individuals—for example, individual polar bears cannot change their genes in order to conform to new environmental conditions.

22.4. Founder effect may explain the difference—that is, the Irish population may have descended from a small group of founders among whom the PKU allele was more common. The allele that causes PKU may be an adaptation in the environment of Ireland, but not in England or Europe—perhaps related to differences in typical diet or exposure to a particular pathogen.

Learning the Basics

1. b **2.** e **3.** b **4.** d **5.** c **6.** e **7.** c, b, a **8.** d **9.** c **10.** d **11.** b **12.** a **13.** d **14.** e **15.** e **16.** e **17.** a **18.** e **19.** c **20.** false

Analyzing and Applying the Basics

1. Here, each set of parentheses designates all of the offspring of a single common ancestor, that is, all species above a branch point. ((((pasture rose, multiflora rose) spring avens) live forever) spring vetch)

2. The specimen has traits that are clearly humanlike (bipedalism) and those that are clearly apelike (small brain, curved fingers, ape voice box)

3. Overall brain size, size and makeup of teeth, and shape of jaw and chin

4. Characteristics associated with bipedalism.

5. If individuals with the expanded mutation had greater fitness than the normal individuals, this mutation would become more common over time as the number of seeds produced by "expanded" individuals that carried the mutation formed a larger and larger percentage of the next generation.

6. The late mutation is unlikely to be selected for since flowering in these plants misses the availability of pollinators. These flowers are unlikely to produce many offspring.

7. The early mutation, because production of flowers in these plants does not overlap with the normal population. There is little gene flow between the two groups of plants.

8. Because they are not on the chromosomes, the genes on mtDNA are not scrambled during reproduction by crossing over, and any genes present on the mtDNA are passed to all offspring at 100% likelihood. The only differences that should occur between people in their mtDNA sequences are those mutations that appeared in their own genealogical line.

9. The Y chromosome is only inherited (by sons) from one parent and does not exchange genetic material very often with the X chromosome. It thus can be used in a similar way to trace evolutionary history through the male line.

Chapter 23

Stop and Stretch

23.1. Carrying capacity is approximately 100 flies. The carrying capacity is likely to go up in response to additions of food. However, the amount of space for flies in the test tube and the amount of waste they can tolerate will also affect the carrying capacity.

23.2. Approximately 75 species. About 25–30% of species would be expected to be lost as a result of the habitat modification described.

23.3. The rattlesnake and badger are likely to be competitors, as are the deer mouse and ground squirrel. This hypothesis could be tested by examining the diets of these animals over time to see if they overlap or by removing one of each pair from the environment to see if the purported competitor increases in number.

Learning the Basics

1. c **2.** b **3.** a **4.** d **5.** d **6.** d **7.** e **8.** c **9.** a **10.** a **11.** d **12.** b **13.** a **14.** b **15.** c **16.** a **17.** a **18.** d **19.** e

Analyzing and Applying the Basics

1. 2,500

2. The first population is likely to eventually be larger because such a large percentage of it is at reproductive age or younger.

3. Its current population size, its distribution (are there several populations or only one), its migration patterns and wintering spots, its habitat and how that habitat is changing to become more or less favorable to the birds, how it breeds (do males and females pair, or does one male defend a harem, which would reduce genetic diversity).

4. If climate changes disrupt plant growth—for instance, favoring certain plants over others—food supplies could go down. And if colder climates mean that more calories are needed to maintain the animal's body temperature, food could become very limited. When climate warms, food supplies may also suffer if drought limits plant growth.

5. The birds are flightless and their nest sites are on the ground. In addition, males apparently are not available to help the female defend the nest from predators.

6. The bird is unlikely to be able to thrive in the wild. Humans will need to prevent the introduction of predators onto their current islands, and it is possible that with a limited number of islands to support them, the population would be too small to prevent genetic drift without the supplemental feeding.

7. If bamboo were to go extinct, humans would likely switch to another plant for the resources it provided. In this case, the species that depend on this other plant would likely become threatened.

Chapter 24

Stop and Stretch

1. Answers to all of these questions depend on local conditions. Local nature centers can be a good place to start to get a lot of these answers.

Learning the Basics

1. e **2.** e **3.** d, a, c, e, b **4.** e **5.** a **6.** c **7.** d **8.** d
9. b **10.** c **11.** a

Analyzing and Applying the Basics

1. The biome here is likely to be tropical rain forest. It is fairly close to the equator and also on the windward side of a mountain range, meaning that it should be quite rainy. Since it is not at high altitude, temperatures should be warm this close to the equator.

2. If the Gulf Stream current stops, warm water from the tropics will not be carried north to Europe to add warmth to the air mass over the continent. It is likely to be much colder in Europe if this happens.

3. Bend and Yakima are on the leeward side of the Cascades, in its rain shadow, while Eugene, Tacoma, and Portland are on the windward side.

4. Brood XIX covers the largest range (although X is close). Brood XIX will emerge in 2011.

5. Brood X inhabits some of the most densely human populated areas of the country. Several of the urban areas in this region have expanded greatly since 1970. However, at the same time, forests in these areas have grown up on abandoned farmland as well. The balance between 1970 and 2001 is likely more loss of cicada habitat than gain, however, so the population was likely lower in 2001.

6. Rivers offered a convenient means for moving goods before railroads and highways became common. In addition, rivers provide a source of freshwater for drinking and irrigation of nearby agricultural lands, and can also be employed to generate hydropower to run milling operations and eventually, electrical plants.

7. 24 acres per person/5.4 Earth landmasses = 4.4 acres per person/single Earth landmass

8. Several are possible: walk or use the bus more; eat lower on the food chain; recycle or reuse materials; consume fewer disposable goods; lower your thermostat in winter, raise it in summer; cut down on air travel; etc.

GLOSSARY

abduction A type of joint-facilitated movement that involves drawing a limb away from the body. (Chapter 6)

ABO blood system A system for categorizing human blood based on the presence or absence of carbohydrates on the surface of red blood cells. (Chapter 8)

abortion Removal of an embryo or fetus from the uterus during pregnancy. (Chapter 18)

acetabulum Cup-shaped socket of hip bone into which the head of the femur fits. (Chapter 6)

acetylcholine A neurotransmitter with many functions, including facilitating muscle movements; thought to be involved in development of Alzheimer's disease. (Chapter 14)

acetylcholinesterase An enzyme in nerve cells that breaks down acetylcholine. (Chapter 14)

acid A substance that increases the concentration of hydrogen ions in a solution. (Chapter 2)

acidosis Low blood pH. (Chapter 11)

acquired immune deficiency syndrome (AIDS) Syndrome characterized by severely reduced immune system function and numerous opportunistic infections. Results from infection with HIV. (Chapter 13)

acrosome An organelle at the tip of the sperm cell containing enzymes that help the sperm penetrate the egg cell. (Chapter 18)

actin A protein found in muscle tissue that, together with myosin, facilitates contraction. (Chapters 5, 6)

action potential Wave of depolarization in a neuron propagated to the end of the axon—also called a nerve impulse. (Chapter 14)

activation energy The amount of energy that reactants in a chemical reaction must absorb before the reaction can start. (Chapter 3)

activator A protein that enhances the transcription of a gene. (Chapter 4)

active site Substrate-binding region of an enzyme. (Chapter 3)

active transport The ATP-requiring movement of substances across a membrane against their concentration gradient. (Chapter 3)

adaptation Trait that is favored by natural selection and increases an individual's fitness in a particular environment. (Chapter 22)

adduction A type of joint-facilitated movement that involves drawing a limb toward the body. (Chapter 6)

Addison's disease Disease caused by low levels of adrenal cortisol. (Chapter 16)

adenosine diphosphate (ADP) A nucleotide composed of adenine, the sugar ribose, and two phosphate groups. (Chapter 3)

adenosine triphosphate (ATP) A nucleotide composed of adenine, the sugar ribose, and three phosphate groups that can be hydrolyzed to release energy. Form of energy that cells can use. (Chapters 3, 4)

adhesion junction A type of adhering junction holding epithelial cells together. (Chapter 5)

adipocyte Fat-storing cell. (Chapter 5)

adipose tissue Fat-storing connective tissue. (Chapter 5)

adolescence The period during human development between puberty and adulthood, during which the brain continues to mature. (Chapter 21)

adrenal gland Either of two endocrine glands, one located atop each kidney, that secrete adrenaline in response to stress or excitement, help maintain water and salt balance, and secrete small amounts of sex hormones. (Chapter 16)

adrenocorticotropic hormone (ACTH) Protein hormone secreted by the anterior pituitary that stimulates the adrenal gland to secrete cortisol. (Chapter 16)

adulthood The period of development where maturation of all organs is complete. (Chapter 21)

aerobic An organism, environment, or cellular process that requires oxygen. (Chapter 3)

aerobic respiration Cellular respiration that uses oxygen as the final electron acceptor. (Chapter 3)

afferent arteriole The blood vessel that enters the glomerular capsule of the nephron. (Chapter 11)

agarose gel A jelly-like slab used to separate molecules on the basis of molecular weight. (Chapter 7)

agglutination Clumping of cells, in particular the reaction that occurs when blood cells react with an antibody to molecules on their plasma membranes. (Chapters 8, 20)

AIDS See acquired immune deficiency syndrome. (Chapter 13)

albumin An abundant protein in plasma with the primary function of keeping the blood isotonic to the body tissues. (Chapter 8)

aldosterone A steroid hormone produced by the adrenal cortex that helps maintain blood pressure by helping to conserve water in the kidneys. (Chapters 11, 16)

alimentary canal Part of the digestive system that forms a tube extending from the mouth to the anus. Also called the digestive tract. (Chapter 7)

allele Alternate versions of the same gene, produced by mutations. (Chapter 17)

allometric growth Variation in relative growth rates of different parts of the body during development. (Chapter 21)

alveoli Sacs inside lungs, making up the respiratory surface in land vertebrates and some fish. (Chapter 10)

Alzheimer's disease Progressive mental deterioration in which there is memory loss along with the loss of control of body functions, ultimately resulting in death. (Chapter 14)

amenorrhea Abnormal cessation of menstrual cycle. (Chapter 3)

amino acid Monomer subunit of a protein. Contains an amino group, a carboxyl group, and a unique side group. (Chapter 2)

amniocentesis A diagnostic test performed during pregnancy in which a small amount of amniotic fluid is removed in order to evaluate the chromosomal number and condition of the developing fetus. (Chapter 21)

amnion The fluid-filled sac in which a developing embryo is suspended. (Chapter 21)

amniotic fluid Fluid that accumulates within the amnion. (Chapter 21)

amygdala Almond-shaped structure in the brain's temporal lobe that helps regulate emotional behaviors. (Chapter 14)

anaerobic respiration A process of energy generation that uses molecules other than oxygen as an electron acceptor. (Chapter 3)

anaphase Stage of mitosis during which microtubules contract and separate sister chromatids. (Chapter 17)

anchorage dependence Phenomenon that holds normal cells in place. Cancer cells can lose anchorage dependence and migrate into other tissues or metastasize. (Chapter 17)

androgen Masculinizing hormone, such as testosterone, secreted by the testes and adrenal glands. (Chapter 18)

anecdotal evidence Information based on one person's personal experience. (Chapter 1)

anemia Illness in which oxygen delivery to cells is too low; may result from deficiencies of red blood cells, hemoglobin, or blood volume. (Chapter 8)

aneurysm A bulge in a blood vessel caused by a weakened vessel wall. (Chapter 9)

angina Chest pain, typically caused by coronary artery disease. (Chapter 9)

angiogenesis Formation of new blood vessels. (Chapter 17)

angioplasty A surgical procedure in which a small balloon is inflated inside a coronary artery in order to flatten atherosclerotic plaques that are restricting blood flow. (Chapter 9)

angiotensin II An enzyme in the blood which when activated causes a number of effects resulting in increased blood pressure. (Chapter 11)

anorexia Self-starvation. (Chapter 3)

antagonistic muscle pair A set of muscles whose actions oppose each other. (Chapter 6)

anterior pituitary gland The portion of the pituitary gland of the brain that secretes growth hormone, luteinizing hormone, follicle-stimulating hormone, and adrenocorticotropic hormone. (Chapter 16)

antibody Protein made by the immune system in response to the presence of foreign substances or antigens. Can serve as a receptor on a B cell or be secreted by plasma cells. (Chapters 8, 12)

anticodon Region of tRNA that binds to an mRNA codon. (Chapter 4)

antidiuretic hormone (ADH) A hormone released from the pituitary gland in response to low blood pressure; it causes increased water reabsorption in the kidneys. (Chapters 11, 16)

antigen Short for antibody-generating substances. A molecule that is foreign to the host and stimulates the immune system to react. (Chapters 8, 12)

antigen-presenting cell (APC) A cell that displays foreign antigens on its cell surface in order to increase an immune response. (Chapter 12)

antigen receptor Proteins in B- and T-cell membranes that bind to specific antigens. (Chapter 12)

antioxidant Certain vitamins and other substances that protect the body from the damaging effects of free radicals. (Chapter 2)

antiparallel Feature of DNA double helix in which nucleotides face "up" on one side of the helix and "down" on the other. (Chapter 2)

anus The end point and outlet of the digestive tract. (Chapter 7)

aorta The major blood vessel emerging from the left ventricle of the heart; delivers oxygenated blood to the systemic arteries. (Chapter 9)

apnea A transient involuntary stoppage of normal breathing. (Chapter 10)

apoptosis Programmed cell death. (Chapter 21)

appendicitis An inflammation of the appendix. (Chapter 7)

appendicular skeleton The part of the skeleton composed of the bones of the hip, shoulder, and limbs. (Chapter 6)

appendix A small pouchlike structure projecting from the large intestine. (Chapter 7)

aquatic Of, or relating to, water. (Chapter 24)

aqueous humor Fluid filling the eye chamber enclosed by the cornea and the lens. (Chapter 15)

arrhythmia Variation in the normal heart rate. (Chapter 9)

artificial heart A pump that replicates the blood-propelling functions of the heart. (Chapter 9)

arteriole Small blood vessel that carries blood away from the arteries toward capillary beds. (Chapter 9)

artery Blood vessels that carry blood from the heart to body tissues or the lungs. (Chapter 9)

artificial pacemaker Electronic device implanted in the chest that provides a regular series of electrical impulses, maintaining heart rhythm. (Chapter 9)

assortative mating Tendency for individuals to mate with someone who is like themselves. (Chapter 22)

asthma A respiratory disease characterized by spasmodic constriction of the air passages in the lungs and overproduction of mucus. Often triggered by air contaminants. (Chapter 10)

astigmatism A defect of vision related to irregularity in the shape of the cornea, preventing light rays from focusing properly on the retina. (Chapter 15)

astrocyte Star-shaped glial cells of the brain that help regulate the composition of fluids in the CNS. (Chapter 5)

asymptomatic Stage in an infection that is characterized by relatively unnoticeable, or absent, symptoms of illness. (Chapter 13)

atherosclerosis Accumulation of fatty deposits within blood vessels; hardening of the arteries. (Chapter 9)

atom The smallest unit of matter that retains the properties of an element. (Chapter 2)

atomic number The number of protons in the nucleus of an atom. Unique to each element, this number is designated by a subscript to the left of the symbol for the element. (Chapter 2)

ATP synthase Enzyme found in the mitochondria that helps synthesize ATP. (Chapter 3)

atrial natriuretic peptide (ANP) Hormone produced by cells in the heart in response to high blood pressure, and which triggers water excretion. (Chapter 11)

atrioventricular (AV) node Structure in the heart that transfers the electrical signal, after a slight delay, from the atria to the ventricles. (Chapter 9)

atrioventricular (AV) valve Heart valve between the atria and the ventricles. (Chapter 9)

atrium An upper chamber of the heart which receives blood from the body or lungs and pumps it to a ventricle. (Chapter 17)

atrophy A wasting or decrease in size of a tissue or organ. (Chapter 5)

auditory canal Passageway from the outer ear to the eardrum. (Chapter 15)

auditory nerve Nerve that transmits signals generated by the auditory receptors in the ear to the brain. (Chapter 15)

auditory receptor Any one of the hair cells in the cochlea of the inner ear that transmits a signal upon being mechanically disrupted by a passing sound wave. (Chapter 15)

auditory tube Channel that allows air or fluid to pass from inner ear into the nasal cavity. Also known as Eustachian tube. (Chapters 10, 15)

autoimmune disease Any disease resulting from an attack by the immune system on normal body cells. (Chapter 12)

autonomic nervous system The branch of the nervous system that regulates involuntary actions. Subdivided into the sympathetic and parasympathetic nervous systems. (Chapter 14)

autosomes Nonsex chromosomes, of which there are 22 pairs in humans. (Chapters 17, 20)

axial skeleton Part of the skeleton that supports the trunk of the body and consists largely of the bones making up the vertebral column or spine and much of the skull. (Chapter 6)

axon Long, wirelike portion of the neuron that ends in a terminal bouton. (Chapter 14)

B lymphocyte (B cell) The type of white blood cell responsible for antibody-mediated immunity. (Chapters 8, 12)

bacterium Single-celled prokaryotic organism. (Chapter 12)

ball-and-socket joints Joints in the hips and shoulders that enable arms and legs to move in three dimensions. (Chapter 6)

barrier method Any of the forms of birth control that prevent sperm and egg from interacting. (Chapter 18)

basal metabolic rate Resting energy use of an awake, alert person. (Chapter 3)

base A substance that reduces the concentration of hydrogen ions in a solution. (Chapter 2)

basement membrane The thin, noncellular layer of an epithelium that attaches to underlying tissues. (Chapter 5)

basilar membrane A structure in the inner ear that flexes in response to movement of sound-generated fluid waves and thus triggers action potentials in hair cells sitting on it. (Chapter 15)

basophil A leukocyte that is involved in the nonspecific immune response and releases histamine. (Chapters 8, 12)

benign Describes a tumor that stays in one place and does not affect surrounding tissues. (Chapter 17)

bias Influence of researchers' or subjects' opinions on experimental results. (Chapter 1)

bile Mixture of substances produced in the liver that aids in digestion by emulsifying fats. (Chapter 7)

bilirubin A pigment that results from the breakdown of hemoglobin, most of which is excreted in bile. (Chapter 8)

binary fission Asexual form of bacterial reproduction. (Chapter 12)

biodiversity Variety within and among living organisms. (Chapter 23)

biogeography The study of the geographic distribution of organisms. (Chapter 22)

biological evolution A change in the characteristics of a population of organisms over the course of generations, as the result of a change in the frequency of particular alleles in the population. See also theory of evolution. (Chapter 22)

biological species concept Definition of a species as a group of individuals that can interbreed and produce fertile offspring but typically cannot breed with members of another species. (Chapter 22)

biology The study of living organisms. (Chapter 1)

biomass The mass of all individuals of a species, or of all individuals on a level of a food web, within an ecosystem. (Chapter 23)

biome A broad ecological community defined by a particular vegetation type (for example, temperate forest, prairie), which is typically determined by climate factors. (Chapter 24)

biopsy Surgical removal of some cells, tissue, or fluid to assay for cancer. (Chapter 17)

bipolar cell A neuron in the retina that transmits signals from rod or cone cells to the optic nerve. (Chapter 15)

birth rate Number of births averaged over the population as a whole. (Chapter 23)

blastocyst An embryonic stage consisting of a hollow ball of cells. (Chapter 21)

blood The combination of cells and liquid that flow through blood vessels in the cardiovascular system; made up of red blood cells, plasma, white blood cells, and platelets. (Chapters 5, 8)

blood clot A mass of the protein fibrin and dead blood cells that forms in the region of blood vessel damage. (Chapter 8)

blood pressure The force of the blood as it travels through the arteries; partially determined by artery diameter and elasticity. (Chapter 9)

blood transfusion The transfer of blood or blood components from one person to the bloodstream of another. (Chapter 8)

blood type Refers to the presence or absence of particular antigens on the surface of red blood cells that differ among individuals. Typically associated with the ABO or Rh blood group system. (Chapter 8)

blood vessel Any one of a number of structures that carry blood in the vascular system throughout the body; arteries, capillaries, and veins. (Chapter 9)

body mass index (BMI) Calculation using height and weight to determine a number that estimates a person's amount of body fat and associated health risks. (Chapter 3)

bone A type of connective tissue consisting of living cells in a matrix rich in collagen and calcium. (Chapter 5)

boreal forest A biome type found in regions with long, cold winters and short, cool summers. Characterized by coniferous trees. (Chapter 24)

brain stem Region of the brain that lies below the thalamus and hypothalamus that governs reflexes and some involuntary functions such as breathing and swallowing. (Chapter 14)

breech Position of a fetus during labor in which part of the lower body (buttocks or feet) is against the cervix. (Chapter 21)

bronchioles The branching air passageways inside the lungs. (Chapter 10)

bronchitis Inflammation of the bronchi and bronchioles in the lungs. (Chapter 10)

bronchus The large air passageway from the trachea into a lung. (Chapter 10)

buffer A substance in a solution that lessens the change in pH. (Chapter 2)

bulbourethral gland Either of the two glands at the base of the penis that secrete acid-neutralizing fluids into semen. (Chapter 18)

bulimia Binge-eating followed by purging. (Chapter 3)

Caesarean section An operation that delivers an infant through an incision in the abdomen. (Chapter 21)

calcitonin A hormone produced by the thyroid that helps lower blood calcium levels. (Chapters 6, 16)

calendar method A method of birth control that relies on using a calendar to predict fertile and nonfertile days of the female menstrual cycle. (Chapter 18)

Calorie A kilocalorie or 1,000 calories. (Chapter 3)

calorie Amount of energy required to raise the temperature of 1 gram of water by 1°C. (Chapter 3)

cancer A disease that occurs when cell division escapes regulatory controls. (Chapters 5, 17)

canine tooth Any of the four pointed teeth located between the incisors and bicuspids. (Chapter 7)

capacitation A change to the chemical characteristics of a sperm cell that makes it able to complete fertilization. (Chapter 21)

capillary The smallest blood vessel of the cardiovascular system, connecting arteries to veins and allowing material exchange across their thin walls. (Chapter 9)

capillary bed A branching network of capillaries supplying a particular organ or region of the body. (Chapter 9)

capsid Protein coat that surrounds a virus. (Chapter 12)

capsule Gelatinous outer covering of bacterial cells that aids in attachment to host cells during an infection. (Chapter 12)

carbohydrate Energy-rich molecule that is the major source of energy for the cell. Consists of carbon, hydrogen, and oxygen in the ratio CH_2O. (Chapter 2)

carcinogen Substance that causes cancer or increases the likelihood of its development. (Chapter 17)

cardiac arrest The sudden cessation of normal heart rhythm. (Chapter 9)

cardiac cycle The cycle of contraction and relaxation that a normally functioning heart undergoes over the course of a single heartbeat. (Chapter 9)

cardiac muscle Muscle that forms the contractile wall of the heart. (Chapter 5)

cardiac veins Blood vessels that drain blood from the heart tissues directly into the right atrium. (Chapter 9)

cardiopulmonary resuscitation (CPR) A technique that involves chest compression and mouth-to-mouth ventilation as an emergency replacement for normal heart and lung function. (Chapter 9)

cardiovascular system The organ system made up of the heart, the blood, and the blood vessels, including arteries, capillaries, and veins. (Chapter 9)

carpal bone Any of the eight small bones of the wrist. (Chapter 6)

carrier Individual who is heterozygous for a recessive disease allele. (Chapter 20)

carrying capacity Maximum population that the environment can support. (Chapter 23)

cartilage Connective tissue found in the skeletal system that is rich in collagen fibers. (Chapter 5)

cataract Changes in the proteins of the lens that make this structure opaque, thus interfering with vision. (Chapter 15)

catalyze To speed up the rate of a chemical reaction. Enzymes are biological catalysts. (Chapter 3)

caudate nucleus Structure within each cerebral hemisphere that functions as part of the pathway that coordinates movement patterns, learning, and memory. (Chapter 14)

cecum The blind pouch forming the beginning of the large intestine. (Chapter 7)

cell body Portion of the neuron that houses the nucleus and organelles. (Chapter 14)

cell differentiation The process during development by which cells mature to become a particular specialized cell type. (Chapter 21)

cell division Process a cell undergoes when it makes copies of itself. Production of daughter cells from an original parent cell. (Chapter 17)

cell migration The orchestrated movement of cells during development. (Chapter 21)

cell wall Tough but elastic structure surrounding bacterial cell membranes. (Chapter 12)

cell-mediated immunity A type of specific immune response carried out by T cells. (Chapter 12)

cellular respiration Metabolic reactions occurring in cells that result in the oxidation of macromolecules to produce ATP. (Chapter 3)

central nervous system (CNS) Includes brain and spinal cord and is responsible for integrating, processing, and coordinating information taken in by the senses. It is the seat of functions such as intelligence, learning, memory, and emotion. (Chapter 14)

centriole A structure in animal cells that helps anchor microtubules during cell division. (Chapters 3, 17)

centromere Region of a chromosome where sister chromatids are attached and to which microtubules bind. (Chapter 17)

cerebellum Region of the brain that controls balance, muscle movement, and coordination. (Chapter 14)

cerebral cortex Deeply wrinkled outer surface of the cerebrum where conscious activity and higher thought originate. (Chapter 14)

cerebrospinal fluid Protective liquid bath that surrounds the brain within the skull. (Chapter 14)

cerebrum Portion of the brain in which language, memory, sensations, and decision making are controlled. The cerebrum has two hemispheres, each of which has four lobes. (Chapter 14)

cervical cap Latex or silicone female birth control device that covers the opening to the uterus. (Chapter 18)

cervix The lower narrow portion of the uterus at the top end of the vagina. (Chapter 18)

chaparral A biome characteristic of climates with hot, dry summers and mild, wet winters and a dominant vegetation of aromatic shrubs. (Chapter 24)

chemical reaction A process by which one or more chemical substances is transformed into one or more different chemical substances. (Chapter 2)

chemoreceptor A type of sensory receptor that responds to the binding of particular chemicals to its plasma membrane. (Chapter 15)

chemotherapy Using chemicals to try to kill rapidly dividing (cancerous) cells. (Chapter 17)

childhood The period of development from infancy to puberty, about ages 2–11. (Chapter 21)

chlamydia A sexually transmitted disease caused by infection by the bacterium *Chlamydia trachomatis*. Often without symptoms, but a primary cause of pelvic inflammatory disease. (Chapter 13)

chondroblast A cell that gives rise to a chondrocyte. (Chapter 6)

chondrocyte A type of cartilage cell that produces collagen and proteoglycans. (Chapter 5)

choroid The vascular region of the eye, found between the sclera and the retina. (Chapter 15)

chorionic villi Fingerlike projections that emerge from the outer sac surrounding a developing embryo, and which will eventually make up part of the placenta. (Chapter 21)

chromosome Subcellular structure composed of a long single molecule of DNA and associated proteins, housed inside the nucleus. (Chapters 4, 17)

chronic obstructive pulmonary disease (COPD) Lung disease, typically consisting of emphysema and chronic bronchitis, that results in labored breathing and poor gas exchange. (Chapter 10)

chronic pain Pain that persists for longer than three months, past the point where it is useful to signal injury or prevent further injury. (Chapter 15)

chyme Partially digested food and enzyme mixture, passed from the stomach to the intestine. (Chapter 7)

ciliary body The muscles and connective tissue that regulate the shape of the lens of the eye. (Chapter 15)

circulatory system The cardiovascular and lymphatic systems, which transport nutrients and waste around the body. (Chapters 8, 9)

circumduction The circular movement of a limb such that the end of the limb delineates an arc. (Chapter 6)

citric acid cycle The series of reactions catalyzed by enzymes located in the mitochondrial matrix that oxidize various substrates to produce ATP and reduced electron carriers. (Chapter 3)

classification system Method for organizing biological diversity. (Chapter 22)

clavicle Either of the two collarbones. Each bone articulates with the sternum and a scapula. (Chapter 6)

cleavage Rapid cell division that occurs early in animal development. (Chapter 21)

climate The average temperature and precipitation as well as seasonality. (Chapter 24)

clinical pregnancy Pregnancy that has progressed to implantation and can be diagnosed by typical clinical tests, such as the test for the presence of human chorionic gonadotropin. (Chapter 21)

clitoris Erectile tissue found in the external genitalia of females that functions in sexual arousal. (Chapter 18)

clonal population Population of identical cells copied from the immune cell that first encounters an antigen. The entire clonal population has the same DNA arrangement, and all cells in a clonal population carry the same receptor on their membrane. (Chapter 12)

cloning Making genetically identical copies of a gene or an organism. (Chapter 4)

CNS See central nervous system. (Chapter 14)

cochlea The spiral-shaped portion of the inner ear containing the auditory receptors. (Chapter 15)

cochlear duct The inner chamber of the cochlea, containing fluid and the structures associated with hearing. (Chapter 15)

cochlear implant An electronic device that replaces the function of the auditory hair cell receptors, sending a crude sound signal to the wearer. (Chapter 15)

codominance Refers to alleles that result in a new protein with a different, but not dominant, activity compared to the normal protein. (Chapter 20)

codon A triplet of mRNA nucleotides. Transfer RNA molecules bind to codons during protein synthesis. (Chapter 4)

cohesion The tendency for molecules of the same material to stick together. (Chapter 2)

collagen An extracellular protein that strengthens connective tissues including skin, bone, and nails. (Chapter 5)

collecting duct Structure in the kidney that receives filtrate from numerous nephrons and transmits filtrate to the renal pelvis. Hormones act on the collecting duct, changing its permeability to water and thus the amount of water retained in the kidney. (Chapter 11)

colon Most of the large intestine, from the cecum following the small intestine to the rectum. (Chapter 7)

columnar epithelium Tissue formed of epithelial cells that are taller than they are wide. (Chapter 5)

combination drug therapy Treatment with at least three different anti-HIV drugs, from two different classes of drugs. The therapy of choice for HIV patients. (Chapter 13)

combined hormone contraceptive Female birth control pill consisting of estrogen and progesterone. (Chapter 18)

community A group of interacting species in the same geographic area. (Chapter 23)

compact bone The hard, outer shell of bones. (Chapter 6)

competition Interaction that occurs when two species of organisms both require the same resources within a habitat; competition tends to limit the size of populations. (Chapter 23)

complement protein A type of protein in the blood with which an antibody-antigen complex can combine in order to kill bacterial cells. Enhances the immune response on many levels. (Chapter 12)

compound A substance consisting of two or more elements in a fixed ratio. (Chapter 2)

condom Latex sheath, in male and female versions, that prevents sperm and egg contact. (Chapter 18)

cones Sensory receptor cells in the eye that respond to specific wavelengths of light, contributing to color vision. (Chapter 15)

confidence interval In statistics, a range of values calculated to have a given probability (usually 95%) of containing the true population mean. (Chapter 1)

congenital syphilis Physiological and morphological changes that occur in children born to women with active syphilis infection. (Chapter 13)

conjoined twins Identical twins that form when the embryo does not separate completely, resulting in two individuals who share organs or tissues. (Chapter 21)

consumer Any organism that relies on the consumption of other organisms in order to survive. (Chapter 23)

contact inhibition Property of cells that prevents them from invading surrounding tissues. Cancer cells may lose this property. (Chapter 17)

contagious Capable of being transmitted by an infected individual to others. (Chapter 12)

continuous variation A range of slightly different values for a trait in a population. (Chapter 19)

contraception Any of the devices or pharmaceuticals used to prevent pregnancy. (Chapter 18)

contraceptive patch An adhesive patch impregnated with hormones that are secreted into the female wearer's body. (Chapter 18)

contraceptive sponge A type of birth control that is inserted into the vagina to block the cervix. (Chapter 18)

control Subject for an experiment who is similar to an experimental subject but is not exposed to the experimental treatment. Used to obtain baseline values to measure effect of an experimental treatment. (Chapter 1)

controlled experiment An experiment that is designed in a way to eliminate nearly all alternative hypotheses, primarily by using a control group. (Chapter 1)

coral reef Highly diverse biome found in warm, shallow saltwater, dominated by the limestone structures created by coral animals. (Chapter 24)

cornea The outer surface of the eye in front of the lens. (Chapter 15)

corpus callosum Bundle of nerve fibers at the base of the cerebral fissure that provides a communication link between the cerebral hemispheres. (Chapter 14)

coronary arteries Blood vessels that supply blood from the aorta to the heart muscle. (Chapter 9)

coronary bypass surgery Surgery that replaces the function of clogged coronary arteries by suturing healthy blood vessels to the artery, bypassing the narrowed region. (Chapter 9)

coronary heart disease (CHD) Disease that results from the dysfunction of coronary arteries, typically caused by vessel narrowing as a result of atherosclerosis. (Chapter 9)

corpus luteum Hormone-producing tissue (the ovarian follicle after ovulation) that makes progesterone and estrogen and degenerates about 12 days after ovulation if fertilization does not occur. (Chapter 18)

correlation Describes a mathematical relationship between two factors. (Chapter 1)

cortex Outer surface of certain organs, such as the adrenal cortex or cerebral cortex. (Chapters 14, 16)

corticotropin-releasing hormone Hypothalamic hormone that stimulates the anterior pituitary to secrete adrenocorticotropic hormone. (Chapter 16)

cortisol Steroid hormone produced by the adrenal cortex that helps regulate carbohydrate metabolism and blood pressure. (Chapter 16)

cortisone Steroid hormone produced by the adrenal glands in response to stress and converted into cortisol. (Chapter 16)

countercurrent exchange The process by which streams of fluid move in parallel but opposite directions, maximizing material, gas, or heat exchange between them. (Chapter 11)

covalent bond A type of strong chemical bond in which two atoms share electrons. (Chapter 2)

cranial nerve Any of the 12 pairs of nerves in humans that arise from the brain stem and connect to muscles and organs of the upper chest and head. (Chapter 14)

cranium The brain-enclosing skull. (Chapter 6)

cross In genetics, the mating of two organisms. (Chapter 19)

crossing over Exchange of some of their portions between members of a homologous pair of chromosomes. (Chapter 17)

crown The part of the tooth that is covered with enamel. (Chapter 7)

cuboidal epithelium Tissue formed of epithelial cells that are cube shaped. (Chapter 5)

cupula Structure at the base of the semicircular canals in the vestibular system where head position and movement is sensed. (Chapter 15)

Cushing's syndrome A disease caused by high levels of cortisol and characterized by central body obesity. (Chapter 16)

cutaneous membrane Epithelial membranes composed of hard, dry skin. (Chapter 5)

cytokinesis Part of the cell cycle during which two daughter cells are formed by the cytoplasm splitting. (Chapter 17)

cytoplasm The entire contents of the cell (except the nucleus) surrounded by the plasma membrane. (Chapter 3)

cytoskeleton A network of tubules and fibers that branch throughout the cytoplasm. (Chapter 3)

cytosol The semifluid portion of the cytoplasm. (Chapter 3)

cytotoxic T cell Immune system cell that attacks and kills virus-infected body cells before the virus has had time to replicate. Releases a chemical that causes the plasma membrane of the target cell to leak. (Chapter 12)

data Information collected by scientists during hypothesis testing. (Chapter 1)

decomposer Organism, typically a bacterium or fungus in the soil, whose action breaks down complex molecules into simpler ones. (Chapter 23)

deductive reasoning Making a prediction about the outcome of a test; "if . . . then" statements. (Chapter 1)

defibrillator A device that delivers a shock to the heart, placing it back into normal rhythm. (Chapter 9)

dehydration A decrease in an organism's required water level. (Chapters 2, 11)

dehydration synthesis Joining monomers to make polymers by removing a hydrogen from one monomer and a hydroxyl from another monomer, yielding water. (Chapter 2)

denatured (1) In proteins, the process where proteins unravel and change their native shape, thus losing their biological activity. (Chapter 2) (2) For DNA, the breaking of hydrogen bonds between the two strands of the double-stranded DNA helix, resulting in single-stranded DNA. (Chapter 20)

dendrite Short extensions of the neuron that receive signals from other cells. (Chapter 14)

density-dependent factor Factor related to a population's size that influences the current growth rate of a population—for example, communicable disease or starvation. (Chapter 23)

density-independent factor Factor unrelated to a population's size that influences the current growth rate of a population—for example, a natural disaster or poor weather conditions. (Chapter 23)

dentin The calcified tissue surrounding the pulp cavity of a tooth. (Chapter 7)

deoxyribonucleic acid (DNA) Molecule of heredity that stores the information required for making all of the proteins required by the cell. (Chapters 2, 17)

dependent variable The variable in a study that is expected to change in response to changes in the independent variable. (Chapter 1)

depolarization Reduction in the charge difference across the neuronal membrane. (Chapter 14)

Depo-Provera Brand name of progesterone shot given every 3 months for birth control. (Chapter 18)

depression Disease that involves feelings of helplessness and despair, and sometimes thoughts of suicide. (Chapter 14)

dermis The layer of skin beneath the epidermis. It contains connective tissue, blood vessels, sweat glands, hair follicles, and nerves. (Chapter 6)

desert The biome found in areas of minimal rainfall. Characterized by sparse vegetation. (Chapter 24)

development All of the progressive changes that produce an organism's body. (Chapter 21)

diabetes mellitus Disorder of carbohydrate metabolism characterized by impaired ability to produce or respond to the hormone insulin. (Chapter 7)

dialysis A process that replaces kidney function, allowing waste to diffuse from the bloodstream across a semipermeable membrane. (Chapter 11)

diaphragm (1) Dome-shaped muscle at the base of the chest cavity. Contraction of this muscle helps draw air into the lungs. (Chapters 5, 10) (2) A birth control device consisting of a flexible contraceptive disk that covers the cervix to prevent the entry of sperm. (Chapter 18)

diaphysis The shaft of a long bone. (Chapter 6)

diastole The stage of the cardiac cycle when the heart relaxes and fills with blood. (Chapter 9)

diastolic pressure The lowest blood pressure in the arteries, occurring during diastole of the cardiac cycle. (Chapter 9)

diffusion The spontaneous movement of substances from a region of their own high concentration to a region of their own low concentration. (Chapter 3)

digestion The breakdown of food in the alimentary canal into forms that can be absorbed into the bloodstream. (Chapter 7)

digestive tract See alimentary canal. (Chapter 7)

diploid Containing homologous pairs of chromosomes ($2n$). (Chapter 17)

distal tubule The last part of a nephron. Attaches to the collecting duct. (Chapter 11)

diuretic Any chemical that promotes water loss from the kidney. (Chapter 11)

diverge In evolution, divergence occurs when gene flow is eliminated between two populations. Over time, traits found in one population begin to differ from traits found in the other population. (Chapter 22)

dizygotic twins Fraternal (nonidentical) twins, which develop from separate zygotes. (Chapter 21)

DNA See deoxyribonucleic acid. (Chapters 2, 17)

DNA fingerprinting Powerful genetic identification technique that takes advantage of differences in DNA sequences between all people other than identical twins. (Chapter 20)

DNA polymerase Enzyme that catalyzes phosphodiester bond formation during DNA synthesis. (Chapter 17)

DNA replication The synthesis of two daughter DNA molecules from one original parent molecule. Takes place during the S phase of interphase. (Chapter 17)

dominant Applies to an allele with an effect that is visible in a heterozygote. (Chapter 19)

dopamine Neurotransmitter in pathways that control emotions and complex movements. (Chapter 14)

double-blind Experimental design protocol when both research subjects and scientists performing the measurements are unaware of either the experimental hypothesis or who is in the control or experimental group. (Chapter 1)

ductus arteriosus A structure present in a fetus that connects the pulmonary artery to the aortic arch. (Chapter 21)

ductus venosus In a fetus, connects the umbilical vein to the inferior vena cava, bypassing the liver. (Chapter 21)

Duffy antigens A protein on the surface of red blood cells in some individuals, and which is used by the malaria parasite to infect the cells. (Chapter 8)

duodenum The initial portion of the small intestine extending from the stomach to the jejunum. (Chapter 7)

ear A special sense organ that contains sensory receptors for hearing and body position. (Chapter 15)

ecological footprint A measure of the natural resources used by a human population or society. (Chapter 24)

ecology Field of biology that focuses on the interactions between organisms and their environment. (Chapter 23)

ecosystem All of the organisms and natural features in a given area. (Chapter 23)

ectoderm The outermost of the three germ layers that arise during animal development. (Chapter 21)

ectopic pregnancy Implantation of an embryo in the walls of an oviduct. Cannot develop successfully and must be surgically removed. (Chapter 21)

effector Muscle, gland, or organ stimulated by a nerve. (Chapter 14)

efferent arteriole The blood vessel leaving the glomerular apparatus and becoming a capillary bed surrounding the nephron. (Chapter 11)

elastin Fibrous protein found in elastic tissues. (Chapter 5)

electrocardiogram (ECG) A trace of the electrical activity of the heart, measured by changes in voltage of the skin. (Chapter 9)

electrolyte A chemical compound that ionizes in solution. (Chapter 2)

electron A negatively charged subatomic particle. (Chapter 2)

electronegative The tendency to attract electrons to form a chemical bond. (Chapter 2)

electron shell An energy level representing the distance of an electron from the nucleus of an atom. (Chapter 2)

electron transport chain A series of proteins in the mitochondrial membrane that move electrons during the redox reactions that release energy to produce ATP. (Chapter 3)

element A substance that cannot be broken down into any other substance. (Chapter 2)

elimination The removal of waste from the body. (Chapter 11)

embolism A blood vessel obstruction, typically a clot, that has traveled from elsewhere in the cardiovascular system. (Chapter 8)

embryo The developmental stage commencing after the first mitotic divisions of the zygote and ending when body structures begin to appear; from about the second week after fertilization to about the ninth week. (Chapters 18, 21)

embryonic disk The structure within a blastocyst that will become the embryo. (Chapter 21)

embryonic stem cell A cell from the embryonic disk that has the potential to become any specialized adult cell. (Chapter 21)

emphysema A lung disease caused by the breakdown of alveoli walls; characterized by shortness of breath and an expanded chest cavity. (Chapter 10)

enamel The hard calcified covering of the crown of a tooth. (Chapter 7)

Endangered Species Act (ESA) U.S. law intended to protect and encourage the population growth of threatened and endangered species. Enacted in 1973. (Chapter 23)

endocrine gland Any of the glands that secrete hormones into the blood. (Chapters 5, 16)

endocytosis The uptake of substances into a cell by a pinching inward of the plasma membrane. (Chapter 3)

endoderm The innermost of the three germ layers that arise during animal development. (Chapter 21)

endometriosis The abnormal occurrence of functional endometrial tissue outside the uterus, resulting in painful menstrual cycles. (Chapter 18)

endometrium Lining of the uterus, shed during menstruation. (Chapter 18)

endoplasmic reticulum (ER) A network of membranes in eukaryotic cells. Rough ER, studded with ribosomes, functions as a workbench for protein synthesis. Smooth ER, devoid of ribosomes, functions in phospholipid and steroid synthesis and in detoxification. (Chapter 3)

end-stage renal disease (ESRD) Failure of kidney function. (Chapter 11)

environmental tobacco smoke (ETS) The tobacco smoke in the air that results from smoldering tobacco on the lit ends of cigarettes and pipes as well as the smoke exhaled by active smokers. (Chapter 10)

enzyme Protein that catalyzes and regulates the rate of metabolic reactions. (Chapter 3)

eosinophil A type of white blood cell that helps the body respond to allergy and asthma. (Chapters 8, 12)

epidemic The rapid and extensive spread of a contagious disease among many individuals. (Chapter 12)

epidemiologist Scientist who attempts to determine who is prone to a particular disease, where risk of the disease is highest, and when the disease is most likely to occur. (Chapter 12)

epidermis The outer, nonvascular layer of skin that covers the dermis. (Chapter 6)

epididymis A coiled tube where sperm are stored, adjacent to the testes. (Chapter 18)

epiglottis Flap that blocks the windpipe so food goes down the pharynx, not into the lungs. (Chapters 7, 10)

epinephrine Adrenal hormone secreted in response to stress that causes increased heart rate and blood pressure. (Chapter 16)

epiphyseal plate In a long bone, cartilage layer that is the site of longitudinal growth. (Chapter 6)

epiphysis Region at the end of a long bone that is separated from the main part of the bone by a layer of cartilage until ossification is complete in early adulthood. (Chapter 6)

epithelial tissue Tightly packed sheets of cells that line organs and body cavities. (Chapter 5)

erythrocyte Red blood cell. (Chapter 8)

erythropoietin A hormone released by the kidney in conditions of low oxygen, and which promotes red blood cell production. (Chapter 8)

esophagus Tube that conducts food from the pharynx to the stomach. (Chapter 7)

essential amino acid One of the eight amino acids that humans cannot synthesize and must obtain from the diet. (Chapter 2)

essential fatty acid One of the fatty acids that animals cannot synthesize and must obtain from the diet. (Chapter 2)

estrogen Feminizing hormones secreted by the ovary in females and adrenal glands in both sexes. (Chapter 16)

estuary An aquatic biome that forms at the outlet of a river into a larger body of water such as a lake or ocean. (Chapter 24)

ethmoid bone The bone between the eye sockets of the cranium. (Chapter 6)

Eustachian tube See auditory tube. (Chapter 10)

evolution See biological evolution; theory of evolution. (Chapter 22)

excretion The filtering of waste from the blood. (Chapter 11)

exhalation The release of air from the lungs. Also called expiration. (Chapter 10)

exocrine gland Any gland that secretes substances through a duct. (Chapter 5)

exocytosis The secretion of molecules from a cell via fusion of membrane-bounded vesicles with the plasma membrane. (Chapter 3)

experiment Contrived situation designed to test specific hypotheses. (Chapter 1)

expiration See exhalation. (Chapter 10)

exponential growth Growth that occurs in proportion to the current total. (Chapter 23)

external respiration Exchange of gases between the body and the environment. Occurs in the lungs. (Chapter 10)

external urethral sphincter A ring of muscle that regulates the release of urine from the urethra. (Chapter 11)

externoreceptor Any sensory receptor that receives signals from the external environment. (Chapter 15)

extinction Complete loss of a species. (Chapter 23)

extraembryonic membrane Any of a number of membranes produced by a fertilized egg which do not become part of the embryo proper, but form protective layers. (Chapter 21)

eye A special sense organ containing photoreceptors and adapted to produce the sense of sight. (Chapter 15)

facilitated diffusion The spontaneous passage of molecules, through membrane proteins, down their concentration gradient. (Chapter 3)

falsifiable Applies to a statement that could potentially be proved false by observations of the measurable universe. (Chapter 1)

fascia Connective tissue that supports and binds together organs. (Chapter 6)

fascicle A bundle of nerve or muscle fibers. (Chapter 6)

fast pain The sharp, prickling pain that results from fast-acting sensory receptors. (Chapter 15)

fat Hydrophobic lipid molecule composed of a three-carbon glycerol skeleton bonded to three fatty acids. (Chapter 2)

fatty acid A lipid consisting of a long chain of hydrocarbons bonded to a carboxyl group. A major component of plant and animal fats. (Chapter 2)

fecal material Waste products of digestion, excreted from the rectum. (Chapter 7)

femur Bone extending from the pelvis to the knee, also called thighbone. (Chapter 6)

fermentation A process that makes a small amount of ATP from glucose without using an electron transport chain. Ethyl alcohol and lactic acid are produced by this process. (Chapter 3)

fertility awareness Using changes in the body, such as changes in cervical mucus, to predict the fertile time of a woman's menstrual cycle. (Chapter 18)

fertilization The fusion of haploid gametes (egg and sperm) to produce a diploid zygote. (Chapter 21)

fetal stem cells Stem cells produced during the fetal period of development. Can be harvested from amniotic fluid. (Chapter 21)

fetus The term used to describe a developing human from the ninth week of development until birth. (Chapter 21)

fever Abnormally high body temperature. (Chapter 12)

fiber The indigestible, structural parts of plants. (Chapter 2)

fibrillation Uncoordinated contractions in the heart. (Chapter 9)

fibrin A protein produced during the process of blood clotting that forms a net to trap and block blood flow from a damaged blood vessel. (Chapter 8)

fibrinogen A protein that circulates in the bloodstream and is converted to the clot-forming fibrin at the end of a clotting cascade. (Chapter 8)

fibroblast A protein-secreting cell found in loose connective tissue. (Chapter 5)

fibula The thinner, outer long bone of the leg, extending from knee to ankle. (Chapter 6)

fimbria Small, thin, fingerlike projections. (Chapter 18)

fissure A deep groove in the brain, such as the one that divides the cerebrum from front to back, into the right and left cerebral hemispheres. (Chapter 14)

fitness Relative survival and reproduction of one variant compared to others in the same population. (Chapter 22)

flagellum (*plural*: flagella) A long cellular projection that aids in motility. (Chapter 12)

flexion Bending a limb so that the angle between bones decreases. (Chapter 6)

follicle Structure in the ovary that contains the developing ovum and secretes estrogen. (Chapter 18)

follicle cell A flattened cell within the single layer that surrounds each primary oocyte. (Chapter 18)

follicle-stimulating hormone (FSH) Hormone secreted by the pituitary gland involved in sperm production, regulation of ovulation, and regulation of menstruation. (Chapters 16, 18)

food chain The linear relationship between trophic levels from producers, to primary consumers, to secondary consumers, and so on. (Chapter 23)

food web The feeding connections between and among organisms in an environment. (Chapter 23)

foramen ovale A hole between the right and left atria in the heart of a developing fetus that allows most blood to bypass the lungs. (Chapter 21)

forest Terrestrial community characterized by the presence of trees. (Chapter 24)

formed elements Cells or cell fragments in blood. (Chapter 8)

fossils Remains of plants or animals that once existed, left in soil or rock. (Chapter 22)

founder effect Type of sampling error that occurs when a small subset of individuals emigrates from the main population to found a new population. Results in differences in the gene pools of source population and the new population. (Chapter 22)

fovea The focal point of the eye. (Chapter 15)

frameshift mutation A mutation that occurs when the number of nucleotides inserted or deleted from a DNA sequence is not a multiple of three. (Chapter 4)

freshwater Water with less than 1,000 milligrams per liter of dissolved solids, primarily salt. (Chapter 24)

frontal bone Upper front portion of the cranium. The forehead. (Chapter 6)

frontal lobe The largest and most anterior portion of each cerebral hemisphere. (Chapter 14)

gallbladder Pear-shaped organ attached to the liver that stores bile and empties into small intestine. (Chapter 7)

gamete Specialized sex cells (sperm and egg in humans) that contain half as many chromosomes as other body cells. (Chapters 17, 18)

gametogenesis The production of gametes. (Chapter 18)

ganglia (*singular*: ganglion) Groups of nerve cell bodies located outside the CNS. (Chapter 14)

ganglion cell Cell in the eye that serves as an interneuron. Ganglion cell axons form the optic nerve. (Chapter 15)

gap junction A gap between adjacent cell membranes containing connections that allow substances to pass from cell to cell. (Chapter 5)

gas pressure The pressure exerted by a volume of gas. (Chapter 10)

gastrin A hormone, secreted by the stomach, that causes gastric juice secretion to increase. (Chapter 16)

gastrula The two-layered, cup-shaped stage of embryonic development. (Chapter 21)

gel electrophoresis The separation of biological molecules on the basis of their size and charge by measuring their rate of movement through an electric field. (Chapter 20)

gene Discrete unit of heritable information about genetic traits. Consists of a sequence of DNA that codes for a specific polypeptide. (Chapter 4)

general senses Senses including temperature, pain, touch, pressure, and body position, or proprioception. The general sensory receptors are scattered throughout the body. (Chapter 15)

genetic code Table showing which mRNA codons code for which amino acids. (Chapter 4)

genetic drift Change in allele frequency that occurs as a result of chance. (Chapter 22)

genetic variation All of the forms of genes, and the distribution of these forms, found within a species. (Chapter 19)

genetically modified organisms (GMOs) Organisms whose genome incorporates genes from another organism; also called transgenic or genetically engineered organisms. (Chapter 4)

genital tubercle Embryonic structure that has the capacity to become either male or female external genitalia. (Chapter 21)

genital warts A sexually transmitted disease caused by infection with one of several types of human papillomavirus. (Chapter 13)

genome Entire suite of genes present in an organism. (Chapters 4, 12)

genotype Genetic composition of an individual. (Chapter 19)

gerontology The branch of medicine that deals with aging and issues faced by the elderly. (Chapter 21)

gland A cell, group of cells, or organ that produces a secretion for use elsewhere in the body (Chapters 5, 16)

glans penis The head of the penis. (Chapter 18)

glaucoma Abnormally high fluid pressure in the eye caused by trapped aqueous humor. May result in blindness. (Chapter 15)

glenoid cavity The hollow head of the scapula into which the head of the humerus fits to form the shoulder joint. (Chapter 6)

global climate change Increases in average temperatures as a result of the release of increased amounts of carbon dioxide and other greenhouse gases into the atmosphere. (Chapter 24)

globulin Any one of a number of proteins in the blood, many of which function in immunity. (Chapter 8)

glomerular capsule The part of a nephron that surrounds the glomerulus and which receives filtrate from the blood. (Chapter 11)

glomerulus A compact cluster of capillaries that release filtrate into the nephron. (Chapter 11)

glottis The opening into the trachea at the upper part of the larynx. (Chapter 10)

glucagon Pancreatic hormone that works opposite insulin to regulate blood glucose levels. (Chapters 7, 16)

glucocorticoid Any of a number of steroid hormones secreted by the adrenal cortex and involved in regulating metabolism and reducing inflammation. The hormone cortisone, activated as cortisol, is a glucocorticoid. (Chapter 16)

glycolysis The splitting of glucose to produce pyruvate, ATP, and NADH. (Chapter 3)

Golgi apparatus An organelle in eukaryotic cells consisting of flattened membranous sacs that modify and sort proteins and other substances. (Chapter 3)

Golgi tendon organ A sensory neuron found in tendons that measures muscle tension and contributes to proprioception. (Chapter 15)

gonad The male and female sex organs; testicles in human males and ovaries in human females. (Chapters 16, 18)

gonadotropin-releasing hormone (GnRH) Hormone produced by the hypothalamus that stimulates the pituitary gland to release FSH and LH, which stimulate the activities of the gonads. (Chapters 16, 18)

gonorrhea A sexually transmitted disease caused by the bacterium *Neisseria gonorrhoeae*. (Chapter 13)

grassland Biome characterized by the dominance of grasses, usually found in regions of lower precipitation. (Chapter 24)

gray matter Unmyelinated axons, combined with dendrites and cell bodies of other neurons that appear gray in cross section. (Chapter 14)

greenhouse effect The retention of heat by carbon dioxide and other greenhouse gases. (Chapter 24)

ground substance An intercellular material in which the cells and fibers of connective tissue are embedded. (Chapter 5)

growth factor Protein that stimulates cell division. (Chapters 16, 17)

growth hormone (GH) Hormone produced by the anterior pituitary that stimulates growth. (Chapter 16)

growth plate The region of a long bone where growth occurs; also called the epiphyseal plate. Located between the epiphysis and diaphysis. (Chapter 6)

gustatory receptor Chemoreceptor in the tongue responsible for the sense of taste. (Chapter 15)

HAART Highly active antiretroviral therapy, a multidrug cocktail that suppresses HIV replication in the body and greatly extends the asymptomatic period of HIV infection. (Chapter 13)

habitat Place where an organism lives. (Chapter 23)

habitat destruction Modification and degradation of natural forests, grasslands, wetlands, and waterways by people; primary cause of species loss. (Chapter 23)

habitat fragmentation Threat to biodiversity caused by humans that occurs when large areas of intact natural habitat are subdivided by human activities. (Chapter 23)

hair cell Sensory receptor in the inner ear responsible for sensing sound waves. (Chapter 15)

haploid Describes cells containing only one member of each homologous pair of chromosomes (*n*); in humans, these cells are eggs and sperm. (Chapter 17)

hard palate The bony anterior roof of the mouth. (Chapter 7)

heart The muscular organ that pumps blood via the vascular system to the lungs and body. (Chapter 9)

heart attack An acute condition during which blood flow is blocked to a portion of the heart muscle, causing part of the muscle to be damaged or die. (Chapter 9)

heart failure A condition in which the heart fails to adequately pump blood to the tissues. (Chapter 9)

heart murmur An abnormal sound of the heart which typically indicates heart valve malfunction. (Chapter 9)

helper T cell Immune system cell that enhances cell-mediated immunity and humoral immunity by secreting a substance that increases the strength of the immune response. Also called T4 cell. (Chapters 12, 13)

hemoglobin An iron-containing protein that carries oxygen in red blood cells. (Chapter 8)

hemophilia Rare genetic disorder caused by a sex-linked recessive allele that prevents normal blood clotting. (Chapters 8, 20)

hemorrhage Blood loss from the vascular system. (Chapter 8)

hemostasis The stoppage of hemorrhage. (Chapter 8)

hepatic portal system The blood vessels that carry blood from capillaries surrounding the digestive organs through a vein and into a capillary bed in the liver. (Chapter 9)

hepatitis Inflammation of the liver, often caused by a viral infection. (Chapter 13)

hepatitis B virus (HBV) A sexually transmitted virus that causes hepatitis, but can be prevented via immunization. (Chapter 13)

heritability The amount of variation for a trait in a population that can be explained by differences in genes among individuals. (Chapter 19)

herpes simplex virus (HSV) Either of two viruses that infect the skin and nervous system, causing recurrent outbreaks of blisters. Often transmitted by intimate contact. (Chapter 13)

heterozygous Genotype consisting of two different alleles for a gene. (Chapter 19)

high-density lipoprotein (HDL) A cholesterol-carrying particle in the blood that is high in protein and low in cholesterol. (Chapter 2)

hinge joint A joint that allows back-and-forth movement. (Chapter 6)

hippocampus Part of the brain, in the temporal lobe, that consists mainly of gray matter and plays a role in memory. (Chapter 14)

histamine A chemical released from mast cells and basophils; acts in the inflammatory response and in allergic reactions. Causes dilation of blood vessels and lowered blood pressure. (Chapter 12)

HIV See human immunodeficiency virus. (Chapter 13)

HIV positive Characterized by the presence of HIV-antibodies in the blood. (Chapter 13)

homeostasis The steady-state condition an organism works to maintain. (Chapters 2, 5)

hominins Humans and earlier humanlike, bipedal species. (Chapter 22)

homologous pair Set of two chromosomes of the same size and shape with centromeres in the same position. Homologous pairs of chromosomes carry the same genes in the same locations but may carry different alleles. (Chapter 17)

homology Similarity in characteristics as a result of common ancestry. (Chapter 22)

homozygous Having two copies of the same allele of a gene. (Chapters 19, 20)

hormone A protein or steroid produced in one tissue that travels through the circulatory system to act on another tissue to produce some physiological effect. (Chapter 16)

human chorionic gonadotropin (HCG) A hormone produced by an embryo early in pregnancy that inhibits degradation of the corpus luteum, maintaining the endometrium. (Chapter 21)

human immunodeficiency virus (HIV) Agent identified as causing the transmission and symptoms of AIDS. (Chapter 13)

human papillomavirus (HPV) Any one of a large number of viruses that cause warts. Several are sexually transmitted, and some are known to cause cervical and anal cancer. (Chapter 13)

humerus The long bone of the arm, extending from the shoulder to the elbow. (Chapter 6)

humoral immunity B-cell-mediated immunity that occurs when a B-cell receptor binds to an antigen. The B cell divides to produce a clonal population of memory cells, and it produces plasma cells. (Chapter 12)

hydrogen atom One negatively charged electron and one positively charged proton. (Chapters 2, 3)

hydrogen bond A type of weak chemical bond in which a hydrogen atom of one molecule is attracted to an electronegative atom of another molecule. (Chapter 2)

hydrophilic Readily dissolving in water. (Chapter 2)

hydrophobic Not able to dissolve in water. (Chapter 2)

hyoid bone A U-shaped bone at the base of the tongue that helps support the tongue. (Chapter 6)

hyperopia Far-sightedness. Condition in which the eye is elongated so that light rays are focused by the lens in front of the retina, resulting in an indistinct image for close objects. (Chapter 15)

hyperpolarized An electrical state in which the inside of the cell is made more negative than the outside of the cell. (Chapter 21)

hypertension High blood pressure. (Chapter 9)

hyperthyroidism The condition of having an overactive thyroid gland. Some symptoms include nervousness, insomnia, and fatigue. (Chapter 16)

hypertonic Having more dissolved solute than the surrounding environment. (Chapter 3)

hypertrophy An atypical enlargement, for example, of a tissue. (Chapter 5)

hypodermis The loose connective tissue layer underlying the dermis. (Chapter 6)

hyponatremia A deficiency of sodium in the blood, leading to water retention in the tissues. (Chapter 11)

hypothalamus Gland that helps regulate body temperature; influences behaviors such as hunger, thirst, and reproduction; and secretes a hormone (GnRH) that stimulates the activities of the gonads. (Chapters 14, 16)

hypothalamus-pituitary-adrenal axis The interactions between the hypothalamus, pituitary, and adrenals that help maintain neuroendocrine homeostasis. (Chapter 16)

hypothesis Tentative explanation for an observation that requires testing to validate. (Chapter 1)

hypothyroidism Condition resulting from overactive thyroid. Symptoms include increased metabolism and weight loss. (Chapter 16)

hypotonic Having less dissolved solute than the surroundings. (Chapter 3)

ileum The end of the small intestine, from the jejunum to the cecum. (Chapter 7)

ilium The flared, upper part of the hip. (Chapter 6)

immortal Property of cancer cells that allows them to divide more times than normal cells. (Chapter 17)

immune response Ability of the body to respond to an infection resulting from increased production of B cells and T cells. (Chapter 12)

immunization Method to produce immunity; typically a vaccination. (Chapter 8)

implantation During development, the process by which a developing embryo becomes attached to the endometrium. (Chapter 21)

in vitro fertilization Fertilization that takes place when sperm and egg are combined in a glass dish or test tube. (Chapter 21)

incisor The four cutting and gnawing teeth at the front of each jaw. (Chapter 7)

incomplete dominance A type of inheritance where the heterozygote has a phenotype intermediate between both homozygotes. (Chapter 20)

independent assortment The separation of homologous pairs of chromosomes into gametes independently of one another during meiosis. (Chapter 19)

independent variable A factor whose value influences the value of the dependent variable, but is not influenced by it. In experiments, the variable that is manipulated. (Chapter 1)

indifferent gonads Embryonic structures with the capacity to become either ovaries or testes. (Chapter 21)

induced fit A change in shape of the active site of an enzyme so that it binds tightly to a substrate. (Chapter 3)

inductive reasoning A logical process that argues from specific instances to a general conclusion. (Chapter 1)

infancy The period of development from birth to about 15 months old. (Chapter 21)

infectious Applies to a pathogen that finds a tissue inside the body that will support its growth. (Chapter 12)

inferior vena cava Major vein returning blood from the lower extremities and abdominal cavity to the right atrium of the heart. (Chapter 9)

inflammatory response A line of defense triggered by a pathogen penetrating the skin or mucous membranes. (Chapter 12)

inhalation The act of drawing in air. Also called inspiration. (Chapter 10)

inner cell mass A cluster of cells in the blastocyst that eventually develops into the embryo. (Chapter 21)

insertion (1) For a muscle, the attachment to the bone moved by the muscle's contraction. (Chapter 6) (2) For a gene, a mutation resulting from addition of a nitrogenous base into the DNA sequence. (Chapter 4)

inspiration See inhalation. (Chapter 10)

insulin A hormone secreted by the pancreas that lowers blood glucose levels by promoting the uptake of glucose by cells and the storage of glucose as glycogen in the liver. (Chapters 7, 16)

insulin-dependent diabetes Type 1 diabetes mellitus, which results from inability to produce insulin. (Chapters 7, 12)

intercellular junction Protein complexes located between cells of an epithelium to hold them together and allow communication between cells. (Chapter 5)

integration Combining multiple neural signals into one response. (Chapter 14)

integumentary system External organ system consisting of skin, hair, nails, and sweat and sebaceous glands and their products. (Chapter 6)

intercostal muscles Muscles between the ribs which, when contracted, increase the size of the thoracic cavity. (Chapter 10)

interferon A chemical messenger produced by virus-infected cells that helps other cells resist infection. (Chapter 12)

internal respiration Gas exchange within the body tissues. (Chapter 10)

internal urethral sphincter A ring of muscle that, when contracted, prevents urine from leaving the bladder. (Chapter 11)

interneuron Neuron located between a sensory and a motor neuron that functions to integrate sensory input and motor output. (Chapter 14)

internoreceptor A sensory receptor that responds to stimuli within the body, such as blood acid levels. (Chapter 15)

interphase Part of the cell cycle when a cell is preparing for division and the DNA is duplicated. Consists of G_1, S, and G_2. (Chapter 17)

introduced species A nonnative species that was intentionally or unintentionally brought to a new environment by humans. (Chapter 23)

involuntary muscle Muscle tissue whose action requires no conscious thought. (Chapter 5)

ion Electrically charged atom. (Chapter 2)

ionic bond A chemical bond resulting from the attraction of oppositely charged ions. (Chapter 2)

iris The colored portion of the eye, consisting of a ring of muscle that contracts or relaxes to regulate the amount of light falling on the retina. (Chapter 15)

ischium The bone forming the lower and back part of the hip bone. (Chapter 6)

isotonic Having the same concentration of dissolved solute as the surroundings. (Chapter 3)

isotope Versions of a chemical element containing the same number of protons but different numbers of neutrons in the nucleus. (Chapter 2)

jejunum The middle portion of the small intestine, between the ileum and duodenum. (Chapter 7)

jaundice Yellow discoloration of the skin and eyes caused by a buildup of bilirubin in the blood. Often indicates liver dysfunction. (Chapter 8)

karyotype Picture of the chromosomes of a cell, with chromosomes arranged in homologous pairs and according to size. (Chapter 17)

keystone species A species that has an unusually strong effect on the structure of the community it inhabits. (Chapter 23)

kidney Major organ of the excretory system, responsible for filtering waste from the blood. (Chapter 11)

kidney stone Crystallized salt that forms a small, hard mass in the kidney. (Chapter 11)

labia majora Paired thick folds of skin that enclose and protect the labia minora of the vulva in females. (Chapter 18)

labia minora Paired thin folds of skin that enclose the urinary and vaginal openings and clitoris in females. (Chapter 18)

labioscrotal swellings Embryonic structures that can differentiate into either the labia majora or the scrotum. (Chapter 21)

labor Strong rhythmic contractions that force a baby from the uterus through the vagina during childbirth. (Chapter 21)

lactation Production of milk to nurse offspring. (Chapter 21)

lacunae Small cavities in bone. (Chapter 6)

lake An aquatic biome that is completely landlocked. (Chapter 24)

laparoscope A thin tubular instrument inserted through an abdominal incision and used to view organs in the pelvic cavity and abdomen. (Chapter 17)

large intestine Portion of the digestive system beginning after the small intestine with the cecum and continuing with the colon, rectum, and anus. It absorbs water and forms feces. (Chapter 7)

laryngitis Inflammation of the larynx, often resulting in changed characteristics of the voice. (Chapter 10)

larynx A portion of the upper respiratory tract made up primarily of stiff cartilage. Also known as the "voice box." (Chapter 10)

latent virus An inactive virus that is integrated into the host genome. (Chapter 12)

lateral geniculate nucleus Structure in the thalamus that is the primary processor of visual information from the optic nerves. (Chapter 15)

lens A transparent structure in the eye that focuses light on the retina. (Chapter 15)

leptin A hormone produced by fat cells that may be involved in the regulation of appetite. (Chapters 3, 16)

leukocyte Any one of a number of white blood cells, all of which function in the immune system. (Chapter 8)

Leydig cells Cells scattered between the seminiferous tubules of the testicles that produce testosterone and other androgens. (Chapter 18)

limbic system A ring of interconnected brain structures involved in emotion, motivation, and behavior. (Chapter 14)

linked genes Genes located on the same chromosome. (Chapter 17)

lipase Any of the enzymes that break down fats. (Chapter 7)

lipid Hydrophobic cellular constituents including fats, phospholipids, and steroids. (Chapter 2)

liver Organ with many functions including the production of bile to aid in the absorption of fats. (Chapter 7)

lobules Subdivisions of the lobes of the liver. (Chapter 7)

logistic growth Pattern of growth seen in populations that are limited by resources available in the environment. A graph of logistic growth over time typically takes the form of an S-shaped curve. (Chapter 23)

loose connective tissue Connective tissue that serves to bind epithelia to underlying tissues and to hold organs in place. (Chapter 5)

low-density lipoprotein (LDL) Cholesterol-carrying substance in the blood that is high in cholesterol and low in protein. (Chapter 2)

lower respiratory tract The larynx, trachea, bronchi, lungs, and alveoli. (Chapter 10)

lungs The primary organ of the respiratory system; the site where gas exchange occurs. (Chapter 10)

lupus An autoimmune disease affecting the connective tissue. Causes fever, joint pain, and affects various organs. (Chapter 12)

luteinizing hormone (LH) Hormone involved in sperm production, regulation of ovulation, and regulation of menstruation. (Chapters 16, 18)

lymph A clear fluid that contains white blood cells and drains from the body tissues into the lymph ducts. (Chapter 9)

lymph ducts Open-ended vessels that collect lymph from the body tissues and return it to the bloodstream. (Chapter 9)

lymph nodes Organs located along lymphatic vessels that filter lymph and help defend against bacteria and viruses. (Chapter 9)

lymphatic system A system of vessels and nodes that return fluid and protein to the blood. (Chapter 9)

lymphocyte White blood cells that make up part of the immune system. (Chapter 12)

lysosome A membrane-bounded sac of hydrolytic enzymes found in the cytoplasm of many cells. (Chapter 3)

macromolecule Large molecule, such as a polysaccharide, protein, or nucleic acid, composed of subunits joined by dehydration synthesis. (Chapter 2)

macrophage Phagocytic white blood cell that swells and releases toxins to kill bacteria. (Chapters 8, 12)

major histocompatibility complex (MHC) A group of genes encoding individual and tissue-specific cell surface markers. (Chapter 12)

malaria An infectious disease caused by a protozoan transmitted to humans by the bite of an infected female mosquito. (Chapter 8)

malignant Describes a tumor that is cancerous, whether it is invasive or metastatic. (Chapter 17)

mandible Bone of the lower jaw. (Chapter 6)

marine Of, or pertaining to, saltwater. (Chapter 24)

marrow Network of soft connective tissue that fills bones and is involved in the production of red blood cells. (Chapter 5)

mass extinction Loss of species that is rapid, global in scale, and affects a wide variety of organisms. (Chapter 23)

mass number The sum of the number of protons and neutrons in an atom's nucleus. (Chapter 2)

mast cell Large connective tissue cell involved in the inflammatory response. (Chapter 12)

matrix (1) In a mitochondrion, the semifluid substance inside the inner mitochondrial membrane, which houses the enzymes of the citric acid cycle. (Chapter 7). (2) In connective tissue, a nonliving substance between cells, ranging from fluid blood plasma to fibrous matrix in tendons to solid bone matrix. (Chapter 5)

maxillae The pair of bones that form the upper part of the jaw. (Chapter 6)

mechanoreceptor Sensory receptor that responds to changes in its shape or that of nearby cells. (Chapter 15)

medulla The center of an organ or gland, such as the kidney or adrenal gland. (Chapter 16)

medulla oblongata Region of the brain stem that is a continuation of the spinal cord and conveys information between the spinal cord and other parts of the brain. (Chapter 14)

medullary cavity The marrow cavity found in the shaft of a long bone. (Chapter 6)

megakaryocyte A large cell found in the bone marrow that produces platelets released into the bloodstream. (Chapter 8)

meiosis Process that sex cells undergo in order to produce gametes. (Chapter 17)

Meissner's corpuscle A sensory receptor that responds to pressure and is found in hairless skin. (Chapter 15)

melanocyte An epidermal cell that produces melanin. (Chapter 6)

melatonin A hormone whose secretion by the pineal gland is inhibited by sunlight. (Chapter 16)

memory Information stored in the brain for later retrieval. (Chapter 14)

memory cell An immune system cell that is part of a clonal population, which is programmed to respond to a specific antigen and to help the body respond quickly if the infectious agent is encountered again. (Chapter 12)

menarche The first occurrence of menstruation at puberty. (Chapter 21)

meninges Membranes that surround the brain and spinal cord. (Chapters 5, 14)

menopause Cessation of menstruation. (Chapter 18)

menstrual cycle Changes that occur in the uterus and depend on intricate interrelationships among the brain, ovaries, and lining of the uterus. (Chapter 18)

menstruation The shedding of the lining of the uterus during the menstrual cycle. (Chapter 18)

Merkel disk Sensory receptors for touch in the skin that respond to pressure. (Chapter 15)

mesoderm The middle of three germ layers that arise during animal development. (Chapter 21)

messenger RNA (mRNA) Complementary RNA copy of a DNA gene, produced during transcription. The mRNA undergoes translation during protein synthesis. (Chapter 4)

metabolic rate Measure of an individual's energy use. (Chapter 3)

metabolism All chemical reactions occurring in the body. (Chapter 3)

metacarpal Any bone of the hand between the wrist and fingers. (Chapter 6)

metaphase Stage of mitosis during which duplicated chromosomes align across the middle of the cell. (Chapter 17)

metastasis When cells from a tumor break away and start new cancers at distant locations. (Chapter 17)

metatarsal Any bone of the foot between the ankle and the toes. (Chapter 6)

microbe Microscopic organism, such as bacteria. (Chapter 12)

microglia Phagocytic cells of the central nervous system. (Chapter 5)

micronutrients Nutrients, such as vitamins and minerals, needed in small quantities. (Chapter 2)

microorganism See microbe. (Chapter 12)

microtubule Protein structure that moves chromosomes around during mitosis and meiosis. (Chapter 17)

microvillus Fine fingerlike projection composed of epithelial cells that function in absorption. (Chapter 7)

micturition reflex Release of urine from the bladder. Also known as urination. (Chapter 11)

midbrain Uppermost region of the brain stem, which adjusts the sensitivity of the eyes to light and the ears to sound. (Chapter 14)

mifepristone Drug that induces abortion of a pregnancy. (Chapter 18)

mineral Inorganic nutrient essential to many cell functions. (Chapter 2)

mineralocorticoid Hormone, such as aldosterone, synthesized by the adrenal gland and involved in water and electrolyte balance. (Chapter 16)

miscarriage The expulsion of a fetus before it is viable, typically in the third through seventh month. (Chapter 21)

missense mutation Point mutation that results in the incorporation of a different amino acid than originally coded for. (Chapter 4)

mitochondrion Organelle in which chemical energy is converted into ATP. (Chapter 3)

mitosis The division of the nucleus that helps produce daughter cells that are genetically identical to the parent cell. (Chapters 5, 17)

mixed nerve A nerve that is composed of sensory and motor fibers. (Chapter 14)

model organisms Nonhuman organisms used in biological studies, including the Human Genome Project, that are easy to manipulate and help scientists understand genes or traits the organism shares with humans. (Chapter 1)

molar Food-grinding tooth. (Chapter 7)

molecule Two or more atoms held together by covalent bonds. (Chapter 2)

monocyte A large circulating white blood cell. (Chapter 8)

monogamous The practice of having a single sexual partner. (Chapter 13)

monomer Individual subunit of a macromolecule. (Chapter 2)

monozygotic twins Identical twins that developed from one fertilized egg. (Chapter 21)

morphogenesis Change in form. During development, transformation from an undifferentiated mass of cells into an embryo with clearly defined parts. (Chapter 21)

motor neuron A neuron that carries information away from the brain or spinal cord to muscles or glands. (Chapters 6, 14)

motor unit A motor neuron and the muscle fibers innervated by it. (Chapter 6)

mRNA See messenger RNA. (Chapter 4)

mucous membrane Mucus-secreting membrane lining body cavities and passages that come in contact with air. (Chapter 5)

Mullerian duct Embryonic ductal structure that proliferates in female embryos to become the oviducts, uterus, cervix, and vagina. The Mullerian duct regresses in male embryos. (Chapter 21)

multiple allelism A gene for which there are more than two alleles in the population. (Chapter 20)

multiple sclerosis Chronic, degenerative nervous system disease caused by breakdown of myelin. (Chapter 12)

multipotent Describes a stem cell able to differentiate into several types of cells, although all are of the same general type. Compare with pluripotent and totipotent. (Chapter 21)

muscle fiber Single cell that aligns in parallel bundles to form muscles. (Chapter 6)

muscle spindle Muscle structure innervated by sensory and motor neurons. (Chapter 15)

muscle tissue Specialized contractile tissue that can conduct electrical impulses. (Chapter 5)

mutation A change to a DNA sequence that may result in the production of altered proteins. (Chapter 4)

mutualism Interaction between two species that provides benefits to both species. (Chapter 23)

myelin sheath Protective layer that coats many axons, formed by supporting cells such as Schwann cells. The myelin sheath increases the speed at which the electrochemical impulse travels down the axon. (Chapter 14)

myocardium The muscle of the heart. (Chapter 9)

myocardial infarction (MI) Damage to or death of heart tissue resulting from cutoff of blood flow to the heart. Also known as heart attack. (Chapter 9)

myofibril Fibril found in muscle cells and composed of thin filaments of actin and thick filaments of myosin. (Chapter 6)

myopia Nearsightedness. A condition in which the eye is foreshortened and light rays focus behind the retina, leading to indistinct images of distant objects. (Chapter 15)

myosin A type of protein that, along with actin, enables muscle cells to contract. (Chapters 5, 6)

nail The hard epidermal tissue on the surface of fingers and toes. (Chapter 6)

nasal bone One of two oblong bones that form the bridge of the nose. (Chapter 6)

nasal cavity Cavity behind the nose and extending from the floor of the braincase to the roof of the mouth. (Chapter 10)

natural experiments Situations with unique circumstances that allow a hypothesis test without prior intervention by researchers. (Chapter 19)

natural killer cell A cell that attacks virus-infected cells or tumor cells without being activated by an immune system cell or antibody. (Chapter 12)

natural selection See theory of natural selection. (Chapter 22)

negative feedback A mechanism of maintaining homeostasis in which the product of the process inhibits the process. (Chapter 5)

neonatal The period of development immediately after birth to about 1 month of age. (Chapter 21)

nephron The functional structure within kidneys, where waste filtration and some urine concentration occurs. (Chapter 11)

nephron loop A long extension found in some nephrons that carries the filtrate through the medulla of the kidney. Formerly known as the loop of Henle. (Chapter 11)

nerve Bundle of neurons; nerves branch out from the brain and spinal cord to eyes, ears, internal organs, skin, and bones. (Chapter 14)

nerve impulse Electrochemical signal that controls the activities of muscles, glands, organs, and organ systems. (Chapter 14)

nerve tract A bundle of myelinated nerve fibers. (Chapter 14)

nervous system Brain, spinal cord, sense organs, and nerves that connect organs and link this system with other organ systems. (Chapter 14)

nervous tissue Tissue composed of neurons and associated cells. (Chapter 5)

neural tube Structure that forms early in development from embryonic ectoderm, and that will become the brain and spinal cord. (Chapter 21)

neuron Specialized message-carrying cells of the nervous system. (Chapters 14, 15)

neuromuscular junction The junction between a nerve fiber and a muscle cell. (Chapter 6)

neurotransmitter One of many chemicals released by the presynaptic neuron into the synapse, which then diffuse across the synapse and bind to receptors on the membrane of the postsynaptic neuron. (Chapters 5, 14)

neutral mutation A genetic mutation that confers no selective advantage or disadvantage. (Chapter 4)

neutron An electrically neutral particle found in the nucleus of an atom. (Chapter 2)

neutrophil A phagocytic white blood cell. (Chapters 8, 12)

nicotinamide adenine dinucleotide (NAD) Intracellular electron carrier. Oxidized form is NAD+; reduced form is NADH. (Chapter 2)

nitrogenous base Nitrogen-containing bases found in DNA (A, C, G, T) and RNA (A, C, G, U). (Chapter 2)

nociceptor A sensory receptor that responds to pain, including chemicals produced by damaged cells or pressure caused by inflammation. Also called pain receptor. (Chapter 15)

node of Ranvier Small indentation separating segments of the myelin sheath. Nerve impulses "jump" successively from one node of Ranvier to the next. (Chapter 14)

nondisjunction The failure of members of a homologous pair of chromosomes to separate from each other during meiosis. (Chapter 17)

nonpolar Unable to dissolve in water. Hydrophobic. (Chapter 2)

nonsense mutation A genetic mutation that changes an amino acid–coding codon to a stop codon. Protein synthesis is prematurely terminated. (Chapter 4)

nonspecific defenses Defense systems against infection that do not distinguish one pathogen from another. Includes the skin, secretions, and mucous membranes. (Chapter 12)

norepinephrine A neurotransmitter and hormone secreted by certain neurons and the adrenal medulla. It activates the sympathetic nervous system. (Chapter 16)

nose Part of the face containing the nostrils and including the olfactory bulb. (Chapter 10)

notochord A long flexible rod that runs through the axis of a vertebrate embryo in the future position of the spinal cord. (Chapter 21)

nuclear envelope The double membrane enclosing the nucleus in eukaryotes. (Chapter 17)

nucleic acid Polymer of nucleotides. DNA and RNA are nucleic acids. (Chapter 2)

nucleoid region The region of a prokaryotic cell where the DNA is located. (Chapter 12)

nucleotide Building blocks of nucleic acids that include a sugar, a phosphate, and a nitrogenous base. (Chapter 2)

nucleus Cell structure that houses DNA; found in eukaryotes. (Chapters 2, 3)

nutrient cycling Process by which nutrients become available to plants. Nutrient cycling in a natural environment relies upon a healthy community of decomposers within the soil. (Chapter 23)

occipital bone Saucer-shaped bone forming the lower, back part of the skull. (Chapter 6)

occipital lobe The posterior lobe of each cerebral hemisphere, containing the visual center of the brain. (Chapter 14)

odorant A chemical that triggers a response from an olfactory receptor. (Chapter 15)

olfactory bulb Organ in the nose containing olfactory receptors. (Chapter 15)

olfactory receptor Sensory chemoreceptor responsible for the sense of smell. (Chapter 15)

oligodendrocyte Central nervous system cell that helps form myelin. (Chapter 5)

oogenesis Formation and development of female gametes, which occurs in the ovaries and results in the production of egg cells. (Chapter 18)

opportunistic infection Diseases that occur only when a weakened immune system allows access. (Chapter 13)

optic chiasm Site in the brain where some of the nerve fibers from the right optic nerve cross to the left side of the brain and vice versa. (Chapter 15)

optic nerve Bundle of nerve fibers exiting the eye and carrying information from the photoreceptors to the brain. (Chapter 15)

oral contraceptive An estrogen and progesterone pill taken to prevent ovulation and conception. (Chapter 18)

organ A specialized structure composed of several different types of tissues. (Chapter 5)

organ system Suite of organs working together to perform a function or functions. (Chapter 5)

organelle Subcellular structure found in the cytoplasm of eukaryotic cells that performs a specific job. (Chapter 3)

organogenesis The developmental process of organ formation from less differentiated cells and tissues. (Chapter 21)

orgasm Peak of sexual excitement. (Chapter 18)

osmosis The diffusion of water across a selectively permeable membrane. (Chapter 3)

ossicle Any one of three tiny bones within the middle ear. (Chapter 15)

osteoblast Bone-forming cell responsible for the deposition of collagen. (Chapter 6)

osteoclast Bone-reabsorbing cell that liberates calcium. (Chapter 6)

osteocyte Highly branched cell found in bone. (Chapters 5, 6)

osteoporosis A condition of weakened bones that elevates the risk of bone breakage. (Chapter 3)

otitis media Infection of the middle ear. (Chapter 10)

otolith organ Structure in the vestibular apparatus containing mechanoreceptors that respond to changes in head movement. (Chapter 15)

oval window A small membranous structure on the exterior surface of the cochlea where movements of the ossicles are transmitted to the fluid in the cochlea. (Chapter 15)

ovarian cycle Development and discharge of the ovarian follicle followed by development and regression of the corpus luteum. (Chapter 18)

ovary One of the paired abdominal structures that produce egg cells and secrete female hormones. (Chapter 18)

overexploitation Threat to biodiversity caused by humans that encompasses overhunting and overharvesting. Overexploitation occurs when the rate of human destruction or use of a species outpaces the ability of the species to reproduce. (Chapter 23)

oviduct Egg-carrying duct that brings egg cells from ovaries to uterus. (Chapter 18)

ovulation Release of an egg cell from the ovary. (Chapter 18)

ovum An egg; the female gamete. Cell produced during oogenesis that receives the majority of the cytoplasmic nutrients and organelles. (Chapter 21)

oxytocin Pituitary hormone that stimulates the contraction of smooth muscle of the uterus during labor and facilitates secretion of milk from the breast during nursing. (Chapter 16)

pain receptor See nociceptor. (Chapter 15)

Pacinian corpuscle A rapidly adapting sensory receptor for touch, found in the skin. (Chapter 15)

pancreas Gland that secretes digestive enzymes and insulin. (Chapter 7)

pancreatic amylase Sugar-digesting enzyme. (Chapter 7)

pancreatic islet Hormone-producing cells of the pancreas. (Chapter 16)

papillae On the tongue, small bumps that contain the taste buds. (Chapter 15)

parasite An organism that benefits from an association with another organism which is harmed by the association. (Chapters 8, 12)

parasympathetic division The division of the nervous system that stimulates digestive secretions, slows the heart, and constricts blood vessels. Its effects are opposite those of the sympathetic division of the nervous system. (Chapter 14)

parathyroid gland One of four endocrine glands located on the thyroid that secrete parathyroid hormone in order to regulate blood calcium levels. (Chapter 16)

parathyroid hormone Hormone secreted by the parathyroid glands that helps regulate blood calcium and phosphorus levels. (Chapters 6, 16)

parietal bone Either of two bones that form the sides and top of the skull. (Chapter 6)

parietal lobe Part of the brain that processes information about touch and is involved in self-awareness. (Chapter 14)

Parkinson's disease (PD) Disease that results in tremors, rigidity, and slowed movements. May be due to faulty dopamine production. (Chapters 14, 21)

partial pressure The pressure exerted by a single gas in a mixture of gases. A measure of the gas's concentration in the mixture. (Chapter 10)

particulates Tiny airborne particles found in smoke and other pollutants. (Chapter 10)

parturition Birth. (Chapter 21)

passive immunity Immunity acquired when antibodies are passed from one individual to another, as from mother to child during breast-feeding. (Chapter 12)

passive transport The diffusion of substances across a membrane with their concentration gradient and not requiring ATP. (Chapter 3)

patella The movable bone at the front of the knee (kneecap). (Chapter 6)

pathogen Any disease-causing organism. (Chapter 12)

pectoral girdle Bones that support the arms including the clavicles, or collarbones, and scapulas. (Chapter 6)

pedigree Family tree that follows the inheritance of a genetic trait for many generations. (Chapter 20)

peer review The process by which reports of scientific research are examined and critiqued by other researchers before they are published in scholarly journals. (Chapter 1)

pelvic girdle Bony arch formed by the ilium, ischium, and pubic bones. (Chapter 6)

pelvic inflammatory disease (PID) Infection of the internal reproductive organs of a woman as the result of an untreated sexually transmitted disease. (Chapter 13)

penis The copulatory structure in males. (Chapter 18)

pepsin Digestive enzyme of the stomach that hydrolyzes proteins. (Chapter 7)

peptide bond Covalent bond that joins the amino group and carboxyl group of adjacent amino acids. (Chapter 2)

perception The recognition and interpretation of sensory stimuli. (Chapter 15)

peripheral nervous system (PNS) Network of nerves outside the brain and spinal cord that links the CNS with sense organs. (Chapter 14)

peripheral vision Visible to the eye outside the central area of focus. (Chapter 15)

peritoneum Serous membrane that lines the abdominal cavity. (Chapter 5)

peristalsis Rhythmic muscle contractions that move food through the digestive system. (Chapter 7)

permafrost Permanently frozen soil. (Chapter 24)

pH scale A measure of the hydrogen ion concentration ranging from 0 to 14. Lower numbers equal higher hydrogen ion concentrations. (Chapter 2)

phagocyte Cell that engulfs food, cell debris, or pathogens. (Chapter 12)

phalange A bone of the finger or toe. (Chapter 6)

pharynx Tube and muscles connecting the mouth to the esophagus; the throat. (Chapters 7, 10)

phenotype Physical and physiological traits of an individual. (Chapter 19)

pheromone A chemical that is released by one animal and modifies the behavior or physiology of another animal. (Chapter 15)

phospholipid One of three types of lipids, phospholipids are components of cell membranes. (Chapter 2)

phospholipid bilayer The membrane that surrounds cells and organelles and is composed of phospholipids (along with proteins and sometimes cholesterol). (Chapter 3)

phosphorylation Addition of a phosphate group, which energizes the molecule it joins. (Chapter 3)

photochemical A chemical that changes conformation when irradiated with a particular wavelength of light. (Chapter 15)

photoreceptor A sensory receptor that responds to exposure to light. (Chapter 15)

pilus A hairlike structure on the surface of a microorganism. (Chapter 12)

pinna External structure of the ear. (Chapter 15)

pineal gland Endocrine gland in the brain that secretes the hormone melatonin. (Chapter 16)

pituitary gland Small gland attached by a stalk to the base of the brain that secretes growth hormone, reproductive hormones, and other hormones. (Chapter 16)

pivot joint A type of joint that allows rotation. (Chapter 6)

placebo Sham treatment, used for comparison with experimental treatment. (Chapter 1)

placenta Organ formed in pregnancy from maternal and embryonic tissues. It secretes hormones to support pregnancy and allows maternal-fetal blood exchange. (Chapter 21)

plasma The liquid portion of blood. (Chapter 8)

plasma cell Cell produced by a clonal population that secretes antibodies specific to an antigen. (Chapter 12)

plasma membrane Phospholipid bilayer that encloses a cell, defining the cell's outer boundary. (Chapter 3)

plasmid Circular piece of bacterial DNA that normally exists separate from the bacterial chromosome and can make copies of itself. (Chapters 4, 12)

platelet Type of blood cell that carries constituents required for the clotting response. (Chapter 8)

pleiotropy The ability of one gene to affect many different traits. (Chapter 20)

pleural membrane One of two membranes that surround the lungs. (Chapters 5, 10)

pluripotent Describes a cell able to differentiate into any cell in the human body. Compare with multipotent, totipotent. (Chapter 21)

pneumonia Inflammation of the lungs. (Chapter 10)

point mutation A change to one nucleotide of a DNA sequence. (Chapter 4)

polar Describes a molecule with regions having different charges; capable of ionizing. (Chapter 2)

polygenic A heritable trait influenced by many genes. (Chapters 19, 20)

polymer General term for a macromolecule composed of many chemically bonded monomers. (Chapter 2)

polymerase chain reaction (PCR) A laboratory technique that allows the production of many identical DNA molecules. (Chapter 20)

polysaccharide A carbohydrate composed of three or more monosaccharides. (Chapter 2)

pond An aquatic biome that is landlocked and is typically smaller than a lake. (Chapter 24)

pons A band of nerve fibers on the brain stem that links the medulla oblongata and the cerebellum with upper portions of the brain. (Chapter 14)

population All individuals of a species in a given area; a subgroup that is somewhat independent from other groups. (Chapters 22, 23)

population bottleneck Dramatic but short-lived reduction in population size followed by an increase in population. (Chapter 22)

population crash Steep decline in number that may occur when a population grows larger than the carrying capacity of its environment. (Chapter 23)

population pyramid A visual representation of the number of individuals in different age categories in a population. (Chapter 23)

positive feedback A relatively uncommon homeostatic mechanism in which the product of a process intensifies the process, thereby promoting change. (Chapter 5)

posterior pituitary gland Part of the pituitary gland that secretes oxytocin and antidiuretic hormone. (Chapter 16)

postsynaptic neuron The neuron that responds to neurotransmitter released from the presynaptic neuron. (Chapter 14)

prairie A grassland biome. (Chapter 24)

precapillary sphincters Bands of muscles surrounding capillaries at the arteriole end of a capillary bed and which, when contracted, cut blood flow to the bed. (Chapter 9)

precipitation Water that turns from vapor in the atmosphere to liquid or solid form and falls to Earth's surface. (Chapter 24)

predation Ecological interaction in which one organism consumes another. (Chapter 23)

prediction Result expected from a particular test of a hypothesis if the hypothesis were true. (Chapter 1)

preembryo The stage of development from fertilization through implantation in the uterus. (Chapter 21)

premolar Baby tooth that will be replaced with a molar. (Chapter 7)

presynaptic neuron The neuron that secretes neurotransmitter into a synapse, transmitting a signal. (Chapter 14)

primary struczture The linear sequence of amino acids of a protein. (Chapter 2)

prion Normally occurring protein produced by brain cells that, when misfolded, causes spongiform encephalopathy; prion is the shortened form of proteinaceous infectious particle. (Chapter 12)

probability Likelihood that an event will or did occur. (Chapter 1)

producer Organism that synthesizes carbohydrates from inorganic carbon; typically via photosynthesis. (Chapter 23)

product The modified chemical that results from a chemical or enzymatic reaction. (Chapter 2)

progesterone Ovarian hormone. High levels have a negative feedback effect on the hypothalamus, causing GnRH secretion to decrease. (Chapter 16)

progesterone-only pill Birth control drug that functions by suppressing ovulation and rendering the female reproductive tract unsuitable for fertilization. (Chapter 18)

prokaryote Type of cell that does not have a nucleus or membrane-bounded organelles. (Chapter 12)

prolactin A hormone produced by the pituitary gland that stimulates the development of mammary glands and lactation. (Chapter 16)

promoter Sequence of nucleotides in DNA to which the polymerase binds to start transcription. (Chapter 4)

prophase Stage of mitosis during which duplicated chromosomes condense. (Chapter 17)

prostate gland A gland in human males that secretes a thin, milky, acid-neutralizing fluid into semen. (Chapter 18)

protein Cellular constituent made of amino acids in a sequence coded for by genes. Proteins play many roles in cells including structural, transport, and enzymatic roles. (Chapter 2)

protein hormone A protein that is secreted in one region and has effects at a different location. (Chapter 16)

protein synthesis Joining amino acids together, in an order dictated by a gene, to produce a protein. (Chapter 4)

proton A positively charged subatomic particle. (Chapter 2)

proximal tubule Portion of a nephron closest to the glomerular apparatus. (Chapter 11)

pseudopodia Cellular extensions of an amoeba, used for feeding and in motility. (Chapter 12)

pubescence The point in human development when male or female hormones are triggered in the body. Males begin to produce sperm at this time, and females begin the menstrual cycle. (Chapter 21)

pubic lice Bloodsucking insects that infest pubic hair. Commonly transmitted sexually. (Chapter 13)

pubis Bones that form the front of the pelvis. (Chapter 6)

pulmonary artery Vessel that delivers deoxygenated blood from the right ventricle to the lungs. (Chapter 9)

pulmonary circuit The path of blood through vessels from the heart, through the lungs, and back to the heart; one of the two parts of the double circulation. (Chapter 9)

pulmonary veins Vessels that deliver oxygenated blood from the lungs to the left atrium of the heart. (Chapter 9)

pulp cavity The central cavity of the tooth including the pulp and the root canal. (Chapter 7)

pulse The volume of blood that passes into the arteries as a result of the heart's contraction. (Chapter 9)

Punnett square Table that lists the different kinds of sperm or eggs parents can produce relative to the gene or genes in question and predicts the possible outcomes of a cross between these parents. (Chapter 19)

pupil Opening in the iris that allows light to strike the retina. (Chapter 15)

pyruvic acid The three-carbon molecule produced by glycolysis. (Chapter 3)

quantitative traits Traits that have many possible values. (Chapter 19)

quaternary structure The level of protein structure characterized by two or more separate polypeptides binding to form a functional protein. (Chapter 2)

radiation therapy Treatment to kill the dividing cells of a tumor by focusing beams of reactive particles on it. (Chapter 17)

radiometric dating Technique that relies on radioactive decay to estimate a fossil's age. (Chapter 22)

radius The shorter and thicker of two forearm bones. (Chapter 6)

random alignment When members of a homologous pair line up randomly with respect to maternal or paternal origin. Occurs during metaphase I of meiosis and increases the genetic diversity of offspring. (Chapter 17)

random assignment Placing individuals into experimental and control groups randomly to eliminate systematic differences between the groups. (Chapter 1)

random fertilization The unpredictability of exactly which gametes will fuse during the process of sexual reproduction. (Chapter 19)

reactant Any starting material in a chemical reaction. (Chapter 2)

reading frame The grouping of mRNAs into three-base codons to specify an amino acid during protein synthesis. (Chapter 4)

receptive field Of a sensory neuron; region of space in which the presence of a stimulus will cause the firing of the neuron. (Chapter 15)

receptor (1) Protein on the surface of a cell that recognizes and binds to a specific chemical signal. (Chapter 16) (2) See sensory receptor. (Chapters 14, 15)

recessive Applies to an allele with an effect that is not visible in a heterozygote. (Chapter 19)

rectum The terminal section of the intestine, ending in the anus. (Chapter 7)

red bone marrow Tissue inside large bones containing blood stem cells. (Chapter 6)

referred pain Pain that is perceived as originating at a site distant from or adjacent to the actual site of injury. Typically indicates damage to an internal organ. (Chapter 15)

reflex Involuntary, automatic response to a stimulus. (Chapter 14)

reflex arc Nerve pathway in which sensory neurons signal motor neurons via the spinal cord, allowing a quick reflex response before the brain can respond. (Chapter 14)

refractory period The period of rest after a neuron or muscle cell fires, before it can fire again. (Chapter 14)

regenerative medicine The branch of medicine dealing with the body's own mechanisms to promote healing, including stem cells. (Chapter 21)

regulatory hormone A hormone that regulates the release of other hormones. (Chapter 16)

releasing hormone A hormone whose main role is to control the release of another hormone. (Chapter 16)

remission The period during which the symptoms of a disease subside. (Chapter 7)

renal cortex Outer region of the kidney, containing most of the nephrons. (Chapter 11)

renal medulla Inner, salty region of the kidney, into which nephron loops may extend. (Chapter 11)

renal pelvis Hollow area in the center of the kidney that collects urine before sending it to the bladder. (Chapter 11)

renin Enzyme released by the kidney in response to low blood pressure and which catalyzes the production of angiotensin. (Chapter 11)

repolarization The restoration of a charge difference across a membrane. (Chapter 14)

reproductive isolation Prevention of gene flow between different biological species due to failure to produce fertile offspring; can include pre- and postmating barriers. (Chapter 22)

residual volume The amount of air that remains in the lungs after maximal exhalation. (Chapter 10)

respiratory center Area of the brain stem that regulates breathing rate. (Chapter 10)

respiratory pump Process that draws blood in the veins toward the heart as a result of pressure differences in the thoracic cavity caused by respiration. (Chapter 9)

respiratory surface Surface across which the body exchanges gas with the air. In humans, the lining of the alveoli of the lungs. (Chapter 10)

respiratory system The organ system involved in gas exchange between an animal and its environment. In humans, the lungs and air passages. (Chapter 10)

resting potential The charge difference between the two sides of a membrane that occurs when a nerve cell is not conducting an impulse. (Chapter 14)

restoration ecology Branch of biology concerned with restoring back to their natural function those habitats and ecosystems damaged by human activity. (Chapter 23)

restriction enzyme A protein catalyst that cleaves DNA at specific nucleotide sequences. (Chapters 4, 20)

reticular formation Extensive network of neurons that runs through parts of the brain stem and regulates consciousness and wakefulness. (Chapter 14)

retina Structure in the eye that contains the photoreceptors and is responsible for our sense of sight. (Chapter 15)

retinal detachment Separation of the retina from the underlying tissues. (Chapter 15)

retrovirus A virus that contains RNA as its genetic material and that uses the enzyme reverse transcriptase to generate DNA upon infecting a cell. (Chapter 13)

reuptake In neurons, the process by which neurotransmitters are reabsorbed by the neuron that secreted them. (Chapter 14)

reverse transcription The production of DNA by transcription of viral RNA. (Chapter 13)

Rh factor Surface molecule found on some red blood cells. (Chapter 8)

rheumatoid arthritis Chronic autoimmune disease resulting in joint inflammation and deformation. (Chapter 12)

rhodopsin Molecule in photoreceptors that changes shape when irradiated by a particular wavelength of light, causing a nerve impulse to be generated. (Chapter 15)

rib cage The enclosure formed by the ribs and the bones to which the ribs attach. (Chapter 6)

ribonucleic acid (RNA) Single-stranded polymer of nucleotides. RNA nucleotides contain the sugar ribose and nitrogenous base uracil. (Chapter 2)

ribosomal RNA (rRNA) A class of RNA that makes up part of the structure of ribosomes. (Chapter 4)

ribosome Subcellular structure that helps translate genetic material into proteins by anchoring and exposing small sequences of mRNA. (Chapters 3, 4)

risk factor Any exposure or behavior that increases the likelihood of disease. (Chapter 17)

river Aquatic biome characterized by flowing water. (Chapter 24)

RNA See ribonucleic acid. (Chapter 2)

RNA polymerase Enzyme that synthesizes mRNA from a DNA template during transcription. (Chapter 4)

rod Photoreceptive cell in the eye that responds to low light levels of a variety of wavelengths. Responsible for detecting movement in low light and peripherally. (Chapter 15)

round window Flexible membrane in the cochlea that releases energy produced by a sound wave back into the middle ear. (Chapter 15)

Ruffini corpuscle A touch receptor in the skin. (Chapter 15)

saccule One chamber of the vestibular apparatus in the inner ear. (Chapter 15)

salts Charged substances that ionize in solution. (Chapter 2)

saltwater Water with a solute concentration of greater than 1,000 milligrams per liter. (Chapter 24)

sample Small subgroup of a population used in an experimental test. (Chapter 1)

sample size Number of individuals in both the experimental and control groups. (Chapter 1)

sampling error Effect of chance on experimental results. (Chapter 1)

sarcolemma A thin membrane surrounding a muscle fiber. (Chapter 6)

sarcomere Any of the repeating units in a myofibril of striated muscle, bounded by Z discs. (Chapter 6)

sarcoplasm The cytoplasm of a striated muscle fiber. (Chapter 6)

sarcoplasmic reticulum Calcium-regulating endoplasmic reticulum found in striated muscle fibers. (Chapter 6)

saturated fat Found in butter and other fats that are solids at room temperature. This type of fat tends to increase blood cholesterol levels. (Chapter 2)

savanna Grassland biome containing scattered trees. (Chapter 24)

scapula Either of the two flat bones forming the shoulder blades. (Chapter 6)

Schwann cells Glial cells that form the myelin sheath along the axons of nerve cells in the peripheral nervous system. (Chapters 5, 14)

scientific method A systematic method of research consisting of putting a hypothesis to a test designed to disprove it, if it is false. (Chapter 1)

scientific theory Body of scientifically accepted general principles that explain natural phenomena. (Chapter 1)

sclera The connective tissue that forms the outer surface of the eye. (Chapter 15)

scrotum The pouch of skin that houses the testes. (Chapter 18)

sebaceous gland A gland in the dermis of the skin that houses the hair follicle and secretes sebum. (Chapter 6)

secondary messenger system A form of cell signaling in which the signaling molecule does not enter the cell. Instead it signals a cascade of events that result in some intracellular change. (Chapter 16)

secondary sex characteristics External physical traits that are often specific to one sex and develop during puberty; for example, breast growth in women and beard growth in men. (Chapter 21)

secondary structure The level of protein structure that consists of alpha helices and beta-pleated sheets maintained by hydrogen bonding. (Chapter 2)

segregation Separation of pairs of alleles during the production of gametes. Results in a 50% probability that a given gamete contains one allele rather than the other. (Chapter 19)

selectively permeable A characteristic of cell membranes in which some substances can cross the membrane and others can't cross unaided. (Chapter 3)

semen Sperm and energy-rich associated fluids. (Chapter 18)

semicircular canals Fluid-containing structures in the inner ear that are responsible for our perception of head movement. (Chapter 15)

semilunar valves Heart valves controlling blood flow from the ventricles into blood vessels leading away from the heart. (Chapter 9)

seminal vesicle Either of two pouchlike glands located on both sides of the bladder that add a fructose-rich fluid to semen prior to ejaculation. (Chapter 18)

seminiferous tubules Highly coiled tubes in the testicles where sperm are formed. (Chapter 18)

senescence The stage of development characterized by decline in function and degradation of structure. (Chapter 21)

sensation The awareness of a sensory stimulus. (Chapter 15)

sense organ A specialized body structure containing sensory receptors. (Chapter 15)

senses Any of the faculties, such as sight, hearing, or touch, by which environmental stimuli are perceived. (Chapter 15)

sensory adaptation The processes that turn down our perception of a sensory stimulus. (Chapter 15)

sensory neuron A neuron that conducts impulses from a sense organ to the central nervous system. (Chapter 14)

sensory receptors Cellular systems that collect information about the environment inside or outside the body and transmit that information to the brain. (Chapters 14, 15)

serous membrane Any of the thin membranes that line body cavities and secrete a watery fluid. (Chapter 5)

septum A wall that divides a chamber. In the nose, the membrane between two halves of the nasal cavity. (Chapter 10)

Sertoli cell A cell, found in the testis, that secretes substances that aid in the development of mature sperm. (Chapter 18)

serum Blood plasma minus clotting factors. (Chapter 8)

sex chromosome Either of the sex-determining chromosomes (X and Y). (Chapters 17, 20)

sex determination Determining the biological sex of an offspring. Humans have a chromosomal mechanism of sex determination in which two X chromosomes produce a female, and an X and a Y chromosome produce a male. (Chapter 20)

sex hormone Any of the steroid hormones that affect development and functions of reproductive structures and secondary sex characteristics. (Chapter 16)

sex-linked gene Any of the genes found on the X or Y sex chromosomes. (Chapter 20)

sexual reproduction Reproduction involving two parents. The result is offspring that have unique combinations of genes. (Chapter 18)

sexual selection Form of natural selection that occurs when a trait influences the likelihood of mating. (Chapter 22)

sexually transmitted disease (STD) Any of a number of diseases caused by a sexually transmitted infection. (Chapter 13)

sexually transmitted infection (STI) Infection by a bacteria, virus, or other organism that is transmitted by sexual activity. (Chapter 13)

shock Collapse of circulatory function, often caused by a dramatic decline in blood pressure. (Chapter 8)

sickle-cell anemia A disease caused by loss of red blood cells under conditions of stress or physical activity as a result of deformation of mutant hemoglobin molecules. (Chapter 8)

signal transduction When a change in a cell or its environment is relayed through various molecules and results in a cellular response. (Chapter 16)

sinoatrial (SA) node Region of the heart muscle that generates an electrical signal which controls heart rate. (Chapter 9)

sinus A cavity or recess in a bone; often refers to the air chambers in the skull. (Chapters 6, 10)

sinusitis Inflammation of the mucous membranes lining the nasal sinuses. (Chapter 10)

sister chromatid Either of the two duplicated, identical copies of a chromosome formed after DNA synthesis. (Chapter 17)

skeletal muscle Striated muscle involved with voluntary movements. (Chapter 5)

skull The bony structure of the head. (Chapter 6)

sliding-filament model The theory that muscles contract when actin filaments slide across myosin filaments, shortening the sarcomere. (Chapter 6)

slow pain Dull ache that is perceived when slow-adapting pain receptors are signaled, usually by chemicals released from damaged tissue. (Chapter 15)

small intestine The narrow, twisting, upper part of the intestine where nutrients are absorbed into the blood. (Chapter 7)

smog Products of fossil fuel combustion in combination with sunlight, producing a brownish haze in still air. (Chapter 24)

smooth muscle Nonstriated, spindle-shaped muscle cells that line organs and blood vessels. (Chapter 5)

sodium bicarbonate ($NaHCO_3$) Soluble compound that functions as antacid. (Chapter 7)

sodium-potassium pump This protein pump in a cell membrane moves sodium out of the cell and potassium into the cell, both against their concentration gradients. (Chapter 14)

soft palate The soft tissue that forms the back of the roof of the mouth. (Chapter 7)

solid waste Garbage. (Chapter 24)

solute The substance that is dissolved in a solution. (Chapter 2)

solution A mixture of two or more substances. (Chapter 2)

solvent A substance, such as water, that a solute is dissolved in to make a solution. (Chapter 2)

somatic cell The body cells in an organism. Any cell that is not a gamete. (Chapter 17)

somatic nervous system The part of the peripheral nervous system associated with voluntary movements. (Chapter 14)

special senses Smell, taste, equilibrium, hearing, and vision. The sensory receptors for these five special senses are found in complex sense organs. (Chapter 15)

species A group of individuals that can breed together to produce fertile offspring and are generally distinct from other species in appearance or behavior. In Linnaeus's classification system, a group in which members have the greatest resemblance. (Chapter 22)

species-area curve Graph describing the relationship between the size of a natural landscape and the relative number of species it contains. (Chapter 23)

specific defense Defense against pathogens that utilizes white blood cells of the immune system. (Chapter 12)

specificity Phenomenon of enzyme shape determining the reaction the enzyme catalyzes. (Chapter 3)

sperm The male gamete. (Chapter 18)

spermatogenesis The production of sperm. (Chapter 18)

spermicide Chemical that kills sperm, often used in concert with other forms of birth control. (Chapter 18)

sphenoid A bone at the base of the anterior cranium. (Chapter 6)

sphincter A ringlike muscle that functions to maintain a constriction. (Chapter 7)

sphygmomanometer Device that measures blood pressure, also known as a blood pressure cuff. (Chapter 9)

spina bifida Birth defect caused by improper closure of the neural tube, resulting in part of the spinal cord appearing outside the vertebral column. (Chapter 21)

spinal cord Thick cord of nervous tissue that extends from the base of the brain through the spinal column. (Chapter 14)

spinal nerves Nerves that branch off from the spinal cord to the rest of the body. (Chapter 14)

spongy bone The porous, honeycomb-like material of the inner bone. (Chapter 6)

squamous epithelium Tissue that lines surfaces and is composed of flattened cells. (Chapter 5)

***SRY* gene** Sex-determining region of Y chromosome. Encodes a protein that stimulates development of the testes. (Chapter 21)

standard error A measure of the variability in a statistical sample; essentially the average distance any single data point is from the mean value. (Chapter 1)

statistical significance Low probability that experimental groups differ simply by chance. (Chapter 1)

statistics Specialized branch of mathematics used in the evaluation of experimental data. (Chapter 1)

stem cell Cell that can divide indefinitely and can differentiate into other cell types. (Chapters 8, 18, 21)

steppe Biome characterized by short grasses, found in regions with relatively little annual precipitation. (Chapter 24)

sternum The breastbone. (Chapter 6)

steroid Any of the fat-soluble hormones including cholesterol, estrogen, and testosterone. (Chapter 16)

stimulus (*plural*: stimuli) Any environmental phenomenon that triggers an action potential in a neuron. (Chapter 15)

stomach A pouchlike organ in the alimentary canal that plays a major role in digesting food. (Chapter 7)

stop codon Any of three mRNA codons that do not code for an amino acid but instead cause the amino acid chain to be released into the cytoplasm. (Chapter 4)

stress response Series of physiological changes that occur in response to physical, emotional, or mental stress, and are triggered by norepinephrine. (Chapter 16)

striated muscle A voluntary muscle made up of elongated, multinucleated fibers. Typically skeletal and cardiac muscle that is distinguished from smooth muscle by the in-register banding patterns of actin and myosin filaments. (Chapter 6)

stroke Acute condition caused by a blood clot that blocks blood flow to an organ or other region of the body. (Chapter 9)

subcutaneous layer Tissue layer below the skin. (Chapter 6)

substrate Any chemical metabolized by an enzyme-catalyzed reaction. (Chapter 3)

sugar-phosphate backbone Series of alternating sugars and phosphates along the length of a DNA or RNA molecule. (Chapter 2)

supernatural Not constrained by the laws of nature. (Chapter 1)

superior vena cava Vessel that collects blood draining from the head and upper extremities and returns it to the right atrium of the heart. (Chapter 9)

surfactant In the lungs, a slippery substance that reduces surface tension in the alveoli, preventing them from collapsing. (Chapter 10)

sweat gland Any of the tubular glands of the skin that secrete perspiration through pores to help regulate body temperature. (Chapter 6)

sympathetic division The division of the autonomic nervous system that becomes active during stress. (Chapter 14)

symptothermal method A method of controlling female fertility that is based on monitoring body temperature and cervical mucus. (Chapter 18)

synapse Gap between neurons consisting of a terminal bouton of the presynaptic neuron, the space between the two adjacent neurons, and the membrane of the postsynaptic neuron. (Chapter 14)

synovial joint A movable joint. (Chapter 6)

synovial membrane A thin membrane that lines the joint capsule of movable joints and secretes fluids that aid in joint movement. (Chapter 5)

syphilis A sexually transmitted disease caused by infection by the bacterium *Treponema pallidum*. (Chapter 13)

systemic circuit Flow of blood from the heart to body capillaries and back to the heart. One of two parts of the double circulation. (Chapter 9)

systole Portion of the cardiac cycle when the heart is contracting, forcing blood into arteries. (Chapter 9)

systolic pressure Force of blood on artery walls when heart is contracting. Highest blood pressure in arteries. (Chapter 9)

T lymphocyte (T cell) Immune system cell that matures in the thymus gland and recognizes and responds to body cells that have gone awry, such as cancer cells or cells invaded by viruses, as well as transplanted tissues and organisms such as fungi and parasitic worms. T cells provide an immune response called cell-mediated immunity. (Chapters 8, 12)

T tubule A tubule that passes from the sarcolemma across the myofibrils in a striated muscle. (Chapter 6)

T_3 (triiodothyronine) A form of thyroid hormone involved in controlling the rate of metabolic processes. (Chapter 16)

T_4 (thyroxine) A form of thyroid hormone, which affects body temperature, growth, metabolic rate, and heart rate. (Chapter 16)

T4 cell See helper T cell. (Chapters 8, 12)

***Taq* polymerase** A catalytic enzyme that can withstand high temperatures and is used in polymerase chain reactions. (Chapter 20)

target cell Cell that responds to regulatory signals such as hormones. (Chapter 16)

target tissue A tissue that responds to a hormonal signal. (Chapter 16)

tarsal Any of the small bones of the ankle. (Chapter 6)

tastant A molecule that binds to a taste cell receptor, eventually triggering a nerve impulse. (Chapter 15)

taste bud A structure that contains gustatory receptors, found on the sides of papillae in the tongue. (Chapter 15)

tectorial membrane A structure in the cochlea. Hair cells are pressed against it when the basilar membrane flexes. (Chapter 15)

telomerase An enzyme that helps prevent the degradation of the tips of chromosomes, active during development and sometimes reactivated during cancer. (Chapters 17, 21)

telomere The ends of a linear chromosome. (Chapter 21)

telophase Stage of cell division during which the nuclear envelope forms around the newly produced daughter nucleus, and chromosomes decondense. (Chapter 17)

temperate forest Biome dominated by deciduous trees. (Chapter 24)

temporal bone Either of the paired bones forming the sides and base of the cranium. (Chapter 6)

temporal lobe Part of the cerebral hemisphere that processes auditory and visual information, memory, and emotion. Located in front of the occipital lobe. (Chapter 14)

terminal bouton Knoblike structure at the end of an axon. (Chapter 14)

tertiary structure A level of protein structure stabilized by many different chemical interactions between side groups of amino acids. (Chapter 2)

testable Applies to a statement that can be evaluated through observations of the measurable universe. (Chapter 1)

testis (*plural*: testes) Either of the paired male gonads, involved in gametogenesis and secretion of reproductive hormones. (Chapter 18)

testosterone Masculinizing hormone secreted by the testes. (Chapters 16, 18)

thalamus Main relay center for sensory impulses between the spinal cord and the cerebrum. (Chapter 14)

theory of common descent Theory that all organisms present on Earth today derive from a single common ancestor. (Chapter 22)

theory of evolution Theory that all species present on Earth today are descendants of a single common ancestor and are the products of millions of years of evolutionary change. See also biological evolution. (Chapter 22)

theory of natural selection Theory that individuals with certain traits have greater survival and reproduction than individuals who lack these traits, resulting in an increase in the frequency of successful alleles and a decrease in the frequency of unsuccessful ones. (Chapter 22)

thermoreceptor Sensory receptor that responds to changes in temperature. (Chapter 15)

thermoregulation The ability to maintain a body temperature within a narrow range. (Chapter 5)

thrombin A molecule in the clotting cascade that helps to trigger the formation of fibrin. (Chapter 8)

thrombosis A blood clot that forms in a vessel and does not travel through the bloodstream. (Chapter 8)

thymopoietin Protein hormone secreted by the thymus that stimulates T-cell development. (Chapter 16)

thymosin Thymus hormone that stimulates T-cell development. (Chapter 16)

thymus An endocrine gland located in the neck region that helps establish the immune system. (Chapter 16)

thyroid gland An endocrine gland in the neck region that stimulates metabolism and regulates blood-calcium levels. (Chapter 16)

tight junction Connection between epithelial cells that reduces their permeability. (Chapter 5)

tibia Long bone of the leg that extends from the knee to ankle (shinbone). (Chapter 6)

tidal volume The amount of air that is exchanged with the environment during each breath. (Chapter 10)

tissue A group of cells with a common function. (Chapter 16)

tonicity The concentration of solute in a solution, which affects the size and shape of cells it is in contact with. (Chapter 3)

tonsil Lymphoid tissue in the throat. (Chapter 7)

tonsillitis Inflammation of the tonsils. (Chapter 10)

totipotent Describes a cell able to specialize into any cell type of its species, including embryonic membrane. Compare with multipotent, pluripotent. (Chapter 21)

trachea Air passage from upper respiratory system into lower respiratory system. Also called "windpipe." (Chapter 10)

transcription Production of an RNA copy of the protein-coding DNA gene sequence. (Chapter 4)

trans fat Contains unsaturated fatty acids that have been hydrogenated, which changes the fat from a liquid to a solid at room temperature. (Chapter 2)

transfer RNA (tRNA) Amino acid–carrying RNA structure with an anticodon that binds to an mRNA codon. (Chapter 4)

transgenic organism Organism whose genome incorporates genes from another organism; also called genetically modified organism (GMO). (Chapter 4)

translation Process by which an mRNA sequence is translated into a sequence of amino acids joined together to produce a protein. (Chapter 4)

trichomoniasis A sexually transmitted disease caused by infection with the protozoan *Trichomonas vaginalis*. (Chapter 13)

tRNA See transfer RNA. (Chapter 4)

trophic level Feeding level or position in a food chain; for example, producers, primary consumers, etc. (Chapter 23)

trophoblast The outer layer of a developing blastocyst that supplies nutrition to the embryo. (Chapter 21)

tropical forest Biome dominated by broad-leaved, evergreen trees; found in areas where temperatures never drop below the freezing point of water. (Chapter 24)

trypsin A digestive system enzyme that hydrolyzes proteins. (Chapter 7)

tubal ligation Female sterilization involving the sealing of the oviducts. (Chapter 18)

tumor Mass of tissue that has no apparent function in the body. (Chapter 17)

tundra Biome that forms under very low-temperature conditions. Characterized by low-growing plants. (Chapter 24)

tympanic membrane Eardrum. Flexible membrane that forms the outer surface of the middle ear. (Chapter 15)

type 1 insulin-dependent diabetes mellitus (IDDM) Early-onset diabetes requiring insulin injections for disease management. (Chapter 7)

type 2 non-insulin-dependent diabetes mellitus (NIDDM) Adult-onset diabetes. (Chapter 7)

ulna The long bone extending from the elbow to the wrist and opposite the thumb. (Chapter 6)

ultrasound An image generated by sound waves bouncing off an object and being picked up by a receiver. Used during pregnancy to evaluate the health of a developing fetus. (Chapter 21)

umbilical cord Structure through which blood vessels travel, connecting the developing embryo and fetus to the placenta. (Chapter 21)

unsaturated fat Fatty acid with many carbon-to-carbon double bonds; liquid at room temperature. (Chapter 2)

upper respiratory tract The nose, mouth, nasal cavities, and throat. (Chapter 10)

urea Nitrogenous waste produced during metabolism and excreted in urine. (Chapter 11)

ureter Tube that delivers urine from a kidney to the bladder. (Chapter 11)

urethra Urine-carrying duct that also carries sperm in males. (Chapters 11, 18)

urine Liquid expressed by the kidneys and expelled from the bladder in mammals, containing the soluble waste products of metabolism. (Chapter 11)

uterus Pear-shaped muscular organ in females that can support pregnancy and that undergoes menstruation when its lining is shed. (Chapter 18)

utricle One of two chambers in the vestibular apparatus of the inner ear. (Chapter 15)

vaccination A preparation of a weakened or killed pathogen, or portion of a pathogen, that will stimulate the immune system of a recipient to prepare a long-term defense (memory cells) against that pathogen. (Chapter 12)

vacuum aspiration A method for removal of the contents of the uterus during abortion or after miscarriage. (Chapter 18)

vagina Muscular canal in females leading from the cervix to the vulva. (Chapter 18)

vaginal ring A hormonal form of birth control that is inserted into the vagina, against the cervix. (Chapter 18)

valence shell The outermost energy shell of an atom. It contains the valence electrons, those most involved in the chemical reactions of the atom. (Chapter 2)

variable A factor that varies in a population or over time. (Chapter 1)

vas deferens (*plural*: vasa deferentia) Either of the two ducts in males that carry sperm from the epididymis to the urethra. (Chapter 18)

vasectomy Surgical sterilization of males during which each vas deferens is cut or blocked to prevent delivery of sperm from testes to urethra. (Chapter 18)

vector An organism that carries a pathogen from one host to another, such as a mosquito carrying West Nile virus. (Chapter 12)

veins Vessels that carry blood from the body tissues back to the heart. (Chapter 9)

ventilation Breathing. Exchange of air with the environment. (Chapter 10)

ventral body cavity Body cavity composed of thoracic, abdominal, and pelvic cavities and located at the front of the body. (Chapter 5)

ventricle (1) Chamber of the heart that pumps blood from the heart to the lungs or systemic circulation. (Chapter 9) (2) In the brain, any of the connecting cavities. (Chapter 14)

venules Small blood vessels that carry blood from capillary beds to veins. (Chapter 9)

vertebra (*plural*: vertebrae) Bone of the spinal column through which the spinal cord passes. (Chapter 6)

vertebral column Also called the spine, the series of vertebrae and cartilaginous disks extending from the brain to the pelvis. (Chapter 6)

vertigo Sensation of dizziness. (Chapter 15)

vestibular apparatus Structure in the inner ear responsible for sensing head movement, especially angular momentum. (Chapter 15)

vestigial traits Features modified to have no, or relatively minor, function compared to the function in other descendants of the same ancestor. (Chapter 22)

villus (*plural*: villi) Small fingerlike projections on the inside of the small intestine that function in nutrient absorption. (Chapter 7)

viral envelope Layer formed around some virus protein coats (capsids) that is derived from the cell membrane of the host cell and may also contain some proteins encoded by the viral genome. (Chapter 12)

virus Infectious intracellular parasite composed of a strand of genetic material and a protein or fatty coating. It can only reproduce by forcing its host to make copies of it. (Chapter 12)

visible light Light of a wavelength that humans can perceive. (Chapter 15)

vital capacity Amount of air that can be exchanged with the environment during the deepest possible breath. (Chapter 10)

vitamin Organic nutrient needed in small amounts. Most vitamins function as coenzymes. (Chapter 2)

vitreous humor Liquid that fills the chamber of the eye between the lens and the retina. (Chapter 15)

vocal cords Cartilaginous strings in the larynx that vibrate to produce vocal sounds. (Chapter 10)

voluntary muscle Muscle normally under conscious control. Mainly skeletal muscle. (Chapter 5)

vulva The outer portion of the female external genitalia including the labia majora, labia minora, clitoris, and vaginal and urethral openings. (Chapter 18)

wastewater Liquid wastes produced by humans. (Chapter 24)

weather Current temperature and precipitation conditions. (Chapter 24)

wetland Biome characterized by standing water, shallow enough to permit plant rooting. (Chapter 24)

white matter Nervous system tissue, especially in the brain and spinal cord, made of myelinated cells. (Chapter 14)

withdrawal A relatively ineffective method of birth control that involves removal of the penis from the vagina prior to ejaculation during intercourse. (Chapter 18)

Wolffian duct Embryonic structure that regresses in females but develops in males, eventually becoming the epididymis, vas deferens, urethra, and associated structures. (Chapter 21)

x-axis The horizontal axis of a graph. Typically describes the independent variable. (Chapter 1)

X-linked gene Any of the genes located on the X chromosome. (Chapter 20)

y-axis The vertical axis of a graph. Typically describes the dependent variable. (Chapter 1)

yellow bone marrow The fat-rich bone marrow at the ends of long bones. (Chapter 6)

Y-linked gene Any of the genes located on the Y chromosome. (Chapter 20)

Z discs The borders of a sarcomere in muscle. (Chapter 6)

zona pellucida The substance surrounding an egg cell that a sperm must penetrate for fertilization to occur. (Chapter 21)

zygomatic bone Either of the paired cheekbones of the human skull. (Chapter 6)

zygote Single cell resulting from the fusion of gametes (egg and sperm). (Chapters 17, 21)

INDEX

Page references followed by *fig* indicate an illustrated figure or photograph; followed by *t* indicate a table.

A

Abduction, 139, 139*fig*
ABO blood system
    genotypes/phenotypes, 490, 490*fig*
    and immune response, 187, 188*t*
Abortion, elective, 459
Acetabulum, 136, 136*fig*
Acetylcholine/acetylcholinesterase, 328
Acid, defined, 34, 50
Acidosis, 263
Acquired immune deficiency syndrome (AIDS). *See* HIV/AIDS
Acrosome, 444, 513, 514
Actin, 113, 142, 142*fig*, 144*fig*, 151
Action potential, 326–328, 327*fig*, 346
Activation energy, 67–68, 79
Active immunity, 289
Active site, 68, 68*fig*, 79
Active transport
    in capillaries, 203
    and concentration gradient, 66, 66*fig*, 79, 258–259, 258*fig*
    in tubular reabsorption, 258–259, 258*fig*, 260, 260*fig*
Adaptations (in natural selection), 551, 554–555, 558
ADD. *See* Attention deficit disorder (ADD)
Adenosine diphosphate (ADP)
    in cellular respiration, 71, 72*fig*
    in muscle contraction, 144–146, 145*fig*
    structure and function, 58, 58*fig*
Adenosine triphosphate (ATP)
    as cell energy source, 57–59, 57*fig*, 58*fig*, 66, 66*fig*, 258–259, 258*fig*
    in muscle contraction, 144–146, 144*fig*, 145*fig*
    synthesis of, 68–73, 69*fig*, 73*fig*, 79
Adhesion junctions, 107, 122
Adipocytes, 110
Adipose tissue, 56–57, 57*fig*, 110–111, 110*fig*, 148
Adolescence (as developmental stage), 526–527
Adrenal cortex, 386, 387, 387*fig*
Adrenal glands, 386–389, 387*fig*, 389*fig*, 396*t*
Adrenaline, 386
Adrenal medulla, 386, 387*fig*, 388
Adrenocorticotropic hormone (ACTH), 389, 389*fig*
Adulthood (as developmental stage), 527–528
Adult stem cells, 512, 528, 531, 533
Aerobic respiration, 69–73
Afferent arteriole, 257
Agarose gel, 502
Agglutination, 187, 187*fig*
Aging, 529–531, 534
Agranulocytes, 183, 183*t*
AIDS. *See* HIV/AIDS
Air pollution, 593, 593*fig*
Albumins, 181
Aldosterone, 260, 260*fig*, 262, 263, 387

Alimentary canal. *See* Digestive system
Alleles. *See also* Genes
    defined, 423
    dominant/recessive, 472–476, 489–490
    and gamete variation, 469–472, 469*fig*
    and Punnett square analysis, 474–476, 474*fig*, 475*fig*, 482
Allergy, 289, 289*fig*
Allometric growth, 527, 527*fig*
Alveoli
    and gas exchange, 238–240, 246
    structure, 230, 231*fig*
    tobacco smoke effect on, 234, 237–238, 237*fig*
Alzheimer's disease (AD), 328, 531, 534
Amenorrhea, 74
Amino acids
    monomers of, 37–38, 38*t*
    and mutations, 88, 90–91, 90*fig*
    in protein structure, 37–40, 38*fig*, 39*fig*, 50, 84
    and ribosome protein synthesis, 61
    and transfer RNA (tRNA), 87–88, 89*fig*, 100
Amniocentesis, 523
Amnion, 517
Amniotic fluid, 517
Amygdala, 340, 340*fig*, 347
Anabolic steroids, 391, 391*fig*
Anaerobic respiration, 73–74, 74*fig*
Anaphase
    in meiosis, 425, 426–427*fig*, 431*fig*
    in mitosis, 413, 414–415*fig*
Anchorage dependence, 418
Androgens, 441
Anecdotal evidence, 17, 22
Anemia
    and malaria infection, 186
    and Rh blood factor, 188, 189*fig*, 194
    sickle cell, 191
Aneurysm, 208, 208*fig*
Angina, 217
Angiogenesis, 418
Angioplasty, 218, 218*fig*
Angiotensin II, 260
Anopheles mosquito, 180, 185–186, 185*fig*
Anorexia, 74–75, 450
Antagonistic muscle pairs, 140, 140*fig*, 151
Anterior pituitary, 387, 388*fig*
Antibodies
    in blood plasma, 181
    and B lymphocytes (B cells), 288–290, 288*fig*, 290*fig*, 292
    and HIV/AIDS, 311, 313
Anticodons, 87, 100
Antidiuretic hormone (ADH), 260, 260*fig*, 262, 387–388
Antigen-presenting cell (APC), 294, 294*fig*
Antigen receptors, 288–291, 288*fig*, 290*fig*, 292

Antigens
    defined, 187, 287
    immune system response to, 287–291, 288*fig*, 290*fig*, 292
Antioxidants, 47–48, 48*t*, 51, 394
Anus, 164, 164*fig*
Aorta, 210, 211*fig*
Apoptosis, 416, 517–518, 533
Appendicular skeleton, 134, 134*fig*, 136–139, 151
Appendix, 163, 164*fig*
Aqueous humor, 370, 370*fig*
Arrhythmia, 216
Arteries
    coronary, 217, 217*fig*
    structure and function, 200, 201*fig*, 221
Arterioles, 202–203, 203*fig*, 205–206, 221
Asexual reproduction *vs.* sexual reproduction, 440
Assortative mating, 556–557, 558
Asthma, 233, 233*fig*
Astigmatism, 370, 372*fig*
Astrocytes, 114
Atherosclerosis, 207, 207*fig*, 217–219, 221
Athleticism and sex differences, 131, 137, 139, 148–149
Atomic number, 28
Atomic structure, 28–29, 50
ATP. *See* Adenosine triphosphate (ATP)
ATP synthase, 71, 71*fig*, 72*fig*, 79
Atrial natriuretic peptide (ANP), 262, 262*fig*, 268
Atrial systole, 214, 215*fig*
Atrioventricular (AV) node, 214, 214*fig*
Atrioventricular (AV) valves, 209*fig*, 210, 214, 215*fig*
Atrium, 209, 209*fig*, 211*fig*
Atrophy, 113, 122
Attention deficit disorder (ADD), 323–324, 330–331, 340, 343–344
Auditory canal, 364, 365*fig*
Auditory nerve, 365–366, 366*fig*
Auditory receptors, 364–367, 365*fig*, 378
Auditory tubes, 229
Autoimmune diseases, 291
Autonomic nervous system, 342–343, 343*fig*, 347
Autosomes, 422*fig*, 423, 494
Axial skeleton, 134–135, 134*fig*, 151
Axons, 324–328, 325*fig*, 327*fig*, 346

B

Bacteria
    antibiotic resistance in, 276
    diseases caused by, 278*t*, 306–308
    and gene cloning, 93–96, 95*fig*
    health benefits of, 277
    reproduction in, 275–276, 275*fig*, 277*fig*
    structure of, 274–275, 275*fig*
Ball-and-socket joints, 139, 139*fig*

I-1

Basal metabolic rate, 74, 393
Base (defined), 34, 50
Basement membrane, 107, 122
Base pairing, 43–44, 44*fig*, 411–412
Basilar membrane, 365–367, 366*fig*
Basophils, 183, 183*t*, 286
Beagle, 540–541
Bell curve, 476*fig*
Beta-pleated sheets, 39, 39*fig*
Bias and experimental design, 9–10, 9*fig*, 21
Bicarbonate ions
    as blood buffer, 263
    and gas exchange, 239–240, 246, 249*fig*
Bile, 43, 167–168
Bilirubin, 167, 189, 194, 255
Binary fission, 275–276, 275*fig*
Biodiversity
    and current extinction rate, 567, 568*fig*, 570, 580
    and Endangered Species Act (ESA), 567
    and global warming, 595
    "hotspots," 575–576, 576*fig*, 581
    and Linnaeus classification, 542, 543*t*
    preservation of, 575–578, 579*t*
    threats to, 568–570, 595
Biogeography, 545–546, 558
Biological classification, 542, 543*t*
Biological evolution. *See* Evolution (process of)
Biomass, 570
Biomes
    aquatic, 590–591, 596
    terrestrial, 586–589, 587*fig*, 596
Biopsy, 419
Bioregional awareness, 585–586
Bipedalism, 547, 547*fig*
Bipolar cells, 373
Birth control
    barrier methods, 453–455, 455*fig*, 456*fig*, 463
    future technology, 460
    hormonal methods, 456, 456*fig*, 463
    method effectiveness, 455*t*
    other methods, 457–459, 457*fig*, 463
Birth defects
    and environmental factors, 517, 518–521, 523*fig*
    screening tests for, 523, 524*t*
    and stem cell treatment, 521, 523–524
Blastocyst, 515, 515*fig*, 517*fig*, 533
"Blind spot," 375, 375*fig*
Blood
    as connective tissue, 110*fig*, 111, 122, 193
    constituents of, 180–184, 180*fig*, 182*fig*, 183*t*, 193
Blood buffers, 34, 263, 263*fig*
Blood calcium, 132, 392, 393*fig*
Blood clotting
    disorders, 190–191
    and plasma proteins, 181
    process of, 189–190, 190*fig*, 194
    and vitamin K, 45, 46*t*
Blood filtration
    by kidneys, 252–259
    by the liver, 167–168
Blood flow. *See* Cardiovascular system
Blood gases
    carbon dioxide, 180–182, 239–240, 240*fig*, 246
    homeostasis of, 186, 187*fig*, 236, 236*fig*
    oxygen, 182, 182*fig*, 239, 239*fig*, 246
Blood glucose
    and glucocorticoids, 387, 389
    and liver, 167

    pancreas regulation of, 169, 169*fig*, 391–392, 391*fig*, 397*t*
Blood pH
    and homeostasis, 34, 180, 236, 236*fig*, 246, 252
    and urinary system, 252, 263, 263*fig*, 268
Blood plasma, 180–181, 180*fig*, 181*fig*, 193
Blood platelets, 184, 189–191, 190*fig*, 194
Blood pressure
    and arteriole diameter, 202–203, 203*fig*, 221
    and blood volume, 214–215, 221
    in capillary beds, 203, 203*fig*
    control of, 205–206
    and glucocorticoids, 387
    and hormone release, 260, 260*fig*, 387
    measurement of, 207–208, 207*fig*
    and urine formation, 257
    in veins, 204
Blood sugar. *See* Blood glucose
Blood transfusions, 186–189
Blood types, 186–189, 188*t*
Blood typing analysis, 489–490
Blood vessels. *See* Cardiovascular system
Blood volume
    and blood pressure, 214–215, 221
    and kidneys, 254, 260, 262, 264, 268
B lymphocytes (B cells), 183*fig*, 183*t*, 184, 288–294, 292*fig*, 295*fig*, 297
Body cavities, 115, 115*fig*, 122, 122*fig*
Body mass index (BMI), 75–77, 76*fig*
Body membranes, 115–116, 116*fig*, 122
Body weight and health, 74–76
Bone
    as connective tissue, 110*fig*, 112–113, 122
    effect of aging on, 530
    growth and development, 130–131, 130*fig*, 131*fig*
    remodeling and repair, 131–134, 132*fig*, 133*fig*
    structure and types of, 128–129, 129*fig*
Bone marrow, 128, 129, 129*fig*, 182*fig*, 184, 289
Bony pelvis, 136–137, 136*fig*
Boreal forest, 588, 588*fig*
Bovine spongiform encephalopathy (BSE). *See* Spongiform encephalopathies
Bowels, 163, 164
Brain
    and attention deficit disorder (ADD), 339, 340, 343
    cardiovascular center in, 205–206
    components of, 336–338, 336*fig*, 337*fig*, 338*fig*
    development, 528
    hemisphere functions, 338
    and limbic system, 340, 340*fig*, 347
    and olfactory information, 362
    and perception, 355–356, 356*fig*, 378
    tobacco smoke effects on, 243
    and visual information, 374–376, 374*fig*, 375*fig*
Brain stem, 338, 338*fig*, 347
Breathing
    control of, 236–237, 236*fig*, 246
    mechanics of, 234–235, 235*fig*, 246
    and tobacco smoke, 237–238, 237*fig*
Breech birth, 525
Bronchi/bronchioles, 229*fig*, 230
Bronchitis, 233
BSE (bovine spongiform encephalopathy). *See* spongiform encephalopathies
Bulbourethral glands, 441
Bulimia, 75

C

Caesarean section, 525
Calcitonin, 132, 392, 393*fig*
Calcium
    in blood, 132, 392, 393*fig*
    and bones, 128, 130, 132, 151, 392, 393*fig*
    and muscle contraction, 142–143, 143*fig*
Calendar method, 457
Calorie *vs.* calorie, 74
Cancer
    detection of, 419–420, 419*fig*
    and human papillomavirus, 309–310
    and inheritance, 417, 429
    and melatonin levels, 394–395
    and mutations, 409, 417
    risk factors for, 408–409
    and tobacco smoke, 227, 233
    treatment of, 420–421
    as unregulated cell division, 406–408, 407*fig*, 416–418
    warning signs, 419*fig*
Candida infection, 305*t*, 306, 318
Capacitation, 513
Capillaries/capillary beds
    in alveoli, 230, 238–240
    structure and function, 200, 201*fig*, 203, 203*fig*, 211*fig*, 221
Capsid, viral, 278, 279*fig*
Capsule, bacterial, 275, 275*fig*
Carbohydrates
    as cell markers, 64, 64*fig*
    digestion of, 56*fig*, 158, 161–162, 162*fig*
    function and types, 35–37, 36*fig*, 37*fig*, 37*t*, 50
Carbon dioxide
    in blood, 180–182, 239–240, 240*fig*, 246
    and cellular respiration, 69–73, 79
    and external respiration, 221, 231*fig*, 234–238, 236*fig*
    and global warming, 593, 594–595, 594*fig*
Carbon molecule, 35, 35*fig*, 50
Carbon monoxide and hemoglobin, 240–241, 246
Carcinogens, 408, 409
Cardiac cycle, 214–217, 214*fig*, 215*fig*
Cardiac muscle, 113, 114*fig*
Cardiac output, 214–215
Cardiac veins, 217, 217*fig*
Cardiopulmonary resuscitation (CPR), 216
Cardiovascular center, 205–206
Cardiovascular disease, 199, 206–209, 207*fig*, 208*t*, 211–213, 217–218
Cardiovascular system. *See also* Lymphatic system
    blood pressure control, 205–206, 207*fig*
    cardiac cycle, 214–217, 214*fig*, 215*fig*
    coronary circulation, 217–218, 217*fig*
    effects of aging on, 530
    hepatic portal system, 210–211, 212*fig*
    maintaining health of, 219
    overview, 200, 210–211, 211*fig*
    vascular system, 200–204, 201*fig*, 203*fig*, 204*fig*
Carotid arteries, and blood pressure, 205
Carpal bones, 137*fig*, 138
Carriers (of recessive traits), 496
Carrying capacity, 565–566, 565*fig*, 580
Cartilage
    and bone repair, 133
    as connective tissue, 110*fig*, 111–112, 122
    effect of aging on, 530

ossification of, 130–131, 130*fig*
in skeletal system, 128–129, 129*fig*, 130, 133
Catalytic agents (enzymes as), 67–68, 68*fig*, 79
Cataracts, 371
Caudate nuclei, 337
Causation *vs.* correlation, 12, 12*fig*
Cecum, 163, 163*fig*, 164*fig*
Cell body (neuron), 324, 325*fig*
Cell cycle
and meiosis, 422–431, 424*fig*, 426–427*fig*, 428*fig*, 430*fig*, 431*fig*
and mitosis, 413–416, 413*fig*, 414–415*fig*, 431*fig*
regulation of, 416–418, 416*fig*
Cell differentiation
and blastocyst production, 515, 533
and gastrulation, 517, 517*fig*, 533
and stem cell therapy, 531
Cell division
and DNA replication, 411–412, 411*fig*
and gametogenesis, 422–424, 443–447, 445*fig*, 447*fig*, 470–472
and meiosis, 422–431, 426–427*fig*, 428*fig*, 430*fig*, 431*fig*
and mitosis, 413–416, 414–415*fig*, 416*fig*, 431*fig*
and oncogenes, 417
overview, 409–412
in preembryonic development, 515
telomere limitation of, 529
Cell junctions, 107, 109*fig*, 122
Cell-mediated immunity, 293–294
Cell membrane. *See* Plasma membrane
Cell migration, 517–518, 533
Cell nucleus. *See also* Meiosis; Mitosis
features of, 43, 60, 60*fig*, 78, 78*fig*
as transcription site, 86, 100
Cells
adenosine triphosphate (ATP) as energy source, 57–59, 57*fig*, 58*fig*, 66, 66*fig*, 258–259, 258*fig*
and diffusion, 64–65, 64*fig*
and osmosis, 65–66, 65*fig*
size of, 59–60, 59*fig*, 60*fig*
Cells, reproductive. *See* Gametes
Cell structures, 60–63, 60*fig*, 61*fig*, 63*fig*
Cellular respiration
citric acid cycle in, 70–71, 71*fig*, 79
electron transport chain in, 71, 72*fig*, 79
fermentation in, 73–74
and gas exchange, 239–240
glycolysis in, 69–70, 70*fig*, 79
and muscle contraction, 144–146
overview, 68–69, 69*fig*, 71, 73*fig*, 79
Cellulose, 36, 36*fig*, 37
Cell wall, bacterial, 275, 275*fig*
Central nervous system (CNS). *See also* Nervous system; Peripheral nervous system
components of, 324*fig*, 334–340, 346–347
formation of, 518–519, 519*fig*
and perception, 355–356, 356*fig*, 378
Centrioles
in cell division, 413, 414–415*fig*, 425, 426–427*fig*
features of, 62, 63*fig*, 79
Centromeres, 410, 413, 414–415*fig*
Cerebellum, 338, 338*fig*, 347
Cerebral cortex, 336–340, 337*fig*
Cerebrospinal fluid, 334, 336
Cerebrum, 336–337, 336*fig*, 346

Cervical cancer, 202, 303, 309–310
Cervical caps, 455, 456*fig*
Cervix, 443
Chaparral, 588, 596
Chemical bonding
in biological molecules, 32*t*
and carbon molecule, 35, 35*fig*, 50
cohesion in, 31, 32*fig*
and dehydration synthesis, 35–36, 50
energy storage in, 56–57, 68
and hemoglobin, 182, 182*fig*
and hydrolysis, 35
and metabolic reactions, 67–68
in protein structure, 37–40
and valence shell, 29–30, 50
Chemical senses, 361–363, 378
Chemoreceptors, 354, 354*t*, 361–363, 361*fig*, 362*fig*
Chemotherapy, 420–421
Childbirth
and hepatitis B, 308
and herpes simplex, 308
process of, 524–526, 525*fig*
Childhood (as developmental stage), 526–527
Chlamydia, 305*t*, 306–307, 306*fig*
Cholesterol
and atherosclerosis, 207, 207*fig*, 217–219, 221
and diet, 11–12, 11*fig*, 15–16
in membrane structure, 64, 64*fig*
and steroid hormones, 384, 392
and stress, 392
structure and function, 42–43, 43*fig*, 50
and tobacco smoke, 243
Chondrocytes, 111, 122
Chorion, 521
Chorionic villi, 521, 522*fig*
Choroid, 370*fig*, 372
Chromatids, 409–410
Chromatin, 60, 409
Chromosomes. *See also* Genes; Homologous pairs
anomalies in, 495–496
degradation of, 418, 529
in gametes, 422–424, 424*fig*, 443, 470–472
as gene carriers, 468, 496–498
as organized DNA, 409–410
and sex determination, 494, 494*fig*
in somatic cells, 422
Chrondroblasts, 130
Chronic obstructive pulmonary disease (COPD), 238
Chronic pain, 360
Chyme, 160
Ciliary body, 370, 371*fig*
Circulatory system. *See also* Cardiovascular system; Lymphatic system
in fetus, 522
overview, 180*fig*, 200*fig*, 221
Circumduction, 139, 139*fig*
Cis configuration, 42, 42*fig*
Citric acid cycle, 70–71, 71*fig*, 79
Classification, biological, 542, 543*t*
Clavicle, 137, 137*fig*
Cleavage (in zygote), 515, 533
Climate, 586–587, 587*t*, 594–595, 596
Climate change, global, 594–595, 594*fig*, 596
Clinical pregnancy (defined), 516
Clitoris, 442
Clonal population, 292
Clotting cascade, 189–190, 190*fig*

CNS. *See* Central nervous system
Cochlea, 364–365, 366*fig*
Cochlear duct, 364–365, 366*fig*
Cochlear implant, 366, 367*fig*
Codominance, 490
Codons, 87, 100
Coenzymes (vitamins as), 45–46, 51, 74
Cohesion, 31, 32*fig*, 50
Collagen, in blood clotting, 189–190
Collagen fibers
in bone tissue, 122, 130, 131
in connective tissue, 109, 111–112
in epithelia, 122, 147
Collecting duct, 256*fig*, 257, 260, 262
Colon, 163, 164*fig*
Columnar epithelium, 107, 122, 122*fig*
Combination drug therapy for HIV/AIDS, 313–315, 314*t*
Common descent, theory of
alternative hypotheses to, 541, 553
and biological classification, 542
and DNA (deoxyribonucleic acid), 545, 545*fig*, 558
evidence from biogeography, 545–546, 558
evidence from embryonic development, 544–545, 545*fig*
evidence from fossil record, 546–549, 558
evidence from homology, 544–545, 544*fig*, 558
as robust theory, 552–553
and theory of evolution, 4, 5*fig*, 540, 540*fig*
Compact bone, 128, 129*fig*, 151
Competition, 571, 574*t*
Complementary pairing (in DNA molecule), 43–44, 44*fig*, 411–412
Complement proteins, 286–287, 287*fig*, 292
Compound (defined), 29
Concentration gradient
and active transport, 66, 66*fig*, 79, 258–259, 258*fig*
in capillary beds, 203, 203*fig*
and facilitated diffusion, 65, 65*fig*
in kidneys, 258*fig*, 259, 261–262, 261*fig*
and passive transport, 64–66, 64*fig*, 79
and placental membrane, 522
and sodium-potassium pump, 66, 326–328, 327*fig*
Condoms, 455, 455*fig*
Cones, 373, 374*fig*
Confidence interval, 15
Conjoined twins, 515
Connective tissue, 108–112, 110*fig*, 122
Conservation International, 575–576
Consumers (in ecology), 569
Contact inhibition, 418
Contagious pathogens (defined), 275
Continuous variation, 476–477
Contraception. *See* Birth control
Control group, 8, 8*fig*, 21
Controlled experiments, 8–9, 21
Coral reefs, 59–591, 590*fig*
Cornea, 369–370, 370*fig*
Coronary arteries, 217, 217*fig*
Coronary bypass surgery, 218
Coronary circulation, 217, 217*fig*, 221
Coronary heart disease (CHD), 218
Corpus callosum, 336*fig*, 337
Corpus luteum, 447*fig*, 449, 450–451, 516–517, 521

Correlation
  in genetics, 477–479, 478 *fig*
  in hypothesis testing, 10–12, 11 *fig*, 12 *fig*, 21
Corticotropin-releasing hormone (CFH), 389, 389 *fig*
Cortisol, 387
Cortisone, 387
Countercurrent exchange, 261–263, 261 *fig*, 268
Covalent bonds
  and amino acid polymers, 35, 38–40, 38 *fig*
  in biological molecules, 32 *t*, 50
  and carbon, 35, 35 *fig*, 50
  and dehydration synthesis, 35
  described, 29, 29 *fig*, 50
  in DNA molecule, 43–44, 44 *fig*, 411–412
Cranial cavity, 115, 115 *fig*
Cranial nerves, 341, 341 *fig*, 347
Cranium, 134, 134 *fig*
Creatine phosphate, and adenosine triphosphate (ATP) synthesis, 144–146, 145 *fig*, 255–256, 256 *fig*
Creatinine, 255–256, 256 *fig*
Creutzfeld-Jakob disease (CJD), 273, 283
Critical habitat designation, 576
Crossing over (in meiosis), 425, 426 *fig*, 428, 428 *fig*
Cuboidal epithelium, 107, 122, 122 *fig*
Cushing's syndrome, 392
Cutaneous membrane, 116
Cystic fibrosis, 230, 232, 473, 473 *fig*
Cytokines, 183–184
Cytokinesis
  in meiosis, 424–425, 426–427 *fig*
  in mitosis, 413, 414–415 *fig*, 416 *fig*
Cytoplasm
  features of, 61, 61 *fig*
  and gamete maturation, 444
  in muscle fibers (cells), 142
  as translation site, 87, 100
Cytoskeleton, 62, 63 *fig*, 79
Cytosol
  features of, 60–61
  glycolysis in, 69–70, 70 *fig*, 79
Cytotoxic T cells, 293–294, 293 *fig*

D

Darwin, Charles
  on Linnaeus' classification, 542, 558
  and natural selection, 550–551
  on species distribution, 545–546
  and theory of evolution, 540–541, 541 *fig*
  on vestigial traits, 544
Decomposers, 574
Deductive reasoning, 6, 21
Defibrillators, 216–217
Dehydration synthesis
  and chemical bonding, 35–36, 35 *fig*, 38 *fig*
  in transcription, 86
  in translation, 87
Denatured DNA molecule (in PCR), 502
Dendrites, 324, 325 *fig*
Dense fibrous connective tissue, 110 *fig*, 111, 122
Density-dependent/density-independent factors, 565
Deoxyribonucleic acid (DNA). *See* DNA (deoxyribonucleic acid)
Dependent/independent variables, 7–8, 11–12, 11 *fig*
Depolarization/polarization (of neurons), 326–328, 327 *fig*, 346
Dermis, 147, 147 *fig*, 151
Deserts, 589, 596
Development after birth, 526–527, 534
Development, embryonic, 516–521, 518 *fig*, 533
Development, fetal, 521–524, 521 *fig*, 533–534
Development, neonatal, 526
Diabetes, 169–171, 170 *fig*, 264, 291, 392, 522
Dialysis, 265, 265 *fig*
Diaphragm (muscle), 115, 115 *fig*, 234
Diaphragms (birth control method), 455, 456 *fig*
Diaphysis, 129, 129 *fig*
Diastole, 214, 215 *fig*
Diastolic pressure, 207–208
Diffusion
  in alveoli, 238–240, 246
  in capillaries, 203, 203 *fig*, 211 *fig*
  in kidneys, 258 *fig*, 259, 261–262, 261 *fig*
  and membranes, 64–65, 64 *fig*
Digestive system
  accessory organs and functions, 166–171, 167 *fig*
  and autonomic nervous system, 342
  digestive tract, 156–166, 156 *fig*, 157 *fig*
  effects of aging on, 530
  hormones produced by, 396
  nutrient breakdown in, 162 *fig*
  secretion regulation, 163, 163 *fig*
  tobacco smoke effects on, 242
Dihybrid crosses, 491–493, 492 *fig*
Diploid cells, 424, 424 *fig*
Disaccharides, 36, 36 *fig*, 37 *t*
Distal tubule, 256 *fig*, 257, 259
Diuretic, 262–263, 262 *fig*
Diversity. *See* Biodiversity
Dizygotic twins, 514
DNA (deoxyribonucleic acid). *See also* Cell division; Genes; Protein synthesis
  in bacteria, 274–275
  in chromosomes, 409–410
  copying by polymerase chain reaction (PCR), 501–502
  and genes, 84, 100, 409–410, 468
  molecular structure, 43–44, 44 *fig*, 410, 410 *fig*
  and mutations, 88, 90–91, 90 *fig*, 91 *fig*, 101, 416–418, 469, 475, 529
  replication, 411–412, 411 *fig*, 416–418
  and skin color, 554, 558
  and theory of common descent, 545, 545 *fig*, 552, 558
  and tobacco carcinogens, 233, 408
  in viruses, 277–278
  vs. RNA (ribonucleic acid), 85 *fig*, 100
DNA fingerprinting, 500–505, 501 *fig*, 502 *fig*, 503 *fig*, 506
DNA polymerase, 411, 501–502
Dominant alleles, 473–476
Dopamine, 328–330
Dorsal cavity, 115, 115 *fig*, 122 *fig*
Double-blind experiments, 9–10, 9 *fig*, 21
Double bonds
  in fats, 41–42, 41 *fig*
  representation of, 29, 29 *fig*
Double helix. *See* DNA (deoxyribonucleic acid)
Ductus arteriosus, 522
Ductus venosus, 522
Duffy antigen, 188–189
Duodenum, 161

E

Ear
  and equilibrium, 367–369, 368 *fig*, 378
  and hearing, 364–367, 365 *fig*, 366 *fig*, 378
Eardrum, 364, 365 *fig*
Echocardiogram, 217 *fig*, 218
Ecological communities, 571–573
Ecological footprints, 592, 596
Ecology (defined), 564
Ecosystems
  defined, 573
  energy in, 569–570, 570 *fig*, 573–574, 580
Ectoderm, 517
Ectopic pregnancy, 443, 516
Effectors, 324, 325 *fig*
Efferent arteriole, 257
Egg production. *See* Oogenesis
Eggs. *See also* Menstrual cycle; Oogenesis
  in fertilization process, 513–514
Ejaculation, 441, 527
Elastin fibers, 109, 122, 147
Electrocardiogram (EDG or EKG), 215–216, 216 *fig*
Electrolytes, 27, 34
Electron cloud, 28–29, 29 *fig*
Electronegativity (in water molecule), 30–31
Electrons
  in atomic structure, 28–29, 29 *fig*
  energy storage in, 68
Electron shell, 29, 29 *fig*
Electron transport chain, 71, 72 *fig*, 79
Elements in humans, 28 *t*
Elimination, 163–165, 252
Embolism, 190
Embolus, 208
Embryo (defined), 513
Embryonic development
  process of, 516–521, 518 *fig*, 533
  and theory of common descent, 544–545, 545 *fig*
Embryonic disk, 517
Embryonic stem cells
  characteristics of, 512, 533
  debate regarding use of, 511–512, 515, 524
  potential uses for, 521, 529, 530–531
  production of, 513–515, 533
Emphysema, 238
Endangered Species Act (ESA), 567, 570, 576
Endocrine glands
  function, 106, 122
  hormones produced by, 384–389, 396–397 *t*
  vs. exocrine glands, 384
Endocrine system
  components of, 385, 386 *fig*
  and homeostasis, 387–389
  and immune system, 395, 396 *fig*
Endocrine tissues (other than endocrine glands), 396–398, 397 *fig*
Endocytosis (defined), 66, 66 *fig*
Endoderm, 517
Endometriosis, 451–452
Endometrium, 443, 450–452
Endoplasmic reticulum (ER)
  features of, 61–62, 62 *fig*, 79
  in muscle, 142, 143 *fig*, 151
Endorphins, 114
Endothelium, vascular, 201 *fig*, 203, 208
Endocrine-disrupting chemicals, 448
End-stage renal disease (ESRD), 265

Energy
    and active transport, 66, 66fig, 258–259, 258fig
    adenosine triphosphate (ATP) as source of, 57–59, 57fig, 58fig
    carbohydrates as source of, 36–37, 50
    and cellular respiration, 68–71, 69fig, 73fig, 79
    in ecosystems, 569–570, 570fig, 573–574, 580
    fats as source of, 40, 42, 50
    and fermentation, 73–74
    measurement of, 74, 74fig
    for muscle contraction, 144–146, 144fig, 145fig
    and natural resources, 591–592
Energy drinks, 27, 33–34, 36–37, 40, 47–49
Environment vs. heredity, 476–481, 483, 529
Enzymes
    in lysosomes, 61
    in metabolic reactions, 67–68, 68fig, 79
    proteins as, 37
    restriction, 94, 95fig, 502, 502fig
    specificity of, 68
    and vitamins, 74
Eosinophils, 183, 183t, 285
Epidemic, 295
Epidemiological studies, 12, 13t
Epidermis, 106, 146–147, 147fig, 151
Epididymis, 441
Epiglottis, 159, 159fig, 173
Epinephrine, 386, 388
Epiphyseal plate, 130, 131fig
Epiphysis, 129, 129fig, 151
Epithelial membranes, 115–116, 122
Epithelial tissues, 106–107, 106fig, 108fig, 122
Equilibrium, and the ear, 367–369, 368fig
Erythrocytes
    as blood constituent, 180fig, 181–182
    flow through capillaries, 203, 203fig
    and malaria infection, 182, 185–186, 186t, 189, 191, 193
    production of, 186, 187fig, 193, 252
    recycling of, 167, 189
Erythropoietin, 186, 187fig, 193, 252, 387
Esophagus, 159, 159fig, 173
Essential amino acids, 37
Essential fatty acids, 40–41
Estrogen
    as contraceptive, 456
    effect on osteoclasts, 132
    in pregnancy, 516–517, 521, 524
    as sex hormone, 391, 446, 449–451, 451fig
Estuaries, 591, 591fig
Ethmoid bone, 134, 134fig
Eukaryotes vs. prokaryotes, 274
Eukaryotic pathogens, 280, 281t, 297, 304, 305t
Eustachian tubes, 228fig, 229
Eutrophication, 570, 590
Evolution (process of)
    and blood types, 188–189
    and color blindness, 375
    and cystic fibrosis, 232
    defined, 540, 558
    and HIV, 313, 313fig
    human, 554–557
    and response to sound, 366–367
    and response to tastants, 362
Evolution (theory of)
    in the classroom, 557
    and common descent, 4, 5fig, 540–549, 558
    debate regarding, 539–540, 541, 553

and natural selection, 4, 549–552
as robust theory, 552–553, 558
Excretion, 252, 255–259, 267–268
Excretory system. See Urinary system
Exhalation, 234–235
Exocrine glands
    function, 106, 122
    vs. endocrine glands, 384
Exocytosis
    defined, 66, 66fig
    neurotransmitter release by, 328
Experimental method. See Hypothesis testing
Experiments
    controlled, 8–10, 8fig, 9fig
    natural, 478–479, 478fig
Expiration, 234–235
Exponential growth
    defined, 564, 565fig, 580
    and populations, 564–565
External respiration, 231fig, 234–240, 234fig, 246
External urethral sphincter, 254
Externoreceptors, 354, 354t
Extinction rates, 567, 568fig, 570, 580
Extinctions
    causes of, 568–570, 580
    consequences of, 571–574
    preventing, 575–578, 579t
Extraembryonic membranes, 515, 516fig, 517, 533
Eye
    focusing by, 369–372, 371fig, 372fig, 378
    photoreceptors in, 372–373
    structure, 369–370, 370fig

F

Facilitated diffusion, 65, 65fig
Falsifiability, 3
Fascia, 142, 142fig
Fascicles, 142, 142fig, 151
Fast pain, 360
Fats
    breakdown of, 56fig, 73, 162fig
    as energy source, 40, 42, 50
    and the liver, 167
    storage of, 56–57, 57fig, 110–111, 110fig
    structure, 40–42, 41fig, 59
Fat-soluble vitamins, 45–46, 46t
Fatty acids, 40–42, 41fig, 42fig, 50
Feces, 163, 252
Feedback. See also Homeostasis; Negative feedback; Positive feedback
    defined, 118
Female reproductive system, 442–443, 442fig, 462
Femur, 136, 136fig
Fermentation
    and anaerobic respiration, 73–74, 74fig
    and muscle contraction, 146
Fertility. See also Infertility
    awareness, 457, 457fig, 458fig
    safeguarding, 460
Fertility control. See Birth control
Fertilization
    and menstrual cycle, 450–451, 462
    in natural vs. assisted reproduction, 513
    process of, 513–514, 514fig
    random, 472
    and spermatogenesis problems, 446
Fetal alcohol syndrome, 523

Fetal circulation, 522, 523fig
Fetal development, 521–524, 521fig, 533–534
Fetal hemoglobin, 522
Fetal stem cells
    characteristics, 512–513, 533
    potential uses for, 523–524
Fetus
    defined, 513
    ossification in, 130
Fever, 287
Fiber, dietary, 37
Fibrillation, 216
Fibrin, 190, 190fig, 194
Fibrinogen, 190
Fibroblasts, 108, 122, 133
Fibrous connective tissue, 110fig
Fibula, 136, 136fig
Filtration
    by the kidney, 252–259
    by the liver, 167–168
Fimbria, 443
Fitness (in natural selection), 551, 558
Fitness waters, 27, 45, 47–49
Flagella, bacterial, 275, 275fig
Flexion, 139, 139fig
Flu. See Influenza
Follicles, ovarian, 446, 450, 451fig
Follicle-stimulating hormone (FSH), 390–391, 444–445, 446fig, 462, 527
Food chain, 569–570, 573–574, 580
Food guide pyramid (USDA), 77fig
Food web, 571–574, 572–573fig
Foramen ovale, 522
Forests and shrublands, 588, 588fig, 596
Formed elements (of blood), 180–184, 180fig
Fossil fuels
    environmental impact, 592, 592fig
    and global warming, 589, 594
Fossilization, 546–547, 546fig
Fossil record
    and extinction rates, 567
    and theory of common descent, 546–549, 558
Founder effect, defined, 556, 558
Four-stage model (Masters and Johnson), 452, 452fig
Fovea, 373
Frameshift mutations, 91, 91fig
Free radicals, 47, 51, 408, 529
Freshwater biomes, 590, 596
Frontal bone, 134, 134fig
Frontal lobe, 336fig, 337, 337fig
FSH (follicle-stimulating hormone), 390–391, 444–445, 446fig, 449–450, 449fig, 462

G

Gallbladder, 168–169, 173
Gamete production. See Gametogenesis
Gametes
    chromosomes in, 422–424, 424fig, 443, 470–472
    defined, 440
    and meiosis, 424–431, 431fig, 443–444, 470–472
    variation in, 425, 470–472, 471fig
Gametogenesis
    and meiosis, 422–424, 443–444, 470–472
    and oogenesis, 446–447, 447fig, 449–450
    overview, 443–444, 462
    and spermatogenesis, 444–446, 445fig
Ganglia, 341

Ganglion cells, 373–374, 374fig
Gap junctions, 107, 122, 515
Gas exchange
    and blood, 239–241, 240fig, 246
    and breathing, 234–238, 236fig
    in the lungs, 238–240, 246
    and smoking, 240–241
Gas pressure, measurement of, 238
Gastric bypass surgery, 165–166, 165fig, 171
Gastric juice, 160
Gastrin, 396
Gastrula/gastrulation, 517, 517fig, 518–519fig
Gel electrophoresis, 502, 502fig
Gene cloning, 93–96, 95fig
Gene expression
    meaning of, 84
    and organogenesis, 519–521
    regulation of, 37, 92–93, 92fig, 93fig, 101
General senses, 354, 356–360, 378
Genes. *See also* Alleles; Chromosomes; DNA (deoxyribonucleic acid); Protein synthesis
    and aging, 529
    and genotypes/phenotypes, 472, 472fig
    and heritability, 476–481, 480fig, 481fig, 483
    linked, 428, 494–500, 497fig
    and quantitative traits, 476–477
    and single-gene (qualitative) traits, 472–476
    and theory of common descent, 545, 545fig
Gene slicing, 94, 94fig, 95fig
Genetically modified organisms (GMOs)
    benefits of, 97–98, 101, 101fig
    debate regarding, 83, 96, 98–99, 99fig, 101
    defined, 95
    examples of, 97–98, 97fig, 98fig
    and U.S. Department of Agriculture, 96, 98–99
Genetic code, 88, 90fig, 100
Genetic crosses
    dihybrid, 491–493, 492fig
    and hemophilia, 497fig
    monohybrid, 489
Genetic disease
    and single-gene traits, 473–476
    vs. infectious disease, 275
Genetic drift, 556, 558, 577, 581
Genetic engineering
    and evolution, 552
    and protein synthesis, 93–96, 101
Genetics
    and correlation, 477–479
    and pedigree analysis, 498–499, 498fig, 500fig
    and probability, 481
    and Punnett square analysis, 474–476, 474fig, 475fig, 482, 491, 492fig
Genetics, Mendelian. *See* Mendelism
Genetic variation
    in human populations, 554–557
    and meiosis, 425, 428–430, 428fig, 430fig, 468–472, 492–493, 493fig
    and natural selection, 4, 550–552, 558
    and population size, 556, 558, 577–578, 581
    and Punnett square analysis, 474–476, 474fig, 475fig, 482
    and quantitative traits, 476–477
    and sexual reproduction, 440
    and single-gene (qualitative) traits, 472–474
Genital herpes, 308
Genitalia. *See* Reproductive organs

Genital ulcer diseases, 308
Genital warts, 303, 306t, 309–310, 309fig
Genome
    defined, 84, 100
    of viruses, 277–278
Genotypes
    and ABO blood system, 490, 490fig
    defined, 472, 472fig
    and Punnett square analysis, 474–476, 474fig, 475fig, 482
    quantitative traits in, 476–477
    single-gene (qualitative) traits in, 472–474
Germ cells, 519
Germ layers, 517
Germ theory, 3–4
Gerontology, 530
Glands
    endocrine, 384–398, 396–397t
    exocrine, 384
    function, 106, 122
Glans penis, 440
Glaucoma, 370
Glenoid cavity, 137, 137fig
Global climate change, 594–595, 594fig, 596
Global warming, 589, 594–595, 595fig
Globulins, 181
Glomerular capsule, 257, 257fig
Glomerular filtration, 256fig, 257, 257fig, 267
Glomerulus, 256fig, 257, 257fig
Glucagon, 169, 169fig, 391–392, 391fig
Glucocorticoids
    effect on thymus, 395
    and melatonin, 394, 395fig
    and prostaglandins, 398
    and stress response, 387, 389, 389fig, 392
Glucose breakdown, 69–70, 70fig
Glucose homeostasis, 167, 169, 169fig, 387, 389, 391–392, 391fig
Glycerol, 40–42, 42, 42fig, 56fig
Glycogen
    and blood-glucose homeostasis, 167, 391–392, 391fig
    food energy stored as, 36, 37fig, 56, 59
    in sarcoplasm, 142
Glycolysis, 69–70, 70fig, 73, 79
GnRH (gonadotropin-releasing hormone), 390–391, 444–445, 446fig, 449–450, 449fig, 462, 527
Golgi apparatus, features of, 62, 62fig, 79
Golgi tendon organs, 357
Gonadotropin-releasing hormone (GnRH), 390–391, 444–445, 446fig, 449–450, 449fig, 462, 527
Gonads
    defined, 440
    development of, 519–521, 520fig
    and endocrine glands, 390–391
    hormones secreted by, 390–391, 396t, 441, 443–445, 446fig, 449–451, 449fig, 451fig
    maturation of, 527
Gonorrhea, 305t, 306–307
Gout, 255fig
Granulocytes, 183, 183t
Grasslands, 588–589, 589fig, 596
Graves' disease, 394, 394fig
Gray matter, 325, 335
Greenhouse effect, 594–595, 594fig
Ground substance, 108, 122
Growth factors, 398, 416, 417

Growth hormone (GH), 93, 389–390
Growth plate, 130
Gustatory receptors. *See* Taste receptors

H

HAART (highly active antiretroviral therapy), 313–315, 314t
Habitat. *See also* Human habitats
    destruction, 568–569, 588–591
    fragmentation, 569–570
    preservation, 575–577
Hair cells. *See* Auditory receptors
Haploid cells, 423–424, 424fig
Hard palate, 157, 158fig
HCG (human chorionic gonadotropin), 451, 516, 533
HDLs (high-density lipoproteins), 43, 243
Hearing
    and the ear, 364–367, 365fig, 366fig
    and perception, 366–367
    and temporal lobe, 336
Heart
    and anorexia, 75
    circulation, 217, 217fig, 221
    cycle, 214–217, 214fig, 215fig
    hormones produced by, 262, 396
    repair of, 211–213, 213fig, 221
    structure and function, 199, 209–210, 209fig, 221
Heart attack, 217
Heart disease, 199, 208t, 211–213, 216–219
Heart failure, 213
Heart-lung machine, 213fig
Heart rate, 120, 120fig, 200, 202fig, 214–217
Helper T cells. *See* T4 cells
Hematocrit (defined), 186
Heme group, 182
Hemoglobin
    affinity for carbon monoxide, 240–241, 246
    as blood buffer, 263
    degradation of, 167, 189, 194, 255
    fetal, 522
    as oxygen transporter, 182, 182fig, 239, 239fig, 246
    and sickle cell anemia, 90, 191
Hemolytic disease, 188
Hemophilia
    as clotting disorder, 190, 491
    as sex-linked, 497, 497fig, 500fig
Hemostasis, 189–190
Hepatic portal system, 210–211, 212fig
Hepatitis B virus (HBV), 305t, 308–309
Heredity *vs.* environment, 476–481, 483, 529
Heritability
    and genes, 476–481, 480fig, 481fig, 483
    and IQ (intelligence quotient), 477–479, 478t
Herpes simplex viruses (HSV), 305t, 308, 308fig
Heterozygous genotype, defined, 472
High-density lipoproteins (HDLs), 43, 243
Highly active antiretroviral therapy (HAART), 313–315, 314t
Hinge joints, 139, 139fig
Hippocampus, 340, 340fig, 347
Histamine, 286, 289
HIV/AIDS
    effect on T4 cells, 294, 310–313, 312fig
    and HIV evolution, 313, 313fig

and HIV positive designation, 311
and HIV reproductive cycle, 311, 312*fig*
immune response to, 311–313, 314*fig*, 318
prevention of, 315–316, 318
transmission of, 306*t*, 315, 315*fig*
treatment of, 313–315, 314*t*, 318
"Hole in the heart," 211–212
Homeostasis
of blood gases, 186, 187*fig*, 236, 236*fig*
and blood glucose, 167, 169, 169*fig*, 387, 389, 391–392, 391*fig*
and blood pH, 34, 180, 236, 236*fig*, 246, 252
and bone remodeling, 131–132, 132*fig*, 151
and calcium, 132, 392, 393*fig*
and capillary blood flow, 203
and endocrine system, 387–389, 391–392, 391*fig*
and erythrocyte production, 186, 187*fig*, 252, 387
and internoreceptors, 354, 354*t*
and lymphocyte testing, 291
as pooled effort, 118–119
and temperature regulation, 119, 342, 347
and urinary system, 252, 263, 263*fig*, 268
Hominins
and bipedalism, 547, 547*fig*
evolution of, 554–557, 558
fossil record, 547–549, 549*fig*
pedigree, 548, 548*fig*
Homologous pairs. *See also* Chromosomes
and crossing over, 425, 426*fig*, 428, 428*fig*
described, 422*fig*, 423, 423–424
independent assortment of, 470–472, 471*fig*, 492–493, 493*fig*
random alignment of, 425, 426*fig*, 428–430, 429*fig*, 430*fig*, 492–493, 493*fig*
segregation of, 470–472
Homozygous genotypes
defined, 472
and small populations, 577–578
Hormonal cycles, 394, 395*fig*
Hormones. *See also* Sex hormones
as chemical messengers, 106, 384–385, 384*fig*, 385*fig*
as contraceptives, 456, 456*fig*
of endocrine glands, 384–395, 396–397*t*
and endocrine-disrupting chemicals, 448, 521
fetal, 524
of other endocrine tissues, 396–398, 397*fig*
placental, 521–522, 524–525, 533
Human body
elements in, 28*t*
levels of organization, 118, 118*fig*
organs and organ systems, 117*t*
Human chorionic gonadotropin (HCG), 451, 516, 533
Human evolution, 554–557, 558
Human habitats. *See also* Habitat
and climate change, 594–595, 596
and energy use, 591–592
and natural resources, 592, 596
and waste production, 592–593, 596
Human immunodeficiency virus(HIV). *See* HIV/AIDS
Human papillomavirus (HPV), 309–310
Human population
and exponential growth, 564–565, 565*fig*
growth time lag, 566–567, 566*fig*
The Human Sexual Response (Masters and Johnson), 452

Humerus, 137, 137*fig*
Humoral immunity, 292, 292*fig*
Huntington's disease, 473, 473*fig*
Hydrogenation, 42, 42*fig*
Hydrogen bonds
in DNA molecule, 43–44, 44*fig*, 410–412
in fats, 40–42, 41*fig*
properties of, 31, 32*t*
in protein structure, 39–40, 39*fig*
in transcription, 86
in water molecule, 31–32, 32*t*, 33*fig*, 50
Hydrogen ions
kidney removal of, 259, 263, 263*fig*
and pH scale, 34, 50
in water molecule, 33-34, 34*fig*, 50
Hydrolysis
of creatine phosphate, 146
in nutrient breakdown, 35, 35*fig*, 162*fig*
Hydrophilic/hydrophobic molecules
defined, 31
in lipids, 40, 42, 42*fig*, 50, 64
and passive transport, 64–65
in vitamins, 45–46
Hyoid bone, 135, 135*fig*
Hyperopia, 370, 372*fig*
Hypertension, 206–207, 221
Hyperthyroidism/hypothyroidism, 394, 394*fig*
Hypertonic solution, 65, 65*fig*, 79
Hypertrophy, 113, 122
Hypodermis, 148
Hyponatremia, 263–264
Hypothalamus
as control center, 337, 346, 387–389, 388*fig*, 393, 396*t*
and gonads, 390–391, 444–445, 446*fig*, 449–450, 462, 527
Hypothalamus-pituitary-adrenal axis, 387–389, 389*fig*
Hypothalamus-pituitary-thyroid axis, 393
Hypotheses, 2–5, 2*fig*, 21
Hypothesis testing
and correlation, 10–12, 21
and experimental design, 8–10, 21
and intelligent design, 541, 553
logic of, 4–7, 6*fig*, 21
and Romanov remains, 505*fig*
and special creation, 541
Hypotonic solution, 65–66, 65*fig*, 79

I

Ileum, 161
Ilium, 136, 136*fig*, 137*fig*
Immortal cells, 418
Immune system. *See also* Lymphatic system
and aging, 409
and antigens, 187, 287–291, 288*fig*, 290*fig*
cell types, 295*t*
HIV/AIDS as disease of, 310–313, 312*fig*
and malaria, 183–186
and melatonin levels, 394–395
nonspecific response, 183–184, 183*t*, 284–287, 294*fig*, 311
specific response, 287–294, 295*fig*
and thymus, 395, 396*fig*
and transfusion reaction, 187, 187*fig*
Immunity, 289, 292–294
Immunization, 192, 309–310, 309*fig*

Immunoglobulins, 288–289
Implantation, 516, 516*fig*
Incomplete dominance, 489, 489*fig*
Independent assortment (in meiosis), 470–472, 471*fig*, 492–493, 493*fig*
Independent/dependent variables, 7–8, 11–12, 11*fig*
Indifferent gonads, 519
Induced fit, 68, 79
Inductive reasoning, 5, 21
Infancy (as developmental stage), 526–527
Infectious agents
transmission of, 283–284, 283*fig*, 297
types of, 274–283, 297
Infectious *vs.* genetic disease, 275
Inferior vena cava, 210
Infertility
female, 443, 450, 451–452
male, 442, 446, 453
Inflammation
as nonspecific immune response, 286, 286*fig*
and prostaglandins, 360, 398
as tobacco smoke effect, 233, 237–238
Influenza, 293–294
Information processing
and chemical senses, 361–363
and general senses, 356–360
and light, 373–376, 374*fig*, 375*fig*, 379
and sensory receptors, 355–356, 356*fig*
and sound, 366–367
Inhalation, 234
Inheritance
and pedigree analysis, 498–499, 498*fig*, 499*fig*, 500*fig*
of quantitative traits, 476–481
and sex-linked genes, 494–500
of single-gene (qualitative) traits, 468–472, 482
Inhibiting hormones, 388
Inner cell mass (of blastocyst), 515
Inner ear, 364, 365*fig*, 366*fig*, 367–368, 368*fig*
Insertion (of a muscle), 140
Inspiration (respiratory), 234
Insulin, 169–170, 391–392, 391*fig*, 522
Insulin-dependent diabetes mellitus (IDDM), 169–170, 291
Integumentary system, 117*t*, 146–149, 151
Intelligence quotient (IQ) and heritability, 477–479, 478*t*
Intelligent design hypothesis, 541, 552, 553
Intercellular junctions, 107, 109*fig*, 122
Intercostal muscles, 234
Interferons, 286
Internal respiration, 234, 234*fig*, 239–240, 239*fig*, 246
Internal urethral sphincter, 253
Interneurons
function, 324–325, 325*fig*, 346
in reflex arc, 342, 342*fig*
as spinal cord components, 335, 335*fig*
Internoreceptors, 354, 354*t*, 356–357
Interphase, 413, 413*fig*, 414–415*fig*, 424–425, 424*fig*
Intervertebral disks, 135, 135*fig*
Intrauterine devices (IUDs), 457, 457*fig*
Introduced species, 570, 574
In vitro fertilization (IVF), 513, 515, 516
Involuntary muscles, 113, 122
Ionic bonds, 29–30, 30*fig*, 32*t*, 34, 40
Ions (defined), 29

IQ (intelligence quotient) and heritability, 477–479, 478*t*
Iris, 370, 370*fig*
Ischium, 136, 136*fig*
Isotonic solution, 65–66, 65*fig*, 79
Isotopes (defined), 29
IUCN (World Conservation Union), 567, 568, 569, 570

J

Jaundice, 189
Jejunum, 161
Johnson, Virginia, 452
Joints and movement, 138–140
"Jury-rigged design," 552

K

Karyotype analysis, 422*fig*, 423, 494
Keystone species, 573, 573*fig*
Kidneys
 and blood pH, 252, 263, 263*fig*
 and blood pressure regulation, 205–206, 260*fig*, 261
 erythropoietin release by, 252
 and excretion, 255–259
 failure of, 264–265, 268
 and salt balance, 263–264
 structure, 253–254, 253*fig*
 and water balance regulation, 259–262, 261*fig*
Kidney stones, 265
Klamath Falls crisis, 563–564, 567, 570, 578

L

Labia majora, 442
Labioscrotal swellings, 519
Labor, 525, 525*fig*
Lactation, 526, 527*fig*
Lactic acid, 73–74, 74*fig*
Lactose, 36
Lacunae, 128, 129*fig*
Lakes and ponds, 590
Landfills, 593, 593*fig*
Laparoscope, 420
Large intestine, 163–165, 164*fig*
Laryngitis, 233
Larynx, 228*fig*, 229, 229*fig*
LASIK surgery, 370
Latent viruses, 279
Lateral geniculate nucleus, 374, 374*fig*
LDLs (low-density lipoproteins), 11–12, 11*fig*, 15–16, 43, 243, 392
Lens, 370–371, 370*fig*, 372*fig*, 378
Leptin, 74, 392
Leukocytes. *See* White blood cells
Leydig cells, 441
Ligaments, 128
Light
 characteristics of, 369, 370*fig*
 focusing of, 369–372, 371*fig*, 372*fig*, 378
 and perception, 373–376, 374*fig*, 379
 and pineal gland, 394
 receptors for, 372–375, 373*fig*
 and temporal lobe, 336
Limbic system, 340, 340*fig*, 347
Linked genes, 428
Linnaeus, Carolus, 542, 554, 558

Lipids
 digestion of, 158, 161, 169
 structure and types of, 40–43
Lipoproteins, 11–12, 11*fig*, 15–16, 43, 243, 392
Liver
 and blood glucose, 391
 and cholesterol, 43
 and insulin-type growth factor (IGF), 398
 structure and function, 167–168, 167*fig*, 168*fig*, 173
 toxin metabolism by, 168
Liver disease, 308–309
Logistic growth, 565, 565*fig*
Long bones, 129*fig*
Loose connective tissue, 108–109, 110*fig*, 122, 148
Low-density lipoproteins (LDLs), 11–12, 11*fig*, 15–16, 43, 243, 392
Lower respiratory tract, 229–230, 229*fig*
Lungs
 and breathing, 234–235, 235*fig*
 gas exchange in, 238–240, 246
 structure, 230, 231*fig*
 tobacco smoke effect on, 233, 234*fig*, 237–238, 237*fig*
Luteinizing hormone (LH), 390–391, 444–445, 446*fig*, 527
Lymph, 205
Lymphatic system, 200, 204–205, 205*fig*, 221, 287*fig*. *See also* Cardiovascular system; Immune system
Lymph ducts, 20, 205, 205*fig*
Lymph nodes, 205*fig*, 287*fig*
Lymphocytes, 287–294
Lysosomes, 61, 61*fig*, 78

M

Macromolecules
 energy storage in, 56–57, 56*fig*
 metabolism of, 69–74, 69*fig*, 73*fig*
 structure and types of, 35–45, 35*fig*, 50
Macrophages, 183, 183*t*, 285, 287, 294, 294*fig*
Macular degeneration, 373
Mad cow disease. *See* Spongiform encephalopathies
Major histocompatibility complex (MHC) proteins, 288
Malaria
 and anemia, 186
 and blood type, 188–189
 body's response to, 182–186, 186*t*, 189, 191, 193
 cerebral, 184, 191
 disease history, 184–185
 efforts to control, 179–180, 192, 194
 forms of, 186*t*
 as vector-borne disease, 180, 185*fig*
Male reproductive system, 440–442, 442*fig*, 462
Mammography, 419*fig*
Mandible, 134, 134*fig*
Marine biomes, 590–591, 596
Mass extinctions, 567, 580
Mass number (defined), 29
Mast cells, 286
Masters, William H., 452
Matrix, in connective tissues, 108, 122
Maturation, 526–528
Maxillae, 134, 134*fig*
Mechanical digestion, 156–158, 160, 173
Mechanoreceptors, 354, 354*t*, 356–359, 357*fig*, 358*fig*, 364–369, 365*fig*

Medulla oblongata, 236, 338, 338*fig*
Medullary cavity, 129, 129*fig*
Megakaryocytes, and platelets, 184
Meiosis
 anomalies in, 495
 in fertilization process, 514
 and gametogenesis, 422–424, 443–447, 445*fig*, 447*fig*, 470–472
 and genetic variation, 425, 428–430, 428*fig*, 430*fig*, 468–472
 stages of, 425–430, 426–427*fig*
 vs. mitosis, 431*fig*
Meissner's corpuscles, 358, 358*fig*
Melanocytes, 147, 151
Melatonin, 394–395, 395*fig*
Membrane, plasma. *See* Plasma membrane
Membranes, body, 115–116, 116*fig*, 122
Memory and limbic system, 340, 347
Memory cells (in immune response), 292–294, 293*fig*
Menarche, 527, 528*fig*
Mendel, Gregor, 472, 472*fig*, 550
Mendelism
 extensions of, 488–491, 506
 and single-gene traits, 472–476
Meningeal membranes, 116, 122
Meninges, 334
Menstrual cycle, 450–452, 451*fig*, 462
Menstruation
 and anorexia, 74
 first occurrence of, 527, 528*fig*
Merkel disks, 358, 358*fig*
Mesoderm, 517
Messenger RNA (mRNA), 86–87, 86*fig*, 89*fig*, 92, 92*fig*, 100
Messengers (in signal transduction), 384
Metabolic rate, 74, 79, 393
Metabolic reactions, 67–68, 68*fig*, 79
Metabolic waste excretion, 250, 252, 255–259
Metabolism
 defined, 67, 79
 of macromolecules, 69–74, 69*fig*, 73*fig*
Metacarpals, 137*fig*, 138
Metaphase
 in meiosis, 425, 426–427*fig*, 431*fig*
 in mitosis, 413, 414–415*fig*, 431*fig*
Metastasis, 407, 407*t*
Metatarsals, 136*fig*, 137
Microbes (defined), 274
Microglia, 114
Micronutrients, 45–49, 46, 46*t*, 47*t*, 51
Microscopes, 59
Microtubules, 62, 413, 414–415*fig*, 425, 426–427*fig*
Microvilli, 162
Micturition reflex, 254–255*fig*
Midbrain, 338, 338*fig*
Middle ear, 364, 365*fig*
Mifepristone, 459
Mineralocorticoids, 387
Minerals
 in bones, 128, 129, 151
 function, 47–48, 47*t*, 48*t*, 51
Missense mutations, 90, 90*fig*
Mitochondria
 in cellular respiration, 69*fig*, 70–71, 71*fig*, 72*fig*, 73*fig*, 79
 features of, 61, 61*fig*, 78

in muscle fibers (cells), 145–146
in sperm, 44
Mitosis
  regulation of, 416–418
  stages of, 413–416, 414–415 fig
  vs. meiosis, 431 fig
Mixed nerves, 341
Model organisms, 10–11, 10 fig, 21
Molecules (defined), 29
Monocytes, 183–185, 183 t
Monohybrid crosses, 489
Monomers (defined), 35
Monosaccharides, 36, 36 fig, 37 t, 50, 56 fig
Monozygotic twins, 515
Morphogenesis, 515
Morula, 515, 517 fig
Motor neurons
  function, 142–143, 143 fig, 324, 325 fig, 335, 346
  in reflex arc, 342, 342 fig
Motor unit, 142, 143 fig
Mouth, structure and function, 157–158, 158 fig
Movement and joints, 138–139
Mucous membranes
  function and types, 116, 122
  as nonspecific defense, 184–185
  in respiratory system, 229–230
  tobacco smoke effect on, 232
Mucus escalator, 230
Mullerian duct, 519, 520 fig
Multiple allelism, 490
Multiple hit model, 418–419
Multiple sclerosis, 291
Multipotent stem cells, 512, 528
Muscle fibers (cells)
  and fermentation, 73–74
  as proprioceptors, 356
  structure and function, 142, 142 fig, 145, 151
Muscle spindles, 356, 357 fig
Muscle tissues, 113, 114 fig, 122
Muscular dystrophy, 146
Mutations
  and aging, 529
  and antibiotic resistance, 276
  and cell cycle regulation, 416–419
  dominant/recessive, 472–476
  and genetic variation, 469–472, 469 fig
  and natural selection, 551–552
  and protein synthesis, 88–91, 90 fig, 101
  types of, 90–91, 90 fig, 91 fig
Mutualism, 571, 574 t
Myelin sheath, 325
Myocardial infarction (MI), 217
Myocardium. See Cardiac muscle
Myofibrils, 142, 142 fig, 151
Myoglobin, 142
Myopia, 370, 372 fig
Myosin, 113, 142, 142 fig, 144 fig, 151

N

Nails, 147
Nasal bone, 134, 134 fig
Nasal cavity, 228–229, 228 fig
Natural experiments, 478–479, 478 fig
Natural killer cells, 286
Natural selection
  and human populations, 554–555, 555 fig, 558
  theory of, 4, 550–552, 558

Negative feedback
  and adrenal hormones, 389
  and blood pressure regulation, 214–215, 221, 260, 260 fig
  in breathing control, 236 fig
  described, 118–119, 119 fig
  and erythropoietin release, 186, 187 fig, 193
  in kidneys, 260, 260 fig
  and menstrual cycle, 450–451, 451 fig
  and ovarian cycle, 449, 449 fig
  and spermatogenesis, 445
  and thyroid hormones, 392–393, 393 fig
Neonatal development, 526
Nephron loop, 256 fig, 257, 261–262, 261 fig
Nephrons, 253, 253 fig, 256–259, 256 fig, 257 fig.258 fig
Nerve cells. See Neurons
Nerve fibers. See Axons
Nerve impulse transmission
  along axons, 325–330, 326 fig, 327 fig, 346
  between neurons, 328–330, 329 fig
Nerves, 325, 341, 341 fig
Nerve tracts, 325, 335
Nervous system. See also Central nervous system; Peripheral nervous system
  and blood pressure control, 205–206
  divisions of, 324, 324 fig
  effects of aging on, 530
  and endocrine system, 386–389
  and recreational drugs, 332–334 t
Nervous system tissues, 114–115, 115 fig, 122, 324–334, 346
Neural tube, 518–519, 519 fig
Neuroendocrine system, 386
Neuroglia, 114, 115 fig, 324, 325
Neuromuscular junction, 142, 143 fig
Neurons
  adaptability of, 528
  in brain, 336
  presynaptic/postsynaptic, 328, 329 fig, 330–331, 331 fig
  in reflex arc, 342
  signal conduction by, 325–329, 327 fig
  signal integration by, 329–330, 330 fig, 346
  in spinal cord, 335
  structure and function, 114, 115 fig, 324–325, 325 fig, 346
Neurotransmitters
  and attention deficit disorder (ADD), 330–331, 331 fig
  in auditory receptors, 365
  and disease, 328–330
  function, 114, 328, 329 fig, 346
  in muscle fibers (cells), 142, 143 fig
  and Parkinson's disease (PD), 531
  in taste cells, 361
Neutral mutations, 91, 91 fig
Neutrons, 28–29
Neutrophils, 183, 183 t, 285
New-variant Creutzfeld-Jakob disease (nvCJD), 274, 283
Nicotinamide adenine dinucleotide (NAD), in glycolysis, 69–71, 70 fig, 71 fig
Nicotine, 241–244, 241 fig, 242 fig, 246
Nitrogenous bases
  in nucleotides, 43–44, 44 fig, 50, 410, 410 fig
  in protein synthesis, 85–88
Nociceptors. See Pain receptors

Nodes of Ranvier, 325, 325 fig
Non-insulin dependent diabetes mellitus (NIDDM), 170–171, 392
Nonpolar molecules (defined), 31
Nonsense mutations, 90
Nonspecific immune response
  and HIV/AIDS, 311
  types of, 284–287
  white blood cells in, 183–184, 183 t, 285–286
Noradrenaline, 386
Norepinephrine, 386, 388
Nose, 228–229, 228 fig
Nuclear envelope
  in cell division, 413, 414–415 fig, 425, 426–427 fig
  as cell structure, 60, 60 fig
Nuclear pores, 60, 60 fig
Nucleases, 92
Nucleic acids, 43–45, 50
Nucleoid region, 275
Nucleolus, 60, 60 fig
Nucleoplasm, 60, 60 fig
Nucleotides, 43–44, 44 fig, 57, 85–86, 410–411, 410 fig, 500–502
Nucleus (in atom), 28–29
Nucleus, cell. See Cell nucleus
Nutrient absorption and vitamins, 45, 46 t
Nutrient breakdown
  in digestion, 56 fig, 162 fig
  as enzyme-catalyzed, 67–68, 68 fig, 158, 160, 162
  and subcellular structures, 61
Nutrient cycling, 573–574, 575 fig, 581

O

Obesity
  and cancer, 409
  and diabetes, 170–171
  effects of, 75, 155
  and genes, 166
Observation and inductive reasoning, 5, 7
Observer bias, 9
Occipital bone, 134, 134 fig
Occipital lobe, 336, 336 fig
Odorants, 362–363
Olfactory bulb, 363
Olfactory receptors, 362–363, 362 fig
Oligodendrocytes, 114
Omega-3 fatty acid, 40–41
Omega-6 fatty acid, 40
Oncogenes, 417
Oocytes, 446, 447 fig, 449–450
Oogenesis, 446, 447 fig, 449–450, 462
Opportunistic infections, 310
Optic chiasm, 374
Optic nerves, 370 fig, 373–375, 374 fig
Oral cavity, structure and function, 157–158, 158 fig
Oral contraceptives, 456, 456 fig
Oral herpes, 308, 308 fig
Organelles, 60–63, 61 fig, 63 fig, 78
Organ formation, 517–521, 518 fig, 533
Organic chemistry (defined), 35, 50
Organogenesis, 517–521, 518 fig, 531, 533
Organs and organ systems
  development of, 517–521, 518 fig
  effects of aging on, 529–530
  and feedback, 118–119
  maintaining fitness of, 119–120

Organs and organ systems (cont.)
  overview, 116–118, 117t, 122
  tobacco smoke effect on, 232–234, 237–238, 240–243
Orgasm, 452
Origin (of muscle), 10
On the Origin of Species (Charles Darwin), 541, 551, 553
Osmosis
  in capillary beds, 203, 203fig, 204
  in kidneys, 258fig, 259, 260–262, 261fig
  and membranes, 65–66, 65fig, 79
Ossicles, 364, 365fig
Ossification, in bone development, 130–131, 130fig
Osteoblasts, 130, 131, 132, 133, 151, 393fig
Osteoclasts, 131, 132, 133, 393fig
Osteocytes, 112, 122, 128, 129fig, 151
Osteoporosis, 75, 112, 132, 132fig
Otitis media, 233
Oval window, 364, 365fig
Ovarian cycle, 446–447, 449
Ovarian follicles, 446, 450, 451fig
Ovaries
  and the menstrual cycle, 450–452, 451fig
  and oogenesis, 446, 447fig, 449–450
  and reproductive anatomy, 440, 443
Overexploitation, 570
Oviducts, 443
Ovulation, 446, 450–452, 451fig
Ovum, 514
Oxcytocin, 524–525, 526
Oxygen
  atomic structure, 29fig
  and the heart, 217–218
  hemoglobin transport of, 182, 182fig, 239, 239fig, 246
Oxygen debt, 73–74
Oxyhemoglobin, 182
Oxytocin, 387–388, 524–525

P

Pacemakers, 216, 216fig
Pacinian corpuscles, 358, 358fig, 359fig
Pain receptors, 354, 354t, 360, 360fig, 369, 378
Pain types, 360
Pancreas
  and blood glucose regulation, 169–170, 169fig, 391–392, 391fig, 397t
  and diabetes, 169–170, 170fig
  digestive secretions, 160, 169, 173
Pancreatic islets, 391–392, 391fig
Papillae, 361, 361fig
Parasites
  defined, 180, 275
  viruses as, 277, 297
Parasites, malaria, 185, 186t
Parasympathetic nervous system, 342, 343fig, 347
Parathyroid hormone (PTH), 132, 392, 393fig, 397t
Parietal bones, 134, 134fig
Parietal lobe, 336–337, 336fig, 337fig
Parkinson's disease (PD), 328–329, 511–512, 531, 534
Partial birth abortion, 459
Partial pressure (defined), 238
Parturition, 524–526, 525fig
Passive immunity, 289
Passive transport, 64–66, 64fig, 79

Patella, 136, 136fig
Pathogens (defined), 275
Pattern recognition, 375–376, 376fig
Pectoral girdle, 137–138, 137fig
Pedigree analysis, 498–499, 498fig, 499fig, 500fig, 504fig, 506
Peer review, 17, 18fig
Pelvic girdle, 136–137, 136fig
Pelvic inflammatory disease (PID), 307
Penis, 440
Pepsin, 160
Pepsinogen, 160
Peptide bonds, in protein structure, 38–40, 38fig
Perception
  and light, 373–376, 374fig, 379
  and odors, 363
  and pain, 360, 361fig
  process of, 355, 356fig
  and sound, 366–367
  tools to enhance, 376
  and triangulation, 367, 367fig, 374
Pericardium, 116, 116fig, 122, 209, 209fig
Periosteum, 129, 129fig, 130, 151
Peripheral nervous system (PNS), 341–343, 347. See also Central nervous system; Nervous system
Peristalsis, 159–160, 165, 173
Peritoneum, 116, 116fig, 122
Permafrost, 589
Permeability (of membranes), 64–65, 64fig, 79
pH. See Blood pH
Phagocytes, 285, 285fig
Phalanges (foot), 136fig, 137
Phalanges (hand), 137fig, 138
Pharynx, 159, 159fig, 173, 228fig, 229, 229fig
Phenotypes
  and ABO blood system, 490, 490fig
  defined, 472, 472fig
  effect of environment on, 476–481, 476fig, 483
  and Punnett square analysis, 474–476, 474fig, 475fig, 482
  quantitative traits in, 476–477
  and sex-linked genes, 494–500
  single-gene (qualitative) traits in, 472–474
Phenotypic ratios, 491, 492fig
Phospholipid bilayer, 64–65, 64fig
Phospholipids
  in plasma membrane, 64, 64fig
  structure, 42, 42fig, 50
Phosphorylation
  and electron transport chain, 72fig, 79
  and energy transfer, 58, 58fig
  in muscle contraction, 144–146, 145fig
Photochemicals, 372–373
Photoreceptors, 354, 354t, 372–374, 374fig
pH scale, 33–34, 34fig, 50, 50fig
Pili, bacterial, 275, 275fig
Pineal gland, 394, 395fig, 397t
Pinna, 364, 365fig, 367
Pituitary gland
  and gonads, 390–391, 444–445, 446fig, 449–450, 449fig
  and growth regulation, 389–390
  hormones secreted by, 393, 396t
  and stress response, 387–389, 388fig
Pivot joints, 139, 139fig
Placebos, 8, 9
Placenta, 521–522, 524, 526, 533
Plasma, blood, 180–181, 180fig, 181fig, 193

Plasma cells, 292
Plasma membrane
  and hormones, 384
  in muscle cells, 142, 143fig, 151
  in sensory receptors, 355
  structure and function, 64–66, 64fig, 66fig, 79, 79fig
  in taste cells, 361
Plasmids, bacterial
  DNA in, 274–275, 275fig
  and gene cloning, 94–96, 95fig, 101
Plasmodium falciparum, 185, 185fig, 191, 192
Plasmodium species, 179, 180, 186t
Plasmodium vivax, 188–189
Platelets, blood, 184, 189–191, 190fig, 194
Pleiotropy, 491
Pleural membranes, 116, 116fig, 122, 230
Pluripotent stem cells, 512, 528
Pneumonia, 233
PNS. See Peripheral nervous system
Point mutations, 90
Polar bodies, 446, 447fig
Polarity (in water molecule), 30–31, 31fig, 50, 50fig
Polarization/depolarization (of neurons), 326–328, 327fig, 346
Pollution, 570, 593
Polygenic traits, 476–477, 489
Polymerase chain reaction (PCR), 501–502, 501fig
Polymers, 35, 35fig, 50
Polypeptides, 38–40, 38fig
Polysaccharides, 36, 36fig, 37t, 50
Polyunsaturated fats, 41. See also Unsaturated fats
Pons, 338, 338fig
Population bottleneck, 556
Population crash, 566, 566fig
Population ecology principles, 564–565
Population pyramids, 566, 566fig
Populations
  and assortative mating, 556–557, 558
  ecologically defined, 564
  ensuring adequate size of, 577–578, 581
  genetic drift in, 556, 558, 577, 581
  growth limits, 564–567, 580
  and natural selection, 552, 554–555, 555fig, 577
  and reproductive isolation, 549–550
  and sexual selection, 555, 555fig
Positive feedback
  and aging, 529
  in breathing control, 236fig
  described, 119, 119fig
  and labor, 525
  and ovarian cycle, 449, 449fig
  and oxytocin, 388
  in urination, 254–255, 255fig
Posterior pituitary, 387–388, 388fig
Prairies, 588–589, 589fig
Precapillary sphincters, 203, 203fig
Predation, 571, 574t
Prediction and scientific process, 6, 21, 376
Preembryo (defined), 513
Preembryonic development, 515, 515fig
Pregnancy
  ectopic, 443, 516
  medical termination of, 459
  and Rh blood factor, 188, 189fig, 194
  and tobacco smoke, 241
Premonitions, 376

Prenatal screening tests, 523, 524t
Presynaptic/postsynaptic neurons, 328, 329fig, 330–331, 331fig
Primary sources, 17–18, 18fig, 22
Primary structure, protein, 38–39, 39fig
Prions. *See also* Spongiform encephalopathies
    as cause of disease, 280–283, 297
    and immune response, 294
    replication of, 283, 283fig
    structure, 280–283, 283fig
Probability (in statistical tests), 15
Producers (in ecology), 569, 573
Products (in chemical reactions), 31
Progesterone
    as contraceptive, 456
    in pregnancy, 516–517, 521
    as sex hormone, 391, 449–451, 451fig
Prokaryotes *vs.* eukaryotes, 274
Prolactin, 390
Promoter, 86, 86fig, 100
Prophase
    in meiosis, 425, 426–427fig, 431fig
    in mitosis, 413, 414–415fig, 431fig
Proprioception, 356–357, 378
Prostaglandins, 398
Prostate gland, 441
Proteases, 92
Protein breakdown, 56fig, 72–73, 162fig, 173, 396
Protein degradation, 93, 93fig
Protein hormones (as class), 384, 384fig
Protein receptors, 384
Proteins
    and active transport, 66, 79
    in blood, 181–182, 182fig
    and cell cycle regulation, 416
    digestion of, 40, 160–161, 169
    in electron transport chain, 71, 72fig, 79
    energy storage in, 56–57
    in nonspecific immune response, 286–287
    and passive transport, 64–65
    in phospholipid bilayer, 64, 64fig
    and sodium-potassium pump, 66, 326–328, 327fig
    structure, 37–40, 38fig, 39fig, 50
    and subcellular structures, 61–62, 79
Proteins, prion, 280–283, 280fig, 283fig
Proteins, self-marker, 288
Protein synthesis. *See also* DNA (deoxyribonucleic acid); Genes
    and mutations, 88–91, 90fig, 91fig
    overview, 84–85, 85fig
    and recombinant proteins, 93–96, 101
    regulation of, 92–93, 92–93fig
    transcription, 86, 86fig, 92, 100
    translation, 87–88, 89fig, 92, 100–101, 100fig
Protons, 28–29, 33
Proto-oncogenes, 417
Protozoans, as infectious agents, 280, 304, 304fig, 305t
Proximal tubule, 256fig, 257–259, 261
Pseudopodia, 285, 285fig
Puberty (as developmental stage), 526–527
Pubescence. *See* Puberty
Pubic lice, 304, 304fig, 305t, 318
Pubis, 136, 136fig
Pulmonary artery, 210
Pulmonary circuit, 210, 211fig
Pulmonary veins, 210, 211fig

Pulse, 200, 202fig
Punnett square analysis, 474–476, 474fig, 475fig, 482, 491, 492fig
Pupil (eye), 370, 370fig
Pyrogens, 287
Pyruvic acid, in cellular respiration, 69–73, 70fig, 71fig, 72fig, 73fig

Q

Q angle, 136–137, 137fig
Quantitative genetics, 476–481
Quantitative traits
    continuous variation in, 476–477
    and heritability, 476–481, 480fig, 481fig, 483
Quarternary structure, protein, 39fig, 40

R

Radiation therapy, 421
Radioactive decay, 547
Radioactive isotopes, 29
Radiometric dating, 547–548, 549fig
Radius (bone), 137fig, 138
Random alignment (in meiosis)
    described, 425, 426fig, 428–430, 429fig, 430fig
    and independent assortment, 470–472, 471fig, 492–493, 493fig
Random assignment (in controlled experiments), 8, 10, 21
Random fertilization, 472
Reactants (defined), 31, 50
Receptive fields, 358, 360, 376
Receptors, protein, 384
Recessive alleles
    defined, 472–476
    and X-linked genes, 496–497
Recombinant bovine growth hormone (rBGH), 93–96
Recombinant genes (defined), 94
Recombinant human growth hormone, 390
Recombinant proteins
    and evolution, 552
    synthesis of, 93–96, 101
Recombination (defined), 93
Rectum, 163, 164fig
Recycling solid waste, 593
Red blood cells. *See* Erythrocytes
Red bone marrow, 128
Referred pain, 350, 350fig
Reflex arc, 342, 342fig
Reflexes, 342, 347, 357, 360
Refractory period, 328
Regenerative medicine, 512
Regulatory hormones, 387–388, 389
Releasing hormones, 388
Remission, 421
Renal cortex, 253, 253fig, 258–259, 258fig, 261–262, 261fig
Renal medulla, 253, 253fig, 257, 261–262, 261fig, 268
Renal pelvis, 253, 253fig
Renin, 260, 260fig, 262
Reproductive cells. *See* Gametes
Reproductive isolation, 549–550, 550fig
Reproductive organs
    development of, 519–521, 520fig
    effects of aging on, 530
    structure of, 442–443, 442fig, 462
Residual volume, 234, 235fig

Respiration (types of), 234, 234fig
Respiratory center, 236, 236fig
Respiratory surface, 238
Respiratory system
    and breathing, 234–237
    and gas exchange, 238–241
    structure and function, 228–230, 228fig
    tobacco smoke effects on, 230–234, 237–238, 240–242
Resting potential, 326, 327fig, 346
Restoration ecology, 578
Restriction enzymes, 94, 95fig, 502, 502fig
Reticular formation
    function, 338–339, 338fig, 347
    and sensory adaptation, 355
Retina, 370–372, 370fig
Retinal detachment, 372
Retrovirus, 311
Reuptake (of neurotransmitters), 328, 329fig, 330–331, 331fig
Reverse transcription, 311
Rheumatoid arthritis, 291
Rh factor, 187–189, 189fig, 194
Rhodopsin, 373
Rhythm method, 457
Ribonucleic acid (RNA). *See* RNA (ribonucleic acid)
Ribosomal RNA (rRNA), 87
Ribosomes
    features of, 60, 61, 61fig, 78
    in protein synthesis, 86fig, 87–88, 100
Ribs/rib cage, 135, 136fig
Ritalin, 330–331, 331fig, 344
Rivers, 590
RNA (ribonucleic acid)
    in protein synthesis, 84–91, 86fig, 87fig, 89fig
    of viruses, 277–278
    vs. DNA (deoxyribonucleic acid), 44, 50, 85fig, 100
RNA polymerase, 86, 86fig, 100
Rods, 373, 374fig
Romanov family
    and DNA fingerprinting, 502–504, 503fig
    and hemophilia allele, 491, 498, 500fig
    and hypothesis testing, 505fig
    pedigree analysis, 504fig
    random alignment illustrated by, 492–493, 493fig
Rotation, 139, 139fig
Rough endoplasmic reticulum, 61–62, 62fig, 78
Round window, 364
Ruffini corpuscles, 358, 358fig

S

Saliva, 158
Salts, 34, 50
Saltwater biomes, 590–591
Sample size, 15–16, 16fig
Sample, statistical, 13–14, 14fig
Sampling error, 14–15, 14fig
Sanitary landfills, 593, 593fig
Sarcolemma, 142, 143fig, 151
Sarcomeres, 142–143, 142fig, 144fig, 151
Sarcoplasm, 142
Sarcoplasmic reticulum, 142, 143fig, 151
Saturated fats, 41–42, 41fig, 43
Savannas, 588
Scapula, 137, 137fig

Schwann cells, 114, 325, 325fig, 326fig
Scientific information evaluation, 17–20, 20t, 22
Scientific method, 2–7, 21, 376, 505fig
Scientific theory
    and intelligent design, 541
    nature of, 3–4, 21
Sclera, 369, 369fig
Screening tests, prenatal, 523, 524t
Scrotum, 441
Sea level rise, 595, 595fig
Seasonal affective disorder (SAD), 394, 395fig
Sebaceous glands, 148
Secondary messenger, cyclic AMP (cAMP) as, 384
Secondary sex characteristics, 527–528
Secondary sources, 17–20, 18fig, 22
Secondary structure, protein, 39, 39fig
Seeing. *See* Vision
Segregation (in meiosis), 470–472
Self-marker proteins, 288, 291, 291fig
Self-reactivity, 291
Semen, 441–442
Semicircular canals, 368, 368fig
Semilunar valves, 209fig, 210, 214, 215fig
Seminal vesicles, 441
Seminiferous tubules, 441, 444
Senescence, 529–531, 534
Sensation, 355
Senses (defined), 354
Sensory adaptation, 355, 359fig, 360, 378
Sensory neurons
    and auditory receptors, 365–366, 365fig
    function, 324, 324–326, 325fig, 346
    and olfactory receptors, 362–363, 362fig
    and photoreceptors, 373–374, 374fig
    receptive field of, 358, 360
    in reflex arc, 342, 342fig
    stimulation of, 355
    and taste receptors, 361
Sensory receptors
    adaptation of, 355, 359fig, 360, 378
    classification of, 354, 354t, 378
    for equilibrium, 367–369, 368fig, 369fig
    for light, 372–376, 374fig, 379
    in nervous system, 324–325
    for pain, 354, 354t, 360, 360fig, 369
    response to stimuli, 355
    for smell, 362–363, 362fig
    and the somatic system, 342
    for sound, 364–367, 365fig
    and spinal nerves, 335
    for taste, 157, 361–362, 361fig
    tobacco smoke effect on, 243
    for touch, 147, 147fig, 358–359, 358fig, 359fig, 378
Serous membranes, 116, 116fig, 122
Sertoli cells, 444
Serum, 181
Sewage treatment, 592–593
Sex cells. *See* Gametes
Sex chromosomes, 422fig, 423, 494, 494fig
Sex determination, 494, 494fig, 506
Sex differences
    and athleticism, 131, 137, 139, 148–149
    in body fat, 148
    in bone, 131, 134, 136, 137, 137fig, 139, 494
    in excretory system, 253–254, 253fig
    measurement of, 128
    in muscle, 143, 146

Sex hormones. *See also* Hormones
    and ovaries, 443, 446, 449–451, 449fig, 451fig
    overview, 391
    in pregnancy, 516–517, 521, 524
    and testes, 441, 444–445, 446fig
Sex-linked genes, 494–500, 497fig
Sexually transmitted diseases (STDs), 304, 305–306t, 443
Sexually transmitted infections (STIs), 304, 304t, 305–306t, 443, 455
Sexual reproduction
    and birth control, 454fig
    and genetic variation, 440
    vs. asexual reproduction, 440
Sexual response, 452–453, 452fig, 462
Sexual selection, 555, 555fig, 558
Sickle-cell anemia, 191
Signal conduction
    by neurons, 325–329, 327fig
    by neurotransmitters, 328, 329fig
Signal integration, 374, 374fig
Signal transduction, 384, 384fig
Single-gene traits, 472–476, 481
Sinoatrial (SA) node, 214, 214fig
Sinuses, 134, 135fig, 229
Sister chromatids, 409–410, 425, 426–427fig, 431fig
Sixth extinction, 567–574, 580–581
"Sixth sense," 353, 376
Skeletal muscle
    activation by motor neurons, 142–143, 143fig, 151
    as connective tissue, 113, 122
    contraction, 142–146, 143fig, 144fig, 145fig, 151
    effects of aging on, 530
    sex differences in, 143, 146
    structure, 114fig, 140–142, 142fig, 151
    types, 140, 141fig
    and venous blood flow, 204, 204fig, 221
Skeletal system, 128–139, 151, 530
Skin
    effects of aging on, 529–530
    as nonspecific defense, 284–285
    pain receptors in, 360, 360fig
    structure, 146–148, 147fig, 151, 151fig
    thermoreceptors in, 359–360, 378
    touch receptors in, 358–359, 358fig, 359fig, 378
Skin color, 554–555, 555fig
Skull bones, 134, 134fig
Sleep/wake cycles, 394, 395fig
Sliding-filament model, 143–144, 144fig
Slow pain, 360
Small intestine, 161–162, 161fig, 173
Smell receptors. *See* Olfactory receptors
Smog, 593, 593fig
Smoking. *See also* Tobacco smoke
    quitting or preventing, 243–244, 244fig
Smooth endoplasmic reticulum, 61–62, 62fig, 79
Smooth muscle
    activation by autonomic nervous system, 342
    features of, 113, 114fig, 122
    and nitric oxide (NO), 206
    in vascular system, 200, 201fig, 202
Sodium-potassium pump, 66, 326–328, 327fig, 346
Sodium transporters (in countercurrent exchange), 262
Soft palate, 157, 158fig, 159fig
Solid waste disposal, 593, 593fig
Solute (defined), 30

Solution
    defined, 30
    and pH scale, 34, 50
Solvent
    defined, 30
    water as, 30–31, 31fig, 50, 50fig
Somatic cells, 422–424, 431fig
Somatic nervous system (of PNS), 342–343, 343fig, 347
Somatic senses, 354
Somatostatin, 391fig, 392
Sound waves
    characteristics of, 364, 364fig, 367
    pathway to auditory receptors, 364–365, 365fig, 378
    reception by auditory receptors, 365–367, 365fig, 366fig, 367fig, 378
Special creation hypothesis, 541, 544, 548
Special senses, 354, 378
Speciation, 549–550, 558
Species
    and biogeography, 545–546, 558
    defined, 549, 558
    ecosystem interactions, 571–573, 574t, 581
    extinction causes, 567–570
    humans as, 554–557, 558
    introduced, 570
    origin of, 549–553, 558
    as resources, 571, 581
    response to global warming, 595
    saving from extinction, 575–578, 579t, 581
Species-area curve, 568, 569fig, 575, 580
Specific immune response
    and HIV/AIDS, 311, 312fig, 313
    white blood cells in, 183t, 184, 287–294
Specificity, enzyme, 68
Specificity, lymphocyte, 288
Speech and cerebral cortex, 337, 337fig
Sperm
    components of, 444, 462
    in fertilization process, 513–514
    and infertility, 442, 446
Spermatogenesis, 441, 444–446, 445fig, 462
Spermicides, 454, 454fig
Sperm production. *See* Spermatogenesis
Sphincter muscles
    and capillary blood flow, 203, 203fig
    and digestion, 159, 173
Sphygmomanometer, 207, 207fig
Spina bifida, 518–519, 520fig
Spinal cord, 114, 115fig, 135, 334–336, 335fig, 346
Spinal nerves, 335, 335fig, 341, 341fig, 342, 347
Spinal reflexes, 342
Spleen, 287fig
Spongiform encephalopathies. *See also* Prions
    cause of, 273–274, 280–283
    and immune response, 294
    prevention of, 295–296
    transmission of, 284
Spongy bone, 128, 129fig, 151
Sports drinks. *See* Energy drinks
Squamous epithelium, 107, 122, 122fig, 146, 203
SRY gene, 519
Standard error, 15
Starch, 36, 37fig, 158, 169
Static model, 541, 544, 548
Statistical significance
    defined, 14–15, 17, 22
    factors influencing, 15–16, 16fig

Statistical tests
  information provided by, 13–15, 15*fig*, 22
  limitations of, 16–17
Statistics (defined), 13
Stem cell lines, 515
Stem cells
  and blood cells, 181, 182*fig*, 193
  and lymphocytes, 289
  potential uses for, 512, 521, 523–524, 528, 529, 530–531
  types of, 512–513
  umbilical cord, 526
Stem cell therapy
  and adult stem cells, 512, 528, 531
  debate regarding, 511–512, 515, 524, 531
  and embryonic stem cells, 512, 521, 529, 530–531
  and fetal stem cells, 523–524
Steppes, 588–589
Sterilization, 457–459, 458*fig*, 459*fig*
Sternum, 135, 136*fig*
Steroid hormones (as class), 384, 385*fig*, 387, 391
Steroids
  anabolic, 391, 391*fig*
  structure, 42–43, 43*fig*, 50
Stimuli (defined), 354
Stomach, 159–160, 160*fig*, 173
Stop codon, 88
Stratified tissue, 106, 122, 122*fig*
Stress response, 385–389, 391, 392, 393, 394
Stretch receptors, 205–206
Striated muscle, 113
Stroke, 209
Stroke volume, 214
Subcellular structures, 60–63, 61*fig*, 63*fig*, 78
Subcutaneous layer, 148
Subject bias, 9, 12
Substrates/substrate binding, 67–68, 68*fig*, 79
Sucrose (as disaccahride), 36, 36*fig*
Sugar-phosphate backbone (in nucleotides), 43–44, 44*fig*, 50, 410, 410*fig*
Superior vena cava, 210
Supernatural, 3, 541
Swallowing, 159, 159*fig*
Sweat and electrolyte loss, 34
Sweat glands, 148
Sympathetic nervous system
  defined, 342–343, 343*fig*, 347
  stress response, 342–343, 386, 392
Symptothermal method, 457
Synapses and neurotransmitters, 328, 329*fig*, 342
Synaptic integration, 329–330, 330*fig*
Synovial joints, 138–140, 138*fig*, 139–140, 139*fig*, 140*fig*
Synovial membranes, 116, 122
Syphilis, 305*t*, 307–308
Systemic circuit, 210, 211*fig*
Systole, 214, 215*fig*
Systolic pressure, 207

T

T_3/T_4 cells, 393, 393*fig*
T4 cells
  HIV/AIDS effect on, 294, 311, 312*fig*, 313
  in immune response, 293–294, 293*fig*
Taq polymerase, 501–502
Target cell, 384

Target tissue, 384
Tarsals, 136*fig*, 137
Tastants, 361–362
Taste buds, 157, 158*fig*, 361–362, 361*fig*, 378
Taste cells, 361, 361*fig*
Taste hairs, 361, 361*fig*
Taste receptors, 157, 158*fig*, 361–362, 361*fig*, 378
Tectorial membrane, 365–366, 366*fig*
Teeth, 157–158, 158*fig*
Telomerase, 418, 529
Telomeres, 529
Telophase
  in meiosis, 425, 426–427*fig*, 431*fig*
  in mitosis, 413, 414–415*fig*, 431*fig*
Temperate forest, 588, 588*fig*
Temperature
  and protein structure, 40, 502
  regulation, 119, 342, 347
  and thermoreceptors, 354, 354*t*, 359–360, 378
Temporal bones, 134, 134*fig*
Temporal lobe, 336, 336*fig*
Tendons, 128, 151
Terminal boutons, 324, 325*fig*
Tertiary structure, protein, 39–40, 39*fig*
Testability
  as hypothesis feature, 2–3
  and theory of evolution, 4, 552–553
Testes
  and sex hormones, 441, 444–445, 446*fig*, 519
  as sperm producers, 440, 441
Testosterone, 391, 444–445, 519
Thalamus, 337, 338*fig*, 346
Thermoreceptors, 354, 354*t*, 359–360, 378
Thermoregulation, 119, 342, 347
Thoracic cage, 135, 136*fig*
Thoracic cavity, 115, 115*fig*
Thrombin, 190
Thrombosis, 190
Thrombus, 208
Thymopoietin, 395
Thymosin, 395
Thymus
  effect of glucocorticoids on, 395
  hormones secreted by, 395, 397*t*
  as lymphatic system organ, 287*fig*
  and T lymphocytes (T cells), 184, 289, 395, 396*fig*
Thyroid hormones, 132, 392–394, 393*fig*, 397*t*
Tibia, 136, 136*fig*
Tidal volume, 234, 235*fig*
Tight junctions, 107, 122
Tissues
  defined, 106, 127
  types of, 106–114
T lymphocytes (T cells), 183*t*, 184, 288–294, 288*fig*, 290*fig*, 293*fig*, 295*fig*, 297
Tobacco addiction, 241–242, 243–244
Tobacco smoke
  and cancer, 408
  constituents of, 230–232, 240–241, 245
  effect on organ systems, 242–243, 242*fig*, 246
  effect on respiratory system, 230–234, 234*fig*, 237–238, 237*fig*, 245–246
  and oxygen deprivation, 240–241
Tongue, 157, 173
Tonicity and osmosis, 65–66, 79
Tonsillitis, 233
Tonsils, 157, 158*fig*

Totipotent cells, 512
Touch receptors, 358–359, 358*fig*, 359*fig*, 378
Toxin metabolism, 168
Trachea, 229*fig*, 230
Traits
  and dihybrid crosses, 491–493–492*fig*
  inheritance of, 468–472, 482
  Mendelian extensions in, 488–491, 489*fig*, 490*fig*
  and natural selection, 550–552, 558
  and pedigree analysis, 498–499, 498*fig*, 499*fig*, 500*fig*
  polygenic, 476–477, 489
  qualitative (single gene), 472–476
  quantitative, 476–481
  and reproductive isolation, 549–550, 550*fig*
  and sexual selection, 555, 555*fig*
Trans configuration, 42, 42*fig*
Transcription, 85*fig*, 86, 86*fig*, 92, 100
Transcription, reverse, 311
Trans fats, 42, 42*fig*
Transfer RNA (tRNA), 87–88, 87*fig*, 89*fig*, 100
Transfusion reaction, 187
Transfusions. *See* Blood transfusions
Transgenic organism, 95
Translation, 85*fig*, 87–88, 89*fig*, 92, 93*fig*, 100–101, 100*fig*
Transport proteins
  and active transport, 66, 79
  in blood plasma, 181
  in electron transport chain, 71, 72*fig*, 79
  and facilitated diffusion, 65, 65*fig*
  in kidneys, 259
  in passive transport, 64–65
  in sodium-potassium pump, 66, 326–328, 327*fig*
Triangulation
  and hearing, 367, 367*fig*
  and seeing, 374
Trichomonas vaginalis, 304, 305*t*, 306, 318
Trichomoniasis, 304, 305*t*, 306, 318
Trophic levels, 569–570, 570*fig*, 573–574
Trophic pyramid, 570, 570*fig*
Trophoblast, 515
Tropical forest, 588, 588*fig*
Tropomyosin, 142
Troponin, 142
T tubules, 142, 143*fig*
Tubal ligation, 458–459, 458*fig*
Tuberculosis, 232
Tubular reabsorption, 256*fig*, 258–259, 258*fig*, 267
Tubular secretion, 256*fig*, 259
Tumors, 407, 416–418
Tumor suppressors, 417
Tundra, 589, 589*fig*, 596
Tympanic membrane, 364, 365*fig*
Type 1 diabetes, 169–170, 291
Type 2 diabetes, 170–171, 392

U

Ulna, 137*fig*, 138
Ultrasound, 523
Umbilical cord, 522, 526
Umbilical cord stem cells, 526
Unsaturated fats, 41–42, 41*fig*
Upper respiratory tract, 228–230, 228*fig*, 230*fig*, 245
Urea, 255
Ureters, 253, 253*fig*
Urethra, 440

Urinary bladder, 253–255, 253*fig*
Urinary system
　and blood pH, 252, 263, 263*fig*
　effects of aging on, 530
　and excretion, 255–259
　overview, 252–255, 253*fig*, 267
　and salt balance, 263–264, 268
　sex differences in, 254, 254*fig*
　and water balance, 252, 259–263, 261*fig*, 268
Urination, 254–255, 255*fig*
Urine
　components of, 255–256
　formation, 256–263
U.S. Department of Agriculture (FDA)
　food guide pyramid, 77*fig*
　GMO regulations, 96, 98–99
U.S. Fish and Wildlife Service, and Klamath Falls crisis, 563
Uterus
　in childbirth, 524–526
　and menstrual cycle, 450–452
　structure, 443

V

Vaccination
　and genital warts, 303, 309–310
　and hepatitis B, 308–309
　and immune response, 292–293
　and influenza, 293–294
　and malaria, 192, 194
Vacuum aspiration, 459
Vagina, 443
Vaginal ring, 456, 456*fig*
Valence shell
　of carbon, 35
　and chemical bonding, 28–29, 35
　defined, 29
Valves, heart
　repair of, 212, 213*fig*, 221
　structure and function, 209*fig*, 210
Variables and experimental method, 7–8, 11–12, 11*fig*
Variation. *See* Genetic variation
Vascular spasm, 189
Vascular system (overview), 200, 201*fig*
Vas deferens, 441
Vasectomy, 458, 458*fig*

Vasoconstriction
　by angiotensin II, 260
　by antidiuretic hormone (ADH), 388
　and blood pressure, 202, 203*fig*, 206
Vasodilation
　and blood pressure, 202, 203*fig*, 206
　histamine produced, 286
Vector-borne transmission, 180, 185*fig*, 283
Veins, 201*fig*, 204, 204*fig*, 210–211, 211*fig*
Veins, cardiac, 217, 217*fig*, 221
Ventilation. *See* Breathing
Ventral cavity, 115, 115*fig*, 122, 122*fig*
Ventricles
　of the brain, 336
　of the heart, 209–210, 209*fig*, 211*fig*, 214, 221
Ventricular systole, 214, 215*fig*
Venules, 201*fig*, 204, 210, 211*fig*
Vertabral column, 135, 135*fig*
Vertebrae, 135, 135*fig*
Vertebral canal, 115*fig*
Vesicles
　defined, 62, 62*fig*
　large molecule transport by, 66, 66*fig*, 79
　and neurotransmitters, 328, 329*fig*
Vestibular apparatus, 367–368, 368*fig*
Vestigial traits, 545, 545*fig*
Villi, 161–162
Viral envelope, 278, 279*fig*
Viruses
　diseases caused by, 280*t*, 305–306*t*
　as parasites, 277, 297
　replication by, 279, 279*fig*
　and sexually transmitted diseases, 308–316, 318
　structure, 277–278, 279*fig*, 297
Visible light, 369, 370*fig*
Vision
　and the cerebrum, 336, 336*fig*
　and color, 373–375, 373*fig*
　and focusing light, 369–372, 371*fig*, 372*fig*, 378
　and perception, 373–376, 374*fig*, 379
　and photoreceptors, 372–373, 373*fig*
Visual cortex, 374–376, 374*fig*, 375*fig*
Vital capacity, 234, 235*fig*
Vitamin C, 6–7, 45, 46*t*
Vitamin D deficiency, 45, 45*fig*
Vitamin K, 45, 46*t*
Vitamins, 45–46, 46*t*, 48*t*, 51

Vitamin waters, 8, 45
Vitreous humor, 370, 370*fig*
Voice and vocal cords, 229–230, 230*fig*
Voluntary muscles, 113
Vulva, 442

W

Wastewater treatment, 592–593
Water
　bottled *vs.* tap, 33, 33*fig*
　molecular structure, 28–30, 29*fig*, 50
　and osmosis, 65–66, 65*fig*, 79
　properties of, 30–33, 31*fig*, 32*fig*, 33*fig*, 50
Water balance and urinary system, 252, 259–263, 261*fig*, 268
Weather, defined, 586
"Web of life," 571–573, 572–573*fig*
Weight-loss surgery, 155
Wetlands, 590, 590*fig*
White blood cells
　in nonspecific immune response, 183–184, 183*t*, 285–286
　in specific immune response, 287–294
　types and functions, 183, 183*t*
White matter, 325, 335
Withdrawal (as birth control method), 454
Wolffian duct, 519, 520*fig*
World Conservation Union (IUCN), 567, 568, 569, 570

X

X chromosome, 494, 494*fig*, 496–497

Y

Y chromosome, 494, 494*fig*, 497–498
Yeast infection, 305*t*, 306, 318
Yolk sac, 517

Z

Z discs, 113, 142, 142*fig*, 144*fig*
Zona pellucida, 513
Zygomatic bones, 134, 134*fig*
Zygote, 424, 424*fig*, 514–515, 533

PHOTO CREDITS

Chapter 1
p. xxviii: Johannes Kroemer/Getty Images; p. 1 (top): David Marsden/Photolibrary.com; p. 1 (middle): Sean Irvine/Photolibrary.com; p. 1 (bottom): Plush Studios/Blend Images/CORBIS; Figure 1.2 (a): Eye of Science/Photo Researchers; Figure 1.2 (b): Tony Ashby/AFP/Getty Images; Figure 1.3 (a): Gary Gaugler/Photo Researchers; Figure 1.3 (b) B. Boonyaratanakornkit & D. S. Clark, G. Vrdoljak/EM Lab, University of CA, Berkeley/Visuals Unlimited; Figure 1.3 (c): Michael Abbey/Visuals Unlimited; Figure 1.3 (d): Jeremy Burgess/Photo Researchers; Figure 1.3 (e): San Juan County, NM/Nature's Images/Photo Researchers; Figure 1.3 (f): Stanley Flegler/Visuals Unlimited; Figure 1.3 (g): Francois Gohier/Photo Researchers; Figure 1.7 (a): Will & Deni McIntyre/Photo Researchers; Figure 1.7 (b): Tom Lynch/Photolibrary; Figure 1.7 (c): Susan Kuklin/Photo Researchers

Chapter 2
p. 26: Ina Fassbender/Reuters/Landov; p. 27 (top): Joe Rimkus Jr./Miami Herald/MCT/Landov; p. 27 (middle): Envision/CORBIS; p. 27 (bottom): Frank Rumpenhorst/dpa/Landov; Figure 2.9: Alex Segre/Alamy; Figure 2.13 (b): Jose Luis Pelaez/Iconica/Getty Images; Figure 2.13 (c): Photo Researchers; Figure 2.23: Jeffrey L. Rottman/CORBIS

Chapter 3
p. 54: Mitchell Sams/Camera Press; p. 55 (top): KRT Photo/Richard Marshall, St. Paul Pioneer Press/Newscom; p. 55 (middle): Christopher La Marca/Redux Pictures; p. 55 (bottom): Jeff J. Mitchell/ Getty Images; Figure 3.2: SPL/Photo Researchers; Figure 3.9: Don Fawcett/Photo Researchers; Figure 3.10: P. Motta & T. Naguro/SPL/Photo Researchers; Figure 3.11: Professors P. Motta & T. Naguro/SPL/Photo Researchers; Figure 3.12: Micrograph and figure adapted from J. A. Lake, *Scientific American* (1981) 245:86. © J. A. Lake; Figure 3.13: Professors P. Motta, S. Makabe, & T. Naguro/SPL/Photo Researchers; Figure 3.14: P. Motta & T. Naguro/SPL/Photo Researchers; Figure 3.15: Don W. Fawcett; Figure 3.16: Jennifer Waters/Photo Researchers; p. 67 (left, middle, right): Sam Singer; Figure 3.34: Suza Scalora/Getty Images

Chapter 4
p. 82: AP Photo; p. 83 (top): Beth Plowes/Proteapix; p. 83 (middle): Courtesy University of Missouri, Columbia; p. 83 (bottom) Paul Grebliunas/Getty Images; Figure 4.18: Pearson Education; Figure 4.19: David Young Wolff/PhotoEdit

Chapter 5
p. 104: Steve Boninil/Getty Images; p. 105 (top): Dominique Faget/AFP/Getty Images; p. 105 (middle): Jupiter Images; p. 105 (bottom): Radius Images/Jupiter Images; Figure 5.5 (a): Ed Reschke/Peter Arnold; Figure 5.5 (b–f): Nina Zanetti, Pearson Science; Figure 5.6 (a): Ed Reschke/Peter Arnold; Figure 5.6 (b–c): Nina Zanetti, Pearson Science; Figure 5.13: Lori Adamski Peek/Getty Images; Figure 5.14: Pearson Science

Chapter 6
p. 126: Steve Burmeister; p. 127 (top) Duluth News Tribune; p. 127 (middle): Scott Halleran/Getty Images/NewsCom; p. 127 (bottom): Gavin Lawrence/Getty Images; Figure 6.1 (a): Andrew Syred/Photo Researchers; Figure 6.1 (b): Prof. P. Motta/University "La Sapienza," Rome/Photo Researchers; Figure 6.2: ISM/Phototake; Figure 6.4: SPL/Photo Researchers; Figure 6.6 (a): Prof. P. Motta/Dept. of Anatomy/University "La Sapienza," Rome/Photo Researchers; Figure 6.6 (b): P. Motta/Photo Researchers; Figure 6.7: SPL/Photo Researchers; Figure 6.23: Don W. Fawcett/Visuals Unlimited; p. 145: ISM/Phototake; Figure 6.29: Catherine Wessel/CORBIS

Chapter 7
p. 154: Reuters/CORBIS; p. 155 (top): Ian Waldie/Getty Images; p. 155: Michael Loccisano/FilmMagic/Newscom; Figure 7.2: Kessel & Dr. Kardon/Visuals Unlimited/Getty Images; Figure 7.6 (b): Ed Reschke/Peter Arnold; Figure 7.6 (c): Walker/Science Photo Library/Photo Researchers; Figure 7.7: Fawcett/Hirokawa/Heuser/SPL/Photo Researchers

Chapter 8
p. 178: Centers for Disease Control; p. 179 (top): Spencer Platt/Getty Images/Newscom; p. 179 (middle): Masamichi Aikawa, M.D./Phototake; p. 179 (bottom): Jon Hrusa/epa/CORBIS; p. 181: Stephen Morrison/epa/CORBIS; Figure 8.8: J.C. Revy/Phototake; Figure 8.10: David Phillips/Visuals Unlimited/Getty Images; p. 191: Oliver Meckes/Nicole Ottawa/Photo Researchers; Figure 8.12 ALEXANDER JOE/AFP/Getty Images/Newscom

Chapter 9
p. 198: Joseph R. Siebert/Custom Medical Stock Photo, Inc.; p. 199 (top): David Joel/Getty Images; p. 199 (middle): Sheila Terry/Photo Researchers; p. 199 (bottom): Jeff Topping/Getty Images; Figure 9.5: Lennart Nilsson/Albert Bonniers Forlag AB; p. 204: ALEX BARTEL/SPL/Photo Researchers; Figure 9.9: Alfred Pasieka/SPL/Photo Researchers; Figure 9.11 CNRI/Photo Researchers; Figure 9.15: Mario Villafuerte/ZUMA Press; Figure 9.16: Wake Forest University Health Sciences; Figure 9.21: Dario Sabljak/Shutterstock; Figure 9.23: Yoav Levy/Phototake; Figure 9.25: Roy Morsch/CORBIS

Chapter 10
p. 226: Reuters/Eric Gaillard; p. 227 (top): AP Photo/Anchorage Daily News, Bob Hallinen; p. 227 (middle): Phanie/Photo Researchers; p. 227 (bottom): Patrick Sheandell O'Carroll/PhotoAlto/Alamy; Figure 10.2: Eye of Science/Photo Researchers; Figure 10.4 (a–b): ISM/Phototake;

Figure 10.6: Ralph Hutchings/Visuals Unlimited; Figure 10.7 (b): Eye of Science/Photo Researchers; Figure 10.8: Janine Wiedel Photolibrary/Alamy; Figure 10.9: Lemoine/Photo Researchers; Figure 10.10: St. Bartholomew's Hospital/Photo Researchers; Figure 10.12 (a): Phanie/Photo Researchers; Figure 10.16: Phanie/Photo Researchers; Figure 10.21: Elmtree Image/Alamy

Chapter 11

p. 250 Bryan Lowry/Alamy; p. 251 (top): Nigel Marple/Getty Images; p. 251 (middle): Carol Hogan/Ironman; p. 251 (bottom): Chris Hondros/Getty Images/Newscom; Figure 11.6: Scott Camazine/Phototake; Figure 11.8: Thomas Frey/epa/CORBIS; Figure 11.10 (a): SPL/Photo Researchers; Figure 11.18: Greg M. Cooper/AP Photo

Chapter 12

p. 272: TEK/SPL/Photo Researchers, Inc.; p. 273 (top): Robert J. Higgins/UC Davis; p. 273 (middle): AP Photo; p. 273 (bottom): SIPA Press; Figure 12.1 (a) (top): Gary Gaugler/Visuals Unlimited; Figure 12.1 (a) (middle): Oliver Meckes/Photo Researchers; Figure 12.1 (a) (bottom): Tina Carvalho/Visuals Unlimited; p. 276: Bananastock/Jupiter Images; Figure 12.4 (a) (top): Linda Stannard, UCT/SPL/Photo Researchers; Figure 12.4 (a) (bottom): CNRI/Science Photo Library/Photo Researchers; p. 281 (a): Oliver Meckes/Nicole Ottawa/Photo Researchers; p. 281 (b): Eye of Science/Photo Researchers; p. 281 (c): SPL/Photo Researchers; p. 281 (d): ISM/Phototake; p. 281 (e): Custom Medical Stock Photo/Newscom; p. 281 (f): Andrew Syred/Photo Researchers; Figure 12.9 (b): Dennis Kunkel/Phototake; Figure 12.11: Lennart Nilsson/Albert Bonniers Forlag AB; Figure 12.18 P. Marazzi/Photo Researchers

Chapter 13

p. 302: Jean-Paul Chassenet/Photo Researchers; p. 303 (top): Kevin Laubacher/Getty Images; p. 303 (middle): Mike Kemp/Getty Images; p. 303 (bottom): Darren Greenwood/Design Pics, Inc./Alamy; Figure 13.1: Chris Fortuna/Getty Images; Figure 13.2 (a): Eye of Science/Photo Researchers; Figure 13.2 (b) David M. Phillips/The Population Control/Photo Researchers; Figure 13.3: David Phillips/Visuals Unlimited/Getty Images; Figure 13.4: SPL/Photo Researchers; Figure 13.5: Custom Medical Stock Photo; Figure 13.6: Steven J. Nussenblatt/Custom Medical Stock Photo; p. 310: Indranil Mukherjee/AFP/Getty Images

Chapter 14

p. 322: Paul Thomas/Getty Images; p. 323 (middle): Victoria Blackie/Getty Images; p. 323 (bottom): Lucidio Studio Inc./CORBIS; Figure 14.4: SPL/Photo Researchers; Figure 14.7 (a): E.R. Lewis, T.E. Everhart, Y.Y. Zeevi/Visuals Unlimited; p. 332 (a): Michael-John Wolfe/Shutterstock; p. 332 (b): David Hoffman/Alamy; p. 332 (c): Sergei Didyk/Shutterstock; p. 332 (d) CORBIS; p. 333 (a): Andrew Burns/Shutterstock; p. 333 (b) George Post/Science Photo Library/Photo Researchers; p. 333 (c): David Buffington/Getty Images; p. 333 (d) Kiselev Andrey Valerevich/Shutterstock; p. 334: Mariano Heluani/iStockphoto; Figure 14.9 (b): Nina Zanetti, Pearson Science; Figure 14.10 (a): Custom Medical Stock Photo

Chapter 15

p. 352: Zuma Press; p. 353 (top): LWA/Dann Tardif/Blend Images/Alamy; p. 353 (middle): Mauricio Anton/Photo Researchers; p. 353 (bottom): Glow Images/Getty Images; Figure 15.1: Leonard Lessin/Photo Researchers; Figure 15.3: Gemstone Images/Alamy; Figure 15.7: Robert Szajkowski/iStockphoto; Figure 15.9 (left) Justin Sullivan/Getty Images; Figure 15.9 (right): AP Photo/Frank Franklin II; Figure 15.10 (top): Prof. P. Motta/University "La Sapienza," Rome/Photo Researchers; Figure 15.10 (middle): Astrid & Hanns-Frieder Michler/Photo Researchers; Figure 15.12: Jim Gensheimer/San Jose Mercury News/KRT/Newscom; Figure 15.15: SPL/Photo Researchers; Figure 15.27: Eye of Science/Photo Reseachers; p. 375: Hartdavis/Science Photo Library/ Photo Researchers; Figure 15.30: NASA

Chapter 16

p. 382: Adrian Sherratt/Alamy; p. 383 (top): Laurence Mouton/Getty Images; p. 383 (middle): Erik Dreyer/Getty Images; p. 383 (bottom): Owen Franken/CORBIS; p. 390: Bettmann/CORBIS; Figure 16.7: Christian Petersen/Getty Images; Figure 16.8: Nina Zanetti, Pearson Science; Figure 16.9: Sharmyn McGraw; Figure 16.12: Pat & Lowe Stevens/Fleshandbones.com; p. 394: John Paul Kay/Peter Arnold, Inc.; Figure 16.14: Laurent/Laeticia/Photo Researchers; Figure 16.17: Time & Life Pictures/Getty Images

Chapter 17

p. 404: SPL/Photo Researchers; p. 405: Jim Whitmer; p. 406: Jim Whitmer; Figure 17.3: Biophoto Associates/Photo Researchers; p. 412 National Cancer Institute; Figure 17.8: Alexey Khodjakov/Photo Researchers Figure 17.9: LookatSciences/Phototake; p. 418 Edelmann/La Villette/Photo Researchers; Figure 17.12: Kings College School of Medicine/Photo Researchers; Figure 17.13: Jim Whitmer; Figure 17.14: CNRI/SPL/Photo Researchers

Chapter 18

p. 438: Stockbyte/Alamy; p. 439 (top): Yorgos Nikas/Getty Images; p. 439 (middle): Virginia Sherwood/ABC News ("© American Broadcasting Companies, Inc."); p. 439 (bottom): AJPhoto/Photo Researchers; Figure 18.3 (b): Eye of Science/Photo Researchers; p. 448: Thinkstock/Jupiter Images; Figure 18.10: Renn Sminky/Creative Digital Vision, Pearson Science; Figure 18.11: Allendale Pharmaceuticals, Inc., 73 Franklin Turnpike, Allendale, NJ 07401; Figure 18.12: Renn Sminky/Creative Digital Vision, Pearson Science; Figure 18.13 (a): Renn Sminkey, Pearson Science; Figure 18.13 (b): Frank LaBua, Pearson Education; Figure 18.14 (a): Don Farrall/Getty Images; Figure 18.14 (b): http://www.ORTHOEVRA.COM; Figure 18.14 (c): Organon, a part of Schering-Plough Corporation; Figure 18.15: Duramed Pharmaceuticals, Inc., a subsidiary of Barr Pharmaceuticals, Inc.; Figure 18.19 (b): Conceptus Incorporated; Figure 18.20: Brenden Smialowski/AFP/Getty Images

Chapter 19

p. 466: Yorgos Nikas/SPL/Photo Researchers; p. 467 (top): ColorBlind Images/Blend Images/Alamy; p. 467 (middle): Elyse Lewin/Getty Images; p. 467 (bottom): Gaetano/CORBIS; p. 472: Getty Images; p. 473 (top): Simon Fraser/Royal Victoria Infirmary/ Photo Researchers, Inc.; p. 473 (bottom): Harvard Brain Tissue Resource Center; p. 476 (top): University of Florida; p. 476 (bottom): Courtesy of New York plastic surgeon, Dr. Darrick E. Antell, www.Antell-MD.com; p. 477 (left): Photodisc/Getty Images; p. 477 (right): Hunter Martin/PGA/Getty Images

Chapter 20

p. 486: FPG/Getty Images; p. 487 (top): Alexander Zemlianichenko/AP Photos; p. 487 (middle): Agence No.7/CORBIS Sygma; p. 487 (bottom): Agence No.7/CORBIS Sygma; Figure 20.6: Andrew Syred/Photo Researchers; p. 495 (top): AP Photo/The Hutchinson News, Lindsey Bauman; pp. 495–496: University of Washington, Dept. of Pathology; Figure 20.9 (a): SPL Science Photo Library/Photo Researchers; Figure 20.9 (b): Spencer Grant/PhotoEdit; Figure 20.9 (c): Ed Kashi/CORBIS; Figure 20.13: David Parker/Photo Researchers; Figure 20.15: Getty Images; Figure 20.17: Anatoly Maltsev/AP Photo

Chapter 21

p. 510 (top): Chuck Kennedy/MCT/ABACAUSE.COM/AP Photos; p. 510: epa/CORBIS; p. 511 (middle): Alex Wong/Newsmakers/Getty Images/Newscom; p. 511 (bottom): LookatSciences/Phototake; Figure 21.1: Yorgos Nikas/Getty Images; Figure 21.4: Lennart Nilsson/Albert Bonniers Forlag AB; p. 518 (top): Specimen 20297, Congenital Anomaly Research Center, Kyoto University; p. 518 (middle): Specimen 4655, Congenital Anomaly Research Center, Kyoto University; p. 518 (bottom): Specimen 10074, Congenital Anomaly Research Center, Kyoto University; Figure 21.9: Biophoto Associates/Science Source/Photo Researchers; Figure 21.11 (top): John Watney/Photo Researchers; Figure 21.11 (middle): James Stevenson/Photo Researchers; Figure 21.11 (bottom): Petit Format/Nestle/Science Source/Photo Researchers; Figure 21.17: Laura Dwight; Figure 21.18: Diane Macdonald/Getty Images; Figure 21.21: AP Photo/Gerald Herbert

Chapter 22

p. 538: James Balog/Getty Images; p. 539 (top): Scala/Art Resource; p. 539 (middle): Dorling Kindersley Media Library; p. 539 (bottom): Phil Schermeister/CORBIS; Figure 22.2: George Richmond/Archiv/Photo Researchers; Figure 22.3 (a) Schafer & Hill/Getty Images; Figure 22.3 (b): Tui De Roy/The Roving Tortoise Nature Photography; Figure 22.6: Richardson, M. K., et al., *Science,* Vol. 280: Pg. 983c, Issue # 5366, May 15, 1998. Embryo from Professor R. O'Rahilly, National Museum of Health and Medicine/Armed Forces Institute of Pathology; Figure 22.8: Harry Taylor/Dorling Kindersley Media Library; Figure 22.12 (a): Barbara Gerlack/Visuals Unlimited; Figure 22.12 (b, left): Henry Ausloos/Animals Animals/Earth Scenes; Figure 22.12 (b, right): Hanan Isachar/CORBIS; Figure 22.12 (b, bottom): Lou Jacobs, Jr./Grant Heilman Photography, Inc.; Figure 22.14: Doug Cheeseman/Peter Arnold; Figure 22.15: Katherine Feng/Globio/Minden Pictures; Figure 22.19: Christof Wermter/zefa/CORBIS

Chapter 23

p. 562: AP Photo; p. 563 (top): Tupper Ansell Blake/USFWS; p. 563 (middle): Stephan Anderson/The Nature Conservancy of Oregon; p. 563 (bottom): AP Photo; Figure 23.3: Wolfgang Kaehler/CORBIS; Figure 23.6: Daisy Gilardini/Danita Delimont Photography; Figure 23.10: Peter Weimann/Animals Animals/Earth Scenes; Figure 23.13: Steven Holt/VIREO; Figure 23.14: U.S. Fish and Wildlife Service

Chapter 24

p. 584: Michael Newman/PhotoEdit; p. 585 (top) Tom & Pat Leeson/Photo Researchers; p. 585 (middle) Wes Thompson/CORBIS; p. 585 (bottom): Helena Bergengren/Nordic Photos/Getty Images; Figure 24.2 (a): Botanica/Jupiter Images; Figure 24.2 (b): Ross M. Horowitz/Getty Images; Figure 24.2 (c): Bryan and Cherry Alexander/Alamy Images; Figure 24.3: Adam Jones/Photo Researchers; Figure 24.4: Stephen Krasemann/NHPA/Photo Researchers; Figure 24.5: Cisca Castelijns/Foto Naturals/Minden Pictures; Figure 24.6: Jeffrey L. Rotman/CORBIS; Figure 24.7: Millard H. Sharp/Photo Researchers; Figure 24.9: Steven Wayne Rotsch; Figure 24.10: James Leynse/CORBIS; Figure 24.11: Stephen Chiang/Getty Images; Figure 24.14: Dave Pape, University of Buffalo